Designs and Drive Control of Electromechanical Machines

Designs and Drive Control of Electromechanical Machines

Editors

Kan Liu
Wei Hu

MDPI
Basel • Beijing • Wuhan • Barcelona • Belgrade • Novi Sad • Cluj • Manchester

Editors
Kan Liu
Hunan University
Changsha
China

Wei Hu
Hunan University
Changsha
China

Editorial Office
MDPI
St. Alban-Anlage 66
4052 Basel, Switzerland

This is a reprint of articles from the Topic published online in the open access journals *Machines* (ISSN 2075-1702), *Energies* (ISSN 1996-1073), *Electronics* (ISSN 2079-9292), and *Applied Sciences* (ISSN 2076-3417) (available at: https://www.mdpi.com/topics/electromechanical_machines).

For citation purposes, cite each article independently as indicated on the article page online and as indicated below:

Lastname, A.A.; Lastname, B.B. Article Title. *Journal Name* **Year**, *Volume Number*, Page Range.

ISBN 978-3-7258-1073-4 (Hbk)
ISBN 978-3-7258-1074-1 (PDF)
doi.org/10.3390/books978-3-7258-1074-1

Contents

About the Editors

Kan Liu

Professor Kan Liu (Senior Member, IEEE) received the B.E. and Ph.D. degrees in automation from Hunan University, Changsha, China, in 2005 and 2011, respectively, and the Ph.D. degree in electronic and electrical engineering from The University of Sheffield, Sheffield, U.K., in 2013. From 2013 to 2016, he was a Post-doctor Research Associate with the Department of Electronic and Electrical Engineering, The University of Sheffield. From 2016 to 2017, he was a Lecturer with the Control Systems Group, Loughborough University. He joined the Hunan University as a Professor of electromechanical engineering in 2017. He is now the Assistant Dean of the College of Mechanical and Vehicle Engineering, Hunan University, and the Director of the Engineering Research Center of Ministry of Education on Automotive Electronics and Control Technology, China. Prof. Liu has pioneered research on parameter identification and state observation of permanent magnet synchronous machines, design of electrical machines for construction and deep-sea machineries, high-power density SiC inverters/converters for electric vehicles and locomotives. He is the inventor of 12 patents and published more than 100 papers in leading journals and conferences, including more than 40 IEEE transactions papers, and 3 Best Paper Awards of International Conference on Electrical Machines and Systems (2019, 2022 and 2023). Prof. Liu was awarded the First Prize of the Science and Technology Progress Award of Hunan Province (2020), the Second Prize of the Natural Science Award of the Chinese Society of Automation (2023), and became a Chief Scientist of National Key Research and Development Program, China, in 2022.

Wei Hu

Assistant Professor Wei Hu (Member, IEEE) received the B.Eng. and the M.Sc. degrees from the Harbin Institute of Technology, Harbin, China, in 2014 and 2016, respectively, and the Ph.D. degree from Zhejiang University, Hangzhou, China in 2020, all in electrical engineering. From 2019 to 2020, he was a Visiting Scholar with the Department of electronic and electrical engineering from Nanyang Technological University, Singapore. From 2020, he joined the College of Mechanical and Vehicle Engineering, Hunan University. His research interest includes the advanced electric machine drives and control for electric transport equipment.

Preface

Being as core components for energy transformation, electrical machines have been widely employed in industrial applications and also widely studied by researchers in both academia and industry. This Reprint presents a collection of state-of-the-art researches which explore the advanced control techniques and design methods in electrical machine drive systems. For example, descriptions of some novel control methods taking into account both inverter and electrical machines are given in this Reprint, as well as some new topologies and structures of electrical machines and power converters. Overall, this Reprint can be a good resource for potential users working in industrial sectors such as wind power, electric vehicles, high-speed train, servo drive systems, robotics and marine propulsion systems.

Kan Liu and Wei Hu

Editors

machines

MDPI

Article

3D Pipe Forming of a New Bending Machine with a 3PUU–3RPS Hybrid Mechanism

Shuai Zhang [1,†], Yusen Li [2,3,†], Zhenming Yue [1], Zhongran Zhang [1], Lianpeng Su [1], Biao Yan [1] and Jun Gao [1,*]

1 School of Mechanical, Electrical and Information Engineering, Shandong University at Weihai, Weihai 264209, China; 202037591@mail.sdu.edu.cn (S.Z.); yuezhenming@sdu.edu.cn (Z.Y.); 201820663@mail.sdu.edu.cn (Z.Z.); 202037577@mail.sdu.edu.cn (L.S.); 202037584@mail.sdu.edu.cn (B.Y.)
2 GAMMA3-INRIA/ICD, University of Technology of Troyes, 10010 Troyes, France; yusen.li@utt.fr
3 Sino-French International Joint Research Laboratory of Computing Technology for Advanced Materials, Shanghai 200240, China
* Correspondence: shdgj@sdu.edu.cn
† These authors contributed equally to this work.

Abstract: To achieve the spatial variable curvature bending process of metal pipes, one 3PUU–3RPS hybrid mechanism designed for the free-bending forming of pipes is presented in this study. Its kinematics model was conducted based on theoretical analysis, and the obtained result was validated through ADAMS simulation. Through the theoretical analysis, the inverse position model of the proposed mechanical construction, which can show the relationship between the motion and the drive of the working platform, was presented. The velocity Jacobian matrix was also obtained and analyzed by establishing the inverse velocity model and inverse acceleration model. In addition, the static stiffness analysis of the proposed mechanical construction was also conducted in ABAQUS. Finally, by investigating its working space, the capability of 3PUU–3RPS mechanism was proved.

Keywords: pipe; free bending; hybrid mechanism; kinematics analysis

Citation: Zhang, S.; Li, Y.; Yue, Z.; Zhang, Z.; Su, L.; Yan, B.; Gao, J. 3D Pipe Forming of a New Bending Machine with a 3PUU–3RPS Hybrid Mechanism. *Machines* **2022**, *10*, 470. https://doi.org/10.3390/machines10060470

Academic Editors: Kan Liu and Wei Hu

Received: 7 May 2022
Accepted: 9 June 2022
Published: 12 June 2022

1. Introduction

The metal pipe not only has good forming property, but also has the characteristics of high strength, high stiffness, and high material utilization [1]. Therefore, the metal pipe has important applications in various industrial fields, such as electric power construction, railway construction, ships, furniture, and other areas of infrastructure. In aerospace, the metal pipe is also widely used in large precision machinery equipment and other high-end areas [2,3]. Therefore, many researchers are studying the precise forming of the metal pipe. H. Li et al. [4] established an AFE-based hybrid analytical-numerical framework for neutral layer shifting (NLS) calculation in tube bending. J. Wu et al. [5] proposed a new comprehensive strategy for the springback control of 3D tubes. J. Ma et al. [6] developed a generalized analytical model to accurately analyze springback in the bending of tubular materials. H. Yang et al. [7] developed a differential heating-based bending method to break the forming limit of difficult-to-form tubular materials. In the past few years, many efforts have been made to predict and compensate for springback [8–12].

However, among the manufacturing processes of metal pipes, the common bending forming methods include roll bending, draw bending, press bending, push bending and rotary-draw bending, etc. [13]. At present, with the developing requirement on the directions of lightweight miniaturization and integration of aerospace products, the space complexity and forming precision of pipeline are greatly improved, and higher requirements for the production of bending pipes are given. In order to meet these requirements, some new forming devices for metal pipes have to been developed [14,15].

To overcome the shortcomings of traditional bending methods, some ways to realize the spatial variable curvature-bending forming have been gradually developed [16].

As a new technology, 3D free-bending technology has the great potential to be suitable for the most tubes and profiles because of its geometrical flexibility and efficiency [17]. Plettke et al. [18] proposed a new method (free bending) which is suitable for arbitrary curve forming in space. The key point of this method is to gradually push the pipe through a bending die which can rotate freely within a certain angle of space. The shape of the bent pipe is controlled by adjusting the rotation of the mobile die. Free bending permits bending of almost all geometries. However, the minimum bending radius is approximately 2.5 times the pipe diameter, and it is also sensitive to the changes in material property.

P. Gantner et al. [19] developed a new bending technique in the hydroforming process chain, which is particularly suitable for the bending profile and tube cross-sections and has advantageous characteristics, such as fast bending speeds and an almost free definable bending geometry with transitionless bend-in-bends and spline bends without re-clamping.

S. Chatti et al. [20] carried out the numerical simulation studies on the torque superposed spatial bending (TSS) [21,22] and three-roll bending [23,24] processes. Based on the roll-bending process, dynamically adjusting the bending profile and superposing torque, controlling forming process parameters are given to achieve the purpose of bending different radii and angles.

H. Goto et al. [25,26] proposed a new type of multifunctional pipe-bending machine. By feeding the pipe into the fixed die and mobile die, adjusting the position of the mobile die can achieve bending. The bending radius is controlled by the relative distance and direction of the mobile die with the fixed die, and the bending angle is controlled by adjusting the feed length of the pipe. The movement of the mobile die is controlled by 6-DOF parallel kinematics (PKM).

Combining serial mechanisms with parallel mechanisms can allow these mechanisms to take full advantages of each other. Therefore, this paper proposes a new type of hybrid mechanism with 3PUU–3RPS structure which can better serve the needs of the spatial variable curvature bending of the metal pipe. The analysis of degrees of freedom, inverse position calculation, the analysis of kinematics and dynamics, stiffness analysis, workspace analysis, and other aspects of the new hybrid mechanism are studied. It turns out that the hybrid mechanism has less cumulative error, higher stiffness, and higher loading capacity than the series mechanism. Compared with the traditional parallel mechanism, the hybrid mechanism is easier to realize real-time control, and its workspace is also increased.

2. The Structural Design of Pipe-Bending Machine

The principle of spatial variable curvature bending forming is shown in Figure 1. This process consists of a fixed die, a mobile die, and a pusher [27,28]. The fixed die acts as a fixed pipe and guide. The mobile die can move and rotate in space and is the main part of this machine. The role of the pusher is feeding pipe. Each parameter is presented in Figure 1 and Table 1. The relationship among parameters is as seen in Equation (1). By this equation, it is found that the bending radius R of the pipe is related to V and u (detailed explanations of these parameters are given in Table 1). The rotation angle of mobile die θ is only for the purpose that the thrust of the mobile die to the pipe is perpendicular to the pipe.

$$\begin{cases} R = \frac{u^2+V^2}{2u} = \frac{u}{2} + \frac{V^2}{2u} \\ \theta = \arcsin\frac{V}{R} \end{cases} \tag{1}$$

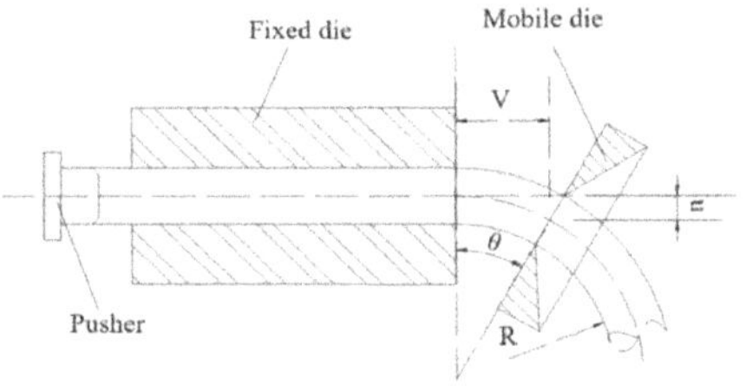

Figure 1. Principle of spatial variable curvature bending forming.

Table 1. Bending parameters of pipe.

R	Bending radius of the pipe
V	Distance between the fixed die and the mobile die
u	Offset between the center line of the fixed die and the center of the mobile die
θ	Angle of rotation of the mobile die

By understanding the principle of spatial variable curvature-bending forming, it is found that the most important part of the pipe-bending machine is the mobile die, which can do the translation and rotation motions in space. A new type of pipe-bending construction is presented in this paper, as is shown in Figure 2. The mobile die of this machine is a kind of 3PUU–3RPS structure. The device is mainly composed of a mounting frame, a pusher, a fixed die, and a mobile die. The mobile die is composed of a 3PUU moving part and a 3RPS rotary part, as shown in Figures 3 and 4.

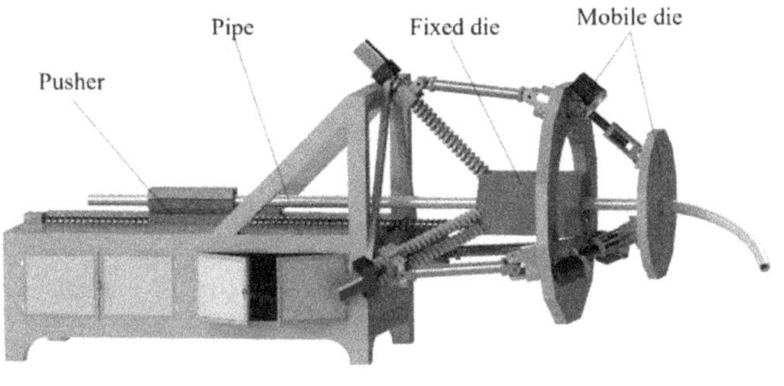

Figure 2. Whole structural design for free bending device.

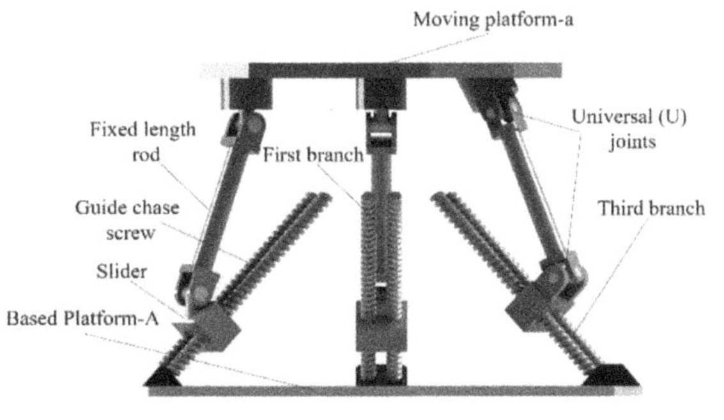

Figure 3. 3PUU moving part.

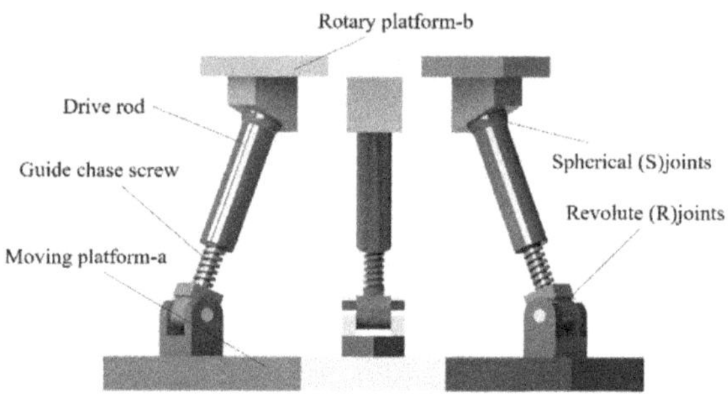

Figure 4. 3RPS rotary part.

The Analysis on Degree of Freedom

The 3PUU moving part proposed in this article has three parallel branched chains which are asymmetrically distributed in space. Each connecting chain is composed of prismatic (P), lower universal, and upper universal joints arranged from the base to the output link. In the first two branches, the axis of the first revolute joint of lower universal (U) joints is perpendicular to the based platform-A, whereas the axis of the second revolute joints of the upper U joint is perpendicular to the moving platform-a. In the third branch, the axis of the first revolute joint of the lower U joint is parallel to the based platform-A, and the axis of the second revolute joints of upper U joint is parallel to the moving platform-a, as is shown in Figure 3.

For the first branched chain, the coordinate system is established as shown in Figure 5. The X-axis and Z-axis are along the two axes of the U joints; the Y-axis is perpendicular to the crosshead plane of the U joints. Therefore, in the initial assembly position, the screw system of the branch chain can be written as follows:

$$\begin{cases} \$_1 = (\begin{matrix} 0 & 0 & 0; & 0 & e_1 & f_1 \end{matrix}) \\ \$_2 = (\begin{matrix} 0 & 0 & 1 & 0 & 0 & 0 \end{matrix}) \\ \$_3 = (\begin{matrix} 1 & 0 & 0; & 0 & 0 & 0 \end{matrix}) \\ \$_4 = (\begin{matrix} 1 & 0 & 0; & 0 & e_4 & f_4 \end{matrix}) \\ \$_5 = (\begin{matrix} 0 & 0 & 1 & d_5 & 0 & 0 \end{matrix}) \end{cases}. \quad (2)$$

To calculate the reciprocal screw of five kinematic screws, all the variables in Equation (2) are imported into the formula $\$^r \circ \$_i = 0$, $(i = 1, 2, 3)$, and the constraint screw of the single branched chain can be obtained as follows:

$$\$_1^r = (0 \quad 0 \quad 0; \quad 0 \quad 1 \quad 0). \quad (3)$$

For the third branched chain, the coordinate system is established as shown in Figure 6. The X-axis and Y-axis are along the two axes of the U joints; the Z-axis is perpendicular to the crosshead plane of the U joints. Therefore, in the initial assembly position, the screw system of branch chain can be written as follows:

$$\begin{cases} \$_1 = (\begin{matrix} 0 & 0 & 0; & 0 & e_1 & f_1 \end{matrix}) \\ \$_2 = (\begin{matrix} 1 & 0 & 0; & 0 & 0 & 0 \end{matrix}) \\ \$_3 = (\begin{matrix} 0 & 1 & 0; & 0 & 0 & 0 \end{matrix}) \\ \$_4 = (\begin{matrix} 0 & 1 & 0; & d_4 & 0 & 0 \end{matrix}) \\ \$_5 = (\begin{matrix} 1 & 0 & 0; & 0 & e_5 & 0 \end{matrix}) \end{cases}. \quad (4)$$

To calculate the reciprocal screw of five kinematic screws, all the variables in Equation (4) are imported into the formula $\$^r \circ \$_i = 0$, $(i = 1, 2, 3)$, and the constraint screw of the single branched chain can be obtained as follows:

$$\$_3^r = (0 \quad 0 \quad 0; \quad 0 \quad 0 \quad 1). \quad (5)$$

In Equations (3) and (5), the first three elements of the constraint spiral are all zero. Thus, these two constraint screws are both constraint couples and perpendicular to the crosshead plane of the respective U joints. The second constraint screw is the same as the first one. Because three couples of branched chains are not parallel and linearly independent of each other, three independent constraints act on the moving platform-a. These three constrained couples constrain three rotational degrees of freedom, and the mobile platform has only three translational degrees of freedom. Therefore, the mobile platform can move along the X-axis, Y-axis, Z-axis and the directions of any linear combination of them.

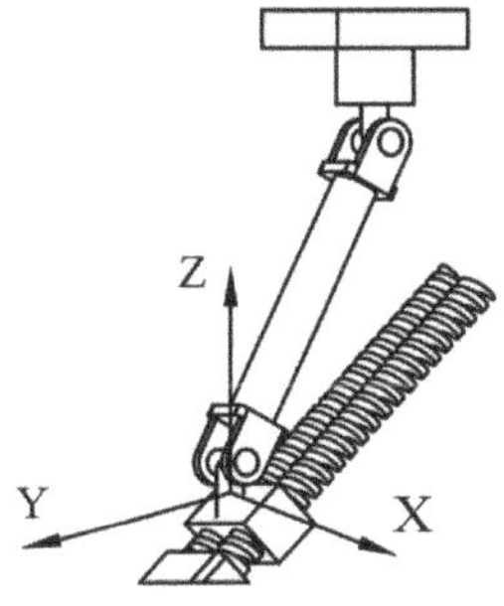

Figure 5. Coordinate system of first branch chain of 3PUU moving part.

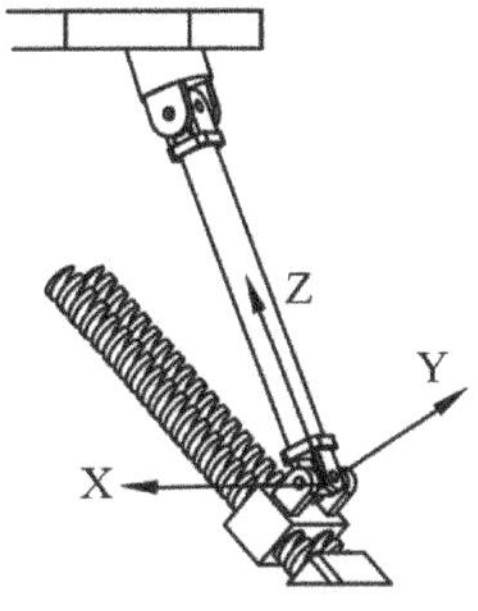

Figure 6. Coordinate system of third branch chain of 3PUU moving part.

The 3RPS rotary part has three parallel branched chains which are symmetrically distributed in space, as shown in Figure 4. Each connecting chain is composed of revolute (R), prismatic (P), and spherical (S) joints. For the first branched chain, the coordinate system is established (Figure 7). The origin is the first kinematic pair. The X-axis is along the axis of this revolute (R) joint, and the Z-axis is perpendicular to the based plane. Thus, the screw system of branch chain can be written as follows:

$$\begin{cases} \$_1 = \begin{pmatrix} 1 & 0 & 0; & 0 & 0 & 0 \end{pmatrix} \\ \$_2 = \begin{pmatrix} 0 & 0 & 0; & 0 & \mathrm{e} & f \end{pmatrix} \\ \$_3 = \begin{pmatrix} 1 & 0 & 0; & 0 & f & -e \end{pmatrix} \\ \$_4 = \begin{pmatrix} 0 & 1 & 0; & -f & 0 & 0 \end{pmatrix} \\ \$_5 = \begin{pmatrix} 0 & 0 & 1; & e & 0 & 0 \end{pmatrix} \end{cases}. \tag{6}$$

In the same way, the constraint screw of the single branched chain can be obtained as follows:

$$\$^r = (1 \quad 0 \quad 0; \quad 0 \quad f \quad -e). \tag{7}$$

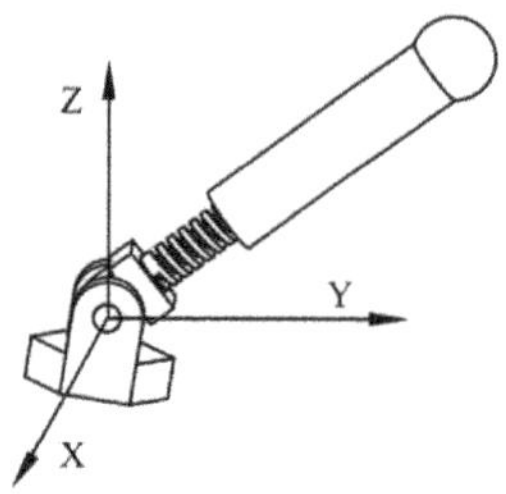

Figure 7. Coordinate system of first branch chain of 3RPS rotary part.

For the 3RPS rotary part, the three chains have a similar constraining force. This force can be applied on the rotary platform-b, located on the center of respective branched chain's sphere joint and parallel to the axis of revolute joint. These three constraining forces are linearly independent and limited to three freedom degrees of the rotary platform-b. Through the analysis, the limited degrees of freedom include two translational degrees and one rotational degree. Thus, the rotary platform can only have a two-dimensional rotation in the plane and a translation along the Z-axis.

Through analysis, the 3PUU-3RPS mobile die has six degrees of freedom in total. The degree of freedom normally can also be calculated by using Equation (8). With the obtained results $M = 6$, the determined degree of freedom can be regarded as consistent with the theoretical obtained results, which also confirms the accuracy of analysis of 3PUU-3RPS system:

$$M = d(n - g - 1) + \sum_{i=1}^{g} f_i + v = 6 \times (15 - 18 - 1) + 30 + 0 = 6. \tag{8}$$

3. Kinematic Analysis

Considering the independence between the 3PUU moving part and the 3RPS rotary part in the 3PUU-3RPS, in coming sections these two parts can be studied and analyzed respectively.

3.1. Inverse Position Model

3.1.1. Inverse Position Model of 3PUU Moving Part

As shown in Figure 8, the based coordinate system is given. In the initial position, the moving coordinate system $o\prime - x\prime y\prime z\prime$ coincides with the based coordinate system $o - xyz$. The direction from the origin O to $O\prime$ can be represented by a vector $\boldsymbol{P}$, and the direction cosine matrix $\boldsymbol{R}$ is represented by Euler angles as is Equation (9), and $C_\phi, C_\theta, C_\psi, S_\phi, S_\theta, S_\psi$ represent $\cos\phi$, $\cos\theta$, $\cos\psi$, $\sin\phi$, $\sin\theta$, $\sin\psi$, respectively.

$${}^{o}_{o\prime}\boldsymbol{R} = \begin{bmatrix} C_\phi C_\theta C_\psi - S_\phi S_\psi & -C_\phi C_\theta S_\psi - S_\phi C_\psi & C_\phi S_\theta \\ S_\phi C_\theta C_\psi + C_\phi S_\psi & -S_\phi C_\theta C_\psi + C_\phi S_\psi & S_\phi S_\theta \\ -S_\theta C_\psi & S_\theta S_\psi & C_\theta \end{bmatrix} \tag{9}$$

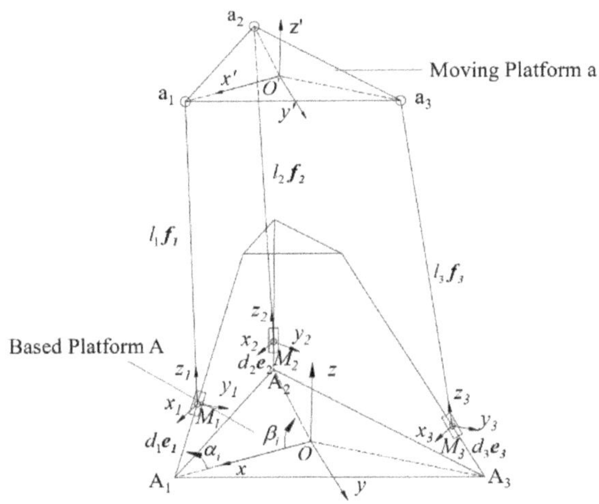

Figure 8. Schematic diagram of 3PUU moving part.

Through the analysis of the degree of freedom, it is found that the 3PUU moving part only has three translational degrees of freedom, which induces $\phi = \theta = \psi = 0$. Thus, the homogeneous transfer matrix ${}^{o}_{o\prime}\boldsymbol{T}$ of the moving platform relative to the based platform can be written as follows:

$$ {}^{o}_{o\prime}\boldsymbol{T} = \begin{bmatrix} 1 & 0 & 0 & \Delta x \\ 0 & 1 & 0 & \Delta y \\ 0 & 0 & 1 & \Delta z \\ 0 & 0 & 0 & 1 \end{bmatrix}, \tag{10} $$

where Δx, Δy, and Δz represent the moving distances of the platform a relative to the based platform-A along three directions of the x-axis, y-axis, and z-axis, respectively.

To conveniently describe the posture of each branch, a moving coordinate system $M_i - x_i y_i z_i$ is established at the center of the U joints which connect the slider and the fixed-length rod, where i represents the connecting chain number (i = 1, 2, 3). The z-axis of the coordinate system is along the direction of the fixed-length rod. The y-axis is determined by the right-hand rule with the z-axis and the z_i-axis. The direction cosine matrix ${}^{o}_{M_i}\boldsymbol{R}$ of the coordinate system relative to the based coordinate system can be written as Equation (11), and the position of the fixed rod in the based coordinate system can be expressed by Equation (12):

$$ {}^{o}_{M_i}\boldsymbol{R} = \begin{bmatrix} C_{\phi i}C_{\theta i} & -S_{\phi i} & C_{\phi i}S_{\theta i} \\ S_{\phi i}C_{\theta i} & C_{\phi i} & S_{\phi i}S_{\theta i} \\ -S_{\theta i} & 0 & C_{\theta i} \end{bmatrix} \tag{11} $$

$$ \boldsymbol{l}_i = {}^{O}_{M_i}\boldsymbol{R} \cdot \boldsymbol{l}_{Mi} = {}^{O}_{M_i}\boldsymbol{R} \begin{bmatrix} 0 \\ 0 \\ l \end{bmatrix} = \begin{bmatrix} lC_{\phi i}S_{\theta i} \\ lS_{\phi i}S_{\theta i} \\ lC_{\phi i} \end{bmatrix}, \ (i = 1,2,3), \tag{12} $$

where $\boldsymbol{P}$ represents the position where the center of the moving platform $o\prime$ in based coordinate system, $\boldsymbol{A}_i$ and $\boldsymbol{a}_i$ represent the position coordinate of point A_i and a_i in the based coordinate system, $\boldsymbol{a}_{oi}$ represents the position where point a_i in the moving coordinate system $o\prime - x\prime y\prime z\prime$, R represents the radius of platform-A, r represents the radius of platform-a, d_i represents the translational distance of the slider, vector $\boldsymbol{e}_i$ represents the unit vector that position of the screw in based coordinate system, $\boldsymbol{l}_i$ represents the position of the fixed-length rod in the based coordinate system, $\boldsymbol{l}_{Mi}$ represents the position of the fixed-length rod in the moving coordinate system $M_i - x_i y_i z_i$, $\boldsymbol{M}_i$ represents the position of point M_i in the based coordinate system. Through geometrical relation, Equations (13)–(18)

can be obtained. Based these formulas, the translational distance of the slider d_i can be calculated by mathematical methods, as Equation (19):

$$\boldsymbol{a}_{oi} = \begin{bmatrix} r\cos\beta_i \\ -r\sin\beta_i \\ 0 \end{bmatrix}, \ (i = 1,2,3) \tag{13}$$

$$\boldsymbol{A}_i = \begin{bmatrix} R\cos\beta_i \\ -R\sin\beta_i \\ 0 \end{bmatrix}, \ (i = 1,2,3) \tag{14}$$

$$\boldsymbol{e}_i = \begin{bmatrix} -\cos\alpha\cos\beta_i \\ \cos\alpha\sin\beta_i \\ \sin\alpha \end{bmatrix}, \ (i = 1,2,3) \tag{15}$$

$$\boldsymbol{M}_i = \boldsymbol{A}_i + d_i\boldsymbol{e}_i, \ (i = 1,2,3) \tag{16}$$

$$\boldsymbol{l}_i = \boldsymbol{\alpha}_i - \boldsymbol{M}_i = \boldsymbol{\alpha}_i - \boldsymbol{A}_i - d_i\boldsymbol{e}_i, \ (i = 1,2,3) \tag{17}$$

$$\boldsymbol{a}_i = {}^{o}_{o\prime}\boldsymbol{T} \cdot \boldsymbol{a}_{oi}, \ (i = 1,2,3) \tag{18}$$

$$d_i = \left(\boldsymbol{a}_i{}^T - \boldsymbol{A}_i{}^T\right)\boldsymbol{e}_i + \sqrt{[(\boldsymbol{a}_i{}^T - \boldsymbol{A}_i{}^T)\boldsymbol{e}_i]^2 - \|\boldsymbol{\alpha}_i - \boldsymbol{A}_i\|^2 + l_i{}^2}, \ (i = 1,2,3). \tag{19}$$

3.1.2. Inverse Position Model of 3RPS Rotary Part

As shown in Figure 9, the x-axis of the based coordinate system points to a_1, the z-axis is perpendicular to the moving platform-a. The $x\prime$-axis of the moving coordinate system points to b_1, the $z\prime$-axis is perpendicular to the rotary platform-b. In the initial position, the moving coordinate system $o\prime - x\prime y\prime z\prime$ coincides with the based coordinate system $o - xyz$. The homogeneous transfer matrix ${}^{o}_{o\prime}\boldsymbol{T}$ of the rotary platform b relative to the moving platform-a can be given as Equation (20). The position of the kinematic pair b_i in the based coordinate system can be shown as Equation (21):

$${}^{o}_{o\prime}\boldsymbol{T} = \begin{bmatrix} C_\phi C_\theta C_\psi - S_\phi S_\psi & -C_\phi C_\theta S_\psi - S_\phi C_\psi & C_\phi S_\theta & \Delta x \\ S_\phi C_\theta C_\psi + C_\phi S_\psi & -S_\phi C_\theta C_\psi + C_\phi S_\psi & S_\phi S_\theta & \Delta y \\ -S_\theta C_\psi & S_\theta S_\psi & C_\theta & \Delta z \\ 0 & 0 & 0 & 1 \end{bmatrix} \tag{20}$$

$$\boldsymbol{b}_i = {}^{o}_{o\prime}\boldsymbol{T} \cdot \boldsymbol{b}_i{}', (i = 1,2,3). \tag{21}$$

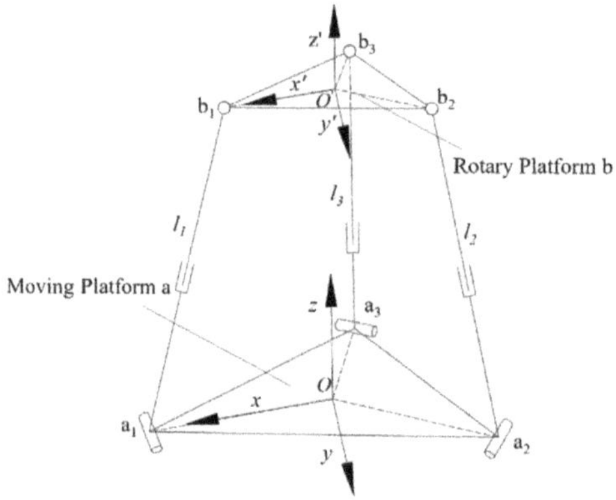

Figure 9. Schematic diagram of 3RPS rotary part.

By geometrical relationship, the coordinate values of point a_i and b_i can be obtained easily in their respective coordinate systems. Through Equation (21), the coordinate values of the hinge point of the platform-b in the based coordinate system can be obtained. The position vector $\boldsymbol{l}_i$ of the drive rod in the based coordinate system can be calculated by Equation (22), and the equation of the inverse position model can be expressed by Equation (23):

$$\boldsymbol{l}_i = \boldsymbol{b}_i - \boldsymbol{a}_i = \begin{bmatrix} l_{ix} & l_{iy} & l_{iz} \end{bmatrix}^T, (i = 1,2,3) \tag{22}$$

$$l_i = \sqrt{l_{ix}{}^2 + l_{iy}{}^2 + l_{iz}{}^2}, (i = 1,2,3). \tag{23}$$

3.2. Inverse Velocity Model and the Inverse Acceleration Model

3.2.1. The Inverse Velocity Model and the Inverse Acceleration Model of 3PUU Moving Part

By geometrical relationship, the position vector $\boldsymbol{P}$ of the center of platform-a can be written by Equation (24). Taking a derivative with respect to time for two sides of Equation (24), the velocity $\boldsymbol{v}$ of the center of the platform-a can be obtained by Equation (25), where $\boldsymbol{\omega}$ represents the angular velocity of the platform-a and $\boldsymbol{\omega}_i$ represents the angular velocity of the fixed-length rod:

$$\boldsymbol{P} = \boldsymbol{A}_i + d_i\boldsymbol{e}_i + l_i - \boldsymbol{\alpha}_i,\ (i = 1,2,3) \tag{24}$$

$$v = \dot{d}_i\boldsymbol{e}_i + \omega_i \times \boldsymbol{l}_i - \omega \times \boldsymbol{\alpha}_i,\ (i = 1,2,3). \tag{25}$$

Calculate the dot product with the directional unit vector of the fixed-length rod $\boldsymbol{f}_i$ for two sides of Equation (25), and through mathematical calculation, the translational speed $\dot{\boldsymbol{d}}$ of the slider can be obtained by Equation (26). The matrix equation of Equation (26) can be expressed as Equation (27):

$$\dot{d}_i = \frac{\boldsymbol{f}_i^T}{\boldsymbol{f}_i^T \cdot \boldsymbol{e}_i}\boldsymbol{v},\ (i = 1,2,3) \tag{26}$$

$$\dot{\boldsymbol{d}} = \boldsymbol{J}_1\dot{\boldsymbol{X}}. \tag{27}$$

In Equation (27), $\dot{\boldsymbol{X}}$ represents the translational speed of the moving platform-a, and $\boldsymbol{J}_1$ represents the velocity Jacobian matrix. They can be expressed as follows:

$$\begin{cases} \dot{\boldsymbol{d}} = \begin{pmatrix} \dot{d}_1 & \dot{d}_2 & \dot{d}_3 \end{pmatrix}^T \\ \dot{\boldsymbol{X}} = \begin{pmatrix} v_x & v_y & v_z \end{pmatrix}^T \end{cases} \tag{28}$$

$$\boldsymbol{J}_1 = \begin{bmatrix} \frac{\boldsymbol{f}_1{}^T}{\boldsymbol{f}_1{}^T \cdot \boldsymbol{e}_1} & \frac{\boldsymbol{f}_2{}^T}{\boldsymbol{f}_2{}^T \cdot \boldsymbol{e}_2} & \frac{\boldsymbol{f}_3{}^T}{\boldsymbol{f}_3{}^T \cdot \boldsymbol{e}_3} \end{bmatrix}^T. \tag{29}$$

To establish the inverse acceleration model, first take a derivative with respect to time for the two sides of Equation (26), as Equation (30). Calculate the dot product with a directional unit vector of the fixed-length rod $\boldsymbol{f}_i$ for two sides of Equation (30), and through mathematical calculation, the translational acceleration $\ddot{\boldsymbol{d}}$ of the slider can be obtained by Equation (31). The matrix equation of Equation (31) can be expressed as Equation (32):

$$\dot{v} = \ddot{d}_i\boldsymbol{e}_i + \dot{\omega}_i \times \boldsymbol{l}_i + \omega_i \times (\omega_i \times \boldsymbol{l}_i),\ (i = 1,2,3) \tag{30}$$

$$\ddot{d}_i = \frac{\boldsymbol{f}_i^T}{\boldsymbol{f}_i^T\boldsymbol{e}_i}\dot{v} - \frac{\boldsymbol{f}_i^T}{\boldsymbol{f}_i^T\boldsymbol{e}_i}\omega_i \times (\omega_i \times \boldsymbol{l}_i),\ (i = 1,2,3) \tag{31}$$

$$\ddot{\boldsymbol{d}} = \boldsymbol{J}_1\ddot{\boldsymbol{X}} + \boldsymbol{K}_1, \tag{32}$$

where, $\ddot{X}$ represents the translational acceleration of the moving platform-a, K_1 represents the second order kinematic influence coefficient. They can be expressed as follows:

$$\begin{cases} \ddot{d} = \begin{pmatrix} \ddot{d}_1 & \ddot{d}_2 & \ddot{d}_3 \end{pmatrix}^T \\ \ddot{X} = \begin{pmatrix} a_x & a_y & a_z \end{pmatrix}^T \end{cases} \tag{33}$$

$$J_1 = \begin{bmatrix} \frac{f_1^T}{f_1^T \cdot e_1} \\ \frac{f_2^T}{f_2^T \cdot e_2} \\ \frac{f_3^T}{f_3^T \cdot e_3} \end{bmatrix}, K_1 = \begin{bmatrix} -\frac{f_1^T}{f_1^T e_1}\omega_1 \times (\omega_1 \times l_1) \\ -\frac{f_2^T}{f_2^T e_2}\omega_2 \times (\omega_2 \times l_2) \\ -\frac{f_3^T}{f_3^T e_3}\omega_3 \times (\omega_3 \times l_3) \end{bmatrix}. \tag{34}$$

As above, the inverse velocity model of the 3PUU moving part is expressed by Equation (27), and the inverse acceleration model is expressed by Equation (32).

3.2.2. Inverse Velocity and Acceleration Models of 3RPS Rotary Part

By mathematical relationship, Equation (35) can be achieved, where $\boldsymbol{l}_i$ represents the relationship between two hinges of the branched chain. Through mathematical calculation, the velocity of the drive rod $\dot{l}_i$ can be obtained by Equations (36) and (37), where $\boldsymbol{v}_{bi}$ represents the velocity vector of the hinge point b_i:

$$l_i^2 = \boldsymbol{l}_i \cdot \boldsymbol{l}_i, (i = 1,2,3) \tag{35}$$

$$\dot{l}_i = \boldsymbol{Q}_i \cdot \boldsymbol{v}_{bi}, (i = 1,2,3) \tag{36}$$

$$\boldsymbol{Q}_i = \frac{\boldsymbol{l}_i}{l_i}, (i = 1,2,3). \tag{37}$$

The velocity vector of the hinge point b_i of the platform-b can also be obtained through the Equation (38), where $\boldsymbol{\omega} = \begin{bmatrix} \omega_x & \omega_y & \omega_z \end{bmatrix}^T$ represents the angular velocity of platform-b, $\boldsymbol{v} = \begin{bmatrix} v_x & v_y & v_z \end{bmatrix}^T$ represents the velocity of the center of platform-b, $\boldsymbol{r}_{bi}$ represents the position vector of hinge point in the moving coordinate system:

$$\boldsymbol{v}_{bi} = \boldsymbol{v} + \boldsymbol{\omega} \times \boldsymbol{r}_{bi}, (i = 1,2,3). \tag{38}$$

Considering $v_x = v_y = \omega_z = 0$, the velocity $\dot{\boldsymbol{P}}$ of the rotary platform-b can be obtained by Equation (39). By mathematical calculation, the velocity $\dot{\boldsymbol{l}}_i$ of the branched chain can be calculated as Equation (40):

$$\dot{\boldsymbol{P}} = \begin{bmatrix} \omega_x & \omega_y & v_z \end{bmatrix}^T \tag{39}$$

$$\begin{cases} \dot{\boldsymbol{l}}_i = \boldsymbol{Q}_i \cdot \boldsymbol{v}_{bi} = \boldsymbol{Q}_i \boldsymbol{G}_{bi} \dot{\boldsymbol{P}} \\ \boldsymbol{G}_{bi} = \begin{bmatrix} \boldsymbol{i} \times \boldsymbol{r}_{bi} & \boldsymbol{j} \times \boldsymbol{r}_{bi} & \boldsymbol{k} \end{bmatrix} \end{cases}, \quad (i = 1,2,3). \tag{40}$$

The matrix equation of Equation (40) can be expressed by Equation (41), where $\dot{q}$ represents the translational speed of the drive rod, $\dot{\boldsymbol{P}}$ represents the translational speed of the rotary platform-b, J_2 represents the velocity Jacobian matrix of 3RPS rotary part. They can be expressed as follows:

$$\dot{q} = J_2 \dot{\boldsymbol{P}}, \tag{41}$$

where $\dot{q}$ and J_2 can be expressed as follows:

$$\dot{q} = \begin{bmatrix} \dot{l}_1 & \dot{l}_2 & \dot{l}_3 \end{bmatrix} \tag{42}$$

$$J_2 = \begin{bmatrix} Q_1 G_{b1} & Q_2 G_{b2} & Q_3 G_{b3} \end{bmatrix}^T \tag{43}$$

By taking the derivation with respect to time for the two sides of Equation (40), and through mathematical calculation, the acceleration of the drive rod $\ddot{l}_i$ can be obtained as Equation (44), where A_{bi} represents the acceleration of the hinge point b_i of rotary platform-b:

$$\ddot{l}_i = Q_i \cdot A_{bi} + \frac{v_{bi} \cdot v_{bi} - \dot{l}_i^2}{l_i}, (i = 1,2,3). \tag{44}$$

The acceleration of the hinge point A_{bi} can also be expressed by Equation (45), where $\varepsilon = \begin{pmatrix} \varepsilon_x & \varepsilon_y & \varepsilon_z \end{pmatrix}^T$ represents the angular acceleration of rotary platform-b and $A = \begin{pmatrix} A_x & A_y & A_z \end{pmatrix}^T$ represents the acceleration of the center of the rotary platform-b:

$$A_{bi} = A + \varepsilon \times r_{bi} + \omega \times (\omega \times r_{bi}), (i = 1,2,3). \tag{45}$$

Due to $A_x = A_y = \varepsilon_z = 0$ in this mechanism, the acceleration A_{bi} of the hinge point b_i can be expressed as Equation (46), where $\ddot{P}$ is the acceleration of the rotary platform-b, H_{bi} can be expressed as Equation (47). Through mathematical calculation, the acceleration $\ddot{l}_i$ can be achieved, as Equation (48):

$$\begin{cases} A_{bi} = G_{bi}\ddot{P} + \dot{P}^T H_{bi} \dot{P} \\ \ddot{P} = \begin{bmatrix} \varepsilon_x & \varepsilon_y & A_z \end{bmatrix}^T \end{cases}, \quad (i = 1,2,3) \tag{46}$$

$$H_{bi} = \begin{bmatrix} i \times (i \times r_{bi}) & j \times (i \times r_{bi}) & 0 \\ i \times (j \times r_{bi}) & j \times (j \times r_{bi}) & 0 \\ 0 & 0 & 0 \end{bmatrix} \tag{47}$$

$$\ddot{l}_i = Q_i^T G_{bi} \ddot{P} + \dot{P}^T \left[Q_i^T * H_{bi} + \frac{1}{l_i} \left(G_{bi}^T G_{bi} - G_{bi} Q_i Q_i^T G_{bi} \right) \right] \dot{P} (i = 1,2,3). \tag{48}$$

The matrix equation of Equation (48) can be expressed by Equation (49), and $\ddot{q}$ represents the acceleration of the drive rod, $\ddot{P}$ represents the acceleration of the rotary platform-b, J_2 represents the velocity Jacobian matrix of 3RPS rotary part, and K_2 represents the second order kinematic influence coefficient of 3RPS rotary part. They can be expressed as follows:

$$\begin{cases} \ddot{q} = J_2 \ddot{P} + \dot{P}^T K_2 \dot{P} \\ \ddot{q} = \begin{bmatrix} \ddot{l}_1 & \ddot{l}_2 & \ddot{l}_3 \end{bmatrix}^T \\ K_{2m:n} = \{ [U_1]_{m:n} \quad [U_2]_{m:n} \quad [U_3]_{m:n} \} \\ [U_i] = Q_i^T * H_{bi} + \frac{1}{l_i} \left\{ G_{bi}^T G_{bi} - G_{bi} Q_i Q_i^T G_{bi} \right\} \end{cases}. \tag{49}$$

As above, the inverse velocity model of the 3RPS rotary part is expressed by Equation (41), and the inverse acceleration model is expressed by Equation (49).

4. Analysis of Numerical Simulation Result

The geometric parameters of the 3PUU-3RPS mobile die is shown in Table 2. R represents the radius of the based platform-a, r represents the radius of the moving platform-a, s represents the radius of the rotary platform-b, l represents the length of the fixed-length rod, and α represents the angle of inclination of the screw in 3PUU moving part.

Table 2. The geometric parameters of 3PUU-3RPS mobile die.

R/m	r/m	s/m	l/m	α /rad
0.5	0.3	0.2	0.35	0.7

4.1. Kinematics Simulation and Calculation Based on ADAMS and MATLAB

In this section, the kinematics simulation was accomplished in ADAMS (Automatic Dynamic Analysis of Mechanical Systems, Newport Beach, CA, USA). At the same time, combined with MATLAB software (Matrix&Laboratory, Dubai, U.A.E), some simulation results and the relevant theoretical analysis results were compared to verify the reliability of the proposed analytic model.

Kinematics simulation is the process of finding the inverse model, that is, giving the output, and then solving the displacement, velocity, and acceleration of each branch. First, a 3D model was built through SolidWorks (Massachusetts, USA). Then, the model was imported into ADAMS, and the constraint of the kinematic pair was added. Finally, a driving force was applied to the end-effector.

Taking the fixed coordinate system $o-xyz$ as the reference system, the motion parameters of the moving platform-a of the 3UPU parallel mechanism can be shown as Equation (50), and the motion parameters of the rotary platform-b of the 3RPS parallel mechanism is as Equation (51). Figures 10 and 11 compare the simulation results with analysis results by the displacement curve, velocity curve, and acceleration curve of each branched chain. The analysis model was validated by the comparison of the simulation results with the calculation results.

$$\begin{cases} x = 0.35 \times \sin(5 \times t) \\ y = 0.35 \times \cos(5 \times t) \\ z = 0 \end{cases} \tag{50}$$

$$\begin{cases} \phi = 0.5 \times \cos(5 \times t) \\ \theta = 0.5 \times \sin(5 \times t) \\ z = 0 \end{cases}, \tag{51}$$

where x, y, and z represent the move function along x-axis, y-axis, and z-axis, respectively and ϕ and θ represent the rotational functions along x-axis and y-axis, respectively.

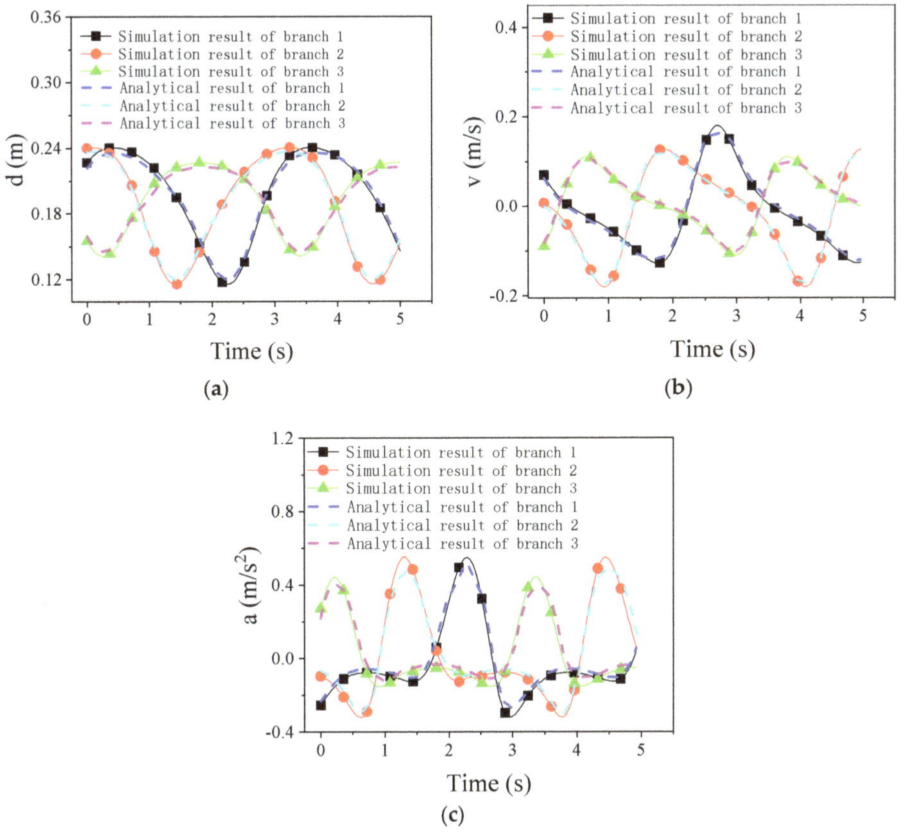

Figure 10. Displacement, velocity, and acceleration curves in the 3PUU moving part. (**a**) Displacement curves of each branched chain in 3PUU moving part; (**b**) velocity curves of each branched chain in 3PUU moving part; (**c**) acceleration curves of each branched chain in the 3PUU moving part.

4.2. Static Stiffness Analysis

During the forming process, the hybrid mechanism normally results in the stress concentration and poor structural stiffness easily. Furthermore, it may affect the dynamic property and positioning accuracy of the mechanism, and even cause mechanical damage. Thus, it is necessary to carry out the static analysis to verify the stress distribution and structural deformation of each component and joint, and also improve and optimize the structural configuration and the rationality.

In this section, finite element simulation software ABAQUS (Paris, France) is used to analyze the stiffness of the whole structure and its branch chains. The geometric models were created and assembled in Solidworks, and then imported into ABAQUS. The FE model is shown in Figure 12 which includes the based platform, moving platform, rotary platform, PUU structure, and RPS structure. The based platform, moving platform, and rotary platform were defined as rigid bodies. The PUU and RPS structures were defined as deformable bodies which were given the material properties of tool steel Cr12Mo1V1, and its specific parameters are given in Table 3. These deformable bodies were meshed by using C3D4 solid elements which were four-node linear tetrahedron elements, and the minimum element size was 1 mm. The interaction was set to general contact, where the slider in the 3PUU structure was tied to the lead screw on the based platform. Because moving parts of the structure are lubricated in practice, the coefficient of friction was identified as 0.1. The analysis step was identified as static general. In addition, the base platform was completely fixed, the mobility and rotation degrees of the 3PUU and 3RPS branch chains were preserved. Through the bending test of an aluminum alloy tube with diameter of 30 mm and wall thickness of 2 mm by using a universal testing machine, it is found that the bending force of the tube is 5 kN [29]. In order to verify the stability and rationality of the designed mechanism, 15 kN force was used to simulate the mechanism. So a static load of 15 kN was applied to the center of the rotary platform-b along four directions as shown in Figure 13. The stress distribution is also shown in Figure 13. The static analysis of each branch chain was also carried out. In addition, the static loads of 8 kN were applied, and the stress distribution was as shown in Figure 14.

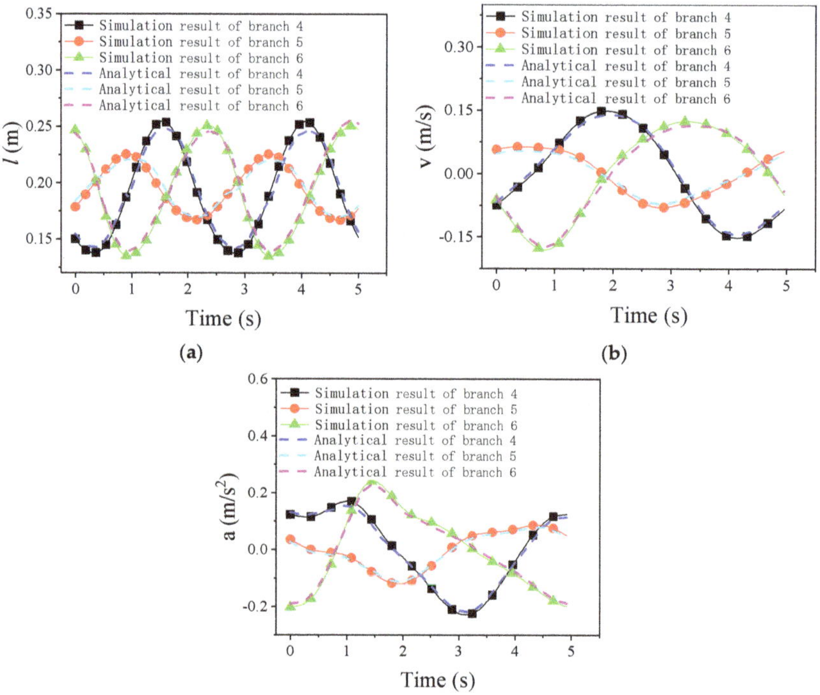

Figure 11. Length, velocity and acceleration curves in the 3RPS rotary part. (**a**) Length curve of each branched chain in 3RPS rotary part; (**b**) velocity curves of each branched chain in the 3RPS rotary part; (**c**) acceleration curves of each branched chain in the 3RPS rotary part.

Table 3. Material parameters of tool steel Cr12Mo1V1.

Material Property	Value
Poisson ratio	0.269
Young's modulus (MPa)	206,000
Yield strength (MPa)	430
Tensile strength (MPa)	780
Density (kg/cm^3)	7.89

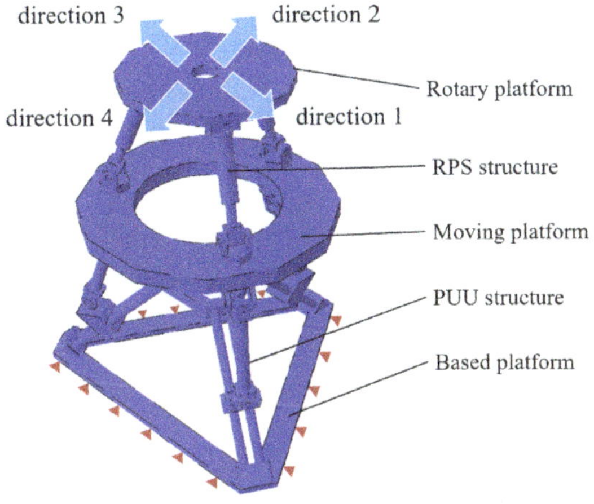

Figure 12. FE modeling of 3PUU-3PRS.

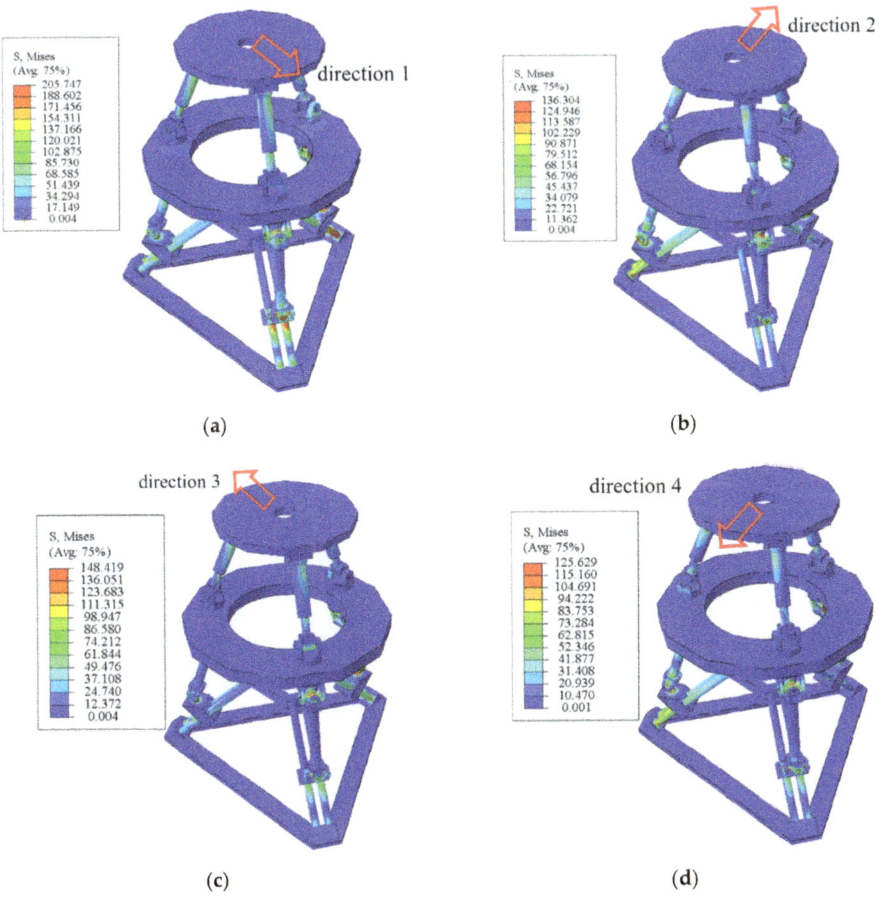

Figure 13. Stress distribution along different directions. (**a**) Stress distribution along the direction-1; (**b**) stress distribution along the direction-2; (**c**) stress distribution along the direction-3; (**d**) stress distribution along the direction-4.

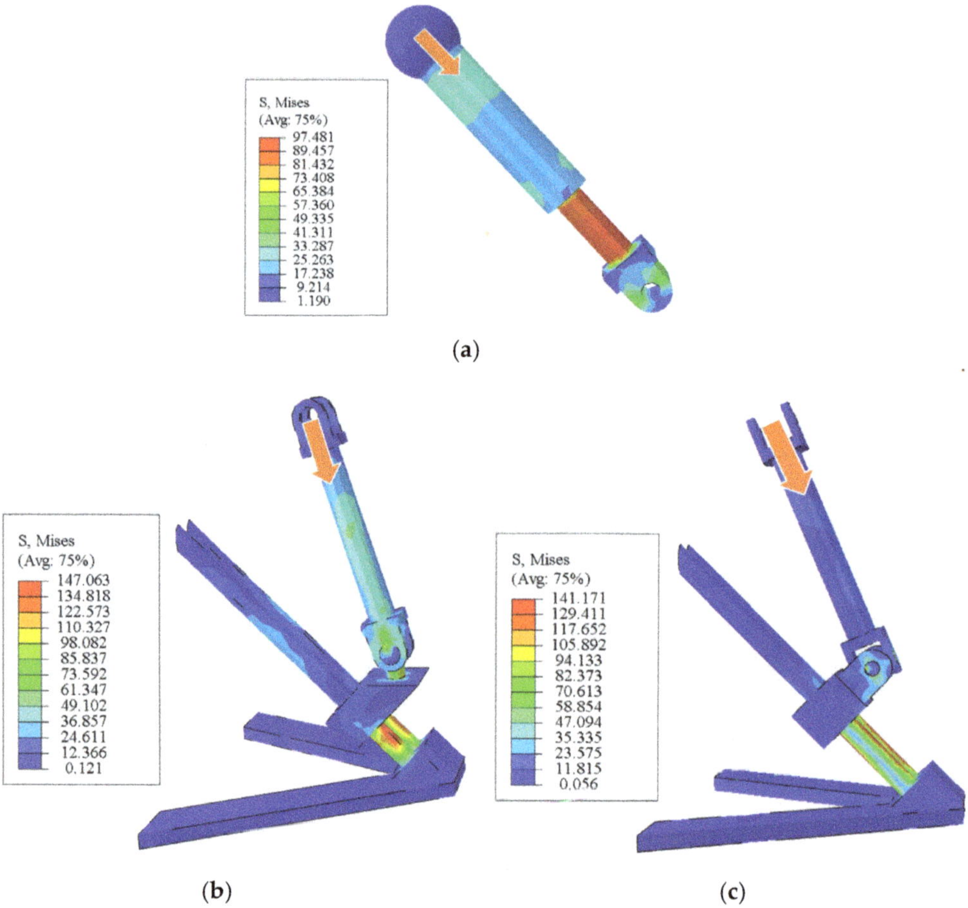

Figure 14. Stress distribution of branch chain. (**a**) Stress distribution of branch chain RPS; (**b**) stress distribution of PUU branch chain-1; (**c**) stress distribution of PUU branch chain-2.

By investigating the whole stress distribution, the maximum stress can be found located at the U-joint of the PUU structure. On the branch chain stress distribution, the maximum stress is located at the lower link of the branch chain. The maximum stress value is 205 MPa, which is far less than the yield strength of tool steel Cr12Mo1V1. It is proved that static stiffness can meet the requirements and the structure design is reasonable.

4.3. Analysis of Workspace

The main factors on the working space of the parallel mechanism are the length of the rod, the angle of rotation of the kinematic pair and the interference of the connecting rods [30]. The limitation about the length of the rod, the translational distance of each slider, and the length of each drive rod must satisfy Equation (52) for this 3PUU-3RPS mobile die:

$$\begin{cases} 0 \le d_i \le d_{\max} \\ 0 \le l_i \le l_{\max} \end{cases}, \qquad (i = 1,\ 2,\ 3). \tag{52}$$

The limitation about the angle of rotation of the kinematic pair of the 3PUU-3RPS mobile die must satisfy Equation (53). Through Equation (12), the angles of rotation ϕ_{Mi} and

θ_{Mi} can be achieved by Equation (54). By geometrical relationship, the angles of rotation ϕ_{ai} and θ_{ai} can be achieved by Equation (55):

$$\begin{cases} 0 \leq \phi_{ai} \leq \phi_{Max} \\ 0 \leq \theta_{ai} \leq \theta_{Max} \\ 0 \leq \phi_{Mi} \leq \phi_{Max} \\ 0 \leq \theta_{Mi} \leq \theta_{Max} \\ 0 \leq \theta_{Bi} \leq \theta_{BMax} \\ 0 \leq \theta_{bi} \leq \theta_{bMax} \end{cases}, (i = 1,2,3) \tag{53}$$

$$\begin{cases} \phi_{Mi} = \arccos\frac{l_{iz}}{l} \\ \theta_{Mi} = \arcsin\frac{l_{ix}}{l_{iz}} \end{cases} \tag{54}$$

$$\begin{cases} \phi_{ai} = \phi_{Mi} \\ \theta_{ai} = \theta_{Mi} \end{cases}. \tag{55}$$

The angles of rotation θ_{Bi} and θ_{bi} in the 3RPS rotary part can be calculated by Equation (56), where vector $\boldsymbol{n}_{bi}$ represents the position vector of the base of spherical (S) joints:

$$\begin{cases} \theta_{bi} = \arccos\frac{\boldsymbol{l}_i \cdot ({}^{o}_{o'}\boldsymbol{T}\boldsymbol{n}_{bi})}{|\boldsymbol{l}_i|} \\ \theta_{Bi} = \arccos\frac{\boldsymbol{l}_i\boldsymbol{B}_i}{|\boldsymbol{l}_i||\boldsymbol{B}_i|} \end{cases}, (i = 1,2,3). \tag{56}$$

For the 3PUU moving part, the maximum rotary angle of U joints is $\pm\pi/4$, and the maximum translational distance of the slider is $d_{\max} = 0.7$ m. For the 3RPS rotary part, the maximum rotary angle of spherical (S) joints is $\pm\pi/4$, the maximum translational distance of slider is $l_{\max} = 0.5$ m. Through the polar coordinate searching method, the workplace of this 3PUU-3RPS mobile die is calculated by using MATLAB and is shown in Figure 15. It is found that the range of the mechanism is 0.3 m in the Z direction, and the range of motion in the X and Y directions is ±0.2 m. Currently, the extreme structure of bending tubes is characterized by small bending radii ($R_d/D < 3$) [31]. According to the results of the previous FE simulation results by Equation (1), it is found that the bending die needs to move 55 mm when the minimum bending radii is 90 mm [32]. Therefore, Equation (1) describes the motion of tube bending that can be performed within the workspace presented in Figure 15.

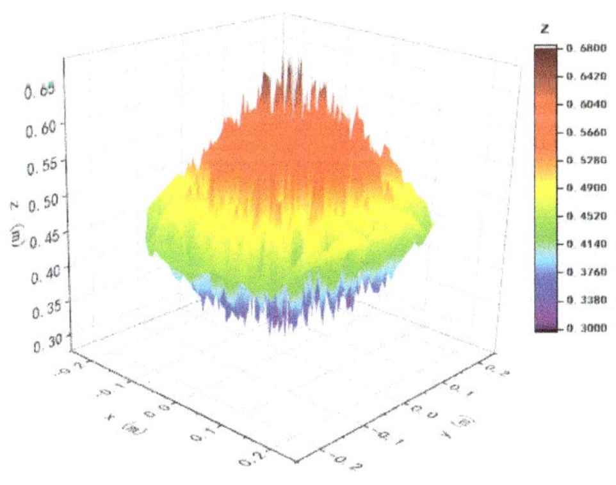

Figure 15. Workplace of 3PUU-3RPS mobile die.

5. Conclusions

In this research, a 3PUU-3RPS hybrid mechanism for the free-bending forming of tubes was presented. The rationality and reliability of the hybrid mechanism are verified by theoretical analysis and simulation. The mechanism can better serve the needs of production activities. The main results are as follows:

1. The degree of freedom of the proposed mechanism was analyzed based on the screw theory. It can meet the requirements of free-bending forming by combining the 3PUU moving part and 3RPS rotary part.
2. In aspects of kinematics, the inverse position model, inverse velocity model and inverse acceleration model were established based on structural characteristics of the 3PUU–3RPS mobile die. Furthermore, the kinematics simulation and static stiffness were accomplished. The corresponding numerical simulation and the relevant theoretical analysis were also conducted to verify the reliability and feasibility of the mechanism.
3. Based on the inverse position model, the working space of this mechanism and the relationship between end-effector and actuator was also presented.

To conclude, this study provides an in-depth understanding of the bending process of tubes by theoretical and simulation analyses. Moreover, the designed 6-DOF hybrid mechanism is appropriate for the spatial variable curvature tube-bending forming, which provides theoretical guidance for building a device to accomplish the free- bending forming of tubes. The next step, the experimental validation, will be carried out in future work.

Author Contributions: Simulation, formal analysis, investigation, writing and reviewing, S.Z. and Y.L.; Funding acquisition, Supervision, Writing—review & editing, Z.Y.; Conceptualization, Z.Z.; Methodology and supervision, L.S.; Validation and supervision, B.Y.; Resources, J.G. All authors have read and agreed to the published version of the manuscript.

Funding: This work was supported by the National Natural Science Foundation of China (No. 51975327 and 52175337).

Institutional Review Board Statement: Not applicable.

Informed Consent Statement: Not applicable.

Data Availability Statement: Not applicable.

Conflicts of Interest: The authors declare no conflict of interest.

References

1. Song, F.; Yang, H.; Li, H.; Zhan, M.; Li, G. Springback prediction of thick-walled high-strength titanium tube bending. *Chin. J. Aeronaut.* **2013**, *26*, 1336–1345. [CrossRef]
2. Wu, J.; Zhang, Z.; Shang, Q.; Li, F.; Hui, Y.; Fan, H. A method for investigating the springback behavior of 3D tubes. *Int. J. Mech. Sci.* **2017**, *131*, 191–204. [CrossRef]
3. Zhou, W.; Shao, Z.; Yu, J.; Lin, J. Advances and Trends in Forming Curved Extrusion Profiles. *Materials* **2021**, *14*, 1603. [CrossRef] [PubMed]
4. Li, H.; Ma, J.; Liu, B.Y.; Gu, R.J.; Li, G.J. An insight into neutral layer shifting in tube bending. *Int. J. Mach. Tools Manuf.* **2018**, *126*, 51–70. [CrossRef]
5. Wu, J.; Zhang, Z. An improved procedure for manufacture of 3D tubes with springback concerned in flexible bending process. *Chin. J. Aeronaut.* **2021**, *34*, 267–276. [CrossRef]
6. Ma, J.; Li, H.; Fu, M.W. Modelling of Springback in Tube Bending: A Generalized Analytical Approach. *Int. J. Mech. Sci.* **2021**, *204*, 106516. [CrossRef]
7. Yang, H.; Li, H.; Ma, J.; Li, G.; Huang, D. Breaking bending limit of difficult-to-form titanium tubes by differential heating-based reconstruction of neutral layer shifting. *Int. J. Mach. Tools Manuf.* **2021**, *166*, 103742. [CrossRef]
8. Zhan, M.; Yang, H.; Huang, L.; Gu, R. Springback analysis of numerical control bending of thin-walled tube using numerical-analytic method. *J. Mater. Processing Technol.* **2006**, *177*, 197–201. [CrossRef]
9. Gu, R.J.; Yang, H.; Zhan, M.; Li, H.; Li, H.W. Research on the springback of thin-walled tube NC bending based on the numerical simulation of the whole process. *Comput. Mater. Sci.* **2008**, *42*, 537–549. [CrossRef]
10. Yang, H.; Li, H.; Zhang, Z.; Zhan, M.; Liu, J.; Li, G. Advances and Trends on Tube Bending Forming Technologies. *Chin. J. Aeronaut.* **2012**, *25*, 1–12. [CrossRef]
11. Wagoner, R.H.; Lim, H.; Lee, M.-G. Advanced Issues in springback. *Int. J. Plast.* **2013**, *45*, 3–20. [CrossRef]
12. Ma, J.; Welo, T. Analytical springback assessment in flexible stretch bending of complex shapes. *Int. J. Mach. Tools Manuf.* **2021**, *160*, 103653. [CrossRef]
13. Staupendahl, D.; Chatti, S.; Tekkaya, A.E. Closed-loop control concept for kinematic 3D-profile bending. *AIP Conf. Proc.* **2016**, *1769*, 150002.
14. Yang, D.Y.; Bambach, M.; Cao, J.; Duflou, J.R.; Groche, P.; Kuboki, T.; Sterzing, A.; Tekkaya, A.E.; Lee, C.W. Flexibility in metal forming. *CIRP Ann.* **2018**, *67*, 743–765. [CrossRef]

15. Zhang, Z.K.; Wu, J.J.; Guo, R.C.; Wang, M.Z.; Li, F.F.; Guo, S.C.; Wang, Y.A.; Liu, W.P. A semi-analytical method for the springback prediction of thick-walled 3D tubes. *Mater. Des.* **2016**, *99*, 57–67. [CrossRef]
16. Zhang, S.; Wu, J.; Wang, Q.; Liang, Z. Research on Springback in Tube Bending Process. *Aeronaut. Manuf. Technol.* **2014**, *454*, 45–50.
17. Murata, M.; Kuboki, T. CNC tube forming method for manufacturing flexibly and 3-Dimensionally bent tubes. In *60 Excellent Inventions in Metal Forming*; Tekkaya, A.E., Homberg, W., Brosius, A., Eds.; Springer: Berlin/Heidelberg, Germany, 2015; pp. 363–368.
18. Plettke, R.; Vatter, P.H.; Vipavc, D. Basics of Process Design for 3D freeform bending. *Steel Res. Int.* **2012**, *S*, 307–310.
19. Gantner, P.; Bauer, H.; Harrison, D.K.; de Silva, A.K.M. Free-Bending—A new bending technique in the hydroforming process chain. *J. Mater. Processing Technol.* **2005**, *167*, 302–308. [CrossRef]
20. Chatti, S.; Hermes, M.; Tekkaya, A.E.; Kleiner, M. The new TSS bending process: 3D bending of profiles with arbitrary cross-sections. *CIRP Ann.* **2010**, *59*, 315–318. [CrossRef]
21. Becker, C.; Staupendahl, D.; Hermes, M.; Chatti, S.; Tekkaya, A.E. Incremental Tube Forming and Torque Superposed Spatial Bending—A View on Process Parameters. *Steel Res. Int.* **2012**, *S*, 415–418.
22. Hudovernik, M.; Kosel, F.; Staupendahl, D.; Tekkaya, A.; Kuzman, K. Application of the bending theory on square-hollow sections made from high-strength steel with a changing angle of the bending plane. *J. Mater. Processing Technol.* **2014**, *214*, 2505–2513. [CrossRef]
23. Strano, M.; Colosimo, B.M.; Castillo, E.D. Improved design of a three roll tube bending process under geometrical uncertainties. *AIP Conf. Proc.* **2011**, *1353*, 35–40.
24. Vatter, P.H.; Plettke, R. Process model for the design of bent 3-dimensional free-form geometries for the three-roll-push-bending process. *Procedia CIRP* **2013**, *7*, 240–245. [CrossRef]
25. Goto, H.; Ichiryu, K.; Saito, H.; Ishikura, Y.; Tanaka, Y. Applications with a new 6-DOF bending machine in tube forming processes. In Proceedings of the JFPS International Symposium on Fluid Power, Toyama, Japan, 15–18 September 2008; The Japan Fluid Power System Society: Tokyo, Japan, 2008; pp. 183–188.
26. Goto, H.; Tanaka, Y.; Ichiryu, K. 3D Tube Forming and Applications of a New Bending Machine with Hydraulic Parallel Kinematics. *Int. J. Autom. Technol.* **2012**, *6*, 509–515. [CrossRef]
27. Kawasumi, S.; Takeda, Y.; Matsuura, D. Precise pipe-bending by 3-RPSR parallel mechanism considering springback and clearances at dies. *Trans. Jpn. Soc. Mech. Eng.* **2014**, *80*, 343.
28. Takeda, Y.; Inada, S.; Kawasumi, S.; Matsuura, D.; Hirose, K.; Ichiryu, K. Kinematic design of 3-RPSR parallel mechanism for movable-die drive mechanism of pipe bender. *Rom. J. Tech. Sci. Appl. Mech.* **2013**, *58*, 71–96.
29. Li, Y.; Li, A.; Yue, Z.; Qiu, L.; Badreddine, H.; Gao, J.; Wang, Y. Springback prediction of AL6061 pipe in free bending process based on finite element and analytic methods. *Int. J. Adv. Manuf. Technol.* **2020**, *109*, 1789–1799. [CrossRef]
30. Xu, H.; Li, J.; Chen, Y.; Zhang, R.; Gao, J. Kinematic Modeling of Spherical Parallel Manipulator with Vectored Thrust Function for Underwater Robot Based on Screw Theory. *Robot* **2016**, *38*, 745–753.
31. Li, H.; Yang, H.; Zhang, Z.Y.; Li, G.J.; Liu, N.; Welo, T. Multiple instability-constrained tube bending limits. *J. Mater. Processing Technol.* **2014**, *214*, 445–455. [CrossRef]
32. Li, Y.; Yue, Z.; Min, X.; Gao, J. Simulation and optimization of the free bending process of aluminum alloy 6061 pipe. *Chin. J. Eng.* **2020**, *42*, 769–777.

energies

MDPI

Article

Design and Research on Electro-Hydraulic Drive and Energy Recovery System of the Electric Excavator Boom

Lin Li [1], Tiezhu Zhang [1,*], Kaiwei Wu [1], Liqun Lu [1], Lianhua Lin [2] and Haigang Xu [2]

1 School of Transportation and Vehicle Engineering, Shandong University of Technology, Zibo 255000, China; lilin20082007@163.com (L.L.); sdlgwkw@163.com (K.W.); luliqunustb@163.com (L.L.)
2 Shandong Shifeng (Group) Co., Ltd., Liaocheng 252800, China; sfdwllc@163.com (L.L.); 13869553168@139.com (H.X.)
* Correspondence: zhangtz@sdut.edu.cn

Abstract: The hydraulic accumulator has the advantages of high power density, fast response, stable operation and high cost performance. However, compared with the electric energy storage method, the hydraulic accumulator has low energy density and large pressure fluctuation while absorbing and discharging energy, which severely limits its application in hydraulic excavators. To improve the potential energy loss of the boom during the lowering process, an electro-hydraulic drive and energy recovery system for excavator booms (EHDR-EEB) based on a battery and accumulator is proposed. As a result, a simulation model of the electro-hydraulic drive and energy management strategy of a 1.6 t pure electric hydraulic excavator is built to investigate the energy regeneration and utilization. The simulation outcomes show that the potential energy recovery rate is as high as 92%. This research on EHDR-EEB makes a significant contribution to the economic improvement of electric hydraulic excavators.

Keywords: electric excavator boom; electro-hydraulic drive and energy recovery system; energy regeneration; energy management strategy

Citation: Li, L.; Zhang, T.; Wu, K.; Lu, L.; Lin, L.; Xu, H. Design and Research on Electro-Hydraulic Drive and Energy Recovery System of the Electric Excavator Boom. *Energies* **2022**, *15*, 4757. https://doi.org/10.3390/en15134757

Academic Editors: Kan Liu and Wei Hu

Received: 17 May 2022
Accepted: 27 June 2022
Published: 28 June 2022

1. Introduction

Currently, the energy crisis and environmental pollution have become increasingly urgent global problems [1]. Construction machinery accounts for about 60% of the carbon dioxide emissions produced by all vehicles [2]. Excavators account for about 80% of the total construction machinery. Moreover, the energy utilization rate of traditional hydraulic excavators is low, and only 20% of the energy of the engine is used by the load [3]. Thus, it is imperative to save energy and develop new energy systems for excavators. Driven by the energy demand, pure electric hydraulic excavators with batteries as energy carriers are a promising solution for energy saving [4].

In terms of the energy potential of excavators, it is more straightforward to recover more energy, realize higher efficiency and lower loss when considering the larger load energy of the boom. Therefore, the recovery and utilization of the electro-hydraulic excavator boom's potential energy can be taken as the research starting point, with the aim of achieving significant energy savings and high efficiency.

1.1. Literature Review

The hydraulic excavator's boom can recover high energy during the falling process. Nevertheless, the existing pure electric excavator lacks the energy recovery link, and a low energy utilization rate still exists during the working process. Research data indicates that the recovery methods can be divided into electric energy storage and accumulator energy storage [5].

The accumulator is an important energy storage element, which has the characteristics of high power-density, low installation cost, and can store and release braking energy in a short time. Caterpillar [6] developed an accumulator-based energy recovery system that has been successfully used on a 50 t hydraulic excavator. The energy consumption is reduced by 37% when the boom rises through the variable pump. Fu et al. [7] proposed a boom hydraulic potential energy recovery system with an accumulator as the energy storage element. The results showed that the potential energy recovery rate of the boom was 22.6%. Lin et al. [8,9] constructed a two-stage idling speed control system based on a hydraulic accumulator, which reduced the energy consumption when the idling speed control was turned off, and further improved the control performance of the brake. Casoli [10] used a hydraulic accumulator as a storage device to recover the potential energy of the boom. The recovered energy was utilized as the mechanical force to operate the motor, thereby reducing the torque required by the internal combustion engine. Tan et al. [11] proposed an energy recovery system with an accumulator to reduce overflow loss in excavator rotary braking. The results showed that the energy recovery of the system reached 53.1% when the excavator completed a working cycle of rotation.

Additionally, the battery is a significant energy storage unit, and scholars have carried out much research utilizing the high energy storage density of batteries. Hitachi Construction machinery launched the first prototype using an electric method to recover the boom potential energy. Yi et al. [12] presented a dynamic programming-based energy management strategy for environmentally friendly electric excavators using supercapacitors. Wang et al. [13] adopted a main valve differential pressure control strategy to improve the operation characteristics of the electric energy recovery system, and the recovery efficiency was increased to 40%. Xia et al. [14] designed an integrated drive and energy recuperation system based on a three-chamber hydraulic cylinder. It reduced the boom energy consumption by 50.1% and power supply peak power by 64.9%. The boom potential energy recovery system designed by Pei [15] combined the energy recovery motor with the generator to ensure that the generator was always in the working state of efficient power generation.

Based on the above research and considering the potential for multi-source hybrid excavator and energy recovery, in this study, a proposed electro-hydraulic drive and energy recovery system of the electric excavator boom is investigated.

1.2. Prevalent Problems

Scholars have conducted much research into energy saving through hydraulic excavator boom energy recovery systems, but these research results are limited to only one kind of excavator power source. However, the following problems generally exist in hydraulic excavators using electric energy storage to recover energy [16–18]: high energy density but low power density; low power density; large weight and volume; and the inability to charge and discharge in a short time. All these problems greatly limit the recovery and utilization of batteries. Furthermore, improper matching of power between the source and load will lead to energy loss. Additionally, the power source cannot quickly adapt to real fluctuations in the load, which results in a low energy utilization rate.

A boom system with accumulator energy storage has the following advantages: relatively few energy-conversion links, high recovery efficiency, and fast charging and discharging speed. Therefore, the accumulator can also be used as an auxiliary power source to reduce installed capacity [19]. However, energy loss is caused by throttling and overflow of the hydraulic system, and it has a low energy density and large pressure fluctuation while absorbing and discharging energy.

Facing severe energy competition, it is urgent to find a suitable energy recovery system for electro-hydraulic excavators to improve energy recovery and efficiency, relieve battery pressure and prolong the life of power system.

1.3. Challenges and the Future

Recently, there are increasing numbers of studies into the energy recovery of the boom in electro-hydraulic excavators, but excavators equipped with both electric storage and accumulator energy storage still face some challenges that cannot be ignored. As an energy storage component, the accumulator plays an important role in low processing cost and high power density, while the battery has the advantages of low power density and high energy storage density. However, the above research is based only on the electrical or hydraulic energy storage mode in unilateral terms. Realizing the effective combination of the two energy storage methods will be the future research direction.

Some researchers have carried out research on the potential energy recovery of electro-hydraulic boom. Gong et al. [20] proposed an electro-hydraulic energy saving system, which can store hydraulic energy and electrical energy. The experimental verification showed that the new system reduced the energy consumption by 17.6% compared with the original system. Chen et al. [21] proposed an electro-hydraulic recovery and utilization system for braking energy of a large inertia slewing mechanism. It combined a hydraulic accumulator and a super capacitor to store energy, and controlled the generator torque to match the pressure of the accumulator, so as to maximize the braking energy recovery efficiency and ensure the slewing stability. Yao et al. [22] used an electro-hydraulic coordinated energy recovery system of a hydraulic excavator to realize the energy reuse when the boom and the slewing mechanism act together, which allowed the boom to lower with an energy recovery rate of 27.23%. Lin et al. [23] proposed two methods for regenerating potential energy of hybrid hydraulic excavators. When the boom was lowered, 41% of the total potential energy could be regenerated, while the recovery efficiency in the motor-generator system was only about 17%.

In the above studies on electro-hydraulic energy recovery and storage, although the research on electro-hydraulic recovery and utilization has improved the economy of excavators, the effects are not significant. In current pure electric-hydraulic excavators, the main power source is the battery, and the working device of the load is the hydraulic system. How to make full use of the advantages of electric energy storage and hydraulic energy storage, and give full play to battery and accumulator in combination still needs to be considered. Therefore, this paper proposes an electro-hydraulic drive and energy recovery system of the electric excavator boom (EHDR-EEB) combining electric energy storage and accumulator energy storage.

1.4. Contribution of This Work

To address the current problems, this article attempts to make the following three significant contributions and improvements to the recent research:

- This paper proposes a new electro-hydraulic drive and energy recovery system for the electric excavator boom, which realizes the mutual conversion of mechanical, electric and hydraulic energy.
- The energy flow of the system, the relationship between the required power flow, and torque is analyzed in order to realize real-time control when the boom is lifting and falling.
- A regular energy management strategy under typical working conditions is proposed, which greatly improves energy utilization and potential energy recovery.

1.5. Organization of This Paper

This paper is organized as follows: Section 2 describes the structure and principle of the EHDR-EEB. Section 3 builds the theoretical calculation and analysis of the system and designs a rule-based energy management strategy. Section 4 performs a simulation analysis to verify the reasonableness of the management strategy. Section 5 summarizes the research of this paper.

2. Design of the EHDR-EEB

2.1. Structure of the EHDR-EEB

A specific electro-hydraulic drive and energy recovery system of an electric excavator boom is shown in Figure 1. The system includes the battery, motor, variable hydraulic pump/motor (VPM), high-pressure accumulator (HPA), low-pressure accumulator (LPA), hydraulic cylinder, pressure cylinder, hydraulic valves(HV) and other components. The VPM is installed between the motor and the LPA to complete the conversion between hydraulic and mechanical energy. The HPA is connected with the VPM, pressure cylinder and hydraulic cylinder through multiple hydraulic valves. The HPA can supply energy to the hydraulic cylinder to realize its lifting function.

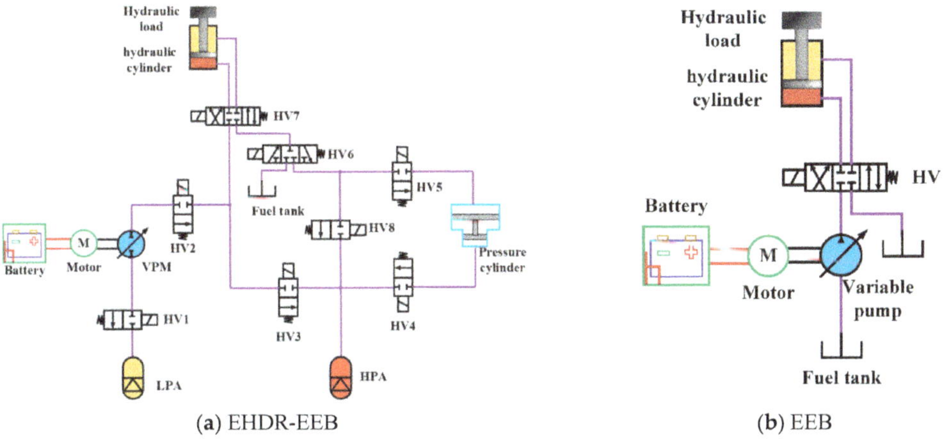

Figure 1. Structure diagram of the EHDR-EEB and EEB: (**a**) EHDR-EEB; (**b**) EEB.

When the boom is lifting, the battery can be used as the power source alone. The electrical energy is converted into mechanical energy by the motor to drive the VPM. Then, the VPM is used as a pump, and the oil is sucked from the LPA and sent to the rodless cavity of the hydraulic cylinder by pressurization, so as to overcome the load and complete lifting. Simultaneously, HPA can be connected directly to the rodless cavity of the hydraulic cylinder, which greatly saves battery power consumption. The gravitational potential energy of the load can be converted into electrical energy and stored in the battery when falling is in progress. It can also recover the hydraulic potential energy without conversion by the HPA. The main function of the LPA is to replenish oil for the HPA. If the LPA pressure is low and the HPA pressure is sufficient, the HPA recharges the LPA. When the HPA pressure is insufficient, the battery can use the motor and VPM to directly charge the HPA to meet the subsequent power requirements of the vehicle.

2.2. Principle and Modes of the EHDR-EEB

According to the system structure and the working characteristics of the hydraulic excavator, the system working mode is divided into five modes (Figure 2):

(1) Electrodynamic drive (ED). In this mode, HV1 and HV2 are open, HV7 is in the right position, and HV6 is in the left position. When the motor works, electrical energy is transformed into hydraulic energy through the battery, motor, and VPM. In this case, the VPM is used as a pump, and the low-pressure oil is sucked from the LPA, then transported to the rodless cavity of the hydraulic cylinder through HV2 and HV7. Finally, the hydraulic cylinder piston overcomes the load gravity under the action of pressure and lifts upward. The oil in the rod cavity of the hydraulic cylinder flows back to the fuel tank through HV7 and HV6.

(2) Hydrodynamic drive (HD). At this time, HV3 is open, HV7 is in the right state, and HV6 is in the left state. The hydraulic oil is delivered from HPA to the rodless cavity of the hydraulic cylinder through HV3 and HV7. The hydraulic cylinder piston lifts upward as described above. The oil in the rod cavity of the hydraulic cylinder flows back to the tank through HV7 and HV6.

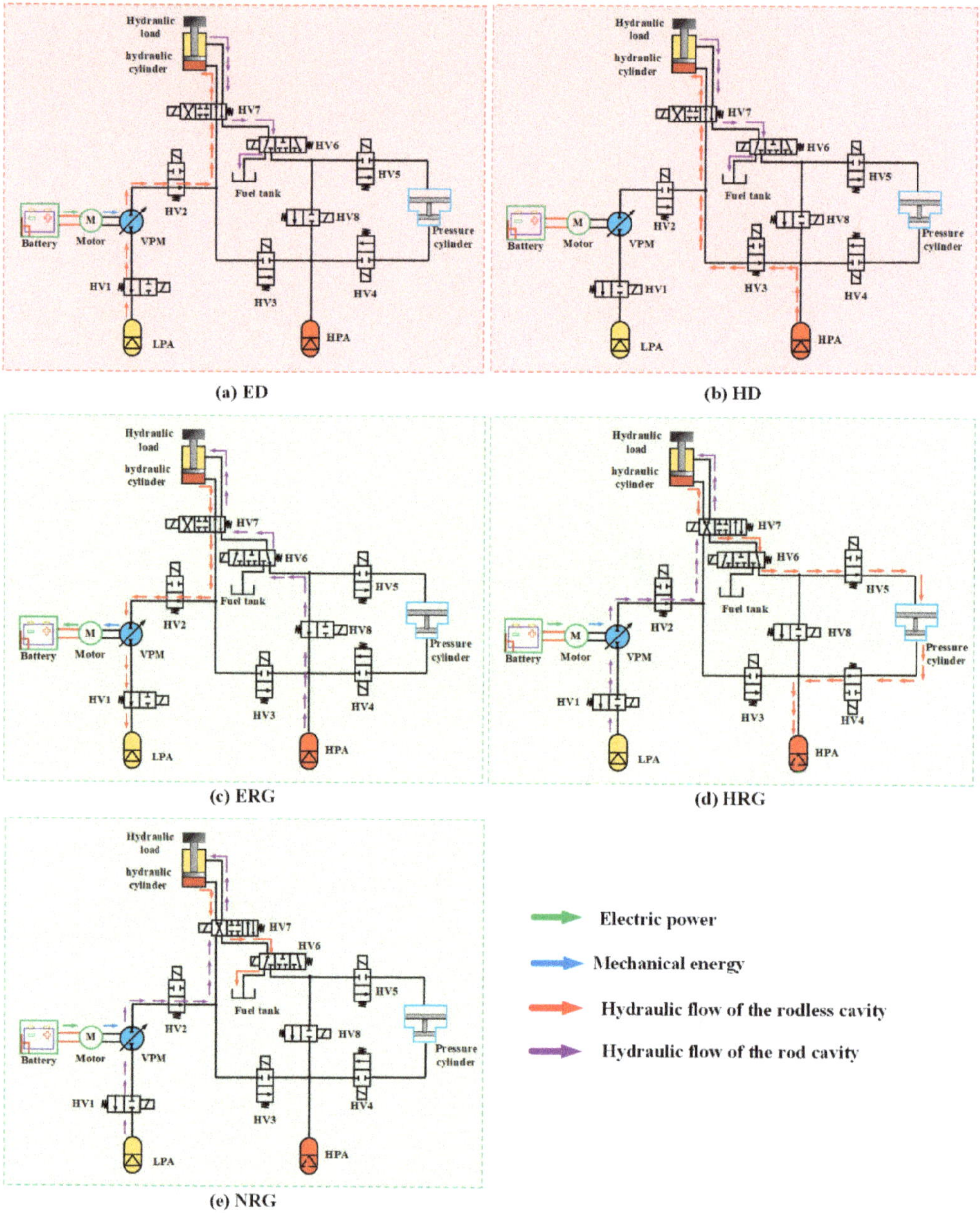

Figure 2. Working principle of the EHDR-EEB: (**a**) ED; (**b**) HD; (**c**) ERG; (**d**) HRG; (**e**) NRG.

(3) Electric regeneration (ERG). During the load falling process, the battery can recover electrical energy relying on the potential energy of the boom. HV1, HV2, HV8 are in the open state, HV7 is in the right position, and HV6 is in the right position. The piston in the hydraulic cylinder descends under the action of boom gravity and the HPA, and pushes the oil in the rodless cavity into the LPA through HV7 and HV2. Meanwhile, the VPM is used as a motor, which converts hydraulic energy into mechanical energy and drives the motor to work. Afterwards, the motor is used as a

generator to output negative torque, convert mechanical energy into electrical energy and store it in the battery.

(4) Hydraulic regeneration (HRG). When the boom falls, the HPA can recover potential hydraulic energy under the gravity of the boom. Now, HV1, HV2, HV4, and HV5 are open, HV7 is in the left state, and HV6 is in the right state. The piston in the hydraulic cylinder descends with the help of the boom gravity and the VPM, and pushes the oil in the rodless cavity into the HPA through HV7, HV6, HV5, the pressure cylinder and HV4. In this case, the hydraulic energy directly charges the HPA without conversion.

(5) No regeneration (NRG). If the battery SOC and the HPA pressure are greater than the set threshold in the falling process, HV2 is turned on, and HV7 and HV6 are in the left position. The piston of the hydraulic cylinder descends. The hydraulic oil enters the rod cavity of the hydraulic cylinder through the LPA, VPM and HV2, while the oil in the rodless cavity enters the fuel tank through HV7 and HV6.

3. Energy Management Strategy Design

This section establishes mathematical modeling of working components: the battery, motor, accumulator and VPM. Subsequently, for excavators equipped with EHDR-EEB, the precise division of the modes and rational design of energy management strategy play an important role in this paper. On this basis, a rule-based energy management strategy is designed, and each operating mode is analyzed.

3.1. Mathematical Modeling

3.1.1. Battery Model

For electric excavators, the battery is the main energy source for the vehicle. The maximum power of the designed battery must meet the maximum power demand of the motor [24]. Meanwhile, the capacity of the battery must meet the needs of the excavator's work equipment and driving needs. Lithium-ion batteries have high specific energy, high specific power density, a long cycle life and superior comprehensive performance [25]. Therefore, this paper adopts lithium-ion battery as the battery.

(1) Battery energy

The battery energy is a measure of the battery's external working ability (kWh), and its stored energy can be expressed as:

$$E_p = V_{bar}C_r/1000 \tag{1}$$

where V_{bar} is the battery rated terminal voltage, (V), and C_r is the battery rated capacity, (Ah).

(2) Battery power

In order to meet the power demand of the excavator, the peak power of the battery needs to meet the peak power of the motor.

$$U_t = E_o - Ri \tag{2}$$

U_t is the battery terminal voltage (V), E_o is the battery electromotive force (V), R is the battery internal resistance (Ω), and i is the discharge current (A).

Then the battery discharge power P_t is:

$$P_t = E_o i - Ri^2 \tag{3}$$

Then the battery theoretical maximum value of the instantaneous power P_{max} can be expressed as:

$$P_{max} = \frac{E_o^2}{4R} \tag{4}$$

Under actual discharge conditions, the following formula is generally used as the power constraint condition of the battery in engineering applications:

$$P_{\max} \geq \frac{P_{\mathrm{mp}}}{\eta} \tag{5}$$

where P_{mp} is the motor peak power (kW), and η is the motor efficiency.

SOC represents the charge state of the battery and is an indicator to measure the degree of battery discharge. The *SOC* state equation is as follows:

$$SOC(t) = SOC_0 - \frac{\int_0^t i(t)dt}{Q_0} \tag{6}$$

where SOC_0 is the initial *SOC* of the battery (%), and Q_0 is the total battery capacity of the battery (mAh).

3.1.2. Motor Model

A permanent magnet synchronous motor is selected for the system. The motor can provide additional peak drive torque or absorb excess boom potential energy. The output mechanical energy of the motor provides rotational kinetic energy for the hydraulic pump/motor. Therefore, the rated power of the motor is the same as that of the hydraulic pump/motor. The motor is coaxially connected to the hydraulic pump/motor, so the speed is consistent. The external load of the boom acts on the rod cavity of the hydraulic cylinder. In the working process of the boom, the rated torque of the motor can be approximated to the average torque of the external load:

$$P_{\mathrm{m}} = \frac{2\pi n_{\mathrm{m}}}{60} \frac{\int_{t_1}^{t_2} T_{\mathrm{load-max}} dt}{t_2 - t_1} \tag{7}$$

where P_{m} is the motor rated power, n_{m} is the motor rated speed, and $T_{\text{load-max}}$ is the maximum value of external load demand torque.

3.1.3. Accumulator Model

The accumulator is a critical component of the hybrid power system. Its parameters include the working pressure, inflation pressure, and accumulator volume, which determine the amount of energy recovery and auxiliary driving energy of the hydraulic regeneration system.

(1) Working pressure of accumulator

The pressure release and energy storage process of the accumulator can be regarded as an adiabatic process. The gas state equation must be satisfied during the storage and release of energy:

$$p_0 V_0^n = p_1 V_1^n = p_2 V_2^n \tag{8}$$

where p_0 is the initial working pressure of the accumulator (Pa), V_0 is the initial volume of the accumulator (L), V_1 is the gas volume when the accumulator pressure is p_1 (L), V_2 is the gas volume when the accumulator pressure is p_2 (L), and n is the gas variability index, assumed to be 1 during the isothermal process, and 1.4 during the adiabatic process.

The higher the maximum working pressure of the accumulator, the more energy can be recovered by the regenerative energy. To ensure as much energy recovery as possible with the accumulator, the minimum working pressure of the accumulator should not be too low. According to the empirical formula [26]:

$$p_{\min} = (0.6 - 0.85) p_{\max} \tag{9}$$

the relationship between the accumulator's lowest pressure p_{min} and highest pressure p_{max} can be obtained.

(2) Inflation pressure

Theoretically, the inflation pressure p_0 of the accumulator should be equal to the minimum working pressure. Still, due to the leakage and charging temperature in the process of energy absorption by the accumulator, the inflation pressure should be less than the minimum working pressure. According to the empirical formula [27], the inflation pressure can be calculated as:

$$p_0 = (0.8 - 0.85)p_{min} \tag{10}$$

(3) The maximum recovery energy and volume of the accumulator

During the HRG process, the volume of the hydraulic accumulator should be able to recover at least the potential energy under medium load, and the upper limit should be set according to the maximum load energy recovered [28]. The maximum energy that the accumulator can recover is expressed as follows:

$$E_{reg} = -\int_{V_0}^{V_1} pdV = -\int_{V_0}^{V_1} p_0(\frac{V_0}{V_1})dV = \frac{p_0 V_0}{n-1}\left[(\frac{p_1}{p_0})^{\frac{n-1}{n}} - 1\right] \tag{11}$$

From the above formula, the initial volume of the accumulator is:

$$V_0 = \frac{(n-1)E_{reg}}{p_0\left[(\frac{p_1}{p_0})^{\frac{n-1}{n}} - 1\right]} \tag{12}$$

In this formula, E_{reg} is the maximum energy recovered by the accumulator, and V_0 is the initial accumulator volume under the inflation pressure, which can be approximately equal to the volume of the accumulator.

3.1.4. Variable Hydraulic Pump/Motor (VPM)

For the bidirectional variable displacement piston pump, the closed volume at the bottom of the plunger will complete the oil suction and discharge for each rotation when the cylinder swings. The instantaneous flow rate of the pump/motor is expressed as follows:

$$Q_i = \frac{\pi d^2}{4} w \cdot R \cdot \tan\alpha \cdot \sin\varphi \tag{13}$$

where d is the plunger diameter, R is the distribution circle radius of the plunger axis in the cylinder, w is the rotational angular velocity of cylinder, α is the swashplate inclination, and φ is the plunger angle.

The VPM is mainly selected in terms of the displacement parameters. During the lifting process, it is used as a motor to output mechanical torque or as a pump in the hydraulic recovery process. The input torque is expressed as follows:

$$T_p = \frac{\beta \cdot V \cdot \Delta p}{20\pi} \tag{14}$$

where T_p is the input torque of the secondary component (Nm), β is the opening of the secondary element swash plate $(-1, 1)$, V is the secondary component displacement (mL/r), and Δp is the working pressure of the secondary element (MPa). The efficiency of the secondary element is regarded as 1.

3.2. *Rule-Based Energy Management Strategy*

3.2.1. Mechanical Model of the Hydraulic Cylinder in Boom

Aiming at the recovery process of gravitational potential energy when the boom is descending, the mechanical model (Figure 3) of the boom hydraulic cylinder is analyzed [29] according to Equation (15):

$$p_2A_2 + F = p_1A_1 + bx' + Mx'' + F_f \tag{15}$$

where p_2 is the cylinder pressure of rod cavity, A_2 is the effective area of the rod cavity, p_1 is the cylinder pressure of the rodless cavity, A_1 is the effective area of the rodless cavity, b is the viscous damping on the piston and load, x' is the piston movement speed, x'' is the piston acceleration, M is the equivalent mass of the piston and load, and F_f is the friction between the hydraulic cylinder and piston.

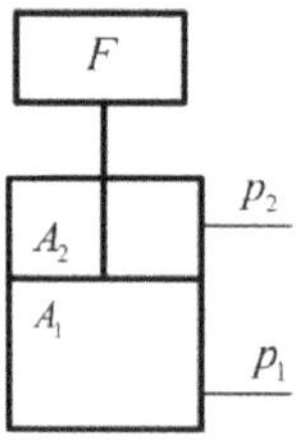

Figure 3. Mechanical model diagram of the hydraulic cylinder.

3.2.2. Demand Flow

The driver operates the joystick to give control commands to the boom [30]. In this paper, the boom is selected to run at a constant speed when it rises and falls.

$$v_{req} = \alpha v_{max} \tag{16}$$

v_{max} is the maximum speed of the boom cylinder, α is the joystick signal, [0, 1], and v_{req} is the required speed of the boom cylinder.

$$q_{req} = v_{req}A_c \tag{17}$$

q_{req} is the required flow of the boom cylinder, and A_c is the piston area of the boom cylinder.

$$P_{req} = p_c q_{req} \tag{18}$$

P_{req} is the demand power of the boom cylinder, and p_c is the cylinder pressure of rodless cavity.

3.2.3. Demand Power and Torque of the Motor

(1) ED

In this mode, all required power is provided by the motor when cylinder is lifting. The power and torque of the motor are expressed as follows:

$$P_{mq} = \frac{P_{req}}{\eta_P} \tag{19}$$

$$T_{mq} = 9550 \cdot \frac{P_{mq}}{n_m} \tag{20}$$

where P_{mq} is the motor's required power, η_P is the motor's efficiency, T_{mq} is the motor's required torque, and n_m is the motor's speed.

(2) HD

In this mode, the motor does not participate in driving, so the motor power and torque are both 0.

(3) ERG

When the boom descends, the gravitational potential energy is converted to electrical energy and stored in the battery, rather than to hydraulic energy stored in the HPA. Meanwhile, the motor is used as a generator.

$$T_{\mathrm{mq}} = \frac{V_{\mathrm{P}} \cdot \Delta p \cdot \beta}{20\pi} \tag{21}$$

β is the opening of the secondary element swash plate $(-1, 1)$, V_{P} is the secondary component displacement (mL/r), and Δp is the working pressure of the secondary element (MPa).

(4) HRG

In this mode, the gravitational potential energy is converted into hydraulic energy and stored in the HPA rather than in the battery when the boom descends. The motor still provides a certain amount of power for the hydraulic cylinder, setting to be 1/5 of the weight of the hydraulic cylinder, where G is the weight of the cylinder.

$$P_{\mathrm{req}} = \frac{G \cdot \Delta x}{5 \cdot \Delta t} \tag{22}$$

$$T_{\mathrm{mq}} = 9550 \cdot \frac{P_{\mathrm{req}}}{n_m \cdot \eta_p} \tag{23}$$

(5) NRG

This working condition generally occurs when the SOC and the pressure of the HPA reach the maximum setting threshold, and there is no energy recovery when the hydraulic cylinder falls. The motor still provides a certain amount of power with the help of the hydraulic cylinder's own weight to lower the boom. Similarly, the power provided is still 1/5 of the self-weight of the hydraulic cylinder. The power and torque required by the motor are the same as the HRG working conditions.

3.2.4. Energy Management Flowchart

Figure 4 shows the control strategy for the lifting and lowering conditions of the boom system, and the strategy can be explained as follows:

Boom lifting process: the boom cylinder sensor receives a driver-operated up signal. First, the HPA determines whether P_{H} is greater than P_{th1}. If $P_{\mathrm{H}} > P_{\mathrm{th1}}$, the system enters HD mode and completes the boom lifting condition. If $P_{\mathrm{H}} \leq P_{\mathrm{th1}}$, it is judged to be the SOC threshold. If SOC > $s1$, the system enters the ED mode to complete the boom lifting condition. If SOC $\leq s1$, the system enters the battery charging mode.

Boom lowering process: the boom cylinder sensor receives a driver-operated descent signal. The HPA judges whether P_{H} is greater than P_{th2}. If $P_{\mathrm{H}} > P_{\mathrm{th2}}$ and SOC < $s2$, the system enters the ERG mode to initiate regenerative braking, and converts the gravitational potential energy of the boom into electrical energy, then recycles it into the battery to complete the lowering condition. If $P_{\mathrm{H}} \leq P_{\mathrm{th2}}$, the system enters the HRG mode, converts the gravitational potential energy of the boom into hydraulic energy and stores it in the HPA. If $P_{\mathrm{H}} > P_{\mathrm{th2}}$, and SOC $\geq s2$, there is no potential energy recovery when boom is lowering.

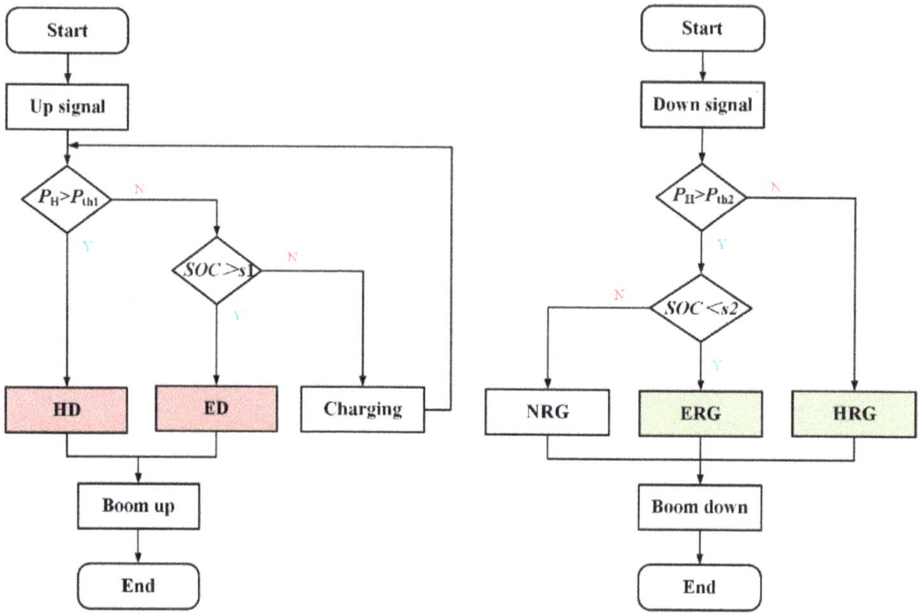

Figure 4. Control strategy of the EHDR-EEB. P_H: the pressure of HPA, P_{th1}: the pressure setting threshold when lifting, P_{th2}: the pressure setting threshold when lowering, s_1: the SOC threshold when lifting, s_2: the SOC threshold when lowering.

4. Simulation Analysis

4.1. Component Selection

The paper examines the recovery of the boom potential energy of an electro-hydraulic excavator during the movement process of the traditional excavator boom, and carries out the recovery and utilization of the boom potential energy. Based on the basic problems existing in the potential energy regeneration system of small hydraulic excavator booms, the design scheme adopted in this paper is as follows. Firstly, the original battery of the electric excavator is used to recover the gravity potential energy of the boom. Secondly, accumulators with high power density are used as auxiliary energy storage elements. The excavator boom system not only has the function of the original excavator boom, but also recycles the potential energy of the boom as much as possible to improve the economy without affecting the operability. Table 1 shows the parameter settings of the small electro-hydraulic excavator. The parameters of the accumulator and hydraulic pump/motor need to be redesigned, while the remaining components remain the same as the original excavator.

Table 1. Hydraulic components and parameters.

Component	Value	Remark and Unit
Hydraulic cylinder	0.055 × 0.03 × 0.8	Piston dia. × rod dia. × stroke length (m)
Hydraulic pump/motor	20	Displacement (mL/r)
Low pressure accumulator	12–18	Working pressure (MPa)
High pressure accumulator	26–38	Working pressure (MPa)
Motor	10	Rated power (kW)
Battery	35	Maximum power (kW)
Load force	50	kN

4.2. Analysis of Boom Motion Characteristics

Figure 5 shows the single-cycle trajectory of the piston in the hydraulic cylinder. The boom full load test is the research object, and the test period is 22 s. To ensure the standardization of the research, the assumptions are as follows:

(1) The reference curve is the motion characteristics of the original boom system, and the actual displacement curve is the EHDR-EEB motion characteristics. The purpose is to compare the EHDR-EEB with the original system and make the new system as close as possible to the kinematic characteristics of the original boom system.

(2) Test cycle time setting: the boom piston starts to move from 3 s; it takes about 3 s to rise to 0.65 m, stays for 3 s, and then descends to the lowest position. After that, it rises to the highest position 0.8 m for 3 s, and stays at the highest position for 3 s. Finally, it takes about 4 s to drop to the lowest position. The test cycle and the movement trajectory of the boom piston are shown in Figure 5.

(3) The position setting of the boom cylinder piston: the piston stops at the maximum position 0.8 m. and the middle position is at about 0.65 M.

(4) During lifting process, it is assumed that the external load remains unchanged. In the falling process, the unloading work has been completed, and the load is zero.

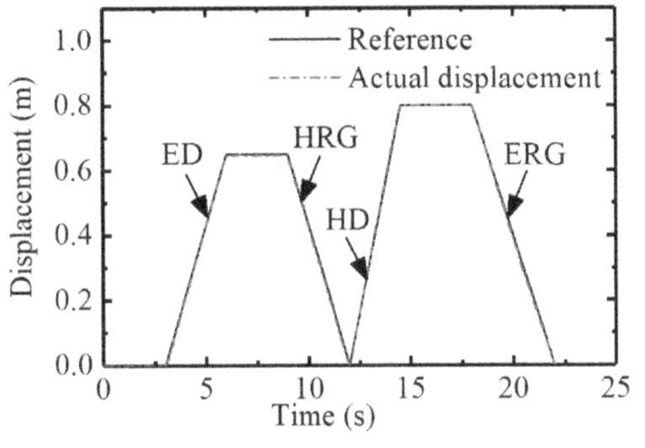

Figure 5. Piston Displacement Curve.

4.3. Operating Energy Consumption Analysis

In this paper, the pressure changes in the rod cavity and the rodless cavity are analyzed in Figure 6. The results illustrate that in the ED stage, the oil pressure in the rodless cavity is about 10 MPa, and the oil pressure in the rod cavity is 4.7 MPa. At this time, the motor drives the VPM to work, and pressurizes the hydraulic oil in the LPA into the rodless cavity of the cylinder to push the piston to move upward and lift the goods against the load gravity. When in balance, the oil in the two chambers remains stationary, but fluctuates slightly. In the HRG stage, the VPM assists in supplying hydraulic oil under the force of the piston and boom. Then the oil in the rodless cavity flows back to the HPA to charge the HPA, which is pressurized by the pressure cylinder. In the HD stage, the HPA supplies oil to the rodless cavity of the hydraulic cylinder. When the piston is in the highest position, the load is unloaded, and the load force suddenly decreases. Therefore, the pressure in the rod cavity drops to 0 in the balance stage of the highest position, and the pressure in the rodless cavity increases to 30 MPa. In the ERG stage, under the action of the HPA, the oil in rod cavity pushes the piston down together with the help of the boom. Accordingly, the pressure in the rodless cavity increases, and the oil enters the LPA under the action of the VPM. In this process, the hydraulic energy is converted into mechanical energy to drive the motor. The motor is used as a generator, and the potential energy of the boom is converted into electrical energy and stored in the battery.

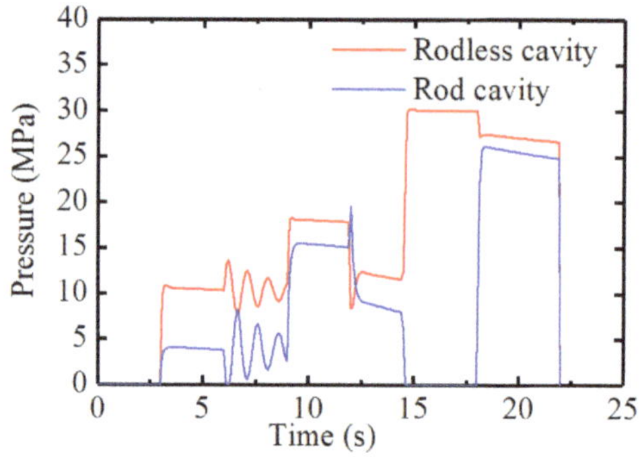

Figure 6. Pressure of the cylinder for the EHDR-EEB.

The pressure curves of the accumulators are shown in Figure 7. The initial pressure of the LPA is 18 MPa, and the initial pressure of the HPA is 32 MPa. In the ED stage, the HPA does not participate in the job, the pressure remains unchanged, and the LPA pumps the oil into the rodless cavity of the hydraulic cylinder by the VPM so the LPA pressure drops. In the HRG stage, the boom potential energy is converted into hydraulic energy stored in the HPA, and the HPA pressure rises to 36 MPa. The LPA participates in this stage, so the LPA pressure drops to 15.8 MPa. In the HD stage, the HPA releases energy to move the piston up, which leads to a pressure drop to 30.2 MPa. The LPA does not work so the pressure remains unchanged. In the ERG stage, the gravitational potential energy of the boom is converted into electrical energy. During this process, the oil in the rodless cavity flows into the LPA through the VPM, and the pressure in the LPA rises to 17.5 MPa. Simultaneously, the HPA assists the boom to move down, and the output hydraulic oil enters the rod cavity. After that, the pressure in the HPA drops to 28 MPa.

From the torque curves of the motor in Figure 8, the following conclusions can be drawn: In the original model (EEB), the motor is only used as a motor which cannot recover energy, and the maximum torque is 150 Nm. In the new model (EHDR-EEB), the motor acts as a motor in the ED, HRG and ERG stages, and outputs positive torque with a maximum torque of 50 Nm. Moreover, in the ERG stage, it is used as a generator to output a negative torque of −30 Nm, convert the kinetic energy transmitted by the motor into electrical energy and store this energy in the battery.

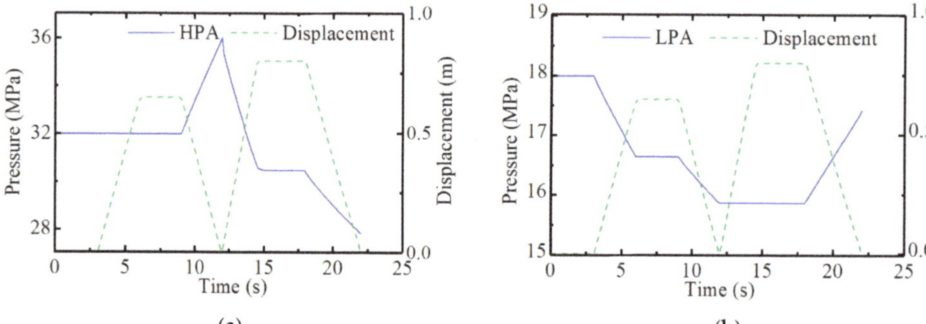

(**a**) (**b**)

Figure 7. Pressure of accumulators: (**a**) HPA; (**b**) LPA.

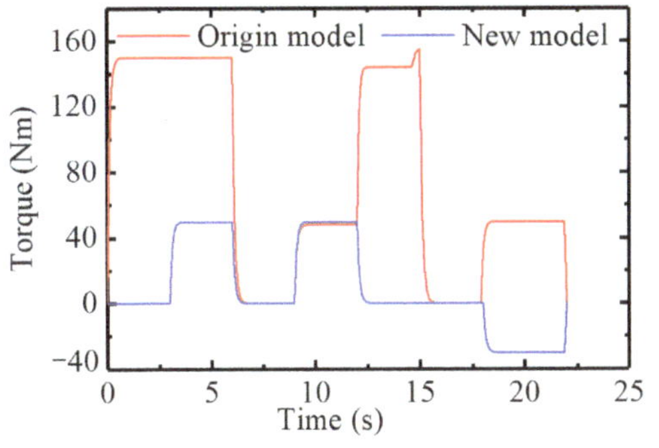

Figure 8. The motor torque comparison chart.

As can be seen in Figure 9, the motor in the original model has jitter problems, and the maximum speed is 1300 r/min, far exceeding the maximum speed of the motor in the new model. Fluctuation occurs when the motor torque suddenly increases from zero to a peak value or drops rapidly to zero in the origin model. The process of changing the speed of the motor from one point to another takes time, and there is a time delay. However, the working time of the boom is short, which results in the phenomenon of rotational speed fluctuation. In the EHDR-EEB, the motor does not work and the speed is zero from 0 to 3 s. The system enters the HRG stage at 9 s. Although the motor participates in work, the variation range of peak torque is very small, and its peak torque is one third that of the original model motor, so it cannot cause fluctuation in speed. Therefore, the new system can compensate the motor speed and torque through the accumulator, eliminate the jitter phenomenon, and reduce the large peak torque during the frequent operation of the motor.

According to the battery SOC curve, it can be seen that the new system has a more obvious energy-saving effect than the original system, and the SOC has increased in Figure 10. As seen in in Figure 11, the maximum power of the VPM in the original model is 19.8 kW, whereas that in the new model is 9 kW. In the ERG stage, the VPM in the new model is used as a motor, producing −8.6 kW to convert the hydraulic energy into electrical energy stored in the battery.

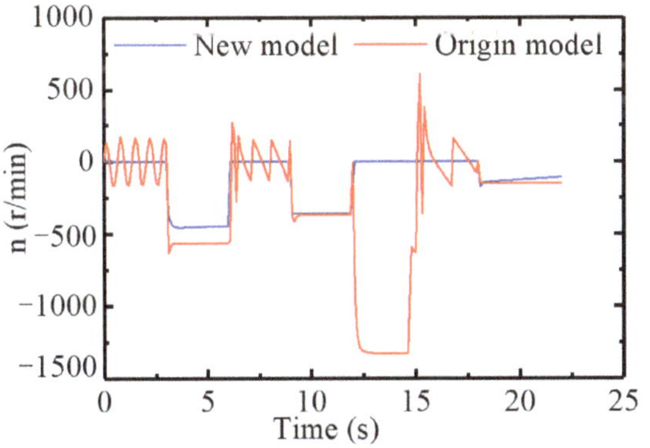

Figure 9. The motor speed comparison chart.

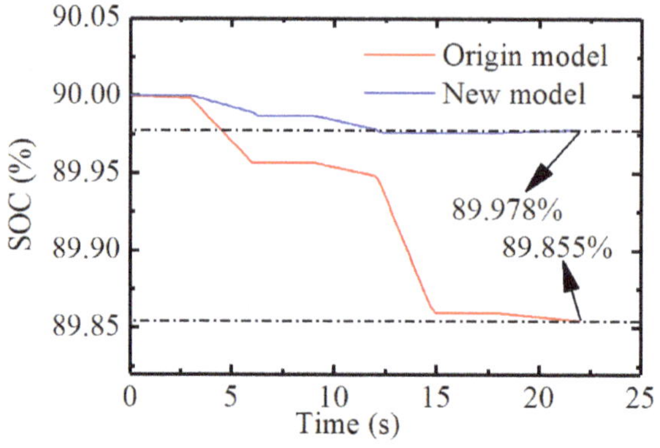

Figure 10. Economic verification results: SOC change.

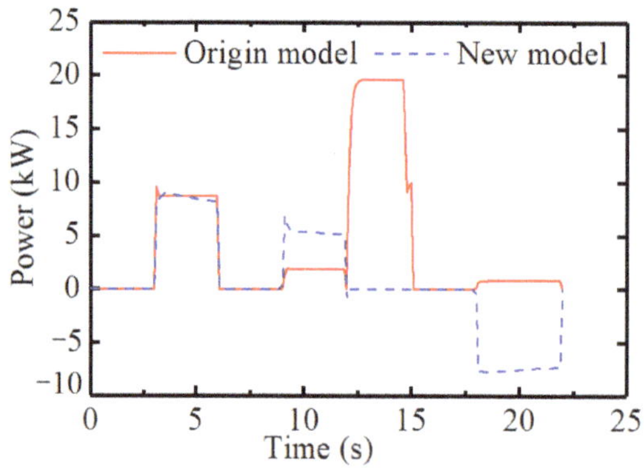

Figure 11. Power curve of the VPM.

Figure 12 represents the distribution of the recovered energy diagram. The boom descends twice during a 22 s period, and its gravitational potential energy is 27,925 J. The hydraulic energy recovered by the HPA is 6579 J in the HRG stage. The energy recovered by the battery is 19,125 J and is converted from hydraulic energy by the VPM and motor in the ERG stage. According to calculations, 92% of the potential energy of the boom can be recovered during the lowering process.

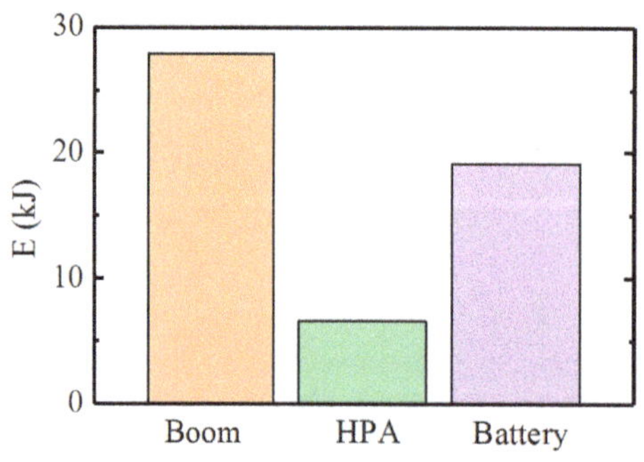

Figure 12. Distribution of the recovered energy during boom lowering.

5. Conclusions

In this paper, a novel electro-hydraulic drive and energy recovery system for the electric excavator boom is proposed. Based on the simulation, the displacement curve of the boom cylinder piston, the pressure curve of the hydraulic cylinder and the accumulator, the motor speed and torque curve, the SOC curve, the VPM power curve, and the recovered energy distribution curve are analyzed in detail during the lowering process. Compared with the original electric excavator, the new system has achieved significant energy savings in the four working modes ED, HD, HRG and EGR. The results show that the potential energy recovery rate is 92%. Most potential energy is stored in the battery by means of electrical energy. In addition, the maximum torque of the motor in the new system is reduced by 66.7% compared with the original system, which greatly reduces the large peak torque when the motor starts and stops frequently, stabilizes the motor speed, and eliminates the motor's chattering problem. This technology provides ideas for the direction of future research and development of energy recovery in electric excavators and a fundamental reference for improvements in the recovery and utilization of boom potential energy.

Author Contributions: Conceptualization, L.L. (Lin Li); methodology, T.Z. and L.L. (Lin Li); validation, T.Z., K.W. and L.L. (Liqun Lu); data curation, L.L. (Lianhua Lin) and H.X.; writing—original draft preparation, L.L. (Lin Li); writing—review and editing, L.L. (Lin Li) and T.Z.; funding acquisition, T.Z. All authors have read and agreed to the published version of the manuscript.

Funding: This research was funded by the International Technology Cooperation projects (No. 2019GHZ016), Innovation Team Project of "Youth Innovation Technology Program" of Shandong Higher Education Institution (No. 2021KJ083), Major science and technology innovation project of Shandong Province of China (No. 20020CXGC010405).

Institutional Review Board Statement: Not applicable.

Informed Consent Statement: Not applicable.

Data Availability Statement: Not applicable.

Conflicts of Interest: The authors declare that they have no known competing financial interest or personal relationships that could have influenced the work reported in this paper.

Nomenclature

EHDR-EEB	Electro-hydraulic drive and energy recovery system of the electric Excavator Boom
EEB	Electric excavator boom
VPM	Variable hydraulic pump/motor
HPA	High pressure accumulator
LPA	Low pressure accumulator
ED	Electrodynamic drive
HD	Hydrodynamic drive
ERG	Electric regeneration
HRG	Hydraulic regeneration
NRG	No regeneration
SOC	State of charge

References

1. Vacca, A. Energy efficiency and controllability of fluid power systems. *Energies* **2018**, *11*, 1169. [CrossRef]
2. Ding, R.; Zhang, J.; Xu, B.; Cheng, M.; Pan, M. Energy efficiency improvement of heavy-load mobile hydraulic manipulator with electronically tunable operating modes. *Energy Convers Manag.* **2019**, *188*, 447–461. [CrossRef]
3. Wu, W. Study on Energy-Saving and Power Matching of Parallel-Hybrid Hydraulic Excavator. Ph.D. Thesis, Southwest Jiaotong University, Chengdu, China, 2015.
4. Chen, M.; Zhao, D. The gravitational potential energy regeneration system with closed-circuit of boom of hydraulic excavator. *Mech. Syst. Signal Process.* **2017**, *82*, 178–192. [CrossRef]
5. Yu, Y.-X.; Ahn, K.K. Energy regeneration and reuse of excavator swing system with hydraulic accumulator. *Int. J. Precis. Eng. Manuf. Green Technol.* **2020**, *7*, 859–873. [CrossRef]

6. Li, J.; Ding, H.; Zhao, J.; Qin, J.; Shi, J.; Wang, Z. Design and simulation analysis of hybrid boom potential energy recovery system for super large hydraulic excavator. *Chin. Hydraul. Pneum.* **2019**, *7*, 33–38.
7. Fu, C.; Zou, G. Boom Potential Energy Recovery of Hydraulic Excavator Working Mechanism. *J. Univ. Jinan (Sci. Technol.)* **2018**, *32*, 4.
8. Ye, Y.; Lin, T.; Ren, H. Research on the potential energy regeneration and utilization in hydraulic based on balance cylinder. *Chin. Hydraul. Pneum.* **2018**, *1*, 71–77.
9. Lin, Y.; Lin, T.; Chen, Q.; Li, Z.; Fu, S.; Ren, H.; Guo, T. Research progress on key technologies of electric construction machinery. *Chin. Hydraul. Pneum.* **2021**, *45*, 1.
10. Casoli, P.; Ricco, L.; Campanini, F.; Bedotti, A. Hydraulic hybrid excavator—Mathematical model validation and energy analysis. *Energies* **2016**, *9*, 1002. [CrossRef]
11. Hao, L.; Fu, K.; Yu, B.; Lou, W.; Ma, G.; Qin, H. Reliability analysis of an incremental digital valve. *Chin. Hydraul. Pneum.* **2020**, *48*, 168–171.
12. Yi, H.; Cha, S. Optimal energy management of the electric excavator using super capacitor. *Int. J. Precis. Eng. Manuf. Green Technol.* **2019**, *8*, 151–164. [CrossRef]
13. Wang, T.; Wang, Q. Efficiency analysis and evaluation of energy-saving pressure compensated circuit for hybrid hydraulic excavator. *Autom. Constr.* **2014**, *47*, 62–68. [CrossRef]
14. Xia, L.; Quan, L.; Ge, L.; Hao, Y. Energy efficiency analysis of integrated drive and energy recuperation system for hydraulic excavator boom. *Energy Convers. Manag.* **2018**, *156*, 680–687. [CrossRef]
15. Zhang, Y. Research on Energy Saving for Hydraulic Excavators Based on Hybrid and Regeneration. Ph.D. Thesis, Zhejiang University, Hangzhou, China, 2006; pp. 27–32.
16. Yuan, X.; Wu, J. Research on the development of pure electric vehicle power battery technology based on patent analysis. In Proceedings of the IOP Conference Series: Earth and Environmental Science, Changsha, China, 18–20 September 2020; Volume 615, p. 012081.
17. Yang, Z.; Ravey, A.; Marion-P'era, M.C. Multi-objective energy management for fuel cell electric vehicles using online-learning enhanced Markov speed predictor. *Energy Convers Manag.* **2020**, *213*, 112821.
18. Parikh, D.; Christensen, T.; Li, J. Correlating the influence of porosity, tortuosity, and mass loading on the energy density of $LiNi_{0.6}Mn_{0.2}Co_{0.2}O_2$ cathodes under extreme fast charging (XFC) conditions. *J. Power Sources* **2020**, *474*, 228601. [CrossRef]
19. Chen, M.; Zhao, D. Research on recoverable potential of hydraulic excavator. *Mach. Tool Hydraul.* **2016**, *44*, 65–68.
20. Gong, J.; Zhang, D.; Liu, C.; Zhao, Y.; Hu, P. Power control strategy and performance evaluation of a novel electro-hydraulic energy-saving system. *Appl. Energy* **2019**, *233*, 724–734. [CrossRef]
21. Chen, Z.; Fu, J.; Zhang, X.; Cheng, H.; Quan, L. Electro-hydraulic recovery system of swing braking energy in hydraulic excavator. *Mach. Tool Hydraul.* **2018**, *46*, 1–8.
22. Yao, M.; Wu, W.; Qin, J.; Sun, L. Study on energy recovery system for electro-hydraulic coordinated hydraulic excavator Compound action. *Mach. Tool Hydraul.* **2018**, *46*, 54–59.
23. Lin, T.; Wang, Q.; Hu, B.; Gong, W. Research on the energy regeneration systems for hybrid hydraulic excavators. *Autom. Constr.* **2010**, *19*, 1016–1026. [CrossRef]
24. Hong, J.; Wang, Z.; Ma, F.; Yang, J.; Xu, X.; Qu, C.; Zhang, J.; Shan, T.; Hou, Y.; Zhou, Y. Thermal runaway prognosis of battery systems using the modified multi-scale entropy in real-world electric vehicles. *IEEE Trans. Transp. Electrif.* **2021**, *7*, 2269–2278. [CrossRef]
25. Yang, J.; Zhang, T.; Zhang, H.; Meng, Z. Research on the starting acceleration characteristics of a new mechanical–electric–hydraulic power coupling electric vehicle. *Energies* **2020**, *13*, 6279. [CrossRef]
26. Yang, J.; Zhang, T.; Hong, J.; Zhang, H.; Zhao, Q.; Meng, Z. Research on driving control strategy and fuzzy logic optimization of a novel mechatronics-electro-hydraulic power coupling electric vehicle. *Energy* **2021**, *233*, 121221. [CrossRef]
27. Hong, J.; Wang, Z.; Chen, W.; Wang, L.; Lin, P.; Qu, C. Online accurate state of health estimation for battery systems on real-world electric vehicles with variable driving conditions considered. *J. Clean. Prod.* **2021**, *294*, 125814. [CrossRef]
28. Geng, Y.; Yang, J.; Quan, L. Power matching of hybrid excavator based on load balance. *Chin. Hydraul. Pneum.* **2017**, *2*, 63–69.
29. Altare, G.; Vacca, A. A design solution for efficient and compact electrohydraulic actuators. *Procedia Eng.* **2015**, *106*, 8–16. [CrossRef]
30. Shi, J.; Quan, L.; Zhang, X.; Xiong, X. Electro-hydraulic velocity and position control based on independent metering valve control in mobile construction equipment. *Autom. Con. Struct.* **2018**, *94*, 73–84. [CrossRef]

energies

MDPI

Review

A Review on Magnet Loss Analysis, Validation, Design Considerations, and Reduction Strategies in Permanent Magnet Synchronous Motors

Samith Sirimanna *, Thanatheepan Balachandran and Kiruba Haran

Department of Electrical Computer Engineering, University of Illinois at Urbana Champaign, Urbana, IL 61801, USA; tb8@illinois.edu (T.B.); kharan@illinois.edu (K.H.)
* Correspondence: sss7@illinois.edu

Abstract: Eddy current losses in magnets are a major consideration in the rotor design of permanent magnet synchronous motors (PMSMs). Stator design choices and the use of modern inverters with high switching frequency introduce harmonics that can contribute to significant losses in the magnets, causing the rotor to heat up. In typical PMSMs, the lack of rotor cooling can cause the magnet's performance to degrade at high temperatures and eventually demagnetize. This review examines a large number of studies analyzing magnet eddy current losses using analytical methods and finite-element analysis. In some of these studies, magnet segmentation is carried out to reduce the losses; however, their loss-reduction effects depend highly on the type of PMSM and the mix of stator harmonics. Magnet segmentation without considering these effects can, in fact, increase the magnet losses, in addition to the extra manufacturing efforts. Multiple design analysis show the influence of rotor–stator geometric features on magnet losses. Although measuring magnet eddy current losses for these motor designs is a tedious task, authors have proposed calorimetric and loss segregation-based techniques to provide validation. This paper addresses magnet loss modeling techniques, PM material considerations, magnet segmentation effectiveness, motor and stator design effects, and experimental validation to inform motor designers about the costs and benefits of rotor designs that minimize rotor losses.

Keywords: magnet eddy current losses; analytical methods; finite-element analysis; loss reduction; segmentation

Citation: Sirimanna, S.; Balachandran, T.; Haran, K. A Review on Magnet Loss Analysis, Validation, Design Considerations, and Reduction Strategies in Permanent Magnet Synchronous Motors. *Energies* **2022**, *15*, 6116. https://doi.org/10.3390/en15176116

Academic Editors: Kan Liu and Wei Hu

Received: 1 August 2022
Accepted: 20 August 2022
Published: 23 August 2022

1. Introduction

High-frequency permanent magnet (PM) motor drives are becoming increasingly available in electric vehicles (EV), aerospace, and other applications which are sensitive to motor weight. Designing these machines encompasses multiple strategies such as distributed vs. concentrated windings on the stator and interior PM, surface PM, and inset PM on the rotor. Independent of design choice, the magnets in the rotor are exposed to a variety of harmonics. They are generated from the slotting effect and inverter switching and are the sources of magnet eddy current losses. Typically, stator slotting produces the first few harmonics, while modern inverters create pulse-width modulation (PWM) switching harmonics that have significantly higher frequencies. The percentage of eddy current losses due to each of these sources may vary depending on stator and rotor design. Popular NdFeB magnets used in most of these applications have a high electrical conductivity that exhibits eddy currents and creates eddy current losses. While the magnitude of this loss component is typically low compared to major losses such as copper, iron, and mechanical losses, and has a lower impact on motor efficiency, excessive losses generated in thermally isolated rotors develop barriers to the continuous operation and lead to thermal magnetic degradation.

Magnet segmentation increases the resistance of eddy current paths and is the prevailing strategy to reduce eddy current losses. Segmentation can be either axial or circumferential. The former is more common in longer stack-length machines and, additionally, benefits the manufacturability of magnets by providing a practical dimensional aspect ratio. The latter is implemented to further curtail the magnet losses. During high-speed operation, when these harmonic frequencies approach the kHz range, typical NdFeB magnet blocks reach their skin depth, such that the eddy currents are concentrated on the magnet edges. Here, the skin depth (δ) is defined as:

$$\delta = \sqrt{\frac{2}{\sigma\omega\mu}} \tag{1}$$

where σ is the material's electrical conductivity, ω is the angular frequency, and μ is the material's permeability. As the frequencies increase to a level that makes skin depth close to half the magnet width, the circuit transitions from a resistance-limited to a reactance-limited circuit, causing the more segmented magnet to generate more losses [1]. Of the two indispensable methods used to model magnet losses, finite-element analysis (FEA) and analytical, FEA better captures the skin effect. While there are several analytical models that incorporate the skin effect, some of them are constrained to a homogeneous or sinusoidal excitation field.

This paper aims to summarize the modeling techniques and various factors affecting magnet losses in PM motors to provide motor designers with design approaches that can minimize rotor magnet losses. Section 2 details the two main mechanisms, hysteresis and eddy currents, that generate magnet losses; methods used in analyzing the eddy current loss; analytical models, and their ability to predict the losses. Section 2.3 describes a simple case study used to investigate the accuracy of a few selected methods. Section 3 describes the PM material properties that influence the losses and provides a comparison between the most common PM materials. Section 4 presents segmentation effects and summarizes findings from the literature. Section 5 investigates the variation of magnet losses in PMSMs due to rotor design and discusses features of available rotor types contributing to magnet eddy current losses. The influence of stator design choices and winding types is outlined in Section 6, while Section 7 summarizes other possible factors that contribute to magnet losses. Section 8 reports on experimental validation mechanisms using calorimetric and loss segregation approaches and their results.

2. Loss Mechanism and Modeling Techniques

2.1. Hysteresis Loss

Two potential loss mechanisms in magnets are hysteresis and eddy current losses. Although the majority of the magnet loss studies in the literature focus only on eddy current losses, rare-earth magnets in PMSMs are electrically conductive metal blocks with permeability slightly higher than air (μ_0); thus, there is the possibility of developing hysteresis loops.

Figure 1a shows the normal curve between the applied magnetic field (H) and the magnetic flux density (B) of a rare-earth PM material. On the original curve, the segment from Br to point C is called the *linear region* where point C is called the *knee point*. After the knee point, the flux density drops gradually and becomes zero when the applied magnetic field is equal to coercivity (Hc). The operating point will be at the intersection of the BH curve and the airgap line (determined by whichever application is being modeled), e.g., point A. In a PMSM, this always occurs inside the second quadrant of the BH curve. If the operating point falls below the knee point, as shown in point D, the magnetic flux density does not return to its original value; rather it follows a new path set out by the recoil line. Consider the operating point represented by point E. During motor operation, a positive and negative magnetic field will be applied around Hm, which will cause the flux to vary along this recoil line. In most designs, this variation will take place within the second quadrant of the BH curve. Thus, it has the possibility of forming a loop, as seen in ferromagnetic materials. This is called a *minor hysteresis loop*. Capturing these minor loops requires special measuring equipment such as a vibrating sample magnetometer or a fluxgate magnetometer. Much of the literature does not report the relatively smaller minor loops (Figure 1b), while some authors have tediously tested this, as shown in Figure 1c. In any case, the relative recoil permeability of sintered NdFeB magnets is close to one, which, according to some studies [2], generates minimal magnet hysteresis losses in a typical PMSM. If the entire magnet is enclosed by the second quadrant, relatively smaller losses are present, as seen from smaller minor loops in Figure 1c.

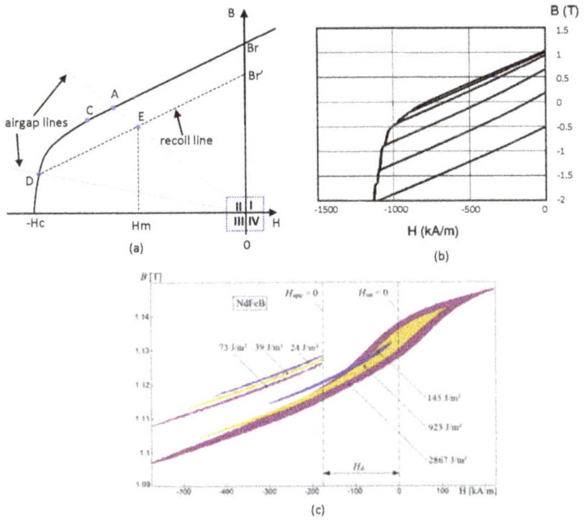

Figure 1. (**a**) Theoretical BH curve of a PM; (**b**) measured recoil curves of a NdFeB sample at 120 °C [3]; (**c**) measured minor hysteresis loops in a magnet at room temperature [4].

The hysteresis losses due to slot harmonics are reported in [5], which provides details on possible scenarios for significant magnet hysteresis loss in PMSM. Changing the sign of magnetic field H is stated as the necessary condition for sizable minor hysteresis loops to form. This causes the operating condition to change from the second to the first quadrant of the BH curve as shown by the larger loops in Figure 1c. FEA modeling of a two-pole high torque PMSM shows that a strong stator field is needed for the operating point to move to the first quadrant of a rare earth magnet [4]. This can easily be avoided with careful consideration at the machine design stage. In summary, hysteresis losses in rare-earth magnets seem to be smaller or avoidable.

2.2. Eddy Current Loss

FEA modeling is frequently used at the machine design stage as well as for loss-component analysis. It can provide a very accurate prediction of PM eddy current losses by capturing both non-homogeneous field distribution and temporal variations for any shape of the excitation waveform. Although, depending on the simulated frequency, the mesh size and time step necessary to capture the skin effect can make the FEA solution too time consuming compared to an analytical model. The solution is obtained by numerically solving the following equations [6]:

$$\nabla \times (\frac{1}{\mu} \times A) = J_a - \sigma\left(\frac{\partial A}{\partial t} + \nabla\phi\right) + \frac{1}{\mu_0}\nabla \times M \quad (2)$$

$$\nabla \cdot \left(\sigma\left(\frac{\partial A}{\partial t} + \nabla\phi\right)\right) = 0 \quad (3)$$

$$V_a = \frac{d\Phi}{dt} + R_a i_a \tag{4}$$

where A and ϕ are the magnetic vector potential and electric scalar potential, respectively, J_a is the armature current density, and M is the PM magnetization. v_a and i_a are the armature voltage and current, respectively, where the R_a and Φ are the armature resistance and flux linkage, respectively. Numerous studies have analyzed the magnet losses using FEA both in 2D and 3D domains [7–10] and hybrid methods have employed both analytical and FEA to increase computational efficiency [11–13].

The 2D analytical methods predicting the magnet eddy current losses can be derived from basic Maxwell equations and PM material characteristics. The governing Helmholtz equation for magnet losses can be derived using

$$B = \mu_m (H + M) \tag{5}$$

$$\nabla \times E = -\frac{\partial B}{\partial t} \tag{6}$$

$$J = \nabla \times H \tag{7}$$

$$J = \sigma_m E \tag{8}$$

$$\nabla \cdot B = 0 \tag{9}$$

where B is the magnetic flux density, E is the induced electric field using Faraday's law, H is the magnetic field, J is the eddy current density in the magnet, and μ_m and σ_m represent magnet permeability and electrical conductivity, respectively. Applying Equations (6) and (7) along with boundary conditions gives

$$\nabla^2 H = \mu_m \sigma_m \frac{\partial H}{\partial t} \tag{10}$$

The skin depth is much larger than the typical magnet block dimensions under relatively lower frequencies. This removes the need for considering magnet eddy current reaction field effects. The majority of the models do not capture this reaction field, and are called *resistance-limited* models [14–26]. At relatively higher frequencies, the skin depth becomes important to the eddy current loss calculations. This causes a significant reaction field on the magnet and, as a result, the magnet losses vs. frequency flatten. A few analytical models that are referred to as *skin-limited* models capture this effect [6,27–36]. Given the high-frequency nature of modern motor drives, skin-limited models are more suitable. Another important factor is the excitation field distribution; this could be spatially homogeneous or non-homogeneous. Some of these models display accurate modeling of eddy current losses with reaction field under a spatially homogeneous field excitation [6,27]. A few studies are available that used analytical methods that capture the non-homogeneity of the excitation field in a 3D domain, both excluding the reaction field [37–39] and including it [40,41].

Accurate modeling of the stator field is equally important to properly model the field excitation source, which is spatially non-homogeneous. Analytical modeling of stator magnetic fields in electric machines has been largely improved in past decades. Stator field modeling techniques for slotless stators are summarized in [42,43]. For the slotted stators, the modeling methods include reluctance networks [44–47], conformal mapping [48,49], and subdomain and exact subdomain methods [50], where the subdomain methods are more commonly used. In most 2D analytical models, only the normal magnetic field component H_y is considered as the source for eddy current loss. This results in eddy currents in the x-z plane, flowing in the x and z directions. In a typical PMSM, the magnet block experiences both normal (H_y) and tangential magnetic field (H_x) due to stator harmonics. Three-dimensional methods capture the effects of the magnetic field in all x, y, and z directions, which creates eddy currents in all x-z, x-y, and y-z planes, as shown in

Figure 2. The depiction of eddy current paths here is only representative of the loop they flow in. The actual depth of penetration of the currents will depend on the magnet block dimensions and operating frequency.

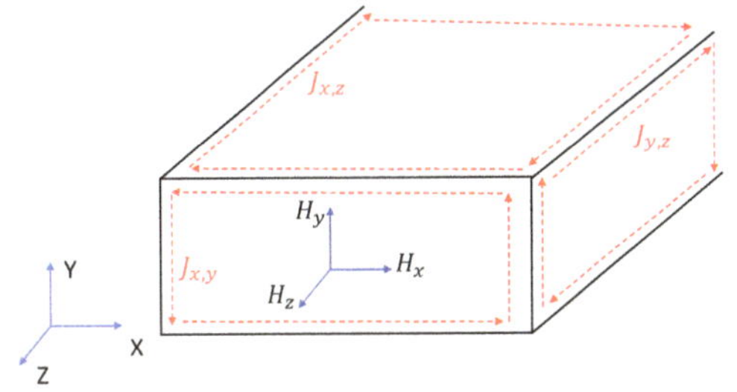

Figure 2. Sources of eddy currents and their paths.

Next, two forms of magnet loss expressions with different levels of complexity will be described. These will be applied in a simple study later in Section 2.3. A basic form of the equation, which considers a homogeneous magnetic field excitation through the magnet block, can be written using the iron-loss equations as [1]:

$$P_v = \frac{\pi^2 f^2 B_m^2 a^2}{6\rho} \tag{11}$$

where f is the excitation frequency, B_m is the mean flux density through the block in the y-direction, a is the magnet width, and ρ is the magnet resistivity. One limitation in this type of modeling is the failure to capture the skin effect, which will cause the losses to keep growing as the frequency increases. As the frequency increases, the magnetic field through the magnet block starts to fringe, which invalidates the homogeneous field assumption.

The 2D analytical model described in [6] finds the field solution to the governing Helmholtz equation:

$$\frac{\partial H_y^2(x,z)}{\partial x^2} + \frac{\partial H_y^2(x,z)}{\partial z^2} = j\omega\mu\sigma H_y(x,y)) \tag{12}$$

and uses

$$J(x,z) = \nabla \times H_y(x,z) \tag{13}$$

to arrive at a spatial distribution of current densities in the x-z plane as

$$J_x(x,z) = \frac{8a\gamma^2 H}{\pi^2} \sum_{n=1}^{\infty} \frac{(-1)^m \lambda_m}{(2m+1)^2 \beta_m} \frac{sinh(\beta_m z)}{cosh(\beta_m b)} cos(\lambda_m x) \tag{14}$$

$$J_z(x,z) = -\gamma H \frac{sinh\gamma x}{cosh\gamma a} + \frac{8a\gamma^2 H}{\pi^2} \sum_{m=1}^{\infty} \frac{(-1)^m \lambda_m}{(2m+1)^2 \beta_m^2} \frac{coshh(\beta_m z)}{cosh(\beta_m b)} cos(\lambda_m x) \tag{15}$$

where γ, λ_m, and β_m are parameters that can be calculated using skin depth, magnet width (a), and the spatial harmonic order (m). Compared to the simple loss equation, this method calculates the positional magnetic field and current density variation to capture the field non-homogeneity on the magnet. Current densities and material conductivity can be used to calculate the loss as in (16).

$$P = \frac{2d}{\sigma} \int_0^a \int_0^b J_x(x,y)^2 + J_z(x,y)^2 dxdy \tag{16}$$

2.3. A Comparison Study

Typical EV motor drives involve inverters with 10–20 kHz switching frequency and can create significant stator current harmonics up to a few times the switching frequency. Thus, it is useful to understand the eddy current losses generated in a magnet block under a wide range of frequency excitation before applying models to the motor application. In order to demonstrate key highlights of this behavior, two magnetic circuits with and without an airgap are used, as shown in Figure 3. These consist of two cut C-cores with a magnet placed in between. The C-core has the cross section dimensions shown in Figure 3, derived from a standard Metglass Amorphous core size to contain the flux path through the magnet block. For the gapped core on left, a gap of 16 mm (8 mm from each C-core) was taken out to place the magnet in the middle of the airgap. This leaves an airgap of 4.7 mm on each side of the magnet. For the ungapped core, only 3.3 mm was taken out from each core, to fit the magnet exactly between them. The magnet block dimensions are taken from a typical hybrid EV motor design [51] as 6.6 mm height, 19.1 mm width, and 30.5 mm depth on the z-axis. The motor in this reference is using sintered NdFeB magnets. Thus, the sample is assumed to have a representative electrical conductivity of 0.66 MS/m and relative permeability of 1.05. For the purpose of comparison, a pure sinusoidal excitation of 100 amp-turns is applied to the excitation coil in each circuit.

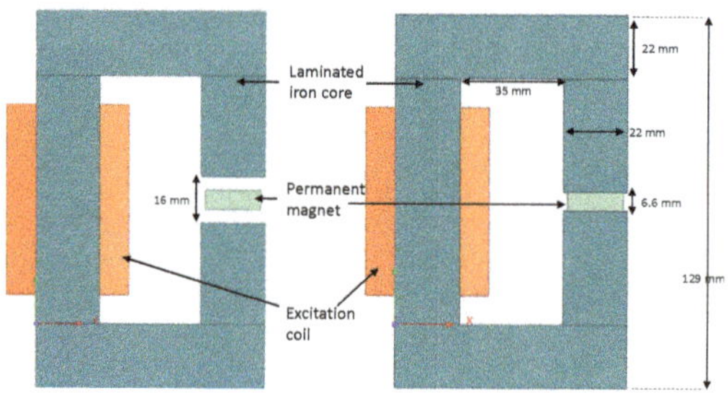

Figure 3. Two-dimensional FEA models: gapped core (**left**) and ungapped core (**right**) to analyze magnet losses.

Figure 4 shows the loss comparison between the simple approach in (11) and the described 2D analytical method in [6] for the gapped-core magnetic circuit. As seen, the simple homogeneous assumption without skin effect is already predicting much higher losses in the frequency range of hundreds of Hz. The plot shows the predicted loss difference in the lower frequency range, where the 2D analytical model has not yet reached the frequencies that would permit skin effect.

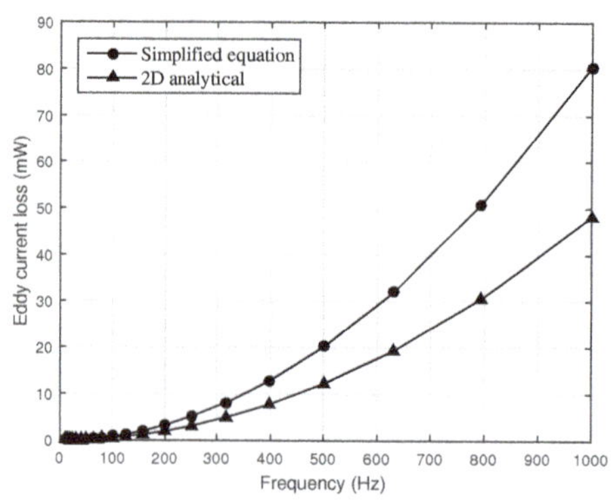

Figure 4. Comparison of loss vs. frequency between analytical models.

Figure 5 gives a comparison of loss prediction between 2D analytical and FEA methods. Both models show the skin effect, which causes the loss rate to decline. At first, the magnet losses increase as the square of the frequency. When the frequency is large enough, the skin effect becomes significant, and the circuit moves from being resistance-limited to reactance-limited [18], causing the loss rate to flatten. Here, the increase is proportional to the square root of the frequency. The exact frequency at which this transition occurs will depend on magnet dimensions and magnetic circuit parameters. Roughly, it initiates around the frequency that makes $\delta = 2a$, where a is the magnet width.

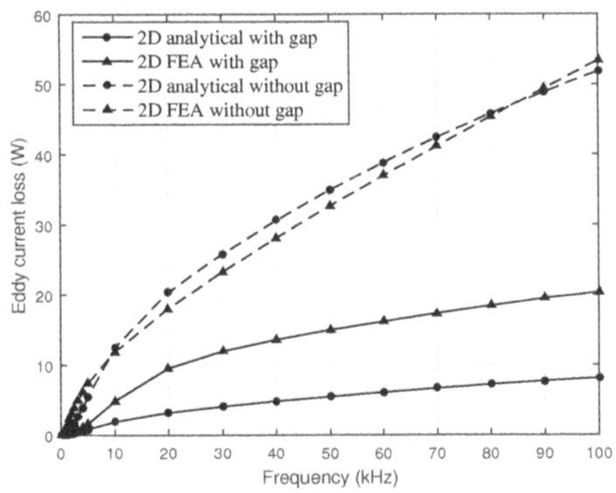

Figure 5. Comparison of loss vs. frequency between analytical and FEA models.

In the ungapped core magnetic circuit, the analytical model and FEA are close to each other. This circuit has a lower magnetic reluctance, which will generate a higher magnetic field and higher losses when the same excitation is applied. However, at higher frequencies, the magnetic circuit with an airgap shows lower losses from the analytical model than from FEA. This is caused by assuming a normal magnetic field (H_y) in the analytical model, a state that is less true than in the ungapped core because of flux fringing. Similarly, depending on the application, analytical models can vary in accuracy, despite faster calculation. A similar comparison is provided in [52], which compares increasingly complex analytical models with the FEA solution, and presents recommendations on which models are more accurate for a given excitation frequency.

3. Material Properties

PM material properties such as electrical conductivity and permeability influence eddy current loss behaviors. Although the NdFeB magnets are the more common choice in most modern PM motors, other magnets such as Sm-Co magnets are preferred in some motor applications, where a higher temperature rating is required. Alnico magnets with inherently lower coercivity are less popular now in motor applications due to the superior rare-earth magnets, although, new grades of Alnico magnets are still used in some motor designs. When it comes to manufacturing, NdFeB magnets have two production methods: sintered and bonded. In the case of a sintered magnet, the pulverized material is heat treated to form the required shape, while in the bonded magnets pulverized particles are mixed with a polymer and formed using injection molding or a similar process. Sintered NdFeB magnets have varying electrical conductivity from around 0.5 MS/m to 0.9 MS/m, depending on the material grade and manufacturing process, while providing a relative permeability of around 1.05. Figure 6 shows the eddy current loss variation of sintered NdFeB magnets based on their conductivity. The same gapped magnetic circuit was used as in Section 2.3. The lowest magnetic conductivity, although it has lower losses at relatively lower frequencies, ends up generating more losses at higher frequencies.

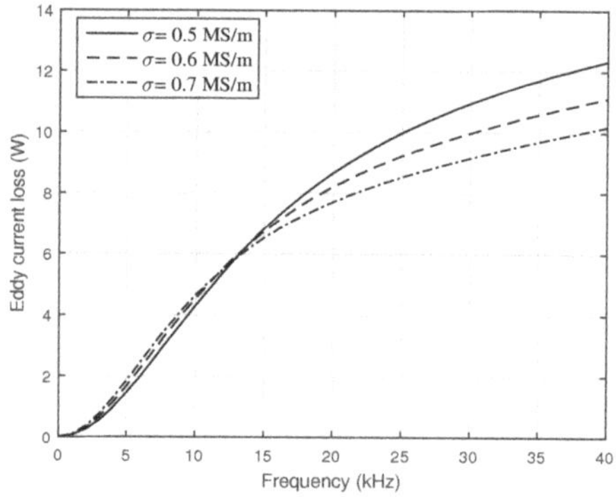

Figure 6. Magnet eddy current losses vs. frequency for sintered NdFeB classes (same isotropic relative permeability of 1.05 assumed).

Next, the loss variation due to the difference in PM permeability is investigated using a few grades of bonded NdFeB magnets. They exhibit a wide range of permeability values while the conductivity is consistently an order of magnitude lower compared to sintered magnets. Figure 7 shows the loss behavior of three different classes of bonded magnets with the same electrical conductivity of 0.01 MS/m and varying permeability. Due to the lower conductivity, the loss saturation takes place at a much higher frequency than in the sintered magnet. Thus, the effect on losses due to permeability variation is minimum within available bonded NdFeB magnet grades.

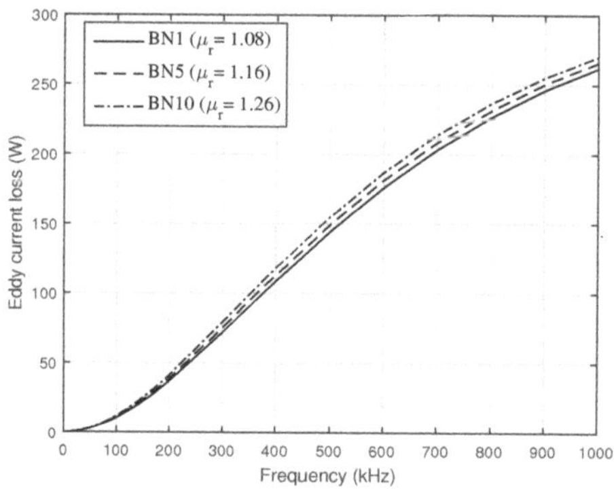

Figure 7. Magnet eddy current losses vs. frequency for bonded NdFeB classes (same conductivity of 0.01 MS/m assumed).

An overall comparison is conducted next. The comparison in Figure 8 shows the eddy current losses of magnet blocks of the same dimensions with different materials subjected to the same current excitation. As shown in Table 1, the electrical conductivity largely varies between the available types of magnet, and the permeability difference is relatively low. As for other familiar magnet types, ceramic magnets are not included here due to the huge electrical resistivity it displays. For a given magnet block size, the material properties result in better loss characteristics for certain frequency ranges. Although the magnet material, manufacturing method, and their grade impact magnet eddy current losses, the selection of material is governed by specifications such as remnant flux density (B_r), maximum energy product (BH_{max}), coercivity (H_c), and temperature ratings. Thus, in most cases, the physical properties of the magnet are typically a lower priority in design consideration.

Table 1. Common magnet types and their properties [53].

Material	μ_r	σ (MS/m)	B_r (T)
NdFeB Bonded (BN10)	1.26	0.01	0.7
Alnico 5	1.5	2	1.0
Sm-Co	1.1	1.16	0.8–1.15
NdFeB Sintered	1.05	0.66	1–1.5

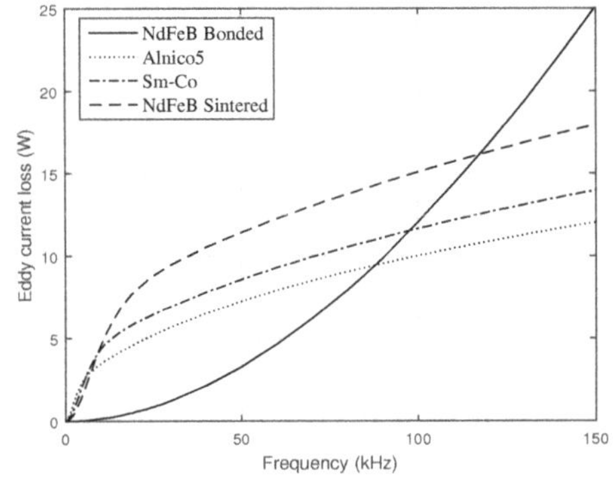

Figure 8. Magnet eddy current losses vs. frequency for a few selected magnet types.

4. Segmentation Effects

Magnet segmentation is the prevailing strategy to cut down magnet eddy current losses in PM machines. The main idea behind segmentation is to build up the effective resistance of eddy current paths, such that the losses will effectively decrease. This method reduces losses when the excitation frequencies are relatively lower and the skin effect has not taken effect. Several analytical models describe how to analyze magnet segmentation in PMSM loss models [1,6,16,23].

First, the focus will be on circumferential segmentation. In a typical PM motor drive with a voltage source inverter (VSI), the excitation harmonics will be voltages, which will naturally dampen the losses corresponding to high-frequency harmonics by reactive impedance. Although, to provide a simplified explanation of circumferential segmentation, the same FEA magnetic circuit from Section 2.3 is used here with constant current excitation. As can be seen from Figure 9, the skin effect reverses the loss reduction expected by circumferential segmentation, also known as the *anomaly of segmentation* [1]. Because of this, the segmentation corresponding to the lowest eddy current loss depends on the frequency of interest. For this particular magnetic circuit, any level of segmentation is effective in reducing losses due to harmonics below 30 kHz. In an EV drive, this frequency range will correspond to the PWM harmonics from the inverter. Usually, the stator harmonics due to the slots will be of much lower frequency and will be effectively reduced by segmentation, although the loss magnitude may be small.

Axial magnet segmentation is frequently found in PMSMs. In a relatively longer stack length motor, the level of axial segmentation can create differences in the eddy current reaction field due to the available eddy current path. This can demonstrate *anomaly of segmentation*, as will be discussed with an example in Section 5. Overall, the effective loss reduction due to magnet segmentation depends on many factors such as the spectrum of harmonics, magnet dimensions, magnetic circuit, and magnet material, and cannot be determined without a detailed look into the particular PMSM drive subsystem and operating conditions [6,14,16,23,54–56].

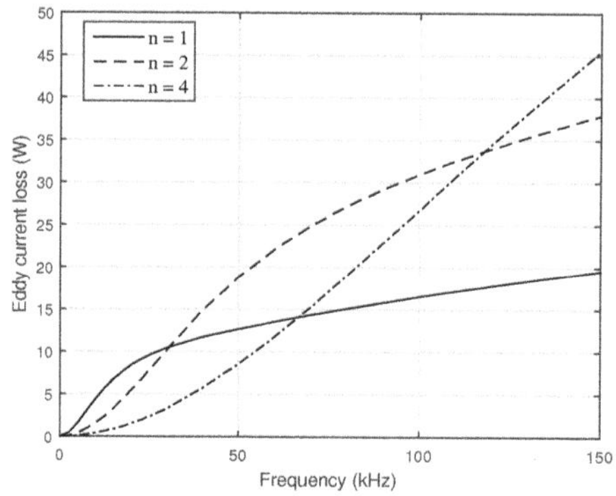

Figure 9. Magnet eddy current losses vs. frequency for different circumferential segmentation.

The contact resistance between the segmented magnets is studied in [57]. It analyzes the component level factors contributing to the effectiveness of segmentation such as the stress between the bonded sintered magnet segments. The results validated that the eddy currents pass through the contact when the compressive force is large, highlighting the importance of manufacturing precision.

5. Rotor Types

There are mainly three rotor types that are used in PM motor designs: surface, interior, and inset. The EV industry is highly focused on interior permanent magnet (IPM) motors due to a variety of reasons including variable speed profiles, requirements for field weakening, and fault tolerance. Some applications that demand constant load profiles, such as aerospace, prefer surface-mount permanent magnet (SPM) motors. Depending on the rotor type, studies show that flux variation on the magnet can vary. FEA results in [58] demonstrate that in low-frequency SPM (<1 kHz), the mid magnet sections experience lower field variation compared to a magnet section facing the airgap. In higher frequency harmonics, the field typically penetrates the entire magnet depth. Due to this reason, the percentage of magnet eddy current losses due to slot harmonics in SPM will be higher than in IPM.

Unlike in SPM, the IPM magnets are buried inside the rotor iron and are not directly exposed to the stator mmf variation by slot harmonics. Although, the PWM harmonics will clearly penetrate the magnets and will create eddy current losses in both cases of IPM and SPM. The analysis in [59] shows the eddy current paths in SPM magnets flow in small loops, which corresponds to the slot openings as shown in Figure 10. This is different in the case of IPM, where the eddy currents flow in a larger loop that covers the entire magnet. While carrier harmonics losses are almost the same for both, the SPM shows significantly higher losses due to slot harmonics at this given loading condition. The authors use an equivalent magnetic circuit to identify the main source of magnet losses in IPM; the main reason being the variation in magnetic path, where the permeance distribution in SPM is almost uniform. Thus, the main source for SPM magnet losses is deduced to be the regional variation of the flux density due to stator slot opening.

An example of total magnetic field decomposition for different rotor types into magnetization flux, armature flux, and reaction flux is presented in [60]. It shows that the ratio of reaction field to total flux is relatively large in IPM. Rotor iron next to the magnet block on each side of IPM makes the skin effect stronger and it causes the *anomaly of segmentation* to occur for the carrier harmonic frequencies. Magnet eddy current loss versus the axial segmentation in Figure 11 shows the continuously decreasing losses for increasing segmentation for both rotor shapes under slot harmonics. Expectedly, the carrier harmonics affect the two rotors differently. While the SPM total magnet losses decrease continuously with axial segmentation, the losses in IPM and inset PM rotors peak at a certain segmentation before it starts to decrease.

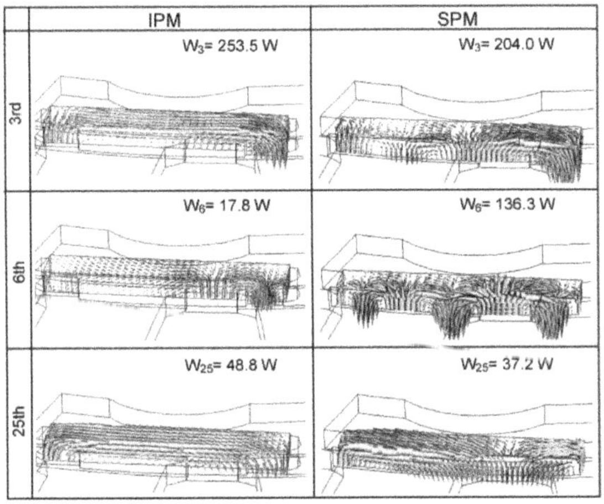

Figure 10. Eddy current paths associated with different harmonics at 300 A and $\beta = 60°$ [59].

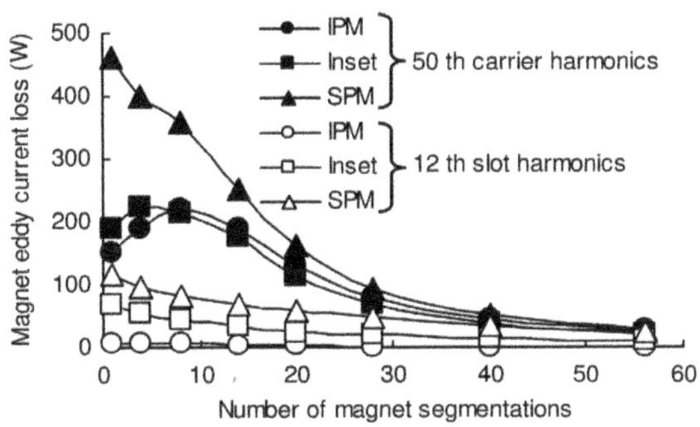

Figure 11. Magnet eddy current loss vs. segmentation for an example slot and carrier harmonic [60].

Although the SPM does not show the anomaly under these particular design and operating conditions, the analysis of SPM magnet eddy current losses in [1] shows the same anomaly in segmentation at slot harmonic frequencies for a locked rotor test.

Magnet eddy current losses are highly sensitive to the stator-rotor geometry. Optimizing the teeth shape to reduce the magnet eddy current losses is discussed in [61]. Various other modifications have been proposed such as flux barriers blocking either d or q axis flux [62] and adding grooves on rotor surface to reduce the harmonic content [63] from stator mmf.

6. Stator and Winding Type

Understanding the magnet loss behavior requires knowledge of the stator winding configuration: concentrated vs. distributed, in addition to the operating frequency, current angle, and loading. A comparison of two identical IPM rotors, each with with a distributed vs. concentrated stator is presented in [59]. As the increasing current angle reduces the operating torque, the magnet eddy current losses vary as it changes the amount of stator flux coupling with magnets. Total magnet eddy current losses are more than ten times larger for concentrated winding compared to the distributed due to large slot harmonics. Only a minor loss difference exists between the two under the carrier harmonics. For the distributed windings, loss due to slot harmonics is much smaller than that from carrier harmonics. The FEA figures show the deep penetration of slot harmonic fields to the rotor in a concentrated machine, due to wide slot pitch, while the same in distributed machine exists at the rotor surface. A comparison of magnet eddy current losses between three winding methods of fractional slot concentrated windings is presented in [15]. It investigates three possible winding methods for a 12 slot/10 pole stator with a surface PM rotor that utilizes the same stator mmf for torque production. The results show the different harmonics contributing to losses in 'all teeth-wound' and 'alternate teeth-wound'. The losses are highest for the 'alternate teeth-wound' design with wider stator teeth. Employing the same slot/pole combination and a new winding pattern involving a different number of turns per coil side, Ref. [64] shows the magnet eddy current losses can be reduced by 80% compared to the previous base design. This is done by minimizing the space harmonics that contribute significantly to the magnet losses in a concentrated winding machine.

Modifications of rotor shapes have been performed to optimize the operation under given winding configurations. An analysis of a fractional slot IPM for loss minimization is presented in [62] with a comparison between four rotor design modifications. A slit in the rotor as a flux barrier in the q-axis generates the lowest magnet losses. The loss behavior and loss reduction effects for two motors with fractional slot numbers are investigated in [14], which analyzes a three-phase external rotor machine and a six-phase fault-tolerant machine. It shows loss reduction effects in a three-phase machine with a fault-tolerant stator than in the six-phase external rotor machine. A novel structure of SPM with damping rings is proposed in [65] to reduce eddy current losses in magnets and rotor sleeves. The trade-off between different slot/pole combinations and the loss reduction effects by multi-layer winding in fractional slot machines are shown in [66].

7. Other Considerations in Loss Reduction

As discussed, the carrier harmonics from the inverter contribute to the magnet eddy current losses, based on the stator–rotor design. The study in [67] proposed modifying the magnet shape along with the employed PWM technique to reduce the harmonic content in the back-emf waveform, which can reduce harmonic content. A similar idea is followed in [63] by adding grooves to the rotor iron while optimizing with the PWM technique to reduce the harmonic content. Variation of magnet and other loss components based on switching frequency and the inverter topology is discussed in [68], which shows the slight loss reduction from two-level to three-level inverter topology. A study was done with a two-level inverter using three types of switching devices [69]: Si IGBT (10 kHz), GaN (30 kHz), and SiC (30 kHz), that shows the harmonics distribution under three cases that cause much lower PM eddy current losses in high switching frequency inverters.

A major goal of reducing magnet losses is to limit rotor heating, which could potentially lead to magnet degradation. The heat transfer and contact resistance between the magnet block and the rest of the rotor are important to make sure the magnets are able to dissipate the heat to the entire magnet quickly without creating a local hotspot [70,71]. Thermal analysis of an IPM is conducted in [72] by considering the contact resistance of magnets. Analysis of the insulation between divided magnet segments is presented in [32], which shows the impact of insulation layer resistivity on magnet loss composition.

8. Validation and Tests

Although a large number of articles discuss the analysis and modeling techniques of magnet eddy current losses in PMSMs, studies with validation results are limited. The methods of validation used involve both calorimetric and loss segregation from electrical measurements. Calorimetric methods use the following simple relationship between the generated loss ($\dot{Q}$), temperature (T) rise of the magnet with time ($\frac{dT}{dt}$), magnet mass (m), and the specific heat capacity of magnetic material (C_p).

$$\dot{Q} = mC_p\frac{dT}{dt} \quad (17)$$

Most of the calorimetric experiments are conducted at the component level, using a simple magnetic circuit to create a high-frequency magnetic field across the magnet [32]. In a well-insulated environment, the initial gradient is assumed to be maintained for a measurable time, during which the gradient is determined. This allows us to clearly see the differences in loss generation under a single frequency excitation across segmented and unsegmented magnets [6,60]. Measurement of magnet temperature, while it is attached to the rotor as reported in [30,73], can be subjected to a few challenges such as the accurate prediction of thermal resistance and rotor-specific heat. A different method is followed in [13,74], where the authors use the back-emf reduction due to heated-up magnets as a primary measure. The entire motor is heated up in an oven and the effects on winding resistance are considered to calculate the voltage constant of the motor. The initial temperature rise of magnets is used to calculate the losses.

The magnetic field of the circuit was measured using a coaxial double coil in [58], where the field input is then used to calculate the losses. Similar field measurements in magnetic circuits were used in [75] to validate losses against 3D FEA results. Both static and rotary tests were conducted using a concentrated winding stator in [76]. The loss calculation was done by subtracting other losses from total input electrical power, although this highly depends on the accuracy of other loss models. Similar measurements with power balance and loss segregation calculations are reported in [19,39].

9. Conclusions

This paper summarized selected work on the analysis, reduction strategies, and validation of magnet eddy current losses in PMSMs, aiming to provide motor designers with considerations for rotor design. Strengths and accuracy of magnet loss modeling methods were discussed. A variety of factors affecting magnet losses in a PMSM, such as material properties, rotor shapes, and winding/stator design are summarized, with simple demonstrations using FEA. The most common magnet loss reduction strategy is magnet segmentation, which is dependent on the excitation frequency and magnet dimensions. Validation of the magnet losses in the literature is done using both calorimetric and loss-segregation methods. While the calorimetric method provides direct measurements of losses dissipated as heat losses, the loss segregation method relies upon the accurate estimation of other loss components of the machine. Overall, the magnet sizing and design in PMSMs should consider the inverter switching frequency, spatial harmonics, and rotor/stator topology, as well as PM material characteristics. Detailed PM rotor designs require weighing all these factors to minimize rotor losses and improve motor performance.

Author Contributions: Conceptualization, methodology, investigation, formal analysis, writing—original draft preparation, S.S.; data curation, writing—review and editing, T.B.; visualization, supervision, project administration, K.H. All authors have read and agreed to the published version of the manuscript.

Funding: Grainger Center for Electric Machinery and Electromechanics (Grainger CEME) at University of Illinois at Urbana-Champaign.

Conflicts of Interest: All authors declare that there is no conflict of interest.

References

1. Huang, W.Y.; Bettayeb, A.; Kaczmarek, R.; Vannier, J.C. Optimization of magnet segmentation for reduction of eddy-current losses in permanent magnet synchronous machine. *IEEE Trans. Energy Convers.* **2010**, *25*, 381–387. [CrossRef]
2. Aoyama, Y.; Miyata, K.; Ohashi, K. Simulations and experiments on eddy current in Nd-Fe-B magnet. *IEEE Trans. Magn.* **2005**, *41*, 3790–3792. [CrossRef]
3. Ruoho, S.; Dlala, E.; Arkkio, A. Comparison of demagnetization models for finite-element analysis of permanent-magnet synchronous machines. *IEEE Trans. Magn.* **2007**, *43*, 3964–3968. [CrossRef]
4. Petrov, I.; Egorov, D.; Link, J.; Stern, R.; Ruoho, S.; Pyrhönen, J. Hysteresis losses in different types of permanent magnets used in PMSMs. *IEEE Trans. Ind. Electron.* **2016**, *64*, 2502–2510. [CrossRef]
5. Pyrhönen, J.; Ruoho, S.; Nerg, J.; Paju, M.; Tuominen, S.; Kankaanpää, H.; Stern, R.; Boglietti, A.; Uzhegov, N. Hysteresis losses in sintered NdFeB permanent magnets in rotating electrical machines. *IEEE Trans. Ind. Electron.* **2014**, *62*, 857–865. [CrossRef]

6. Yamazaki, K.; Fukushima, Y. Effect of eddy-current loss reduction by magnet segmentation in synchronous motors with concentrated windings. *IEEE Trans. Ind. Appl.* **2011**, *47*, 779–788. [CrossRef]
7. Sergeant, P.; Van den Bossche, A. Segmentation of magnets to reduce losses in permanent-magnet synchronous machines. *IEEE Trans. Magn.* **2008**, *44*, 4409–4412. [CrossRef]
8. Wang, Y.; Ma, J.; Liu, C.; Lei, G.; Guo, Y.; Zhu, J. Reduction of magnet eddy current loss in PMSM by using partial magnet segment method. *IEEE Trans. Magn.* **2019**, *55*, 8105105. [CrossRef]
9. Kim, D.M.; Kim, J.H.; Lee, S.G.; Park, M.R.; Lee, G.H.; Lim, M.S. Estimation Method for Rotor Eddy Current Loss in Ultrahigh-Speed Surface-Mounted Permanent Magnet Synchronous Motor. *IEEE Trans. Magn.* **2020**, *57*, 8103205. [CrossRef]
10. Hemeida, A.; Sergeant, P.; Vansompel, H. Comparison of methods for permanent magnet eddy-current loss computations with and without reaction field considerations in axial flux PMSM. *IEEE Trans. Magn.* **2015**, *51*, 8107511. [CrossRef]
11. Wu, X.; Wrobel, R.; Mellor, P.H.; Zhang, C. A computationally efficient PM power loss mapping for brushless AC PM machines with surface-mounted PM rotor construction. *IEEE Trans. Ind. Electron.* **2015**, *62*, 7391–7401. [CrossRef]
12. Zhang, P.; Sizov, G.Y.; He, J.; Ionel, D.M.; Demerdash, N.A. Calculation of magnet losses in concentrated-winding permanent-magnet synchronous machines using a computationally efficient finite-element method. *IEEE Trans. Ind. Appl.* **2013**, *49*, 2524–2532. [CrossRef]
13. Gerlach, T.; Rabenstein, L.; Dietz, A.; Kremser, A.; Gerling, D. Determination of eddy current losses in permanent magnets of SPMSM with concentrated windings: A hybrid loss calculation method and experimental verification. In Proceedings of the 2018 Thirteenth International Conference on Ecological Vehicles and Renewable Energies (EVER), Monte Carlo, Monaco, 10–12 April 2018; pp. 1–8.
14. Atallah, K.; Howe, D.; Mellor, P.H.; Stone, D.A. Rotor loss in permanent-magnet brushless AC machines. *IEEE Trans. Ind. Appl.* **2000**, *36*, 1612–1618.
15. Ishak, D.; Zhu, Z.; Howe, D. Eddy-current loss in the rotor magnets of permanent-magnet brushless machines having a fractional number of slots per pole. *IEEE Trans. Magn.* **2005**, *41*, 2462–2469. [CrossRef]
16. Toda, H.; Xia, Z.; Wang, J.; Atallah, K.; Howe, D. Rotor eddy-current loss in permanent magnet brushless machines. *IEEE Trans. Magn.* **2004**, *40*, 2104–2106. [CrossRef]
17. Ding, X.; Mi, C. Modeling of eddy current loss in the magnets of permanent magnet machines for hybrid and electric vehicle traction applications. In Proceedings of the 2009 IEEE Vehicle Power and Propulsion Conference, Dearborn, MI, USA, 7–10 September 2009; pp. 419–424.
18. Jumayev, S.; Merdzan, M.; Boynov, K.; Paulides, J.; Pyrhönen, J.; Lomonova, E. The effect of PWM on rotor eddy-current losses in high-speed permanent magnet machines. *IEEE Trans. Magn.* **2015**, *51*, 8109204. [CrossRef]
19. Balamurali, A.; Lai, C.; Mollaeian, A.; Loukanov, V.; Kar, N.C. Analytical investigation into magnet eddy current losses in interior permanent magnet motor using modified winding function theory accounting for pulsewidth modulation harmonics. *IEEE Trans. Magn.* **2016**, *52*, 8106805. [CrossRef]
20. Wu, L.; Zhu, Z.; Staton, D.; Popescu, M.; Hawkins, D. Analytical model for predicting magnet loss of surface-mounted permanent magnet machines accounting for slotting effect and load. *IEEE Trans. Magn.* **2011**, *48*, 107–117. [CrossRef]
21. Wu, L.; Zhu, Z.; Staton, D.; Popescu, M.; Hawkins, D. Analytical modeling and analysis of open-circuit magnet loss in surface-mounted permanent-magnet machines. *IEEE Trans. Magn.* **2011**, *48*, 1234–1247. [CrossRef]
22. Ede, J.D.; Atallah, K.; Jewell, G.W.; Wang, J.B.; Howe, D. Effect of axial segmentation of permanent magnets on rotor loss in modular permanent-magnet brushless machines. *IEEE Trans. Ind. Appl.* **2007**, *43*, 1207–1213. [CrossRef]
23. Wang, J.; Atallah, K.; Chin, R.; Arshad, W.; Lendenmann, H. Rotor eddy-current loss in permanent-magnet brushless AC machines. *IEEE Trans. Magn.* **2010**, *46*, 2701–2707. [CrossRef]
24. Bellara, A.; Bali, H.; Belfkira, R.; Amara, Y.; Barakat, G. Analytical prediction of open-circuit eddy-current loss in series double excitation synchronous machines. *IEEE Trans. Magn.* **2011**, *47*, 2261–2268. [CrossRef]
25. Wu, L.; Zhu, Z.; Staton, D.; Popescu, M.; Hawkins, D. Analytical model of eddy current loss in windings of permanent-magnet machines accounting for load. *IEEE Trans. Magn.* **2012**, *48*, 2138–2151. [CrossRef]
26. Amara, Y.; Reghem, P.; Barakat, G. Analytical prediction of eddy-current loss in armature windings of permanent magnet brushless AC machines. *IEEE Trans. Magn.* **2010**, *46*, 3481–3484. [CrossRef]
27. Mukerji, S.K.; George, M.; Ramamurthy, M.; Asaduzzaman, K. Eddy currents in solid rectangular cores. *Prog. Electromagn. Res. B* **2008**, *7*, 117–131. [CrossRef]
28. Pyrhonen, J.; Jussila, H.; Alexandrova, Y.; Rafajdus, P.; Nerg, J. Harmonic loss calculation in rotor surface permanent magnets—New analytic approach. *IEEE Trans. Magn.* **2012**, *48*, 2358–2366. [CrossRef]
29. Dubas, F.; Espanet, C.; Miraoui, A. Field diffusion equation in high-speed surface mounted permanent magnet motors, parasitic eddy-current losses. In Proceedings of the ELECTROMOTION, Lausanne, Switzerland, 27–29 September 2005; pp. 1–6.
30. Zhu, Z.; Ng, K.; Schofield, N.; Howe, D. Improved analytical modelling of rotor eddy current loss in brushless machines equipped with surface-mounted permanent magnets. *IEE Proc.-Electr. Power Appl.* **2004**, *151*, 641–650. [CrossRef]
31. Markovic, M.; Perriard, Y. Analytical solution for rotor eddy-current losses in a slotless permanent-magnet motor: The case of current sheet excitation. *IEEE Trans. Magn.* **2008**, *44*, 386–393. [CrossRef]
32. Yamazaki, K.; Shina, M.; Miwa, M.; Hagiwara, J. Investigation of eddy current loss in divided Nd–Fe–B sintered magnets for synchronous motors due to insulation resistance and frequency. *IEEE Trans. Magn.* **2008**, *44*, 4269–4272. [CrossRef]

33. de la Barriere, O.; Hlioui, S.; Ahmed, H.B.; Gabsi, M. An analytical model for the computation of no-load eddy-current losses in the rotor of a permanent magnet synchronous machine. *IEEE Trans. Magn.* **2013**, *52*, 8103813. [CrossRef]
34. Dubas, F.; Rahideh, A. Two-dimensional analytical permanent-magnet eddy-current loss calculations in slotless PMSM equipped with surface-inset magnets. *IEEE Trans. Magn.* **2013**, *50*, 54–73. [CrossRef]
35. Nuscheler, R. Two-dimensional analytical model for eddy-current loss calculation in the magnets and solid rotor yokes of permanent magnet synchronous machines. In Proceedings of the 2008 18th International Conference on Electrical Machines, Vilamoura, Portugal, 6–9 September 2008; pp. 1–6.
36. Deng, F.; Nehl, T.W. Analytical modeling of eddy-current losses caused by pulse-width-modulation switching in permanent-magnet brushless direct-current motors. *IEEE Trans. Magn.* **1998**, *34*, 3728–3736. [CrossRef]
37. Chen, L.; Wang, J.; Nair, S.S. An analytical method for predicting 3-D eddy current loss in permanent magnet machines based on generalized image theory. *IEEE Trans. Magn.* **2015**, *52*, 8103311. [CrossRef]
38. Nair, S.S.; Wang, J.; Chen, L.; Chin, R.; Manolas, I.; Svechkarenko, D. Prediction of 3-D high-frequency eddy current loss in rotor magnets of SPM machines. *IEEE Trans. Magn.* **2016**, *52*, 8107910. [CrossRef]
39. Zhu, Z.; Meng, Z. 3D analysis of eddy current loss in the permanent magnet coupling. *Rev. Sci. Instrum.* **2016**, *87*, 074701. [CrossRef]
40. Mirzaei, M.; Binder, A.; Funieru, B.; Susic, M. Analytical calculations of induced eddy currents losses in the magnets of surface mounted PM machines with consideration of circumferential and axial segmentation effects. *IEEE Trans. Magn.* **2012**, *48*, 4831–4841. [CrossRef]
41. Sahu, R.; Pellerey, P.; Laskaris, K. Eddy current loss model unifying the effects of reaction field and non-homogeneous 3-D magnetic field. *IEEE Trans. Magn.* **2020**, *56*, 7508404. [CrossRef]
42. Rahideh, A.; Korakianitis, T. Analytical calculation of open-circuit magnetic field distribution of slotless brushless PM machines. *Int. J. Electr. Power Energy Syst.* **2013**, *44*, 99–114. [CrossRef]
43. Pfister, P.D.; Perriard, Y. Slotless permanent-magnet machines: General analytical magnetic field calculation. *IEEE Trans. Magn.* **2011**, *47*, 1739–1752. [CrossRef]
44. Song, Z.; Liu, C.; Chen, Y.; Huang, R. Air-gap Permeance and Reluctance Network Models for Analyzing Vibrational Exciting Force of In-wheel PMSM. *IEEE Trans. Veh. Technol.* **2022**, *71*, 7122–7133. [CrossRef]
45. Mukundan, S.; Dhulipati, H.; Tjong, J.; Kar, N.C. Parameter determination of PMSM using coupled electromagnetic and thermal model incorporating current harmonics. *IEEE Trans. Magn.* **2018**, *54*, 8110505. [CrossRef]
46. Hemeida, A.; Sergeant, P. Analytical modeling of surface PMSM using a combined solution of Maxwell–s equations and magnetic equivalent circuit. *IEEE Trans. Magn.* **2014**, *50*, 7027913. [CrossRef]
47. Tariq, A.R.; Nino-Baron, C.E.; Strangas, E.G. Iron and magnet losses and torque calculation of interior permanent magnet synchronous machines using magnetic equivalent circuit. *IEEE Trans. Magn.* **2010**, *46*, 4073–4080. [CrossRef]
48. Abbaszadeh, K.; Alam, F.R. On-load field component separation in surface-mounted permanent-magnet motors using an improved conformal mapping method. *IEEE Trans. Magn.* **2015**, *52*, 5200112. [CrossRef]
49. Alam, F.R.; Abbaszadeh, K. Magnetic field analysis in eccentric surface-mounted permanent-magnet motors using an improved conformal mapping method. *IEEE Trans. Energy Convers.* **2015**, *31*, 333–344. [CrossRef]
50. Pfister, P.D.; Yin, X.; Fang, Y. Slotted permanent-magnet machines: General analytical model of magnetic fields, torque, eddy currents, and permanent-magnet power losses including the diffusion effect. *IEEE Trans. Magn.* **2015**, *52*, 8103013. [CrossRef]
51. Burress, T.; Coomer, C.; Campbell, S.; Seiber, L.; Marlino, L.D.; Staunton, R.; Cunningham, J. *Evaluation of the 2007 Toyota Camry Hybrid Synergy Drive System*; Technical Report; Oak Ridge National Lab. (ORNL): Oak Ridge, TN, USA, 2008.
52. Zhang, H.; Wallmark, O. Limitations and constraints of eddy-current loss models for interior permanent-magnet motors with fractional-slot concentrated windings. *Energies* **2017**, *10*, 379. [CrossRef]
53. Standard Specifications For Permanent Magnet Materials, No. 0100-00. Available online: https://www.datasheetarchive.com/whats_new/ba4f284c073012d3e2490a1b41e0e98c.html (accessed on 31 July 2022).
54. Yamazaki, K.; Abe, A. Loss investigation of interior permanent-magnet motors considering carrier harmonics and magnet eddy currents. *IEEE Trans. Ind. Appl.* **2009**, *45*, 659–665. [CrossRef]
55. Yamazaki, K.; Watari, S. Loss analysis of permanent-magnet motor considering carrier harmonics of PWM inverter using combination of 2-D and 3-D finite-element method. *IEEE Trans. Magn.* **2005**, *41*, 1980–1983. [CrossRef]
56. Ou, J.; Liu, Y.; Liang, D.; Doppelbauer, M. Investigation of PM eddy current losses in surface-mounted PM motors caused by PWM. *IEEE Trans. Power Electron.* **2019**, *34*, 11253–11263. [CrossRef]
57. Takahashi, N.; Shinagawa, H.; Miyagi, D.; Doi, Y.; Miyata, K. Analysis of eddy current losses of segmented Nd–Fe–B sintered magnets considering contact resistance. *IEEE Trans. Magn.* **2009**, *45*, 1234–1237. [CrossRef]
58. Fukuma, A.; Kanazawa, S.; Miyagi, D.; Takahashi, N. Investigation of AC loss of permanent magnet of SPM motor considering hysteresis and eddy-current losses. *IEEE Trans. Magn.* **2005**, *41*, 1964–1967. [CrossRef]
59. Yamazaki, K.; Fukushima, Y.; Sato, M. Loss analysis of permanent magnet motors with concentrated windings-variation of magnet eddy current loss due to stator and rotor shapes. In Proceedings of the 2008 IEEE Industry Applications Society Annual Meeting, Edmonton, AB, Canada, 5–9 October 2008; pp. 1–8.
60. Yamazaki, K.; Shina, M.; Kanou, Y.; Miwa, M.; Hagiwara, J. Effect of eddy current loss reduction by segmentation of magnets in synchronous motors: Difference between interior and surface types. *IEEE Trans. Magn.* **2009**, *45*, 4756–4759. [CrossRef]

61. Yamazaki, K.; Kitayuguchi, K. Teeth shape optimization of surface and interior permanent-magnet motors with concentrated windings to reduce magnet eddy current losses. In Proceedings of the 2010 International Conference on Electrical Machines and Systems, Incheon, Korea, 10–13 October 2010; pp. 990–995.
62. Choi, G.; Jahns, T.M. Reduction of eddy-current losses in fractional-slot concentrated-winding synchronous PM machines. *IEEE Trans. Magn.* **2016**, *52*, 8105904. [CrossRef]
63. Chaithongsuk, S.; Takorabet, N.; Kreuawan, S. Reduction of eddy-current losses in fractional-slot concentrated-winding synchronous PM motors. *IEEE Trans. Magn.* **2015**, *51*, 8102204. [CrossRef]
64. Dajaku, G.; Xie, W.; Gerling, D. Reduction of low space harmonics for the fractional slot concentrated windings using a novel stator design. *IEEE Trans. Magn.* **2013**, *50*, 8201012. [CrossRef]
65. Wu, L.; Qu, R.; Li, D. Reduction of rotor eddy-current losses for surface PM machines with fractional slot concentrated windings and retaining sleeve. *IEEE Trans. Magn.* **2014**, *50*, 8205704. [CrossRef]
66. Reddy, P.B.; El-Refaie, A.M.; Huh, K.K. Effect of number of layers on performance of fractional-slot concentrated-windings interior permanent magnet machines. *IEEE Trans. Power Electron.* **2014**, *30*, 2205–2218. [CrossRef]
67. Chaithongsuk, S.; Takorabet, N.; Meibody-Tabar, F. On the use of pulse width modulation method for the elimination of flux density harmonics in the air-gap of surface PM motors. *IEEE Trans. Magn.* **2009**, *45*, 1736–1739. [CrossRef]
68. Sato, D.; Itoh, J.i. Total loss comparison of inverter circuit topologies with interior permanent magnet synchronous motor drive system. In Proceedings of the 2013 IEEE ECCE Asia Downunder, Melbourne, VIC, Australia, 3–6 June 2013; pp. 537–543.
69. Korta, P.; Kundu, A.; Balamurali, A.; Iyer, L.V.; Schlager, G.; Kar, N.C. A Novel Hybrid Modelling Approach Towards Comprehensive Drive Cycle Analysis of Si, SiC, and GaN based Electric Motor Drives. In Proceedings of the IECON 2019-45th Annual Conference of the IEEE Industrial Electronics Society, Lisbon, Portugal, 14–17 October 2019; Volume 1, pp. 6331–6336.
70. Kuenzler, M.; Werner, Q.; Schaefer, U. Multidisciplinary Analysis of Permanent Magnet Machines Considering Thermal Contact Resistance. In Proceedings of the 2020 International Conference on Electrical Machines (ICEM), Gothenburg, Sweden, 23–26 August 2020; Volume 1, pp. 940–946.
71. Driesen, J.; Belmans, R.J.; Hameyer, K. Finite-element modeling of thermal contact resistances and insulation layers in electrical machines. *IEEE Trans. Ind. Appl.* **2001**, *37*, 15–20. [CrossRef]
72. Joo, D.; Cho, J.H.; Woo, K.; Kim, B.T.; Kim, D.K. Electromagnetic field and thermal linked analysis of interior permanent-magnet synchronous motor for agricultural electric vehicle. *IEEE Trans. Magn.* **2011**, *47*, 4242–4245. [CrossRef]
73. Zhao, N.; Zhu, Z.; Liu, W. Rotor eddy current loss calculation and thermal analysis of permanent magnet motor and generator. *IEEE Trans. Magn.* **2011**, *47*, 4199–4202. [CrossRef]
74. Malloy, A.C.; Martinez-Botas, R.F.; Lampérth, M. Measurement of magnet losses in a surface mounted permanent magnet synchronous machine. *IEEE Trans. Energy Convers.* **2014**, *30*, 323–330. [CrossRef]
75. Fratila, R.; Benabou, A.; Tounzi, A.; Mipo, J.C. A combined experimental and finite element analysis method for the estimation of eddy-current loss in NdFeB magnets. *Sensors* **2014**, *14*, 8505–8512. [CrossRef] [PubMed]
76. Liu, D.; Jassal, A.; Polinder, H.; Ferreira, J. Validation of eddy current loss models for permanent magnet machines with fractional-slot concentrated windings. In Proceedings of the 2013 International Electric Machines & Drives Conference, Chicago, IL, USA, 12–15 May 2013; pp. 678–685.

electronics

MDPI

Article

Modeling and Simulation of Magnetic Balance Current Sensor Based on Magnetic Scalar Potential Volume Integral

Dacheng Ni [1,2], Derong Luo [1], Shoudao Huang [1,*] and Yang Lv [2]

[1] College of Electrical and Information Engineering, Hunan University, Changsha 410082, China
[2] CRRC Zhu Zhou Institute Co., Ltd., Zhuzhou 412000, China
* Correspondence: hsd1962@hnu.edu.cn

Abstract: In this paper, a field–circuit combined simulation method, based on the magnetic scalar potential volume integral equation (MSP-VIE) and its fast algorithms, are proposed for the transient simulation and nonlinear distortion analysis of the magnetic balance current sensor. The magnetic part of the sensor is modeled and simulated by MSP-VIE with the field matrices extracted by the method of moments. By directly implementing the magnetic field equation in the circuit, these field matrices can be regarded as equivalent circuit parameters of the magnetic part, to construct the corresponding SPICE model. Finally, the field–circuit combined model of the entire sensor is unified in a circuit, so with the SPICE solver, the transient simulation is accomplished in the time domain. Moreover, aiming at the time-consuming problem, this paper presents a corresponding fast method to accelerate the simulation. The comparison of measurement and simulation demonstrates that the proposed method not only realizes the transient simulation of the whole sensor, but also simulates some hidden performance details; thus, it can be applied to practical engineering, to guide and test the early design of the product.

Keywords: closed loop; magnetic; field–circuit simulation

Citation: Ni, D.; Luo, D.; Huang, S.; Lv, Y. Modeling and Simulation of Magnetic Balance Current Sensor Based on Magnetic Scalar Potential Volume Integral. *Electronics* **2022**, *11*, 3008. https://doi.org/10.3390/electronics11193008

Academic Editor: Ahmed Abu-Siada

Received: 22 July 2022
Accepted: 13 September 2022
Published: 22 September 2022

1. Introduction

Magnetic balance currents have the advantages of wide measurement range, fast response speed, high measurement accuracy, good linearity, and operating frequency bandwidth [1]. In recent years, the research has focused on system-on-chip [2], thermal drift [3], magnetic core structure [4] etc. In different application environments, sensors needs to be designed with different performance, accordingly. Moreover, due to the complexity of the application environment, the sensor will have some faults when it is disturbed by the outside world, so it is necessary to analyze the cause of the breakdown. However, it is difficult to test a circuit in the actual working conditions. Thus, modelling and simulation is of great significance for design and improvement. In many situations, such as checking the hidden danger, testing conjecture etc., it is very convenient to have a model.

The sensor is composed of the magnetic part (magnetic core, coil) and the circuit part (Hall element, circuit board), so the overall performance is closely related to the two parts [4]. Because of the closed-loop working principle, the magnetic part and the circuit part must be considered comprehensively and simulated jointly. Consideration should not only be given to: the shape of the magnetic core; the magnetic properties of the magnetic core material; the winding mode of the coil; and the positional relationship between the coil and the magnetic core; but also the output characteristics of the magnetic part, which should match the circuit part.

Ultimately, the sensor needs closed-loop simulation and field–circuit combined simulation. Aiming at this problem, it is necessary to establish an equivalent circuit model, considering various factors of magnetic core. Some analysis methods and solutions have been proposed in relevant papers. Pankau J [5] regarded the magnetic parts of the sensor as a single-phase transformer, and then adopted the equivalent circuit model to realize the high-frequency modeling and accomplish the field–circuit combined simulation. Pejovic P [6] and Ai X [7] took the sensor as a current transformer and established the equivalent circuit model through the transformation function, and accordingly conducted transient analysis on the low-frequency characteristics and sensitivity.

Additionally, Sixdenier utilized finite element simulation to obtain magnetic field parameters, such as magnetic flux, and magnetic flux leakage of the magnetic core, and constructing the magnetic equivalent circuit, based on these module units, ultimately implemented the magnetic circuit modeling and field–circuit combined analysis [8]. To a certain degree, these methods can be used to implement the overall analysis of the sensor, but in fact, they have some drawbacks. These models are simplified a lot and it is difficult to consider the core structure, coil position and other factors in a simple equivalent model. Moreover, some papers have reported sensor simulation and optimization methods using finite element software [9,10]. The circuit element model in the finite element analysis is not accurate, and due to many grids in the air, the closed circuit analysis of magnetic circuit and electric circuit is difficult to converge.

There are some more accurate magnetic equivalent circuit models outside the field of magnetic balance sensors, such as magnetic shielding effectiveness in Rogowski coils [11], transformers [12,13], and switched reluctance motors [14]. However, due to the differences in structure and working-principle circumstances, these models can only be used in their respective fields.

Different from the above traditional methods, [15] reported a transient analysis method based on magnetic field integral equation, to research the fluxgate sensor. The main idea in the paper is to use magnetic scalar potential volume integral equation (MSP-VIE) to extract the magnetic field equation and matrix data, which serve as the equivalent circuit parameters, to construct the corresponding SPICE model, and finally utilize the SPICE solver to realize the field–circuit combined simulation.

This paper further improves and realizes the matrix parameter extraction and the circuit modeling based on MSP-VIE, and applies it to a closed-loop simulation of a magnetic balance current sensor. In addition, this paper also develops a fast method of solving the time-consuming problem in the SPICE simulation. The simulation and measurement results show that this method can not only implement closed-loop simulation and field–circuit combined simulation, but also simulate some hidden performance details, by accurately considering the magnetic core. This method can be applied to practical engineering, in order to guide, optimize and test the product design.

2. Theory and Equation

2.1. Working Principle of Magnetic Balance Current Sensor

As shown in Figure 1, the measured current is loaded on the straight wire and passes through the magnetic core. The sensor is composed of a magnetic core, circuit, Hall generator and compensation coil on the core. The Hall element, placed in the air gap of the magnetic core, detects the magnetic induction intensity and drive; the electric circuit then supplies a secondary current (Is) to the coil, which creates a flux equal in amplitude, but opposite in direction, to the flux created by the primary current.

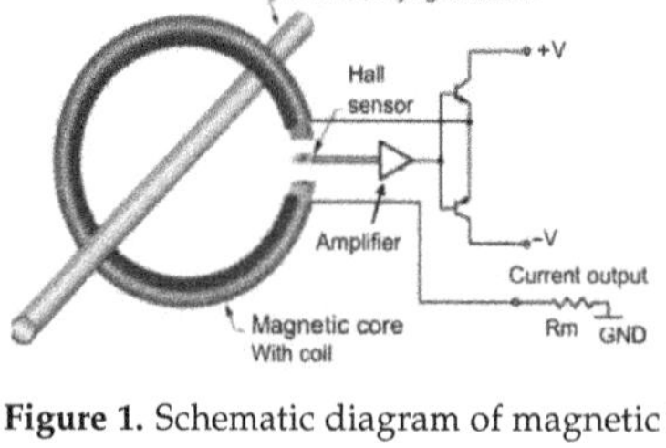

Figure 1. Schematic diagram of magnetic balance current sensor.

The secondary winding acts as a current transformer at higher frequencies. At low frequencies, the sensor operates using the Hall generator. At higher frequencies the secondary coil operates as a current transformer, providing a secondary output current again, defined by the turns ratio and converted to a voltage by the measuring resistor.

To give an order of magnitude, the typical number of secondary turns is NS = 1000 ... 5000 and the secondary current is usually between IS = 25 ... 300 mA, although it could be as high as 2 A. The unique design of closed-loop transducers provides an excellent bandwidth.

2.2. Magnetic Scalar Potential Volume Integral Equation

To realize the field–circuit combined simulation for the entire sensor, the magnetic part (coil, magnetic core and straight wire) should be processed first. Considering the low-frequency working condition, this paper uses a static magnetic field as an approximation.

In the excitation of the magnetic field, generated by external constant current source, magnetic materials will be magnetized. The magnetization behavior law is quantitatively defined by:

$$\mathbf{M}(\mathbf{r}) = \chi(\mathbf{r},\mathbf{H})\mathbf{H}(\mathbf{r}) \tag{1}$$

where $\mathbf{M}(\mathbf{r})$ is the magnetization, $\mathbf{H}(\mathbf{r})$ is the magnetic field at point of coordinates $\mathbf{r}$, and $\chi(\mathbf{r},\mathbf{H})$ is the nonlinear magnetic susceptibility.

At any point in space, the total magnetic field $\mathbf{H}(\mathbf{r})$ is a vector sum of the reduced magnetic field $\mathbf{H}_d(\mathbf{r})$ generated by the magnetic material and the magnetic field source $\mathbf{H}_0(\mathbf{r})$ created by the external current:

$$\mathbf{H}(\mathbf{r}) = \mathbf{H}_0(\mathbf{r}) + \mathbf{H}_d(\mathbf{r}) \tag{2}$$

In the source-free region, the magnetic field is irrotational, so Equation (2) can be expressed by magnetic scalar potential as below [16]:

$$\phi(\mathbf{r}) + \frac{1}{4\pi}\int_{V'} \chi(\mathbf{r}',\mathbf{H})\frac{\nabla\phi(\mathbf{r}')\cdot(\mathbf{r}-\mathbf{r}')}{|\mathbf{r}-\mathbf{r}'|^3}dV' = \phi_0(\mathbf{r}) \tag{3}$$

where $\phi_0(\mathbf{r})$ is the source potential of $\mathbf{H}_0(\mathbf{r})$ and $\phi(\mathbf{r})$ is the total scalar potential of $\mathbf{H}(\mathbf{r})$, satisfying $\mathbf{H}_0(\mathbf{r}) = -\nabla\phi_0(\mathbf{r})$, $\mathbf{H}(\mathbf{r}) = -\nabla\phi(\mathbf{r})$. The integral domain V' denotes the magnetic core.

2.3. Solution with Method of Moments

To solve Equation (3), the magnetic material region is discretized with a linear tetrahedron, and the total scalar potential $\phi(\mathbf{r})$ is approximated with first-order nodal shape function:

$$\phi \approx \sum_{j=1}^{N} \phi_j N_j \tag{4}$$

where ϕ_j and N_j denote the unknown and shape function associated to the j-th mesh node, respectively. N is the number of total nodes.

Then the collocation method at mesh nodes is used to match Equation (4), and, finally, it leads to the solution of a system of algebraic equations:

$$[\mathbf{I}+\mathbf{A}]\phi = \phi_0 \tag{5}$$

$$A_{ij} = \frac{1}{4\pi}\sum_m \int_{V'_m} \chi_m(\mathbf{r}',\mathbf{H})\frac{\nabla N_j(\mathbf{r}')\cdot(\mathbf{r}_i-\mathbf{r}')}{|\mathbf{r}_i-\mathbf{r}'|^3}dV'_m \tag{6}$$

where $\mathbf{I}$ denotes the unit matrix and $\mathbf{A}$ is the impedance matrix, with each element calculated by Formula (6). A_{ij} is the impact of mesh elements with j-th node on i-th node potential. $\int_{V'_m} dV'_m$ expresses the volume integral at the m-th mesh element and M is the number of total elements.

The excitation potential on the right side of the Equation (5) can be obtained through Biot-Savart's law (7) and path integral (8):

$$\mathbf{H}_0(\mathbf{r}) = \frac{1}{4\pi}\int_{V'} \frac{\vec{J}(\mathbf{r}') \times (\mathbf{r} - \mathbf{r}')}{|\mathbf{r} - \mathbf{r}'|^3} dV' \tag{7}$$

$$\phi_0(\mathbf{r}) = \int \mathbf{H}_0(\mathbf{r}) \cdot d\mathbf{r} \tag{8}$$

The magnetic source field $\mathbf{H}_0(\mathbf{r})$ include $\mathbf{H}_a$ and $\mathbf{H}_s$, which are induced by the measured current and the compensation coil current, $\phi_0(\mathbf{r})$ can be expressed as follows:

$$\phi_0(\mathbf{r}) = \int d\mathbf{r} \cdot (\mathbf{H}_a(\mathbf{r}) + \mathbf{H}_s(\mathbf{r})) \tag{9}$$

Within the tetrahedron region, the gradient of first-order nodal shape function is constant, which means the impedance matrix $\mathbf{A}$ can be separated into three parts:

$$\mathbf{A} = -\mathbf{D} \cdot \chi_{\mathbf{m}} \cdot \mathbf{G} \tag{10}$$

where $\chi_{\mathbf{m}}$ is a diagonal matrix comprising the magnetic susceptibility in each mesh element, it is a nonlinear part. The matrix $\mathbf{D}$ and $\mathbf{G}$ are independent of the magnetic material, they are the invariable parts. The element of them is expressed by:

$$\mathbf{D}_{im} = \frac{1}{4\pi}\int_{V'_m} \frac{\mathbf{r}_i - \mathbf{r}'}{|\mathbf{r}_i - \mathbf{r}'|^3} dV'_m \tag{11}$$

$$\mathbf{G}_{mj} = -\nabla N_j,\ r_j \in V_m \tag{12}$$

Furthermore, after normalizing and decomposing the excitation potential, the matrix equation can be obtained as follows:

$$[\mathbf{I} - \mathbf{D} \cdot \chi_{\mathbf{m}} \cdot \mathbf{G}]\phi = \phi_0 = \mathbf{C}I_a + \mathbf{K}I_s \tag{13}$$

where I_a represents the amplitude of the straight wire current, I_s represents the amplitude of the coil current, $\mathbf{C}$ and $\mathbf{K}$ are the corresponding excitation vectors, respectively. From (7)–(9), the elements of them are expressed by:

$$C_i = \int d\vec{r} \cdot \frac{1}{4\pi S_a} \int \frac{\vec{n}_a(\vec{r}') \times (\vec{r}_i - \vec{r}')}{\left|\vec{r}_i - \vec{r}'\right|^3} dV' \tag{14}$$

$$K_i = \int d\vec{r} \cdot \frac{1}{4\pi S_s} \int \frac{\vec{n}_s(\vec{r}') \times (\vec{r}_i - \vec{r}')}{\left|\vec{r}_i - \vec{r}'\right|^3} dV' \tag{15}$$

where V' is the volume of current, $\vec{n}_a(\vec{r}')$ is the direction of the measured current, $\vec{n}_s(\vec{r}')$ is the direction of the coil current, I_s/S_s is the compensation coil current density, I_a/S_a is the measured current density. In this paper, the measured current under test is equivalent to the ideal line current and the compensation coil is equivalent to the volume current model.

The second part of (13) is the reduced potential, and at i-th node ϕ_{di} is:

$$\phi_{di} = \sum_m \mathbf{D}_{im} \cdot \mathbf{M}_m \tag{16}$$

$$\mathbf{M}_m = \chi_m \sum_j \mathbf{G}_{mj} \cdot \phi_j \tag{17}$$

In particular, the modified Born iteration method is utilized to solve the nonlinear equation. The iterative formula is:

$$\left[\mathbf{I} - \mathbf{D}\cdot\chi_{\mathbf{m}}^{(k-1)}\cdot\mathbf{G}\right]\boldsymbol{\phi}^{(k)} = \boldsymbol{\phi}_0 \tag{18}$$

In the iterative process, the initial value of $\chi_{\mathbf{m}}^{(0)}$ and $\boldsymbol{\phi}^{(0)}$ should be given at the same time. Depending on the magnetization curve of the magnetic core and the iterative equation, the next iteration value can be obtained immediately.

With computer programming, these parameter matrices of the magnetic part can be extracted by MSP-VIE.

3. Spice Modeling Based on MSP-VIE

MSP-VIE and method of moments construct the full-wave electromagnetic model of the magnetic part and extract the matrix. The parameter matrices **D**, **G**, **C**, **K** in (13) can be regarded as the equivalent circuit parameters, to establish the equivalent circuit model.

The flow chart of the modeling and simulation is shown in Figure 2. The parameter matrices **D**, **G**, **C**, **K** of the magnetic part can be extracted by MSP-VIE. Then the obtained parameters, as the equivalent circuit parameters of the magnetic component, are compiled to the netlist, as described in Section 3.1. Section 3.1 describes the equivalent relationship between the magnetic scalar potential equation and the Kirchhoff voltage equation corresponding to the netlist in Section 3.1. Section 3.2 describes the equivalent circuit of magnetic components considering parasitic parameters. With the netlist in Section 3.1, the magnetic field and circuit can be completed in the same platform: SPICE.

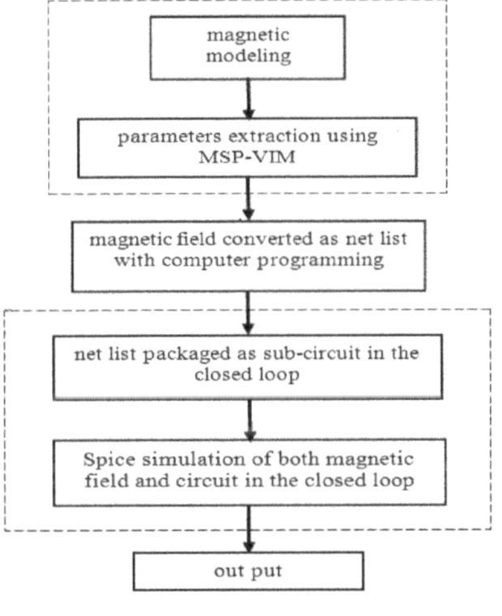

Figure 2. The flow chart of the proposed method.

3.1. Circuit Realization of MSP-VIE

Once the geometry of the coil and core were known, the value of the coefficient matrices **D**, **G**, **C**, **K** in (13) were determined and calculated by MSP-VIE. The magnetic scalar potential (Equation (13)) in the field was then equivalent to the Kirchhoff voltage equation in the circuit. Figure 3 shows the circuit implementation form of the i-th sub-equation in (13). The magnetic scalar potential in the field space can be treated as the node potential in the circuit. The voltage value across the resistor in Figure 3 is the total scalar potential at the i-th node. The voltage at node npi is the source potential at the i-th node $\phi_0(\mathbf{r})$, which is realized with a current-controlled voltage source. The voltage at node nni is the source potential at the i-th node $\phi_d(\mathbf{r})$, which is realized with a voltage-controlled voltage source. The inputs are the measured current I_a and compensation coil current

I_s. The compensation coil current I_s is obtained from the voltage node of the sampling resistor R_m in Figure 1. For convenience, the measured current I_a is also obtained from the voltage node.

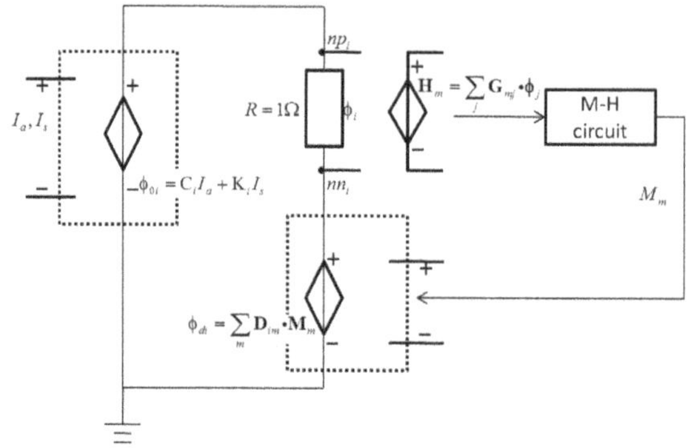

Figure 3. Circuit diagram of the i-th MSP-VIE sub-equation.

Because the magnetic fields inside the magnetic core counteract each other in the magnetic balance state, the magnetic core is usually unsaturated. Therefore, the magnetic field inside the core is very small, which means the hysteresis loss and eddy current loss are fairly tiny. Moreover, a laminated core will be used to reduce the eddy current in a practical application. Only the permeability changing with frequency, caused by eddy currents, is considered in this paper; the eddy current hysteresis loss is not considered. The M-H behavior of the nonlinear magnetic core is described by the magnetization model [17]. However, unlike [17], the frequency effect is considered, taking into account the difference in permeability caused by changing frequency.

3.2. Parasitic Parameters Model

There are parasitic parameters in the magnetic part of the sensor, which should be taken into account in the simulation. The circuit with parasitic parameters for the magnetic core and coil is shown in Figure 4a. R and C are the parasitic resistance and capacitance of the coil, respectively, and the mutual inductance of the coil is described by the inductance, L.

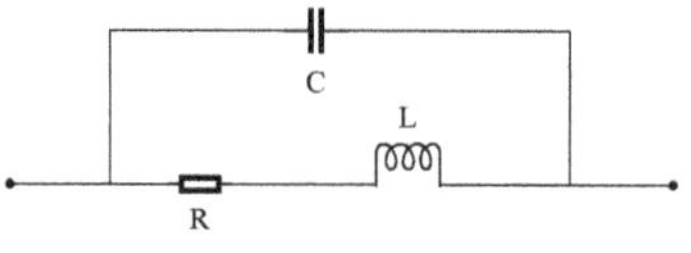

(**a**)

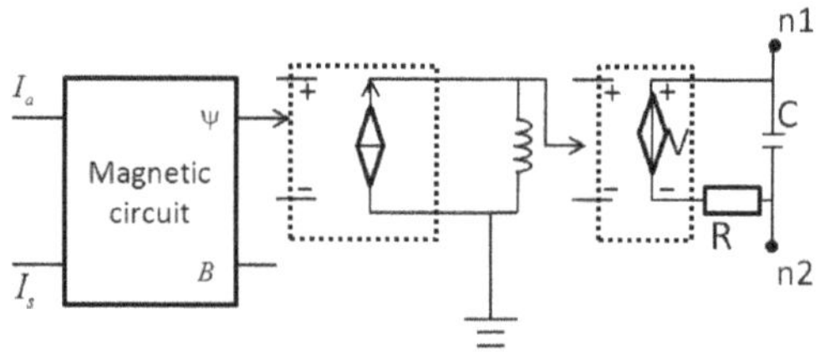

(**b**)

Figure 4. Parasitic parameters circuit. (**a**) Magnetic core and coil diagram; (**b**) Details of the mutual inductance.

The parasitic resistance and the parasitic capacitance can be calculated using the empirical formula. The electromagnetic induction effect of the coil inductance is equivalent to the time differential of the magnetic flux. The average magnetic flux in the magnetic core is computed as follows:

$$\begin{aligned}\psi &= N\int_S \mathbf{B}\cdot d\vec{S} = \frac{N\int_L \int_S \mathbf{B}\cdot d\vec{S}dl}{L} = \frac{N\int_L \int_S \mathbf{B}\cdot \hat{t}dV}{L} \\ &= \frac{N}{L}\sum_{m=1}^{M} \mathbf{B}_m\cdot \vec{t}_m \Delta V_m \\ &= \frac{N}{L}\sum_{m=1}^{M} \mu_0(\mathbf{M}_m + \mathbf{H}_m)\cdot \vec{t}_m \Delta V_m \end{aligned} \tag{19}$$

where L is the coil length, N is coil turns, M the number of tetrahedral elements of the magnetic core surrounded by the coil, $\hat{t}$ is the unit tangential vector of the magnetic core, ΔV_m is the volume of the m-th tetrahedral element. $\mathbf{B}_m$, $\mathbf{M}_m$ and $\mathbf{H}_m$ are, respectively, the magnetic induction strength, magnetization strength and magnetic field strength of the m-th tetrahedral element.

The voltage of the inductor in Figure 4a is obtained by the circuit in Figure 4b. The left part in Figure 4b is the magnetic component model in Section 3.1. The middle part in Figure 4b is the differential circuit, composed of a voltage-controlled current source and inductance. By calling upon the magnetic component model, the induced voltage V is obtained through the differential circuit. I_a is the measured current and I_s is the compensation coil current. B is the magnetic induction intensity at the air gap. ψ is the average magnetic flux.

3.3. SPICE Netlist of Magnetic Circuit Model

One MSP-VIE sub-equation corresponds to one sub-circuit. The netlist for the sub-circuit corresponding to the i-th node (Figure 3) is shown in Figure 5. If the traditional method is applied to generate the netlist, a large circuit for n sub-equations (13) needs to be drawn in Workview. Through programming, using 'ofstream' head file and loop statements to change the node number, netlists for all sub-equations in (13) were written to the output file.

```
*Net list for circuit in Fig.3
EPhi0(i)  np(i)  0  POLY(2)  ncable  0  ncoil  0  0  C(i)  K(i)
R(i)        np(i)       nn(i)  1
EPhid(i)  nn(i)  0  POLY(3n)
+mxnl  0          mxn2 0      ...mxnn  0
+mynl  0          myn2 0      ...mynn  0
+mznl  0          mzn2 0      ...mznn  0
+0
+Dx(il)              Dx(i2)        ...Dx(in)
+Dy(il)              Dy(i2)        ...Dy(in)
+Dz(il)              Dz(i2)        ...Dz(in)

*Net list for H and M
EHx(m)  nhx(m)  0  POLY(4)
+np(i) nn(i)          np(j) nn(j)     np(k) nn(k)     np(1) nn(1)
+0   Gx(i)            Gx(j)           Gx(k)           Gx(1)
EHy(m)  nhy(m)  0  POLY(4)
+np(i) nn(i)          np(j) nn(j)     np(k) nn(k)     np(1) nn(1)
+0 Gy(i)              Gy(j)           Gy(k)           Gy(1)
EHz(m)  nhz(m)  0  POLY(4)
+np(i) nn(i)          np(j) nn(j)     np(k) nn(k)     np(1) nn(1)
+0  Gz(i)             Gz(j)           Gz(k)           Gz(1)
Xmh  nhx  nhy  nhz  mxn  myn  mzn  M-H-circuit
```

Figure 5. Netlist for the i-th node circuit (Figure 3).

4. Fast Methods of Field–Circuit Combined Simulation

Compared with the simple equivalent models, the model in Section 3 considers the core structure, coil position and other factors in the equivalent circuit model, with the parameter matrices **D**, **G**, **C**, **K** in MSP-VIE regarded as the equivalent circuit parameters. At the same time, these many parameters also result in complicated models, and simulation-time consumption.

A well-established method in electromagnetic analysis, when symmetry exists, half, quarter or eighth models are widely used, to simplify the model and save computation time, because the number of elements is reduced. Another fast method has been developed, as follows:

Because of the sparsity characteristic of the integral equation in the far area, there are a large number of small components in the corresponding matrix elements. The value of matrix elements represents the coupling relationship between sub-circuits, so it is feasible to utilize the matrix sparsity to reduce the circuit complexity and simplify the magnetic SPICE model.

The basic sparse strategy is 'remove the small and preserve the big', that is, to retain the large matrix elements and discard the small matrix elements. The size of the element value is judged by a sparse factor SF (man-made setting). If the ratio of the element in the **D** matrix to the maximum value is greater than the SF, the element is retained, otherwise, the element is removed.

$$\mathbf{D} = \begin{bmatrix} D_{11} & \cdots & D_{1j} & \cdots \\ \vdots & \ddots & \vdots & \\ D_{i1} & \cdots & D_{ij} & \cdots \\ \vdots & & \vdots & \end{bmatrix} \begin{cases} SF > \left| \frac{D_{ij}}{D_{imax}} \right|, \text{ removed} \\ SF \leq \left| \frac{D_{ij}}{D_{imax}} \right|, \text{ reserved} \end{cases} \tag{20}$$

With symmetry in the magnetic field and matrix sparsity, the number of matrix parameters in the circuit is reduced; therefore, the SPICE model and netlist is simplified accordingly.

5. Numerical Example

5.1. Model Description

The detailed model for Figure 1 is shown below in Figure 6. R_m is the sampling resistance. The circuit between node n1 and node n2 is the magnetic core, with coil, which is described in Figure 4. According to the different application environments, the actual magnetic cores are various. This paper takes the silicon steel core shown in Figure 7 as an example, to test the proposed method. The outer diameter is 45 mm, the inner diameter is 35mm, the height is 5mm, and the air gap is 1.5 mm. The number of magnetic core meshes is 233. The number of compensation coil turns is 5000.

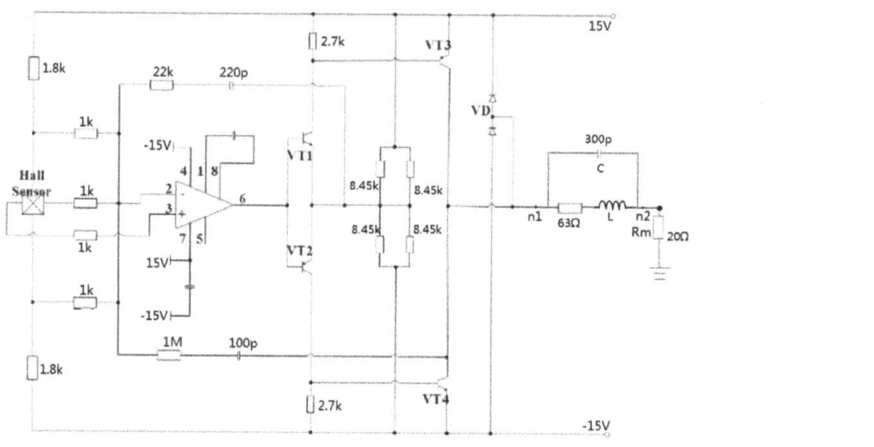

Figure 6. Circuit schematic for sensor.

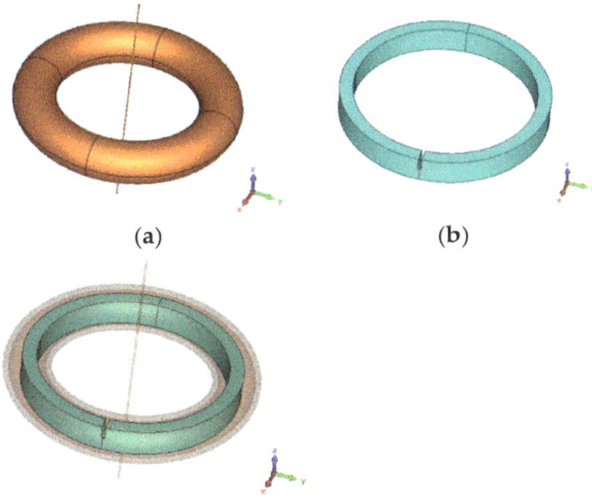

(**a**) (**b**)

(**c**)

Figure 7. Numerical model of magnetic part. (**a**) Coil and straight wire; (**b**) Magnetic core with air gap; (**c**) Position relationship between coil and core.

5.2. Simulation Results

Based on the SPICE model and netlist file, the joint simulation of the entire sensor can be implemented in SPICE.

Figure 8 is the experimental device. The current flowing through the feedback coil can be obtained by measuring the voltage of the sampling resistor R_m. Figure 9 shows the result (output current of the sensor) comparison between measurement and simulation, where the measured currents have different frequencies and amplitudes.

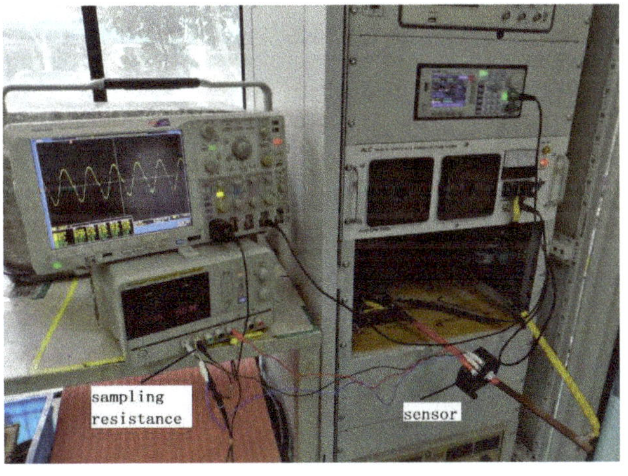

Figure 8. Experimental device.

As shown in Figure 9, the proposed method can be used to carry out the field–circuit combined simulation for the sensor and obtain the transient output current for different measured input currents, which indicates that the method dynamically simulates the magnetic balance working process and gives the time-domain response. There are some differences between simulation and practice. This is because the parasitic capacitance of the coil is a distributed parameter, and it is not very accurate to use one capacitance description.

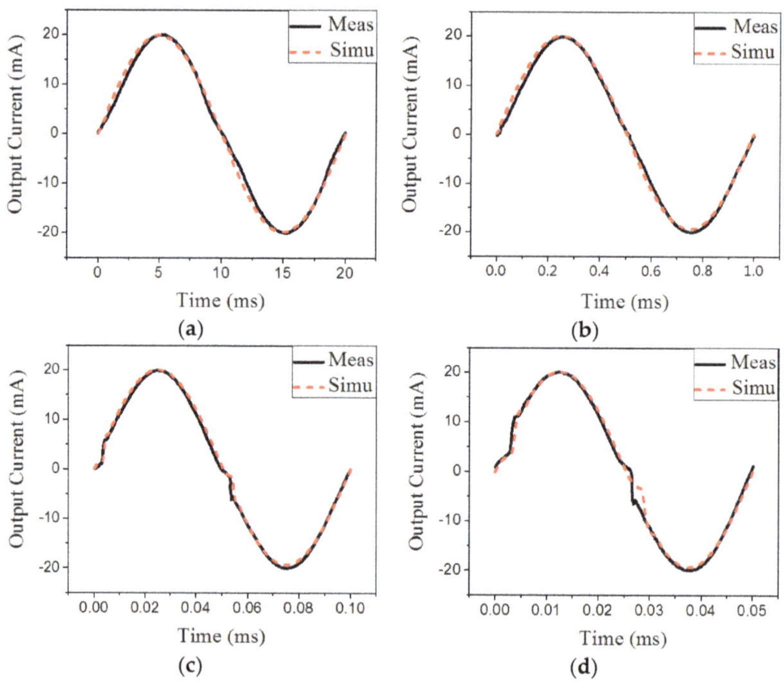

Figure 9. Comparison between simulation and measurement. (**a**) 100 A & 50 Hz; (**b**) 100 A & 1 KHz; (**c**) 100 A & 10 KHz. (**d**) 100 A & 20 KHz.

Figure 10 shows the simulation waveform for distortion analysis. As the frequency of the measured signal increases, the induced voltage increases and exceeds the power-supply voltage, resulting in n1 voltage peak shaving (Figure 10a). Coupled with the influence of parasitic capacitance, the current in the sampling resistor (Figure 10b) is distorted.

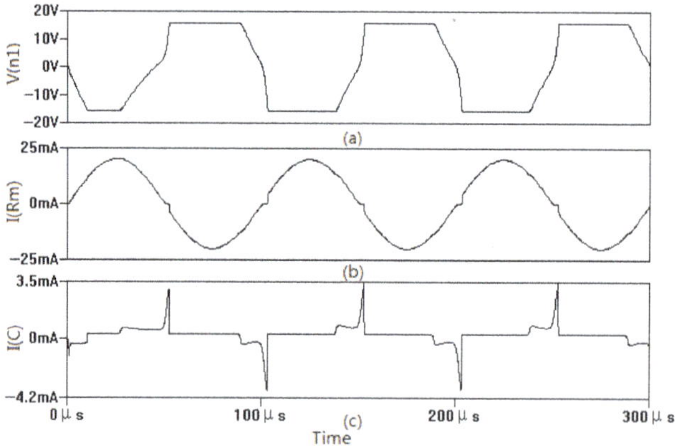

Figure 10. Simulation waveform at 100 A & 10 KHz for distortion analysis. (**a**) Voltage at node n1. (**b**) Current flowing through the sampling resistor. (**c**) Current flowing through the parasitic capacitance.

5.3. Fast Method Validation

Some numerical simulations were completed to evaluate the proposed fast method.

In the matrix sparsity method, the sparse factors were set as 0 (representing no sparse), 0.1, 0.01, and 0.001, respectively, for the SPICE simulation, and the measured current inputs were set as 100 A & 20 KHz. The results were as follows in Figure 11:

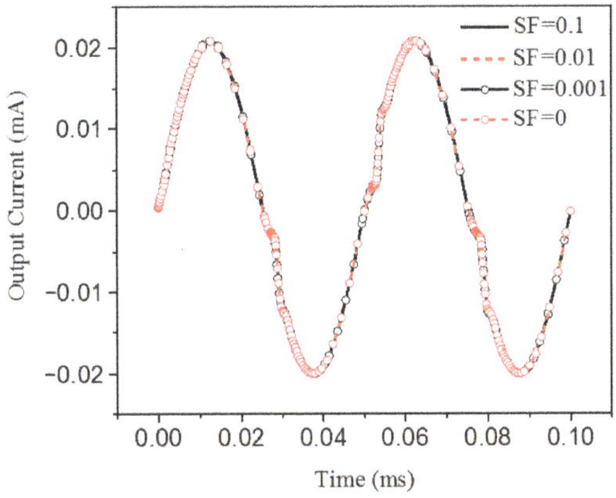

Figure 11. Simulation results with different SF.

The comparison of simulation time under different sparse factors is displayed in the following Table 1. The numerical results show that the matrix sparsity greatly improves the SPICE simulation speed, with high accuracy.

Table 1. Simulation consumption with different SF.

SF	SPICE Netlist Size (kb)	Simulation Time (s)
0.1	5047	620.06
0.01	16,836	3690.46
0.001	42,381	12,380.85
0	71,836	18,205.75

6. Conclusions

In this paper, a field–circuit combined simulation method, based on the magnetic scalar potential volume integral equation (MSP-VIE), is proposed, which discretizes the magnetic part of the sensor, by the electromagnetic field numerical method, and realizes field–circuit combination through solving the magnetic field equation in the circuit. The matrix extracted from MSP-VIE is put into the circuit as the equivalent circuit parameters of the magnetic part, and the magnetic scalar potential equation in the field is equivalent to the Kirchhoff voltage equation in the circuit, thus finishing the magnetic SPICE modeling. In addition, this paper presents a fast method with which to accelerate the SPICE simulation, which makes the method more convenient in practical applications.

Simulation and measurement results reveal that this field–circuit combined method can dynamically simulate the working process of the magnetic balance current sensor and produce the time-domain transient response, as well as simulate some hidden performance details, by accurately considering the magnetic core. All of these factors make this method suitable for use in design, as well as product examination in practical engineering.

Author Contributions: Conceptualization, D.N.; methodology, D.N.; software, D.L.; validation, S.H. and Y.L.; formal analysis, D.N., D.L., S.H. and Y.L.; investigation, D.N.; resources, S.H.; data curation, D.L.; writing—original draft preparation, Y.L.; writing—review and editing, Y.L.; visualization, D.L.; supervision, S.H.; project administration, S.H. All authors have read and agreed to the published version of the manuscript.

Funding: This research received no external funding.

Data Availability Statement: Not applicable.

Conflicts of Interest: The authors declare no conflict of interest.

References

1. Liu, Z. Overview of hall current sensor optimal design. In Proceedings of the 2020 IEEE International Conference on High Voltage Engineering and Application (ICHVE), Beijing, China, 6–10 September 2020.
2. Liu, B.; Sun, Y.; Ding, Y.; Cao, P.; Liu, A.; Ong, S.Y.; Tiong, M.; Cheng, G.; Islam, M.N.; Jain, R.; et al. Low-power and high-sensitivity system-on-chip hall effect sensor. In Proceedings of the 2017 IEEE Sensors, Glasgow, UK, 29 October–1 November 2017.
3. Xu, C.; Liu, J.-G.; Zhang, Q.; Xu, C.; Yang, Y. Investigation of the thermal drift of open-loop Hall Effect current sensor and its improvement. In Proceedings of the IEEE International Workshop on Applied Measurements for Power Systems (AMPS), Aachen, Germany, 23–25 September 2015.
4. Ivan, Y.; Sen, M.; Balabozov, I.; Kostov, I. Influence of magnetic concentrator on Hall effect based current sensor. *COMPEL Int. J. Comput. Math. Electr.* **2018**, *37*, 1481–1488.
5. Pankau, J.; Leggate, D.; Schlegel, D.; Kerkman, R.; Skibiniski, G. High frequency modeling of current sensors. In Proceedings of the 14th Annual Applied Power Electronics Conference and Exposition, (Cat. No. 99CH36285), Dallas, TX, USA, 14–18 March 1999.
6. Pejovic, P.; Stankovic, D. Transient analysis of compensated Hall-effect current transducers. In Proceedings of the 1997 21st International Conference on Microelectronics, Nis, Yugoslavia, 14–17 September 1997.
7. Ai, X.; Bao, H.; Song, Y.H. Novel method of error current compensation for Hall-effect-based high-accuracy current transformer. *IEEE Trans. Power Del.* **2005**, *20*, 11–14. [CrossRef]
8. Sixdenier, F.; Raulet, M.A. Current sensor modeling with a FE-Tuned MEC: Parameters identification protocol. *IEEE Sens. J.* **2012**, *12*, 859–863. [CrossRef]
9. Qiu, J.; Liu, J.; Zhang, Q.; Lin, J. Simulation and optimization of conductor structural parameters of free-space hall-effect current sensor. In Proceedings of the 2014 IEEE International Workshop on Applied Measurements for Power Systems Proceedings (AMPS), Aachen, Germany, 24–26 September 2014.
10. Sen, M.; Balabozov, I.; Yatchev, I.; Ivanov, R. Modelling of current sensor based on hall effect. In Proceedings of the 2017 15th International Conference on Electrical Machines, Drives and Power Systems (ELMA), Sofia, Bulgaria, 1–3 June 2017.
11. Ayhan, B.; Ucak, C. Improved magnetic circuit model for magnetic shielding effectiveness in rogowski coil. *IEEE Trans. Magn.* **2020**, *56*, 1–10. [CrossRef]
12. Luo, M.; Dujic, D.; Allmeling, J. Leakage flux modeling of multiwinding transformers for system-level simulations. *IEEE Trans. Power Electron.* **2018**, *33*, 2471–2483. [CrossRef]
13. Wei, Z.; Rao, X.; He, X.; Wang, Y.; Qiu, A. Circuit simulation with nonlinear magnetic core of a new linear transformer driver dtage. *IEEE Trans. Plasma Sci.* **2019**, *47*, 4084–4090. [CrossRef]
14. Grebennikov, N.; Talakhadze, T.; Kashuba, A. Equivalent magnetic circuit for switched reluctance motor with strong mutual coupling between phases. In Proceedings of the 2019 26th International Workshop on Electric Drives: Improvement in Efficiency of Electric Drives (IWED), Moscow, Russia, 30 January–2 February 2019.
15. Li, J.; Liu, S.; Guo, B.; Hou, X.; Yang, S.; Qiu, S. Method for modelling of fluxgate sensor including the geometry influence. *IET Sci. Meas. Technol.* **2014**, *8*, 214–219. [CrossRef]
16. Carpentier, A.; Chadebec, O.; Galopin, N.; Meunier, G.; Bannwarth, B. Resolution of Nonlinear Magnetostatic Problems with a Volume Integral Method Using the Magnetic Scalar Potential. *IEEE Trans. Magn.* **2013**, *49*, 1685–1688. [CrossRef]
17. Brachtendorf, H.G.; Eck, C.; Laur, R. Macromodeling of hysteresis phenomena with SPICE. In *Circuits and Systems II: Analog and Digital Signal Processing*; IEEE: New York, NY, USA, 1997; Volume 44, pp. 378–388.

machines

MDPI

Communication

Electromagnetic and Mechanical Analysis and Measurements of Interior Permanent Magnet Motors Based on Equivalent Magnetic Circuit Method

Young-Keun Lee [1], Tae-Kyoung Bang [2], Seong-Tae Jo [1], Yong-Joo Kim [3], Kyung-Hun Shin [4,*] and Jang-Young Choi [1,*]

1 Department of Electrical Engineering, Chungnam National University, Daejeon 34134, Korea
2 Advanced Brake System Engineering Cell, Hyundai Mobis, Yongin-si 16891, Korea
3 Department of Smart Agriculture Systems, Chungnam National University, Daejeon 34134, Korea
4 Department of Power System Engineering, Chonnam National University, Jeonnam 59626, Korea
* Correspondence: kshin@chonnam.ac.kr (K.-H.S.); choi_jy@cnu.ac.kr (J.-Y.C.)

Abstract: This paper is about the magnetic field analysis of an interior permanent magnet motor (IPM motor) by using the equivalent magnetic circuit method (EMC method), which requires a small amount of computation time compared with the finite element method (FEM). IPM motors have a specific shape of rotor in which the permanent magnets are embedded. Therefore, in the bridge region, the magnetic saturation is generated due to the shape of the permanent magnet, which affects the magnetic flux density distribution in the air gap and the characteristics of the IPM motor. Thus, to design an IPM motor, the magnetic saturation effects should be considered along with the rotor shape. In addition, because the rotor of the IPM rotates at a high speed directly in connection with the load, the stress generated from the rotor must be stably distributed. Consequently, according to the rotor shape characteristics of the IPM, the stress is concentrated in the thin bridge region during high-speed rotation. When the stress generated in the bridge region exceeds the yield stress of the rotor iron core material, the bridge part is destroyed. Therefore, it is important to analyze the stress that occurs in the rotor during high-speed rotation in the rotor design stage of the IPM. In this study, we analyzed the magnetic field characteristics of an IPM motor using its equivalent magnetic circuit while considering the magnetic saturation in the bridge region. The stability of the rotor was determined by presenting a safety factor based on the maximum stress generated at the rotor for each speed. We derived the stator natural frequency to evaluate the resonance possibility between the electrical frequency and the stator natural frequency. Finally, the validity of the constructed equivalent magnetic circuit was verified by comparing the results with those obtained via the FEM analysis and experiments.

Keywords: equivalent magnetic circuit; magnetic field analysis; permanent magnet; interior permanent magnet motor; stress analysis

Citation: Lee, Y.-K.; Bang, T.-K.; Jo, S.-T.; Kim, Y.-J.; Shin, K.-H.; Choi, J.-Y. Electromagnetic and Mechanical Analysis and Measurements of Interior Permanent Magnet Motors Based on Equivalent Magnetic Circuit Method. *Machines* **2022**, *10*, 915. https://doi.org/10.3390/machines10100915

Academic Editors: Kan Liu and Wei Hu

Received: 6 September 2022
Accepted: 5 October 2022
Published: 9 October 2022

1. Introduction

Through the development of permanent magnet manufacturing technology, the use of permanent magnet electric motors having the advantages of high efficiency and high output power is increasing. These permanent magnet motors can be divided into two types depending on the shape of the rotor: surface-mounted permanent magnet (SPM) motors and interior permanent magnet (IPM) motors. Because the permanent magnet is attached to the surface of the rotor in SPM motors, it is necessary to fabricate a non-magnetic-material tube for preventing the magnet from scattering during high-speed operation. On the other hand, IPM motors could prevent the scattering of permanent magnets during high-speed operation because the permanent magnet is embedded in the rotor. However, placing the permanent magnet in the rotor causes it to have a magnetic polarity due to the differences in reluctance. Therefore, IPM motors could make better use of the magnetic and reluctance torques during operation time, and, through field weakening control strategies, make use of their wide output range and their advantages at high-speed operation [1,2]. The analysis methods for permanent magnet motors can be divided into the analytical method, the finite element method (FEM), and the equivalent magnetic circuit (EMC) method. The analytical method consists in analyzing the magnetic field characteristics of the motor by solving its differential equations using Maxwell's equations [3–6]. It is able to derive high accuracy and relatively quick analysis results compared with FEM. In order to analyze the magnetic field, we should solve the equation derived by Maxwell equation, which has a high level of difficulty. In the case of FEM, it is easy to analyze through commercial tools. However, in order to obtain the accuracy of analytical results, it is necessary to have a large number of element divisions, which leads to a long analysis time. Therefore, for having reliable results, the FEM analysis model needs to appropriate the number of elements depending on the experience of the analyzer. The EMC method does not consist in performing analyses through complex mathematical equations but can be equally used to analyze the magnetic circuit without taking into consideration the shape of the machine while considering the magnetic saturation of the magnetic circuit [7]. In addition, if the saturation of the rotor bridge region is not considered, the magnitude of the back-EMF derived through the air-gap magnetic flux density becomes inaccurate. This leads to difficulty in deriving detailed dimensions of the rotor and stator during the initial design of the IPM. Therefore, considering the proposed saturation region, the EMC method in this paper contributes to the fast characteristic analysis and the initial design of the bar-type IPM.

Furthermore, the IPM should identify the mechanical characteristics owing to the rotor structure in which the permanent magnet is embedded in the rotor iron core. Therefore, although the electromagnetic characteristics are excellent, if the maximum stress exceeds the yield stress of the rotor core, the core may be damaged during high-speed operation. The mechanical stability was evaluated by comparing the stress distribution and magnitude of the maximum stress generated by the rated speed. In addition, when the resonant frequency of the stator and rotor is close to the electromagnetic excitation frequency, the vibration increases further. For this reason, it is necessary to analyze the structural characteristics of the motor, namely the natural frequency and the mode shape. Therefore, this paper performed the magnetic field characteristics of an IPM motor, which were analyzed using the EMC method, and the rotor stability was determined by analyzing the stress generated by the rotor when rotating at the rated speed. Furthermore, by deriving the stator mode frequency, we evaluated the resonance. Finally, its feasibility was verified by comparison with the results of the FEM analysis.

2. Mechanical and Magnetic Characteristics Analysis Using the Equivalent Magnetic Circuit Method Considering the Magnetic Saturation of Bridge Region

2.1. The Analysis Model and Assumptions

The actual model and the analysis model used for our EMC analysis are shown in Figures 1 and 2, respectively.

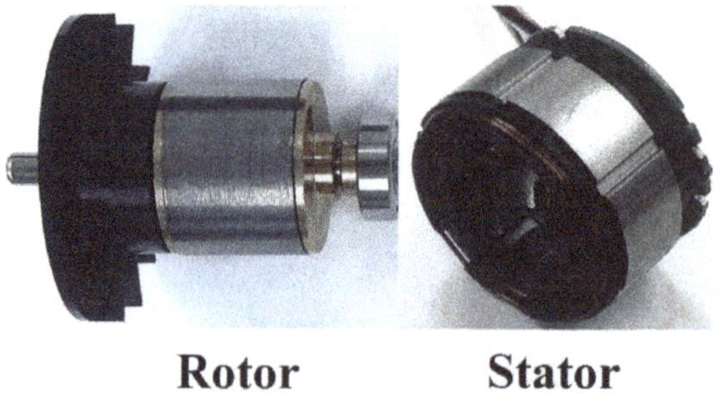

Figure 1. The actual model used for the equivalent magnetic circuit analysis.

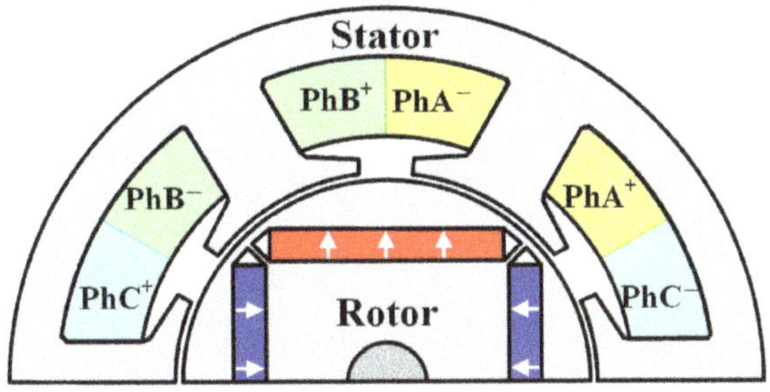

Figure 2. The shape of analysis model.

They consisted of an IPM motor with a combination of four poles and six slots with concentrated winding. For our analysis, some simplifying assumptions were required for using the EMC method. First, the magnetic permeability of the iron core was assumed to be infinite. Secondly, the magnetic saturation was ignored, except for the magnetic saturation at the rib and bridge regions. Thirdly, the demagnetization of the permanent magnet was not considered. Fourthly, the end effect in the axial direction was not considered. Fifthly, the slot effect was ignored. The rotor geometry of the analysis model used consisted of a barrier, a bridge, a pole piece, a rib, and a permanent magnet, while having the shape of a typical IPM motor. The analysis model used for the EMC method after applying the simplifying assumptions and its equivalent magnetic circuit according to the rotor components are shown in Figures 3 and 4, respectively. Each of the resistors represents the reluctance of a component of the motor according to our assumptions.

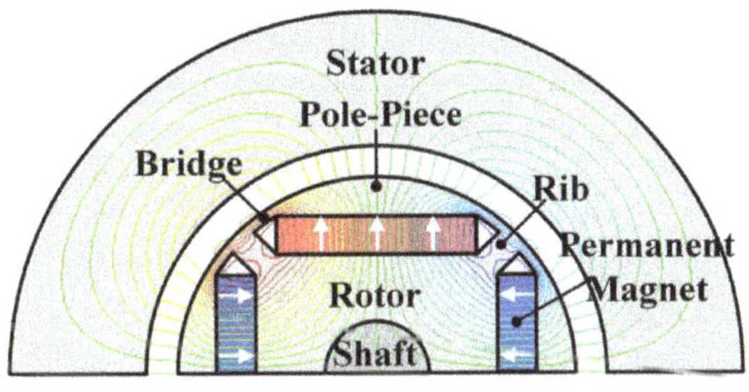

Figure 3. The shape of analysis model with assumptions.

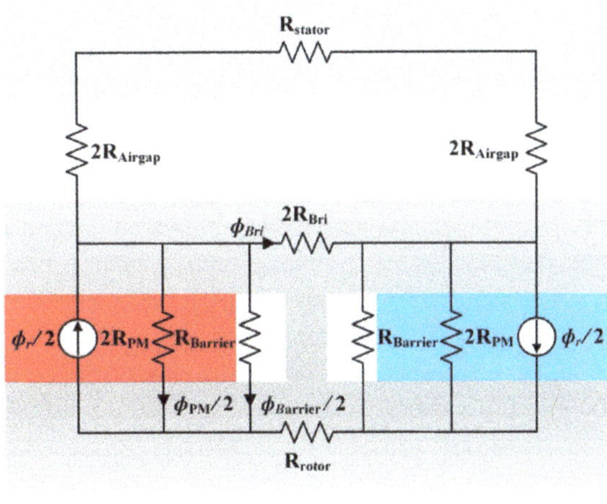

Figure 4. The magnetic circuit constructed for Equivalent Magnetic Circuit Method.

2.2. The Magnetic Characteristic Analysis

To analyze the magnetic field characteristics using the EMC method, a circuit analysis has to be performed. In this case, because the reluctances and magnetic fluxes are represented by resistances and currents in the equivalent electrical circuit, respectively, their definition should be explained. Figure 5 shows the concept of reluctance in the bridge region. If the angle of the bridge region from the end of the barrier to the pole piece region increases, the magnetic saturation becomes more intense as it approaches the permanent magnet, so that the magnitude of the reluctance of the bridge region increases from the end of the barrier to the permanent magnet. In order to consider the magnetic saturation of the bridge region, the reluctance of the bridge region consisted of the reluctance units, as shown Figure 5. These units of reluctance on the bridge region can be expressed as follows in Equations (1)–(6).

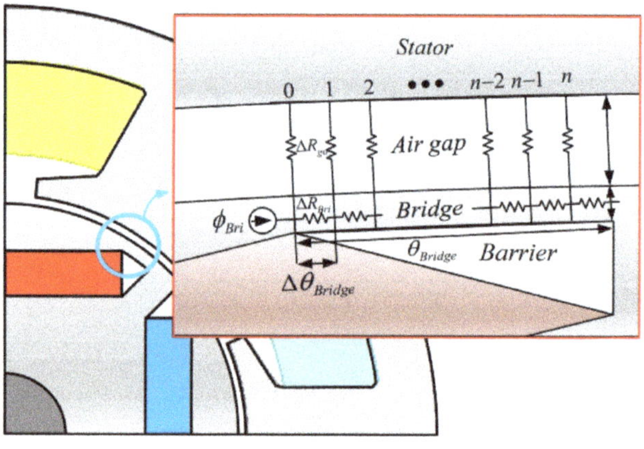

Figure 5. The concept of the reluctance to consider the magnetic saturation of the bridge region of IPM rotor.

For calculating the reluctance units of the bridge region, the nonlinear magnetic permeability should be considered, which can be derived via curve-fitting methods. There are parameters that need to be defined before deriving the reluctances and magnetic fluxes using the equations.

Table 1 and Figure 6 show the parameter values required to derive the rotor's reluctance and magnetic flux. An appropriate thickness of the bridge should be chosen so as to prevent breakage due to inertia during high-speed operation. The thickness of the bridge depends on the size of the device, but is usually chosen to be within 0.6 mm. Therefore, the formula for the magnetic flux of the equivalent magnetic circuit can be summarized as shown in Equations (1)–(3).

$$\phi_r = B_r \cdot A_{PM} = B_r \cdot W_{PM} \cdot L_{stk} \tag{1}$$

$$\phi_{Bri} = B_{sat} \cdot A_{Bri} = B_{sat} \cdot T_{Bri} \cdot L_{stk} \tag{2}$$

$$\frac{\phi_{Airgap}}{2} = \frac{\phi_r}{2} - \left(\frac{\phi_{PM}}{2} + \frac{\phi_{Barrier}}{2} + \phi_{Rib}\right) \tag{3}$$

Here, A_{PM} is the area of the permanent magnet, A_{Bri} is the area of the bridge, and B_{sat} is the saturation flux density at the bridge. The reluctance of each region can be expanded as shown in Equations (4)–(6).

$$R_{Airgap} = \frac{g_0}{\mu_0 \cdot A_{Airgap}} \tag{4}$$

$$R_{PM} = \frac{H_{PM}}{\mu_0 \cdot \mu_r \cdot A_{PM}} = \frac{H_{PM}}{\mu_0 \cdot \mu_r \cdot W_{PM} \cdot L_{stk}} \tag{5}$$

$$R_{Barrier} = \frac{H_{Barrier\_Avg}}{\mu_0 \cdot W_{Barrier\_Avg} \cdot L_{stk}} \tag{6}$$

Here, R_{Airgap} is the reluctance of the air gap, R_{PM} is the reluctance of the permanent magnet, $R_{Barrier}$ is the reluctance of the barrier, $H_{Barrier\_Avg}$ is the height of the barrier, and $W_{Barrier\_Avg}$ is the width of the barrier. In order to derive the magnetic flux density distribution in the air gap, the area at the center of the gap should be calculated as follows. In our case, we considered the reluctances of the air gap having the same angle. Thus, the reluctance of the bridge region can be expressed by Equations (7)–(9) [8–10].

$$A_{Airgap} = \alpha_p \cdot \frac{2\pi \cdot (R_{si} - g_0/2)}{2p} \cdot L_{stk} \tag{7}$$

$$\Delta R_{go} = \frac{g_0}{\mu_0 \cdot (R_{so} \cdot \Delta\theta_{Bridge} \cdot \pi/180) \cdot L_{stk}} \tag{8}$$

$$\Delta R_{Bri} = \frac{(R_{ro} \cdot \Delta\theta_{Bridge} \cdot \pi/180)}{\mu_0 \cdot \mu_r \cdot T_{Bridge} \cdot L_{stk}} \tag{9}$$

$$\lambda_a = k_w \cdot N_{ph} \cdot \phi_{airgap} \cos(\theta_e) \tag{10}$$

$$e_a = -\frac{d\lambda_a}{dt} \tag{11}$$

Through the results of the magnetic flux density distribution at the air gap, the magnetic flux of one phase winding can be expressed in the form of its fundamental wave. The magnitude of the back-EMF can be derived through the linkage flux of one phase winding. For this purpose, the basic waveform of the back-EMF can be derived from Equations (10) and (11). Here k_ω is the winding coefficient and θ_e is the electrical angular degree. Figure 7 shows the comparison results between the waveform of the back-EMF derived through the finite element analysis and the back-EMF derived through the proposed EMC method.

Through Figure 7, it was confirmed that the air-gap flux density of the EMC method considering the saturation of the bridge region presented in this paper coincides with the air-gap flux density derived from the finite element analysis.

Table 1. Parameters for the equivalent magnetic circuit method for considering the magnetic saturation in the bridge region.

Parameter	Value	Unit
Pole number: $2 \cdot p$	4	-
Height of PM: H_{PM}	2	mm
Length of Air gap: $g_{Air-gap}$	0.5	mm
Thickness of Bridge: T_{Bri}	0.8	mm
Length of Stack: L_{stk}	15	mm
Outer Radius of Stator: R_{so}	48	mm
Pole-Arc: P_{arc1}	45	deg
Magnet-Arc: P_{arc3}	33.75	deg
Bridge-Arc: θ_{Bridge}, $(P_{arc2} - P_{arc3})$	6.21	deg

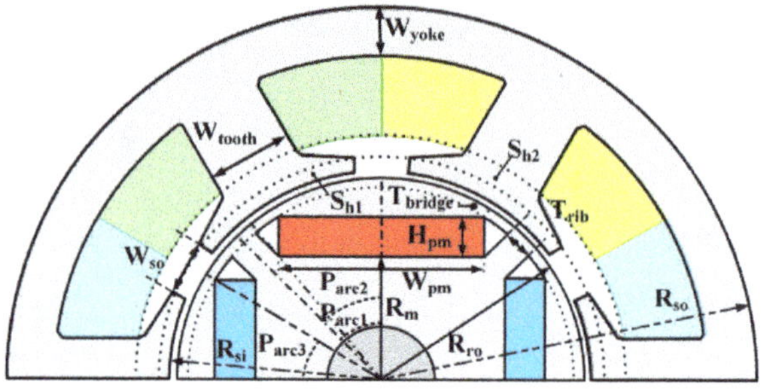

Figure 6. The parameter of the equivalent magnetic circuit method for the magnetic saturation of the bridge region.

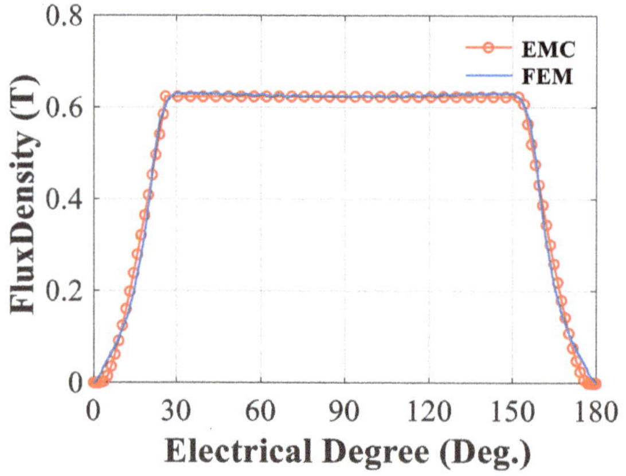

Figure 7. The magnetic flux density distribution of the equivalent magnetic circuit method considering the magnetic saturation of the bridge region.

2.3. *Rotor Stress Analysis*

In the IPM, the design of a bridge to secure the stability of the rotor against high-speed rotation is important, and the width is determined by the stress-limiting condition of the rotor core. Considering only the electromagnetic characteristics, it is recommended to narrow the width of the bridge in each barrier layer to minimize the leakage of the magnetic flux remaining in the rotor as well as maximize the saliency ratio [11,12]. However, if the width of the bridge is designed to be narrow, and the driving speed of the device increases according to the requirements, the bridge of the rotor may be damaged by the centrifugal force generated by the rotation of the motor. Therefore, a wider width and better consideration of the mechanical strength can prevent damage to the rotor [10]. In the motor designer's aspect, it is essential in the design stage to select the appropriate bridge width through electromagnetic and structural analysis of the rotor, as well as to determine whether the rotor core is damaged at operating speed through stress distribution. In this study, the stability was identified based on the safety factor derived from the maximum stress generated from the rotor, and the safety rate was derived using Equation (12). Table 2 show the mechanical property of core and permanent magnet of rotor.

$$Safety\_Factor = \frac{Tensile\ Yield\ Strength\ [\text{MPa}]}{Maximum\ Stress\ [\text{MPa}]} \geq 1.5 \quad (12)$$

Table 2. Rotor core and magnet material properties.

Items	Value		Unit
	35PN440	N35SH	
Density	7700	7600	Kg/m^3
Young's modulus	195	160	GPa
Poisson's ratio	0.25	0.24	-
Tensile Yield strength	273	80	MPa

The safety factor, which was determined to be appropriate to operate within the operating speed of the motor, was selected as 1.5. Moreover, as illustrated in Figure 8, it was derived through rotor structure analysis with a safety rate of approximately 43,000 rpm. Therefore, in the case of the IPM motor in this study, it was considered that there was no damage to the rotor core at the rated speed.

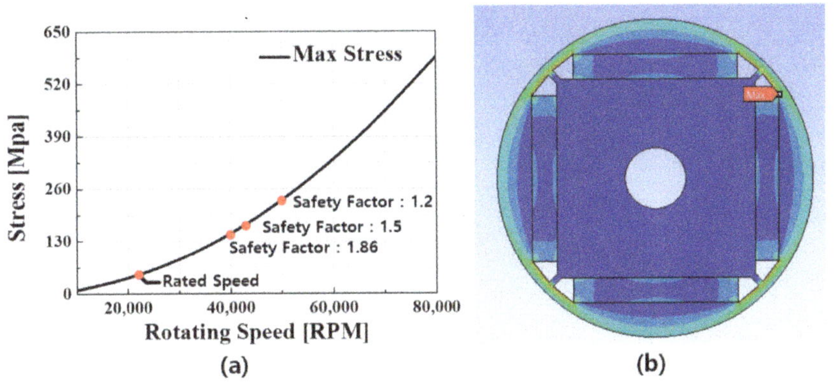

Figure 8. Stress analysis result according to rotating speed and distribution. (**a**) Max stress plot. (**b**) Stress distribution at the rotor core.

2.4. Stator Modal Analysis

As a method for predicting the natural frequency of the stator, the natural frequency and mode shape were predicted by viewing the stator as a simple ring and considering the stator teeth and coil as additional mass. In addition, the stator teeth were considered as cantilever beams connected to the stator and an analysis model was selected. The stator of the motor is a major component of the motor and directly receives the electromagnetic excitation force and has the greatest structural influence on the vibration and noise generation of the motor, such as the resonance frequency being within the frequency range of interest. Figure 9 shows the mode analysis results of the stator core designed using the finite element analysis. The natural frequencies in the 2nd, 3rd and 4th modes are 2800.8 Hz, 7056.7 Hz, and 1120.3 Hz, respectively. From the analysis result, it can be considered that there is no effect due to resonance because there is no corresponding operating range for the electric frequency of the motor and the natural frequency of the stator.

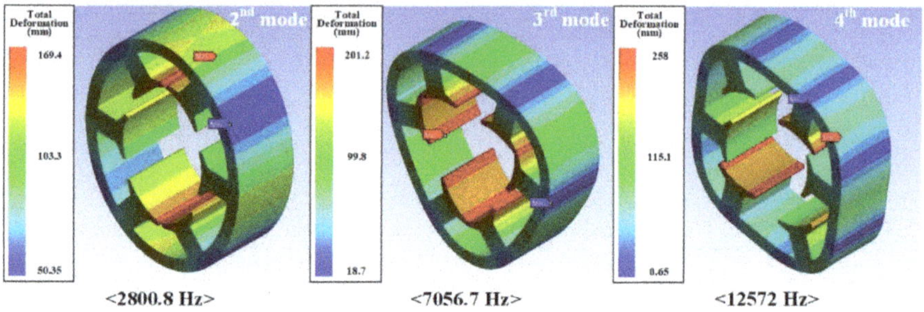

Figure 9. Stator modal analysis results: 2nd mode, 3rd mode, 4th mode.

3. Experimental Validation

Experimental Validation of Analysis Results

A model with the design initially presented was constructed to compare the results of our analysis of the model initially designed (obtained via the equivalent magnetic circuit method presented in this paper) with the measured results and to verify the validity of the design method.

An experimental model was made for the verification of our analysis results. 35PN440 was used as the iron core material for the rotor and stator, and the permanent magnet used was made of N35SH material. Figure 10 shows the measurement system for the no-load back-EMF experiment. No-load back-EMF is the voltage of the motor that occurs when the rotor of each phase of the motor is open and the rotor is rotated at an arbitrary speed through the input driver. In this study, no-load back-EMF experiments were performed at 200 rpm steps from 600 rpm to 1400 rpm.

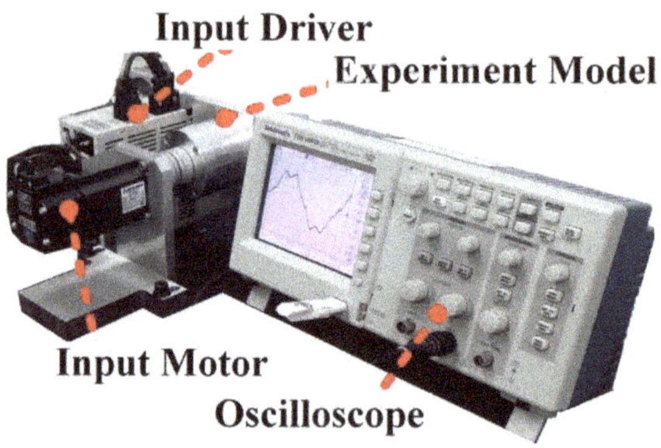

Figure 10. The experiment system for validation of the equivalent magnetic circuit method through measuring the back-EMF.

The measured values were compared with our analytical results using the equivalent magnetic circuit method and the FEM. Figure 11 shows the results of a no-load back-EMF measurement, showing a varying back-EMF and back-EMF constants according to the motor speed. It can be seen that the results of the FEM analysis and the measured waveforms are in good agreement due to the effects of the stator slot shape and the magnetic saturation phenomenon.

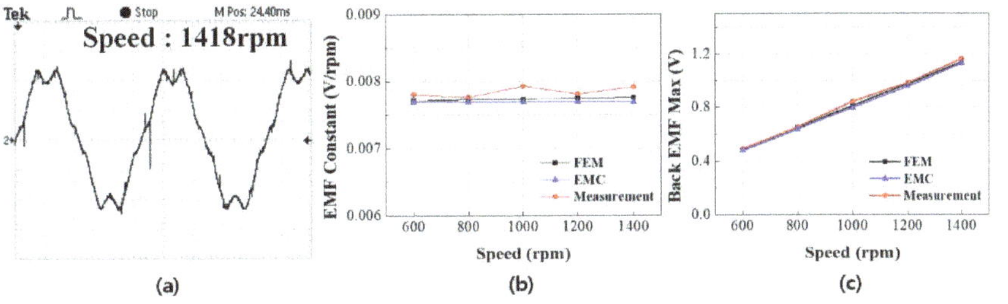

Figure 11. The results of the back-EMF measurement. (**a**) experiment result, (**b**) back-EMF constant, and (**c**) back-EMF max value result.

Because of the influence of harmonics in the experiment, the results of the measurements, those of the equivalent magnetic circuit, and those obtained via the FEM analysis show differences, but these are lower than 5%.

Therefore, it can be seen that the errors of the FEM analysis and the error in the values obtained via the equivalent magnetic circuit did not exceed 5%. Figure 12 shows a graph comparing the changes in the back-EMF values. The errors of the back-EMF values obtained with the proposed method can be seen by comparing them with the analytical and measurement results of the back-EMF experiments.

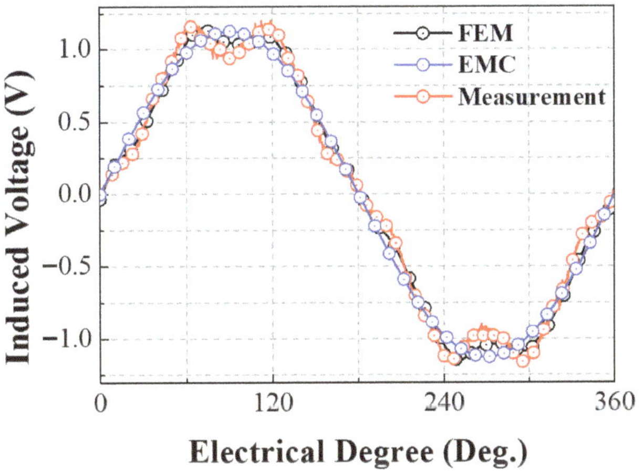

Figure 12. The results of the back-EMF measurement.

4. Conclusions

This paper is about a magnetic characteristics analysis using the equivalent magnetic circuit method while considering the magnetic saturation. The analysis was performed considering the saturation in the bridge region. The validities of the equivalent magnetic circuit model and the proposed analytical method considering the saturation of the bridge region were verified by comparing their results with the FEM results. The experimental results of the back-EMF measurement of the proposed analytical model were compared with the results of the FEM analysis using the equivalent magnetic circuit method. We confirmed that the error between the analytical results obtained via the equivalent magnetic circuit method and the results of the FEM analysis is within 5%. In addition, the safety factor was derived based on the maximum stress generated in the rotor according to each speed through analysis of the rotor structure, and the mechanical stability at the rated speed was identified through the safety factor. Based on the results, it is possible to design a rotor structure using the EMC method proposed in this study, and the IPM mechanical stability can be determined through structural analysis. Finally, the resonance between the electrical frequency and the natural frequency of the stator was determined through the stator modal analysis, and the effect on the vibration and noise caused by the resonance was analyzed. This process and methodology can contribute to the design of IPM rotors.

Author Contributions: K.-H.S.: conceptualization, review and editing; J.-Y.C.: conceptualization, review and editing; Y.-K.L.: analysis, original draft preparation; T.-K.B.: experiment, motor control algorithm and co-simulation; S.-T.J.: experiment and motor control algorithm; Y.-J.K.: experiment and motor control algorithm. All authors have read and agreed to the published version of the manuscript.

Funding: This work was supported by the Korea Institute of Planning and Evaluation for Technology in Food, Agriculture, Forestry (IPET) through the Eco-friendly Power Source Application Agricultural Machinery Technology Development Program, funded by the Ministry of Agriculture, Food and Rural Affairs (MAFRA) (322047-5). This work was supported by the National Research Foundation of Korea (NRF) grant funded by the Korea government (MSIT). (No.2020R1A2C1007353).

Institutional Review Board Statement: Not applicable.

Informed Consent Statement: Not applicable.

Data Availability Statement: The data presented in this study are available on request from the corresponding author.

Conflicts of Interest: The authors declare no conflict of interest.

References

1. Gerada, D.; Mebarki, A.; Brown, N.L.; Gerada, C.; Cavagnino, A.; Boglietti, A. High-Speed Electrical Machines Technologies, Trends, and Developments. *IEEE Trans. Ind. Electr.* **2014**, *61*, 2946–2959. [CrossRef]
2. Kim, H.K.; Hur, J. Dynamic Characteristic Analysis of Irreversible Demagnetization in SPM- and IPM-Type BLDC Motors. *IEEE Trans. Ind. Appl.* **2012**, *53*, 982–990. [CrossRef]
3. Lee, T.Y.; Seo, M.K.; Kim, Y.J.; Jung, S.Y. Motor Design and Characteristics Comparison of Outer-Rotor-Type BLDC Motor and BLAC Motor Based on Numerical Analysis. *IEEE Trans. Ind. Appl.* **2016**, *26*, 5205506. [CrossRef]
4. Yu, J.S. Optimal Design of Interior Permanent Magnet Synchronous Motor considering Magnetic Saturation and Flux Weakening Control Capability. Master's Thesis, Chungnam National University, Daejeon, Korea, 2014.
5. Lubin, T.; Mezani, S.; Rezzoug, A. Exact Analytical Method for Magnetic Field Computation in the Air Gap of Cylindrical Electrical Machines Considering Slotting Effects. *IEEE Tran. Magn.* **2010**, *46*, 1092–1099. [CrossRef]
6. Zhu, Z.Q.; Howe, D.; Chan, C.C. Improved Analytical Model for Predicting the Magnetic Field Distribution in Brushless Permanent-Magnet Machines. *IEEE Trans. on Magn.* **2002**, *38*, 229–238. [CrossRef]
7. Bianchi, N.; Bolognani, S.; Luise, F. Analysis and design of a PM Brushless Motor for high-speed operations. *IEEE Trans. Energy Convers.* **2014**, *20*, 629–637. [CrossRef]
8. Kim, W.H.; Kim, K.C.; Kim, S.J.; Kang, D.W.; Go, S.C.; Lee, H.W.; Chun, Y.D.; Lee, J. A Study on the Optimal Rotor Design of LSPM Considering the Starting Torque and Efficiency. *IEEE Tran. Magn.* **2005**, *45*, 1808–1811. [CrossRef]
9. Kim, K.C.; Koo, D.H.; Hong, J.P.; Lee, J. A Study on the Characteristics Due to Pole-Arc to Pole-Pitch Ratio and Saliency to Improve Torque Performance of IPMSM. *IEEE Tran. Magn.* **2007**, *43*, 2516–2518. [CrossRef]
10. Lee, J.-J.; Lee, J.; Kim, K.-S. Design of a WFSM for Electric Vehicle Based on a Nonlinear Magnetic Equivalent Circuit. *IEEE Trans. Appl. Supercon.* **2018**, *28*, 5206304. [CrossRef]
11. Shin, K.H.; Choi, J.Y.; Cho, H.W. Characteristic Analysis of Interior Permanent-Magnet Synchronous Machine with Fractional-Slot Concentrated Winding Considering Nonlinear Magnetic Saturation. *IEEE Trans. Appl. Supercond.* **2016**, *26*, 5200404. [CrossRef]
12. Cao, H.; Wang, S.; Huang, S. Performance Analysis of IPMSM Based on Stress Calculation of Iron Core. In Proceedings of the 2017 IEEE Transportation Electrification Conference and Expo, Asia-Pacific, Harbin, China, 7–10 August 2017.

machines

Article

Research on Parameter Design and Control Method for Current Source Inverter–Fed IM Drive Systems

Pengyun Song, Yanghui Liu and Chao Liu *

College of Electronical Engineering, Southwest Minzu University, Chengdu 610225, China
* Correspondence: 202031006203@stu.swun.edu.cn

Abstract: In the current source inverter (CSI)-fed inductor motor (IM) drive system, DC-bus inductance and AC-filter capacitance have a direct impact on the dynamic response speed, power quality and power density. In addition, due to the addition of filter capacitors, the mathematical model formed by the IM and capacitor is a third-order system, which increases the difficulty of parameter tuning for the control loop. To solve the above problems in the CSI, a DC-bus inductance design method based on current response speed and the ripple of the DC-link current and an AC-filter capacitance design method based on current utilization and filtering characteristics are presented. Then, the analytical expressions between the open-loop cut-off frequency, phase margin and the PI controller parameters of the current loop and speed loop are derived. Finally, an experimental platform is established to validate the proposed method.

Keywords: inductor motor (IM); current source inverter (CSI); parameter design; tuning; control loop

Citation: Song, P.; Liu, Y.; Liu, C. Research on Parameter Design and Control Method for Current Source Inverter–Fed IM Drive Systems. *Machines* **2022**, *10*, 922. https://doi.org/10.3390/machines10100922

Academic Editors: Kan Liu and Wei Hu

Received: 18 August 2022
Accepted: 28 September 2022
Published: 10 October 2022

1. Introduction

Inverters can be divided into voltage source inverter (VSI) and current source inverter (CSI) according to the characteristics of the power supply in the DC side [1,2]. For a long time, VSI has been a research hotspot and has been widely used in various AC drive and grid-connected applications [3,4]. However, the current ripple on the DC side of VSI is large, and it can only operate under the buck mode; deadtime is needed to prevent DC power supply from passing through [5]. Compared with VSI, CSI can boost the voltage and operate without short circuit protection [6]. Due to the presence of filter capacitors on the AC side, the voltage of the stator side of the motor has good waveform quality [7], so in many work situations, CSI is more suitable than VSI [8,9].

Although the CSI has the above advantages, due to some inherent defects, the application scope of CSI in motor drive systems is restricted. Instead of a constant current source, DC-link current is generated from a DC-bus inductor with a DC voltage source, and the DC-bus inductor is charged or discharged under different switching states, so the DC-link current of traditional CSI cannot be controlled [10,11]. Meanwhile, the CSI is unable to operate under the buck mode; otherwise, the DC-link current will increase continuously [12]. To achieve the control of DC-link current and a wide range of output voltage of CSI, many novel CSI topologies are proposed. In Refs. [13,14], a PWM current source rectifier is used as the front stage to control the DC-link current, but this topology is only suitable for the occasion of AC-DC-AC. In Ref. [15], a quasi-Z-source current source converter is added into the DC side, but the difficulty of control increases greatly with the addition of too many passive components. In Ref. [11], the bypass IGBT was paralleled on the DC-link inductor through the control of the switch device. Under the buck operation, the continuous increase of the DC-link current is effectively restrained. Compared to the above topologies, in Refs. [16,17], a bidirectional DC chopper is added in the front stage of the CSI; this topology is simple and consists of only a pair of switching tubes and diodes, and the DC-link current control can be achieved under the operations of boost, buck, and deceleration by regulating the switching state of the bidirectional DC chopper. Therefore, this paper takes the topology of CSI with bidirectional DC chopper as the research object.

In the AC side of CSI, due to the presence of filter capacitors, the mathematical model formed by the motor and capacitor is a third-order system, which increases the design difficulty of the control system of the inductor motor (IM). In recent years, the vector control technologies for the CSI-fed drive system have been studied by some scholars. In Ref. [13], the references of stator current in the d-q axis are obtained by the controllers of rotor flux and speed, respectively; then, the references of the output current of the inverter bridge in the d-q axis are calculated according to the steady state equation of the IM. However, this control strategy is more sensitive to the accuracy of the IM parameters. In Ref. [18], a small signal mathematical model of a CSI-fed permanent magnet synchronous motor is constructed through the state space averaging method; a speed control system based on duty cycle regulation is proposed, but the steady-state performance of this method is not ideal, and the stator current has obvious harmonic distortion. In Ref. [19], various decoupling methods and the active damping method are presented and compared, and a control structure of voltage inner loop and current outer loop are proposed. In addition to the method of decoupling and controlling, DC-bus inductance and AC-filter capacitance also have a direct impact on the dynamic response speed, power quality, and power density [20]; however, research on parameter design methods for DC-bus inductance and AC-filter capacitance is rarely reported.

In this paper, the design principles of DC-bus inductance and AC-filter capacitance are firstly studied. Then, to reduce the number of the control loops, a vector control technology without a voltage loop is proposed, and the analytical expressions of the control parameters are derived. Finally, under various conditions, experimental validation of the proposed method is carried out.

2. Topology and Mathematical Model

2.1. Topology of CSI-Fed IM Drive System with Bidirectional DC Chopper

To realize the decoupling control of the DC-link current and IM, a bidirectional DC chopper is added between the DC voltage source and DC-link inductor; the topology is shown in Figure 1. In the figure, u_{dc} is the DC input voltage; i_{dc} is the current of DC-link inductor L_{dc}; S_1–S_6 are six switching tubes of the inverter bridge; D_1–D_6 are the diodes connected in series with S_1–S_6; C_a, C_b and C_c are the three-phase filter capacitors in the AC side; i_{at}, i_{bt} and i_{ct} are the three-phase output currents of the inverter bridge; i_{as}, i_{bs} and i_{cs} are the stator currents of the IM; u_{as}, u_{bs} and u_{cs} are the three-phase voltages of the filter capacitors. The bidirectional DC chopper is composed of the diodes D_a, D_b and the switching tubes S_a, S_b.

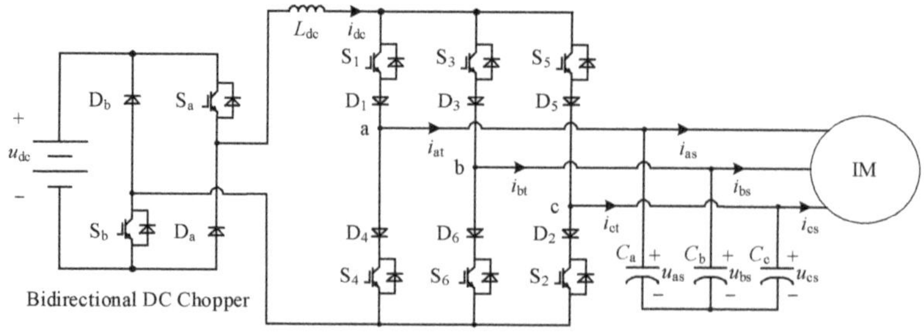

Figure 1. The topology of CSI-fed IM drive system with bidirectional DC chopper.

When the control method proposed in Ref. [17] is adopted, the circuit of DC input side can be regarded as a controlled current source. In the process of mathematical modeling and control system design for the AC side of CSI, the influence on the DC-link current can be ignored, and the topology of the CSI-fed IM drive system with bidirectional DC chopper can be simplified as seen in the circuit shown in Figure 2.

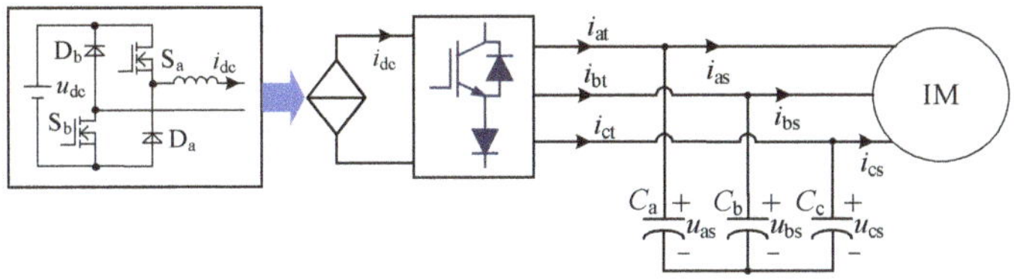

Figure 2. The equivalent circuit of CSI-fed IM drive system with bidirectional DC chopper.

2.2. Mathematical Model of Filter Capacitors and IM

In the synchronous rotating frame (d-q axis) based on rotor flux orientation, the IM voltage equation can be expressed as

$$\begin{cases} u_{ds} = \sigma L_s \frac{di_{ds}}{dt} + i_{ds} R_s - \sigma \omega_s L_s i_{qs} + \frac{L_m d\psi_r}{L_r dt} \\ u_{qs} = \sigma L_s \frac{di_{qs}}{dt} + i_{qs} R_s + \sigma \omega_s L_s i_{ds} + \omega_s \frac{L_m \psi_r}{L_r} \end{cases} \tag{1}$$

where i_{ds} and i_{qs} are the stator currents in the d-q axis; u_{ds} and u_{qs} are the voltages of filter capacitor in the d-q axis; ψ_r is the rotor flux; σ, R_s, L_m, L_r and w_s are the flux leakage coefficient, stator resistor, mutual inductance, rotor inductance and synchronous angular frequency, respectively.

According to Figure 2, the current equation in three-phase stationary coordinate frame can be expressed as

$$\begin{cases} C\frac{du_{as}}{dt} = i_{at} - i_{as} \\ C\frac{du_{bs}}{dt} = i_{bt} - i_{bs} \\ C\frac{du_{cs}}{dt} = i_{ct} - i_{cs} \end{cases} \tag{2}$$

where C represents the value of filter capacitors. Clarke transform and Park transform are carried out on Formula (3); in turn, the current equation in the d-q axis can be obtained and expressed as

$$\begin{cases} C\frac{du_{ds}}{dt} = i_{dt} - i_{ds} + \omega_s C u_{qs} \\ C\frac{du_{qs}}{dt} = i_{qt} - i_{qs} - \omega_s C u_{ds} \end{cases} \tag{3}$$

where i_{dt} and i_{qt} are the output current of the inverter bridge in the d-q axis. The rotor flux equation in the d-q axis is expressed as

$$T_r \frac{d\psi_r}{dt} + \psi_r = L_m i_{ds} \tag{4}$$

where ψ_r and L_m are the rotor flux and excitation inductor of IM, respectively, and T_r is the rotor time constant. The equations of torque and speed are expressed as

$$\begin{cases} T_e = n_p \frac{L_m}{L_r} \psi_r i_{qs} \\ J\frac{d\omega_r}{dt} = T_e - T_L \end{cases} \tag{5}$$

where J is the moment of inertia, ω_r is the mechanical speed, and T_e and T_L are the electromagnetic torque and load torque, respectively.

In summary, the mathematical model of filter capacitors and IM in the AC side of CSI can be represented by Equation (1) to Equation (5).

3. Parameter Design of DC-Link Inductance and AC-Filter Capacitance

3.1. Selection Range of DC-Link Inductance

A strong output impedance and constant current source characteristics of the DC side in CSI can be obtained by a larger DC-link inductance. However, the current regulation becomes slower as the inductance increases, which leads to overmodulation and generates low-order current harmonics. Therefore, it is necessary to determine a selection range of DC-link inductance.

In the start-up stage, the DC-link current is too low to provide the required current for the IM, so the zero switching state is adopted, and L_{dc} is charged by u_{dc}; the equivalent circuit is shown in Figure 3a. The state equation of i_{dc} is expressed as

$$L_{dc}\frac{di_{dc}}{dt} = u_{dc} \tag{6}$$

where T_{up_max} is the maximum allowable charging time, making sure that the charging time for i_{dc} to increase from zero state to maximum DC-link current i_{dc_max} is less than T_{up_max}. L_{dc} should satisfy the following condition:

$$L_{dc} < \frac{u_{dc}T_{up\_max}}{i_{dc\_max}} \tag{7}$$

According to the conclusion in Ref. [21], the current fluctuation Δi_{dc} can be calculated when the IM is powered independently by L_{dc}, which contains two active switching states in one carried period. For example, in sector I, the equivalent discharge circuits are shown in Figure 3b,c, and Δi_{dc} is expressed as

$$\Delta i_{dc} = \frac{3T_sUI\cos\theta}{2L_{dc}i_{dc}} \tag{8}$$

where T_s is the carried period, equal to 100 μs; U is the fundamental amplitude of the stator voltage; I is the fundamental amplitude of the stator current; θ is the phase deviation between stator voltage and stator current. In other sectors, the voltage of L_{dc} and the switching state are shown in Table 1, and the expression of Δi_{dc} is consistent with that of sector I, as shown in Equation (8). Making sure that Δi_{dc} is less than the maximum ripple Δi_{dc_max} under θ = 1, L_{dc} should satisfy the follow condition:

$$L_{dc} > \frac{3m_{imax}m_{umax}T_su_{dc}}{2\Delta i_{dc\_max}} \tag{9}$$

where m_{imax} and m_{umax} are the maximum modulation ratio and boost ratio of CSI, respectively. Combining Equations (7) and (9), the selection range of L_{dc} is expressed as

$$L_{dc} \in (\frac{3m_{imax}m_{umax}T_su_{dc}}{2\Delta i_{dc\_max}}, \frac{u_{dc}T_{up\_max}}{i_{dc\_max}}) \tag{10}$$

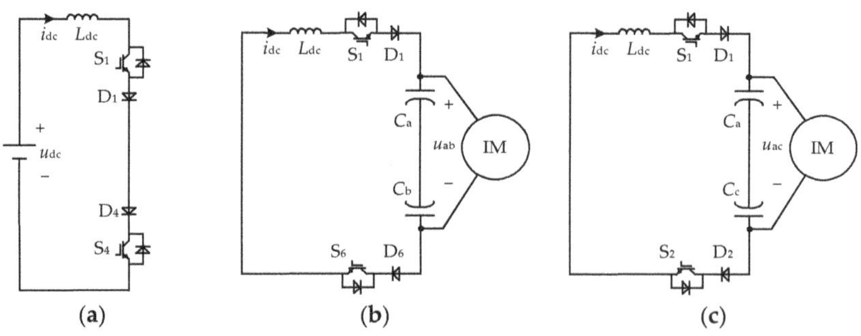

Figure 3. The equivalent circuits of charging and discharging for L_{dc}. (**a**) Equivalent charging circuit; (**b**) equivalent discharging circuit when S_1 and S_6 are turned on; (**c**) equivalent discharging circuit when S_1 and S_2 are turned on.

Table 1. The voltage of L_{dc} and switching state in sectors II–VI.

Sector	On-State Switch	Voltage of L_{dc}
II (active vector 1)	S_1 and S_2	u_{ac}
II (active vector 2)	S_2 and S_3	u_{bc}
III (active vector 1)	S_2 and S_3	u_{bc}
III (active vector 2)	S_3 and S_4	u_{ba}
IV (active vector 1)	S_3 and S_4	u_{ba}
IV (active vector 2)	S_4 and S_5	u_{ca}
V (active vector 1)	S_4 and S_5	u_{ca}
V (active vector 2)	S_5 and S_6	u_{cb}
VI (active vector 1)	S_5 and S_6	u_{cb}
VI (active vector 2)	S_6 and S_1	u_{ab}

The parameters are set as follows: $m_{imax} = 1$, $m_{umax} = 1$, $u_{dc} = 24$ V, $\Delta i_{dc\_max} = 1$ A, $i_{dc\_max} = 50$ A and $T_{up\_max} = 20$ ms. Substituting the above parameters into Equation (10), the minimum value and maximum value of DC-bus inductance are 3.6 mH and 9.6 mH, respectively.

3.2. Selection Range of AC-Filter Capacitance

The IM is represented by the T-type equivalent circuit; the phase-a circuit model of IM and the capacitor is shown in Figure 4, where $L_{s\sigma}$ and $L_{r\sigma}$ are the stator leakage inductor and rotor leakage inductor, respectively; R_r is the rotor resistor; I_{ac} is the current vector of C_a; and s_r is the slip ratio.

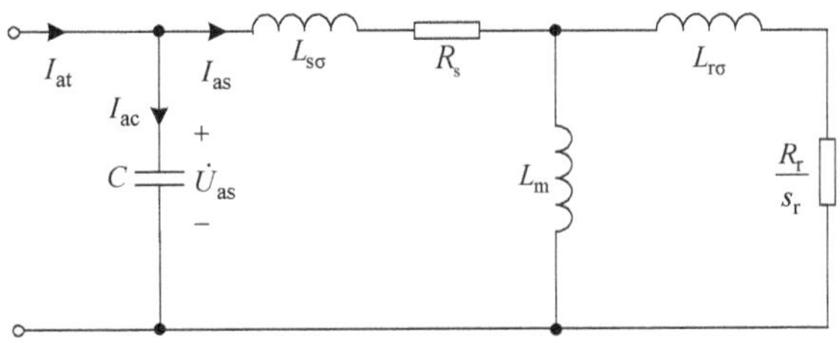

Figure 4. T-type equivalent circuit of IM with filter capacitor.

The vector diagram composed of I_{at}, I_{as}, I_{ac} and U_{as} is shown in Figure 5, and the following relationship can be obtained:

$$I_{at} = I_{ac} + I_{at} = j\omega_s C U_{as} + I_{as} \tag{11}$$

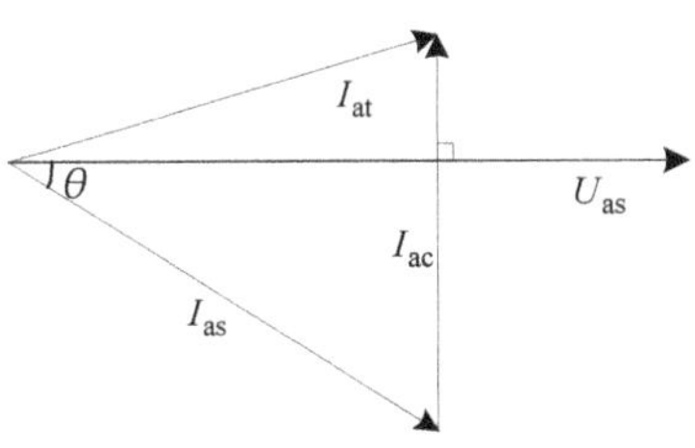

Figure 5. The vector diagram.

In order to maximize the AC current utilization rate, the following equation should be satisfied:

$$I_{at} \leq I_{as} \tag{12}$$

According to the geometric relationship in Figure 5, I_{ac} should be satisfied as follows:

$$I_{ac} \leq I_{as} \sin\theta \tag{13}$$

Substituting Equation (11) into Equation (13), the expression of maximum capacitance C_{max} can be obtained as

$$C \leq C_{max} = \frac{2I_{as} \sin\theta}{\omega_s U_{as}} \tag{14}$$

The mathematical relationship of U_{as}, I_{as} and sinθ can be obtained according to the T-type equivalent circuit of IM shown in Figure 4, and Equation (14) can be rewritten as

$$C \leq C_{max} = \frac{2(\omega_s L_{s\sigma} + A/B)}{\omega_s \sqrt{(R_s + A/B)^2 + (\omega_s L_{s\sigma} + D/B)^2}} \tag{15}$$

where the specific forms of A, B and D are expressed as

$$\begin{cases} A &= \left(\frac{R_r R_m}{s_r} - \omega_s^2 \sigma L_r L_m\right)\left(\frac{R_r}{s_r} + R_m\right) \\ &+\left(\frac{R_r \omega_s L_m}{s_r} + R_m \omega_s \sigma L_r\right)(\omega_s \sigma L_r + \omega_s L_m) \\ B &= \left(\frac{R_r}{s_r} + R_m\right)^2 + (\omega_s \sigma L_r + \omega_s L_m)^2 \\ D &= \left(\frac{R_r \omega_s L_m}{s_r} + R_m \omega_s \sigma L_r\right)\left(\frac{R_r}{s_r} + R_m\right) \\ &-\left(\frac{R_r R_m}{s_r} - \omega_s^2 \sigma L_r L_m\right)(\omega_s \sigma L_r + \omega_s L_m) \end{cases} \tag{16}$$

From Equations (15) and (16), C_{max} is a function of s_r, and ω_s is set as the rated angular frequency of 100π rad/s. Substituting the IM parameters listed in Table 2 into Equations (14) and (15), the relationship curve between C_{max} and s_r can be obtained, which is shown in Figure 6. It is obvious that a minimum value occurs in C_{max}, so the AC-filter capacitance should be less than 2.37 mF.

Table 2. The parameters of the IM.

Parameters	Description	Value
R_s (Ω)	Stator resistance	0.07
R_r (Ω)	Rotor resistance	0.05
L_s (mH)	Stator inductance	4.51
L_r (mH)	Rotor inductance	4.63
L_m (mH)	Mutual inductance	4.38
σ	Flux leakage coefficient	0.088
P_n (kW)	Rated power	1.2
U_n (V)	Rated voltage	48
f_n (Hz)	Rated frequency	50
J (kg·m^2)	Rotational inertia	0.001
n_p	Number of pole pairs	2
R_s (Ω)	Stator resistance	0.07

Next, the influence mechanism on harmonic characteristics of the stator current caused by AC-filter capacitance will be discussed. The T-type equivalent circuit of IM is a third-order system; it is difficult to derive the transfer function and mathematical relationship between the capacitance and bandwidth. Since the high-order harmonic contents near the switching frequency of i_{ds} (i_{qs}) are same as the stator currents in three-phase stationary coordinate, according to Equations (1) and (2), the equivalent circuits of the IM and AC-filter capacitor in the d-q axis based on rotor flux orientation are established, as shown in Figure 7. i_{dt} and i_{qt} are composed of a DC component and several high-order harmonics components near the switching frequency. The equivalent controlled current sources, voltage sources, and back EMFs in Figure 7 cannot amplify or attenuate harmonics, so the suppression effect of high-order harmonics is determined by R_s, σL_s and C, and the transfer function $G_{it2is}(s)$ from i_{dt} (i_{qt}) to i_{ds} (i_{qs}) can be expressed as

$$G_{it2is}(s) = \frac{1}{\sigma L_s C s^2 + R_s C s + 1} \quad (17)$$

and the oscillation angular frequency ω_n can be expressed as

$$\omega_n = \frac{1}{\sqrt{\sigma L_s C}} \quad (18)$$

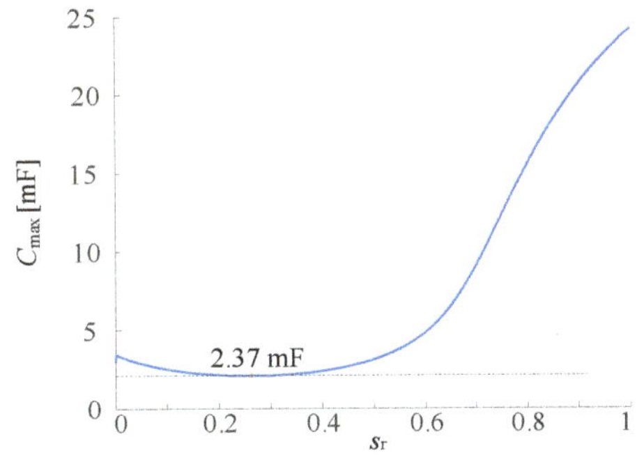

Figure 6. The relationship curve between C_{max} and s_r.

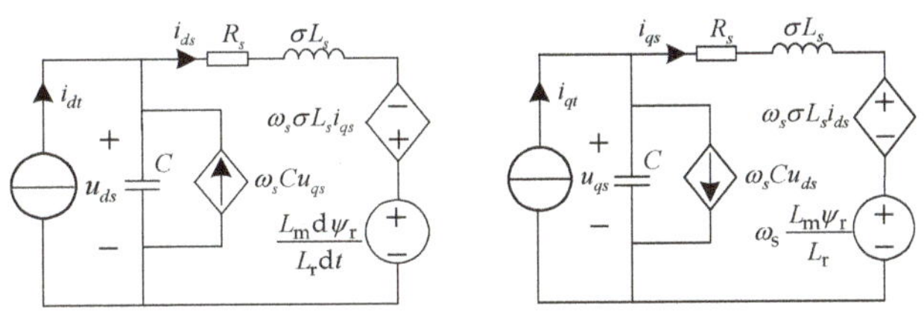

(a) (b)

Figure 7. Second order equivalent circuit of the IM and AC-filter capacitor in the d-q axis: (**a**) d-axis; (**b**) q-axis.

Substituting the value of R_s and L_s into Equation (16), the Bode plot of $G_{it2is}(s)$ is drawn under different AC-filter capacitance, which is shown in Figure 8. It can be seen that $G_{it2is}(s)$ has a resonance spike, and the cutoff frequency decreases with the increase of C. In order to obtain the high quality stator current, $G_{it2is}(s)$ should effectively suppress the harmonic components in i_{at} near the switching frequency f_s, so the oscillation frequency should be far below f_s. This paper stipulates that the oscillation frequency is less than $f_s/2$; according to Equation (17), AC-filter capacitance should be satisfied as

$$C > \frac{1}{\sigma L_s \pi^2 f_s^2} \quad (19)$$

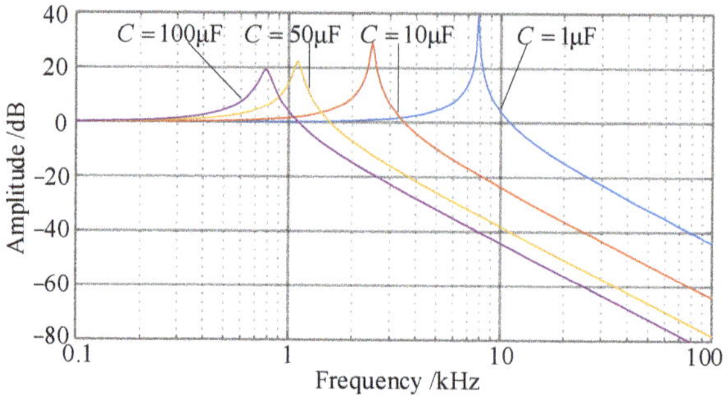

Figure 8. The Bode plot of $G_{\mathrm{it2is}}(s)$ under different AC-filter capacitance.

Substituting the IM parameters listed in Table 2 into Equation (19), the minimum value of C is 2.5 μF.

4. Design Principle for Vector Control Strategy of CSI-Fed IM Drive System

According to Section 1, the mathematical model formed by the IM and capacitor is a third-order system; when the three closed-loop control strategy is adopted, more voltage sensors are needed, and the dynamic response to torque is slower than the VSI-fed IM drive system. In this section, a novel vector control strategy without a voltage inner loop for the CSI-fed IM drive system will be proposed, which includes current loop, flux loop and speed loop.

4.1. Analysis and Parameter Design for Current Loop

The switching delay and digital control delay are equivalent to a first-order transfer function, which is expressed as

$$G_{\mathrm{d}}(s) = \frac{1}{1 + T_{\mathrm{s}}s} \tag{20}$$

Due to the presence of high-frequency noise in the stator current sampling signal, a filter circuit is adopted in this paper, and its equivalent transfer function is

$$G_{\mathrm{cf}}(s) = \frac{1}{1 + T_{\mathrm{cf}}s} \tag{21}$$

where T_{cf} is the filter time constant, equal to 6×10^{-5} in this paper.

Ignoring the coupling terms in Equations (1) and (3), such as $\omega_{\mathrm{s}}\sigma L_{\mathrm{s}}i_{\mathrm{qs}}$, $\omega_{\mathrm{s}}\sigma L_{\mathrm{s}}i_{\mathrm{ds}}$, $\omega_{\mathrm{s}}Cu_{\mathrm{qs}}$ and $\omega_{\mathrm{s}}Cu_{\mathrm{ds}}$, in the d-q axis, the control diagram of the current loop without stator voltage sampling and control is shown in Figure 9, where k_{cp} and τ_{ci} are the proportional coefficient and integral coefficient of PI controller for adjusting the stator current, i^*_{dt} and i^*_{qt} are the reference value of output current of the inverter bridge in the d-q axis, i^*_{ds} and i^*_{qs} are the reference value of the stator currents in the d-q axis.

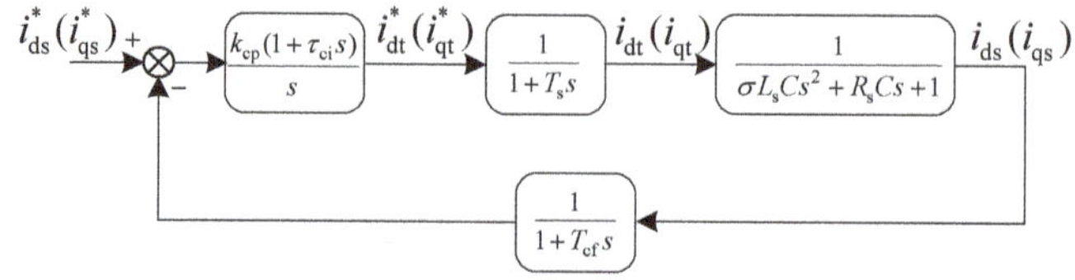

Figure 9. The control diagram of the current loop.

In this paper, the target cut-off frequency and phase margin of current loop are represented as ω_{cc} and φ_{cc}. According to Figure 9, Q_{c_amp}, the amplitude of the current open-loop at ω_{cc}, can be derived as follows:

$$Q_{c\_amp} = \sqrt{\frac{1}{[1+(T_s\omega_{cc})^2][1+(T_{cf}\omega_{cc})^2][(1-\sigma L_s C\omega_{cc}^2)^2 + R_s^2C^2\omega_{cc}^2]}} \tag{22}$$

φ_{c_phase}, the phase of the current open-loop at ω_{cc}, can be derived as follows:

$$\varphi_{c\_phase} = -\arctan(T_s\omega_{cc}) - \arctan(T_{cf}\omega_{cc}) - \arctan(\frac{R_sC\omega_{cc}}{1-\sigma L_sC\omega_{cc}^2}) \tag{23}$$

To ensure that the phase angle margin of the current loop with the PI controller is greater than φ_{cc}, τ_{ci} should be satisfied as follows:

$$\tau_{ci} \geq \frac{\tan(\varphi_{cc} - \varphi_{c\_phase} - \frac{\pi}{2})}{\omega_{cc}} \tag{24}$$

To ensure that the amplitude of the current loop with the PI controller is equal to 1, k_{cp} should be satisfied as follows:

$$k_{cp} = \frac{\omega_{cc}}{Q_{c\_amp}\sqrt{1+\tau_{ci}^2\omega_{cc}^2}} \tag{25}$$

In order to achieve fast dynamic response and the effective suppression of high-order harmonics in the output current of the inverter bridge, ω_{cc} is set to 2000π rad/s, which is one-tenth of the switching frequency. Meanwhile, to obtain a better stability, φ_{cc} is set to $\pi/4$ rad. Substituting the parameters listed in Table 2 into Equation (22) to Equation (25), k_{cp} and τ_{ci} can be calculated as 0.98 and 22.15, respectively. The Bode plot of the current closed-loop transfer function $G_i(s)$ is shown in Figure 10; the bandwidth of $G_i(s)$ is 1kHz, and the amplitude remains 0 dB when the frequency is less than 1kHz. When the frequency is greater than 1 kHz, the amplitude decays with a slope close to −60 dB/dec, which can effectively suppress the high-order harmonics generated in the switching process. This explains that a good steady-state and dynamic performance can be achieved by the PI parameter optimization proposed in this paper. In order to facilitate the subsequent parameter design of the external loop control system, $G_i(s)$ is simplified to a second-order transfer function and expressed as

$$G_i(s) = \frac{3.13\times10^5}{1.65\times10^{-3}s^2 + 37.1s + 3.13\times10^5} \tag{26}$$

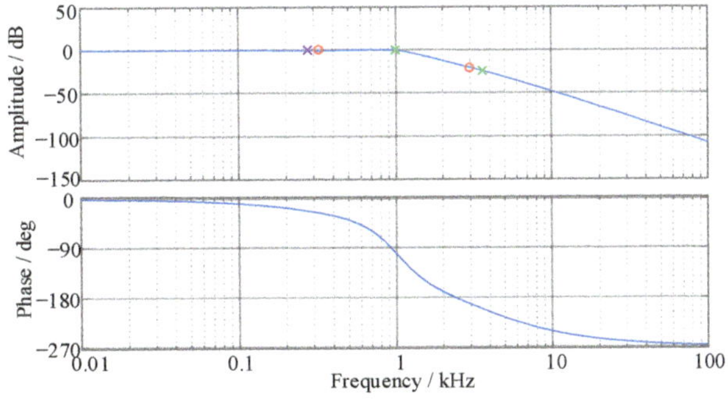

Figure 10. Bode plot of transfer function of current closed loop.

4.2. Analysis and Parameter Design for Flux Loop

A PI controller is adopted to adjust rotor flux ψ_r and is combined with Equation (4); the diagram of the flux loop with a PI controller is shown in Figure 11. $k_{\psi p}$ and $\tau_{\psi i}$ are the proportional coefficient and integral coefficient of the PI controller.

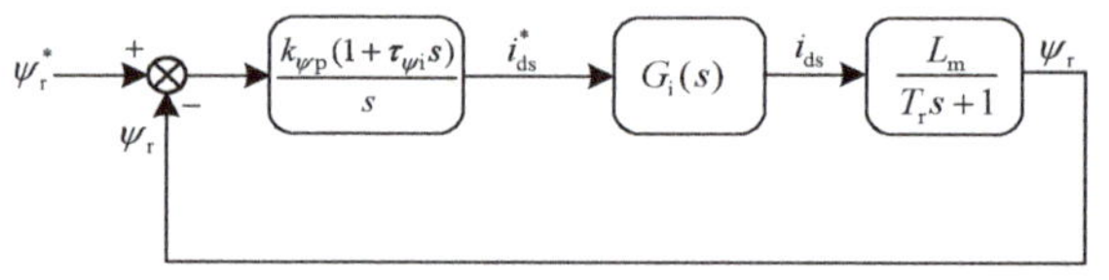

Figure 11. The control diagram of the flux loop.

To simplify the design method of the control parameters, zero-pole cancellation is adopted, and $\tau_{\psi i}$ is set to T_r. Then, the root locus plot is utilized to determine $k_{\psi p}$, which is shown in Figure 12. It is obvious that the condition for the stability of the flux loop is as follows:

$$k_{\psi P} \in (0, 6.21/L_m) \tag{27}$$

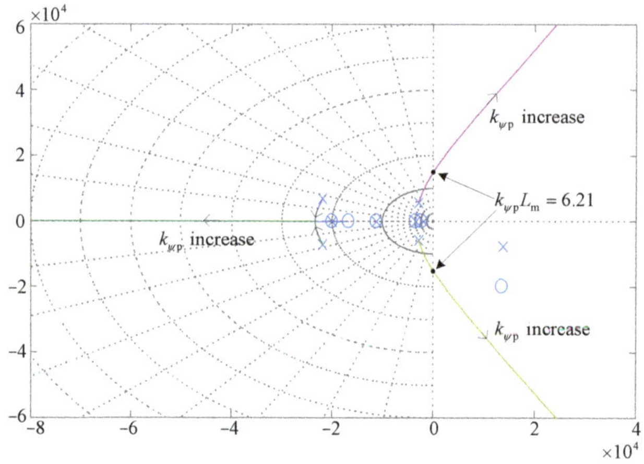

Figure 12. Root locus plot of rotor flux loop.

Finally, $k_{\psi p}$ is obtained by the Sisotool box of Matlab, where $k_{\psi p}$ is equal to $3.15/L_m$; the Bode plot of flux open-loop is shown in Figure 13. The cut-off frequency is 600 Hz, and the phase margin is 60°.

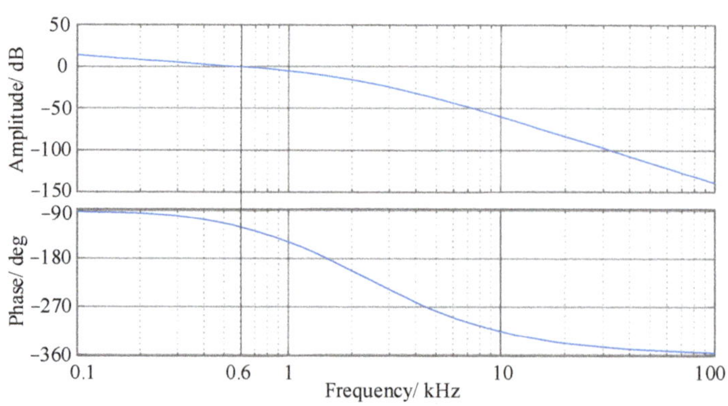

Figure 13. Bode plot of rotor flux open loop.

4.3. Analysis and Parameter Design for Speed Loop

The speed of IM is sampled by an orthogonal encoder, and the sampling frequency is half of f_s; the transfer function of speed sampling and delay is expressed as

$$G_{sf}(s) = \frac{1}{1 + 2T_s s} \tag{28}$$

In accordance with Equations (5) and (6), the block diagram of the speed loop under the PI controller is shown in Figure 14, where k_{sp} and τ_{si} are the proportional coefficient and integral coefficient of the PI controller for adjusting the stator current, respectively. ω_s^* is the rated angular frequency. It is noted that the speed loop contains variable ψ_r, so the cut-off frequency of the speed open loop should be less than that of the flux open loop, and ψ_r is set as 0.05 Wb. The open-loop transfer function of speed $G_{s\_open}(s)$ can be expressed as

$$G_{s\_open}(s) = \frac{3n_p L_m \psi_r k_{sp} G_i(s)(1 + \tau_{si})}{2JL_r s^2(1 + 2T_s s)} \tag{29}$$

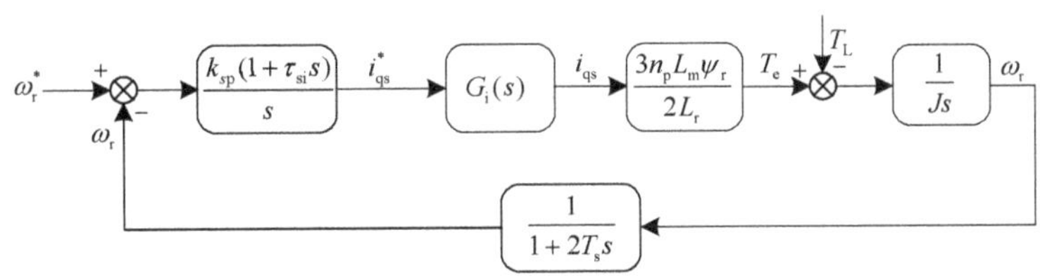

Figure 14. The control diagram of the speed loop.

The close-loop transfer function of speed $G_{s\_close}(s)$ is expressed as

$$G_{s\_close}(s) = \frac{G_{s\_open}(s)(1 + 2T_s s)}{1 + G_{s\_open}(s)} \tag{30}$$

The target cut-off frequency and phase margin of the speed loop are represented as ω_{ss} and φ_{ss}. Using the same method as the current PI controller, the analytical formulas of ω_{ss} and φ_{ss} are derived as follows:

$$\begin{cases} \tau_{si} = \dfrac{\tan(\varphi_{ss} - \varphi_{s\_phase} - \frac{\pi}{2})}{\omega_{ss}} \\ k_{sp} = \dfrac{\omega_{ss}}{Q_{s\_amp}\sqrt{1+\tau_{si}^2\omega_{ss}^2}} \end{cases} \tag{31}$$

where Q_{s_amp} and φ_{s_phase} are the amplitude and phase of the speed open loop at ω_{ss} before the PI controller, respectively. k_{sp} and τ_{si} are obtained by substituting the equation ω_{ss} = 200π rad/s and φ_{ss} = π/6 rad into Equation (31). The specific expression of $G_{s\_close}(s)$ is as follows:

$$G_{s\_close}(s) = \frac{30.89s^4 + 8.5 \times 10^5 s^3 + 9.34 \times 10^9 s^2 + 2.93 \times 10^{13} s + 2.93 \times 10^{11}}{5.425 \times 10^{-10} s^7 + 2.716 \times 10^{-5} s^6 + 0.6s^5 + 7045s^4 + 4.29 \times 10^7 s^3 + 1.01 \times 10^{11} s^2 + 2.93 \times 10^{13} s + 2.93 \times 10^{11}} \tag{32}$$

The Bode plots of open-loop and closed-loop transfer functions are shown in Figure 15. The cut-off frequency of the speed open-loop transfer function is 100 Hz, and the phase margin is 30°. The closed-loop transfer function has a good following characteristic in the bandwidth of 40 Hz. The unit step response of the speed closed-loop transfer function is shown in Figure 16; the rise time is only 2 ms, the overshoot is little, and no steady-state error occurs.

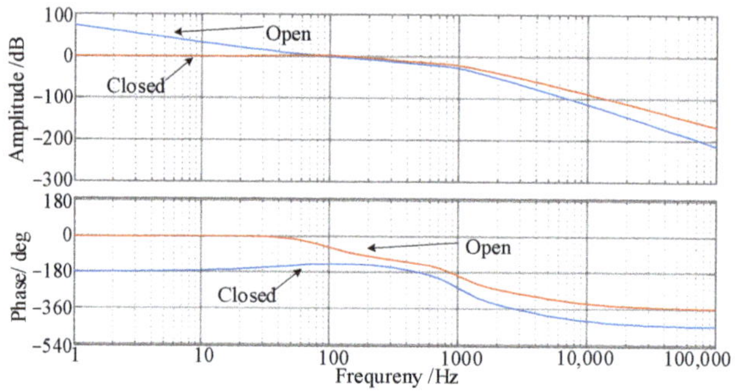

Figure 15. Bode plot of speed open loop and closed loop.

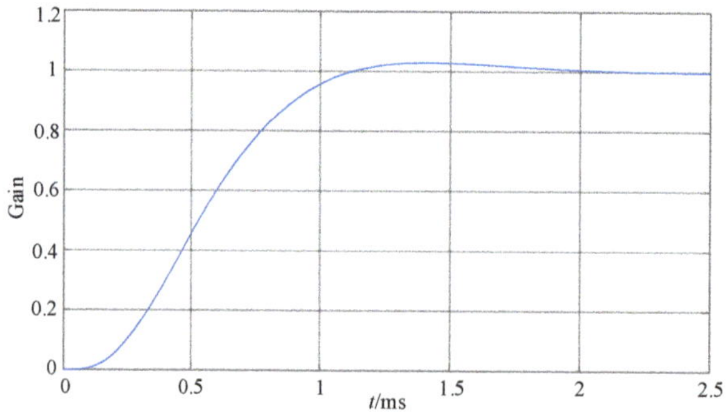

Figure 16. Step response plot of speed closed loop.

5. Simulation Results

The simulation model of the CSI-fed IM drive system with bidirectional DC chopper is built on MATLAB Simulink, including the power module, control module and modulation module. The parameters of IM adopted are consistent with Table 2, and other simulation parameters are shown in Table 3.

Table 3. Simulation parameters.

Parameters	Value
DC-link inductor L_{dc}	4 mH
Filter capacitor C_a, C_b, C_c	66 μF
DC-bus voltage u_{dc}	20 V
Switching frequency	10 kHz
Sampling frequency	10 kHz

5.1. Load Acceleration Condition

The value of load torque is set to 3 N·m, the initial reference speed is 1500 rpm, and the reference speed is 3000 rpm when the time is 0.8 s. The simulation waveforms of stator phase a current, stator currents in the d-q axis, electromagnetic torque, motor speed and rotor flux in the whole simulation process are shown in Figure 17.

To improve DC-link current utilization and reduce losses and harmonic distortion, according to the MTPA method proposed in Ref. [17], the references of rotor flux and DC-link current are adjusted with the change of operating point; the specific principle is as follows:

A. In the dynamic process, in order to provide greater torque, the reference DC-link current is set to 50 A, which is the maximum current of the DC-link inductor.

B. In the steady state, the references of rotor flux ψ_r^* and DC-link current i_{dc}^* are adjusted according to the load torque and speed, which are shown in Table 4.

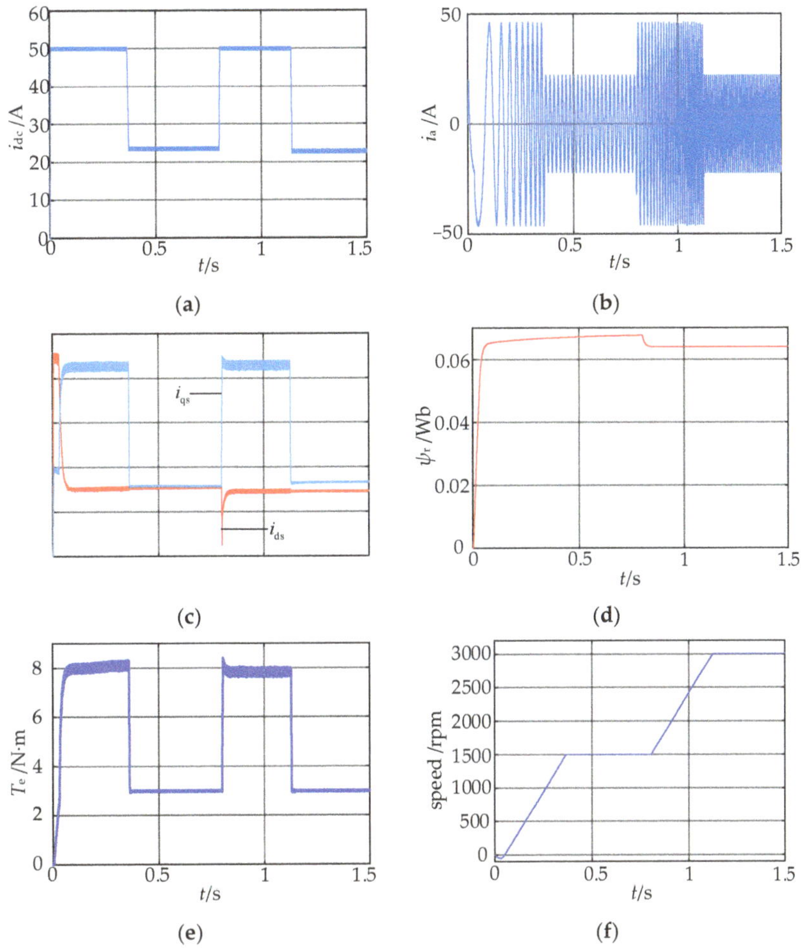

Figure 17. Simulation waveform under load torque of 3 N·m. (**a**) i_{dc}; (**b**) i_a; (**c**) i_d and i_q; (**d**) ψ_r; (**e**) T_e; (**f**) Speed.

Table 4. The reference of rotor flux and DC-link current under different load torque.

T_e (N·m)	Speed (rpm)	ψ_r^* (Wb)	i_{dc}^* (A)
0	1500	0.03	11.5
0	900	0.031	12
1	1500	0.039	14.8
2	1500	0.055	20.9
3	1500	0.068	23.5
3	3000	0.064	22.8

The whole simulation process is divided into three stages.

Start-up stage: in order to make the motor speed reach the reference speed as soon as possible, the hysteresis control strategy of DC-link current is adopted to regulate i_{dc} to 50 A, which is the max reference, so as to meet the demand of current. The optimal reference value of rotor flux ψ_r^* is 0.068 Wb. After 0.05 s, ψ_r^* reaches the reference value, i_{ds} is around 15 A, i_{qs} is around 43 A, and T_e is 8 N·m. After 0.4 s, the speed is adjusted to 1500 rpm, and T_e is reduced to 3 N·m.

Steady-state operation stage at 1500 rpm: i_{dc} is regulated to 23.5 A, as shown in Figure 17a. It can be seen from Figure 17e that when i_{dc} is reduced to 23.5 A, the ripple of torque is significantly reduced.

Second acceleration stage: at the time of 0.8 s, the value of step speed is 3000 rpm. i_{dc} reach to 50 A again, which is the same as the start-up stage. When ψ_r^* is adjusted to 0.064 Wb, i_{ds} is reduced immediately, ψ_r follows the given value quickly, i_{qs} and T_e are the same as the start-up. Finally, at 1.08 s, the speed reaches to 3000 rpm.

5.2. Deceleration Condition with No Load

This section simulates the deceleration condition of the motor with no load. When the IM speed reaches the rated speed of 1500 rpm, the reference speed is changed to 900 rpm at 0.6 s. The simulation waveforms of electromagnetic torque, speed, DC-link current and output power of DC power supply are shown in Figure 18, respectively. When the speed decelerates, i_{dc} is around 30 A. The electromagnetic torque plays a braking role and reaches to −2.3 N·m, and the speed reaches stability again after 0.28 s. The DC power supply intermittently absorbs the energy feedback from the motor decelerating process, and the output power is less than zero, so the energy feedback is realized.

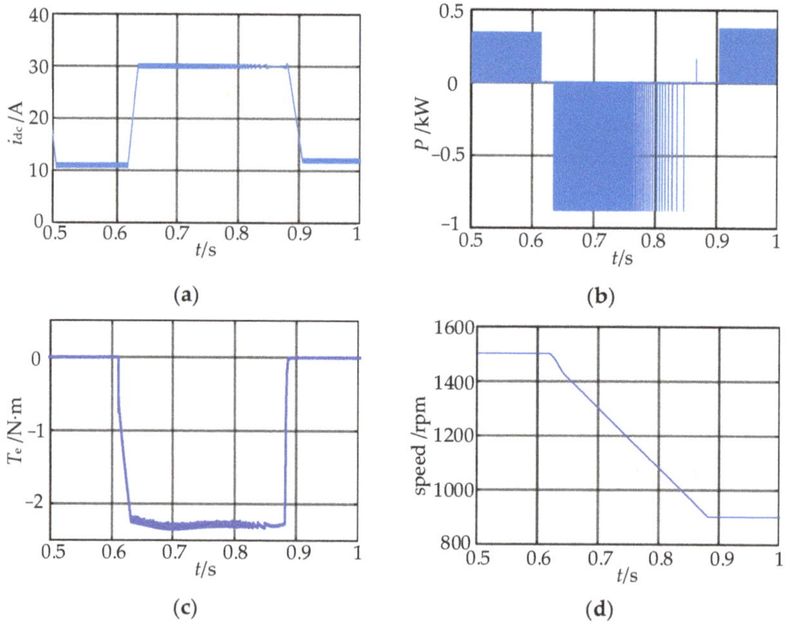

Figure 18. Simulation waveform under no-load deceleration. (**a**) i_{dc}; (**b**) output power of DC power supply; (**c**) T_e; (**d**) speed.

5.3. Load Disturbance Condition

Figure 19 are the simulation waveforms of load disturbance, when the speed remains at 1500 rpm, the load torque is reduced from 3 N·m to 1 N·m at 0.1 s, and the speed controller makes the electromagnetic torque drop rapidly to 1 N·m, where the speed remains unchanged. At the same time, the ψ_r and i_{dc} are reduced to the optimal value of 0.039 Wb and the minimum value of 14.8 A, respectively. When the load torque is increased to 2 N·m at 0.3 s, the speed is stabilized at 1500 rpm after 0.16 s, and the maximum error of speed is only 14 rpm. The ψ_r and i_{dc} are adjusted to 0.055 Wb and 20.9 A, respectively.

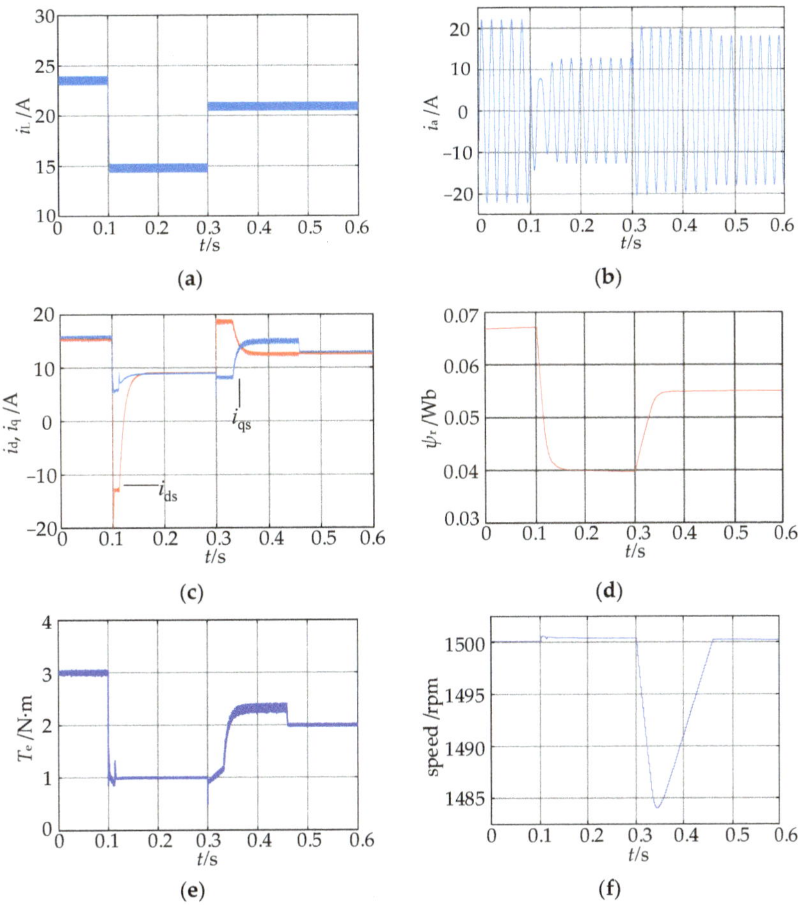

Figure 19. Simulation waveforms under the step perturbation of load torque. (**a**) i_L; (**b**) i_a; (**c**) i_d and i_q; (**d**) ψ_r; (**e**) T_e; (**f**) speed.

6. Experimental Results

In order to verify the feasibility of parameter design for DC-link inductance and AC-filter capacitance and the effectiveness of the vector control strategy proposed in this paper, the experimental platform was established, as shown in Figure 20; the experimental parameters are shown in Table 5. The value of DC-link inductance and AC-filter capacitance are satisfied with the selection range specified in Section 3. Because the motor speed, rotor flux, and stator d-q axis current cannot be measured by the oscilloscope, in order to facilitate debugging as well as the observation and storage of these signals, this paper develops a host computer program on the GUI platform of MATLAB to realize the data communication with DSP.

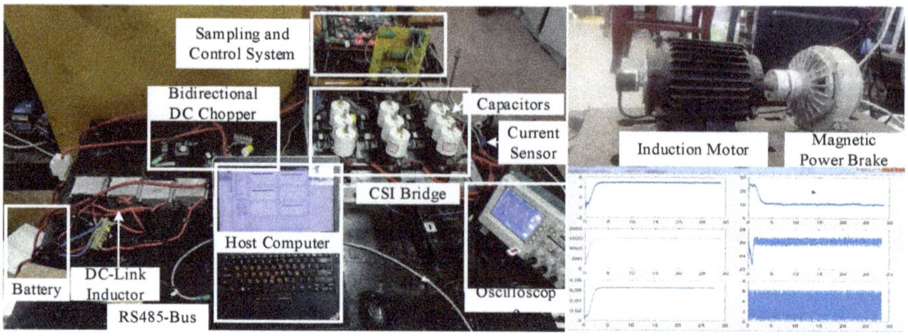

Figure 20. Prototype for CSI-fed IM drive system with bidirectional DC chopper.

Table 5. Models and parameters of the experimental platform.

Category	Part Number	Parameters
DC-link inductor L_{dc}	custom-made	4 mH
Filter capacitor C_a, C_b, C_c	MKP1847610354P4	66 µF
DC-bus voltage u_{dc}	Battery	24 V
Current sensors	LEM LA28-P	/
Diode modules	RM300HA-24F	240 V/300 A
IGBT modules	PM400HSA120	400 V/120 A
DSP	TMS320F28335	/
Encoder	OIH, 2500C/T	/

Next, the experimental results will be carried out under the three operation conditions: acceleration with load, deceleration with no-load and change of load.

6.1. *Performance under Acceleration with Load of 3 N·m*

Experimental waveforms of accelerated conditions are shown in Figure 21, under the load of 3 N·m. The experimental process consists of two stages. The reference speed changes from 1500 rpm to 3000 rpm, in order to reduce the harmonics and losses, and to improve current utilization, the references of i_{dc} and ψ_r are also adjusted according to the calculation method for the minimum DC-link current and the optimal flux proposed in Ref. [17]. In the dynamic process, to obtain a larger torque and reduce the dynamic response time, the reference i_{dc} is set to 50 A. In the first stage, the reference speed is set to 1500 rpm, and the reference i_{dc} and ψ_r are set to 23.5 A and 0.068 Wb, respectively. In the second stage, the reference speed is set to 3000 rpm, and the reference i_{dc} and ψ_r are set to 22.5 A and 0.065 Wb, respectively. From Figure 21, the rise time of i_{dc} increasing from 0 A to 50 A is only 15 ms, and the ripple of i_{dc} is less than 0.5 A, which demonstrates that the parameter design method of the DC-link inductor is feasible. In two stages, the speed and rotor flux can quickly track the references, and the THD of the stator current is less than 2%.

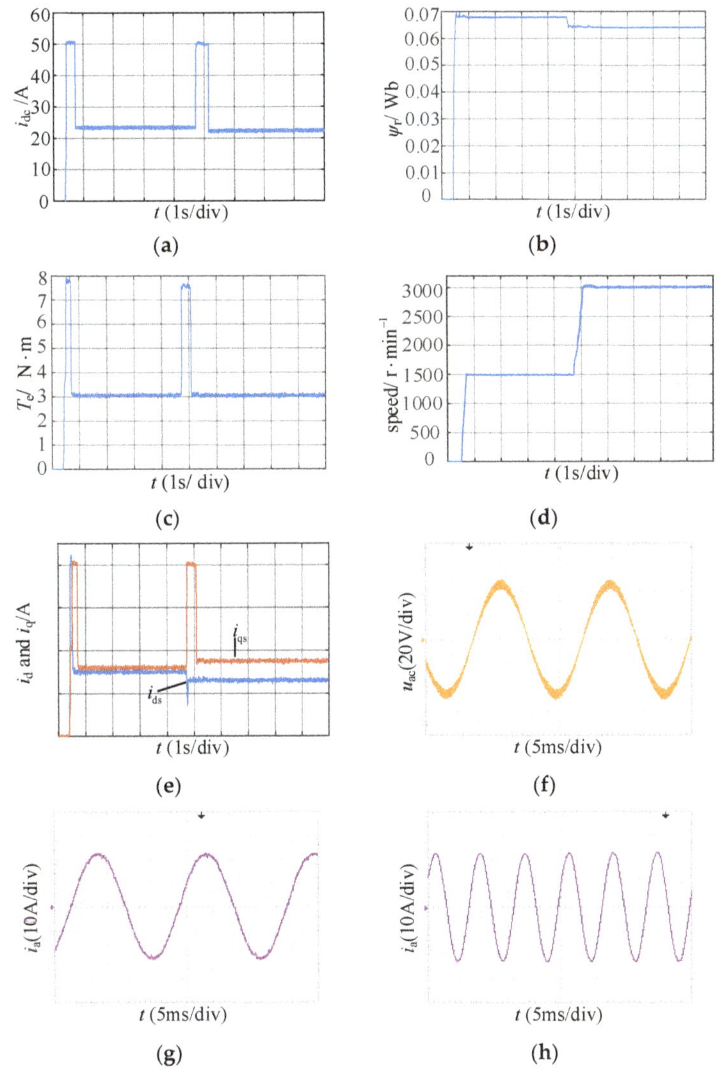

Figure 21. Experimental waveforms under acceleration with load of 3 N·m: (**a**) DC-link current; (**b**) Rotor flux; (**c**) Torque; (**d**) Speed; (**e**) Stator current in d-q axis; (**f**) Stator voltage of phase-a; (**g**) Stator current of phase-a under speed of 1500 rpm; (**h**) Stator current of phase-a under speed of 3000 rpm.

6.2. Performance under Deceleration with No Load

The experimental waveforms of the reference speed decreasing from 1500 rpm to 900 rpm with no load are shown in Figure 22. Unlike with the operational condition where the load is 3 N·m, the amplitude of the stator voltage is 18.5 V, which is less than u_{dc}. i_{dc} remains at 12 A with little ripple in the steady state, and during the dynamic process of speed decreasing from 1500 rpm to 900 rpm, the reference of i_{dc} is set to 30 A, and the mechanical energy from IM is recovered by the DC voltage source. Due to a fast dynamic and little ripple of T_e, dynamic and steady performance of speed can be obtained, the rotor flux can quickly track the references, and the stator current has very little harmonic distortion.

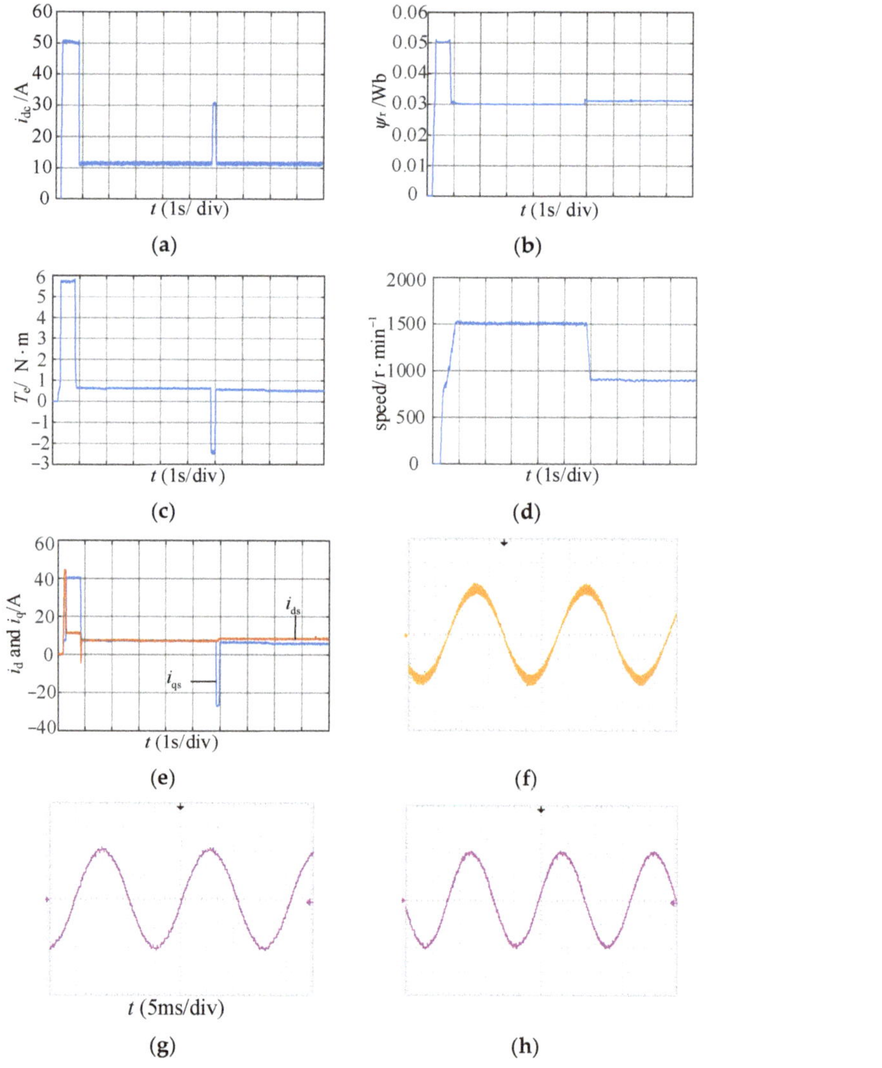

Figure 22. Experimental waveforms under deceleration with no load: (**a**) DC-link current; (**b**) Rotor flux; (**c**) Torque; (**d**) Speed; (**e**) Stator current in d-q axis; (**f**) Stator voltage of phase-a under speed of 900 rpm; (**g**) Stator current of phase-a under speed of 1500 rpm; (**h**) Stator current of phase-a under speed of 900 rpm.

6.3. Performance under Different Loads

The reference speed is set to 1500 rpm. The IM starts with the load of 3 N·m, then the load is suddenly reduced to 1 N·m, and finally the load is increased to 2 N·m. The experimental waveforms are shown for the reference speed decreasing from 1500 rpm to 900 rpm with no load are shown in Figure 23. Since the torque can be quickly adjusted when the sudden change occurs on the load, the fluctuation of the speed is less than 3%. Meanwhile, i_{dc} and ψ_r can quickly track the references.

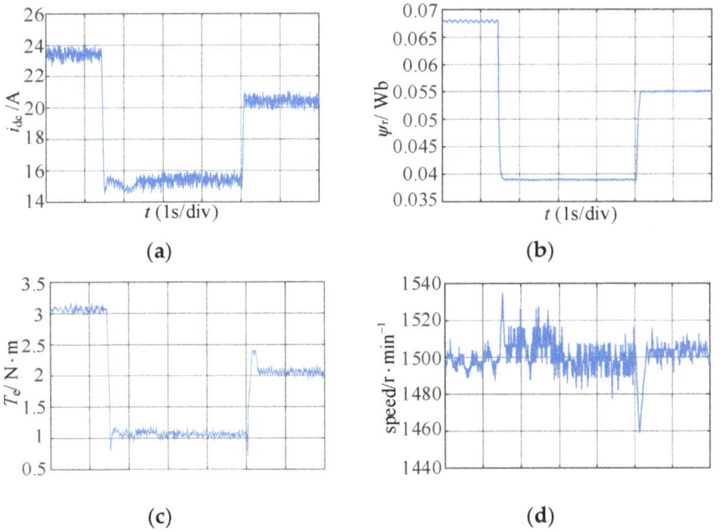

Figure 23. Experimental waveforms under different loads: (**a**) DC-link current; (**b**) Rotor flux; (**c**) Torque; (**d**) Speed.

7. Conclusions

In order to improve the performance of the current source inverter–fed inductor motor drive system, this paper presents a novel design principle of DC-bus inductance and AC-filter capacitance, and a vector control technology without voltage loop. The feasibility and correctness are verified by the simulation and experimental results under various operating conditions. The main contributions can be summarized as follows:

(1) Under consideration of the dynamic response speed and current ripple, the range expression of the DC-link inductance is derived.
(2) Under consideration of the DC-link current utilization and harmonic suppression, the range expression of the AC-filter inductance is derived.
(3) The simulation and experimental results show that the fast dynamic performance of speed and flux with no overshoot and a stator current with low harmonic distortion can be obtained by the vector control strategy without voltage loop for current source inverter–fed inductor motor drive systems. The analytical expressions of the control parameters derived in this paper are also suitable for other IM parameters.

Author Contributions: Conceptualization and methodology, P.S.; simulation, Y.L.; validation and experiment, P.S. and C.L.; writing and supervision, P.S.; review and editing, C.L. funding acquisition, P.S. All authors have read and agreed to the published version of the manuscript.

Funding: This work was supported in part by the Southwest Petroleum University City-School Cooperation Strategic Project under Grant 18SXHZ0043.

Data Availability Statement: Not applicable.

Conflicts of Interest: The authors declare that there is no conflict of interest regarding the publication of this paper.

References

1. Lorenzani, E.; Migliazza, G.; Immovilli, F.; Buticchi, G. CSI and CSI7 current source inverters for modular transformerless PV inverters. *Chin. J. Electr. Eng.* **2019**, *5*, 32–42. [CrossRef]
2. Wang, W.; Gao, F.; Yang, Y. An eight-switch five-level current source inverter. *IEEE Trans. Power Electron.* **2018**, *34*, 8389–8404. [CrossRef]

3. Zhou, S.; Liu, K.; Hu, W.; Chen, Y.; Zhang, D.; Huang, Q.; Tong, Q.; Zhang, Q. Harmonic-Separation-Based Direct Extraction and Compensation of Inverter Nonlinearity for State Observation Control of PMSM. *IEEE Access* **2021**, *9*, 142028–142045. [CrossRef]
4. Dong, Z.; Liu, C.; Song, Z.; Liu, S. Suppression of Dual-Harmonic Components for Five-Phase Series-Winding PMSM. *IEEE Trans. Transp. Electrif.* **2022**, *8*, 121–134. [CrossRef]
5. Zhao, L.; Chen, Z.; Wang, H.; Li, L.; Mao, X.; Li, Z.; Zhang, J.; Wu, D. An Improved Deadbeat Current Controller of PMSM Based on Bilinear Discretization. *Machines* **2022**, *10*, 79. [CrossRef]
6. Zhao, J.; Lu, H. Study on a new control strategy based on three phase current inverter. In Proceedings of the 2017 Eighth International Conference on Intelligent Control and Information Processing (ICICIP), Hangzhou, China, 3–5 November 2017; pp. 151–156.
7. Rajeev, M.; Agarwal, V. Single phase current source inverter with multiloop control for transformerless grid—PV interface. *IEEE Trans. Ind. Appl.* **2018**, *54*, 2416–2424. [CrossRef]
8. Chen, D.; Jiang, J.; Qiu, Y.; Zhang, J.; Huang, F. Single-stage three-phase current-source photovoltaic grid-connected inverter high voltage transmission ratio. *IEEE Trans. Power Electron.* **2016**, *32*, 7591–7601. [CrossRef]
9. Xing, L.; Wei, Q. Series-connected current source inverters with less switches. *IEEE Trans. Power Electron.* **2019**, *35*, 5553–5556. [CrossRef]
10. Xu, Y.; Wang, Z.; Liu, P.; Wu, J.; Cheng, M. A new control method for current-source inverter fed motor drive system without additional DC-link current regulator. In Proceedings of the 2020 IEEE 9th International Power Electronics and Motion Control Conference (IPEMC2020-ECCE Asia), Nanjing, China, 29 November–2 December 2020; pp. 382–387.
11. Miao, Y.; Liu, H.; Wang, H.; Wang, M. The modulation and control strategy of storage inductance current for single-phase current source inverter. *Trans. China Electrotech. Soc.* **2018**, *33*, 1227–1237.
12. Wang, W.; Gao, F.; Yang, Y.; Blaabjerg, F. Operation and modulation of H7 current-source inverter with hybrid SiC and Si semiconductor switches. *IEEE J. Emerg. Sel. Top. Power Electron.* **2017**, *6*, 387–399. [CrossRef]
13. Li, Y.W.; Pande, M.; Zargari, N.R.; Wu, B. DC-link current minimization for high-power current-source motor drives. *IEEE Trans. Power Electron.* **2008**, *24*, 232–240.
14. Ding, L.; Li, Y.W.; Zargari, N.R. Discrete-time SMO sensorless control of current source converter-fed PMSM drives with low switching frequency. *IEEE Trans. Ind. Electron.* **2020**, *68*, 2120–2129. [CrossRef]
15. Yang, S.; Peng, F.Z.; Lei, Q.; Inoshita, R.; Qian, Z. Current-fed quasi-Z-source inverter with voltage Buck–Boost and regeneration capability. *IEEE Trans. Ind. Appl.* **2010**, *47*, 882–892. [CrossRef]
16. Xu, Y.; Wang, Z.; Liu, P.; He, J. A Soft-Switching Current-Source-Inverter-Fed Motor Drive With Reduced Common-Mode Voltage. *IEEE Trans. Ind. Electron.* **2021**, *68*, 3012–3021. [CrossRef]
17. Miao, Y.; Liu, H.; Zhang, W.; Liu, P. Optimal control strategy of storage inductor current and rotor flux for current source inverter motor drive system. *Proc. CSEE* **2021**, *39*, 2757–2767.
18. Migliazza, G.; Buticchi, G.; Carfagna, E.; Lorenzani, E.; Madonna, V.; Giangrande, P.; Galea, M. DC current control for a single-stage current source inverter in motor drive application. *IEEE Trans. Power Electron.* **2021**, *36*, 3367–3376. [CrossRef]
19. Lee, H.-J.; Jung, S.; Sul, S.-K. A current controller design for current source inverter-fed AC machine drive system. *IEEE Trans. Power Electron.* **2013**, *28*, 1366–1381. [CrossRef]
20. Fuchs, F.W.; Kloenne, A. DC link and dynamic performance features of PWM IGBT current source converter induction machine drives with respect to industrial requirements. In Proceedings of the 4th International Power Electronics and Motion Control Conference (IPEMC), Xi'an, China, 14–16 August 2004; Volume 3, pp. 1393–1398.
21. Miao, Y.; Liu, H.; Hua, Z.; Liu, J.; Cheng, Z. Control strategy for three-phase current source inverter based on optimal given value of DC storage inductance current. *Trans. China Electrotech. Soc.* **2019**, *34*, 349–362.

energies

Article

Robust Sensorless Control of Interior Permanent Magnet Synchronous Motor Using Deadbeat Extended Electromotive Force Observer

Seung-Taik Kim, In-Sik Yoon, Sung-Chul Jung and Jong-Sun Ko *

Department of Electronic and Electrical Engineering, Dankook University, 152, Jukjeon-ro, Suji-gu, Yongin-si 16890, Korea
* Correspondence: jsko@dankook.ac.kr

Abstract: This paper proposes a novel and robust method of sensorless speed control using a deadbeat observer for an interior permanent magnet synchronous motor (IPMSM). The proposed sensorless speed control method uses a deadbeat observer to estimate the extended electromotive force (EEMF) in a rotational coordinate system. The estimated EEMF is used in the IPMSM velocity estimation algorithm. The deadbeat EEMF observer (DEEMFO) shows greater robustness compared to the reconstructor, which estimates the EEMF by simply recalculating the voltage equation. Unlike a reconstructor, DEEMFO has a feedback component, so it can compensate for errors due to uncertainty in motor parameters and errors due to parameter fluctuations that may occur during use. By simulating and experimenting with speed, load torque, and parameter fluctuations, it is proved to be more robust and precise than the reconstructor. The simulation is performed with MATLAB/Simulink, and the experiments were carried out using a DSP TMS320F28335 and a motor-generator set (M-G Set). The simulation and experiment results show the reliability and precision of the proposed sensorless control method.

Keywords: interior permanent magnet synchronous motor; sensorless speed control; extended electromotive force; deadbeat observer

Citation: Kim, S.-T.; Yoon, I.-S.; Jung, S.-C.; Ko, J.-S. Robust Sensorless Control of Interior Permanent Magnet Synchronous Motor Using Deadbeat Extended Electromotive Force Observer. *Energies* **2022**, *15*, 7568. https://doi.org/10.3390/en15207568

Academic Editors: Kan Liu and Wei Hu

Received: 17 September 2022
Accepted: 9 October 2022
Published: 13 October 2022

1. Introduction

IPMSM is widely applied in various driving fields that require high performance, power factor and efficiency, such as industrial facilities and electric vehicles. IPMSM has the advantages that it generates a high torque even with a small rotor size by generating magnetic torque as well as a reluctance torque due to its saliency [1,2]. To efficiently drive an AC motor such as the IPMSM, rotor speed and position information are required. A sensor such as an optical encoder or resolver is required to obtain this information. However, the installation and maintenance costs increase due to the use of the sensor. Furthermore, the problems with the sensor resolution and a bad installation environment cause issues due to errors and disturbances. Therefore, the research on sensorless control without those sensors has been conducted.

One of several methods of sensorless control is using electromotive force. It has a relatively simple structure and shows stable performance over a wide operating range [3–6]. However, when the AC motor is operating at a low speed, the generated back EMF is small, so it is difficult to properly estimate speed [7]. The electromotive force observer method includes a model reference adaptive system [8], extended Kalman filter [9], and sliding mode observer [10]. The second method is the high-frequency signal injection method. Its advantage is that speed and position can be estimated through the injected high frequency even at an extremely low operating speed [11–14]. However, there are losses and noise, additional components are needed for high frequency injection, and it is difficult to find and apply a suitable high frequency signal. Recently, in order to solve the limitations of each method, various methods such as I/f control for low-speed region control [15], a nonlinear extended state observer [16], and a modified model reference adaptive system were being studied [17]. In particular, the sensorless method using electromotive force also has limitations in that it requires some approximation to be used in IPMSM. To solve this, a sensorless control method using an EEMF model that does not require approximation has been proposed in the stationary frame [18], and rotating frame [19].

The paper proposes a robust and innovative sensorless control method for IPMSM. The proposed method uses a deadbeat observer to estimate the EEMF in a rotating coordinate system. A deadbeat observer is easy to pole placement and can accurately estimate parameters using large gains [20]. The DEEMFO is more robust compared to the reconstructor, which estimates the EEMF by simply recalculating the voltage equation. Unlike a reconstructor, DEEMFO has a feedback component, so it can compensate for errors due to uncertainty in motor parameters and errors due to parameter fluctuations that may occur during use.

In this paper, the definition of EEMF and the reconstruction method for estimating it are explained first. The proposed DEEMFO, which solves the limitations of the reconstructor, is described in detail. After that, the simulation and experimental results are presented under conditions of no-load, speed change, load change, and parameter variation. The simulation and experiment results show the reliability and precision of the proposed sensorless control method.

2. The EEMF Estimation Algorithm

2.1. Mathematical Model of IPMSM

Figure 1 shows stationary and rotating reference frames. The α-β frame is a stationary frame fixed to the stator winding. The *d*-*q* rotating reference frame is also known as a synchronous reference frame. In this coordinate system, the d-axis indicates the N pole direction of the permanent magnet. The q-axis indicates a direction that is 90 degrees in front of the *d*-axis by the electrical angle [21]. The IPMSM voltage equations for the *d*-*q* rotation reference frame are expressed as (1) and (2):

$$\begin{bmatrix} v_d \\ v_q \end{bmatrix} = \begin{bmatrix} R_s + pL_d & -\omega L_q \\ \omega L_q & R_s + pL_d \end{bmatrix} \begin{bmatrix} i_d \\ i_q \end{bmatrix} + \begin{bmatrix} 0 \\ E_{ex} \end{bmatrix} \quad (1)$$

$$E_{ex} = \omega\left[(L_d - L_q)i_d + \lambda_{pm}\right] - (L_d - L_q)(pi_q) \quad (2)$$

where v_d, v_q are the values of stator voltage; i_d, i_q are the values of stator current; L_d, L_q are the values of stator inductance in the *d*-*q* frame. R_s is the stator resistance, ω is the rotor angular speed, λ_{pm} is the magnet flux linkage and p is the differential operator. The second term in (1) is called EEMF as (2) [19].

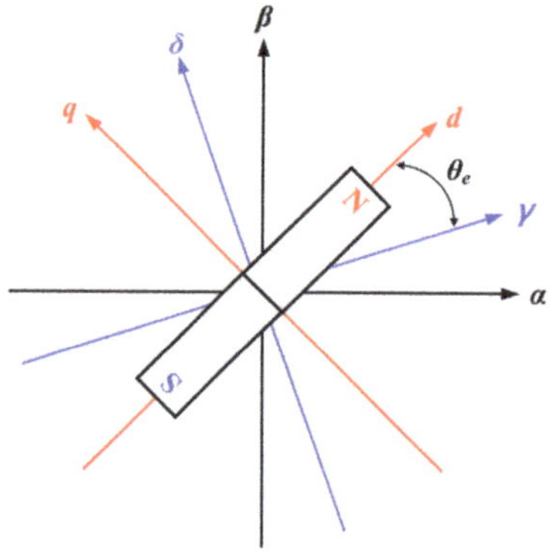

Figure 1. Reference frames for IPMSM Control.

In a sensorless system, speed and position are estimated. The estimated position is located on the γ-δ rotating frame, which differ from the actual d-q frame by the position error θ_e. Therefore, the voltage equations in the γ-δ reference frame can be expressed as (3) and (4):

$$\begin{bmatrix} v_\gamma \\ v_\delta \end{bmatrix} = \begin{bmatrix} R_s + pL_d & -\omega L_q \\ \omega L_q & R_s + pL_d \end{bmatrix} \begin{bmatrix} i_\gamma \\ i_\delta \end{bmatrix} + \begin{bmatrix} e_\gamma \\ e_\delta \end{bmatrix} \quad (3)$$

$$\begin{bmatrix} e_\gamma \\ e_\delta \end{bmatrix} = E_{ex} \begin{bmatrix} -\sin\theta_e \\ \cos\theta_e \end{bmatrix} + (\hat{\omega} - \omega) L_d \begin{bmatrix} -i_\delta \\ i_\gamma \end{bmatrix} \quad (4)$$

where v_γ, v_δ are the values of the stator voltage; i_γ, i_δ are the values of stator current; e_γ, e_δ are the values of EEMF in the γ-δ frame; ^ denotes the estimated value; R_s is the stator resistance; ω is the rotor angular speed; λ_{pm} is the magnet flux linkage and p is the differential operator.

2.2. EEMF Estimation Using Reconstructor

Assuming that the error of the estimated velocity is small and can be considered 0, and the derivative of EEMF is zero, the EEMF can be estimated by the reconstructor using the state space Equations (5) and (6) [19]. This method can estimate the EEMF by simply reconstructing the voltage Equation (7):

$$p \begin{bmatrix} i_\gamma \\ e_\gamma \end{bmatrix} = \frac{1}{L_d} \begin{bmatrix} -R_s & -1 \\ 0 & 0 \end{bmatrix} \begin{bmatrix} i_\gamma \\ e_\gamma \end{bmatrix} + \frac{1}{L_d} \begin{bmatrix} 1 \\ 0 \end{bmatrix} (v_\gamma + \omega L_q i_\delta) \quad (5)$$

$$p \begin{bmatrix} i_\delta \\ e_\delta \end{bmatrix} = \frac{1}{L_d} \begin{bmatrix} -R_s & -1 \\ 0 & 0 \end{bmatrix} \begin{bmatrix} i_\delta \\ e_\delta \end{bmatrix} + \frac{1}{L_d} \begin{bmatrix} 1 \\ 0 \end{bmatrix} (v_\delta - \omega L_q i_\gamma) \quad (6)$$

$$\begin{bmatrix} \hat{e}_\gamma \\ \hat{e}_\delta \end{bmatrix} = \begin{bmatrix} v_\gamma + \hat{\omega} L_q i_\delta - (R_s + pL_d) i_\gamma \\ v_\delta - \hat{\omega} L_q i_\gamma - (R_s + pL_d) i_\delta \end{bmatrix} \quad (7)$$

where $\hat{e}_\gamma$, $\hat{e}_\delta$ are the values of estimated EEMF and $\hat{\omega}$ is estimated speed.

From Equation (7), the derivative term allows the reconstructor to estimate the EEMF even at low-speed operation if the current changes [18]. However, when the current changes rapidly due to changes in speed and load, the differential value of the current is very large and may cause system instability. Therefore, a low-pass filter is required. In addition, when applied to a digital system, if the current value is not measured because it does not match the sampling period when the current changes, the amount of change cannot be reflected in the calculation. Furthermore, in Equation (7) motor parameters are used that may be uncertain or changed during operation. If the parameter used for calculation in the reconstructor has a different value from the actual motor parameter, the calculated EEMF will have a different value from the actual value. If the calculated EEMF is used, the estimated speed value will of course have an error. However, it cannot compensate for parameter errors, so precise and robust control is not possible.

2.3. Proposed Deadbeat EEMF Observer

In this paper, we propose the novel EEMF estimation method using deadbeat observer. In the proposed DEEMFO, the EEMF estimation algorithm by the observer can solve the problem caused by the current differential term and compensate for the motor parameter variation by the feedback term. Through this feedback term, the difference between the actual value and the estimated value is multiplied by the observer gain, which has the effect of gradually reducing the error, even if the motor parameters are uncertain or there is variation. To ensure that the time required to calculate the EEMF is shorter than the entire system response time and to quickly compensate for errors, a deadbeat observer is desirable. As a result, DEEMFO is more robust to parameter fluctuations by reducing the error through the observer gain by receiving current feedback. In addition, the DEEMFO observer gain can be easily obtained and the system can quickly and stably use a large observer gain. We reduced the size of the matrix for the calculation time, by dividing it along the γ-δ frame.

The observer formulas are expressed as Equations (8)–(13):

$$p\begin{bmatrix}\hat{i}_\gamma \\ \hat{e}_\gamma\end{bmatrix} = \frac{1}{L_d}\begin{bmatrix}-R_s & -1 \\ 0 & 0\end{bmatrix}\begin{bmatrix}\hat{i}_\gamma \\ \hat{e}_\gamma\end{bmatrix} + \frac{1}{L_d}\begin{bmatrix}1 \\ 0\end{bmatrix}v_{\gamma 1} + \boldsymbol{ek}_\gamma \tag{8}$$

$$p\begin{bmatrix}\hat{i}_\delta \\ \hat{e}_\delta\end{bmatrix} = \frac{1}{L_d}\begin{bmatrix}-R_s & -1 \\ 0 & 0\end{bmatrix}\begin{bmatrix}\hat{i}_\delta \\ \hat{e}_\delta\end{bmatrix} + \frac{1}{L_d}\begin{bmatrix}1 \\ 0\end{bmatrix}v_{\delta 1} + \boldsymbol{ek}_\delta \tag{9}$$

$$v_{\gamma 1} = v_\gamma + \hat{\omega} L_q i_\delta \tag{10}$$

$$v_{\delta 1} = v_\delta - \hat{\omega} L_q i_\gamma \tag{11}$$

$$\boldsymbol{ek}_\gamma = \begin{bmatrix}ek_{\gamma 1} \\ ek_{\gamma 2}\end{bmatrix}\left(i_\gamma - \begin{bmatrix}1 & 0\end{bmatrix}\begin{bmatrix}\hat{i}_\gamma \\ \hat{e}_\gamma\end{bmatrix}\right) \tag{12}$$

$$\boldsymbol{ek}_\delta = \begin{bmatrix}ek_{\delta 1} \\ ek_{\delta 2}\end{bmatrix}\left(i_\delta - \begin{bmatrix}1 & 0\end{bmatrix}\begin{bmatrix}\hat{i}_\delta \\ \hat{e}_\delta\end{bmatrix}\right) \tag{13}$$

where $ek_{\gamma 1}$, $ek_{\gamma 2}$, $ek_{\delta 1}$, and $ek_{\delta 2}$ are observer gain.

The discrete observer can be expressed as Equations (14) and (15):

$$\begin{bmatrix}\hat{i}_\gamma[k+1] \\ \hat{e}_\gamma[k+1]\end{bmatrix} = \boldsymbol{A}_d\begin{bmatrix}\hat{i}_\gamma[k] \\ \hat{e}_\gamma[k]\end{bmatrix} + \boldsymbol{B}_d v_{\gamma 1} + \boldsymbol{EK}_\gamma\left(i_\gamma[k] - \hat{i}_\gamma[k]\right) \tag{14}$$

$$\begin{bmatrix}\hat{i}_\delta[k+1] \\ \hat{e}_\delta[k+1]\end{bmatrix} = \boldsymbol{A}_d\begin{bmatrix}\hat{i}_\delta[k] \\ \hat{e}_\delta[k]\end{bmatrix} + \boldsymbol{B}_d v_{\delta 1} + \boldsymbol{EK}_\delta\left(i_\delta[k] - \hat{i}_\delta[k]\right) \tag{15}$$

where $\boldsymbol{A}_d$, $\boldsymbol{B}_d$ are the system matrix and the input matrix. $\boldsymbol{EK}_\gamma$, $\boldsymbol{EK}_\delta$ are the observer gain matrices.

Since the rank of the observer ability matrix is 2, the observer is observable. This paper proposes and uses DEEMFO by applying the deadbeat observer as an observer to estimate the EEMF. Since the deadbeat observer places the desired pole arrangement at the origin (0, 0) of the unit concentric circle in the z-domain as shown in Figure 2, observer gain can be easily obtained and the system can quickly and stably use a large observer gain. The observer gain matrices $\boldsymbol{EK}_\gamma$ and $\boldsymbol{EK}_\delta$ can be calculated by the pole placement using Ackermann's formula [22]:

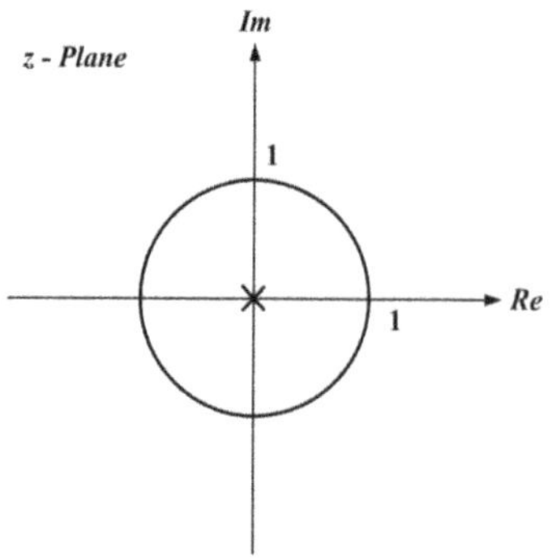

Figure 2. The z-plane for the deadbeat EEMF observer pole placement.

Figure 3 shows the block diagram of DEEMFO on the γ-axis. The δ-axis also has the same configuration. In this system, the observer gain matrices $\boldsymbol{EK}_\gamma$, $\boldsymbol{EK}_\delta$ are equal to $\boldsymbol{EK}_d$, where $\boldsymbol{C}_d$ is the output matrix. With voltage, current, stator inductance L_q and estimated speed as inputs, EEMF and current are estimated by DEEMFO.

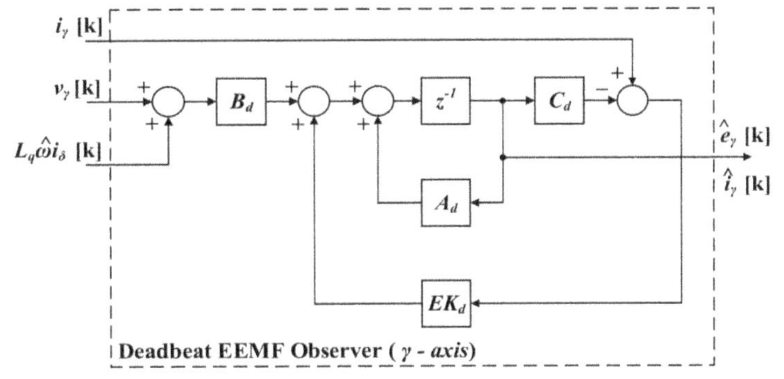

Figure 3. Block diagram of DEEMFO.

Since the difference between the real current i_γ, i_δ and the estimated current $\hat{i}_\gamma$, $\hat{i}_\delta$ is multiplied by the observer gain matrix EK_d, there is a feedback term to compensate for the parameter error and variation effects. After EEMF $\hat{e}_\gamma$, $\hat{e}_\delta$ is estimated by the DEEMFO, the position error $\hat{\theta}_e$ is estimated by Equation (16):

$$\hat{\theta}_e = tan^{-1}\left(-\frac{\hat{e}_\gamma}{\hat{e}_\delta}\right) \tag{16}$$

Figure 4 shows the block diagram of the rotor speed and position estimation algorithm. The estimated position error $\hat{\theta}_e$ is used to estimate the rotor speed. By compensating the estimated position error $\hat{\theta}_e$ to be 0 using the PI compensator, the rotor speed is obtained and the rotor position is estimated by its integration.

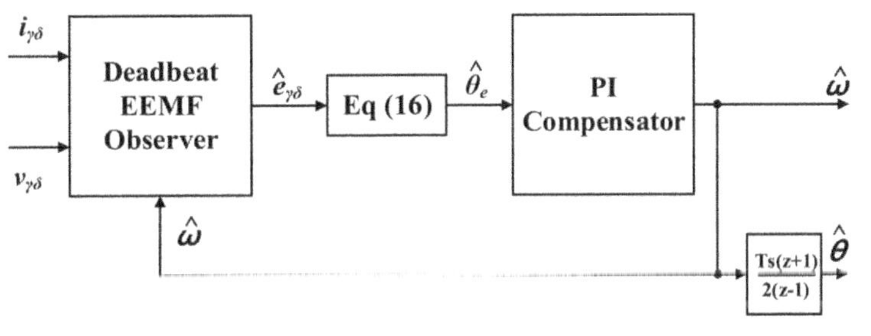

Figure 4. Block diagram of the rotor speed and position estimation algorithm.

Figure 5 represents the proposed overall system block diagram including maximum torque per ampere (MTPA) control for efficient IPMSM control and pulse width modulation (PWM) control along with the proposed speed estimation algorithm. When the speed command ω^* is an input of the speed controller, a torque command T_{com} is output to drive the IPMSM at the appropriate speed. The torque command T_{com} is used as an input of the MTPA and outputs the command current i^*_{dq}. The voltage commands v^*_{dq} obtained by the current controller (PI controller) are converted to three-phase and input into the pulse width modulation (PWM) inverter. The real stator current i_{dq} and the voltage commands v^*_{dq} are used for the DEEMFO input. By applying the speed and position estimation algorithm, the system of Figure 5 performs the sensorless control of the IPMSM.

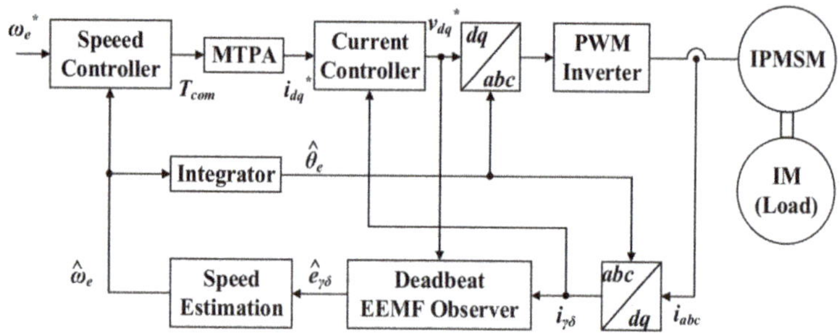

Figure 5. Block diagram of the sensorless speed control system using the proposed deadbeat EEMF Observer.

3. Simulation

3.1. Set-Up

The simulations are performed by MATLAB/Simulink. In the simulation part, the results are analyzed using DEEMFO and compared with the reconstructor. The IPMSM parameters applied to the simulation and experiment are shown in Table 1.

Table 1. IPMSM Parameters.

Parameter	Value	Unit
Rated power	4.0	kW
Rated speed	3500	r/min
Rated torque	12	Nm
Rated voltage	380	V
Rated Current	10	A
Stator resistance (R_s)	0.332	Ω
Inductance d-axis (L_d)	9.91	mH
Inductance q-axis (L_q)	10.93	mH
Flux linkage	0.118	Wb
Pole pairs	5	-

Figure 6 shows the configuration of the proposed IPMSM sensorless control with the use of the DEEMFO simulator. The simulator includes DEEMFO, speed estimation algorithm, speed controller, MTPA operation, current controller, and space vector PWM (SVPWM) inverter. It was designed in the same way as shown in Figure 5. The DEEMFO sampling frequency is 5 kHz, the current controller is 5 kHz, and the speed controller is 0.5 kHz. The switching frequency of the SVPWM inverter has been set to 5 kHz.

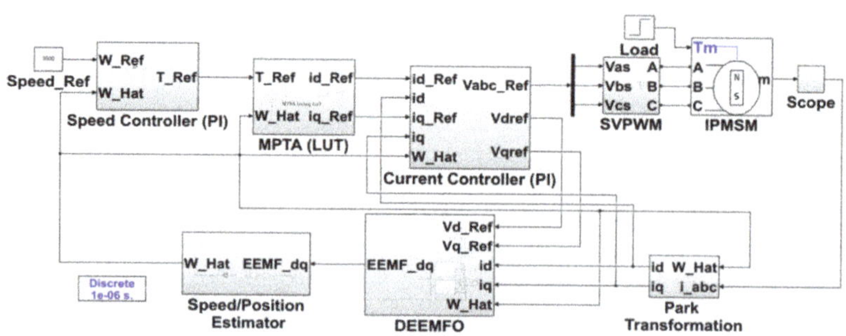

Figure 6. MATLAB/Simulink simulator for IPMSM sensorless control.

Robust and precise sensorless speed control requires accurate rotor speed and position estimation even if the speed or load torque increases or decreases rapidly or parameters are uncertain or changed. To verify this, the simulation was compared and verified under four conditions. First, the control results were compared using the reconstructor and the proposed DEEMFO under no-load conditions; second, when there is a speed command change from 3000 rpm to 3500 rpm; and third, when the load torque increases from 0 Nm to 6 Nm, which is half of the rated torque. As the most important condition, the results were compared by changing the stator resistance R_s and the stator inductance L_d, L_q by 30%, assuming there was a parameter variation or if there are errors in the parameter values in the datasheet.

3.2. Simulation Result

Figure 7 shows the simulation results under no-load conditions using the reconstructor and the proposed DEEMFO. It shows the actual rotor speed ω_r estimated speed $\hat{\omega}_r$, reference speed ω_r^*, stator current $\hat{i}_\gamma$, $\hat{i}_\delta$, and the error as the difference between the actual speed/position and the estimated speed/position.

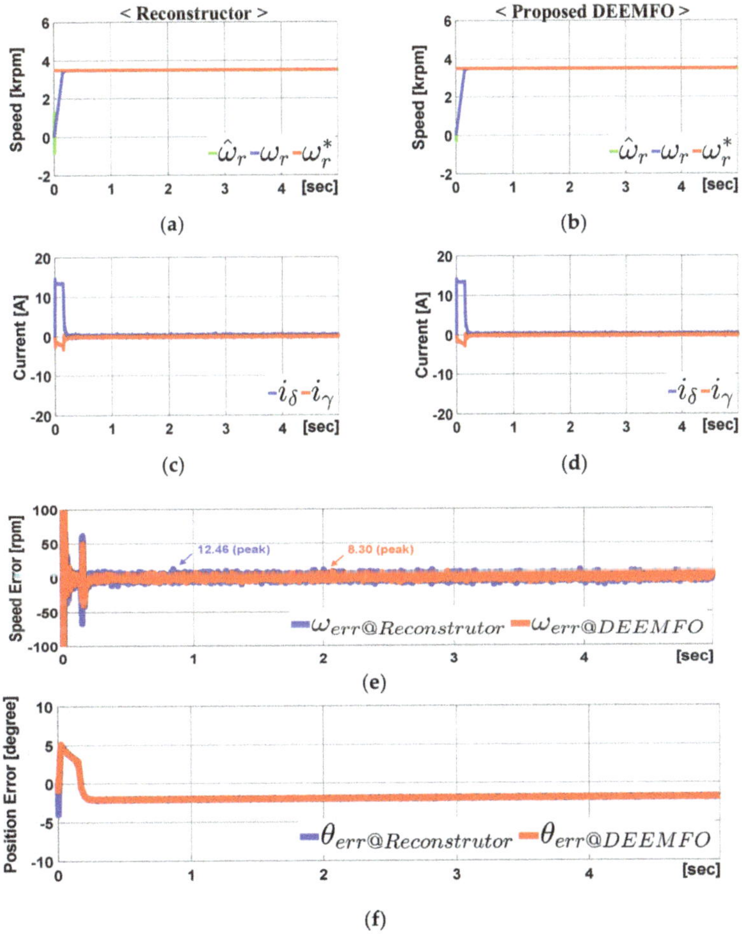

Figure 7. Simulation results under no load condition. (**a**) reference, real, and estimated speed with reconstructor; (**b**) reference, real, and estimated speed with DEEMFO; (**c**) $\gamma - \delta$ axis currents with reconstructor; (**d**) $\gamma - \delta$ axis currents with DEEMFO; (**e**) comparison of speed errors in reconstructor and DEEMFO; (**f**) comparison of position errors in reconstructor and DEEMFO.

When using the reconstructor, it peaks at around 12.46 rpm at a steady state, and when applying the DEEMFO, it shows a peak value of about 8.30 rpm and the speed error of around 4 rpm is reduced. This means an error improvement of approximately 33.4%.

Figure 8 summarizes the results under the condition that the speed command increased from 3000 rpm to 3500 rpm. When the speed command increases at 2.5 s, a speed error ω_{err} occurs. At this time, the speed error when using the reconstructor is 737.5 rpm, and the speed error when using the proposed DEEMFO is 396.2 rpm. This means an error improvement of approximately 46.3%.

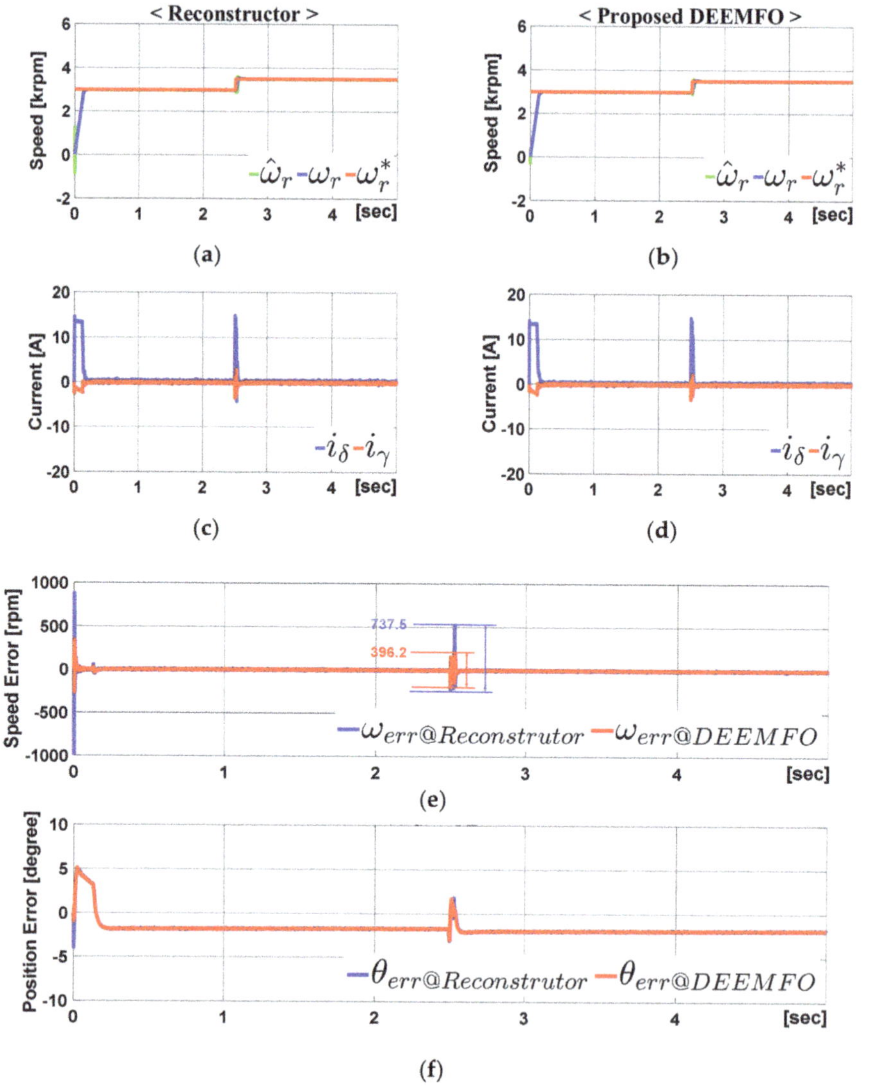

Figure 8. Simulation results under the condition that the speed reference increased from 3000 rpm to 3500 rpm. (**a**) reference, real, and estimated speed with reconstructor; (**b**) reference, real, and estimated speed with DEEMFO; (**c**) $\gamma - \delta$ axis currents with reconstructor; (**d**) $\gamma - \delta$ axis currents with DEEMFO; (**e**) comparison of speed errors in reconstructor and DEEMFO; (**f**) comparison of position errors in reconstructor and DEEMFO.

Figure 9 summarizes the simulation results provided for the load torque increased from 0 Nm to 6 Nm. It exhibits excellent control performance due to its high and accurate speed. Figure 9 shows also the speed and position error. In the reconstruction method, a speed error ω_{err} is 79.88 rpm. In contrast, meanwhile, in DEEMFO there is a speed error ω_{err} of about 71.84 rpm. This means an error improvement of about 10.1%.

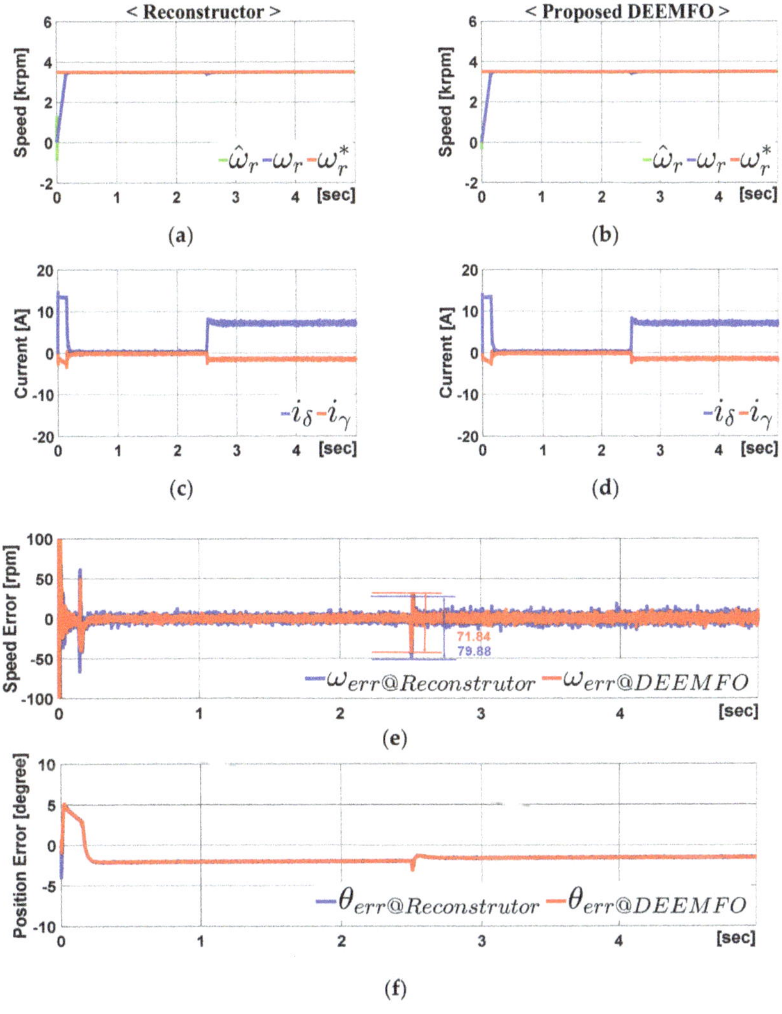

Figure 9. Simulation results under the condition of increasing the load torque from 0 Nm to 6 Nm. (**a**) reference, real, and estimated speed with reconstructor; (**b**) reference, real, and estimated speed with DEEMFO; (**c**) $\gamma - \delta$ axis currents with reconstructor; (**d**) $\gamma - \delta$ axis currents with DEEMFO; (**e**) comparison of speed errors in reconstructor and DEEMFO; (**f**) comparison of position errors in reconstructor and DEEMFO.

Figure 10 represents a summary of the simulation results, provided that the IPMSM parameters changed by 30%. The speed error ω_{err} of the reconstructor is greater than the proposed DEEMFO. The detail value is shown in Figure 10. The speed error ω_{err} is approximately 825.6 rpm when the reconstructor is used. However, when applying DEEMFO, the speed error ω_{err} is around 298.7 rpm. This means an error improvement of about 63.8%.

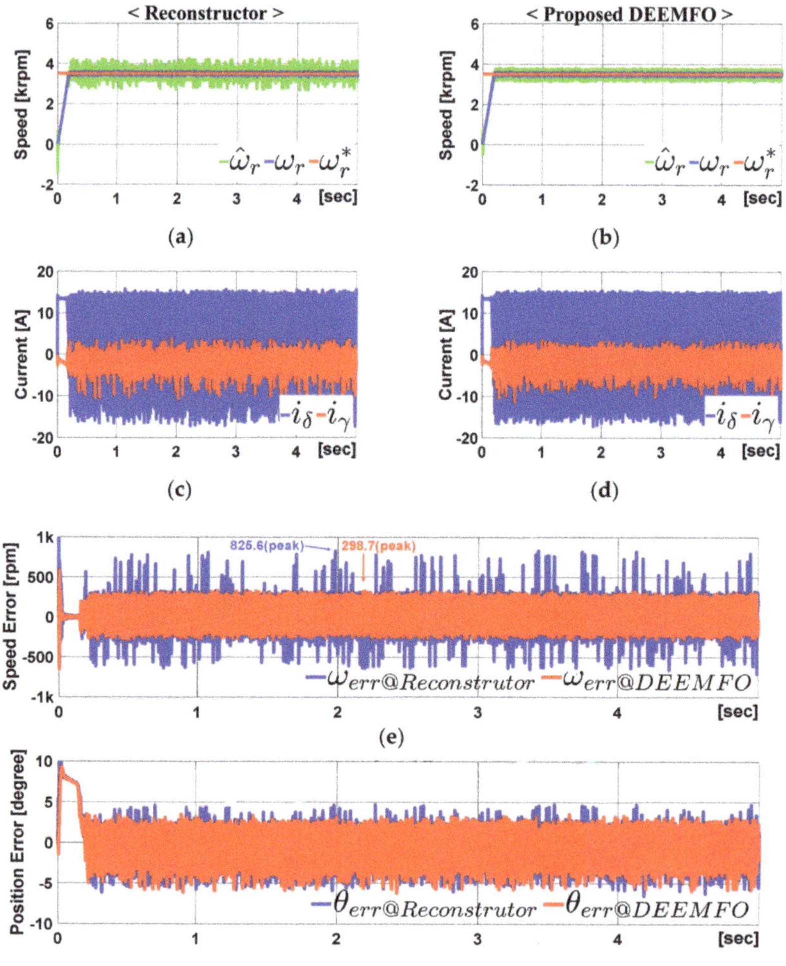

(f)

Figure 10. Simulation results under the condition that the parameter changed by 30%. (**a**) reference, real, and estimated speed with reconstructor; (**b**) reference, real, and estimated speed with DEEMFO; (**c**) $\gamma - \delta$ axis currents with reconstructor; (**d**) $\gamma - \delta$ axis currents with DEEMFO; (**e**) comparison of speed errors in reconstructor and DEEMFO; (**f**) comparison of position errors in reconstructor and DEEMFO.

The control performance was compared by simulating it according to several parameter variations. Table 2 shows whether the sensorless speed control of the reconstructor and DEEMFO is possible when the motor parameter varies from 0.72 pu to a maximum of 1.79 pu. The reconstructor becomes a stable sensorless control only when the electrical parameters (R_s, L_d, L_q) of the actual motor change from 0.94 pu to 1.54 pu. However, when the proposed DEEMFO is applied, stable sensorless control is possible, even if the parameters range from 0.73 pu to 1.79 pu. If the actual motor parameter and the parameter used to estimate the EEMF are different, it can be analyzed that the error is large because the difference cannot be compensated for, because there is no internal feedback. However, if DEEMFO is applied, relatively quick and robust control can be achieved.

Table 2. Speed control result according to parameter change.

Parameter Variation	Reconstructor	DEEMFO
$(R_s, L_d, L_q) \times 0.72$	X	X
$(R_s, L_d, L_q) \times 0.73$	X	O
⋮	X	O
$(R_s, L_d, L_q) \times 0.94$	O	O
⋮	O	O
$(R_s, L_d, L_q) \times 1.00$	O	O
⋮	O	O
$(R_s, L_d, L_q) \times 1.54$	O	O
$(R_s, L_d, L_q) \times 1.55$	O	O
⋮	X	O
$(R_s, L_d, L_q) \times 1.78$	X	O
$(R_s, L_d, L_q) \times 1.79$	X	X
$(R_s, L_d, L_q) \times 0.72$	X	X
$(R_s, L_d, L_q) \times 0.73$	X	O
⋮	X	O

4. Experiment

4.1. Set-Up

Figure 11 shows the M-G set wiring diagram for the experiment. The DSP TMS320F38225 controller, M-G Set, and measuring equipment are connected. Three-phase power is applied to the IPMSM through the power module. The encoder, an IPMSM position sensor, is connected to the controller to measure the actual position, but this position information is not used for actual control as it is sensorless control. The 5.5 kW induction motor (IM) for applying load torque operates via a commercial driver. The control system consists of the controller and easyDSP to output parameters such as position, speed, voltage command, and current command from the IPMSM and a torque sensor/display for checking speed and torque. A 12-bit oscilloscope was connected to obtain a waveform as a result of the experiment. The experimental devices are shown in Figures 12 and 13.

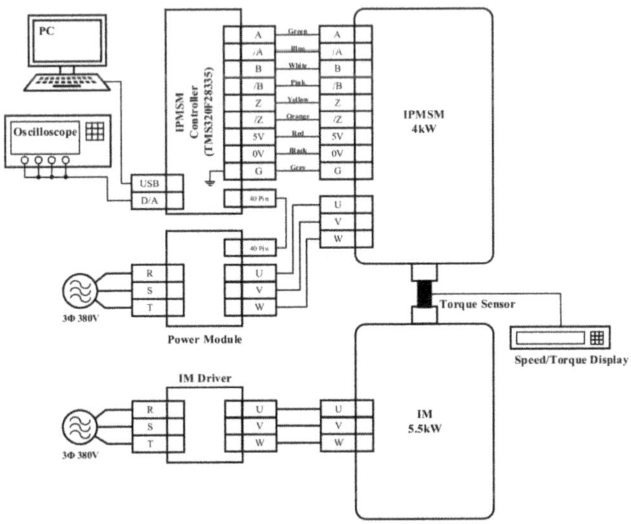

Figure 11. M-G set wiring diagram for experiment.

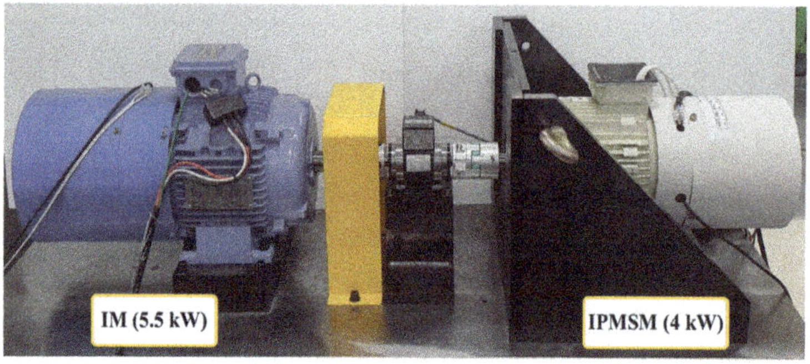

Figure 12. M-G set for the experiment.

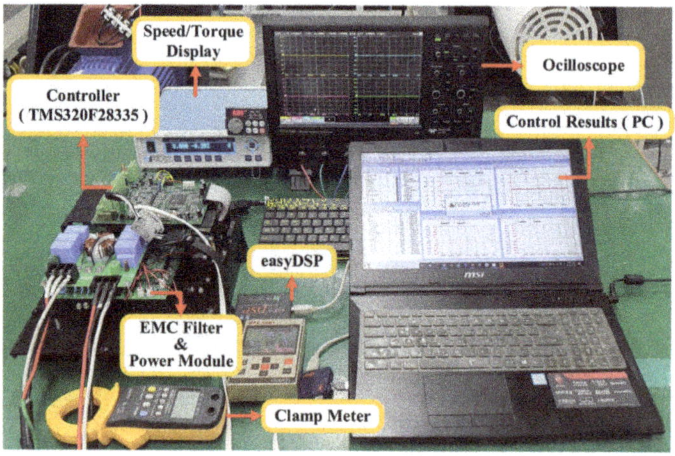

Figure 13. Device configuration for experiments.

4.2. Experiment Result

The experiment was conducted under four conditions as in the simulation. Figure 14 shows the speed obtained in the IPMSM sensorless speed control experiment under no load conditions. It can be seen that the speed error ω_{err} is smaller in the results when using DEEMFO than in the reconstruction method. The experimental results were analyzed to be the same as the simulation results.

Figure 15 shows the results when changing the speed from 3000 rpm to 3500 rpm. The reconstructor has a speed error ω_{err} with a peak-to-peak value of 50.4 rpm. However, using DEEMFO, the speed error ω_{err} reaches a peak-to-peak value of 40.3 rpm. It shows an error improvement of about 20.0%.

Figure 16 represents the results when the load torques were changed from 0 Nm to 6 Nm. In the experiment, the load torque does not increase rapidly due to the limitation of the induction motor applying the load torque, so the speed error is small compared to the simulation. However, even as the load torque increases, the speed error ω_{err} is smaller when the DEEMFO is used compared to the conventional system. The speed error ω_{err} was measured in the control method by the conventional method, reaching a peak-to-peak value of 53.7 rpm. However, the experiment results obtained via the proposed DEEMFO showed the peak-to-peak error equal to 41.6 rpm. It shows an error improvement of about 22.5%.

Figure 17 shows the results of the experiment carried out with the assumption that the electrical parameter of IPMSM increased by 30%. Since it is difficult to directly change the parameters of the IPMSM, we experimented by applying a factor of 1.3 to the nominal parameter. In the case of IPMSM electrical parameters' variation, the peak-to-peak error of 34.6 rpm was measured in the control method using the reconstructor. However, in the control method using the proposed DEEMFO, the peak-to-peak error equals to 21.6 rpm. It shows an error improvement of about 37.6%.

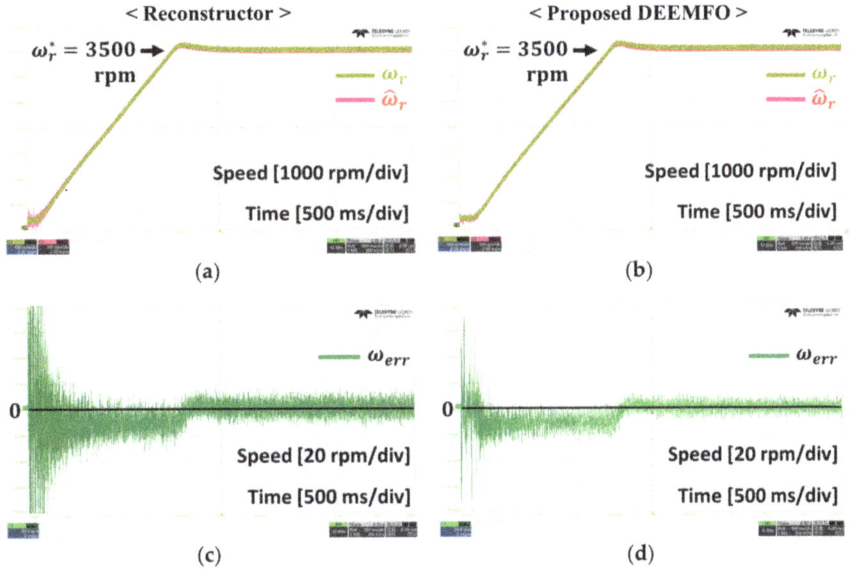

Figure 14. Summary of the experiment results under no load condition. (**a**) reference, real, and estimated speed with reconstructor; (**b**) reference, real, and estimated speed with DEEMFO; (**c**) speed error in reconstructor; (**d**) speed error in DEEMFO.

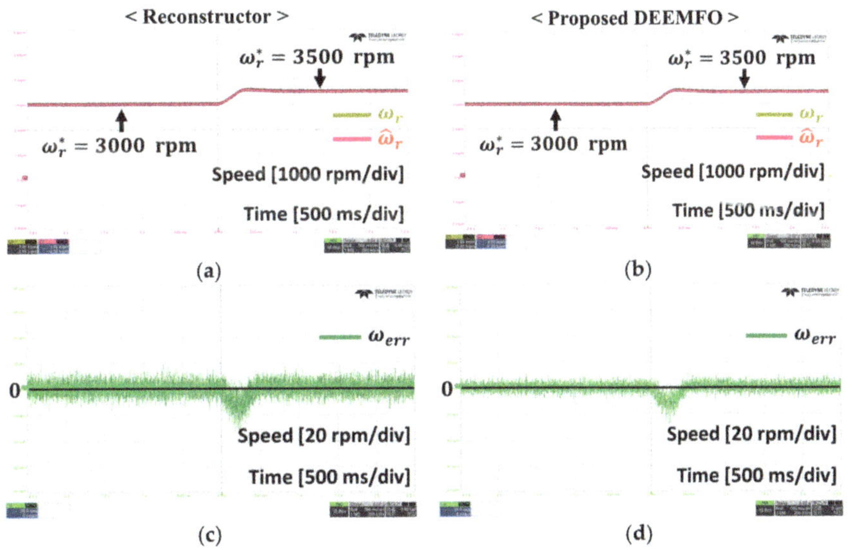

Figure 15. Summary of the experiment results on the condition of increasing the speed reference from 3000 rpm to 3500 rpm. (**a**) reference, real, and estimated speed with reconstructor; (**b**) reference, real, and estimated speed with DEEMFO; (**c**) speed error in reconstructor; (**d**) speed error in DEEMFO.

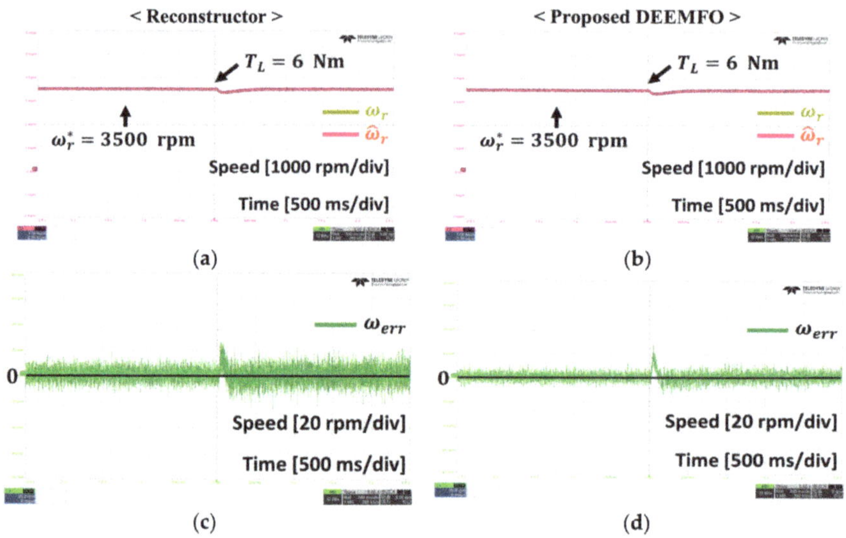

Figure 16. Summary of the experiment results under the condition of increasing the load torque from 0 Nm to 6 Nm. (**a**) reference, real, and estimated speed with reconstructor; (**b**) reference, real, and estimated speed with DEEMFO; (**c**) speed error in reconstructor; (**d**) speed error in DEEMFO.

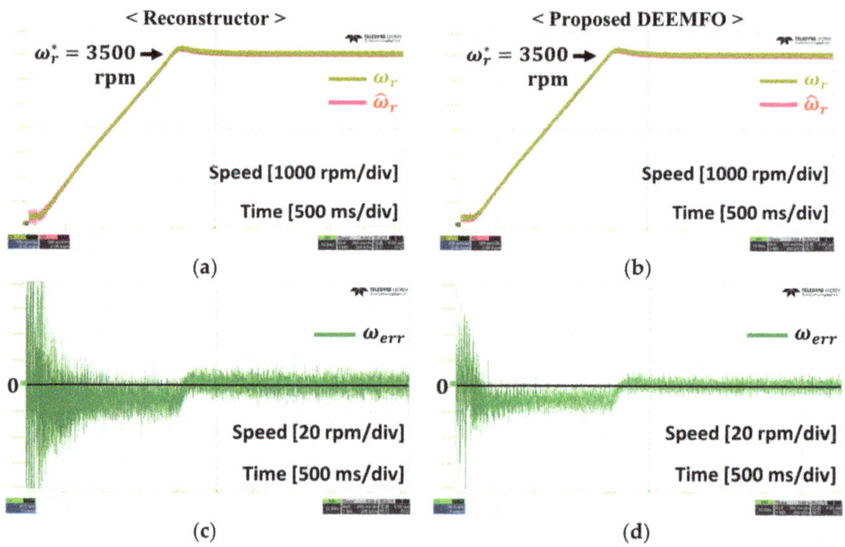

Figure 17. Summary of the experiment results provided for the parameters of IPMSM changed by 30%. (**a**) reference, real, and estimated speed with reconstructor; (**b**) reference, real, and estimated speed with DEEMFO; (**c**) speed error in reconstructor; (**d**) speed error in DEEMFO.

Through experiments under various conditions, it was experimentally verified that the proposed control method using DEEMFO is more accurate than the control method using the conventional method. By more accurately estimating the actual speed, it is possible to obtain control performance with more stable and robust dynamic characteristics even in terms of speed, load torque fluctuation, and parameters' variation.

5. Conclusions

In this paper, by estimating the EEMF using the deadbeat observer for sensorless speed control of IPMSM, it was shown that the speed control is robust and precise in various situations. DEEMFO was proposed to overcome the problem of the current differential term of the reconstructor and the limitation of being vulnerable to parameter fluctuations because the EEMF estimation relies only on simple formulas. DEEMFO can reduce the error between the actual value and the estimated value by compensating through the feedback term. Therefore, the sensorless speed control using the proposed DEEMFO has the advantage of being more precise, robust, and easy to apply because the observer gain can be easily obtained.

Through simulations and experimental results, the proposed DEEMFO shows a better speed control performance than the reconstructor under four conditions. As a result of analyzing the speed error of the reconstructor and DEEMFO, the speed error was improved by at least 10%. To be more precise, in the no-load condition, the error improvement effect of about 33.4% and about 20% compared to the reconstructor was proven in simulation and experiment, respectively. Error improvements of about 46.3% and about 20.0% in the simulation and experiment under the speed fluctuation condition and about 10.1% and 22.5% under the load fluctuation condition were proven, respectively. In particular, in the presence of parameter variations and errors, DEEMFO improves error by about 63.8% in simulation and about 37.6% in experiment.

Research will be conducted to improve operation in the low-speed range and to apply it to IPMSM of various capacities so that it can be used universally and stably. The IPMSM sensorless speed control method using DEEMFO is expected to be applicable to electric vehicles that have a lot of load fluctuations and motor parameters that can vary depending on weather and usage environment. It can also be used in pumps that are located in wet and hot basements and have a high risk of sensor failure.

Author Contributions: Conceptualization, S.-T.K. and I.-S.Y.; Formal analysis, S.-T.K. and I.-S.Y.; Investigation, S.-T.K.; Project administration, J.-S.K.; Software, S.-T.K.; Supervision, S.-C.J. and J.-S.K.; Validation, S.-C.J. and J.-S.K.; Writing—original draft, S.-T.K. All authors have read and agreed to the published version of the manuscript.

Funding: The work was supported in part by the Department of Electronics and Electrical Engineering, through the Research-Focused Department Promotion Project as a part of the University Innovation Support Program for Dankook University in 2021, and in part by a Korea Institute for Advancement of Technology grant funded by the MOTIE "The Competency Development Program for Industry specialist" (Foster R&D specialist of parts for eco-friendly vehicle (xEV), under Grant P0017120).

Conflicts of Interest: The authors declare no conflict of interest.

References

1. Zhu, S.; Hu, Y.; Liu, C.; Jiang, B. Shaping of the Air Gap in a V-Typed IPMSM for Compressed-Air System Applications. *IEEE Trans. Magn.* **2021**, *57*, 8103705. [CrossRef]
2. Yang, Y.; Sandra, M.C.; Rong, Y.; Berker, B.; Anand, S.; Hossein, D.; Ali, E. Design and Comparison of Interior Permanent Magnet Motor Topologies for Traction Applications. *IEEE Trans. Transp. Electrif.* **2017**, *3*, 86–97. [CrossRef]
3. Lin, F.-J.; Hung, Y.-C.; Chen, J.-M.; Yeh, C.-M. Sensorless IPMSM drive system using saliency back-EMF-based intelligent torque observer with MTPA control. *IEEE Trans. Ind. Inform.* **2014**, *10*, 1226–1241.
4. Wang, G.; Yang, R.; Xu, D. DSP-Based control of sensorless IPMSM drives for wide-speed-range operation. *IEEE Trans. Ind. Electron.* **2012**, *60*, 720–727. [CrossRef]
5. Genduso, F.; Miceli, R.; Rando, C.; Galluzzo, G.R. Back EMF Sensorless-Control Algorithm for High-Dynamic Performance PMSM. *IEEE Trans. Ind. Electron.* **2010**, *57*, 2092–2100. [CrossRef]
6. Mobarakeh, B.N.; Tabar, F.M.; Sargos, F.M. Mechanical sensorless control of PMSM with online estimation of stator resistance. *IEEE Trans. Ind. Appl.* **2004**, *40*, 457–471. [CrossRef]
7. Sun, J.; Zhao, J.; Tian, L.; Song, Y.; Liu, Y. Bandwidth and Audible Noise Improvement of Sensorless IPMSM Drives based on Amplitude Modulation Multi-random Frequency Injection. *IEEE Trans. Power Electron.* **2022**, *37*, 14126–14140. [CrossRef]

8. Pal, A.; Das, S.; Chattopadhyay, A.K. An Improved Rotor Flux Space Vector Based MRAS for Field-Oriented Control of Induction Motor Drives. *IEEE Trans. Power Electron.* **2018**, *33*, 5131–5141. [CrossRef]
9. Yin, Z.; Li, G.; Zhang, Y.; Liu, J.; Sun, X.; Zhong, Y. A Speed and Flux Observer of Induction Motor Based on Extended Kalman Filter and Markov Chain. *IEEE Trans. Power Electron.* **2017**, *32*, 7096–7117. [CrossRef]
10. Wang, G.; Li, Z.; Zhang, G.; Yu, Y.; Xu, D. Quadrature PLL-Based High-Order Sliding-Mode Observer for IPMSM Sensorless Control with Online MTPA Control Strategy. *IEEE Trans. Energy Convers.* **2013**, *28*, 214–224. [CrossRef]
11. Zhang, G.; Wang, G.; Yuan, B.; Liu, R.; Xu, D. Active Disturbance Rejection Control Strategy for Signal Injection-Based Sensorless IPMSM Drives. *IEEE Trans. Transp. Electrif.* **2018**, *4*, 330–339. [CrossRef]
12. Tang, Q.; Shen, A.; Luo, X.; Xu, J. IPMSM sensorless control by injecting bidirectional rotating HF carrier signals. *IEEE Trans. Power Electron.* **2018**, *33*, 10698–10707. [CrossRef]
13. Kim, S.-Y.; Ha, I.J. A new observer design method for HF signal injection sensorless control of IPMSMs. *IEEE Trans. Ind. Electron.* **2008**, *55*, 2525–2529.
14. Bolognani, S.; Oboe, R.; Zigliotto, M. Sensorless full-digital PMSM drive with EKF estimation of speed and rotor position. *IEEE Trans. Ind. Electron.* **1999**, *46*, 184–191. [CrossRef]
15. Song, Z.; Yao, W.; Lee, K.; Li, W. An Efficient and Robust I-f Control of Sensorless IPMSM With Large Startup Torque Based on Current Vector Angle Controller. *IEEE Trans. Power Electron.* **2022**, *37*, 15308–15321. [CrossRef]
16. Zhang, Y.; Yin, Z.; Bai, C.; Wang, G.; Liu, J. A Rotor Position and Speed Estimation Method Using an Improved Linear Extended State Observer for IPMSM Sensorless Drives. *IEEE Trans. Power Electron.* **2021**, *36*, 14062–14073. [CrossRef]
17. Sun, X.; Zhang, Y.; Tian, X.; Cao, J.; Zhu, J. Speed Sensorless Control for IPMSMs Using a Modified MRAS With Gray Wolf Optimization Algorithm. *IEEE Trans. Electrif.* **2022**, *8*, 1326–1337. [CrossRef]
18. Chen, Z.; Tomita, M.; Doki, S.; Okuma, S. An extended electromotive force model for sensorless control of interior permanent-magnet synchronous motors. *IEEE Trans. Ind. Electron.* **2003**, *50*, 288–295. [CrossRef]
19. Morimoto, S.; Kawamoto, K.; Sanada, M.; Takeda, Y. Sensorless control strategy for salient-pole PMSM based on extended EMF in rotating reference frame. *IEEE Trans. Ind. Appl.* **2002**, *38*, 1054–1061. [CrossRef]
20. Ko, J.S.; Lee, J.H.; Chung, S.K.; Youn, M.J. A robust digital position control of brushless DC motor with deadbeat load torque observer. *IEEE Tran. Ind. Elec.* **1993**, *40*, 512–520. [CrossRef]
21. Bose, B.K. *Modern Power Electronics and AC Drives*; Prentice-Hall Inc.: Upper Saddle River, NJ, USA, 2002.
22. Astrom, K.J.; Wittenmark, B. *Computer-Controlled Systems: Theory and Design*, 3rd ed.; Prentice Hall: Hoboken, NJ, USA, 1996.

machines

MDPI

Article

Fertilization Control System Research in Orchard Based on the PSO-BP-PID Control Algorithm

Chang Wan [1,2,3,†], Jiawei Yang [1,†], Ling Zhou [2,3], Shuo Wang [1], Jie Peng [4] and Yu Tan [1,*]

1 College of Engineering, China Agricultural University, Beijing 100107, China
2 Key Laboratory of Modern Agricultural Engineering, Xinjiang Uygur Autonomous Region, Alar 843300, China
3 College of Mechanical and Electrical Engineering, Tarim University, Alar 843300, China
4 College of Agriculture, Tarim University, Alar 843300, China
* Correspondence: tanyu@cau.edu.cn
† These authors contributed equally to this work.

Abstract: In order to improve the precision of the variable-rate fertilization system in orchards, this paper conducted a simulation by MATLAB and experimental research based on a variable-rate fertilization experiment platform. The variable-rate fertilization experimental platform was mainly composed of a power supply, DC motors, a PPC-15A1 on-board computer that contains a PCI8932 PC-DAQ, speed sensors, fertilizer dischargers, and a NAV60 module that can receive Beidou Navigation Satellite System positioning data. According to the fertilizer application mechanism of an external grooved wheel fertilizer applicator, the control system model of the variable-rate fertilization driven by the DC motor for orchards was established. A BP neural network adaptive PID controller based on particle swarm optimization (PSO) was proposed to improve the control precision of the system. The step response simulation results by MATLAB show that the overshoot of the BP-PID controller optimized by the PSO algorithm (PSO-BP-PID) is 12.7%, and the adjustment time is 0.557 s. The variable-rate fertilization experiments were conducted, in which the control system was tested by using the PSO-BP-PID controller. The variable fertilizer seeder control system of the Chinese national standard was adopted to evaluate the performance indexes of the system, such as the range of fertilizer amount adjustment, the response time of fertilizer amount adjustment, and the control precision of fertilizer amount. In the variable rate fertilization experiments, the average fertilization errors, respectively, are 1.16% and 1.07%, under the conditions of changing the target fertilization amount and the vehicle speed. The test results are consistent with the simulation results, and the variable-rate fertilization performance parameters are improved.

Keywords: variable-rate fertilization; BP neural network; PSO algorithm; PID controller

Citation: Wan, C.; Yang, J.; Zhou, L.; Wang, S.; Peng, J.; Tan, Y. Fertilization Control System Research in Orchard Based on the PSO-BP-PID Control Algorithm. *Machines* **2022**, *10*, 982. https://doi.org/10.3390/machines10110982

Academic Editors: Kan Liu and Wei Hu

Received: 7 October 2022
Accepted: 25 October 2022
Published: 27 October 2022

1. Introduction

Variable-rate fertilization can reduce the harm caused by excessive chemical fertilization and fertilization cost. With the development of precision agriculture, variable-rate fertilization technology has achieved a series of achievements. The key technology for map-based variable-rate application of fertilizer lies in fertilizer should be applied precisely at each location within the field according to a prescription map. Meng, Z.J. et al. proposed a kind of algorithm for a real-time application map identifier relying on machine location [1]. In order to solve the problems of low precision of variable-rate fertilization and uneven fertilization flow of field liquid fertilizer applicator, Bai, J.B. et al. designed a control system of variable-rate fertilization of liquid fertilizer based on beetle antennae, and a search algorithm was proposed [2]. Nitrogen (N) plays a key role in the optimization between corn grain yield and its economic return. Martins, R.N. et al. aimed to evaluate the technical feasibility of a portable chlorophyll meter for recommending an economic dose of N under variable-rate using different threshold values of the nitrogen sufficiency index (NSI) [3].

Variable-rate application (VRA) of agricultural inputs is essential for optimum crop yield. Alameen, A.A. et al. developed a control system for variable-rate application (VRA) of granular fertilizers and evaluated its performance [4]. In order to solve the fertilizer layered problem caused by the different densities of mixed pellet fertilizer, An, X.F. et al. developed a four-element variable-rate fertilization control system [5].

According to the calculation method of fertilization amount, variable-rate fertilization can be divided into the following two types. The first calculates the fertilizer amount according to the variable-rate fertilization chart. Meng, Z.J. et al. developed a variable-rate fertilization operation system based on the prescription chart. The system realized closed-loop control of the hydraulic motor by receiving the fertilization amount control signal from the airborne control terminal in real-time, combining with the current operating speed and width to achieve the purpose of variable-rate fertilization [6]. The second calculates the fertilization amount based on the information on soil nutrients or crops obtained by real-time sensors. Yang, S. et al. proposed the target fertilization method for the Lycium barbarum orchard in China based on the infrared photoelectric sensors and realized the target fertilization [7].

The smaller the discharge error of variable-rate fertilization is, the closer the fertilization amount is to the fertilizer prescription-based soil nutrient distribution. Focusing on the control precision of fertilization amount and response time of fertilization amount adjustment, scholars have carried out control strategy research to improve the system performance. Huai, B.F. et al. used a PID controller based on a BP neural network in variable-rate fertilization systems, and the error of fertilization in the field experiment was 2.52% [8]. Wang, X.C et al. designed a variable topdressing control system for winter wheat based on the fuzzy PID, and the error of fertilizer amount in the field experiment was 9.07% [9]. A solid fertilizer application control system based on incremental PID was designed by Zhang, J.C., and the error of fertilizer amount in the field experiment was 2.84% [10]. Yuan, Q.C. et al. used a PID algorithm to design a hierarchical variable fertilizer control system for orchards, and the error of fertilizer amount in the field experiment was 6.20% [11]. Sun, Y.J. et al. realized variable-rate fertilization by using the discrete incremental PID control algorithm, and the error of fertilizer amount in the field experiment was 3.41% [12].

Precision fertilization can increase yield and quality in an orchard. Precision fertilization can reduce environmental pollution such as soil hardening and provide green pollution-free fruit. In this paper, a control system of variable-rate fertilization in orchards was designed. The positioning information was obtained through the Beidou satellite navigation and positioning system (BDS), and the adaptive PID control strategy of back-propagation (BP) neural network optimized by particle swarm optimization (PSO) was put forward according to the prescription map. The simulation and experimental results show that the control strategy improves the fertilization accuracy of the variable-rate fertilization system. The PSO-BP-PID algorithm in this paper provides a feasible control algorithm for the orchard variable-rate fertilization system according to the needs of chemical fertilizer.

2. Materials and Methods

2.1. Materials

2.1.1. Orchard Fertilization Process

Figure 1 is the schematic diagram of the orchard trenching and fertilization operation. According to the prescription chart, the DC motor is used as the driving element of fertilization, and the control process of variable-rate fertilization is as follows. The longitude and latitude information are collected by GNSS, the target fertilization amount is obtained according to the prescription chart, the current vehicle speed is obtained by the speed sensor, the control system automatically calculates the theoretical speed of the fertilizer discharging motor, and changes the speed of the DC motor, thus changing the fertilizer discharge amount, and realizing variable-rate fertilization control. When fertilizing, the furrow opener ditches near the root of the fruit tree, and the controller controls the fertilizer distributor to discharge fertilizer according to the requirement of the prescription chart.

The fertilizer falls into the gully through the fertilizer guide pipe, and then the soil mulcher covers the soil to complete the variable-rate fertilization operation.

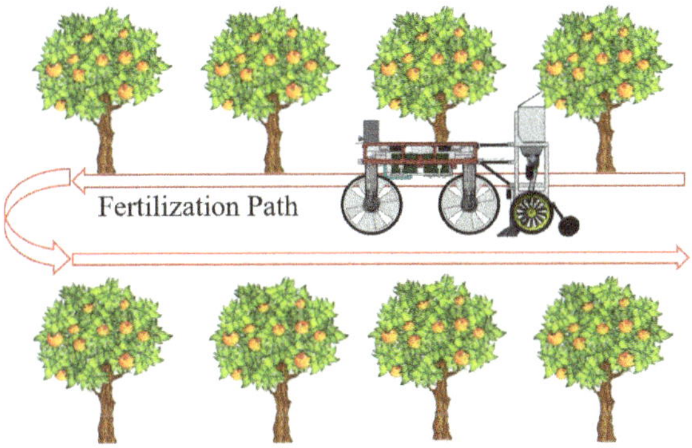

Figure 1. Schematic diagram of fertilization operation in an orchard.

2.1.2. Variable-Rate Fertilization Experiment Platform

As shown in Figure 2, in order to verify the variable-rate fertilization control algorithm, a variable-rate fertilization experimental platform is set up in this paper, which is mainly composed of a power supply, DC motors, a PPC-15A1 on-board computer that contains a PCI8932 PC-DAQ, speed sensors, fertilizer dischargers, and a NAV60 module which can receive BDS (Beidou Navigation Satellite System) positioning data.

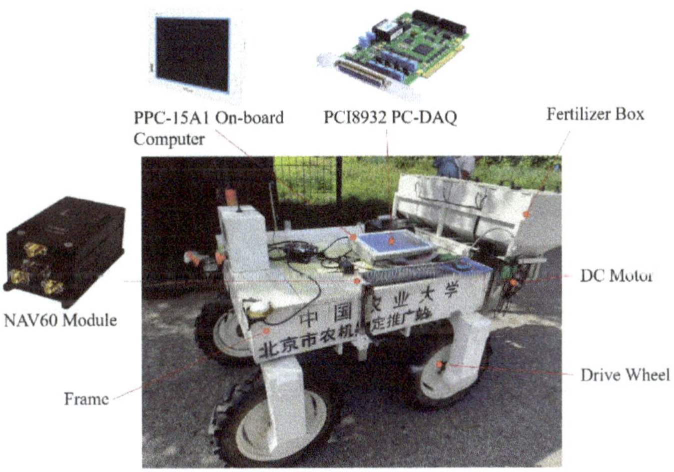

Figure 2. Fertilization experiment platform.

As shown in Figure 3, the on-board computer is equipped with a 15-inch human-computer interactive touch screen, which receives BDS positioning data through the RS232 serial port, collects speed sensor data through the PCI data acquisition card, drives the DC motor, increases the torque through the reducer to drive fertilizer discharge, and configures fertilizer application operating parameters through the on-board computer.

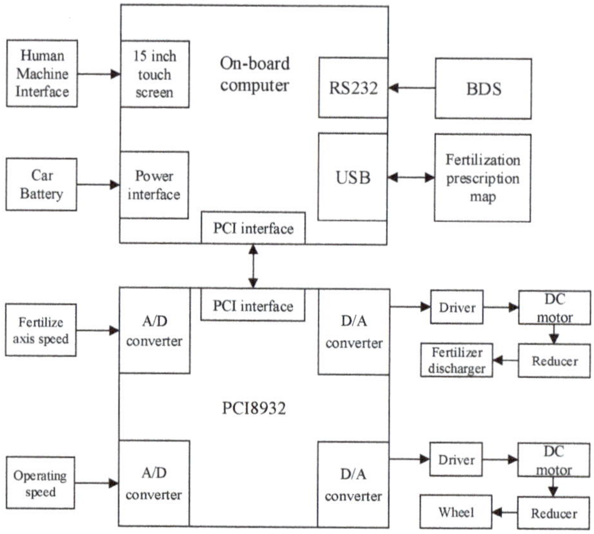

Figure 3. Principle of fertilization platform.

The on-board computer software for the variable-rate fertilization (VRF) system of the orchard is developed by LabVIEW programming, which has a man-machine interface and has the functions of importing fertilization prescription maps, displaying positioning information, displaying fertilization prescription maps in real-time, moving track, and real-time information of the fertilization amount, etc. (Figure 4). Labview realizes the PID algorithm of the BP neural network based on PSO optimization by calling the external DLL link library.

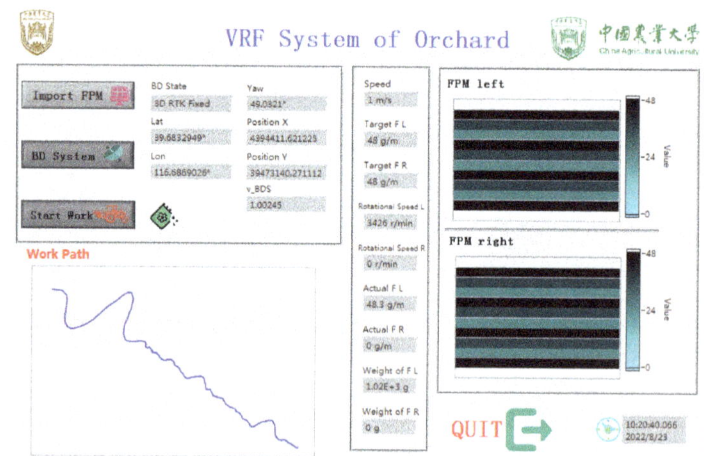

Figure 4. Man-machine Interface.

2.2. *Methods of Model Construction and Calibration Test*

2.2.1. Mathematical Model of Rotating Speed and Fertilizer Discharge

According to the principle of the fertilizer application mechanism of the external grooved wheel fertilizer applicator, under the condition of a certain opening, the amount of fertilizer discharged is determined by the rotation speed of the fertilizer application shaft [13]. The formula for calculating the rotating speed of the fertilizer discharge mechanism is shown in Equation (1).

$$n = \frac{125Qv}{27qN} \tag{1}$$

where Q is the target fertilization amount, g/m; n is the target speed of the motor, r/min; v is the operating speed, m/s; q is the amount of fertilizer discharged by the fertilizer discharger per turn, and g/r; N is the number of fertilizer dischargers.

According to the operating mechanism in this paper and the actual situation of the experiment, a fertilizer discharger is used in the experiment; that is, N is 1; Simplify Equation (1), and the simplified result is shown in Equation (2).

$$n = \frac{125Qv}{27q} \tag{2}$$

It can be known from Equation (2) that the factors that affect the rotation speed of the fertilizer discharging motor include the target fertilizer amount Q, the operation speed v, and the fertilizer discharging amount q per revolution. The target fertilization is obtained according to the prescription chart; the operating speed is obtained by the encoder at the wheel. To realize the precise control of fertilizer application, the key lies in the calibration of fertilizer application q per turn [14].

2.2.2. Calibration Method of Fertilization at Different Speeds

In the experiment, urea, compound fertilizer, and polyethylene plastic particles (simulated fertilizer) were calibrated at different rotating speeds, and the amount of fertilizer discharged per revolution was obtained under different rotational speeds. The rotating speed was increased by 500 r/min from 0 to 3500 r/min, and the rotating speed was kept for one minute. The weight of fertilizer discharged was measured to obtain the actual amount of fertilizer discharged. During the experiment, the fertilizer is easy to melt after being exposed to the air, which affects the weighing after the experiment. Except for the calibration test, polyethylene plastic particles are used to replace the fertilizer in the subsequent tests. SPSS software was used to analyze the bivariate correlation between the rotation speed of different fertilizers and the actual fertilizer discharged, and the p-value and R-value were obtained, with the p-value representing the significant coefficient and the R-value representing the Pearson coefficient. We then performed a linear fit and the fitting coefficient q is obtained from Equation (3).

$$M = qn \tag{3}$$

M is the actual amount of fertilizer discharged, g/min; q is the amount of fertilizer discharged per rotation, g/r; n is the speed, r/min.

2.2.3. Rotating Speed Model of Fertilizer Discharging Motor

In order to facilitate the simulation experiment, the transfer function of the speed control of the fertilizer discharging motor in the variable-rate fertilization system is constructed. In this paper, the control model of medium-precision variable-rate fertilization takes the theoretical fertilization amount as the input of the controller, and the controller sends the electric signal to the DC motor drive module after calculation. The DC motor controls the rotation speed of the fertilizer discharging shaft, and the output is the rotation speed of the motor. The motor rotation speed is fed back to the controller through the encoder, and the controller realizes closed-loop control. The block diagram of the variable-rate fertilization control system is shown in Figure 5.

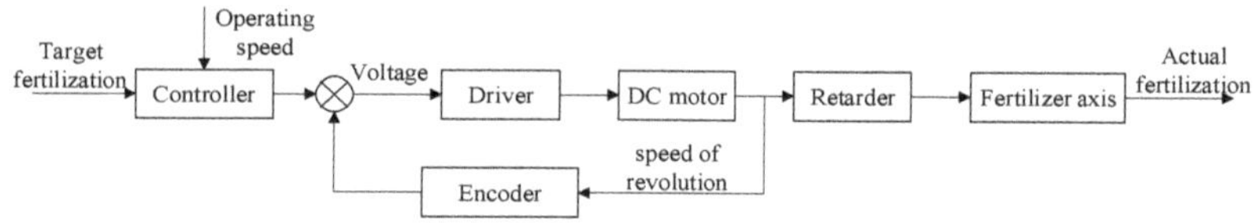

Figure 5. Variable-rate fertilization system in an orchard control block diagram.

The fertilizer discharging motor is a DC brushless motor of Beijing Times Chaoqun Company, and the transfer function of the fertilizer discharging motor is established according to Figure 5. In a DC motor, the control signal is the voltage input and the output is the motor shaft speed. The signal control circuit of a DC motor includes armature circuit balance, electromagnetic induction of the motor rotor, and torque balance of the motor shaft. See Equation (4) for the balance equation

$$\begin{cases} u_a(t) - E = R_a \cdot (i(t) + T_e \frac{di(t)}{dt}) \\ E = k_e \cdot n \\ i(t) - i_l(t) = \frac{T_m}{R_a} \cdot \frac{dE}{dt} \end{cases} \quad (4)$$

where t is time, s; $u_a(t)$ is the voltage of the input DC motor, V; E is the electromotive force of the motor, V; $i(t)$ is armature current, A; T_e is the electromagnetic time constant; R_a is armature resistance, Ω; k_e is the back EMF coefficient; n is the motor speed, rad/s; $i_l(t)$ is the load current, and A; T_m is the electromechanical time constant.

Laplace transform is performed in Equation (4), and the open-loop transfer function of a DC motor is obtained under no-load conditions, as shown in Equation (5).

$$G_k(s) = \frac{N(s)}{U_a(s)} = \frac{k_e}{T_m T_e s^2 + T_m s + 1} \quad (5)$$

According to the control model and each link function, the closed-loop unit negative feedback control transfer function of the variable-rate fertilization control system in this paper is expressed as follows.

$$G_b(s) = \frac{G(s)}{1 + G(s)} = \frac{1.823 \times 10^5}{s^2 + 32.33s + 237.3} \quad (6)$$

where $G_b(s)$ is the closed-loop transfer function of fertilizer discharging motor speed.

According to Equation (6), the characteristic equation is as shown in Equation (7).

$$D(s) = s^2 + 32.33s + 237.3 = 0 \quad (7)$$

According to the Routh–Hurwitz stability criterion, if the coefficients of the characteristic equation of the second-order system are positive, the system is stable.

During the simulation, the transfer function of the rotation speed of the fertilizer discharging motor is used to build the fertilizer discharging system model in Simulink. The PID controller, fuzzy PID controller, and improved BP neural network adaptive PID controller are, respectively, used to control the model.

2.3. *BP Neural Network PID Controller Design Based on Particle Swarm Optimization*

2.3.1. Principle of Particle Swarm Optimization Algorithm

The basic idea of Particle Swarm Optimization (PSO) is to randomly initialize a group of particles without volume and mass. Each particle is a feasible solution to the optimization problem, and the quality of particles is evaluated by a predetermined fitness function. Each particle flies in the feasible solution space, and its direction and distance are determined by a velocity variable. The current optimal particle will be tracked through the particles, and the global optimal solution will be finally found through iterative search [15–18].

In each iteration, the particle will track its optimal solution and the optimal solution of the whole population. The formula for the particle swarm optimization algorithm is shown in Equation (8).

$$\begin{cases} v_{i+1} = \omega \cdot v_i + c_1 r_1 (P_i - x_i) + c_2 r_2 (G_i - x_i) \\ x_{i+1} = x_i + v_{i+1} \end{cases} \tag{8}$$

where i is the particle serial number; v_i is the velocity of the ith particle; x_i is the position of the ith particle; ω is the inertia weight; c_1 and c_2 are learning factors; r_1 and r_2 are random numbers within the range of [0, 1]; P_i is the optimal position of the individual; and G_i is the optimal position of the population.

For inertia weight ω, when ω is small, it is conducive to the local search ability of particles, and the accuracy of solving the problem is relatively high. When ω is large, it is conducive to the global search ability of particles, and to a certain extent it can avoid falling into local optimization. The inertia weight ω decreases by 0.4 from 0.9 linearly, which will generally achieve good optimization results. Therefore, the current inertia weight is calculated according to Equation (9) during iteration.

$$\omega = \omega_{start} - \frac{\omega_{start} - \omega_{end}}{iter_{\max}} \times iter \tag{9}$$

where ω_{start} is the initial inertia weight; ω_{end} is the ending inertia weight; *iter* is the number of iterations; and $iter_{\max}$ is the maximum number of iterations.

2.3.2. Design of the BP Neural Network PID Controller Based on PSO Optimization

The BP neural network is a neural network using the backpropagation algorithm. By combining the BP neural network algorithm with PID control, an adaptive PID controller based on the BP neural network can be designed, which improves the poor control effect of PID in nonlinear systems. The initial weights of the traditional BP neural network are generally generated by using random numbers. After adjusting the network connection weights through backpropagation, it is easy to fall into the local optimal solution, which cannot achieve the optimal control effect. However, the BP neural network optimized by the PSO algorithm can search in a larger space, avoiding the problem of the local optimal solution to a certain extent. The structure diagram of the PSO-BP neural network PID parameter self-tuning control is shown in Figure 6.

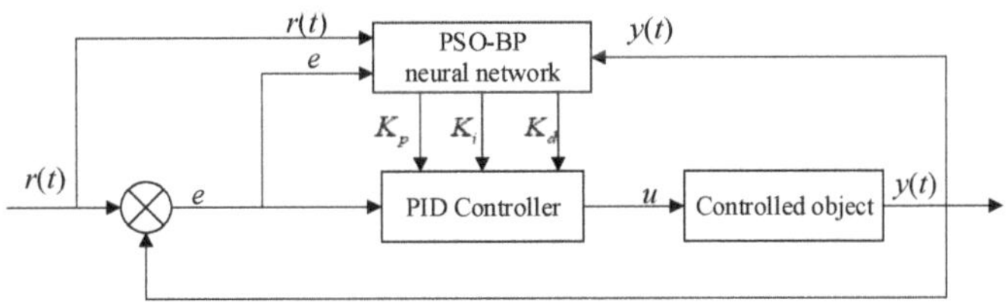

Figure 6. A PID control system diagram based on the PSO-BP neural network.

The learning process of the BP neural network has two parts: forward propagation and backpropagation. The input signal is transmitted to the output layer through the processing of the hidden layer through forwarding propagation. If the expected result cannot be obtained at the output layer, the error signal will return along the original path, and the error signal will be minimized by modifying the weights between the layers. The key to solving nonlinear problems in neural networks is the activation function. If there is no activation function, no matter how many layers of neural networks, the output is a linear combination of inputs, and the nonlinear activation function can effectively solve the problem. There are several types of activation functions: relu activation function, sigmoid activation function, tanh activation function, etc. Since PID parameters must be nonnegative, the sigmoid activation function is selected in this paper. As shown in Equation (10), the mapping range of the output value of the activation function is between 0 and 1.

$$f(x) = \frac{1}{1+e^{-x}} \tag{10}$$

In this paper, the BP neural network structure is 3-5-3. With three layers of input including expected value *rin*, actual value *yout*, and deviation *e*, the hidden layer has five layers, and the output layer has three layers: *Kp*, *Ki*, *Kd*. The value of the learning rate η is 0.5. The weight between the input layer and the hidden layer is wi, and the weight between the hidden layer and the output layer is wo.

In the forward propagation process, the network input layer input, hidden layer input, hidden layer output, output layer input, and output layer output are calculated by sampling the expected value *rin* (*k*), actual value *yout* (*k*), and deviation *e* (*k*) of the system at time *k*. The output of the input layer is:

$$O_j^{(1)} = x(j), j = 1,2,3 \tag{11}$$

The input and output of the hidden layer are:

$$net_i^{(2)}(k) = \sum_{j=0}^{4} w_{ij}^{(2)} O_j^{(1)}(k) \tag{12}$$

$$O_i^{(2)}(k) = f(net_i^{(2)}(k)), i = 1,2,\ldots,5 \tag{13}$$

The input and output of the output layer are:

$$net_l^{(3)}(k) = \sum_{0}^{4} w_{li}^{(3)} O_i^{(2)}(k) \tag{14}$$

$$O_l^{(3)}(k) = f(net_l^{(3)}(k)), l = 1,2,3 \tag{15}$$

The sigmoid function is selected as the activation function of the hidden layer and the output layer, as shown in Equation (10). Adjust the weight value w of the output layer and the hidden layer according to the gradient descent method. The performance index function is selected as:

$$E_k = \frac{1}{2}[r(k) - y(k)]^2 \tag{16}$$

The adjustment amount of hidden layer weight is:

$$\Delta w_{ij}^{(2)}(k) = -\eta \varphi_i^{(2)} O_j^{(1)}(k) \tag{17}$$

Including: $\varphi_i^{(2)} = f'(net_i^{(2)}(k)\sum_{j=1}^{2} \varphi_l^{(3)} w_{li}^{(3)}(k)), i = 1,2,\ldots,5.$

The adjustment amount of the weight value of the output layer is:

$$\Delta w_{li}^{(3)}(k) = -\eta \varphi_l^{(3)} O_i^{(2)}(k) \tag{18}$$

Including: $\varphi_l^{(3)} = e(k)\text{sgn}(k) f'(net_l^{(3)}(k)), l = 1,2,3.$

When the performance index *E*(*k*) is less than the set value, it means that the three parameters output by the BP neural network has met the requirements of the PID controller. And the learning ends. On the contrary, let *k* = *k* + 1 and continue learning.

However, the randomness of the selection of the initial weights of the BP neural network will lead to slow convergence speed and will fall into the local optimal solution, making the whole system unable to work in the optimal state. Particle swarm optimization (PSO) is a global search algorithm with fast convergence speed and a simple algorithm. The initial weights used to optimize BP neural network can better avoid falling into the local optimal solution. It is possible to use the PSO algorithm to train the BP neural network. Firstly, the weights in the network are coded, and each weight is coded into a vector. In this paper, the structure of the BP neural network is 3-5-3, and there are 30 parameters to be optimized, that is, each particle vector is 30 dimensions. Set the particle swarm size to 20 and the maximum number of iterations to 100. Stop training when the optimal solution in the sample is less than the set fitness value or reaches the maximum number of iterations. The fitness function is shown in Equation (19).

$$g_x = \frac{1}{2}[r(k+1) - y(k+1)]^2 = \frac{1}{2}e(k+1) \tag{19}$$

The initial weights *wi* and *wo* of the BP neural network are obtained by using particle swarm optimization (PSO) training, and then the BP neural network adjusts the weights through error backpropagation to adaptively adjust the three PID parameters. The design flow chart of the BP neural network PID controller based on PSO optimization is shown in Figure 7.

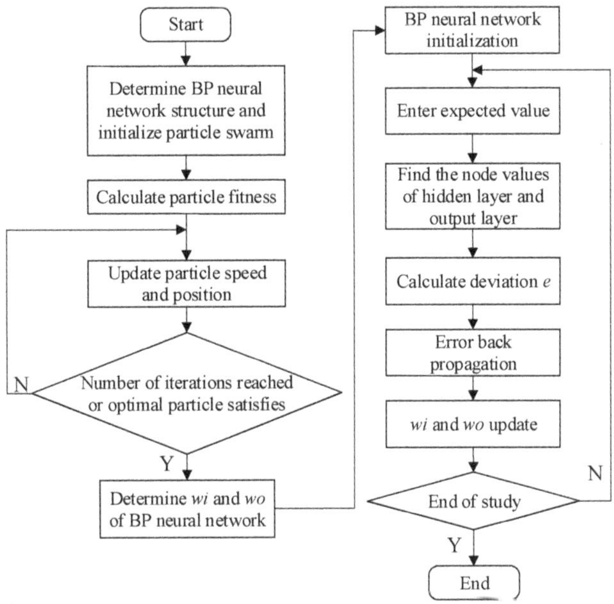

Figure 7. Flow chart of the BP neural network optimization based on particle swarm optimization algorithm.

2.4. Methods of Simulation Test and Variable Rate Fertilization Test

2.4.1. Methods of Simulation Test

In industrial process control, the proportion of PID control strategy in the actual system exceeds 90%. The most difficult problem in the practical application of PID controllers lies in the selection of proportion *Kp*, integral *Ki*, and differential *Kd* coefficients. The fuzzy PID controller can modify the proportion coefficient, integral coefficient, and differential coefficient online according to the deviation *e* and deviation change rate *ec* as well as the the fuzzy rules to meet the requirements for control parameters in different *e* and *ec*, but the quantization factor and scale factor in the controller are still obtained according to the empirical formula [19–23]. The BP neural network is a relatively mature neural network, which can be used to solve nonlinear problems. The BP neural network optimized by particle swarm optimization (PSO) can also effectively avoid the result falling into local optimization. Using the optimized BP neural network can improve the performance of the PID controller [24,25].

According to the transfer function of the fertilizer discharging motor, the traditional PID controller model, fuzzy PID control model, BP neural network, and the PID control models are based on the particle swarm optimization algorithm, which is built using Simulink. The step response simulation test is carried out for the fertilizer discharge motor.

The initial values of *Kp*, *Ki*, and *Kd* in the controller are obtained according to Ziegler Nichols (ZN) method, and the setting formula is shown in Table 1. Neglect the effect of integral and differential, change *Kp* to make the step response in the critical oscillation, and obtain the critical coefficient *Ku* = 0.005, and the oscillation period *Tu* = 0.0016. According to Table 1, *Kp*, *Ki*, and *Kd* are 0.003, 0.075, and 0.000288, respectively. The final values adjusted on this basis are 0.003, 0.2, and 0.0006.

Table 1. ZN Critical Scaling Method Parameter Tuning.

PID Controller	*Kp*	*Ti*	*Td*
P	0.5*Ku*	/	/
PI	0.45*Ku*	0.85*Tu*	/
PID	0.6*Ku*	0.5*Tu*	0.12*Tu*

$Ki = Kp \times T/Ti$, $Kd = Kp \times Td/T$, $T = 0.02$.

The input and output fuzzy rules in the fuzzy controller are set according to expert experience. The fuzzy subset of input and output variables is divided into seven blocks, namely {NB, NM, NS, ZO, PS, PM, PB}. The membership function selects the trimf function. The quantification factor is obtained according to empirical Equations (20) and (21).

$$K_e = \frac{1}{e} \tag{20}$$

$$K_{ec} = \frac{2}{e \times ec} \tag{21}$$

where, *e* is the absolute value of the maximum error in the response process, and *ec* is the absolute value of the maximum error change rate in the response process.

A PID controller based on the PSO-BP neural network cannot be built directly by using Simulink. In this section, the BP-PID controller is simulated by the S function, and the BP-PID controller is compiled through M code in the S function. The initial weights *wi* and *wo* of the BP neural network are obtained through offline training of the particle swarm optimization algorithm.

The BP-PID controller is established in Simulink using the initial weight optimized by the PSO algorithm through the S function, and the step response time domain simulation of the fertilizer discharge motor is carried out. The process is shown in Figure 8. The S function is initialized to seven inputs and four outputs, and the sampling time *T* is set to 0.02 s.

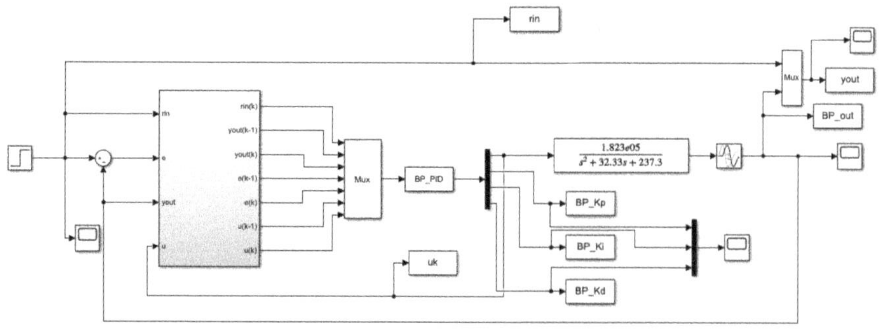

Figure 8. Simulation flow chart of the PSO-BP-PID controller.

In the simulation, the traditional PID controller, fuzzy PID controller, and PSO-BP-PID controller are also tested for anti-interference. During the simulation process, an interference signal with a duration of 0.5 s and a size of 0.01 V is added in 2 s to test the anti-disturbance capability of the three controllers.

2.4.2. Methods of the Variable-Rate Fertilization Test

In order to verify the control performance of the optimized BP neural network PID controller in the variable-rate fertilization system, a variable-rate fertilization experiment was conducted, in which the control system was tested by using the optimized BP neural network PID controller. The variable fertilizer seeder control system (Chinese standard 35487-2017) is adopted to evaluate the performance indexes of the system, such as the range of fertilizer amount adjustment, the response time of fertilizer amount adjustment, and the control precision of fertilizer amount. As shown in Figure 9, to facilitate the weighing of fertilizer, the fertilizer directly falls into the black cloth through the fertilizer guide pipe.

Figure 9. Variable-rate fertilization test.

(1) Response test of fertilizer discharge motor speed.

The response time of the fertilizer discharge adjustment is expressed by measuring the response time of motor speed. During the test, a single fertilizer discharge motor is taken as the object, the feedback speed of the encoder is taken as the actual speed, the target speed is set to increase from 0 to 3500 r/min, and then decrease from 3500 r/min to 0. The motor speed is then recorded.

(2) Fertilizer discharge accuracy test of fertilizer discharger.

During the test, a single fertilizer discharger is taken as the object, polyethylene plastic particles (simulated fertilizer) are used for the test, and the target fertilization and vehicle speed are separately taken as variables for the test. During the test, black havelock is laid on the ground to facilitate the collection and weighing of simulated fertilizer. The actual fertilization is calculated according to Equation (22).

$$Q_2 = \frac{G}{x} \tag{22}$$

where G is the weight of fertilizer discharged, g; x is the driving distance, m.

The error of fertilizer discharge is calculated according to Equation (23).

$$\delta = \left(\frac{Q_1 - Q_2}{Q_1}\right) \times 100\% \tag{23}$$

where δ is the control error of fertilizer discharge, %; Q_1 is the target fertilization, g/m; Q_2 is the actual amount of fertilization, g/m.

The coefficient of variation is calculated according to Equation (24).

$$CV = \frac{SD}{Mean} \times 100\% \tag{24}$$

where CV is the coefficient of variation, %; SD is the standard deviation of multiple groups of tests during weighing; and $Mean$ is the average value of multiple groups of tests during weighing.

According to the upper computer interface of the variable-rate fertilization system, the monitoring fertilization amount is obtained, and the formula for calculating the monitoring error is as Equation (25).

$$\gamma = \left(\frac{Q_3 - Q_2}{Q_3}\right) \times 100\% \tag{25}$$

where γ is the monitoring error of fertilizer discharge, %; Q_2 is the actual amount of fertilization, g/m; and Q_3 is the monitored amount of fertilization, g/m.

1. Target fertilization as a variable.

With the target amount of fertilizer discharged as a variable, import the prescription map into the on-board computer. During the driving process, the prescription map changes the target amount of fertilizer discharged every 20 m, which is 48 g/m, 24 g/m, and 36 g/m in sequence. During the test, the traveling speed of the four-wheel operation platform is maintained at 1 m/s. The fertilizer discharging motor directly drops the fertilizer into the black havelock and weighs the collected fertilizer. After repeated tests, record the average value. Calculate the actual fertilization, coefficient of variation, fertilizer discharge error, and other parameters according to Equations (22)–(25).

2. Vehicle speed as a variable.

With the vehicle speed as the variable, the fertilization amount is set at 48 g/m, and the change speed is 0.6, 0.8, and 1.0 m/s every 20 m when the four-wheel operation platform is running. Collect and weigh the fertilizer discharged from the fertilizer discharger. Repeat the test three times and take the average value. Calculate the actual fertilization, coefficient of variation, error of fertilizer discharge amount, and other parameters according to Equations (22)–(25).

3. Results

3.1. Results of the Calibration Test

The calibration test results of urea, compound fertilizer, and polyethylene plastic particles (simulated fertilizer) at different rotational speeds are shown in Figure 10. The bivariate correlation analysis results of SPSS are shown in Table 1.

According to Table 2, the p-value of the three fertilizers is far less than 0.01, and the R-value is close to 1, that is, the actual fertilizer discharged by the three fertilizers is linearly and positively correlated with the rotation speed.

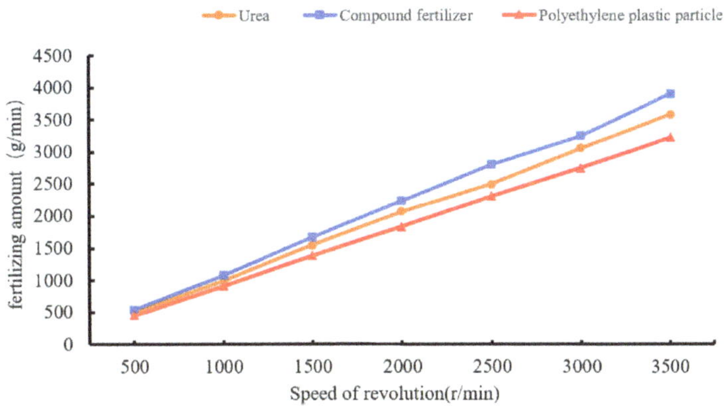

Figure 10. Actual fertilizer discharge amount at different rotating speeds.

Table 2. Correlation analysis between different fertilizer rotation speeds and fertilizer discharge.

Fertilizer Type	p-Value	R-Value
Urea	3.4×10^{-9}	0.99968
Compound fertilizer	1.08×10^{-8}	0.99950
Polyethylene plastic particles	1×10^{-10}	0.99997

According to the linear correlation between the rotating speed and the actual fertilizer amount, the linear regression equation is fitted. According to Equation (3), the fitting coefficient is the fertilizer amount per turn. According to the fitting result, q values of urea, compound fertilizer, and polyethylene plastic particles are 1.016, 1.104, and 0.919 respectively.

3.2. Results of Simulation Test

3.2.1. Results of the PSO-BP-PID Step Response Test

According to the structure of the BP neural network, the particle dimension is 30. By setting the particle swarm size to 20, the inertia weight w decreases from 0.9 to 0.4, the number of iterations is set to 100, and the minimum fitness is set to 1×10^{-5}. The initial weights wi and wo of the BP neural network are obtained through offline training of the particle swarm optimization algorithm, as shown in Equations (26) and (27).

$$wi = \begin{bmatrix} -0.516 & 0.856 & -0.046 \\ 0.802 & 0.542 & -0.212 \\ -1.458 & 0.186 & 0.257 \\ 0.389 & 0.426 & 0.079 \\ 0.557 & -0.333 & 0.559 \end{bmatrix} \quad (26)$$

$$wo = \begin{bmatrix} -0.829 & -1.178 & -0.639 & -2.172 & -2.155 \\ -0.711 & -0.863 & -2.233 & -1.275 & -0.921 \\ -0.957 & -1.376 & -0.432 & -1.563 & -1.142 \end{bmatrix} \quad (27)$$

The time domain response curve is shown in Figure 11. The overshoot is 12.7%, the rise time is 0.105 s, and the adjustment time is 0.557 s. There is no steady-state error. The adaptive values of Kp, Ki, and Kd are shown in Figure 12. According to the simulation results, the PSO-BP-PID controller can adaptively adjust the parameters of Kp, Ki, and Kd, greatly improving the control performance.

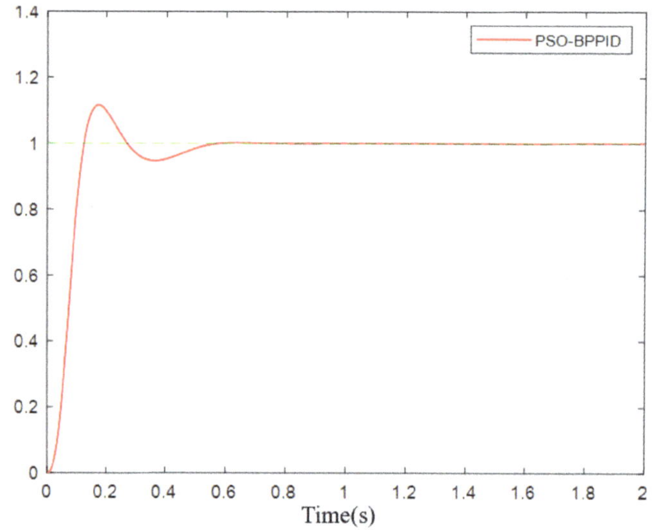

Figure 11. Time domain simulation curve of the PSO-BP-PID controller step response.

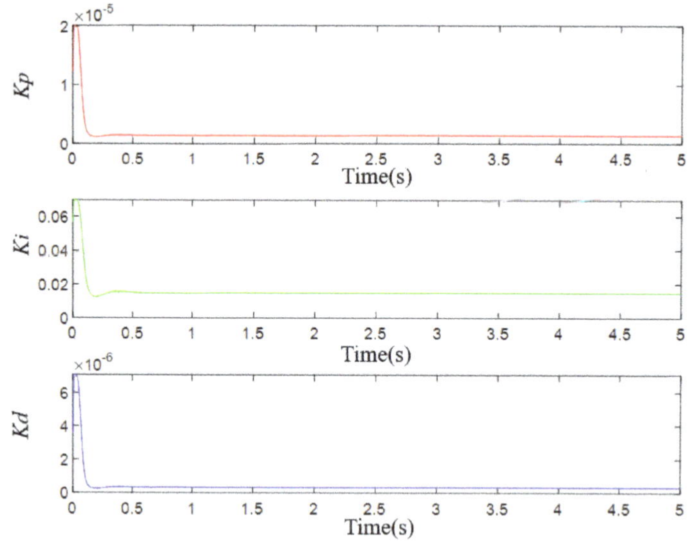

Figure 12. The adaptive curve of step response parameter of the PSO-BP-PID controller.

3.2.2. Results of Step Response Contrast Test

The PID controller and fuzzy PID controller are built in Simulink to simulate the step response of the fertilizer discharging motor, and compared with the optimized BP neural network PID controller. The time domain response curve is shown in Figure 13. The performance index comparison of time domain response is shown in Table 3. Through comparison, it can be found that the performance of the fuzzy PID and the PSO-BP-PID controllers is better than that of traditional PID controllers, which can reduce the adjustment time to a certain extent, and the PSO-BP-PID controller can reduce the overshoot to a certain extent.

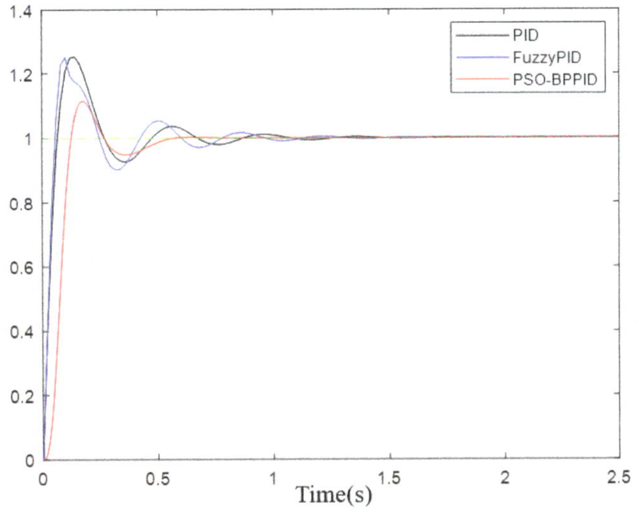

Figure 13. Comparison of time domain response curves.

Table 3. Step response performance index comparison.

Controller	Overshoot σ	Rise Time t_r	Adjustment Time t_s
Traditional PID	25.4%	0.058 s	1.555 s
Fuzzy PID	25.4%	0.053 s	1.443 s
PSO-BP-PID	12.7%	0.105 s	0.557 s

3.2.3. Results of Immunity Comparison Test

In the simulation, the traditional PID controller, fuzzy PID controller, and the PSO-BP-PID controller are also tested for anti-interference. During the simulation process, an interference signal with a duration of 0.5 s and a size of 0.01 V is added in 2 s to test the anti-disturbance capability of the three controllers. The time domain response curve of the three added disturbances is shown in Figure 14. The comparison of anti-disturbance performance indexes can be compared in Table 4, The traditional PID controller and fuzzy PID controller have a poor anti-disturbance ability, the system recovery speed is slow, and the PSO-BP-PID controller has a strong anti-disturbance ability. After the disturbance, the system recovery speed is fast and the fluctuation after the disturbance is minimal.

Table 4. Comparison of anti-disturbance performance indexes.

Controller	Disturbance Recovery Time t_i	Maximum Fluctuation μ
Traditional PID	3.654 s	6.8%
Fuzzy PID	3.923 s	8.1%
PSO-BP-PID	2.849 s	3.2%

In conclusion, the adaptive PID controller based on the BP neural network optimized by the PSO algorithm has the best control performance and anti-disturbance ability. The variable-rate fertilization control system adopts the improved BP neural network adaptive PID controller.

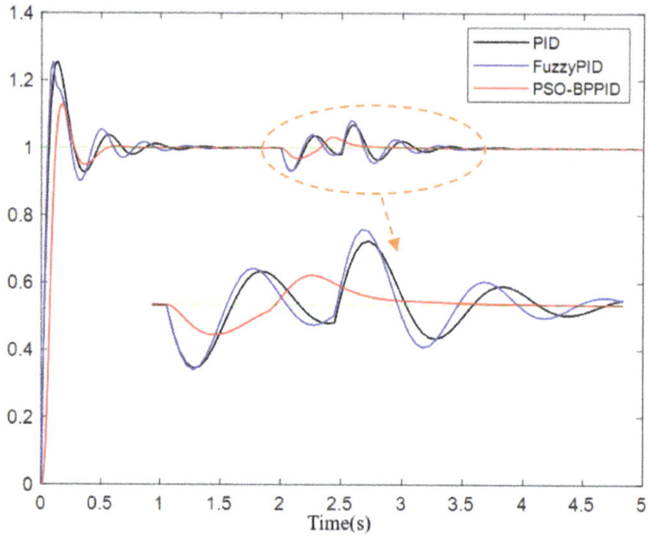

Figure 14. Anti-disturbance simulation curves.

3.3. *Results of the Variable-Rate Fertilization Test*

3.3.1. Result of Response Test of Fertilizer Discharge Motor Speed

The curve shown in Figure 15 is to obtain the motor response time. See Tables 5 and 6 for the test results.

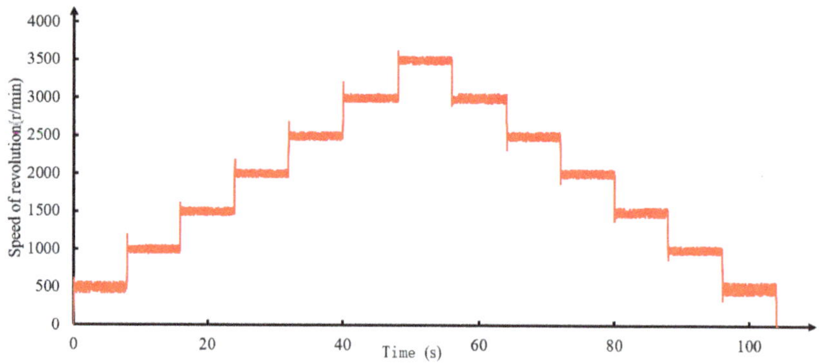

Figure 15. Speed response curve.

Table 5. Response test for increasing speed of dc motor of single fertilizer apparatus.

Rotating Speed (Increasing) r/min	0 to 500	500 to 1000	1000 to 1500	1500 to 2000	2000 to 2500	2500 to 3000	3000 to 3500
Response time s	0.28 s	0.24 s	0.2 s	0.14 s	0.18 s	0.22 s	0.24 s

Table 6. Response test of dc motor speed decline of single fertilizer apparatus.

Rotating Speed (Decreasing) r/min	3500 to 3000	3000 to 2500	2500 to 2000	2000 to 1500	1500 to 1000	1000 to 500	500 to 0
Response time s	0.22 s	0.22 s	0.18 s	0.16 s	0.18 s	0.2 s	0.26 s

According to Tables 5 and 6, the average response time of fertilizer discharge adjustment is 0.21 s, and the adjustment range is 500–3500 r/min. The adjustment range of a fertilizer discharge is 16−52 g/m converted from Equation (1) according to different operating speeds.

3.3.2. Result of Fertilizer Discharge Accuracy Test of Fertilizer Discharger

(1) Result of target fertilization as a variable experiment

Figure 16 shows target fertilization as a variable experiment. According to Table 7, the variable-rate fertilization system can achieve variable-rate fertilization by adjusting the rotation speed of the fertilizer discharge axis according to the prescription map input. The average coefficient of variation of fertilizer discharge of the variable-rate fertilization system under different target fertilizer discharge amounts is 1.93%, the average value of fertilizer discharge error is 1.16%, and the average value of monitoring error is 2.63%. Compared with the traditional PID and fuzzy PID controllers in the references, the accuracy of the controller is improved by 1.68~7.91 percentage points, and the accuracy of the BP-PID controller is improved by 1.36 percentage points.

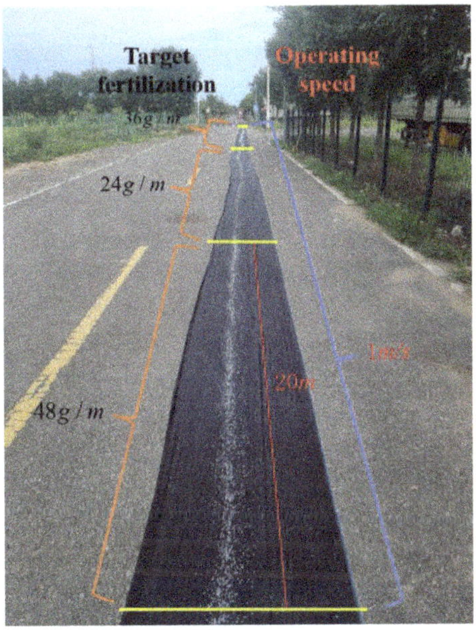

Figure 16. Precision test of fertilizer discharge under different amounts.

Table 7. Fertilizer discharge precision test of variable-rate fertilization system under different fertilizer discharge amounts.

Operating Speed m/s	Target Fertilization g/m	Actual Fertilization g/m	Coefficient of Variation %	Fertilizer Discharge Error %	Monitoring Fertilization g/m	Monitoring Error %
1.0	24	23.472	2.83	2.20	24.52	4.47
	36	35.784	1.34	0.60	36.448	1.86
	48	48.320	1.63	0.67	49.068	1.55

(2) Result of vehicle speed as a variable experiment

Figure 17 shows vehicle speed as a variable experiment. According to Table 8, the variable-rate fertilization system can adjust the rotation speed of the fertilizer discharging shaft according to different vehicle speeds to achieve variable-rate fertilization. The average coefficient of variation of fertilizer discharge amount of variable-rate fertilization system at different speeds is 1.05%, the average value of fertilizer discharge error is 1.07%, and the average value of monitoring error is 1.22%. Compared with the traditional PID controller and fuzzy PID controller in the reference, the precision of the controller is increased by 1.77 to 8 percentage points, and the precision of the BP-PID controller is increased by 1.45 percentage points.

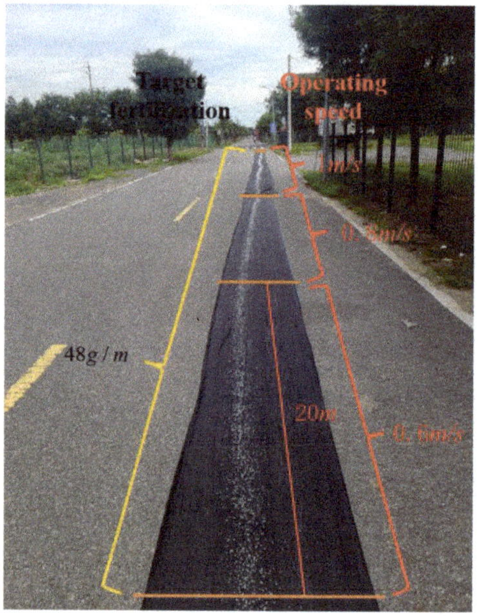

Figure 17. Fertilizer discharge accuracy test under different vehicle speeds.

Table 8. Fertilizer discharge accuracy test of variable-rate fertilization system at different speeds.

Target Fertilization g/m	Operating Speed m/s	Actual Fertilization g/m	Coefficient of Variation %	Fertilizer Discharge Error %	Monitoring Fertilization g/m	Monitoring Error %
	0.6	47.432	1.06	1.18	48.252	1.69
48	0.8	47.396	1.22	1.25	48.344	1.96
	1.0	48.372	0.88	0.78	48.868	1.02

In the precision test of fertilizer discharge, only polyethylene plastic particles (simulated fertilizer) were used as test materials. The variable-rate fertilization control performance of urea, compound fertilizer, and other fertilizers needs further experimental analysis.

4. Discussion

In this paper, the PSO-BP-PID controller is designed and verified by simulation and variable-rate fertilization experiments. In the simulation test, the PSO algorithm was used to optimize the initial weight of the BP neural network, which made the adaptive performance of the BP-PID controller better, that is, the overshoot of the PSO-BP-PID controller in the test was reduced, and the anti-interference performance was improved. The experiment of variable-rate fertilization proves that the PSO-BP-PID controller can complete variable-rate fertilization with small errors.

PSO is a global optimization algorithm, which can be used for offline iteration. Previous researchers used this algorithm to optimize the initial parameters of each controller, including the PID controller and fuzzy PID controller. The initial parameters of the PID controller and fuzzy PID controller are few and can be obtained by trial and error, but the weights of the BP neural network are too large to be obtained by trial and error. In this case, the performance of the BP neural network can be greatly improved by using a PSO algorithm for optimization.

Because the parameters of the PSO-BP-PID controller are adaptively adjusted, the controller is also suitable for nonlinear systems. Since the orchard fertilization work requires ditching, the resistance is large. Most of the fertilization machinery is powered by hydraulic systems. The controller can still effectively control the nonlinear hydraulic motors and other controlled objects.

5. Conclusions

Through the mathematical model test of rotation speed and fertilizer discharge amount, it is proved that the rotation speed is a linear positive correlation with the actual fertilizer discharge amount, further proving the feasibility of controlling the fertilizer discharge amount by controlling the rotation speed of the fertilizer discharge motor, and calibrating the fertilizer discharge amount per rotation through the regression coefficient. This establishes the transfer function of the fertilizer discharge motor.

The initial weights of the BP neural network are optimized by particle swarm optimization algorithm, and the adaptive PID controller of the BP neural network is designed. By comparing with the traditional PID controller and fuzzy PID controller through simulation tests, it can be concluded that the optimized BP neural network adaptive PID controller can reduce overshoot, reduce adjustment time, and have a strong ability to resist disturbance. The overall performance is better than other controllers.

The variable-rate fertilization experiment shows that the PSO-BP-PID control system conducts better fertilization performance according to the prescription chart and vehicle speed. The average adjustment response time of fertilizer discharge is 0.21 s, and the adjustment range of a fertilizer discharge is 16–52 g/m according to different operating speeds. The result of target fertilization as variable experiment show that the average fertilizer discharge error is 1.16%, and the result of vehicle speed as variable experiment show that the average fertilizer discharge error is 1.07%. The above results of fertilization control accuracy meet the national standard, and are 1.36 to 8 percentage points higher than the existing fertilizer discharge accuracy. The results provide a feasible scheme for the control strategy of variable-rate fertilization system in orchards.

Author Contributions: Conceptualization, J.Y. and C.W.; methodology, J.Y. and Y.T.; software, J.Y.; validation, J.Y., S.W. and Y.T.; formal analysis, J.Y.; investigation, J.Y.; resources, C.W., L.Z. and Y.T.; data curation, J.Y. and S.W.; writing—original draft preparation, J.Y. and C.W.; writing—review and editing, J.Y., J.P. and C.W.; supervision, Y.T. and C.W.; funding acquisition, C.W., L.Z. and Y.T. All authors have read and agreed to the published version of the manuscript.

Funding: This paper was supported by the following projects: the Innovation and Development Support Program for Key Industries in Southern Xinjiang supported by Xinjiang Production and Construction Corps in China (Grant No. 2020DB003), the Key Program of Open Project from Key Laboratory of Modern Agricultural Engineering of Xinjiang Uygur Autonomous Region in China (Grant No. TDNG20180101), Alar City Science and Technology Project of the First Division of Xinjiang Production and Construction Corps in China (Grant No. 2020ZB03).

Institutional Review Board Statement: Not applicable.

Informed Consent Statement: Not applicable.

Data Availability Statement: Not applicable.

Acknowledgments: The author would like to thank Beijing Zhong Nong Lv Tong Co., Ltd. for providing the test site.

Conflicts of Interest: The authors declare no conflict of interest. The funders had no role in the design of the study; in the collection, analyses, or interpretation of data; in the writing of the manuscript; or in the decision to publish the results.

References

1. Meng, Z.J.; Zhao, C.J.; Fu, W.Q.; Ji, Y.X.; Wu, G.W. Prescription Map Identification and Position Lag Calibration Method for Variable Rate Application of Fertilizer. *Trans. Chin. Soc. Agric. Mach.* **2011**, *42*, 204–209.
2. Bai, J.B.; Tian, M.; Li, J.Q. Control System of Liquid Fertilizer Variable-Rate Fertilization Based on Beetle Antennae Search Algorithm. *Processes* **2022**, *10*, 357. [CrossRef]
3. Martins, R.N.; Pinto, F.A.; Moura, A.D.; Siqueira, W.C.; Villar, F.M. Nitrogen variable rate fertilization in corn crop prescribed by optical sensor. *J. Plant Nutr.* **2020**, *43*, 1681–1688. [CrossRef]
4. Alameen, A.A.; Al-Gaadi, K.A.; Tola, E.K. Development and performance evaluation of a control system for variable rate granular fertilizer application. *Comput. Electron. Agric.* **2019**, *160*, 31–39. [CrossRef]
5. An, X.F.; Fu, W.Q.; Wei, X.L.; Cong, Y.; Wang, P. Evaluation of Four-element Variable Rate Application of Fertilization Based on Maps. *Trans. Chin. Soc. Agric. Mach.* **2017**, *48*, 66–70.
6. Meng, Z.J.; Zhao, C.J.; Liu, H.; Huang, W.Q.; Fu, W.Q.; Wang, X. Development and performance assessment of map-based variable rate granule application system. *J. Jiangsu Univ.* **2009**, *30*, 338–342.
7. Yang, S.; Zhai, C.Y.; Long, J.; Zhang, B.; Li, H.Z. Wolfberry tree dual-model detection method and orchard target-oriented fertilization system based on photoelectric sensors. *Int. J. Agric. Biol. Eng.* **2018**, *11*, 65–73. [CrossRef]
8. Huai, B.F.; Zhang, C.S.; Zhang, P.X.; Han, J.; Wang, X.; Zhuang, W.D. BP-PID Control Method in Control System of Variable Rate Fertilizer Application. *J. Heilongjiang Bayi Agric. Univ.* **2015**, *27*, 95–98.
9. Wang, X.C.; Chen, M.; Sun, G.X.; Zhang, Y.; Zhang, Y.N. Design and test of control system on variable fertilizer applicator for winter wheat. *Trans. Chin. Soc. Agric. Eng.* **2015**, *31*, 88–92.
10. Zhang, J.C.; Yan, S.C.; Ji, W.Y.; Song, B.G.; Zheng, P. Precision Fertilization Control System Research for Solid Fertilizers Based on Incremental PID Control Algorithm. *Trans. Chin. Soc. Agric. Mach.* **2021**, *52*, 99–106.
11. Yuan, Q.C.; Xu, L.M.; Niu, C.; Ma, S.; Yan, C.G.; Zhao, S.J. Design and Test of Layered Variable Rate Fertilizer Discharge Control System for Organic Fertilizer Deep Applicator. *Trans. Chin. Soc. Agric. Mach.* **2020**, *51*, 195–202.
12. Sun, Y.J.; Shen, J.X.; Dou, Q.Q.; Li, Q.L.; Chen, G.; Sun, Y.T. Design and Test of Monitoring System of Notillage Planter Based on Cortex-M3 Processor. *Trans. Chin. Soc. Agric. Eng.* **2018**, *49*, 50–58.
13. An, X.F.; Fu, W.Q.; Wang, P.; Wei, X.L.; Li, L.W.; Meng, Z.J. Development of Variable Rate Fertilization Control System Based on Matching Fertilizer Line and Seed Line of Wheat. *Trans. Chin. Soc. Agric. Eng.* **2019**, *50*, 96–101.
14. Zhang, J.Q.; Liu, G.; Hu, H.; Huang, J.Y. Development of bivariate fertilizer control system via independent control of fertilizing unit. *Trans. Chin. Soc. Agric. Eng.* **2021**, *37*, 38–45.
15. Ren, H.J.; Hou, B.; Zhou, G.; Shen, L.; Wei, C.; Li, Q. Variable Pitch Active Disturbance Rejection Control of Wind Turbines Based on BP Neural Network PID. *IEEE Access* **2020**, *8*, 71782–71797. [CrossRef]
16. Luo, Q.Y.; Li, J.X.; Zhang, H. Drag coefficient modeling of heterogeneous connected platooning vehicles via BP neural network and PSO algorithm. *Neurocomputing* **2022**, *484*, 117–127. [CrossRef]
17. Ibrahim, B.; Abdelkader, H.; Patrice, W. Advanced Control of Doubly Fed Induction Generator for Wind Power Systems: Optimal Control of Power Using PSO Algorithm. *Appl. Mech. Mater.* **2022**, *905*, 29–42.
18. Chavoshian, M.; Taghizadeh, M.; Mazare, M. Hybrid Dynamic Neural Network and PID Control of Pneumatic Artificial Muscle Using the PSO Algorithm. *Int. J. Autom. Comput.* **2020**, *17*, 428–438. [CrossRef]
19. Chen, M.; Lu, W.; Wang, X.C.; Sun, G.X.; Zhang, Y.; Pan, F. Design and Experiment of Optimization Control System for Variable Fertilization in Winter Wheat Field Based on Fuzzy PID. *Trans. Chin. Soc. Agric. Mach.* **2016**, *47*, 71–76.
20. Song, X.; Li, H.; Chen, C.; Xia, H.; Zhang, Z.; Tang, P. Design and Experimental Testing of a Control System for a Solid-Fertilizer-Dissolving Device Based on Fuzzy PID. *Agriculture* **2022**, *12*, 1382. [CrossRef]
21. Yao, Y.F.; Chen, X.G.; Ji, C.; Chen, J.C.; Zhang, H.; Pan, F. Design and experiments of the single driver for maize precision seeders based on fuzzy PID control. *Trans. Chin. Soc. Agric. Eng.* **2022**, *38*, 12–21.
22. Nuchkrua, T.; Leephakpreeda, T. Fuzzy Self-Tuning PID Control of Hydrogen-Driven Pneumatic Artificial Muscle Actuator. *J. Bionic Eng.* **2013**, *10*, 329–340. [CrossRef]
23. Li, K.; Boonto, S.; Nuchkrua, T. On-line Self Tuning of Contouring Control for High Accuracy Robot Manipulators under Various Operations. *Int. J. Control Autom. Syst.* **2020**, *18*, 1818–1828. [CrossRef]
24. Zhang, M.L.; Zhang, Y.J.; He, X.L.; Gao, Z.J. Adaptive PID Control and Its Application Based on a Double-Layer BP Neural Network. *Processes* **2021**, *9*, 1475. [CrossRef]
25. Pei, G.J.; Yu, M.; Xu, Y.H.; Ma, C.; Lai, H.H.; Chen, F.K.; Lin, H. An Improved PID Controller for the Compliant Constant-Force Actuator Based on BP Neural Network and Smith Predictor. *Appl. Sci.* **2021**, *11*, 2685. [CrossRef]

applied sciences

MDPI

Article

Adapting Neural-Based Models for Position Error Compensation in Robotic Catheter Systems

Toluwanimi O. Akinyemi [1,2,3], Olatunji M. Omisore [1,4,*], Xingyu Chen [1,4], Wenke Duan [1], Wenjing Du [1], Guanlin Yi [1] and Lei Wang [1,4,*]

1 Research Center for Medical Robotics and MIS Devices, Shenzhen Institutes of Advanced Technology, Chinese Academy of Sciences, Shenzhen 518055, China
2 University of Chinese Academy of Sciences, Beijing 100049, China
3 Shenzhen College of Advanced Technology, University of Chinese Academy of Sciences, Shenzhen 518055, China
4 CAS Key Laboratory for Health Informatics, Shenzhen Institutes of Advanced Technology, Chinese Academy of Sciences, Shenzhen 518055, China
* Correspondence: omisore@siat.ac.cn (O.M.O.); wang.lei@siat.ac.cn (L.W.)

Featured Application: Robot-assisted surgery has simplified minimally invasive treatments by providing surgeons with more enhanced skill, precision, and control of surgical instruments for microscale operations than is possible with conventional methods. A use case is robot-assisted percutaneous coronary intervention, which utilizes a robotic catheter system to remove lesions in the coronary arteries. Typically, the RCS provides discrete, repetitive, and steady motion to endovascular tools for tooltip translation from an insertion site to the blocked coronary arteries while ensuring vessel damage avoidance. Therefore, this study aims to enhance master–slave motion accuracy toward the realization of autonomous navigation during robot-assisted cardiac interventions.

Abstract: Robotic catheter systems with master–slave designs are employed for teleoperated navigation of flexible endovascular tools for treating calcified lesions. Despite improved tool manipulation techniques, patient safety and lowering operative risks remain top priorities. Therefore, minimizing undesirable drifts and imprecise navigation of flexible tools during intravascular catheterization is essential. In the current master–slave designs, finite displacement lag between position command and actual navigation action at the slave device affects smooth catheterization. In this study, we designed and developed a compact 2-DOF robotic catheter system and characterized the influence of displacement step values, velocity, and motion gap on the position error at the slave device. For uniform and varying motion commands from the master platform, the results indicate that the overall position error increases with the distance traveled and the displacement step values, respectively. Hence, we proposed using recurrent neural networks—long short-term memory and gated recurrent unit controllers to predict the slave robot's position and appropriate compensation value per translation step. An analysis of in-silico studies with CoppeliaSim showed that the neural-based controllers can ensure uniform motion mapping between the master–slave devices. Furthermore, we implemented the models within the RCS for a catheterization length of 120 mm. The result demonstrates that the controllers suitably aid the slave robot's stepwise displacement. Thus, the neural-based controllers help match the translational motion and precise tool navigation by the slave robotic device. Therefore, the neural-based controllers could contribute to alleviating patients' safety concerns during robotic interventions.

Keywords: robot-assisted catheterization; position error control; teleoperation; learning-based systems; deep learning

Citation: Akinyemi, T.O.; Omisore, O.M.; Chen, X.; Duan, W.; Du, W.; Yi, G.; Wang, L. Adapting Neural-Based Models for Position Error Compensation in Robotic Catheter Systems. *Appl. Sci.* **2022**, *12*, 10936. https://doi.org/10.3390/app122110936

Academic Editors: Kan Liu and Wei Hu

Received: 22 September 2022
Accepted: 25 October 2022
Published: 28 October 2022

1. Introduction

Vascular diseases, such as aneurysms and stenosis, adversely affect blood flow to vital organs and other body tissues. This condition has been categorized as atherosclerosis when it affects the coronary, carotid, or peripheral arteries [1]. As a systemic and inflammatory vascular process, atherosclerosis causes plaque aggregation within the arterial lumen. The presence of atherosclerotic lesions in the coronary arteries, referred to as coronary artery disease (CAD), interrupts the flow of oxygen-rich blood to the heart muscle, thereby eliciting symptoms such as numbness, angina, and myocardial infarction [2]. As a result, CAD is a significant cause of global disease burden and accounts for over 80% of sudden cardiac deaths [3]. Conventionally, interventionalists require large incisions into the chest cavity for procedures such as graft bypass around clogged arteries to reroute blood flow to the heart. However, percutaneous coronary intervention (PCI) has been widely adopted for the diagnosis and treatment of CAD as it offers patients reduced perioperative morbidities and faster recovery times [4–7].

In PCI procedures, intravascular catheterization is a vital step in which patients' vessels are cannulated with thin and flexible endovascular tools, viz. guidewire, catheter, and rapid exchange devices. This routine procedure involves tool navigation from an entry port, typically on the radial, femoral, or brachial vessels, to noticeable lesions in the coronary arteries. Intravascular catheterization is performed with trained interventionalists applying fine motor skills [7,8]. This includes axial operations such as translational (push/pull) and rotational movements applied proximally, while the movement transmits distally to navigate endovascular tools within the patient's vasculature. However, precise manipulation of endovascular tools in a dynamic cardiac environment such as a beating heart and fragile vessels persists as a challenge in *cath labs*. Hence, interventionalists undergo extensive training and evaluation sessions for safe tool manipulation during PCI. Unfortunately, this limits the availability of operators with certified proficiency for manual catheterization relative to the recent high global prevalence rate of vascular diseases such as CAD [8–11].

Robotic catheter systems (RCSs) have been introduced into *cath labs* to reduce PCI operational risks such as radiation exposure hazards and orthopedic injuries to interventionalists. Robotic-assisted PCI (R-PCI) is performed with the RCS having a unique master and slave platform to ensure the surgeons' safety while offering remote tool manipulation [12]. Globally, CorPath® 200 and CorPath® GRX (Siemens Healthineers, Walpole, MA, USA) are FDA-approved commercial products developed to facilitate precise navigation of flexible endovascular tools with enhanced dexterity and a reduced learning curve for R-PCIs [13,14]. In addition, CorPath® GRX offers higher functionalities such as guided control and improved device exchanges. However, to achieve smooth motion during procedures, the CorPath systems translate endovascular tools with a displacement step of 1 mm or a maximum speed of 12 mm/s. Hence, its motion accuracy for higher displacement step values is unsubstantiated [15]. Despite the evidence supporting the safety and feasibility of RCSs for R-PCIs, they are currently available in only a few hospitals globally. Furthermore, it is projected that RCSs with higher autonomy can aid better patient outcomes, and thus it is deemed an area requiring further research for technological improvements [16,17].

RCS prototypes with master–slave position-tracking capabilities for safe and responsive tool navigation at different displacement step values were developed in several studies. For instance, Al-Ahmad et al. [18] proposed a catheter drive system with sleeve-based grippers using pneumatic actuation. The clamping strategy ensures an equivalent force distribution along the catheter while providing continuous motion. However, further validation of the catheter driver is required. Similarly, in [19], the authors integrated the master–slave components into a singular unit. The slave platform replicates an operator's action on the master handle via a linear actuator, while a clutch mechanism allows the device to reposition the master unit for continuous catheter motion. Nevertheless, single-unit systems impede remote operation, which requires a distinct master and slave unit.

Furthermore, Cha et al. [20] developed a master–slave robotic platform with two degrees of freedom (2-DOF) for catheter and guidewire continuous motion during R-

PCI. However, the system utilized a user-defined scaling factor to achieve master–slave position accuracy during procedures. Likewise, the researchers in ref. [21] evaluated the design of a 2-DOF RCS system for teleoperated catheter navigation. The tool manipulator relies on a linear drive mechanism and a slip ring framework to translate and rotate the catheter within a vascular phantom by reproducing the operator's axial and radial motion at the master unit. Nonetheless, the system required a constant scaling factor for master–slave position mapping. Motion scaling is a complementary motion control strategy used in telerobotics. However, it holds better potential in teleoperated systems with high network latency, such as the Da Vinci robot [22]. The endovascular RCS developed by Sankaran et al. [23] endeavors to replicate the interventionalist's natural dexterity, requiring a minimal learning curve. The master unit has a quick-release mechanism with friction wheels to advance or retract the catheter or guidewire. The RCS utilized an adaptive input shaper and the proportional-integral-derivative (PID) controller for position control at the master–slave ends. Overall, these existing studies utilized motion scaling or conventional controllers to track the accuracy of master–slave positions during robot-assisted procedures. Nevertheless, tuning conventional controllers could be simplistic and imprecise resulting in a loss of performance. However, to enhance operational safety during robotic interventions, some other studies have developed different approaches, such as position control [24–26], motion scaling or compensation [20,21,27,28], and force feedback [29,30]. Nonetheless, there is limited proof of their stability and robustness for position control tasks that involve flexible endovascular tools. Furthermore, manufacturing constraints and miniature size requirements complicate the integration of force sensors at the tool's tip.

The choice of the slave robot's drive mechanism and motion control strategy is an essential criterion for safe tool navigation during R-PCI. As shown in Figure 1, the conventional drive mechanisms include (1) ball screws, a belt drive, and roller screws and (2) clamping and translation mechanisms, while magnetic actuation helps to manipulate a robotic catheter's tip. For instance, gripping or ball screw actuation provides the necessary thrust for the catheter's translational motion. However, these actuation designs subject the endovascular tools and RCS to backlash and hysteresis, motion nonlinearities, and uncertain disturbances. These are due to frictions, inertial forces, or temporal interruptions when restoring the master's device to its starting position during translation operations. While backlash in endovascular tool motion has been addressed [27,31,32], nonlinearity in RCSs is yet to receive any attention. The latter causes control complexities, such as motion lags in slave robots during catheterization [27]. Hence, existing systems operate at lower displacement step values.

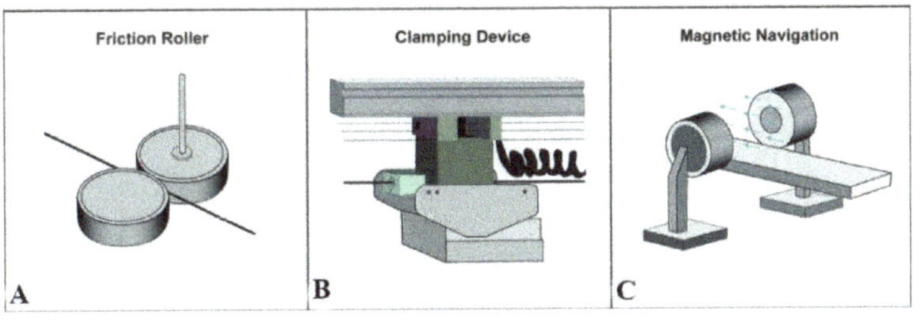

Figure 1. Catheter driver mechanisms using (**A**) friction roller, (**B**) gripping and translation, (**C**) magnetic actuation. Adapted with permission from Ref. [33]. © 2021, American Society of Neuroradiology.

Modeling the slave robot's motion dynamics to characterize uncertain disturbances, e.g., hysteresis in the under-actuated RCS, is challenging and imprecise [34]. In addition, this could be a vital underlying factor hindering robot autonomy in R-PCIs. In response to the complexity of modeling surgical robots, data-driven approaches based on the black box model have become increasingly popular. Omisore et al. [32] employed a nonlinear autoregressive neural network to estimate the hysteresis effect on a uni-directional coronary catheter using its kinematic parameters. The researchers in ref. [35] utilized the long short-term memory network (LSTM) to model the hysteresis of a uni-directional robotic catheter and to predict its pose. Cursi et al. [36] used a feed-forward neural network to model the kinematics of a tendon-driven surgical tool and to compensate for the motor's backlash. However, these approaches apply to steerable catheters or continuum robots [35,36]. In R-PCI, the standard coronary catheter is a non-steerable flexible tool without links or joints. Similarly, RCSs are mainly under-actuated, and physics-based modeling serves little benefit for position control. Furthermore, the existing studies have not utilized a data-driven approach to compensate for position error in such under-actuated RCSs.

This study aims to improve master–slave position tracking capabilities using a recurrent neural network, implemented and evaluated for real-time prediction and error compensation in a slave robot during catheterization. We hypothesized that a learning-based control system would help to optimize the slave robot's precision and response to motion commands received from the master platform. Thus, the catheter drive mechanism could navigate the flexible endovascular tools with higher precision and better minimize position errors that may arise from unknown disturbances and nonlinearities during catheterization. We characterized position errors using a set of actuation commands in a self-developed RCS and trained two recurrent neural network models to learn master–slave motion mapping and to predict the slave robot position for given input variables. Therefore, we evaluated the feasibility and performance of the LSTM and gated recurrent unit (GRU) based controllers. The main contributions of this study are:

(1) The design and development of a custom 2-DOF RCS for endovascular navigation with specification analysis of its main components.
(2) The experimental characterization of the slave robot's responses to uniform and varying motion commands.
(3) An open-loop control model with validation of the neural-based controllers for position prediction and error compensation in the slave robot using in-silico and in-RCS experiments.

The remainder of the paper follows an outline where we present the design of the custom RCS in Section 2, while the characterization of the axial translational position error and the backlash gap within the RCS is presented in Section 3. In addition, we present the neural-based model proposed for the prediction of the slave robot position in Section 4. Furthermore, we highlight the performance analysis of the models based on in-silico and in-RCS analyses in Section 5. Finally, Section 6 highlights the discussion, conclusions, and future work.

2. Design of the Robotic Catheter System

We present the functional design of the RCS and its main components in this section. The teleoperated system comprises a master and slave platform for endovascular tool remote manipulation during R-PCIs. The master–slave platform has dimensions of 45 cm × 19 cm × 13 cm in length, width, and height, and weighs 15.5 Kg and 26.0 Kg, respectively. It is worth mentioning that the current prototype is an improvement on our previous versions [12,26,27]. Furthermore, the designed mechanism ensures that the interventionalist requires a minimal learning curve to operate the RCS for intraluminal navigation. We present an outline of the detailed description of the master–slave platform below.

2.1. *Design of the Master Robotic Device*

By encoding the operator's natural hand movements, the master device (Figure 2) communicates appropriate signals to the slave platform (typically at the patient's bedside) to perform the axial translation and radial movement of flexible tools during R-PCI. The bi-axial DOF design consists of an incremental encoder, rotary collector, encoder module, etc., as shown in Figure 2a,b. Incremental encoders provide excellent feedback regarding speed and distance by generating pulses corresponding to their axial displacement across the electromagnetic handle. The signals are transmitted to the encoder module, where they are aggregated, and the corresponding position command is sent to the slave robot platform. This includes forward commands that drive the slave robot to advance an endovascular tool and pull movements for tool retraction during procedures. Similarly, the encoder (EKT58 series, JADE Optoelectronic Company, Wuxi, China) has a resolution of 5000 cycles per revolution (CPR) and a measurement frequency of 100 KHz. It generates corresponding pulses from clockwise or anticlockwise twists of the rotary collector by an operator, which are converted into discrete angular values to enable precise rotational movement of an endovascular tool by the slave robot.

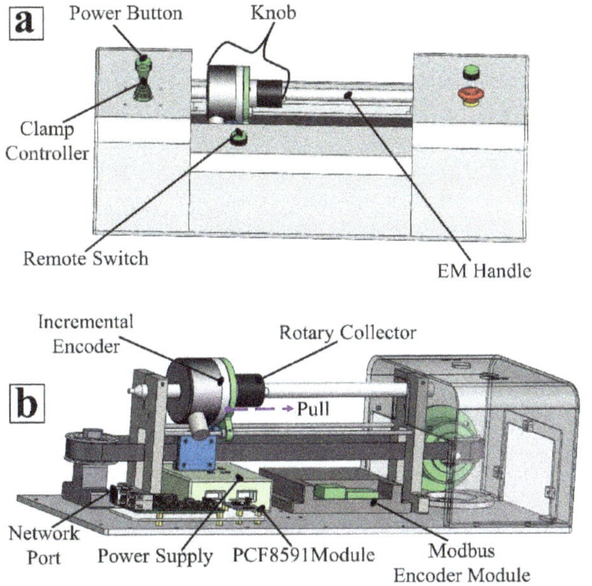

Figure 2. The design and assembling of the master platform. (**a**) The master device's outer casing and exterior components, and (**b**) the inner view and main components of the master robotic device.

In addition, the master device allows for simultaneous translational and rotational movements. Typically, this movement includes rotating on advance or retract operations generated by turning the rotary collector after moving the encoder along the handle or vice versa. This unique movement has been proven as a way interventionalists skillfully navigate guidewire or catheters through complex paths within the heart coronaries [14]. Furthermore, the clamp controller coordinates the slave's tool-grasping fixtures to maintain a fixed pose of the flexible tool, thus preventing tool slippage and potential risks from blood vessel puncture during procedures. The stroke length of the master device is 120 mm. However, to achieve an infinite translational motion, the remote switch (Figure 2a) triggers after a full stroke for either forward or backward drive of the cylindrical knob towards the starting position without corresponding movement at the slave side. On the other hand, when the remote switch turns off, a displacement of the cylindrical knob advances or retracts the tool further at the slave side. Hence, the realization of an infinite motion for a given intravascular procedure. The encoded motion information at the master side is transmitted to the slave device via a wireless network using a transmission control protocol/internet protocol (TCP/IP) address. Additionally, a 12-volt DC-powered rechargeable lithium-ion battery, as shown in Figure 2b, is used to power the master device. At the same time, a DC boost module ensures the realization of varying step-up and step-down voltage-to-current during a given procedure, thus offering high efficiency across input and output voltage ranges.

2.2. Design of the Slave Robotic Device

As shown in Figure 3a,b, the slave-side robotic device comprises a 2-DOF mechanism geared by stepper motors and a linear actuator. The slave platform implements the actual navigation of the endovascular tool based on input commands received from the master unit.

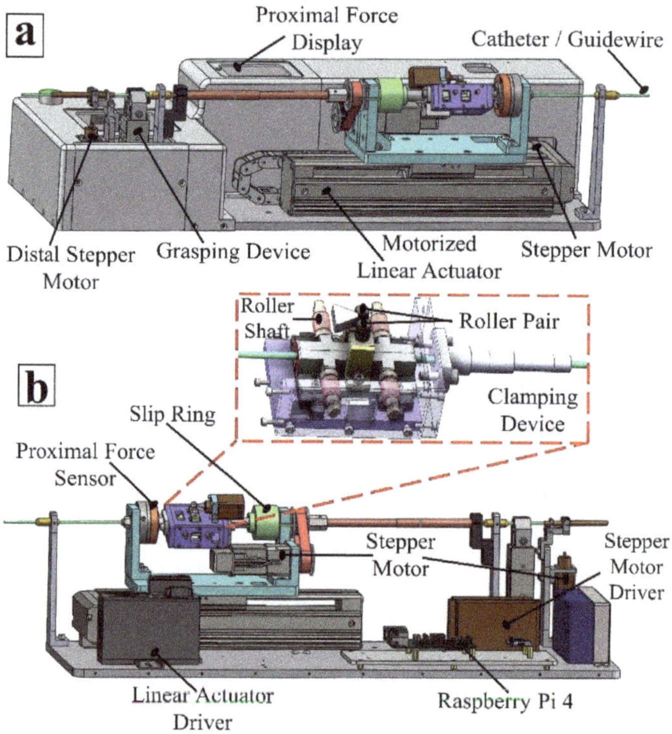

Figure 3. The design and assembling of the slave robot platform. (**a**) The exterior components of the slave platform, and (**b**) the clamping mechanism and interior components of the slave robotic device.

For this custom setup, the slave platform utilizes a motorized linear actuator (EAS4RXE015, Oriental motor, Tokyo, Japan) with a width and height of 58.4 mm × 60 mm, a stroke length of 150 mm (accuracy: ±0.02 mm), and a resolution of 5000 CPR, as shown in Figure 3a. The actuator employs a ball screw mechanism for its axial drive and is powered by a DC-input stepper motor (AZM24AK, Oriental motor, Tokyo, Japan), resolution: 1000 CPR, a motor driver (SR series, MOONS', Shanghai, China), and a linear actuator driver (AZD-KD, Oriental motor, Tokyo, Japan), as shown in Figure 3b. In addition, two limit sensors placed 15 mm from both ends of the linear actuator prevent the sliding plate from going beyond the terminal ends of the axial stage. Thus, one-to-one position mapping exists between the master–slave devices, with an active stroke length of 120 mm. The sliding plate houses the proximal force sensor, slip ring, and clamping mechanism. A rotary stepper motor attached to the sliding plate drives the belt and pulley system for rotational movement of the endovascular tool as controlled by the master.

In addition, as shown in Figure 3b, a miniature linear stepper motor (8k2105, SAMSR motor, Shenzhen Melike Technology Co., Ltd., Shenzhen, China) positioned above the clamping device turns a roller pair, controlling the opening and closing of the clamp via a lateral shift of the clamping controller at the master side. When opened, an operator manually steers the guidewire or catheter through the clamp, the guide sheath, and the grasping device to position the flexible tool at the entrance of a peripheral access port, from where the master–slave device operates the endovascular tool for intravascular the catheterization procedure. The clamp achieves a fixed grasp of the endovascular tool during axial and rotary motion when held in a locked position via four smaller roller shafts, as shown in Figure 3b. The clamping force between the device and the tool, as measured by the proximal sensor (KD24s, ME-Meßsysteme GmbH, Hennigsdorf, Germany) with a resolution of 0.002 N and a range of 2 N, is fed back to the operator during procedures. Input-to-output motion and imaging feedback during procedures could aid the development of state-based navigation in RCSs using learning-based models [10,37].

Furthermore, as shown in Figure 3, the distally positioned stepper motor facilitates adjustment of the guide sheath angle to optimize the tool's navigation during catheterization procedures. A DC 24V 10,000-mAh rechargeable lithium-ion battery powers the slave device for a mean of 270 min, while a Raspberry PI 4 (Raspberry Pi Foundation, South Cambridgeshire, UK) single-board computer with 8 GB RAM is used for data processing and transfer between master and slave devices.

2.3. Communication Modality

The master–slave platform facilitates data exchange via an open-source communication protocol from Modbus (Schneider Automation Inc., Rueil-Malmaison, France) and a wireless network protocol as shown in Figure 4. The RCS intercommunicates through the Modbus remote terminal unit, based on the RS485 standard, with a controller area network (CAN) and expansion board that allows transmitting/receiving data between Raspberry PI and other devices via RS485/CAN functions. Typically, the slave robot executes a given action and performs an error check before feedback of the completed action to the master device for subsequent commands. An 8-bit analog-to-digital (A/D) and digital-analog (D/A) converter chip, PCF8591 module (NXP Semiconductors, Eindhoven, Netherlands), aids A/D–D/A sampling in Raspberry PI via an inter-integrated circuit communication interface. Both linear and rotary data are intra-processed by the system module within the master–slave Raspberry PI and transmitted to the corresponding motor drivers, which actuate the linear or rotary motors for an axial or rotary drive of the catheter/guidewire.

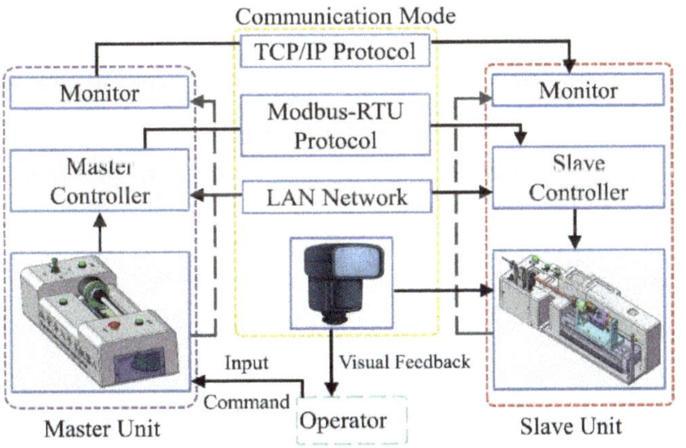

Figure 4. The master–slave robotic platform communication scheme.

The interventionalist relies on sensory feedback such as visual feedback and intercommunication between the master–slave devices for precise tool navigation. The system employs an onboard control system wherein the slave device receives control actions via preprogrammed instructions in a Raspberry PI operating system and thus could perform routine tool movements with discrete motion commands.

3. Characterization of Position-Tracking Error in Robotic Catheter Systems

Smooth and safe tool manipulation requires developing a precise position-tracking ability in the RCS presented above. Position tracking involves iterative geometric mapping between the axial commands (viz. linear or angular displacement) issued via the master device and the actual response from the slave device. This suitably helps to maintain steady translation and rotation movements between the master and slave platforms for robot-assisted catheterization. First, we characterized the RCS parts to establish the presence of motion lag or backlash gaps between the master and slave platforms. Existing studies hypothesized sources of motion lags in RCSs to be from actuated gears, belts, and pulley configurations in linear systems and that become evident during axial translations [21,37]. Practically, this could cause nonlinear but observable marked differences between the motion applied at the master device and the actual cannulation achieved with the slave side. Although new technologies in linear drive systems seek to produce low-backlash devices, none can eliminate the occurrence of position error when using the master–slave setup for R-PCI.

To characterize this situation in our RCS, we performed several experiments to observe the tracking errors between the master–slave platforms. This includes navigation trials performed with varying conditions, i.e., by issuing different position values at the master side and other trials under constant displacement. In all trials, higher-order motion data were derived from the position and velocity outputs from the in-built encoders' readings.

3.1. Experimental Observation of the Master–Slave Position and Velocity Based on an Operator Hand Motion

To quantify the occurrence of position-tracking errors during axial translations with our custom RCS, we segmented the master's stroke into several discrete points to have the RCS navigate a well-controlled path length each time. An operator was made to navigate the master through 30, 45, 70, 100, and 120 mm lengths. Each navigation trial was performed five times, and the average displacement data were recorded for the navigation length. In this procedure, the operator randomly moved the cylindrical knob in the master device from a starting position to each discrete point. This triggers the slave robot to translate along the linear path. Ideally, the master–slave positions should align from the start to the end of a given translational motion with an equal magnitude of distance traveled at both ends. However, we observed that distinct position errors existed in the trials. From the statistical analysis outlined in Table 1, the mean and standard deviation of the slave device's position error for each discrete point indicate that the tracking error increases as the length of the path navigated increases. Similarly, based on the operator's hand motion, the average resultant errors (ARE) for 70 and 120 mm were 10.26 and 18.61 mm, respectively, as shown in Figure 5A,C, while the percentage resultant error (ARE/distance) varied from 14.6 to 19.9%.

Table 1. Average position error of the slave robotic device for given distances.

Distance (mm)	Error (mm)	Resultant Error (%)
30	3.84 ± 2.03	19.90
45	6.38 ± 2.30	14.83
70	7.43 ± 3.72	14.66
100	10.67 ± 4.30	15.11
120	13.14 ± 6.11	15.51

The mean velocity of the master–slave devices obtained when distances of 70 and 120 mm were navigated are shown in Figure 6A,B. The interval between the master–slave velocities was greater in Figure 6A. Relating this to Figure 5A, it indicates that the motion gap increases with the velocity gap in the master–slave platforms for varying translational motion but not necessarily with the catheterization distance. Thus, this elucidates why Figure 5A (70 mm) has a more significant motion gap than Figure 5C (120 mm). In addition, the magnitude of the interval between consecutive points during a catheterization stroke is proportional to the motion gap experienced by the slave robot. Again, we have observed that for a shorter distance (30–70 mm), a novice operator is more likely to move the master knob faster (and at a higher velocity) than for a longer distance (100–120 mm). This could be why fewer motion gaps occurred between the master and slave trajectories when longer distances were moved. Nonetheless, we found that the magnitude of the resultant position error increases with catheterization length.

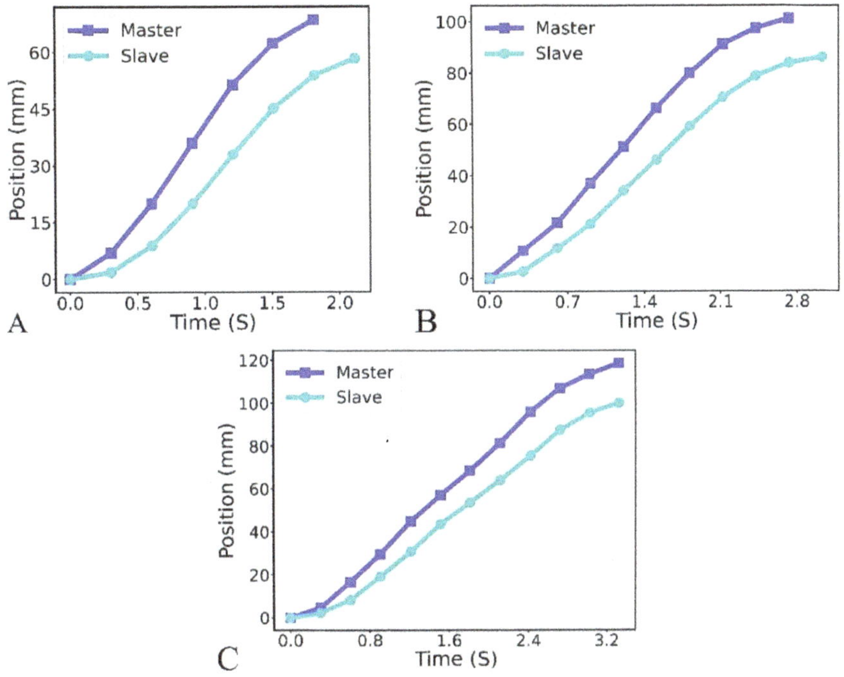

Figure 5. Master–slave position trajectories based on an operator's hand motion for a distance of (**A**) 70 mm, (**B**) 100 mm, and (**C**) 120 mm.

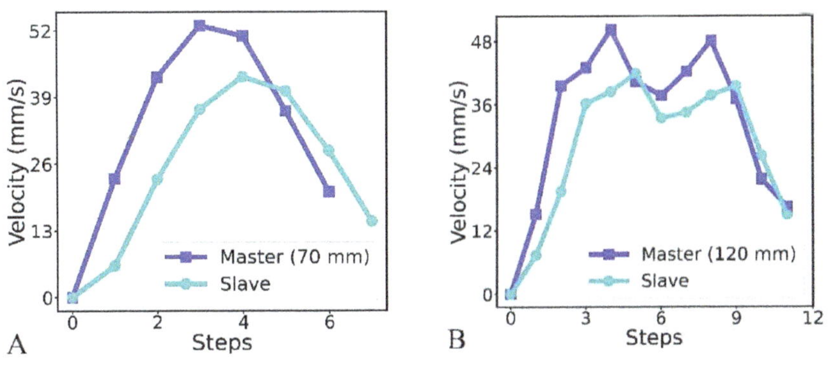

Figure 6. Master–slave velocity based on an operator's hand motion for a distance of (**A**) 70 mm and (**B**) 120 mm.

3.2. Experimental Observation of Master–Slave Position and Velocity under Uniform Translational Motion

The motion lag study was also performed for position and velocity under uniform translational motion. For this, uniform translational motion in steps of 1–10 mm was issued to the slave robot to quantify the tracking error. In addition, each trial was performed five times for each displacement step value to trigger navigation, while the robotic devices' mean position and velocity output were recorded. In Figure 7A–C, we present the distance traveled by both master and slave platforms when displacement values of 3, 5, and 10 mm were, respectively, set. In Figure 7A, the slave device closely matched the displacement of the master platform for a translation step of 3 mm. In contrast, the slave device response was below the input command per step when actuated with a uniform displacement of 5 mm, as shown in Figure 7B. However, above 5 mm, a more visible gap exists such that it extends to the proportion shown in Figure 7C when operated under a displacement step value of 10 mm. Overall, the accruing error influences the resultant slave positions, as shown in Figure 7B,C.

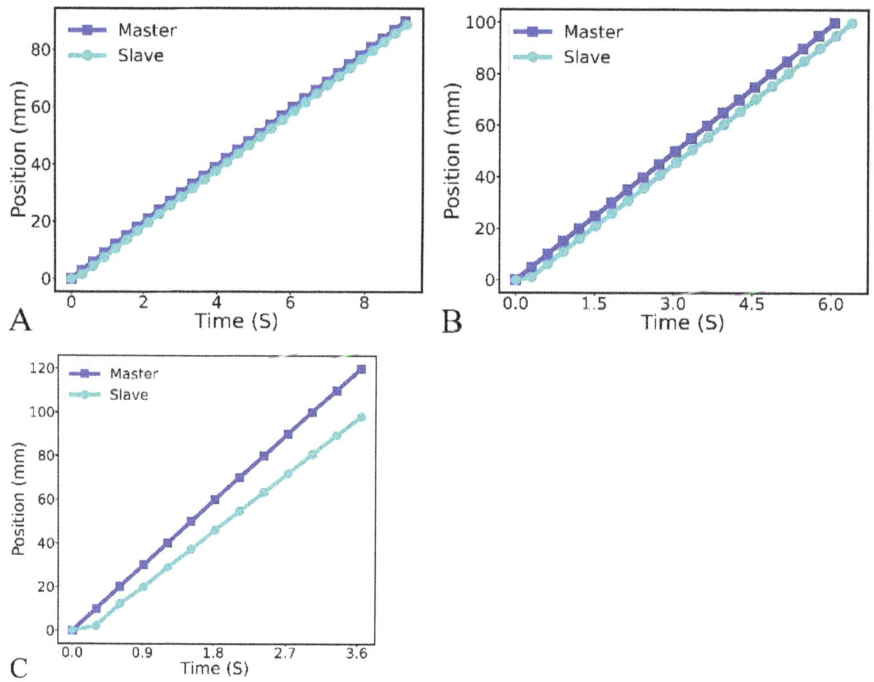

Figure 7. Master–slave position trajectories based on uniform displacement values of (**A**) 3 mm, (**B**) 5 mm, and (**C**) 10 mm.

In addition, the velocity trajectories of the master and slave platforms for 5 and 10 mm are presented in Figure 8A,B, respectively. In Figure 8A, the variation of the slave robot's velocity was within the threshold of the motion gap observed in Figure 7B. This could infer that the slave's position tracking is smooth for small displacement step values ($\leq$ 3 mm). However, limiting an operator to this axial displacement range during catheterization could increase the procedural duration and induce boredom in interventionalists, especially when navigating through vessels with a larger diameter. This includes endovascular tool movement from within the radial vessels to the aorta. However, smaller displacement step values are essential for safe tool movement beyond the aorta and within the coronary arteries. Furthermore, this could allow the operator to choose between different displacement step values depending on the vessel diameter and the catheterization path while a steady translational motion could be achieved.

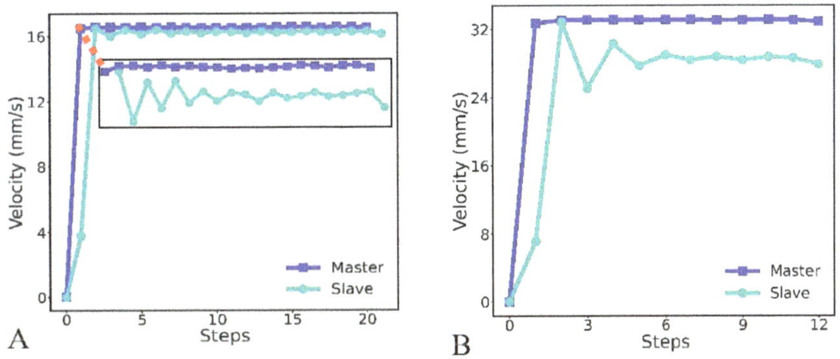

Figure 8. Master–slave velocity trajectories based on uniform displacement values of (**A**) 5 mm and (**B**) 10 mm. Inset shows the enlarged view of the master-slave velocity trajectories for a uniform displacement of 5 mm.

Figure 8B shows a noticeable difference when the slave robot actuates under a higher velocity, such as applying >30 mm/s. Similarly, this correlates with the evident position-tracking error in Figure 7C. Thus, it is remarked that a velocity of 16 mm/s (or below) at the master side could suitably maintain a minimal position error during master–slave robotic catheterization. Comparing Figures 5–8, our study suggests that both position and velocity contribute to the hysteresis observed at the slave-side robotic device during catheterization. In conclusion, Figures 5–8 indicate that the nonlinearities and uncertainties could be higher and unbounded when applying variable displacement values. This typical scenario usually happens during robot-assisted tool catheterization. We, therefore, proposed a neural-based system for learning the dynamics of the master–slave setup to aptly determine and compensate for position error occurrences in the RCS.

4. Recurrent Neural Network Modeling

In this section, we utilized variants of recurrent neural network models, such as the LSTM and GRU networks, to learn typical motion dynamics when operating the master–slave platform. The trained network varieties can estimate the slave robot's displacement from given input motion variables. The estimated value and an appropriate compensating value are then transmitted to the slave robotic device to achieve uniform motion mapping with the master platform. This data-driven approach requires experimental data to pre-train recurrent neural networks. Thus, we conducted a series of controlled studies similar to those described in Section 3, and describes the data processing procedures in this section.

4.1. Data Pre-Processing

As highlighted in Section 3, analysis of the position and velocity illustrates better position tracking between the master and slave platforms when operated with constant displacement values.

Thus, we carried out phantom studies for these displacement values to acquire the experimental data for training the neural-based models. To ensure consistency in all data tuples, we fixed the start and end points for all the navigation trials that we performed. Motion commands from the master device emanate from custom code written in Python and executed online via the Raspberry controller. The data acquired are first pre-processed by removing empty and incomplete records that occasionally occur at the slave side. With a total of 188 experimental runs conducted, we obtained 7500 data sequences with exactly 1500 data sequences selected for each 1–5 mm discrete motion. Subsequently, each step's master–slave motion data were divided randomly into training (70%) and test (30%) datasets.

The displacement and velocity from the master side in addition to the ratio of velocity between the master–slave devices are taken as input variables into the network to estimate the slave robot positions. The training and test datasets for the master–slave positions are shown in Figure 9. The slave robot's response to discrete input commands is shown in Figure 9b. This typifies the occurrence of random values (errors) increasing proportionally with the magnitude of the translational motion. That is, between 1–5 mm, specifically, from 3 mm, there were increasing data points, where the output position of the slave robot were below the input command, thus symbolizing a noisy data. This is a typical response of the slave robot to higher displacement step values without position control or compensation. Furthermore, this also indicates the cause of jerks and vibrations in the slave robot during typical master-slave movements in the absence of position control methods.

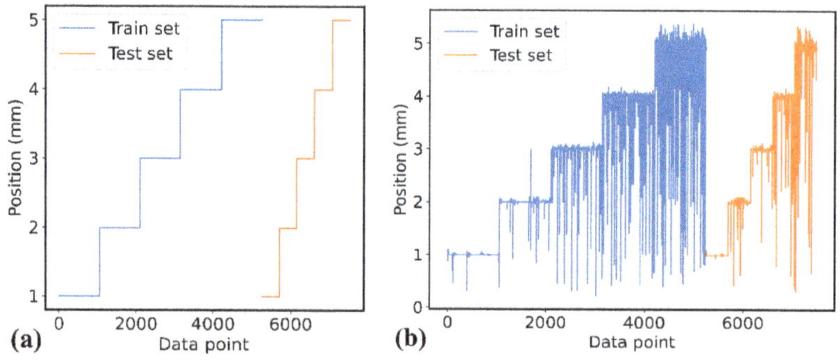

Figure 9. Training and test data split for (**a**) master position (as part of the input variable) and (**b**) the slave position (target variable).

4.2. Recurrent Neural Network

The LSTM network consists of a stack of LSTM units, each equipped with a cell and three gates—a forget gate, an input gate, and an output gate that regulate information flow across the cells. An LSTM unit typically processes information in accordance with the following equations [38]. Thus, the model output at time-step t is given by $h_{t \forall t=1,2,\ldots,k}$ as in Equation (1). Where σ is the logistic function with a range of zero and one that determines the ratio of information flow through each gate. The forget gate ($f^{(t)}$) controls what information is retained from the previous cell state ($C^{(t-1)}$), whereas the input gate ($i^{(t)}$) combines with the input-modulating gate ($C_i^{(t)}$)—which uses a hyperbolic tangent activation (tanh) function to update the cell state to ($C^{(t)}$). In addition, the output gate ($o^{(t)}$) sieves the information within the new cell state ($C^{(t)}$), using both the logistic and tanh functions to regulate its output, and is transferred to the hidden state ($h^{(t)}$). W_* and b_* represent the weight and bias parameter matrix in each gate, respectively. Combining the gating mechanisms enables the LSTM to extract relevant information from an input sequence.

$$\begin{aligned} f^{(t)} &= \sigma\left(W_f\left[h^{(t-1)},\ x^{(t)}\right] + b_f\right) \\ i^{(t)} &= \sigma\left(W_i\left[h^{(t-1)},\ x^{(t)}\right] + b_i\right) \\ C_i^{(t)} &= \tanh\left(W_c\left[h^{(t-1)},\ x^{(t)}\right] + b_c\right) \\ C^{(t)} &= f^{(t)} * C^{(t-1)} + i^{(t)} * C_i^{(t)} \\ o^{(t)} &= \sigma\left(W_o\left[h^{(t-1)}, x^{(t)}\right] + b_o\right) \\ h^{(t)} &= o^{(t)} * \tanh\left(C^{(t)}\right) \end{aligned} \tag{1}$$

In contrast, the stacked GRU units comprise a structure where each unit is equipped with two gates—a reset gate $\left(r^{(t)}\right)$ and update gate $\left(z^{(t)}\right)$. The update gate regulates what information flows into the memory using logistic function within the interval (0, 1), similar to LSTM. This controls the similarity between the previous state and the new state. The reset gate determines what percentage of the previous state is retained and how to compute the candidate hidden state ($h^{\sim(t)}$). An elementwise operation of the update gate with the previous state ($h^{(t-1)}$) and the candidate hidden state result in the final output of the model $h^{(t)}$ is calculated as shown in the following equations below [39].

$$
\begin{aligned}
r^{(t)} &= \sigma\left(\mathrm{W_r}\left[h^{(t-1)},\ x^{(t)}\right] + \mathrm{b_r}\right) \\
z^{(t)} &= \sigma\left(\mathrm{W_z}\left[h^{(t-1)},\ x^{(t)}\right] + \mathrm{b_z}\right) \\
\widetilde{h}^{(t)} &= \tanh\left(\mathrm{W}_h\left[r^{(t)} * h^{(t-1)},\ x^{(t)}\right] + b_h\right) \\
h^{(t)} &= z^{(t)} * h^{(t-1)} + \left(1 - z^{(t)}\right) * \widetilde{h}^{(t)}
\end{aligned}
\tag{2}
$$

4.3. Model Training and Evaluation

The RNN models were implemented in Python using the Keras library and the TensorFlow platform. The LSTM model has an input layer (N $\times$ 3 $\times$ 3), two hidden layers with 32 units each, and a fully connected layer, which outputs a predicted value using the rectified linear unit (ReLU) activation function, as shown in Figure 10a. The LSTM unit can determine important features from an input sequence and keep such information in its memory. To train the model, experimental data for different step values were stacked together with feature scaling performed on the data. The model was trained with a batch size of 32 for 90 epochs, while 20% of the training set was used for validation. The learning rate was set to 1×10^{-4} with a time step of three and the Adam optimizer algorithm employed. Similarly, the GRU model has an input layer, two hidden layers with 16 neurons each, and a fully connected layer with the ReLU activation function, as shown in Figure 10b. The learning rate and batch size were $1{\times}10^{-4}$ and 24, while the model was trained for 20 epoch with the chosen Adam optimizer algorithm.

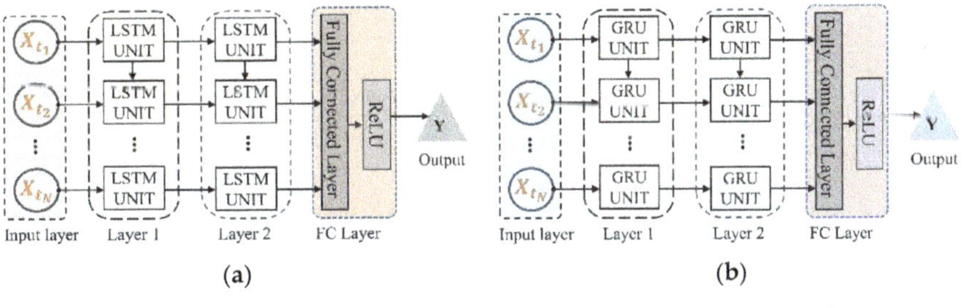

Figure 10. A stacked 2-layer architecture for slave robot's position prediction and compensation using (**a**) LSTM model and (**b**) GRU model.

Furthermore, the description of the two quantitative metrics used to evaluate the models' performance is stated below, with their corresponding values given in Table 2.

Table 2. MAE and MSE for the LSTM and GRU models.

Model	MAE (mm)	MSE (mm^2)
LSTM	0.0825	0.0958
GRU	0.0906	0.0964

1. Mean Absolute Error: The mean absolute error indicates the average sum of the absolute difference between the actual slave robot position (y_i) and the predicted position (x_i), as given in Equation (3).

$$\mathrm{MAE} = \frac{1}{\mathrm{N}} \sum_{\mathrm{i}=1}^{\mathrm{N}} |\mathrm{y_i} - x_i| \tag{3}$$

1. Mean Square Error: The mean square error is the average of the squared error values between the model's prediction and the slave robot response, represented by Equation (4).

$$\mathrm{MSE} = \frac{1}{\mathrm{N}} \sum_{\mathrm{i}=1}^{\mathrm{N}} (\mathrm{y_i} - \hat{\mathrm{y}})^2 \tag{4}$$

Based on the error values (in Table 2), the RNN models have comparative performance. The models' training, validation, and prediction performances are shown in Figure 11a–d. Figure 11 indicates that the validation loss for each model was less than 2% under each category, while the models aptly predicted the intended slave displacement for each set-value, i.e., 1–5 mm. The LSTM model has a longer training time compared with the GRU model, which utilized the least number of neurons and epochs. Thus, we suggest that either of the models could be deployed for position prediction and error compensation to eliminate motion lag between the master–slave platform during robot-assisted tool navigation.

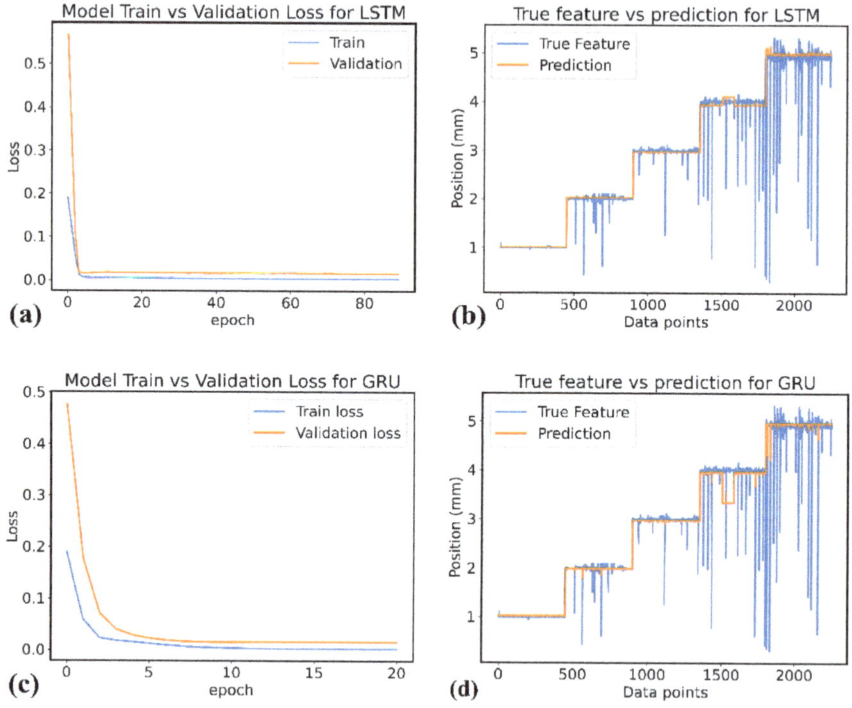

Figure 11. The performance of the RNN models: (**a**) training and validation loss for the LSTM model, (**b**) prediction on the test set based on the LSTM model, (**c**) training and validation loss for the GRU model, (**d**) prediction on the test set based on the GRU model.

5. Experimental Study and Results

We performed in-silico and hardware-based experiments to validate the performance of the RNN-based models when deployed within the slave robot. In addition, we implemented an open-loop control using the models' prediction for error compensation to achieve identical master–slave position mapping for uniform and varying input commands.

5.1. In-Silico Modeling and Evaluation

The in-silico experiments performed in CoppeliaSim Edu (Coppelia Robotics Ltd., Zurich, Switzerland) involve the conversion of the slave robot's 3D CAD model to a unified robot description format (URDF), which is then assembled within the simulation scene, as shown in Figure 12. We parameterized the static and dynamic properties of the robot to mimic the translation mode of the slave robot. To facilitate the position control of the device, a prismatic joint and the global positioning system (GPS) provided feedback about the device's position and orientation. To evaluate the RNN model's performance, a uniform displacement command (5 mm) is transmitted to the slave robot, while a random value (rand_val) represented by Equation (5) generates the uncertainties resulting in stepwise error. For each step, the RNN model predicts the slave robot's response based on input motion variables prior to the robot's translation. Afterwards, the difference between the model's prediction and the actual robot displacement is added to the initial slave robot's movement with the final error computed per step. This ensures that the slave robot reproduces exact position commands received from the master platform, thereby eliminating accruing errors within the slave robot along its traveling distance.

$$\text{rand\_val} = \sigma \times n_p.\,\text{random.normal}(0,\ 1) \tag{5}$$

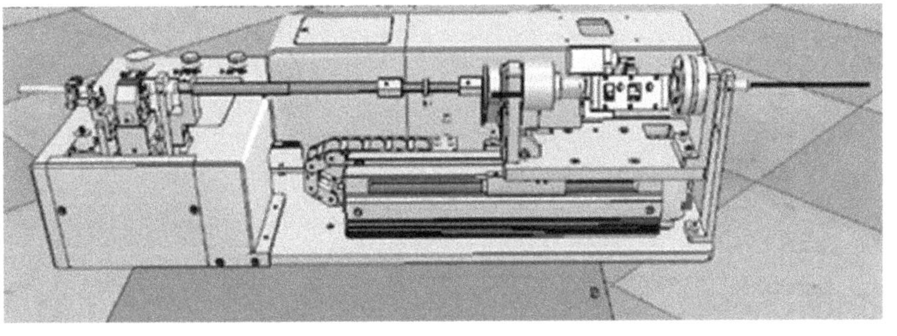

Figure 12. The slave device assembly and translational motion simulation in CoppeliaSim.

5.2. In-Silico Results and Evaluation

The in-silico analyses of the GRU and LSTM-based models are presented in Figure 13. In Figure 13a, a comparison of the slave robot's original and final response for a traveling distance of 150 mm with a scalar value (σ) of 1.0 mm is shown. Although the original error rose to 40% (2.0 mm) of the input command for some steps, the minimum and maximum final errors were 0.001 mm and 0.1 mm (2% of the input command) per step. The figure shows that the GRU-based controller could predict the expected slave robot response and then utilize this to regress the initial position error. Similarly, in Figure 13b, we present the performance of the LSTM model when the slave robot actuates under a uniform motion for a distance of 150 mm with a scalar value of 1.5 mm. The model predicted values indicate that the LSTM-based controller makes a good prediction of the slave robot response for a typical displacement of 5.0 mm, with a resultant final error of 1.25 mm. Compared with Table 1, the neural-based models ensure higher precision and accuracy of master–slave position-tracking capability. Overall, the models are suitable for position prediction and error compensation within RCSs.

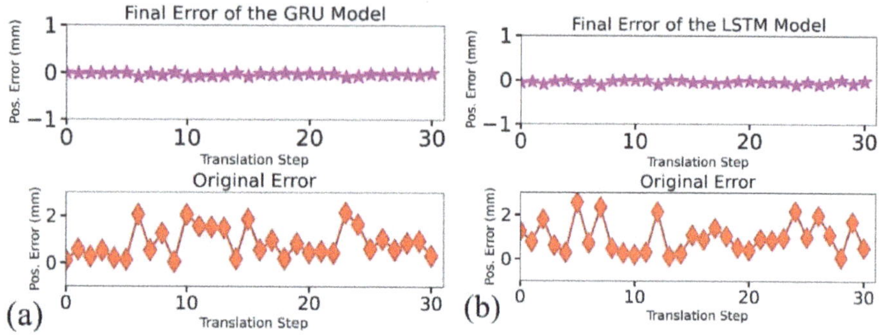

Figure 13. The original position error of the slave robotic device is shown in the lower panel compared with (**a**) GRU-based and (**b**) LSTM-based predictions for a distance of 150 mm with a translational step of 5 mm.

5.3. In-RCS Results and Evaluation

To validate the real-time usability of the learning-based models in the RCS, we utilized the Jetson AGX Orin Development Kit (NVIDIA, Santa Clara, CA, USA) for the slave robot operation, and evaluated the model's performance under uniform and varying input commands using the control block diagram presented in Figure 14. For a typical master's displacement (M_p), velocity (M_v), and the velocity ratio (V_r), the RNN model makes an appropriate prediction (RNN_p) and performs an error check. In addition, the slave robot receives a compensated displacement value (that incorporates the difference between the model's prediction and the input command). Afterwards, the slave robot actuates for the given step and the difference between the master and slave position (S_p) is computed as the final error (motion lag) for that step. In Figure 15, we present the experimental setup for the evaluation of the above control approach within the RCS using the neural-based controllers.

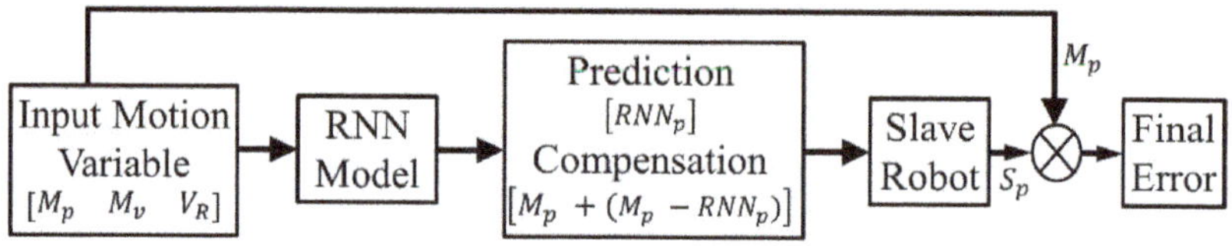

Figure 14. Control block with error compensation for the slave robot.

Figure 15. The hardware setup for experimental validation of the neural-based models within the RCS using Jetson Orin as the slave device's processor.

The average performance of the GRU model for varying displacement commands (3–5 mm) and a traveling distance of 120 mm by the slave robot is presented in Figure 16.

The choice of 3–5 mm aligns with the experimental observation of an increasing motion gap between the master and slave robot under uniform translational motion as presented in Figure 7A,B. The figure shows the positive impact of the real-time compensating values for input displacement such as 5 mm, thus maintaining the slave robot's smooth translational motion with a minimum and maximum error of 0.014 and 0.04 mm, respectively, as shown in Figure 16. This infers the neural-based model's suitability for prediction and compensation tasks within the slave robot during catheterization and the realization of an equivalent master–slave position trajectory for displacement steps above 1 mm.

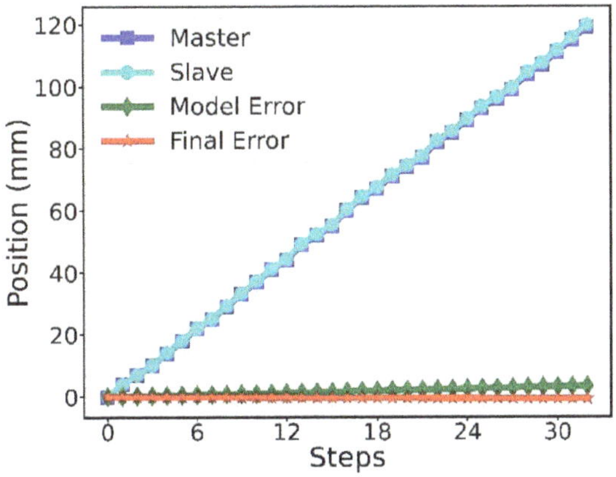

Figure 16. The master–slave position trajectory for a distance of 120 mm under varying input commands with the GRU model prediction and error compensation within the slave robot.

6. Discussion

This study further supports our approach to real-time detection and elimination of the slave robot's position errors during catheterization. We demonstrated that the control approach has the potential to create steady master–slave motion and contribute to procedural safety during R-PCI procedures. In addition, a comparative analysis of existing control approaches for master–slave position tracking shows that the neural-based controllers have a much lower maximum error compared with MLP-PID, ADRC, and FSMC for in-RCS experiments, as indicated in Table 3. Although, RL-PID has the best performance, the control method was only verified with simulation experiments. Overall, we have presented a data-driven control approach for master–slave position tracking, which leverages on neural network capability for robust features learning and representation from training examples, thus yielding a higher precision, accuracy, and lower percentage error.

Table 3. Performance evaluation of the proposed method with existing studies.

Authors	Control Method	Experiment Type	Max. Error (mm)
Ma et al. [24]	MLP-PID	Simulation/in-RCS	1.50
Omisore et al. [10]	RL-PID	Simulation	0.003
Zhou et al. [40]	ADRC	RCS	0.87
Wang et al. [28]	FSMC	Simulation	0.65
This study	LSTM/GRU	Simulation/in-RCS	0.04

Note MLP-PID—multi-layer perceptron-based PID controller; RL-PID—reinforcement learning-based PID controller; ADRC—active disturbance rejection control; FSMC—fuzzy sliding mode controller.

Although single-board computers facilitate easy mobility of the robotic device, we experienced hardware limitations deploying RNN models within Raspberry PI for real-time error predictions. This study suggests the need for more powerful single-board computers. The recent launch of Jetson AGX Orin seems timely; however, large-scale validation is still needed. In addition, machine-learning methods using intelligent agents such as reinforcement learning would be operable within the RCS for state-based navigation or task-specific autonomous control. Furthermore, adding reliable proximal and distal force navigation data would enable learning the master–slave robot kinematics as well as tool dynamics within the vasculature, which could lead to the development of a robust model with the potential realization of closed-loop control during R-PCI. At present, acquiring operator-based motion data is time-consuming and laborious. This is because fewer displacement steps are covered within a full stroke under varying motion (Figure 7A–C) compared with uniform translational motion (Figure 9a–c). However, such data contain a higher degree of nonlinearities, thus are closer to real-world characterization of the slave robot's position error. While intraluminal navigation safety is at the top of patients' safety concerns during minimally invasive procedures, recent investigations show potential for task-specific autonomy for tool navigation during R-PCIs.

In this study, we described the development of a custom-built master–slave RCS to navigate endovascular tools. We further analyzed the causative factors for position errors at the slave side of the robotic system using the displacement values at both master and slave sides, their velocity, and backlash gaps. Thus, we proposed and validated using learning-based control models for online error compensation in the RCS. The RNN models were trained with master–slave forward motion profiles recorded from the RCS. Furthermore, this study focused on the characterization and control of translational motion in the RCS. However, a hybrid of translational and rotary motion is essential for semi-autonomous navigation. To achieve this, high-resolution rotary encoders would be essential. Therefore, further studies will seek to incorporate high-resolution rotary encoders and consolidate the characterization and control of rotary motion. This would provide for the development of a robust learning-based controller for translational and rotary motion during R-PCI. In conclusion, this study has shown the feasibility and self-adaptability of neural-based controllers with verification for RCSs towards achieving smooth translational motion during minimally invasive treatment of CAD.

Author Contributions: Conceptualization, O.M.O. and T.O.A.; methodology, T.O.A.; software, T.O.A. and G.Y.; validation, T.O.A., W.D. (Wenke Duan) and X.C.; formal analysis, T.O.A., O.M.O. and W.D. (Wenjing Du); investigation, T.O.A. and O.M.O.; resources, W.D. (Wenke Duan) and X.C.; data curation, T.O.A.; writing—original draft preparation, T.O.A.; writing—review and editing, O.M.O. and T.O.A.; visualization, T.O.A. and L.W.; supervision, L.W. and O.M.O.; project administration, L.W. and W.D (Wenjing Du); funding acquisition, L.W. All authors have read and agreed to the published version of the manuscript.

Funding: This research was funded in parts by the National Key Research and Development Program of China under Grant 2019YFB1311700, the National Natural Science Foundation of China under Grants (#U1713219, #61950410618), the Shenzhen Natural Science Foundation under Grant JCYJ20190812173205538, and in part by the CAS PIFI Fellowship.

Institutional Review Board Statement: Not applicable.

Informed Consent Statement: Not applicable.

Data Availability Statement: Not applicable.

Conflicts of Interest: The authors declare no conflict of interest.

References

1. Virani, S.S.; Alonso, A.; Aparicio, H.J.; Benjamin, E.J.; Bittencourt, M.S.; Callaway, C.W.; Carson, A.P.; Chamberlain, A.M.; Cheng, S.; Delling, F.N.; et al. Heart Disease and Stroke Statistics-2021 Update A Report from the American Heart Association. *Circulation* **2021**, *143*, E254–E743. [CrossRef]
2. Gatto, L.; Prati, F. Subclinical atherosclerosis: How and when to treat it? *Eur. Hear. J. Suppl.* **2020**, *22*, E87–E90. [CrossRef]
3. Cohn, W.E.; Frazier, O.H.; Mallidi, H.R.; Cooley, D.A. *Coronary Artery Disease, Cardiovascular Medicine*; Willerson, J.T., Holmes, D.R., Jr., Eds.; Springer: London, UK, 2015; p. 657. [CrossRef]
4. Akinyemi, T.O.; Omisore, O.M.; Duan, W.; Lu, G.; Al-Handerish, Y.; Han, S.; Wang, L. Fiber Bragg Grating-Based Force Sensing in Robot-Assisted Cardiac Interventions: A Review. *IEEE Sens. J.* **2021**, *21*, 10317–10331. [CrossRef]
5. Watson, T.J.; Ong, P.J.L.; Tcheng, J.E. *Primary Angioplasty: A Practical Guide*; Springer: Singapore, 2018.

6. Bergman, P.; Blacker, S.J.; Kottenstette, N.; Saber, O.; Sokhanvar, S. Robotic-Assisted Percutaneous Coronary Intervention. In *Handbook of Robotic and Image-Guided Surgery*; Elsevier: Amsterdam, The Netherlands, 2020; pp. 341–362.
7. Akinyemi, T.O.; Omisore, O.M.; Duan, W.; Lu, G.; Du, W.; Alhanderish, Y.; Li, Y.; Wang, L. Development of a Millinewton FBG-Based Distal Force Sensor for Intravascular Interventions. In Proceedings of the 16th International Conference on Control, Automation, Robotics and Vision (ICARCV), Shenzhen, China, 13–15 December 2020; IEEE: New York, NY, USA, 2021; pp. 511–515. [CrossRef]
8. Kalra, A.; Bhatt, D.L.; Kleiman, N.S. A 24-Month Interventional Cardiology Fellowship: Learning Motor Skills Through Blocked Repetition. *JACC Cardiovasc. Interv.* **2017**, *10*, 210–211. [CrossRef]
9. Lee, C.W.J.; Rao, A. Current training in Cardiac devices–a Cardiology trainee perspective: A questionnaire survey. *MedEdPublish* **2017**, *6*, 130.
10. Omisore, O.M.; Akinyemi, T.; Duan, W.; Wang, L. A Novel Sample-efficient Deep Reinforcement Learning with Episodic Policy Transfer for PID-Based Control in Cardiac Catheterization Robots. October 2021. Available online: http://arxiv.org/abs/2110.14941 (accessed on 1 June 2022).
11. Kinnaird, T.; Kwok, C.S.; Kontopantelis, E.; Ossei-Gerning, N.; Ludman, P.; de Belder, M.; Andreson, R.; Mamas, M.A. Incidence, determinants, and outcomes of coronary perforation during percutaneous coronary intervention in the United Kingdom Between 2006 and 2013. *Circ. Cardiovasc. Interv.* **2016**, *9*, e003449. [CrossRef]
12. Omisore, O.M.; Duan, W.; Akinyemi, T.; Han, S.; Du, W.; Alhanderish, Y.; Wang, L. Design of a Master-Slave Robotic System for Intravascular Catheterization during Cardiac Interventions. In Proceedings of the International Conference on Control, Automation, Robotics and Vision (ICARCV), Shenzhen, China, 13–15 December 2020; IEEE: New York, NY, USA, 2021; pp. 996–1000.
13. Weisz, G.; Metzger, D.C.; Caputo, R.P.; Delgado, J.A.; Marshall, J.J.; Vetrovec, G.W.; Reisman, M.; Waksman, R.; Granada, J.F.; Novack, V.; et al. Safety and feasibility of robotic percutaneous coronary intervention: PRECISE (percutaneous robotically-enhanced coronary intervention) study. *J. Am. Coll. Cardiol.* **2013**, *61*, 1596–1600. [CrossRef]
14. Smitson, C.C.; Ang, L.; Pourdjabbar, A.; Reeves, R.; Patel, M.; Mahmud, E. Safety and feasibility of a novel, second-generation robotic-assisted system for percutaneous coronary intervention: First in-human report. *J. Invasive Cardiol.* **2018**, *30*, 152–156. [CrossRef]
15. CorPath GRX System Operator's Manual. Available online: https://www.roboticpci.com/assets/Documents/150-07702_Rev_C_CorPath_GRX_Operator\T1\textquoterights_Manual.pdf (accessed on 31 August 2022).
16. Dupont, P.E.; Nelson, B.J.; Goldfarb, M.; Hannaford, B.; Menciassi, A.; O'Malley, M.K.; Simaan, N.; Valdastri, P.; Yang, G.Z. A decade retrospective of medical robotics research from 2010 to 2020. *Sci. Robot.* **2021**, *6*, 8017. [CrossRef]
17. Akinyemi, T.O.; Omisore, O.M.; Duan, W.; Lu, G.; Al-Handerish, Y.; Han, S.; Wang, L. Towards a Fiber Bragg Grating-Based Force Sensing in Robot-Assisted Cardiac Interventions. *IEEE Sens. Lett.* **2022**, *21*, 5000104. [CrossRef]
18. Al-Ahmad, O.; Ourak, M.; Smits, J.; Jeanquart, S.; Deserranno, N.; Bernhard, F.; Kassahun, Y.; Yu, B.; Vander Poorten, E. Development of an Innovative Sleeve-Based Robotic Catheter Driver. In Proceedings of the 8th Joint Workshop on New Technologies for Computer/Robot Assisted Surgery, London, UK, 10–11 September 2018; pp. 2–3.
19. Payne, C.J.; Rafii-Tari, H.; Yang, G.-Z. A force feedback system for Endovascular catheterization. In Proceedings of the International Conference on Intelligent Robots and Systems, Vilamoura-Algarve, Portugal, 7–12 October 2012; IEEE: New York, NY, USA, 2012; pp. 1298–1304.
20. Cha, H.J.; Yi, B.J.; Won, J.Y. An assembly-type master-slave catheter and guidewire driving system for vascular intervention. *Proc. Inst. Mech. Eng. H* **2017**, *231*, 69–79. [CrossRef] [PubMed]
21. Thakur, Y.; Bax, J.S.; Holdsworth, D.W.; Drangova, M. Design and performance evaluation of a remote catheter navigation system. *IEEE Trans. Biomed. Eng.* **2009**, *56*, 1901–1908. [CrossRef] [PubMed]
22. Richter, F.; Orosco, R.K.; Yip, M.C. Motion scaling solutions for improved performance in high delay surgical teleoperation. In Proceedings of the International Conference on Robotics and Automation Montreal, Montreal, QC, Canada, 20–24 May 2019; IEEE: New York, NY, USA, 2019; pp. 1590–1595.
23. Sankaran, N.K.; Chembrammel, P.; Siddiqui, A.; Snyder, K.; Kesavadas, T. Design and Development of Surgeon Augmented Endovascular Robotic System. *IEEE Trans. Biomed. Eng.* **2018**, *65*, 2483–2493. [CrossRef] [PubMed]
24. Ma, X.; Zhou, J.; Zhang, X.; Zhou, Q. Development of a Robotic Catheter Manipulation System Based on BP Neural Network PID Controller. *Appl. Bionics Biomech.* **2020**, *2020*, 8870106. [CrossRef] [PubMed]
25. Guo, J.; Jin, X.; Guo, S.; Du, W. Study on the tracking performance of the vascular interventional surgical robotic system based on the fuzzy-PID controller. In Proceedings of the International Conference on Mechatronics and Automation (ICMA), Takamatsu, Japan, 6–9 August 2017; IEEE: New York, NY, USA, 2017; pp. 29–34.
26. Omisore, O.M.; Han, S.; Ren, L.; Wang, L. A fuzzy-PD model for master-slave tracking in teleoperated robotic surgery. In Proceedings of the Biomedical Circuits and Systems Conference, Shanghai, China, 17–19 October 2016; IEEE: New York, NY, USA, 2017.
27. Omisore, O.M.; Han, S.P.; Ren, L.X.; Wang, G.S.; Ou, F.L.; Li, H.; Wang, L. Towards Characterization and Adaptive Compensation of Backlash in a Novel Robotic Catheter System for Cardiovascular Interventions. *IEEE Trans. Biomed. Circuits Syst.* **2018**, *12*, 824–838. [CrossRef]
28. Wang, H.; Chang, J.; Yu, H.; Liu, H.; Hou, C.; Lu, H. Research on a Novel Vascular Interventional Surgery Robot and Control Method Based on Precise Delivery. *IEEE Access* **2021**, *9*, 26568–26582. [CrossRef]

29. Yu, H.; Wang, H.; Chang, J.; Niu, J.; Wang, F.; Yan, Y.; Tian, H.; Fang, J.; Lu, H. A Novel Vascular Intervention Surgical Robot Based on Force Feedback and Flexible Clamping. *App. Sci.* **2021**, *11*, 611. [CrossRef]
30. Guo, S.; Song, Y.; Yin, X.; Zhang, L.; Tamiya, T.; Hirata, H.; Ishihara, H. A novel robot-assisted endovascular catheterization system with haptic force feedback. *IEEE Trans. Robot.* **2019**, *35*, 685–696. [CrossRef]
31. Hasanzadeh, S.; Janabi-Sharifi, F.; Keenan, P. Backlash characterization and position control of a robotic catheter manipulator using experimentally-based kinematic model. *Mechatronics* **2017**, *44*, 94–106. [CrossRef]
32. Omisore, O.M.; Han, S.; Zhou, T.; Al-Handarish, Y.; Du, W.; Ivanov, K.; Wang, L. Learning-based Parameter Estimation for Hysteresis Modeling in Robotic Catheterization. In Proceedings of the 41st IEEE Conference of Engineering in Medicine and Biology Society, Berlin, Germany, 23–27 July 2019; IEEE: New York, NY, USA, 2019; pp. 5399–5402.
33. Beaman, C.B.; Kaneko, N.; Meyers, P.M.; Tateshima, S. A review of robotic interventional neuroradiology. *Am. J. Neuroradiol.* **2021**, *42*, 808–814. [CrossRef]
34. Bai, W.; Cursi, F.; Guo, X.; Huang, B.; Lo, B.; Yang, G.Z.; Yeatman, E.M. Task-Based LSTM Kinematic Modeling for a Tendon-Driven Flexible Surgical Robot. *IEEE Trans. Med. Robot. Bionics* **2022**, *4*, 339–342. [CrossRef]
35. Wu, D.; Zhang, Y.; Ourak, M.; Niu, K.; Dankelman, J.; Vander Poorten, E. Hysteresis Modeling of Robotic Catheters Based on Long Short-Term Memory Network for Improved Environment Reconstruction. *IEEE Rob. Autom Lett.* **2021**, *6*, 2106–2113. [CrossRef]
36. Cursi, F.; Bai, W.; Yeatman, E.M.; Kormushev, P. Model Learning With Backlash Compensation for a Tendon-Driven Surgical Robot. *IEEE Rob. Autom. Lett.* **2022**, *7*, 7958–7965. [CrossRef]
37. Choi, J.; Park, S.; Kim, Y.H.; Moon, Y.; Choi, J. A vascular intervention assist device using bi-motional roller cartridge structure and clinical evaluation. *Biosensors* **2021**, *11*, 329. [CrossRef] [PubMed]
38. Józefowicz, R.; Zaremba, W.; Sutskever, I. An Empirical Exploration of Recurrent Network Architectures. In Proceedings of the 32nd International Conference on Machine Learning, ICML, Lille, France, 6–11 July 2015; JMLR: Massuchesetts, MA, USA; pp. 2342–2350.
39. Cho, K.; Van Merriënboer, B.; Bahdanau, D.; Bengio, Y. On the properties of neural machine translation: Encoder-decoder approaches. *arXiv* **2014**, arXiv:1409.1259.
40. Zhou, W.; Guo, S.; Guo, J.; Meng, F.; Chen, Z. ADRC-Based Control Method for the Vascular Intervention Master–Slave Surgical Robotic System. *Micromachines* **2021**, *12*, 1439. [CrossRef]

applied sciences

MDPI

Article

Fuzzy PID Control of the Three-Degree-of-Freedom Parallel Mechanism Based on Genetic Algorithm

Zhifang Zhu [1,2], Yuanjie Liu [2], Yuling He [2], Wenhao Wu [2], Hongzhou Wang [3], Chong Huang [4] and Bingliang Ye [1,*]

1 Faculty of Mechanical Engineering, Zhejiang Sci-Tech University, Hangzhou 310018, China
2 Jiangxi Province Key Laboratory of Precision Drive and Control, Nanchang Institute of Technology, Nanchang 330099, China
3 Jiangxi Institute of Mechanical Science, Nanchang 330002, China
4 College of Intelligent Manufacturing, Jiangxi Technical College of Manufacturing, Nanchang 330095, China
* Correspondence: zist_ybl@zstu.edu.cn; Tel.: +86-13336060776

Abstract: It is necessary to upgrade and transform the sorting equipment in the industrial production line. In order to improve production efficiency and reduce labor intensity, a high-speed lightweight parallel mechanism control system for the high-speed sorting and packaging of light items was studied. A fuzzy PID controller based on genetic algorithm (GA) optimization is proposed according to the nonlinear and strong coupling characteristics of the parallel mechanism (PM) control system. The inverse kinematic analysis was conducted to map the workspace trajectory tracking problem to the joint space. It was transformed into the trajectory planning and solving problems in the joint space. The motion trajectory was obtained utilizing quintic polynomial interpolation. Finally, the servo control system model was established, and the PID control parameters were optimized and self-tuned by the GA. They were applied to the fuzzy PID controller for simulation experiments. The simulation results showed that the GA-optimized fuzzy PID control system compared with the fuzzy PID control system had a 23.39% shorter rise time, 22.32% less regulation time, and 7.18% less steady-state error. The control system had a good dynamic and steady-state performance.

Keywords: parallel mechanism; genetic algorithm; fuzzy PID controller; self-tuned

Citation: Zhu, Z.; Liu, Y.; He, Y.; Wu, W.; Wang, H.; Huang, C.; Ye, B. Fuzzy PID Control of the Three-Degree-of-Freedom Parallel Mechanism Based on Genetic Algorithm. *Appl. Sci.* **2022**, *12*, 11128. https://doi.org/10.3390/app122111128

Academic Editors: Vincent A. Cicirello and Gaetano Zizzo

Received: 3 October 2022
Accepted: 1 November 2022
Published: 2 November 2022

1. Introduction

In the packaging industry of industrial production, it is usually necessary to complete the classification, inspection, packaging, and other operations of light and small objects at a faster speed to improve work efficiency. It is difficult to complete the precise operation with high strength and high speed for a long time using the traditional manual operation method. Compared with the series mechanism [1,2], the parallel mechanism (PM) has high stiffness, small dead weight load ratios, a strong bearing capacity, high precision, and a compact structure. The PM is suitable for applications with a small working space and large load strength, and it is widely used in machine tool processing, aircraft manufacturing, and medical treatment. Artificial intelligence is used to store mathematical models and operation experience in the computer with the development of computer technologies. The established control system model aims to control the whole mechanical system easily.

Proportional-integral-derivative (PID) controllers are still widely used in the industrial process control. Engineers can tune these three gains through experience or simple principles such as classical tuning rules proposed by Ziegler-Nichols [3]. There are many factors in the control system of PMs, such as uncertainty, nonlinearity, and external disturbance. The conventional PID controller has some problems, such as poor parameter settings and poor adaptability to variable working conditions. The combination of fuzzy control and PID control theory can solve these problems.

Li [4] proposed a novel fuzzy logic controller (FLC) for the gap between the time response and the rule base. It performs well in both transient and steady states without using multiple decision tables. Carvajal [5] presented a new fuzzy PID control method for nonlinear systems that are structurally difficult to model. Najafizadeh [6] used two kinds of fuzzy inference engines to construct an adaptive fuzzy PID controller, which achieves a fast convergence time and high performance. Zhou [7] proposed an orthogonal fuzzy PID control method to control the manipulator, which improves system accuracy and reduces the oscillation process near the steady state. Phu [8,9] studied some qualitative properties of fuzzy PID control systems in fuzzy number space. HuKuhara differentiability and fuzzy second-order differential equations are used to solve the multi-boundary problem, which proves the existence and uniqueness of the solution of the differential equation. Liu [10] proposed a cascading predictive fuzzy PID (FPID) controller with weight and used the fastest descent method to calculate weight and improve the accuracy of trajectory tracking.

The quantization factor and scale factor affect the control effect of the fuzzy controller. Traditional selection methods are mostly based on expert experience and industrial knowledge, which will make the control effect unsatisfactory. Therefore, the optimization algorithm is introduced to optimize PID control parameters quickly and accurately using its global optimization ability. Tsai [11] proposed a novel adaptive PID control method—using predictive control and outputting recursive fuzzy wavelet neural networks to process a set of nonlinear digital delayed dynamic systems. Pelusi [12–14] previously studied the use of the GA and neuro-fuzzy techniques to design optimal control systems. The results can be used as benchmarks to compare with the proposed design. Purnama [15] compared various controllers. The PID controller optimized by the GA has a shorter rise time, and smaller steady-state error, but higher theoretical complexity. The proposed fuzzy PID controller was applied to the servo control system [16], showing that the fuzzy PID controller optimized by a GA has good speed control and anti-jamming ability. Chao [17,18] proposed that the membership function should be adjusted by nonlinear factors, which greatly improves the GA and verifies its feasibility. Moran [19,20] used the manual tuning PID and GA PID for comparative control experiments on DC electric machines. The genetic Algorithm PID can obtain more suitable PID parameters, but the system responds slowly.

Vijaya [21] used a fuzzy PID speed controller based on a GA to control a permanent magnet synchronous motor. The multi-carrier PWM is used for analysis, which can achieve the required speed faster than the conventional PID controller. Dogruer [22] optimized the fuzzy PID controller by a GA to improve the robustness of the voltage regulator. Alouache [23] found that the fuzzy PID controller optimized by a GA controls the mobile robot for trajectory tracking in the case of interferences with good control effects.

Therefore, the optimization algorithm in the traditional fuzzy PID control was introduced in the work. The global optimization ability and parallel ability of the GA were used to optimize PID control parameters. Thus, the robustness of the control system and the trajectory tracking accuracy of the PM were greatly improved.

2. Model of the Three-Translation PM

2.1. Introduction to PMs

The three-translation PM in the work was composed of a static platform, a moving platform, three composite branch chains with the same structure, and a parallelogram closed-loop subchain with variable rod lengths. Three composite branch chains were evenly distributed on the static platform with an included angle of 120°. Each composite branch chain was composed of an active arm and a variable-length parallelogram closed-loop subchain connected by a rotating pair. The variable-length parallelogram closed-loop subchain was composed of connectors, and the sliding rod was formed by connecting the moving pair. The active arm and the variable-length parallelogram closed-loop subchain were always kept at 90°, and the parallelogram closed-loop subchain was connected with the moving platform by a spherical pair. Driving motors were fixed on the static platform, which made the machine have good motion performance. Figure 1 shows the PM structure.

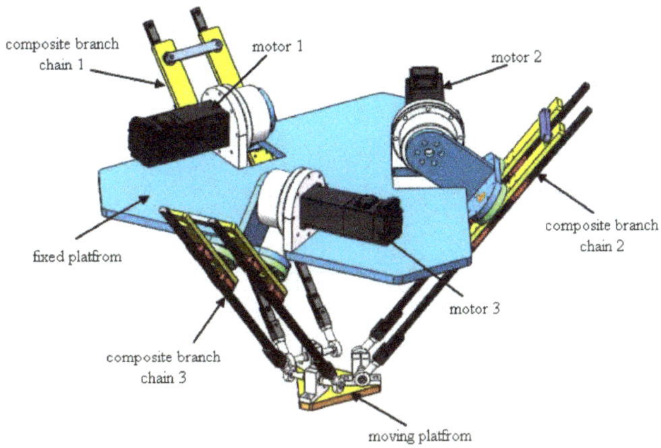

Figure 1. PM structure.

2.2. Coordinate System Establishment

Each driven arm of the PM was a parallelogram, and the distance between the two sliding rods was unchanged; therefore, the structure could be simplified (see Figure 2). Three points, L1, L2, and L3, were the vertices of the equilateral triangle, and R was its circumscribed circle radius. The angle between it and the X-axis was α_i $(i = 1, 2, 3)$ with the circle center as the origin, the direction pointing to the circle center as the X-axis, and the normal plane direction of the regular triangle as the Z-axis. Three points, N_1, N_2, and N_3, were the vertices of an equilateral triangle whose circumscribed circle radius was r.

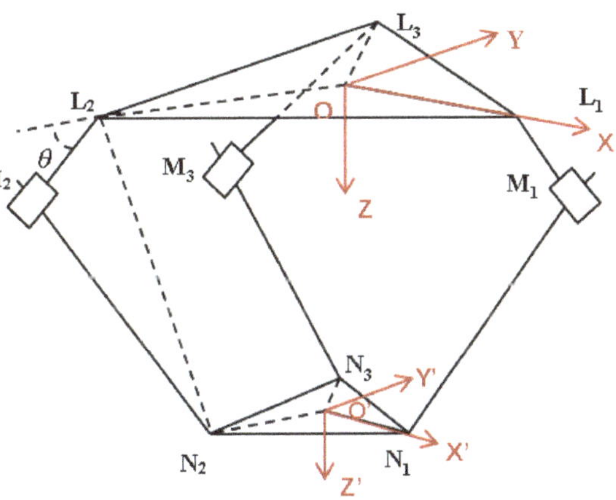

Figure 2. Established coordinate system.

The circle center was taken as the origin, and the three coordinate axes at the initial position were parallel to the basic coordinate system. Dynamic coordinate system O′-X′Y′Z′ was established. l_b is the length of the active arm L_iM_i, and l_{ai} is the length of the slave arm M_iN_i. θ_i is the angle between the static platform and the active arm as well as the input parameter, and i = 1, 2, and 3. The center coordinate of the moving platform is O′(x,y,z), where x, y, and z are output parameters.

2.3. Position Inverse Solution

According to the geometric relationship $L_iM_i \perp M_iN_i$, the simplified mechanism from the Pythagorean theorem is shown in Equation (1):

$$\overline{LN}^2 = \overline{LM}^2 + \overline{MN}^2 \tag{1}$$

where point coordinate L_i:$L_i = R\begin{bmatrix} cos\alpha_i \\ sin\alpha_i \\ 0 \end{bmatrix}$; point coordinate N_i:$N_i = \begin{bmatrix} x + rcos\alpha_i \\ y + rsin\alpha_i \\ z \end{bmatrix}$; point coordinate M_i : $M_i = \begin{bmatrix} (l_b cos\theta_i + R)cos\alpha_i \\ (l_b cos\theta_i + R)sin\alpha_i \\ l_b sin\theta_i \end{bmatrix}$; θ_i is the rotation angle of the active arm; and i = 1, 2, and 3. The calculation results from known coordinates are shown in Equation (2):

$$\begin{cases} \overline{LN}^2 = [(x + rcos\alpha_i) - Rcos\alpha_i]^2 + [(x + rsin\alpha_i) - Rsin\alpha_i]^2 + z^2, \\ \overline{LM}^2 = l_b^2, \\ \overline{MN}^2 = [(x + rcos\alpha_i) - (l_b cos\theta_i + R)cos\alpha_i]^2 + [(x + rsin\alpha_i) - (l_b cos\theta_i + R)sin\alpha_i]^2 \\ \quad + (z - l_b sin\theta_i)^2 \end{cases} \tag{2}$$

$\overline{LN}^2$, $\overline{LM}^2$, and $\overline{MN}^2$ are substituted into Equation (1) to obtain the inverse solution of the mechanism. The inverse solution of the mechanism is shown in Equation (3):

$$\theta_i = 2\arctan(t_i) \tag{3}$$

3. Trajectory Planning

The trajectory planning of the PM involves the path and temporal relationship of its motion. The time-varying position of the moving platform must be located within the workspace defined by mechanical boundaries. Additionally, there are the maximum velocity and acceleration in the physical limit range. The activated-joint angle calculated by the inverse kinematics constitutes the motion trajectory from the starting position to the target position.

3.1. Coordinate Space Trajectory Planning

According to the requirement of the actual task, the motion trajectory of the PM was designed. A door-shaped trajectory was introduced to meet the needs of picking and placing operations in the industry.

P_0 is the picking point and P_5 is the placing point (see Figure 3). The trajectory was composed of three straight lines and two arcs. The transition arcs at both ends could avoid sudden changes in speeds and acceleration.

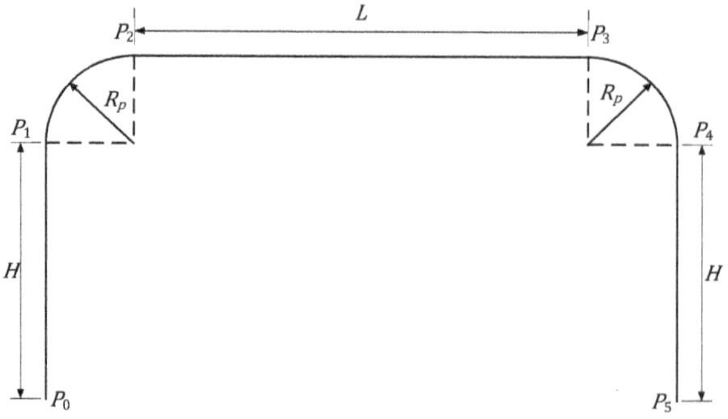

Figure 3. Door-shaped trajectory.

Assuming that the door trajectory is in the XOY plane, and $P_0(x_0, 0, z_0)$ is the initial position, the straight line and circular arc in Figure 3 were interpolated, and the door-shaped trajectory was obtained, as shown in Equation (4):

$$\begin{cases} (x, y, z) = (x_0, 0, z_0 - s(t_i)) & t_0 < t_i \leq t_1 \\ (x, y, z) = (x_0 + r_p - r_p cos\varphi_1, 0, z_0 - h - r_p sin\varphi_1) & t_1 < t_i \leq t_2 \\ (x, y, z) = (x_0 + r_p + s(t_i) - 0.5\pi r_p - h, 0, z_0 - h - r_p) & t_2 < t_i \leq t_3 \\ (x, y, z) = (x_0 + r_p + l + r_p sin\varphi_2, 0, z_0 - h - r_p cos\varphi_2) & t_3 < t_i \leq t_4 \\ (x, y, z) = \left(x_0 + 2r_p + l, 0, z_0 - s_f + s(t_i)\right) & t_4 < t_i \leq t_5 \end{cases} \quad (4)$$

$\varphi_1 = (s(t_i) - h)/r$ and $\varphi_2 = (s(t_i) - h - (1/2)\pi r_p - l)/r_p$,$t_1 \ldots t_5$ correspond to the arrival time at $P_1 \ldots P_5$, respectively. t_i is any time in the cycle.

3.2. Joint Space Trajectory Planning

The fifth-order polynomial for interpolation was used as shown in Equation (5) to plan the PM trajectory, ensure the smooth movement of each joint, and prevent the sudden change in acceleration.

$$\begin{cases} s(t) = a_0 + a_1 t + a_2 t^2 + a_3 t^3 + a_4 t^4 + a_5 t^5 \\ v(t) = a_1 + 2a_2 t + 3a_3 t^2 + 4a_4 t^3 + 5a_5 t^4 \\ a(t) = 2a_2 + 6a_3 t + 12a_4 t^2 + 20a_5 t^3 \end{cases} \quad (5)$$

where s, v, and a represent the path length, velocity, and acceleration, respectively; a_0, a_1, ..., and a_5 are the undetermined coefficients of the polynomial; t is the time.

The following boundary constraints are specified to ensure the smooth operation of the PM. Equation (6) is the specified boundary constraint condition:

$$\begin{gathered} s(0) = 0, v(0) = 0, a(0) = 0 \\ s\left(t_f\right) = s_f, v\left(t_f\right) = 0, a\left(t_f\right) = 0 \end{gathered} \quad (6)$$

where t_f represents the terminal time of the trajectory and s_f is the total length of the trajectory planning path.

Undetermined coefficients can be solved by substituting the above boundary conditions. The trajectory planning Equation (7) can be obtained as:

$$\begin{cases} s(t) = \frac{10s_f}{t_f^3}t^3 - \frac{15s_f}{t_f^4}t^4 + \frac{6s_f}{t_f^5}t^5 \\ v(t) = \frac{30s_f}{t_f^3}t^2 - \frac{60s_f}{t_f^4}t^3 + \frac{30s_f}{t_f^5}t^4 \\ a(t) = \frac{60s_f}{t_f^3}t - \frac{180s_f}{t_f^4}t^2 + \frac{120s_f}{t_f^5}t^3 \end{cases} \quad (7)$$

4. PM Control System

The servo control system of the PM is a complex, nonlinear, and strong coupling system. Since the parameters in the system are time varying, only relevant approximate models can be established, which hinders control. Therefore, the combination of a genetic algorithm and fuzzy PID controller has the reliability of PID control, the robustness of fuzzy control, rapid adjustments, and the global optimization of the GA.

4.1. Fuzzy PID Controller

The PID controller is a feedback element commonly used in industrial control [24]. The input value can be adjusted according to the feedback value and the difference value so that the system is more accurate and stable. PID algorithms can be divided into positional PID and incremental PID.

$$u(k) = K_P e(k) + K_I \sum_{i=0}^{k} e(i) + K_D[e(k) - e(k-1)] \quad (8)$$

$$\Delta u(k) = K_P[e(k) - e(k-1)] + K_I e(k) + K_D[e(k) - 2e(k-1) + e(k-2)] \quad (9)$$

Equations (8) and (9) are place style and increment style, respectively. $e(K)$ represents the error. Fuzzy logic control (FLC) has the advantage of using human brains to solve problems, with its core divided into the fuzzification interface, fuzzy rule base, fuzzy decision, and defuzzification. Fuzzy inferences include membership function and rule table, and the knowledge base is obtained from expert experience.

Fuzzy PID (FPID) combines Fuzzy control and PID control, with a simple structure and strong self-adaptability [25]. The fuzzy controller takes error E as the input. The output is the modified values of PID parameters ΔK_p, ΔK_i, and ΔK_d through quantization factors, fuzzy control rules, and scale factors. K_e and K_{ec} are quantization factors; K_{up} is the scale factor for ΔK_p; K_{ui} the scaling factor for ΔK_i; and K_{ud} the scaling factor for ΔK_d. Figure 4 shows the block of the fuzzy PID structure. The expression of the fuzzy PID controller is shown in Equation (10).

$$\begin{cases} K_p = K_{p0} + \Delta K_p \\ K_i = K_{i0} + \Delta K_i \\ K_d = K_{d0} + \Delta K_d \end{cases} \quad (10)$$

where K_{p0}, K_{i0}, and K_{d0} represent the proportional, integral, and differential initial coefficients in the traditional PID controller, respectively. Based on the fuzzy rule table, the fuzzy inference results show that ΔK_p, ΔK_i, and ΔK_d are the change values of proportion, integral, and differential coefficients, respectively.

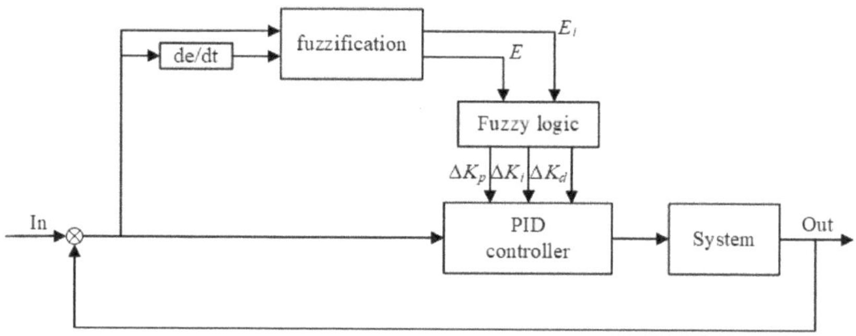

Figure 4. Fuzzy PID controller.

The corresponding membership function is generated by the fuzzy toolbox of mathematical software. Correctly constructing the membership function is one of the keys to using fuzzy control. Figure 5 shows the quantization domain and fuzzy subset. The fuzzy subset with a sharp shape of the membership function has high resolutions and a high control sensitivity. On the contrary, the shape of the membership function curve is relatively flat, with good stability.

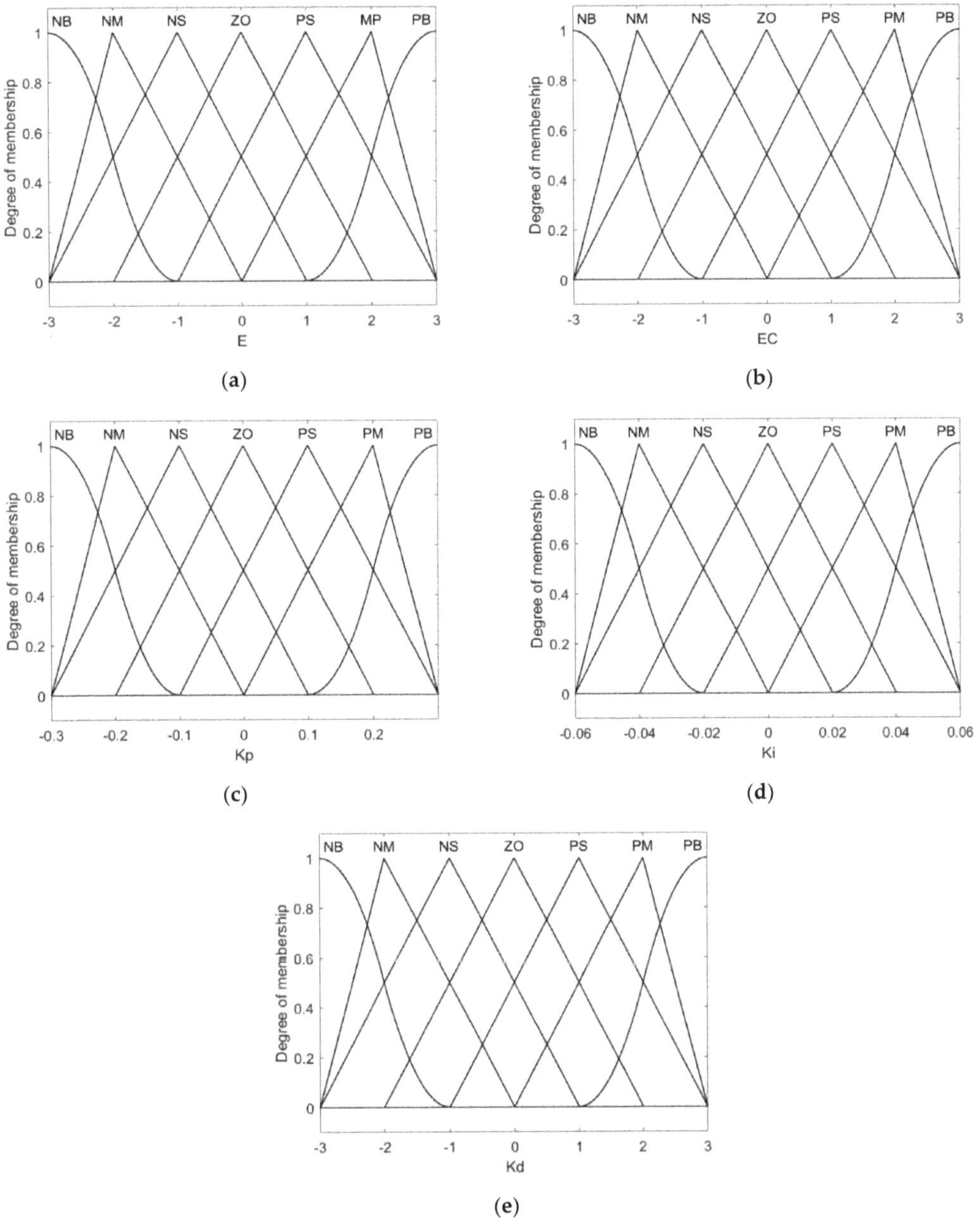

Figure 5. Membership function: (**a**) error; (**b**) changes in errors; (**c**–**e**) K_p, K_i, and K_d membership functions for fuzzy self-tuning PID, respectively.

The relationship between the three input parameters of the PID controller and the general tuning principles summed up by expert experience are as follows:

(1) The absolute value of the input error is large. The K_i value is zero; the larger the K_p value, the smaller the K_d value taken simultaneously.

(2) The absolute value of the input error is the median. K_i takes an appropriate value, and K_p should take a small value. The K_d value significantly affects the system.

(3) The absolute value of the input error is small. K_i and K_p values should be large, and the K_d value depends on the absolute value of the change rate of the input error. When the change rate is small, K_d takes an intermediate value; otherwise, K_d takes a small value.

Tables 1–3 show fuzzy rules. The input fuzzy variable derivative of position errors has seven linguistic variables, namely NB (negative big), NM (negative medium), NS (negative small), ZO (zero), PS (positive small), PM (positive medium), and PB (positive big). Three points, ΔK_p, ΔK_i, and ΔK_d, are defined as the fuzzy set in the defuzzification process.

Table 1. Fuzzy rule of ΔK_p.

E \ EC	NB	NM	NS	ZO	PS	PM	PB
NB	PB	PB	PM	PM	PS	ZO	ZO
NM	PB	PB	PM	PS	PS	ZO	NS
NS	PM	PM	PM	PS	ZO	NS	NS
ZO	PM	PM	PS	ZO	NS	NM	NM
PS	PS	PS	ZO	NS	NS	NM	NM
PM	PS	ZO	NS	NM	NM	NM	NB
PB	ZO	ZO	NM	NM	NM	NB	NB
NB	PB	PB	PM	PM	PS	ZO	ZO

Table 2. Fuzzy rule of ΔK_i.

E \ EC	NB	NM	NS	ZO	PS	PM	PB
NB	NB	NB	NB	NM	NM	NS	ZO
NM	NM	NB	NB	NM	NS	NS	ZO
NS	NS	NB	NM	NS	NS	ZO	PS
ZO	ZO	NM	NM	NS	ZO	PS	PM
PS	PS	NM	NS	ZO	PS	PS	PM
PM	PM	ZO	ZO	PS	PS	PM	PB
PB	PB	ZO	ZO	PS	PM	PM	PB
NB	NB	NB	NB	NM	NM	NS	ZO

Table 3. Fuzzy rule of ΔK_d.

E \ EC	NB	NM	NS	ZO	PS	PM	PB
NB	NB	PS	NS	NB	NB	NB	NM
NM	NM	PS	NS	NB	NM	NM	NS
NS	NS	ZO	NS	NM	NM	NS	NS
ZO	ZO	ZO	NS	NS	NS	NS	NS
PS	PS	ZO	ZO	ZO	ZO	ZO	ZO
PM	PM	PB	NS	PS	PS	PS	PS
PB	PB	PB	PM	PM	PM	PS	PS
NB	NB	PS	NS	NB	NB	NB	NM

4.2. Optimization of Fuzzy PID Parameters by the GA

A GA is an optimization method for finding the optimal solution to a problem based on Darwin's theory of biological evolution [26–28]. Traditional fuzzy PID has great differences in selecting rule tables and membership functions, and it is difficult to generalize according to expert experience. The quantization factor and scale factor of the fuzzy PID controller are optimized by the excellent global optimization ability of the GA. The structure diagram of fuzzy PID optimized by genetic algorithm is shown in Figure 6, and the algorithm flow chart is shown in Figure 7:

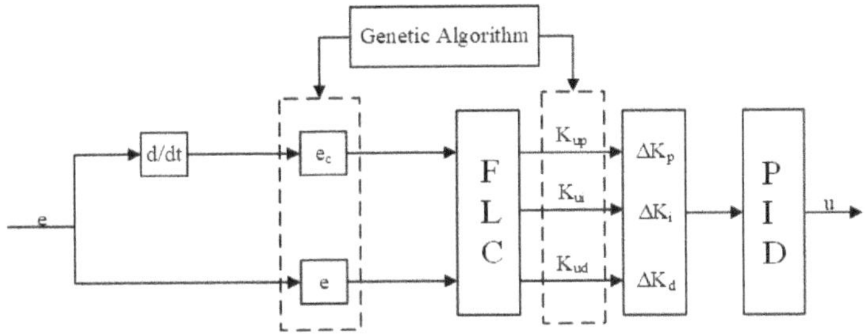

Figure 6. Structure of the GA to optimize fuzzy PID.

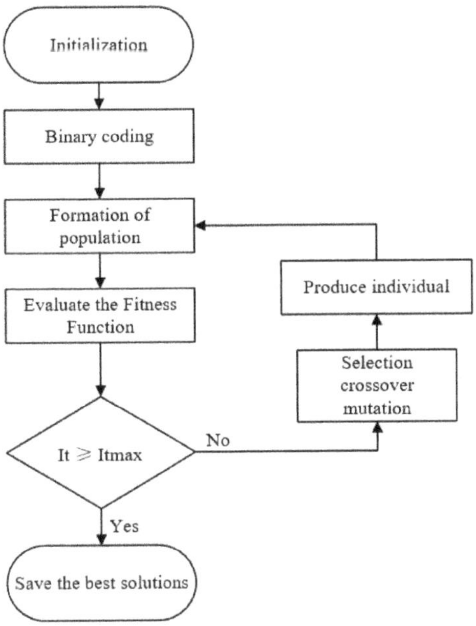

Figure 7. Flow of the GA.

$\{e,\ e_c,\ K_p,\ K_i,\ K_d\}$ are the genes of individuals using the continuous GA to optimize the fuzzy PID controller. Each gene is coded in decimal notation. Fitness function Equation (11) is used to calculate the fitness values of each individual [29]. Equation (12) calculates the probabilities of each individual to be selected [30]:

$$F_i = \frac{1}{\int_0^{\infty} |e_i(t)| dt} \tag{11}$$

$$P_i = \frac{F_i}{\sum_{i=1}^{n} Fi} \tag{12}$$

The genetic mode is set to the high probability of crossover and the low probability of mutation and is iterated for 100 generations as the termination condition.

5. Model Building and Simulation Analysis

The Alternating Current (AC) servo motor system is a typical nonlinear controlled object, and it is necessary to comprehensively consider the characteristics of the uncertain system of the AC servo motor. It is assumed that the magnetic circuit is not saturated and the magnetic field is sinusoidal. The eddy current loss and hysteresis loss are ignored. Let $L_d = L_p = L$, and the state Equations (13) and (14) in the d-q coordinates are as follows:

$$G(s) = \frac{\omega_m(s)}{I_m(s)} = K_{pi}K_fK_u/JLFF(s) \tag{13}$$

$$F(s) = S^2 + (R_s + K_{pi}K_i)S/L + K_fK_u/JL \tag{14}$$

where K_f, R_s, K_u, and K_i represent the torque coefficient, stator winding, potential coefficient, and current feedback coefficient, respectively.

Transfer function $G(s)$ characterizes the dynamic characteristics of the complex system under no moment of inertia and torque interferences. The transfer function is shown in Equation (15):

$$G(s) = \frac{47.5}{s} - \frac{48}{s+6} + \frac{0.5}{s+534} \tag{15}$$

According to the transfer function of the AC servo motor and parameters obtained by the GA, the simulation model is established in mathematical simulation software. The fuzzy PID control system is taken as an example. According to the influence of PID controller parameters on the control effect, combined with fuzzy controller, a fuzzy PID controller is designed (see Figure 8 for the fuzzy PID control model). The three terms used in the block diagram (K_p, K_i, and K_d) are gains for the proportional controller, integral controller, and derivative controller. The values of K_p, K_i, and K_d were 69.25, 0.993, and 10.5, respectively. The values of K_e, K_{ec}, and K_u were 3, 0. 1, and 0. 1, respectively. The controller dynamically adjusts PID parameters through the GA during the system operation.

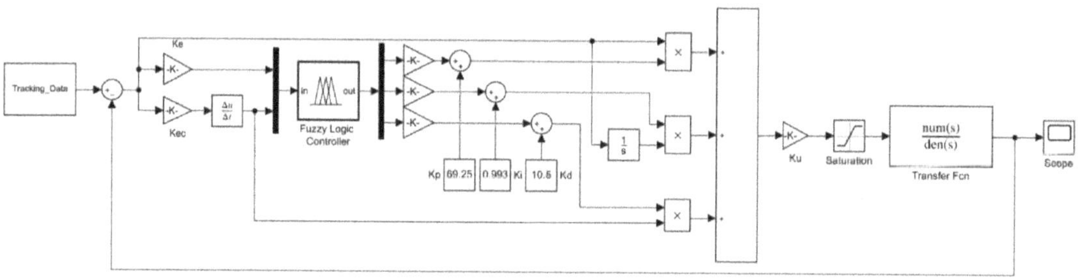

Figure 8. Blocks of the fuzzy self-tuning PID control algorithm.

6. Analysis of Simulation Results

According to the determined structure and control method of the servo control system, a simulation experiment was carried out on the position control of the servo motor combined with the engineering tuning method. The superiority of the control strategy designed in the work was verified by comparing the control effect of the fuzzy PID (FPID) control strategy and genetic algorithm optimized fuzzy PID (GAFPID) control strategy on the position signal tracking of the servo control system.

6.1. Iterative Analysis of GAs

According to the parameters of the setting algorithm of the fuzzy PID control system structure, the population size and the iteration number were set to 30 and 100, respectively. The crossover probability and mutation probability were 0.9 and 0.1, respectively. $K_p = 100$, $K_i = 100$, and $K_d = 10$ at the initial stage.

Figure 9 shows that it converged at the 52nd iteration, while traditional calculations often require thousands of iterations. The traditional method of calculation mentioned here is trial and error. Trial and error is a method of setting parameters empirically. In the closed-loop control system, the adjustment was carried out in the order of K_p, K_i and K_d. While adjusting the parameters, the process was observed until the requirements were met. The optimal objective function value and the optimal matching parameter value of the fuzzy PID controller could be obtained at the end of the iteration. The parameter values of the fuzzy PID controller obtained by the GA were substituted into the simulation model of the servo control system. The experimental analysis of step characteristics, sinusoidal characteristics, and joint trajectory tracking was carried out.

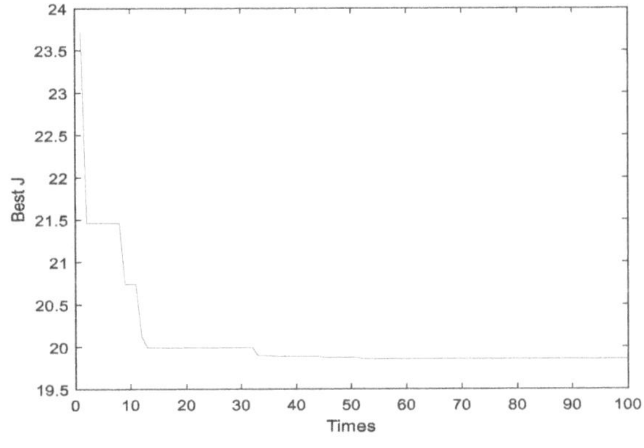

Figure 9. Iterative curve of the GA.

6.2. Simulation Analysis of Step Characteristics

PID control, fuzzy PID control, and PID control optimized by the GA were applied to the control system under the condition that the input was an 8° step signal simulating the load. The sampling time was set to 0.01 s. The dynamic and static indices of the system under different control strategies were analyzed (see Figure 10 for simulation results).

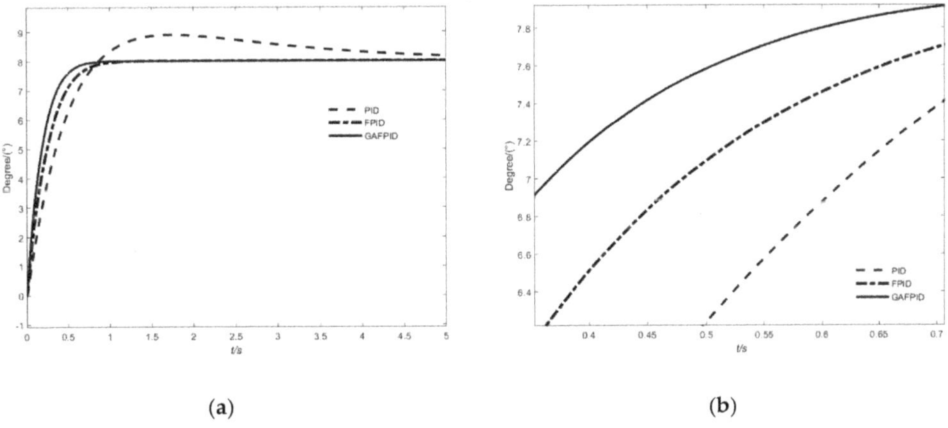

Figure 10. (**a**) Step response; (**b**) local enlarged view.

Table 4 shows the simulation results of the step response. The rise time of the system under the control of a fuzzy PID controller optimized by the GA was reduced by about 37.91% compared with the classical PID control under the loaded step response. It was about 23.39% shorter than the fuzzy PID control. The adjustment time of the system controlled by the fuzzy PID controller optimized by the GA was about 32.46% shorter than that of the PID control and about 22.32% shorter than that of the fuzzy PID control. The steady-state error of the system under genetic fuzzy PID control was 88.67% less than that of the classical PID control, and 7.18% less than that of the fuzzy PID control in the steady-state response.

Table 4. Simulation results of the step response.

The Controller Type	Setting Time/(s)	Rise Time/(s)	Steady-State Error/(°)
PID	0.752	0.612	0.1713
FPID	0.654	0.496	0.0209
GAFPID	0.508	0.380	0.0194

6.3. Sinusoidal Characteristic Simulation Analysis

The input was a 12° sinusoidal signal at 2 Hz to simulate the on-load condition. Four different control strategies were applied to the control system. The sampling time was set to 0.01 s. The maximum amplitude error and the maximum phase error of the system under different control strategies were compared. Figure 11 shows the simulation results.

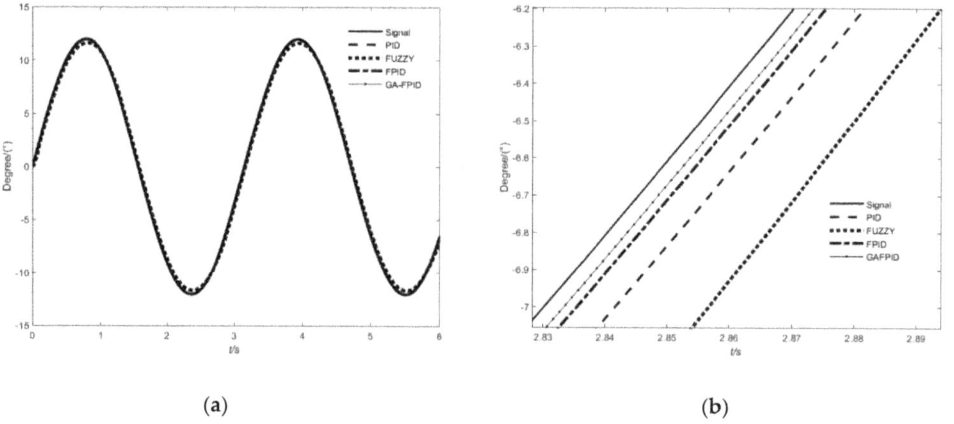

Figure 11. Sinusoidal input signal at 2 Hz: (**a**) tracking response; (**b**) local enlarged view.

Table 5 shows the simulation data of frequency response under loading. The sinusoidal signal at 2 Hz and 12° was input under loading. The system under the genetic fuzzy PID control was 32.30% less than the fuzzy PID control system, 80.01% less than the pure fuzzy control system, and 60.03% less than the classical PID control system in terms of the maximum amplitude error. The system under the fuzzy PID control optimized by the GA was 33.43% less than the fuzzy PID, 86.90% less than the pure fuzzy control system, and 67.29% less than the classical PID in terms of the maximum phase error.

Table 5. Simulation data of the frequency response under loading.

The Controller Type	Maximum Phase Error/(°)	Maximum Magnitude Error/(°)
PID	0.573	0.2614
FUZZY	1.146	0.6527
FPID	0.344	0.1263
GAFPID	0.229	0.0855

6.4. Input-Joint-Trajectory Simulation Analysis

The controllers of the three joints were successively optimized, and the trajectory planning was carried out in a Cartesian coordinate system. The end effector was moved along the gate-shaped trajectory, and the inverse kinematics were used to map the operation space to the joint space. The trajectories planned by interpolating quintic polynomials were used as the input. The simulation time was set to 3 s. The traditional PID control, fuzzy control, fuzzy PID control, and fuzzy PID control strategies optimized by GAs were used to control the PM. The trajectory tracking capabilities of the four control methods were compared.

Figures 12–14 show the trajectory tracking of the three joints, respectively. The average absolute error was about 0.0327 using the GAFPID control strategy for Q1 joint motion. The enlarged partial image presents that the joint trajectory based on the GAFPID control strategy was the closest to the ideal trajectory, with the optimal control effect.

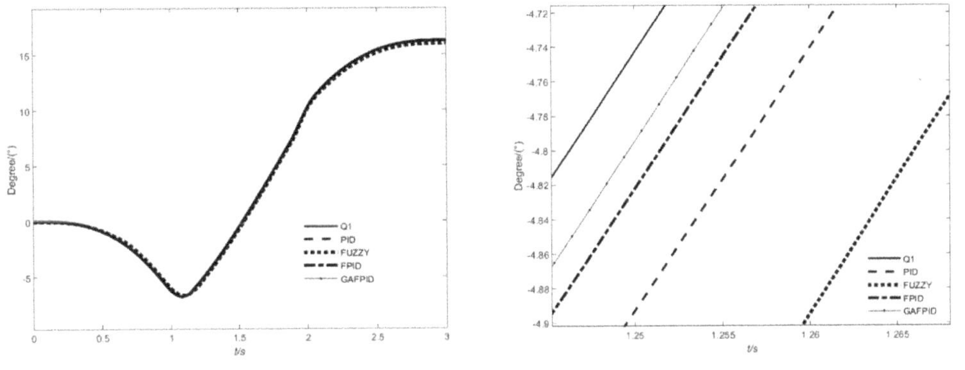

(a) (b)

Figure 12. Joint trajectory Q1: (**a**) tracking response; (**b**) local enlarged view.

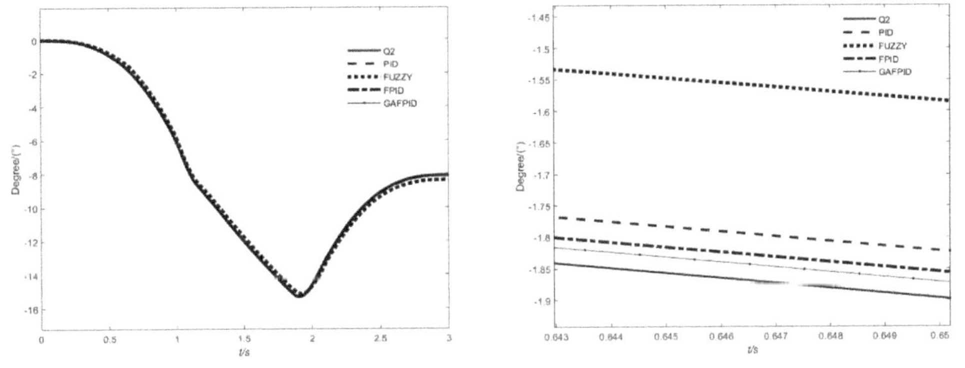

(a) (b)

Figure 13. Joint trajectory Q2: (**a**) tracking response; (**b**) local enlarged view.

According to Tables 6 and 7, the minimum integral time absolute error (ITAE) value of the joint trajectory is based on the GAFPID control strategy. The dynamic response overshoot was small and the adjustment time was short. The integral absolute error (IAE) index was the smallest, which means that small deviations in the control system can be suppressed.

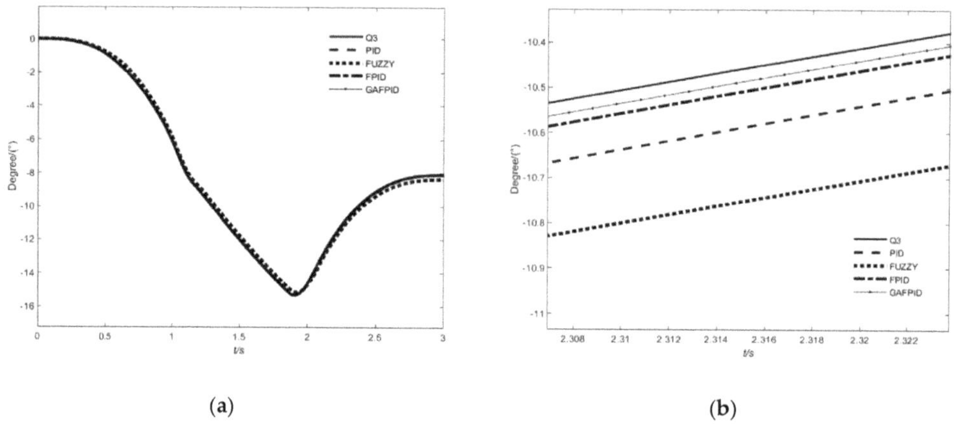

Figure 14. Joint trajectory Q3: (**a**) tracking response; (**b**) local enlarged view.

Table 6. ITAE values.

The Controller Type	Joint Q1	Joint Q2	Joint Q3
PID	0.0655	0.057	0.057
FUZZY	0.1568	0.1667	0.1667
FPID	0.0388	0.0319	0.0319
GAFPID	0.0265	0.0197	0.0197

Table 7. IAE values.

The Controller Type	Joint Q1	Joint Q2	Joint Q3
PID	0.0436	0.042	0.042
FUZZY	0.1094	0.1196	0.1196
FPID	0.0237	0.0235	0.0235
GAFPID	0.0147	0.0144	0.0144

7. Conclusions

(1) The PM is a highly nonlinear, strongly coupled, and time-varying control system. It is difficult to obtain the ideal control effect with the traditional control method. In the trajectory tracking control of 3-DOF PM, the PID control was combined to improve the control accuracy. Without relying on the exact mathematical model of the controlled object, the fuzzy control method was used to approximate the nonlinear system. Through the combination of the fuzzy controller and PID controller, a fuzzy PID controller was designed, which could self-tune PID parameters online in the control process. In view of the problem that the design of the fuzzy control system needs expert knowledge and experience to obtain the best control effect, the fuzzy PID controller was optimized by a genetic algorithm.

(2) The traditional PID control, fuzzy control, fuzzy PID control, and fuzzy PID control optimized by the GA were carried out to simulate signal-tracking and to verify the effectiveness of the optimized controller. The simulation showed that the servo system under the fuzzy PID control based on the GA had significantly improved dynamic response characteristics and steady-state accuracy, with a better trajectory-tracking effect. The ideal control performance of the PM in trajectory tracking control depended on the design of the control system. Through the reasonable selection and design of the control strategy, the control deviation of the joint angle and angular velocity of the PM could be reduced. Thus, the response speed and stability of the system were improved, which is of great significance to the application of PM.

(3) The optimization of other intelligent algorithms and the combination of each other's advantages will be further considered in the future to greatly adapt to more complex regulatory processes. The neural network fuzzy control was considered. Neural networks can adequately approximate arbitrary complex nonlinear relationships. It can learn and adapt to the dynamic characteristics of uncertain systems. The adaptive fuzzy control system was introduced, which can identify system parameters and adjust control parameters online. The expert intelligent self-tuning PID controller was used to recognize the system error pattern when the closed-loop system was disturbed. The key problems to be solved in the above control strategy were online parameter identification, parameter adaptive adjustment, the establishment of an expert system, and the combination of a neural network and fuzzy control. The above needs to be studied in further research work.

Author Contributions: Conceptualization, Z.Z. and Y.L.; methodology, Y.H.; software, W.W. and Y.L.; validation, Y.L., H.W., Z.Z. and C.H.; formal analysis, Y.L.; investigation, Z.Z.; resources, H.W.; data curation, Z.Z., Y.L. and B.Y.; writing—original draft preparation, Y.L. and Y.H.; writing—review and editing, Z.Z. and Y.L.; visualization, B.Y.; supervision, H.W. and C.H.; project administration, B.Y.; funding acquisition, Z.Z. and B.Y. All authors have read and agreed to the published version of the manuscript.

Funding: This research was funded by the Science Funding in Education Department of Jiangxi province grant number GJJ201929, GJJ214704 and GJJ181463, and the Key Scientific and Technological Support Topics in Jiangxi Province under Grant no. 20203ABC28W016.

Institutional Review Board Statement: Not applicable.

Informed Consent Statement: Not applicable.

Data Availability Statement: Not applicable.

Conflicts of Interest: The authors declare no conflict of interest.

Abbreviation

Abbreviation	Wording
PM	Parallel mechanism
PID	Proportional-integral-derivative
FLC	Fuzzy logic controller
FPID	Fuzzy PID controller
GA	Genetic algorithm
DC	Direct current
AC	Alternating current
GAFPID	The genetic algorithm optimized fuzzy PID
ITAE	Integral time absolute error
IAE	Integral absolute error

References

1. Kong, X.; Gosselin, C.M.; Richard, P.-L. Type synthesis of parallel mechanisms with multiple operation modes. In Proceedings of the International Design Engineering Technical Conferences and Computers and Information in Engineering Conference, Philadelphia, PA, USA, 10–13 September 2006; pp. 1037–1046. [CrossRef]
2. Huang, Y.; Lu, Q.; Wang, H.; Liu, J.; Li, Z.; Zou, X.; Zhan, X. Kinematics Analysis and Simulation of a Novel 3T Parallel Mechanism. *Math. Probl. Eng.* **2022**, *2022*, 3424012. [CrossRef]
3. Ziegler, J.G.; Nichols, N.B. Optimum settings for automatic controllers. *Trans. ASME* **1942**, *64*, 759–768. [CrossRef]
4. Li, H.-X.; Gatland, H. A new methodology for designing a fuzzy logic controller. *IEEE Trans. Syst. Man Cybern.* **1995**, *25*, 505–512. [CrossRef]
5. Carvajal, J.; Chen, G.; Ogmen, H. Fuzzy PID controller: Design, performance evaluation, and stability analysis. *Inf. Sci.* **2000**, *123*, 249–270. [CrossRef]
6. Najafizadeh Sari, N.; Jahanshahi, H.; Fakoor, M. Adaptive fuzzy PID control strategy for spacecraft attitude control. *Int. J. Fuzzy Syst.* **2019**, *21*, 769–781. [CrossRef]

7. Zhou, H.; Chen, R.; Zhou, S.; Liu, Z. Design and analysis of a drive system for a series manipulator based on orthogonal-fuzzy PID control. *Electronics* **2019**, *8*, 1051. [CrossRef]
8. Phu, N.D.; Hung, N.N.; Ahmadian, A.; Senu, N. A new fuzzy PID control system based on fuzzy PID controller and fuzzy control process. *Int. J. Fuzzy Syst.* **2020**, *22*, 2163–2187. [CrossRef]
9. Phu, N.D.; Hung, N.N. Some solving methods for a fuzzy multi-point boundary value problem. *Soft Comput.* **2020**, *24*, 483–499. [CrossRef]
10. Liu, Y.; Fan, K.; Ouyang, Q. Intelligent traction control method based on model predictive fuzzy PID control and online optimization for permanent magnetic maglev trains. *IEEE Access* **2021**, *9*, 29032–29046. [CrossRef]
11. Tsai, C.-C.; Yu, C.-C.; Tsai, C.-T. Adaptive ORFWNN-based predictive PID control. *Int. J. Fuzzy Syst.* **2019**, *21*, 1544–1559. [CrossRef]
12. Pelusi, D. Optimization of a fuzzy logic controller using genetic algorithms. In Proceedings of the 2011 Third International Conference on Intelligent Human-Machine Systems and Cybernetics, Hangzhou, China, 26–27 August 2011; pp. 143–146. [CrossRef]
13. Pelusi, D. Designing neural networks to improve timing performances of intelligent controllers. *J. Discret. Math. Sci. Cryptogr.* **2013**, *16*, 187–193. [CrossRef]
14. Pelusi, D.; Mascella, R. Optimal control Algorithms for second order Systems. *J. Comput. Sci.* **2013**, *9*, 183–197. [CrossRef]
15. Purnama, H.S.; Sutikno, T.; Alavandar, S.; Subrata, A.C. Intelligent control strategies for tuning PID of speed control of DC motor: A review. In Proceedings of the 2019 IEEE Conference on Energy Conversion (CENCON), Yogyakarta, Indonesia, 16–17 October 2019; pp. 24–30. [CrossRef]
16. Singh, A.; Giri, V. Design and analysis of DC motor speed control by GA based tuning of fuzzy logic controller. *Int. J. Eng. Res. Technol.* **2012**, *1*, 1–6.
17. Chao, C.-T.; Sutarna, N.; Chiou, J.-S.; Wang, C.-J. An optimal fuzzy PID controller design based on conventional PID control and nonlinear factors. *Appl. Sci.* **2019**, *9*, 1224. [CrossRef]
18. Chao, C.-T.; Sutarna, N.; Chiou, J.-S.; Wang, C.-J. Equivalence between fuzzy PID controllers and conventional PID controllers. *Appl. Sci.* **2017**, *7*, 513. [CrossRef]
19. Flores-Morán, E.; Yánez-Pazmiño, W.; Barzola-Monteses, J. Genetic algorithm and fuzzy self-tuning PID for DC motor position controllers. In Proceedings of the 2018 19th International Carpathian Control Conference (ICCC), Szilvasvarad, Hungary, 28–31 May 2018; pp. 162–168. [CrossRef]
20. Morán, M.E.F.; Viera, N.A.P. Comparative study for DC motor position controllers. In Proceedings of the 2017 IEEE Second Ecuador Technical Chapters Meeting (ETCM), Salinas, Ecuador, 16–20 October 2017; pp. 1–6. [CrossRef]
21. Vijaya, S.; Bharathiraja, P.; Nithyanandam, J. A Performance Comparison of Conventional and New Seven Level Inverter Topology Fed PM Synchronous Motor Using GA Based Fuzzy PID Speed Controller. In Proceedings of the 2018 International Conference on Computation of Power, Energy, Information and Communication (ICCPEIC), Chennai, India, 28–29 March 2018; pp. 275–279. [CrossRef]
22. Dogruer, T.; Can, M.S. Design and robustness analysis of fuzzy PID controller for automatic voltage regulator system using genetic algorithm. *Trans. Inst. Meas. Control* **2022**, *44*, 1862–1873. [CrossRef]
23. Alouache, A.; Wu, Q. Genetic algorithms for trajectory tracking of mobile robot based on PID controller. In Proceedings of the 2018 IEEE 14th International Conference on Intelligent Computer Communication and Processing (ICCP), Cluj-Napoca, Romania, 6–8 September 2018; pp. 237–241. [CrossRef]
24. Ang, K.H.; Chong, G.; Li, Y. PID control system analysis, design, and technology. *IEEE Trans. Control Syst. Technol.* **2005**, *13*, 559–576. [CrossRef]
25. Tang, K.-S.; Man, K.F.; Chen, G.; Kwong, S. An optimal fuzzy PID controller. *IEEE Trans. Ind. Electron.* **2001**, *48*, 757–765. [CrossRef]
26. Forrest, S. Genetic algorithms. *ACM Comput. Surv.* **1996**, *28*, 77–80. [CrossRef]
27. Holland, J.H. Genetic algorithms. *Sci. Am.* **1992**, *267*, 66–73. [CrossRef]
28. Sivanandam, S.; Deepa, S. Genetic algorithms. In *Introduction to Genetic Algorithms*; Springer: Berlin/Heidelberg, Germany, 2008; pp. 15–37. [CrossRef]
29. Mahfud, S.; Derouich, A.; El Ouanjli, N.; Mossa, M.A.; Motahhir, S.; El Mahfoud, M.; Al-Sumaiti, A.S. Comparative Study between Cost Functions of Genetic Algorithm Used in Direct Torque Control of a Doubly Fed Induction Motor. *Appl. Sci.* **2022**, *12*, 8717. [CrossRef]
30. Lin, C.-L.; Jan, H.-Y.; Shieh, N.-C. GA-based multiobjective PID control for a linear brushless DC motor. *IEEE/ASME Trans. Mechatron.* **2003**, *8*, 56–65. [CrossRef]

machines

MDPI

Article

Characteristics Analysis and Optimization of the Structural Parameters of PTRB

Shicheng Zheng [1], Yongling Fu [1], Deyi Wang [2], Junlin Pan [1], Ziyu Liu [3] and Juan Chen [1,*]

[1] School of Mechanical Engineering and Automation, Beihang University, Beijing 100191, China
[2] National Key Laboratory of Science and Technology on Aerospace Intelligence Control, Beijing Aerospace Automatic Control Institute, Beijing 100854, China
[3] System Integration General Laboratory, Beijing General Institute of Electronic Engineering, Beijing 100039, China
* Correspondence: chen.juan@buaa.edu.cn

Abstract: Bearings, as integral parts and components in machinery and equipment, play a critical role in system performance. In this paper, the theoretical models of basic axial load rating, basic radial load rating, and friction torque for the planetary thread roller bearing (PTRB) were derived. Then, a detailed comparison was made between the performance indexes of a PTRB prototype and common high-quality bearing products to demonstrate the advantages of PTRB in volume, bearing capacity, and use method. Next, the effects of different structural parameters on the performance indexes of PTRB were analyzed. On this basis, a multi-objective optimization design was carried out on PTRB under axial load and radial load separately with basic load rating and friction torque as optimization objects to obtain PTRB with superior comprehensive performance. Designers could, in consideration of their specific needs and optimized design results, select the structural parameters of PTRB. Furthermore, a dedicated performance test system was researched and developed, which verified the good static bearing capacity and impact resistance of the PTRB prototype. Within the range of dynamic load rating, the test data were slightly larger than the simulation data for the friction torque, manifesting a relatively high degree of precision of the theoretical model.

Keywords: planetary thread roller bearing; load rating; friction torque; characteristics analysis; structure optimization; experimental tests

Citation: Zheng, S.; Fu, Y.; Wang, D.; Pan, J.; Liu, Z.; Chen, J. Characteristics Analysis and Optimization of the Structural Parameters of PTRB. *Machines* **2022**, *10*, 1051. https://doi.org/10.3390/machines10111051

Academic Editors: Kan Liu and Wei Hu

Received: 10 October 2022
Accepted: 6 November 2022
Published: 9 November 2022

1. Introduction

The performance of parts and components has a direct impact on that of machinery and equipment. Advancements in technology and industrial upgrading gave birth to high-performance equipment, which places greater demands on the performance of bearings. Traditional roller bearings, as core parts and components in electro-mechanical systems, have developed a relatively complete research methodology on performance after multiple rounds of iteration and optimization. To obtain bearings with better performance, researchers mainly enhance the performance of roller bearings by technical means such as developing new materials [1–3], reinforcing surface treatment [4–6], improving machining accuracy, or making structural innovations [7].

In terms of structural innovation, Tian developed a kind of sliding–rolling compound bearing by combining rolling friction pairs with sliding friction pairs. The bearing has a relatively high rotating speed and improved bearing capacity and impact resistance. Tian et al. also analyzed the bearing performance through finite element analysis and contrast simulation experiments [8,9]. Lu conducted a theoretical analysis and fexperimental study on the bearing capacity and contact pattern of sliding–rolling bearings. The results demonstrate that sliding–rolling bearings have excellent heavy-load characteristics and good low-rate start capacity; however, their friction torque is significantly greater than traditional roller bearings [10]. Afterward, Lu improved the structure of sliding–rolling bearings by replacing solid and hollow rollers with spiral elastic rollers and analyzed the bearing performance through theoretical calculation and finite element simulation. The results show that the modified structure can further improve the bearing capacity of sliding–rolling bearings [11,12]. Yao et al. presented an elastic composite cylindrical roller bearing by embedding polytetrafluoroethylene (PTFE) material into a hollow cylindrical roller. This kind of bearing has better capacity against the damage caused by bending fatigue because of the reduced bending stress on the roller's inner wall, so the dynamic performance of the bearing is superior to that of a solid bearing. When the degree of filling of the roller is 55%, the bearing has the optimum fatigue life; when the degree of filling is 40% or 50%, the maximum amplitude of the bearing is smaller than that of a solid bearing [13,14]. Md et al. took dynamic capacity and elasto-hydrodynamic minimum film thickness as objective functions, then carried out multi-objective robust optimization along with design variables of bearing pitch diameter, ball diameter, and the number of rolling elements, which definitely enhanced the performance of the deep-groove ball bearing [15]. Pavel used a computational model of lubrication and heat transfer to optimize the thrust bearing. The prototype bearing designed using this approach reduced the mechanical losses of the entire turbocharger by approximately 20%; at the same time, there was no significant decrease in the bearing ability [16].

PTRB is a kind of roller bearing with a new structure [17]. The author has described its outstanding bearing characteristics in another paper [17]. Due to its unique structure and operating principle, it has a larger load rating than a ball bearing of equal volume, which has important research value and application prospects. However, there is still room for improvement in its comprehensive performance. Judging from the contact pattern of the roller with the raceway, PTRB belongs to the ball bearing. Meanwhile, given that its unique annular zero-lead thread structure has similarities to the traditional thread connection, the research methodology adopted by researchers on roller bearings [18–21] and thread connection [22–25] serves as a good reference for the performance analysis of PTRB. However, since PTRB has a novel and complicated structure, theoretical models and structure optimization models specific to the structural characteristics of PTRB should be built based on Hertz contact theory to improve its comprehensive performance further. In addition, a test system is needed to conduct experimental tests on the performance of PTRB.

In Section 2, preliminary work and PTRB's structure are briefly introduced. The theoretical models of basic load rating and friction torque are built in Section 3. In Sections 4 and 5, the PTRB prototype's performance and the effects of structural parameters on the performance indexes of PTRB are analyzed, respectively. A multi-objective optimization design is conducted on PTRB in Section 6 to enhance its comprehensive performance further. In Section 7, a dedicated performance test system is researched and developed to test the friction torque, static bearing capacity, and impact life of the PTRB prototype. Finally, the research results are summarized in Section 8.

2. Introduction to the Structure

Since the operating principle of PTRB has been described in detail in the author's previous paper [17], this paper only gives a brief introduction. As shown in Figure 1, PTRB consists of a threaded roller, an inner ring, an outer ring, a planet carrier, and a steel wire retaining ring. Multiple evenly spaced rings of annular threads (zero-lead threads) are machined on the threaded roller, inner ring, and outer ring. The thread teeth of the inner ring and outer ring are in a plane or double-sided arc form, and the threaded roller's thread teeth are in a double-sided arc form, with an arc radius smaller than the other two. The contact pair formed by mutual contact between the annular thread of the inner ring and that of the threaded roller is defined as an inner ring-threaded roller contact pair (ITP), and the contact pair formed by mutual contact between the annular thread of the outer ring and that of the threaded roller is defined as an outer ring-threaded roller contact pair (OTP). Each contact pair can be regarded as the mutual contact between an equivalent steel ball and a plane or arc surface.

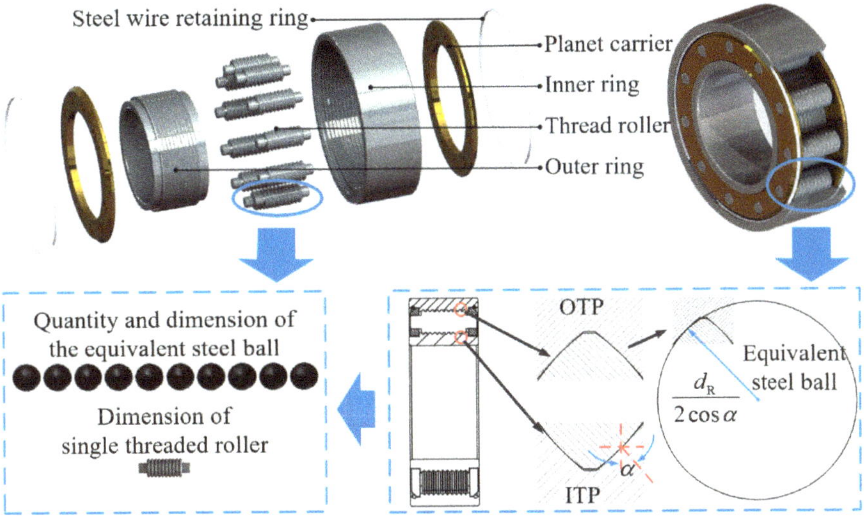

Figure 1. Schematic diagram of PTRB structure composition and equivalent steel ball.

The specific structural parameters of the PTRB prototype studied in this paper are shown in Table 1. The contact radius of threaded roller R_{rr} and the pitch diameter of threaded roller d_R of the PTRB prototype were 5.1 mm and 6 mm, respectively. Each threaded roller was equipped with 10 annular threads. When PTRB was under axial load, each threaded roller could be equivalent to 10 equivalent steel balls with a diameter of $d_R/\cos\alpha$ = 8.5 mm; when PTRB was under radial load, each threaded roller could be equivalent to 20 equivalent steel balls with a diameter of 8.5 mm [17]. The equivalent steel balls of PTRB outstripped those of an identically sized ball bearing in size and quantity, giving PTRB a higher bearing capacity. That is to say, PTRB was smaller than a normal bearing given the same bearing capacity.

Table 1. Specific structural parameters of PTRB prototype.

Structure Parameter	Value	Structure Parameter	Value
Pitch diameter of a threaded roller set D_{pw} (mm)	42.5	Inner diameter of PTRB d_0 (mm)	30
Pitch diameter of threaded roller d_R (mm)	6	Outside diameter of PTRB D_0 (mm)	55
Pitch diameter of inner ring d_I (mm)	36.5	Width H_0 (mm)	18
Pitch diameter of outer ring d_O (mm)	48.5	Number of threaded rollers Z(1)	11
Contact radius of threaded roller R_{rr} (mm)	5.1	Number of thread teeth per roller n(1)	10
Contact radius of inner ring r_I (mm)	∞	Contact angle α (°)	45
Contact radius of outer ring r_O (mm)	∞	Tooth dedendum h (mm)	0.24
Thread bottom thickness of roller B_R (mm)	1.05	Thread ridge thickness s (mm)	0.6
Thread bottom thickness of inner ring B_I (mm)	1.08	Elastic modulus E (MPa)	212,000
Thread bottom thickness of outer ring B_O (mm)	1.08	Poisson's ratio u(1)	0.29
Distance between adjacent threads P (mm)	1.2		

The deformation by force of each main bearing component of PTRB and the calculation method of load distribution have been described in detail in the author's previous paper [23]. It was because of Hertz contact deformation, deformation of the thread teeth, and compressive deformation of the shaft that the load might be unevenly distributed when PTRB is under axial load. The axial load distribution coefficients of ITP and OTP were denoted as f_{Iai} and f_{Oai}, and the maximum axial load distribution coefficients of ITP and OTP were denoted as f_{Iamax} and f_{Oamax}, respectively. When PTRB rotated, the position of the threaded roller kept changing, so the loads exerted on the threaded roller at each position of the distribution circle were different in both direction and value. For this reason, the load might be unevenly distributed when PTRB was under radial load as well. The radial load distribution coefficient of PTRB was denoted as f_{ri}, and the maximum radial load distribution coefficient of PTRB was denoted as f_{rmax} [26,27].

3. Calculation Models of the Performance Parameters

3.1. Theoretical Model of Basic Static Load Rating

According to the requirements of the static load rating of roller bearings in relevant standards, in most applications, a total permanent deformation amount of 0.0001 times the roller's diameter is allowed at the center of the contact point, which will not cause harmful effects in the follow-up running of the bearing. Experiments show that all ball bearings other than the self-aligning ball bearing, including PTRB, would generate 4200 MPa stress at the center of the contact point where the maximum load was applied under the impact of the basic static load rating [28].

According to the Hertz theory, the maximum contact stress of the contact point exists at the contact ellipse center. That is, the contact stress of each contact point of PTRB should satisfy the following conditions:

$$\sigma_{\max} = \frac{3F_{nimax}}{2\pi ab} \leq 4200\text{MPa} \tag{1}$$

where a and b are the long/short semi-axis length of the contact pair, F_{nimax} is the maximum value in the load force of ITP and OTP.

So, the basic static axial load rating of PTRB was [17]:

$$C_{0a} = \frac{Zn\sin\alpha}{f_{amax}}\left\{4200 \times \frac{2\pi}{3} m_{aX} m_{bX} [\frac{3}{(\Sigma\rho)_X}\frac{1-u^2}{E}]^{2/3}\right\}^3 \tag{2}$$

where m_{aX} and m_{bX} is the correlation coefficient of the contact pair, which can be calculated by the contact characteristics. $(\Sigma\rho)_X$ is the sum of principal curvatures of ITP or OTP, Z is the number of threaded rollers, n the number of thread teeth per roller, and f_{amax} is the maximum axial load distribution coefficient. E and u are the elastic modulus and Poisson's ratio of the threaded roller, inner ring, and outer ring, respectively.

GCr15, through quenching and tempering, was used for the main bearing components of PTRB. The chemical composition of GCr15 is shown in Table 2.

Table 2. The chemical composition of GCr15 [29].

Mass Fraction/%						
C	Si	Mn	Cr	Mo	P	S
0.95~1.05	0.15~0.35	0.25~0.45	1.40~1.65	≤0.10	≤0.025	≤0.015

When PTRB was under axial load, the thread teeth of the annular threads were vulnerable spots. To prevent the rupture of the thread teeth at the roots, the strength of tooth roots should meet appropriate requirements. So, a strength check was conducted on the tooth roots of the annular threads based on the strength checking formula for thread connection:

$$\sigma_b = \frac{6hZF_{nimax}\sin\alpha}{\pi d_X B_X^2} \leq [\sigma_b] = 348\text{MPa} \tag{3}$$

where B is the thread bottom thickness, d is the pitch diameter, and X can be R, I, or O. The relevant parameters are listed in Table 1.

The shear strength of the annular thread tooth roots should meet the following formula:

$$\tau_c = \frac{ZF_{nimax}\sin\alpha}{\pi d_X B_X} \leq [\tau_c] = 208.8\text{MPa} \tag{4}$$

The basic static radial load rating of PTRB was:

$$C_{0r} = \frac{2n\cos\alpha}{f_{rmax}}\left\{4200 \times \frac{2\pi}{3} m_{aX} m_{bX} [\frac{3}{(\sum\rho)_X}\frac{1-u^2}{E}]^{2/3}\right\}^3 \tag{5}$$

where f_{rmax} is the maximum radial load distribution coefficient.

When PTRB was under radial load, each bearing thread tooth carried the same normal load on both sides, which offset the bending moment and shear stress at the thread tooth roots. So, there was no need to check the strength of the thread tooth roots.

3.2. Theoretical Model of Basic Dynamic Load Rating

According to the requirements of the dynamic load rating of roller bearings in relevant standards, when the basic rating life is one million revolutions, the constant axial load carried by the roller bearing is the basic dynamic axial load rating, and the constant radial load is the basic dynamic radial load rating. Based on the theory of the maximum dynamic shear stress, Lundberg G and Palmgren A presented the theoretical formulas for calculating basic dynamic axial load rating C_{Da} and basic dynamic radial load rating C_{Dr} of single-row ball bearings, which have been included in the ISO standard [30–32].

C_{Da} and C_{Dr} of PTRB were:

$$C_{Da} = \begin{cases} b_m f_{ca}(\cos\alpha)^{0.7} Z^{\frac{2}{3}} (\frac{d_R}{\cos\alpha})^{1.8} \tan\alpha & , \frac{d_R}{\cos\alpha} \leq 25.4 \\ 3.647 b_m f_{ca}(\cos\alpha)^{0.7} Z^{\frac{2}{3}} (\frac{d_R}{\cos\alpha})^{1.4} \tan\alpha & , \frac{d_R}{\cos\alpha} > 25.4 \end{cases} \tag{6}$$

$$C_{Dr} = \begin{cases} b_m f_{cr}(\cos\alpha)^{0.7} Z^{\frac{2}{3}} (\frac{d_R}{\cos\alpha})^{1.8} & , \frac{d_R}{\cos\alpha} \leq 25.4 \\ 3.647 b_m f_{cr}(\cos\alpha)^{0.7} Z^{\frac{2}{3}} (\frac{d_R}{\cos\alpha})^{1.4} & , \frac{d_R}{\cos\alpha} > 25.4 \end{cases} \tag{7}$$

Dynamic load coefficient f_c was [32]:

$$f_{ca} = 0.9 \times 98.07\left(1 - \frac{\sin\alpha}{3}\right)\frac{\gamma^{0.3}(1-\gamma)^{1.39}}{(1+\gamma)^{\frac{1}{3}}}\left[\frac{2R_{rr}r_I\cos\alpha}{d_R(r_I - R_{rr})}\right]^{0.41}\left[1 + \left(\frac{1-\gamma}{1+\gamma}\right)^{\frac{17.2}{3}}\right]^{-0.3} \tag{8}$$

$$f_{cr} = 0.9 \times 39.91\frac{\gamma^{0.3}(1-\gamma)^{1.39}}{(1+\gamma)^{\frac{1}{3}}}\left[\frac{2R_{rr}r_I\cos\alpha}{d_R(r_I - R_{rr})}\right]^{0.41}\left\{1 + \left[1.04\left(\frac{1-\gamma}{1+\gamma}\right)^{1.72}\right]^{\frac{10}{3}}\right\}^{-0.3} \tag{9}$$

Based on the specific structure and movement relation of PTRB, the relation between structural coefficient γ and stress number of cycles β was:

$$\begin{cases} \beta_I = \frac{Z(D_{pw}+d_R)}{2D_{pw}} = \frac{Z}{2}(1+\gamma) \\ \beta_O = \frac{Z(D_{pw}-d_R)}{2D_{pw}} = \frac{Z}{2}(1-\gamma) \end{cases} \tag{10}$$

The basic dynamic load rating of i rows of common ball bearings that carried single-direction load was:

$$C_P = (Z_1 + \cdots + Z_i)[(\frac{Z_1}{C_{D1}})^{\frac{10}{3}} \cdots + (\frac{Z_i}{C_{Di}})^{\frac{10}{3}}]^{-0.3} = C_D i^{0.7} \tag{11}$$

where C_D is the dynamic load rating of the single thread.

When PTRB was under radial load, the calculation method for multiple rows of common ball bearings could be applied. However, when PTRB was under axial load, the load exerted on different contact pairs of the same threaded roller might be unevenly distributed. Therefore, the basic dynamic axial load rating C_a and the basic dynamic radial load rating C_r were obtained as follows according to the specific structure of PTRB:

$$C_a = f_{amax}^{-1} n^{0.7} C_{Da} \tag{12}$$

$$C_r = (2n)^{0.7} C_{Dr} \tag{13}$$

3.3. Theoretical Models of Friction Torque and Operating Efficiency

Besides the basic load rating, friction torque and operating efficiency are also important performance indexes of the bearing. Based on the friction torque generation mechanism and the methods for calculating the friction torque of traditional ball bearings, the author has derived the formulas for calculating friction torque caused by elastic hysteresis M_e, friction torque caused by spin sliding M_s, friction torque caused by differential sliding M_d, and friction torque caused by viscous resistance of lubricant M_v in another paper [17] that are specific to the operating principle and structure of PTRB. Furthermore, in a sound, dynamic balancing state, no additional force was needed from the planet carrier to correct the position of each threaded roller on the circumference when the threaded roller revolved. There was only a small contact force between the two components, so the threaded roller could drive the planet carrier to rotate smoothly through its optical axes on both sides. The sliding friction caused by the planet carrier and the threaded roller was denoted as M_h [33].

The schematic diagram of each friction torque is shown in Figure 2. Among all friction torques, M_e, M_s, and M_s were determined by the external load, structural performance, and material characteristics; M_h was determined by the structural performance and bearing speed; M_v was not only associated with the external load, structural performance, and material characteristics, but also affected by the bearing speed and lubricant type.

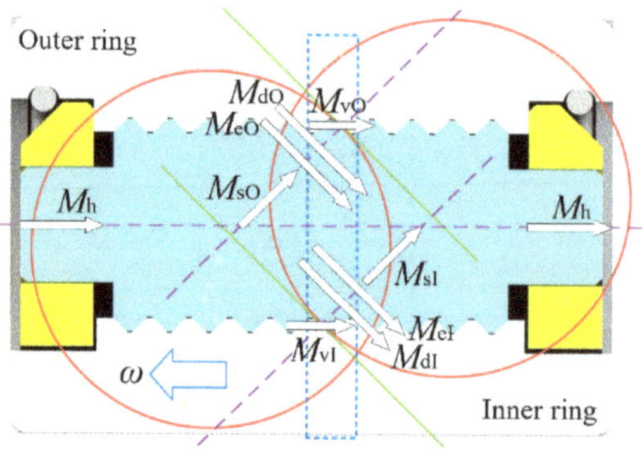

Figure 2. Schematic diagram of each friction torque.

In combination with Figure 2, the friction torques of ITP and OTP, i.e., M_{Ii} and M_{Oi}, were:

$$\begin{cases} M_{Ii} = M_{eIi}\cos\alpha + M_{sIi}\sin\alpha + M_{dIi}\cos\alpha + M_{vIi} \\ M_{Oi} = M_{eOi}\cos\alpha + M_{sOi}\sin\alpha + M_{dOi}\cos\alpha + M_{vOi} \end{cases} \tag{14}$$

When PTRB was under axial load, the total friction torque M_{PTRBa} and the friction coefficient μ_{PTRBa} were:

$$M_{PTRBa} = Z\sum_{i=1}^{n} M_{Ii} + Z\sum_{i=1}^{n} M_{Oi} + M_h \mu_{PTRBa} = \frac{2M_{PTRBa}}{d_0 F_{Pa}} \tag{15}$$

When PTRB was under radial load, the total friction torque M_{PTRBr} and the friction coefficient μ_{PTRBr} were:

$$M_{PTRBr} = 2n\sum_{i=1}^{Z_e} M_{Ii} + 2n\sum_{i=1}^{Z_e} M_{Oi} + M_h\mu_{PTRBr} = \frac{2M_{PTRBr}}{d_0 F_{Pr}} \quad (16)$$

When the torque at the input side was M_D, the operating efficiency of PTRB could be expressed as follows:

$$\eta = (1 - \frac{M_{PTRB}}{M_{PTRB} + M_D}) \times 100\% \quad (17)$$

4. Analysis of the PTRB Prototype's Performance

According to Formulas (1)–(13), for the PTRB prototype, C_{0a} and C_{0r} were 38.36 kN and 19.93 kN, while C_a and C_r were 33.20 kN and 30.26 kN.

The lubricant chosen for PTRB is lubricating oils for total loss systems L-AN. The pressure–viscosity coefficient is 0.0246 mm^2/N, and the dynamic viscosity under normal pressure is 0.04 MPa•s [34]. Under an external load of 0–30 kN and a bearing speed of 100–4000 rpm, the variation of friction torque of the PTRB prototype with external load and bearing speed is shown in Figure 3. The friction torque increased with the increase in external load or bearing speed. Within the range of basic dynamic load rating, M_{PTRBa} ranged from 0 Nm to 1.045 Nm, and M_{PTRBr} ranged from 0 Nm to 1.389 Nm. When the external load was 30 kN, with the bearing speed increasing from 100 rpm to 4000 rpm, M_{PTRBa} increased from 0.715 Nm to 1.045 Nm, and M_{PTRBr} increased from 1.042 Nm to 1.389 Nm.

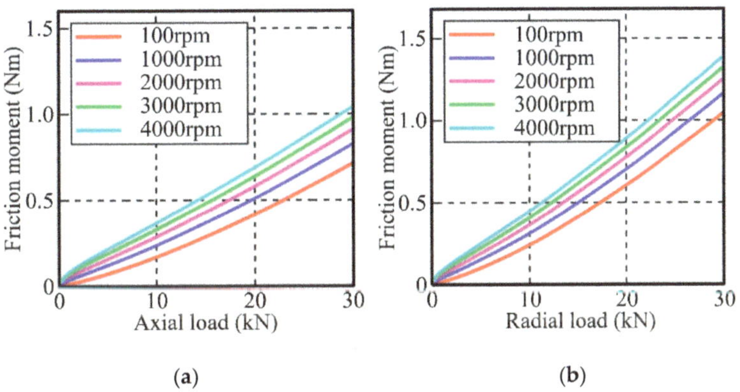

Figure 3. Variation of friction torque of PTRB prototype with external load and bearing speed. (**a**) Variation of friction torque with axial load and bearing speed. (**b**) Variation of friction torque with radial load and bearing speed.

In the machinery field, bearing models are often selected according to the fatigue life of the bearing. When a constant load of 40–100% of the basic dynamic load rating is exerted on the bearing, the normal rotation turns of the bearing range from 15.6 million to 1.0 million, which can meet the requirements of the equipment on the fatigue life of the bearing. Therefore, in this paper, the friction coefficient of the PTRB prototype under an external load of 10–30 kN was analyzed. When the bearing speed was 100–4000 rpm, the variation of friction coefficient of the PTRB prototype with external load and bearing speed is shown in Figure 4. The friction coefficient increased with the increase in bearing speed, wherein μ_{PTRBa} ranged from 0.0011 to 0.0024, and μ_{PTRBr} ranged from 0.0016 to 0.0031.

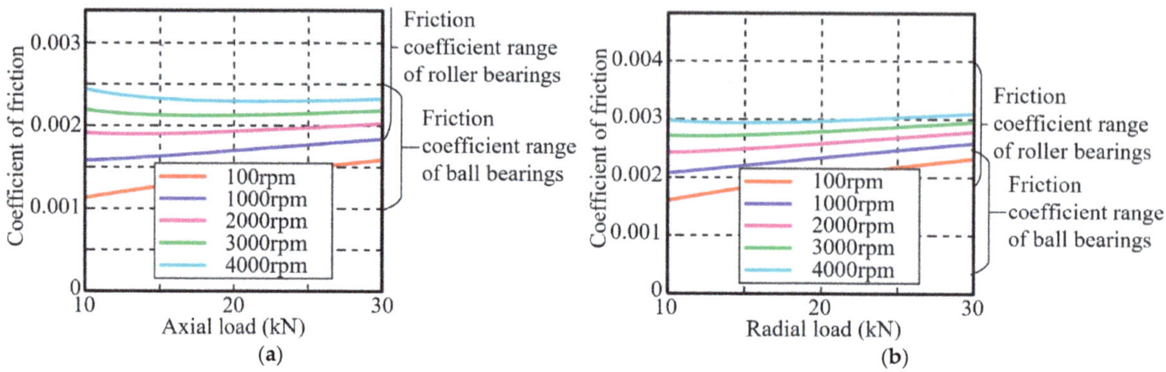

Figure 4. Variation of friction coefficient of PTRB prototype with external load and bearing speed. (**a**) Variation of friction coefficient with axial load and bearing speed. (**b**) Variation of friction coefficient with radial load and bearing speed.

The performance indexes of the PTRB prototype and common high-quality bearing products were compared, as shown in Table 3.

Table 3. Performance indexes of PTRB prototype and common high-quality bearing products [35,36].

Bearing Type	D_0 (mm)	d_0 (mm)	H_0 (mm)	C_{0a} (kN)	C_{0r} (kN)	C_a (kN)	C_r (kN)
PTRB prototype	55	30	18	38.36	19.93	33.20	30.26
Single-row angular contact ball bearing NSK	62	30	16	—	14.80	—	22.50
Double-row angular contact ball bearing NSK	55	30	26	—	20.50	—	24.60
Deep-groove ball bearing SKF	62	30	16	—	12.90	—	23.40
Single-direction thrust ball bearing SKF	52	30	16	51.00	—	25.50	—
Double-direction thrust ball bearing SKF	62	30	34	73.50	—	35.10	—
Single-row tapered roller bearing SKF	62	30	17.25	—	44.00	—	50.00
Single-row tapered roller bearing SKF	62	30	34.5	—	88.00	—	85.70
Cylindrical roller bearing SKF	55	30	13	—	17.30	—	17.90
Unidirectional cylindrical roller thrust bearing SKF	52	30	16	134.0	—	50.00	—

(1) Compared with ball bearings and roller bearings that had similar volume parameters and could carry axial load only, C_a of the PTRB prototype was superior to that of the single-direction thrust ball bearing and the double-direction thrust ball bearing but inferior to that of the unidirectional cylindrical roller thrust bearing. C_{0a} of the PTRB prototype was inferior to that of the above three types of bearings. However, the above three types of bearings could not carry radial load and needed to be used in combination with other types of bearings, thus oversizing the overall bearing assembly. PTRB could carry axial–radial combined load or bidirectional axial load without being used cooperatively with other bearings.

(2) Compared with ball bearings and roller bearings that had similar volume parameters and could carry radial load only, both C_r and C_{0r} of the PTRB prototype were superior to those of the deep-groove ball bearing and the cylindrical roller bearing, and the PTRB prototype was also able to carry axial load.

(3) Compared with ball bearings that had similar volume parameters and could carry axial–radial combined load, the basic static and dynamic load ratings of the PTRB prototype

were superior to those of the single-row angular contact ball bearing and the double-row angular contact ball bearing.

(4) Compared with roller bearings that had similar volume parameters and could carry axial–radial combined load, the basic static and dynamic load ratings of the PTRB prototype were inferior to those of the single-row tapered roller bearing and the double-row tapered roller bearing. However, the single-row tapered roller bearing should be used in combination with other bearings, and the double-row tapered roller bearing had a large axial dimension.

Among the common high-quality bearing products in Table 3, the friction coefficient of ball bearings ranged from 0.0010 to 0.0025, and the friction coefficient of roller bearings ranged from 0.0018 to 0.0040. The friction coefficient range of the PTRB prototype under axial load was 0.0011–0.0024, falling into the friction coefficient range of ball bearings. The friction coefficient range of the PTRB prototype under radial load was 0.0016–0.0031, falling into the friction coefficient range of ball and roller bearings.

According to the above analysis, the PTRB prototype had certain advantages over the common high-quality bearing products, such as small volume, high bearing capacity, no need to be used in pairs or combinations, and the ability to carry axial–radial combined load and bidirectional axial load. The PTRB with optimized structural parameters would be equipped with better performance and more outstanding advantages.

5. Effects of Structural Parameters on the Performance Indexes of PTRB

The effects of different structural parameters on the performance of the PTRB prototype were simulated and analyzed based on the theoretical models of load rating and friction torque. Conditions provided for the simulation and analysis are as follows: the dimension was unchanged, the axial load or radial load was 20 kN, the bearing speed was 4000 rpm, and the shear strength safety coefficient of the thread tooth roots was 2.

The value range of each structural parameter was set as follows:

(1) Since PTRB can carry both axial load and radial load, the value range of the contact angle was set to 30–60° by referring to the angular contact ball bearing.

(2) With the dimension of PTRB being constant, the number of thread teeth per roller was inversely proportional to the distance between adjacent threads. Therefore, a large number of thread teeth would narrow the distance between adjacent threads, resulting in high manufacturing costs and the risk of sliding wire in the bearing process; and a small number of thread teeth would reduce the load rating of PTRB. In light of the above, the value range of the number of thread teeth per roller was set to 8–16, and the corresponding distance between adjacent threads was set to 1.6–0.8 mm.

(3) To ensure that the contacting dimension of the contact ellipse meets the Hertz theory, the contact radius of the threaded roller should be larger than $d_R/2\cos\alpha$, and the maximum value should be one order of magnitude less than the contact radius of the inner and outer rings. For ease of analysis, the value range of the contact radius of the threaded roller was set to 4.5–7.5 mm.

(4) With the dimension of PTRB being constant, the pitch diameter of the threaded roller was associated with that of the inner ring and outer ring. The larger the pitch diameter of the threaded roller was, the smaller the pitch diameter of the inner ring and outer ring was. Considering the pitch diameter of the threaded roller should be no larger than $2R_{rr}\cos\alpha$ required by the Hertz theory and the strength requirement, the value range of the pitch diameter of the threaded roller was set to 4.8–7.2 mm.

(5) The maximum number of threaded rollers was determined by the installation method of PTRB [23], and the minimum value was 3. Given that a small number of threaded rollers might weaken the operating stability and bearing capacity of PTRB, the value range of the number of threaded rollers was set to 7–11.

The parameter improvement of PTRB can be classified into three types: load rating increases while friction torque keeps unchanged or slightly increases; friction torque decreases while load rating keeps unchanged or slightly decreases; load rating increases while friction torque decreases.

Through simulation analysis, the performance indexes of PTRB monotonically increased or decreased with the increase in the structural parameter values. Due to space limitations, the effect of structural parameters on performance indexes was expressed in figures and tables in this paper. The effect of different structural parameters on load rating is shown in Figure 5 and Table 4, and the effect of different structural parameters on friction torque is shown in Figure 6 and Table 5.

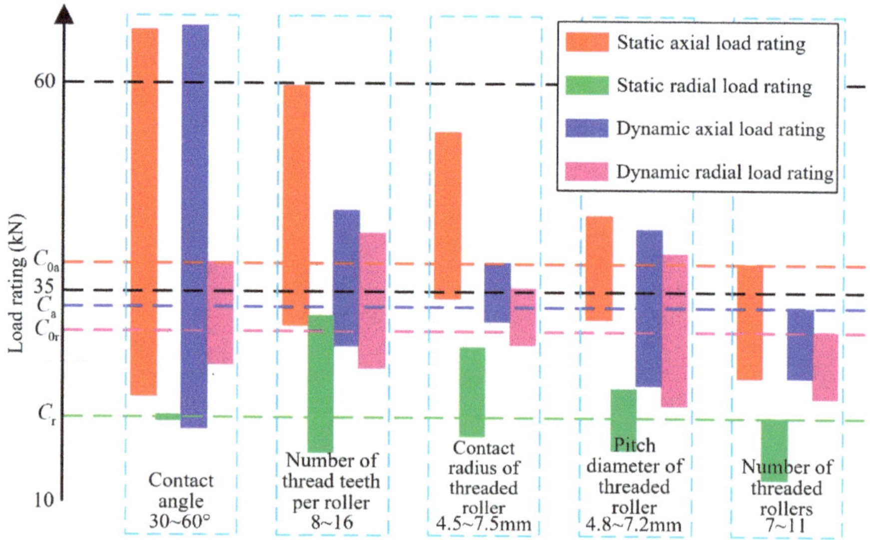

Figure 5. Influence of different structural parameters on load rating.

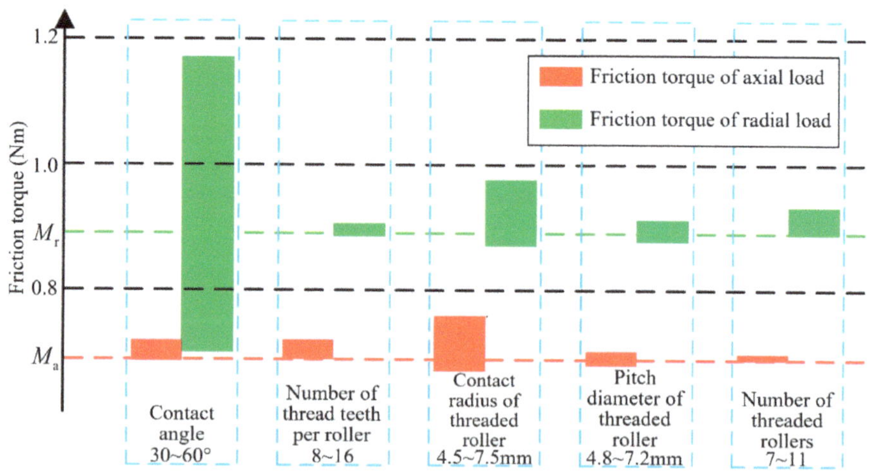

Figure 6. Influence of different structural parameters on friction torque.

Table 4. Influence of different structural parameters on load rating.

Structure Parameter	Value of Parameter	Static Axial Load Rating (kN)	Static Radial Load Rating (kN)	Dynamic Axial Load Rating (kN)	Dynamic Radial Load Rating (kN)
Contact angle	30~60°	22.49~66.27	19.60~20.19	18.63~66.79	26.31~38.44
Number of thread teeth per roller	8~16	30.98~59.65	15.87~32.14	28.53~44.75	25.89~42.05
Contact radius of threaded roller	4.5~7.5 mm	34.24~54.05	17.72~28.40	31.52~38.48	28.79~35.42
Pitch diameter of threaded roller	4.8~7.2 mm	31.77~44.12	16.16~23.41	23.88~42.50	21.45~39.62
Number of threaded rollers	7~11	24.85~38.36	12.67~19.93	24.87~33.07	22.39~30.26

Table 5. Influence of different structural parameters on friction torque.

Structure Parameter	Value of Parameter	Friction Torque of Axial Load (Nm)	Friction Torque of Radial Load (Nm)
Contact angle	30–60°	0.688~0.719 (↑)	0.701~1.170 (↑)
Number of thread teeth per roller	8~16	0.690~0.711 (↑)	0.906~0.886 (↓)
Contact radius of threaded roller	4.5~7.5 mm	0.672~0.760 (↑)	0.874~0.974 (↑)
Pitch diameter of threaded roller	4.8~7.2 mm	0.681~0.703 (↑)	0.878~0.912 (↑)
Number of threaded rollers	7~11	0.699~0.689 (↓)	0.932~0.890 (↓)

It could be seen from the analysis of figures and tables that:

(1) With the increase in the structural parameter values, the static and dynamic load ratings increased, leaving space for improvement, wherein the variation of the contact angle had the most significant effect on C_{0a} and C_a, the variation of the number of thread teeth per roller had the most significant effect on C_{0r}, and the variation of the pitch diameter of the threaded roller had the most significant effect on C_r.

(2) The friction torque increased with the increase in the structural parameter values, with exceptions that M_{PTRBa} decreased with the increase in the number of threaded rollers and M_{PTRBr} decreased with the increase in the number of threaded rollers or the number of thread teeth per roller. The variation of the contact angle had the most significant effect on M_{PTRBr}, and the effect of other structural parameters on friction torque was less than 0.1 Nm.

(3) With the increase in the number of threaded rollers, the friction torque decreased, and the static and dynamic load ratings increased. Assuming that the mass was neglected, the maximum number of threaded rollers should be taken to obtain the optimum comprehensive performance.

(4) With the increase in four parameters, i.e., contact angle, number of thread teeth per roller, contact radius of the threaded roller, and pitch diameter of the threaded roller, the static load rating and the dynamic load rating increased, and so did the friction torque, causing conflict in the performance improvement of PTRB.

(5) Beyond the structural parameters given in Tables 4 and 5, the variation of the pitch diameter of a threaded roller set had a minimal effect on load rating and friction torque, which could be neglected.

6. PTRB's Parameter Optimization

Targeting the problem that there was a conflict between the variation of the structural parameters and the comprehensive performance improvement of PTRB, a multi-objective optimization design was conducted on PTRB under 20 kN and 4000 rpm in this section.

The maximum number of threaded rollers was always taken, and the contact angle, number of thread teeth per roller, contact radius of the threaded roller, and pitch diameter of the threaded roller were deemed as design variables:

$$x = [\alpha, n, R_{rr}, d_R]^T \tag{18}$$

The maximum load rating and the minimum friction torque were set as optimization objects:

$$\begin{cases} f_{a1}(x) = -C_{0a} \\ f_{a2}(x) = -C_a \\ f_{a3}(x) = M_a \end{cases} \quad \text{or} \quad \begin{cases} f_{r1}(x) = -C_{0r} \\ f_{r2}(x) = -C_r \\ f_{r3}(x) = M_r \end{cases} \tag{19}$$

The constraint conditions and value ranges were defined according to Section 5, so the details are not described in this section.

Through the multi-objective optimization design, the axial performance optimization result of PTRB is shown on the left side of Figure 7, and the radial performance optimization result of PTRB is shown on the right side of Figure 7. The friction torque increased with the increase in load rating. Two non-inferior solutions were selected from the two figures in Figure 7, and the structural parameters and performance indexes are listed in Table 6.

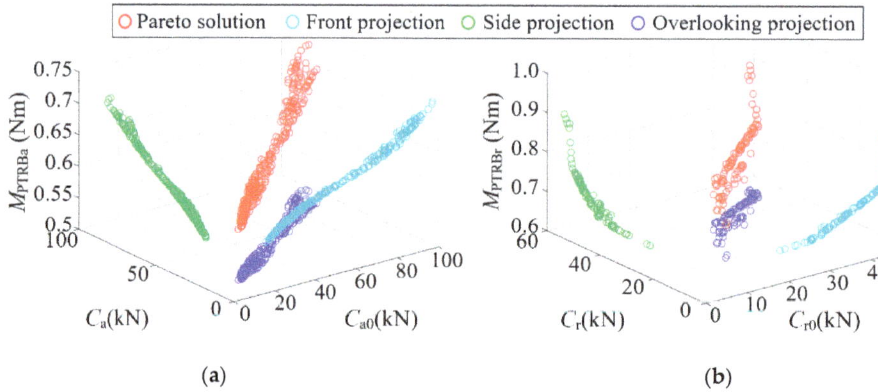

Figure 7. Axial and radial performance optimization results of PTRB. (**a**) Axial performance optimization results. (**b**) Radial performance optimization results.

Table 6. Structure parameter and Performance index of PTRB optimization results.

Structure Parameter	x_{a1}	x_{a2}	x_{r1}	x_{r2}	PTRB Prototype	Performance Index	x_{a1}	x_{a2}	x_{r1}	x_{r2}	PTRB Prototype
P (mm)	0.83	0.95	0.83	0.92	1.2	C_{0a} (kN)	97.4	40.9	96.1	38.4	38.36
d_R (mm)	6.96	5.92	7.06	6.37	6.0	C_a (kN)	78.9	37.1	75.7	33.3	33.20
R_{rr} (mm)	6.64	4.51	7.50	4.54	5.1	μ_{PTRBa} (10^{-3})	2.4	2.1	2.5	2.1	2.3
α (°)	53.5	47.6	52.5	39.3	45	C_{0r} (kN)	40.4	20.0	44.8	24.5	19.93
n(1)	15	11	15	13	10	C_r (kN)	59.0	31.1	53.4	35.6	30.26
						μ_{PTRBr} (10^{-3})	3.6	3.0	3.1	2.6	3.0

Comparative analysis suggests that taking the x_{a1} point where the load rating was large among the axial performance optimization results, though μ_{PTRBa} at x_{a1} was 4% larger than that of the PTRB prototype, its axial and radial load ratings were twice those of the PTRB prototype. Similarly, taking the x_{r1} point where the load rating was large among the radial performance optimization results, though μ_{PTRBr} at x_{r1} was 3% larger than that of the PTRB prototype, its axial and radial load ratings were also twice those of the PTRB prototype. Such performance optimization result was suitable for the equipment having a high requirement for the bearing capacity. At the points x_{a2} and x_{r2}, where the friction torque and friction coefficient were small, though the load rating was slightly increased compared with the PTRB prototype, μ_{PTRBa} and μ_{PTRBr} were lowered by 9% and 13%, respectively. Such performance optimization result was suitable for the equipment having a high requirement for the friction coefficient of the bearing.

Therefore, the designers could, in consideration of their specific needs and optimized design results, select the structural parameters of PTRB in a targeted manner to enhance the design efficiency.

7. PTRB's Performance Test

7.1. Research and Development of a Dedicated Performance Test System

A dedicated performance test system was researched and developed according to the theoretical calculating data and relevant research for testing the performance of the PTRB prototype. The system could provide, either in an axial or radial direction of PTRB, a constant force of no less than 50 kN, an impact force of no less than 35 kN, a rotating speed of 4000 rpm, and a sweep-frequency motion with a rotating angle of ±30° and a frequency of 2 Hz. As shown in Figure 8, the mechanical part of the performance test system includes a motor, load components, friction torque, a radial life test assembly, an axial life test assembly, a mounting bracket, and a heating/cooling assembly, wherein the heating/cooling assembly was an externally connected independent air heater/cooler, which was used to heat or cool the PTRB by outputting hot or cool air to maintain the temperature of PTRB at 20 °C–60 °C. The material object is not shown in the figure.

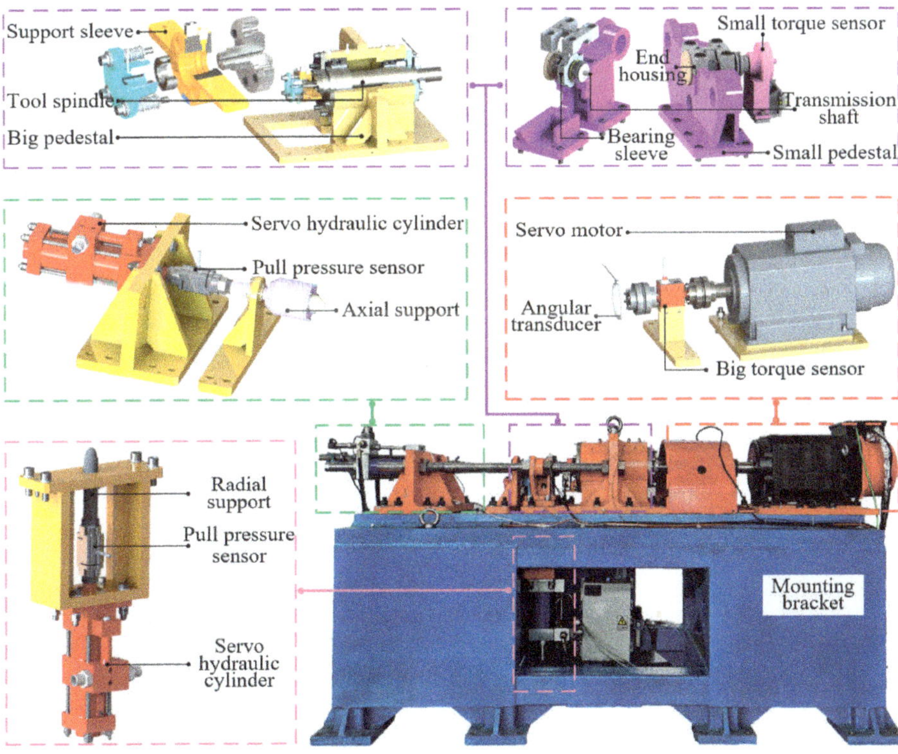

Figure 8. Mechanical scheme of dedicated performance test system.

The test and control part of the performance test system includes motion control, signal acquisition, and data processing. Since the motion complexity of the servo motor and the servo-hydraulic cylinder were not highly demanded in the test, the closed-loop control method is not described in detail hereunder. As for signal acquisition and data processing, firstly, the data acquisition and control system was used to acquire the temperature, tension and pressure, torque, rotating speed, and acceleration signal, followed by the data reading, processing, and storage of the upper computer. Finally, the results were displayed in the upper computer interface programmed by LabVIEW in tables, average values, and curves.

The control system and upper computer interface are shown in Figure 9. The real-time interface of the upper computer is used to display the shock pulse value of PTRB in real time and monitor its health status. It also displays the PTRB friction torque, the load applied to PTRB, and the motor rotation speed in real time, as well as the outer ring temperature. If the PTRB temperature, system friction torque, or load data are abnormal, the buzzer alarm will be triggered. The lookup interface of the upper computer is used to check the test data and images of the temperature sensor, tension and pressure sensor, acceleration sensor, torque sensor, and angle sensor in any time period.

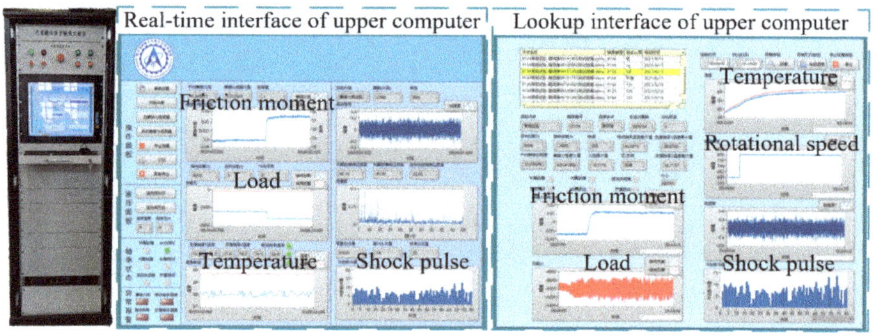

Figure 9. Control system and upper computer interface.

7.2. Performance Test and Analysis of the PTRB Prototype

7.2.1. Preparations

Before the test, all components of PTRB were cleaned and dried with kerosene to ensure the cleanness of PTRB and the frock clamp, as shown in Figure 10. After the assembly of PTRB, the inner and outer rings were rotated manually. If the rotation was smooth without being stuck, the bearing could be defined as a test object, and then a small amount of grease was evenly added, or oil was dropped for lubrication during the test; otherwise, the bearing could not be tested. Then, the inner ring of PTRB was installed on the transmission shaft, and the outer ring of PTRB was installed in the frock clamp, both in an interference fit manner. Running-in was conducted on PTRB under 2 kN, 500 rpm, and 10,000 r. Relatively stable and accurate test data should be guaranteed during the test process.

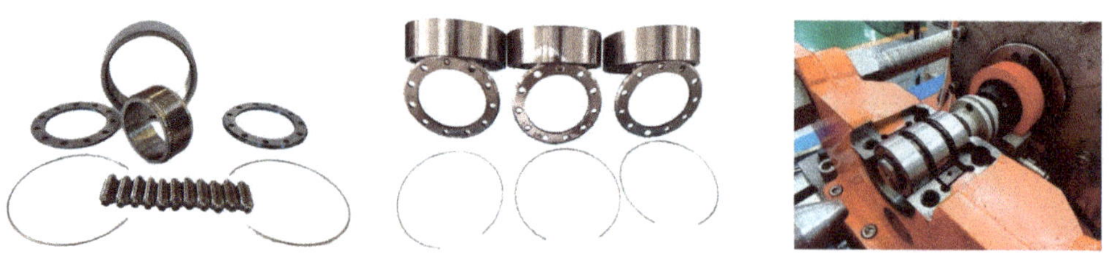

(**a**) (**b**)

Figure 10. Cleaning and installation of PTRB. (**a**) Cleaning of PTRB. (**b**) Installation of PTRB.

7.2.2. Friction Torque Test

The temperature of PTRB was maintained at 40 °C using the heating/cooling assembly. Based on this, a friction torque test was carried out on the PTRB prototype under an external load of 5 kN, a bearing speed of 800 rpm, and a single test duration of 60 s. The M_{PTRBa} and M_{PTRBr} curves obtained are shown in Figure 11, respectively. By averaging the test values in a stable segment after repeated measurements, the M_{PTRBa} and M_{PTRBr} of the PTRB prototype obtained in this state were 0.146 Nm and 0.167 Nm, respectively.

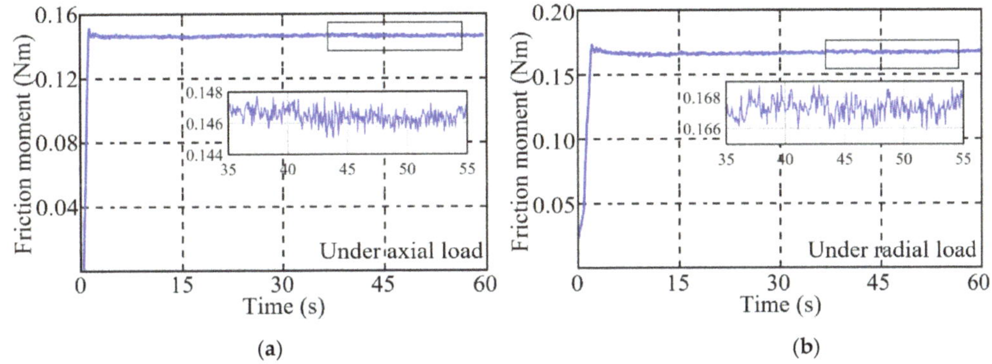

Figure 11. Friction torque curve under axial load and radial load. (**a**) Friction torque curve under axial load. (**b**) Friction torque curve under radial load.

To analyze the effect of the variation of external load on the friction torque of PTRB, a friction torque test was carried out on the PTRB prototype under an external load of 1–30 kN and a bearing speed of 600 rpm. The value of the friction torque in a stable segment was averaged after repeated measurements, and the simulation and test data are listed in Table 7. The test data were fitted and compared with the theoretical simulation results, as shown in Figure 12. Under the same bearing speed, the variation trend of the test data and the simulation data with the external load was consistent, and an approximately linear increase was seen in friction torque with the increase in external load. Under any external load, the test data were slightly greater than the simulation data, with a deviation of no more than 0.1 Nm. Under an external load of 20 kN and a bearing speed of 600 rpm, μ_{PTRBa} and μ_{PTRBr} were 0.0019 and 0.0025, both falling into the range of theoretical calculation.

Table 7. Effect of load on friction torque.

Load (kN)		**1**	**5**	**10**	**15**	**20**	**25**	**30**
Friction moment under axial load (Nm)	test data	0.058	0.143	0.257	0.408	0.573	0.723	0.885
	simulation data	0.028	0.103	0.212	0.338	0.476	0.626	0.786
	deviation	0.030	0.040	0.045	0.070	0.097	0.097	0.099
Friction moment under radial load (Nm)	test data	0.061	0.163	0.339	0.536	0.738	0.977	1.196
	simulation data	0.032	0.132	0.286	0.465	0.667	0.885	1.121
	deviation	0.029	0.031	0.053	0.071	0.071	0.092	0.075

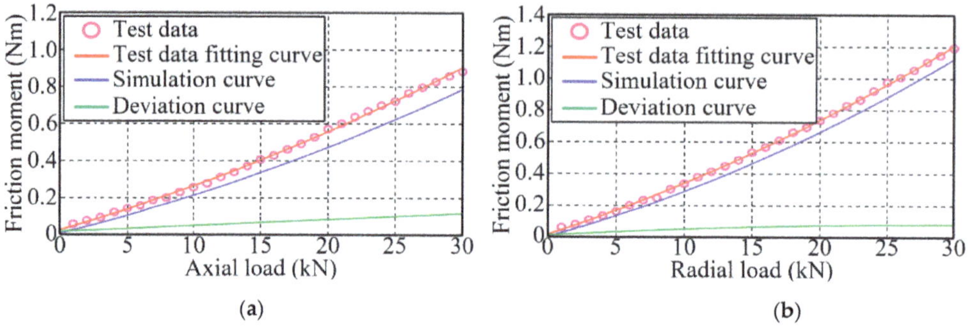

Figure 12. Effect of load on friction torque. (**a**) Effect of axial load on friction torque. (**b**) Effect of radial load on friction torque.

To analyze the effect of the variation of bearing speed on the friction torque of PTRB, a friction torque test was carried out on the PTRB prototype under an external load of 2 kN/5 kN/8 kN and a bearing speed of 600 rpm–2000 rpm. The simulation data and the test data are listed in Table 8. A comparison was conducted between the test data and the simulation data, as shown in Figures 13 and 14. Under the same external load, the variation trend of the test data and the simulation data with the bearing speed was consistent, and an approximately linear increase was seen in friction torque with the increase in bearing speed. Under any bearing speed and external load, the test data were slightly greater than the simulation data, with a deviation of no more than 0.05 Nm.

7.2.3. Static Bearing Capacity Test

The shock pulse method is a common way to judge the health status of roller bearings. As shown in Figure 15, when the shock pulse ranged from 0 to 20 dB, it indicates that the bearing was running in a healthy state; when the shock pulse ranged from 20 to 35 dB, it indicates that the bearing suffered from early damage; when the shock pulse ranged from 35 to 60 dB, it indicates that the bearing was obviously damaged [37–39].

Table 8. Effect of rotating speed on friction torque.

Rotational Speed (rpm)		**600**	**800**	**1000**	**1200**	**1400**	**1600**	**1800**	**2000**
Friction moment under axial load (Nm)	test data (2 kN)	0.076	0.081	0.083	0.086	0.091	0.094	0.098	0.101
	simulation data (2 kN)	0.046	0.052	0.057	0.063	0.067	0.072	0.077	0.081
	test data (5 kN)	0.143	0.146	0.154	0.159	0.166	0.170	0.175	0.182
	simulation data (5 kN)	0.103	0.112	0.121	0.129	0.136	0.143	0.150	0.157
	test data (8 kN)	0.215	0.224	0.233	0.238	0.250	0.256	0.264	0.273
	simulation data (8 kN)	0.167	0.178	0.189	0.199	0.208	0.217	0.226	0.234
Friction moment under radial load (Nm)	test data (2 kN)	0.078	0.082	0.087	0.091	0.095	0.100	0.103	0.108
	simulation data (2 kN)	0.055	0.061	0.067	0.072	0.077	0.082	0.087	0.091
	test data (5 kN)	0.161	0.167	0.176	0.182	0.187	0.194	0.202	0.208
	simulation data (5 kN)	0.132	0.142	0.151	0.159	0.167	0.174	0.181	0.188
	test data (8 kN)	0.249	0.255	0.266	0.275	0.282	0.288	0.300	0.308
	simulation data (8 kN)	0.221	0.233	0.244	0.255	0.264	0.274	0.283	0.291

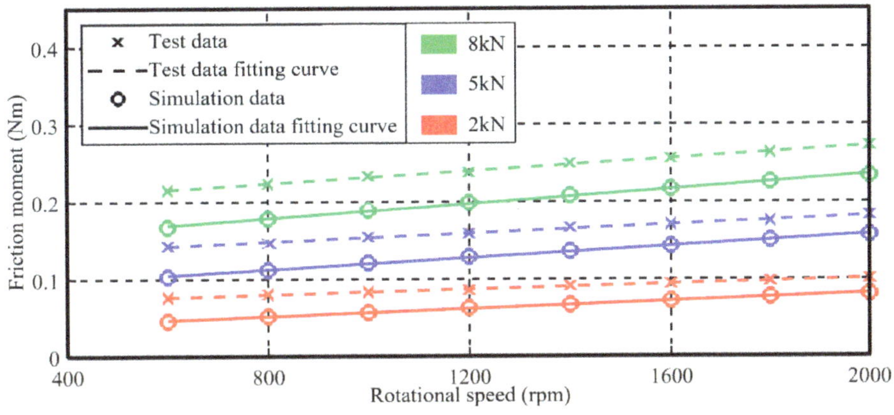

Figure 13. Influence of bearing speed on friction torque under axial load.

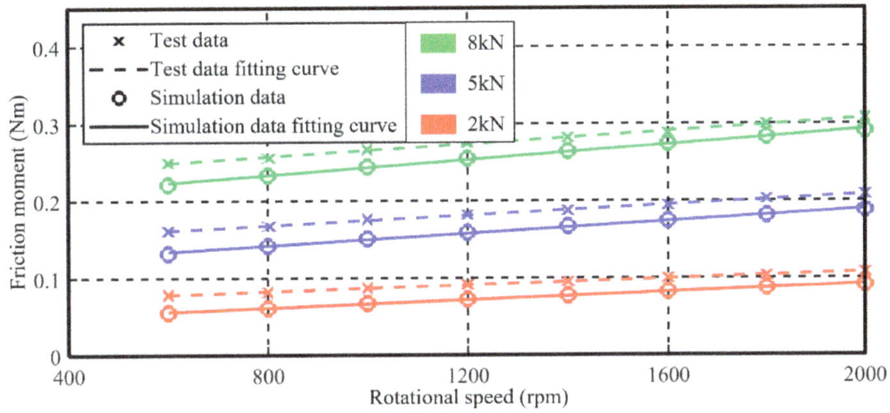

Figure 14. Influence of bearing speed on friction torque under radial load.

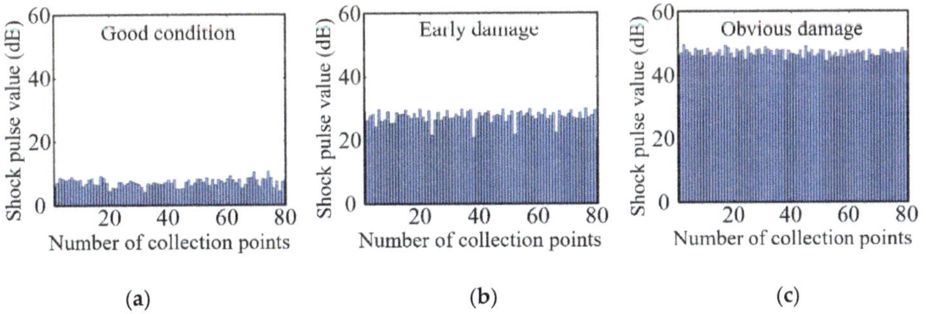

Figure 15. Relationship between shock pulse value and bearing health status. (**a**) Good condition. (**b**) Early damage. (**c**) Obvious damage.

The shock pulse method was used to judge the health status of the PTRB prototype in this paper. With the bearing speed maintained at 0, a static bearing capacity test was carried out on the PTRB prototype under a single test duration of 3 h and an axial load of 38 kN or a radial load of 20 kN. After a single test, a friction torque test was conducted on the PTRB prototype under an external load of 5 kN and a bearing speed of 800 rpm. Meanwhile, the shock pulse value of the bearing was monitored.

After multiple tests, the friction torque and the shock pulse value of the PTRB prototype had no significant changes, meeting the requirements of the relevant standards on the basic static load rating. That is, the basic static axial load rating of the PTRB prototype was no less than 38 kN, and the basic static radial load rating was no less than 20 kN.

7.2.4. Impact Life Test

It is known that the ball bearing of the same size as the PTRB prototype cannot satisfy the requirements of the large mountain walking machine platform on impact life. For testing the impact resistance of the PTRB prototype, an impact life test was conducted on the PTRB prototype under an impact load of 20 kN, a rotating angle of $\pm 30°$, a frequency of 2 Hz, and 100,000 impact times. Before and after the impact life test, friction torque tests were conducted on the PTRB prototype under an external load of 10 kN and a bearing speed of 600 rpm. Meanwhile, the shock pulse value of the bearing was monitored.

Multiple measurement results show that before the impact life test, M_{PTRBa} and M_{PTRBr} were 0.257 Nm and 0.339 Nm, and the shock pulse value was approximately 10 dB. After the impact life test, a slight increase was seen in the dynamic friction torque compared with that before the test, M_{PTRBa} and M_{PTRBr} were 0.329 Nm and 0.416 Nm, and the shock pulse value was approximately 27 dB. It indicates that the bearing suffered from early damage rather than complete damage. Hence, the PTRB prototype exhibited an impact resistance superior to the common high-quality bearing products and could preliminarily meet the requirements of the large mountain walking machine platform.

To sum up, within the range of dynamic load rating, the test data were slightly larger than the simulation data for friction torque, with a deviation of no more than 0.1 Nm, manifesting a relatively high degree of precision of the theoretical model. The deviation might be attributed to machining error and nonlinear friction torque. When the bearing was required to carry a dynamic load of 20 kN-30 kN, the system would obtain great driving force, and the deviation of 0.1 Nm would have a minimal effect on the system efficiency. When the PTRB prototype was used as an electro-mechanical actuator in combination with a planetary roller screw pair of the same power level, the efficiency was nearly 97%, and the deviation was no more than 0.5%. Though the PTRB prototype had a slightly larger friction torque than the ball bearing, it exhibited better static bearing capacity and impact resistance life. In light of the good comprehensive performance of the PTRB prototype, the researchers could conduct a targeted optimization design according to their specific needs to obtain better comprehensive performance.

8. Conclusions

In this paper, an all-around study was carried out on PTRB with a unique structure and strong bearing capacity. The main efforts and important conclusions can be summarized as follows:

1. Based on the contact characteristics and the theoretical model of load distribution, the theoretical models of load rating and friction torque for PTRB under axial load or radial load were derived, respectively, according to the relevant standards. Then, an analysis was conducted on the performance of the PTRB prototype, with C_{0a}, C_a, C_{0r}, and C_r being 38.36 kN, 33.20 kN, 19.93 kN, and 30.26 kN. Under an external load of 0 kN–30 kN and a bearing speed of 100 rpm–4000 rpm, μ_{PTRBa} and μ_{PTRBr} ranged from 0.0011 to 0.0023 and 0.0016 to 0.0031, respectively. The PTRB prototype had certain advantages over the common high-quality bearing products, such as small volume, high bearing capacity, no need to be used in pairs or combinations, and the ability to carry axial–radial combined load and bidirectional axial load.
2. The effects of different structural parameters on the performance of PTRB were simulated and analyzed with the dimension being constant. Within the value range of the structural parameters, the basic load rating and the friction torque of PTRB monotonically increased or decreased with the increase in the structural parameter values, and there was a certain space for improvement for the basic load rating. The

maximum number of threaded rollers was taken to obtain the optimum comprehensive performance, and the effect of the variation of the pitch diameter of a threaded roller set on PTRB performance could be neglected. There was a conflict between the variation of four parameters and the performance improvement of PTRB, i.e., contact angle, number of thread teeth per roller, contact radius of threaded roller, and pitch diameter of threaded roller.

3. A multi-objective optimization design was carried out on PTRB under axial load and radial load separately with basic load rating and friction torque as optimization objects to improve its comprehensive performance further. When the bearing capacity of PTRB was highly demanded, the optimization results selected could improve the axial and radial load ratings to 200% of the PTRB prototype, and the increase in friction torque was no more than 4%. When the friction coefficient of PTRB was required to be small, the optimization results selected could reduce μ_{PTRBa} and μ_{PTRBr} by 9% and 13%, respectively, and also slightly increase the load rating. The multi-objective optimization and the selection of optimization results could effectively improve the design efficiency of PTRB.
4. A dedicated performance test system was built to verify $C_{0a} \nless 38$ kN and $C_{0r} \nless 20$ kN of the PTRB prototype. Moreover, the PTRB prototype could meet the requirement of the large mountain walking machine platform on the PTRB's impact life of more than 100,000 times under an impact load of 20 kN. Within the range of load rating, the deviation between the test data and the simulation data for the friction torque was less than 0.1 Nm. When PTRB was used in combination with a planetary roller screw pair of the same power level, the efficiency was nearly 97%, and the deviation was no more than 0.5%. The performance of the PTRB prototype and the correctness of the theoretical model were verified through tests. In the follow-up work, further optimization should be carried out on PTRB according to the specific needs to raise the comprehensive performance of PTRB to a higher level.

Author Contributions: Methodology, S.Z. and J.C.; Software, S.Z. and J.P.; Validation, Y.F. and D.W.; Investigation, S.Z. and Z.L.; Data curation, D.W., Y.F. and Z.L.; Writing—original draft preparation, S.Z. and D.W.; Writing—review and editing, S.Z. and J.C.; Supervision, Y.F.; project administration, Y.F.; All authors have read and agreed to the published version of the manuscript.

Funding: This research received no external funding.

Data Availability Statement: Not applicable.

Conflicts of Interest: The authors declare no conflict of interest.

References

1. Zhou, X.C.; Zhou, F.; Huang, J.; Ji, J.; Zhong, M.J.; Yan, Z.M.; Zheng, B. Friction and wear properties of a new plastic rubber water-lubricated bearing material. In Proceedings of the 2017 4th International Conference on Transportation Information and Safety (ICTIS), Banff, AB, Canada, 8–10 August 2017; pp. 1022–1027.
2. Wang, F.F.; Zhou, C.G.; Zheng, L.J.; Zhang, H. Improvement of the corrosion and tribological properties of CSS-42L aerospace bearing steel using carbon ion implantation. *Appl. Surf. Sci.* **2017**, *392*, 305–311. [CrossRef]
3. Gu, L.; Wang, L.Q.; Qi, Y.L.; Su, W.; Qian, Z.D.; Zheng, Y.Q. Study on Performance of ultra-low temperature and High Speed Hybrid Ceramic Bearing. *Harbin Gongye Daxue Xuebao* **2004**, *2*, 157–159.
4. Zhang, C.W. Study on Preparation and Bearing Characteristics of Solid Lubricating Coating for Ceramic Ball Bearing. Master's Thesis, Harbin Institute of Technology, Shenzhen, China, 2009.
5. Stewart, S.; Ahmed, R. Rolling Contact Fatigue of Surface Coatings—A Review. *Wear* **2002**, *253*, 1132–1144. [CrossRef]
6. Zaretsky, E.V. Rolling bearing steels—A technical and historical perspective. *Mater. Sci. Tech-Lond.* **2012**, *28*, 58–69. [CrossRef]
7. Lars, K.; Matthias, M.; Rainer, S.; Christoph, B.; Oliver, H.; Ludwig, S. Static and Dynamic Behavior Of A Superconducting Magnetic Bearing Using Ybco Bulk Material. *IEEE. Trans. Appl. Supercond.* **2007**, *17*, 2079–2082.
8. Tian, H.P. Roller Sliding Composite Bearing. 1999. Available online: https://d.wanfangdata.com.cn/patent/ChJQYXRlbnROZXdTMjAyMjAzMjMSDENOOTkxMTY0OTAuMxoIbzZhNmwxZHg%3D (accessed on 1 September 2022).
9. Tian, H.P.; Zhang, F.R.; Yang, C.M. Research on Compound Bearing of Sliding and Rolling. *Prospect. Eng.* **2003**, *1*, 60–62.
10. Lu, L.M.; Li, F.; Li, Z.H. Research on Friction Torque of Roller Bearing Based on ADAMS. *J. Mech. Electr. Eng.* **2020**, *37*, 1026–1031.

11. Lu, L.M.; Qin, Y.J. Bearing Performance Analysis of Spiral Elastic Cylindrical Roller of Roller Bearing. *Nanchang Univ. Xuebao* **2014**, *36*, 46–51.
12. Lu, L.M.; Qin, Y.J. Finite Element Analysis of Dynamic Characteristics of Roller Bearing using ABAQUS. *Mod. Manuf. Eng.* **2015**, *5*, 7–11.
13. Yao, Q.S.; Yang, W.; Yu, J.H.; Zhang, R. Structural Design of Elastic Composite Cylindrical Roller Bearing. *China Mech. Eng.* **2012**, *23*, 2899–2902.
14. Yao, Q.S.; Xiang, L.; Li, C.; Yu, J.H. Dynamic Characteristics Analysis of a New Type Cylindrical Roller Bearing. *Vib. Eng. Xuebao* **2020**, *33*, 734–741.
15. Md, S.A.; Rajiv, T.; Twinkle, M. Multi-Objective Robust Optimization of Deep Groove Ball Bearings Considering Manufacturing Tolerances Based on Fatigue and Wear Considerations. *J. Tribol-T. ASME* **2022**, *144*, 022301.
16. Pavel, N.; Martin, J.; Jiri, V. Evolutionary Optimisation of the Thrust Bearing Considering Multiple Operating Conditions in Turbomachinery. *Int. J. Mech. Sci.* **2021**, *195*, 106240.
17. Zheng, S.C.; Fu, Y.L.; Wang, D.Y.; Pan, J.L.; Li, L.J.; Wang, J.B. A Novel Planetary Thread Roller Bearing: Design and Analysis of Load Characteristic. *J. Mech. Design.* **2021**, *143*, 064501. [CrossRef]
18. Rauert, T.; Herrmann, J.; Dalhoff, P.; Sander, M. Fretting fatigue induced surface cracks under shrink fitted main bearings in wind turbine rotor shafts. *Procedia Struct. Integr.* **2016**, *2*, 3601–3609. [CrossRef]
19. Yang, K.; Wang, Y.W.; Ma, C.Y.; Zhang, G.; Ren, Z.M.; Zhang, Z.Q. Influence of Structural Design Parameters on Fatigue Life of High-Speed Iron Bearing. *Ind. Contrl. Comput.* **2019**, *32*, 108–110+113.
20. Zhang, J.H.; Fang, B.; Hong, J.; Zhu, Y.S. Effect of preload on ball-raceway contact state and fatigue life of angular contact ball bearing. *Tribol. Int.* **2017**, *114*, 365–372. [CrossRef]
21. Pan, X.J.; Dong, S.J.; Mu, S.F.; Zhao, X.X. Study on Load Distribution and Fatigue Life of Flexible Bearing in Harmonic Reducer under Assembly Deformation. *Jixie Qiangdu* **2020**, *42*, 453–458.
22. Blinov, D.S.; Kolobov, A.Y.; Dikun, E.V. On Unevenness of Load Distribution between Mating Thread Ridges of Parts of Roller Screw Gears. *AIP Conf. Proc.* **2021**, *2318*, 140002.
23. Huang, H.Y.; Wang, K.Y.; Zhao, P.H. Simulation and Analysis of Variable Pitch Thread Based on Workbench. *Coal Mine Mach.* **2016**, *37*, 162–165.
24. Zhou, X.H.; Sun, Y.S.; Zhang, E.W. Analysis of axial load distribution of drive screw based on finite element method. *Mach. Des. Manuf.* **2008**, *1*, 16–18.
25. Wang, W.; Marshek, K.M. Determination of load distribution in a threaded connector with yielding threads. *Mech. Mach. Theory.* **1996**, *31*, 229–244. [CrossRef]
26. Zheng, S.C.; Fu, Y.L.; Pan, J.L.; Li, L.J.; Wang, D.Y. An in-depth study on load distribution characteristics of the planetary threaded roller bearing. *Math. Probl. Eng.* **2020**, *2020*, 9593051. [CrossRef]
27. *GB/T 192-2003/ISO 68-1:1998*; Basic Tooth Shape of Common Thread. Standardization Press of China: Beijing, China, 2003.
28. *GB/T 4662-2003/ISO 76:1987*; Rated Static Load of Rolling Bearing. Standardization Press of China: Beijing, China, 2003.
29. *GB/T 18254-2016*; High-Carbon Chromium Bearing Steel. Standardization Press of China: Beijing, China, 2017.
30. *GB/T 6391-2010/ISO 281:2007*; Rated Dynamic Load and Rated Life of Rolling Bearing. Standardization Press of China: Beijing, China, 2010.
31. *ISO/TR 1281-1:2008*; Rolling Breaings—Explanatory Notes on ISO 281—Part 1: Basic Dynamic Load Rating and Basic Rating Life. International Organization for Standardization: Geneva, Switzerland, 2008.
32. Wan, C.S. *Analysis Method of Rolling Bearing*; China Machine Pres: Beijing, China, 1987; pp. 44–298.
33. Gao, C. Dynamic Prediction and Thermal Characteristics Analysis of Friction Torque of Grease Lubricated Angular Contact Ball Bearings. Master's Thesis, Lanzhou Ligong University, Lanzhou, China, 2016.
34. *GB 443-89*; Lubricating Oils for Total Loss Systems L-AN. Standardization Press of China: Beijing, China, 1989.
35. SKF. Product Samples of Ball Bearings. 2022. Available online: https://www.skf.com/cn/products/rolling-bearings (accessed on 1 October 2022).
36. NSK. Product Samples of Ball Bearings. 2022. Available online: https://www.nsk.com/products/ballbearing/ (accessed on 1 October 2022).
37. Xue, G.H.; Zhao, X.Y.; Ji, X.D. Research on Condition Monitoring of Rolling Bearing Based on Impact Pulse Method. *Coal Eng.* **2013**, *45*, 119–121.
38. Yang, C.H.; Qian, Y.X. Application of Vibration Monitoring of Rolling Bearing in Equipment Management. *Manag. Technol. SME* **2015**, *7*, 165.
39. Liu, Z. Experimental Study on the Performance of Planetary Threaded Roller Bearing. Master's Thesis, Beihang University, Beijing, China, 2022.

electronics

MDPI

Article

Application of Hardware-in-the-Loop Simulation Technology in the Development of Electro-Hydraulic Servo System Control Algorithms

Quan Liang *,†, Jun Gao ‡, Feihong Liu ‡, Kelei Wang, Haiyang Zhang, Zhike Wang and Donghai Su

Mechanical Engineering Department, Shenyang University of Technology, Shenyang 110870, China
* Correspondence: liangquan@sut.edu.cn; Tel.: +86-180-0402-6017
† Current address: Shenyang Economic and Technological Development Zone, 111 Shenliao West Road, Shenyang 110870, China.
‡ These authors contributed equally to this work.

Abstract: In this paper, we present a method to identify paramaters for controlling electro-hydraulic servo systems in a real-time environment. With the aim of addressing the problem that it is difficult to accurately obtain the state space equation parameters of the physical entity of the electro-hydraulic servo system, we introduce an online identification theory (recursive least squares method) for identifying said parameters of the state space model in a valve-controlled symmetrical cylinder system. After accurately obtaining the parameters of the system, nonlinear control of the valve-controlled symmetrical cylinder system is carried out using a backstepping algorithm. In order to verify the actual effect of the online identification algorithm and backstepping algorithm, a hardware-in-the-loop (HIL) simulation platform for the valve-controlled symmetrical cylinder system is built in a Linux real-time system, and the real-time performance of the system is evaluated, which demonstrates that the platform can be reliably applied for subsequent system identification and backstepping verification. The results of the HIL simulation test demonstrate that the online identification algorithm and backstepping control method developed in this paper are effective and reliable.

Keywords: electro-hydraulic servo system; system identification; backstepping control; hardware-in-the-loop simulation

Citation: Liang, Q.; Gao, J.; Liu, F.; Wang, K.; Zhang, H.; Wang, Z.; Su, D. Application of Hardware-in-the-Loop Simulation Technology in the Development of Electro-Hydraulic Servo System Control Algorithms. *Electronics* **2022**, *11*, 3850. https://doi.org/10.3390/electronics11233850

Academic Editors: Kan Liu and Wei Hu

Received: 27 October 2022
Accepted: 12 November 2022
Published: 22 November 2022

1. Introduction

Electro-hydraulic servo (EHS) systems and EHS actuators form one of the fundamental drive systems applied in industrial processes and engineering practices. Compared with electrical servo systems, EHS systems typically have higher power to weight ratios, fast and smooth start-up characteristics, and higher power storage capabilities [1,2]

The ability of EHS actuators to perform their powerful control functions is heavily dependent on the control algorithms. Typically, EHS systems, or the mechanisms they drive, exhibit strong nonlinearity [3–6]. Thus, nonlinear control algorithms are particularly suitable for EHS actuators [7–14], of which the performance of the backstepping method is the most prominent [15–19]. However, the backstepping method is highly dependent on the dynamic model of the controlled system and its parameters, which are usually difficult to obtain accurately and requires the use of adaptive control algorithms [20–28] or identification methods to obtain the exact parameters of the system [14].

In the past few decades, numerous scholars have carried out research work focused on parameter identification and nonlinear control algorithms for EHS systems. In [29], some practical issues concerning the identification of EHS actuators are discussed, with the use of discrete time linear models. Hydraulic wind power transfer systems exhibit a highly nonlinear dynamic influenced by system actuator hysteresis and disturbances from wind speed and load torque. A system identification approach to approximate such a nonlinear dynamic is presented in [30]. Wos [31] presents selected issues of position force control of EHS system using adaptive methods. This kind of measure extends the capabilities of control systems that only use position measurements. In [32], experimental work is presented on nonrecursive identification of an EHS actuator system that is represented by a discrete time model in open-loop configuration. A least squares method is used to estimate the unknown parameters of the system based on autoregression with an exogenous input (ARX) model. Jin [33] proposes a novel double-layered network scheme for accurate and efficient identification of hydraulic cylinder. Paper [34,35] presents experimental online identification of an electro-hydraulic system represented by a discrete time model. A recursive least squares (RLS) method is used to estimate the unknown parameters of the system based on autoregression using an exogenous input (ARX) model. Paper [36] presents the system identification process conducted on an industrial EHS actuator system. Ghazali [37] presents system identification of EHS actuator system using a nonrecursive or offline technique, where the mathematical model is determined first by considering fixed supply pressure and load. Lee [38] proposed a system identification algorithm of an EHS for a feedforward position control method. Cologni [39] proposed a global parametric identification procedure that has been performed to identify all unknown parameters.

The common point of the above studies is that, on the one hand, the model parameters of the electro-hydraulic servo system are obtained by means of the system identification algorithm; on the other hand, the parameters obtained by the system identification are applied to the control algorithm in order to obtain improved performance.

However, since both the system identification algorithm and the backstepping control method are complex, debugging the algorithm directly on the physical entity of the controlled object is risky and inefficient. Therefore, this paper makes use of hardware-in-the-loop (HIL) simulation technology to debug the system identification algorithm and the backstepping control algorithm in order to remedy the problem of the difficult development and debugging of control algorithms. Past studies on HIL simulation of EHS have focused on electro-hydraulic braking systems [40–46]. In addition, there are examples in the literature of HIL simulation for testing of electro-hydraulic fuel control unit in a jet engine application [47]. However, none of the aforementioned articles describes the method of building a HIL simulation platform in a Linux real-time environment, nor do they describe the debugging of system identification and control algorithms with the help of a HIL simulation platform. In this paper, HIL simulation technology is applied to the development and debugging of an online identification algorithm and backstepping control algorithm, which reduces the difficulty and risk associated with programs development and improves the efficiency.

The paper then introduces the basic principle of system identification and the control algorithm combined with nonlinear backstepping control and the method for its implementation; then, the nonlinear backstepping control algorithm for valve-controlled symmetric cylinder (VCSC) is derived; finally, the method of building the HIL simulation platform based on a Linux real-time system is introduced, and the identification and backstepping control method is carried out using this method. Finally, the HIL simulation platform based on Linux real-time system is presented and used to conduct identification and backstepping control experiments.

2. Electro-Hydraulic Servo System

The HIL simulation technique is reliant on the accurate mathematical description of the system dynamic characteristics model. The object of HIL simulation in this paper is an EHS system. Therefore, we first introduce the method for derivation of the mathematical model of the EHS system.

The typical actuator of the EHS system is the servo hydraulic cylinder or motor, where the former mainly executes the linear motion, and the latter mainly executes the rotary servo motion. Compared with a linear electrical motor, the former has higher power density ratio and faster response speed, so it is irreplaceable in some instances.

From a representative point of view, a valve-controlled double piston rod hydraulic cylinder is selected as a typical model for introduction, and the dynamic model of the valve-controlled hydraulic motor and cylinder is similar. In this study, it is assumed that the EHS cylinder is vertically mounted with the inertial load placed at one end of the piston rod, and a schematic diagram is shown in Figure 1. This EHS system mainly consists of a servo valve and a hydraulic cylinder and also includes pressure sensors for measuring displacement, external load, and two chambers of the hydraulic cylinder. The data obtained from these sensors will be used in the system identification and backstepping control algorithm presented in this paper.

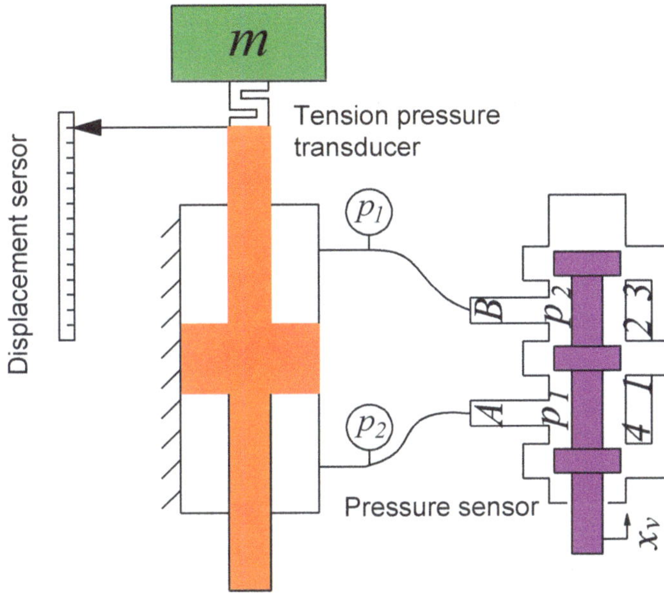

Figure 1. Schematic diagram of electro-hydraulic servo system for VCSC.

2.1. Servo Valve

A servo valve is the most important control element in the EHS system, which amplifies the electrical signal and converts it into high power density hydraulic energy to drive the actuator to externally perform the work. The dynamic characteristics of the servo valve largely affect the overall performance of the EHS system, for which it is usually difficult to build a detailed mathematical model. For simplicity, the model of servo valve in this study is considered as first-order system, that is, the spool valve displacement x_v is related to the voltage input u by a first-order system given by

$$\dot{x}_v = -\frac{1}{\tau_v}x_v + \frac{k_v}{\tau_v}u \tag{1}$$

where τ_v and k_v are the time constant and gain of the servo valve respectively.

2.2. Load Model

In this paper, it is assumed that the EHS cylinder drives the inertial load to complete the motion, and the EHS cylinder and its load are vertically arranged. This way, the system will always be subjected to an external force in the vertical direction equal to the gravitational force applied to the inertial load, so it is necessary to establish the differential equation of the dynamic characteristics of the inertial load. According to Newton's second law, the dynamics of the inertia load can be described by [48]

$$m\ddot{x} = p_L A_c - b_c \dot{x} - k_c x - F_L \tag{2}$$

where x and m represent the displacement and the mass of the load, respectively. $p_L = p_1 - p_2$ is the load pressure of the cylinder, p_1 and p_2 are the left and right chamber pressure, respectively, A_c is the active area of the cylinder, and b_c represents the combined coefficient of the modeled damping and viscous friction forces on the load and the cylinder rod. k_c is the spring stiffness of the external load, which in this case is mainly determined by the elastic deformation of the tension pressure sensor and the spring stiffness of the attached mechanical components. F_L is the external load force. In this paper, since the hydraulic cylinder is assumed to be vertically arranged, the external load force is gravity, that is, $F_L = mg$, where g is the acceleration due to gravity.

2.3. Actuator

Not ignoring the effect of leakage flows in the cylinder and servo valve, the actuator (or the cylinder) dynamics can be written as [48,49]

$$\frac{V_t}{4\beta_e}\dot{p}_L = -A_c\dot{x} - C_l p_L + Q_L \tag{3}$$

where V_t is the total volume of the two-chamber cylinder and hoses between it and the servo valve, β_e is the effective bulk modulus, C_l is the coefficient of the total internal leakage of the cylinder due to pressure, and Q_L is the load flow. Q_L is related to the spool valve displacement of the servo valve by

$$Q_L = C_d w x_v \sqrt{\frac{p_s - \text{sgn}(x_v) p_L}{\rho}} \tag{4}$$

where C_d is the discharge coefficient, w is the spool valve area gradient for cylindrical slide valves, $w = \pi d_v$, d_v is the spool diameter, and p_s is the supply pressure of the system.

In the load flow formula, the function sgn$(*)$ is used to distinguish the positive and negative directions of the hydraulic cylinder movement sign function. Its value is

$$\text{sgn}(*) = \begin{cases} 1, * \geq 0 \\ -1, * < 0 \end{cases} \tag{5}$$

However, the existence of this function causes the load flow formula to be discontinuous, which will cause the mathematical problem of derivative discontinuity, whether for system identification or nonlinear control. Therefore, without loss of generality, the above discontinuous function is replaced by the hyperbolic tangent [49], which is defined as

$$\text{sgn}(*) = \tanh(k_a *) \tag{6}$$

where k_a is a proportional coefficient, which absolutely determines the slope of the hyperbolic tangent function.

Define the state variables $x = [x_1, x_2, x_3, x_4]^T \triangleq [x, \dot{x}, p_L, x_v]^T$. The system can be expressed in state space form as

$$\begin{aligned}
\dot{x}_1 &= x_2 \\
\dot{x}_2 &= -\frac{k_c}{m}x_1 - \frac{b_c}{m}x_2 + \frac{A_c}{m}x_3 - \frac{1}{m}F_L \\
\dot{x}_3 &= -\frac{4\beta_e A_c}{V_t}x_2 - \frac{4\beta_e C_l}{V_t}x_3 + \frac{4\beta_e}{V_t}C_d w\sqrt{\frac{1}{\rho}}\sqrt{|p_s - \tanh(k_a x_4)x_3|}x_4 \\
\dot{x}_4 &= -\frac{1}{\tau_v} + \frac{k_v}{\tau_v}u \\
y &= x_1
\end{aligned} \tag{7}$$

The above mathematical model is characterized by the use of function tanh instead of function sgn such that the discontinuous mathematical model is transformed into a continuous model. The physical characteristics of the valve-controlled cylinder system are correctly simulated, which provides the precondition for subsequent development of the HIL simulation system.

3. System Identification

To obtain increased performance of EHS systems, it is usually necessary to accurately obtain the model system parameters. Due to the common EHS actuator operating mainly through the throttle action of the servo valve to achieve servo control, the mathematical relationship between the pressure difference before and after the throttle orifice of the servo valve and the flow through the servo valve is nonlinear (square root). At the same time, the EHS system parameters, such as the elastic modulus of the oil and the flow coefficient of the throttle port, are easily affected by external factors and time-varying, which makes the dynamic model of the system and its key parameters difficult to obtain under normal circumstances.

For some control algorithms, such as PID control, it is not necessary to obtain accurate parameters of the system; for other algorithms, such as the nonlinear backstepping method, accurate parameters of the system are usually required to obtain better control results than with the PID control effect. More importantly, for the ordinary PID algorithm, when the parameters of the controlled system change due to some reason, the control algorithm has no ability to adapt to this change. If the parameters of system can be obtained online in real time, even if some parameters change, the control algorithm can adapt to this and then achieve the best control effect.

Therefore, the EHS system is dependent on the following two factors to obtain a good control effect:

- To be able to obtain accurate mathematical model and parameters of the system;
- if there is a change in parameters over time, it is necessary to obtain the changing parameter values in real time to modify the control algorithm.

Successfully addressing the above two factors depends on the system identification technology.

The purpose of system identification is to calculate the unknown parameters of the system by measuring the input and output data of the dynamic characteristics of the model with the help of mathematical algorithms. System identification is usually divided into offline and online. The advantage of the former is that the algorithm is simple. The disadvantage of the latter is the need to acquire a sufficient set of experiment test data of the system, which may require a long time and large storage space. More importantly,to achieve a better control effect, the parameter values obtained by offline identification can only be applied to the control algorithm after the identification work is completed. The online identification technology can obtain the key parameters of the system online and in real time such that the control algorithm relying on the dynamic characteristics of the system parameters can achieve better control results, such as in the case of the nonlinear method. Online identification saves time in data collection and improves the model accuracy and reliability. Moreover, any changes in the system structure and components will be taken into consideration in the model without the need for data collection.

Online identification can be defined as an algorithm that estimates dynamic mathematical model parameters using data measured in a real-time environment. In this paper, the dynamic model is the estimated state space function of the electro-hydraulic servo system.

The most basic method of system identification is the least squares method. The problem to be solved by least squares is to find $\theta_i, i = 1, \ldots, n$ according a series of data for known $x_i, i = 1, \ldots, n$ and y , i.e.,

$$\mathbf{y} = [x_1, \ldots, x_n] \begin{bmatrix} \theta_1 \\ \vdots \\ \theta_n \end{bmatrix}$$

written as a matrix of the form

$$\begin{bmatrix} y_1 \\ \vdots \\ y_k \end{bmatrix} = \begin{bmatrix} \phi_1^T \\ \vdots \\ \phi_k^T \end{bmatrix} \boldsymbol{\theta} \tag{8}$$

where

$$\phi_i^T = \begin{bmatrix} x_1^i & \cdots & x_n^i \end{bmatrix} \in \mathbb{R}^{1\times n}$$
$$\boldsymbol{\theta} = \begin{bmatrix} \theta_1 \\ \vdots \\ \theta_n \end{bmatrix} \in \mathbb{R}^{n\times 1}$$

The right-hand side of the Equation (8) represents input observations for the ith set of data, and the left-hand side represents the output observations for the ith set of data, where k represents the data with k sets of observations.

If we define

$$\boldsymbol{\Phi}_k = \begin{bmatrix} \phi_1^T \\ \vdots \\ \phi_k^T \end{bmatrix} \in \mathbb{R}^{k\times n}, \mathbf{Y}_k = \begin{bmatrix} y_1 \\ \vdots \\ y_k \end{bmatrix} \in \mathbb{R}^{k\times 1}$$

then the algorithm of least squares is

$$\hat{\boldsymbol{\theta}}_k = \left(\boldsymbol{\Phi}_k^T \boldsymbol{\Phi}_k\right)^{-1} \boldsymbol{\Phi}_k^T \mathbf{Y}_k \tag{9}$$

If the data are continuously transferred online, it is quite resource and memory consuming to keep using least squares for solving, so a recursive form is needed to ensure the online updating of $\hat{\boldsymbol{\theta}}_k$. The formula for $\hat{\boldsymbol{\theta}}_k$ is provided in the literature [34]

$$\begin{aligned} \varepsilon_k &= y_k - \phi_k^T \hat{\boldsymbol{\theta}}_{k-1} \\ \hat{\boldsymbol{\theta}}_k &= \hat{\boldsymbol{\theta}}_{k-1} + K_k \varepsilon_k \\ \mathbf{K}_k &= \mathbf{P}_k \phi_k \\ \mathbf{P}_k &= \mathbf{P}_{k-1} - \frac{\mathbf{P}_{k-1}\phi_k \phi_k^T \mathbf{P}_{k-1}}{1 + \phi_k^T \mathbf{P}_{k-1}\phi_k} \end{aligned} \tag{10}$$

In this paper, Equation (10) is used for online system identification. The state space expression for VCSC was derived in the previous section. For online system identification, the state space equation of the system is rewritten as

$$\begin{aligned} \dot{x}_1 &= x_2 \\ \dot{x}_2 &= -\theta_{21}x_1 - \theta_{22}x_2 - \theta_{23}F_L + \theta_{24}x_3 \\ \dot{x}_3 &= -\theta_{31}x_2 - \theta_{32}x_3 + \theta_{33}\sqrt{|p_s - \tanh(k_a x_4)x_3|}\,x_4 \\ \dot{x}_4 &= -\theta_{41}x_4 + \theta_{42}u \end{aligned} \tag{11}$$

where

$$\begin{aligned} &\theta_{21} = \tfrac{k_c}{m}, \theta_{22} = \tfrac{b_c}{m}, \theta_{23} = \tfrac{1}{m}, \theta_{24} = \tfrac{A_c}{m} \\ &\theta_{31} = \tfrac{4\beta_e A_c}{V_t}, \theta_{32} = \tfrac{4\beta_e C_l}{V_t}, \theta_{33} = \tfrac{4\beta_e}{V_t} C_d w \sqrt{\tfrac{1}{\rho}} \\ &\theta_{41} = \tfrac{1}{\tau_v}, \theta_{42} = \tfrac{k_v}{\tau_v} \end{aligned} \tag{12}$$

We assume that we do not know the parameters of the dynamic characteristics of the servo valve; therefore, parameter identification is conducted for the whole system, and the matrix form of the state space equations of the system that has to be identified is

$$\begin{bmatrix} \dot{x}_2 \\ \dot{x}_3 \\ \dot{x}_4 \end{bmatrix} = \begin{bmatrix} -x_1 & -x_2 & -F_L & x_3 & 0 & 0 & 0 & 0 & 0 \\ 0 & 0 & 0 & 0 & -x_2 & -x_3 & \sqrt{|p_s - \tanh(k_a x_4)x_3|}x_4 & 0 & 0 \\ 0 & 0 & 0 & 0 & 0 & 0 & 0 & -x_4 & u \end{bmatrix} \begin{bmatrix} \theta_{21} \\ \theta_{22} \\ \theta_{23} \\ \theta_{24} \\ \theta_{31} \\ \theta_{32} \\ \theta_{33} \\ \theta_{41} \\ \theta_{42} \end{bmatrix}$$

The above dynamic characteristic equation is rewritten as

$$\mathbf{y} = \mathbf{\Psi}^T \mathbf{\Theta} \tag{13}$$

where

$$\begin{aligned} \mathbf{y} &= \begin{bmatrix} \dot{x}_2 & \dot{x}_3 & \dot{x}_4 \end{bmatrix}^T \\ \mathbf{\Psi}^T &= \begin{bmatrix} -x_1 & -x_2 & -F_L & x_3 & 0 & 0 & 0 & 0 & 0 \\ 0 & 0 & 0 & 0 & -x_2 & -x_3 & \sqrt{|p_s - \tanh(k_a x_4)x_3|}x_4 & 0 & 0 \\ 0 & 0 & 0 & 0 & 0 & 0 & 0 & -x_4 & u \end{bmatrix} \\ \mathbf{\Theta} &= \begin{bmatrix} \theta_{21} & \theta_{22} & \theta_{23} & \theta_{24} & \theta_{31} & \theta_{32} & \theta_{33} & \theta_{41} & \theta_{42} \end{bmatrix} \end{aligned}$$

The corresponding online identification algorithm can be developed based on Equations (10) and (13).

4. Nonlinear Backstepping Control Method

As can be seen from Equation (7), the state space expression is a nonlinear system, and the backstepping method is needed to obtain a better control effect. The basic principle of the backstepping method is to construct a virtual control quantity equal to the order according to the order of the state space equation, that is, to construct feedback linearization for each subequation of the state space equation such that the nonlinear equation becomes linear and closed-loop control is performed. If more accurate parameter values of system state space expression can be obtained, the control effect of backstepping method is better than that of a traditional control strategy, such as PID.

When the servo valve is considered as the first inertia link, the state space expression of the SCSC system is the fourth order, so the backstepping control algorithm should be carried out in four steps.

The first step of backstepping. Set $i = 1$, define

$$e_1 = x_1 - x_{1,d}$$

Take the derivative of the above equation, then

$$\begin{aligned} \dot{e}_1 &= \dot{x}_1 - \dot{x}_{1,d} \\ &= x_2 - \dot{x}_{1,d} \end{aligned}$$

Define a Lyapunov function as

$$V_1 = e_1^2/2$$

and the derivative of the Lyapunov function is

$$\dot{V}_1 = e_1\dot{e}_1 = e_1(x_2 - \dot{x}_{1,d})$$

when

$$x_2 = x_{2,d} = \dot{x}_{1,d} - k_1 e_1$$

The derivative of the Lyapunov function is less than 0, i.e.,

$$\dot{V}_1 = e_1\dot{e}_1 = e_1(x_2 - \dot{x}_{1,d}) = -k_1 e_1^2 < 0$$

The first equation of state space is stable.
The second step of backstepping. Set $i = 2$, define

$$e_2 = x_2 - x_{2,d}$$

Take the derivative of the above equation, then

$$\begin{aligned} \dot{e}_2 &= \dot{x}_2 - \dot{x}_{2,d} \\ &= -\theta_1 x_2 - \theta_2 F_L + \theta_3 x_3 - \dot{x}_{2,d} \end{aligned}$$

Define Lyapunov function as

$$V_2 = V_1 + e_2^2/2$$

The derivative of the above Lyapunov function is

$$\begin{aligned} \dot{V}_2 &= \dot{V}_1 + e_2\dot{e}_2 \\ &= \dot{V}_1 + e_2[-\theta_1 x_2 - \theta_2 F_L + \theta_3 x_3 - \dot{x}_{2,d}] \end{aligned}$$

if

$$x_3 = x_{3,d} = \frac{-k_2 e_2 + \dot{x}_{2,d} + \theta_1 x_2 + \theta_2 F_L}{\theta_3}$$

then

$$\begin{aligned} \dot{V}_2 &= \dot{V}_1 + e_2\dot{e}_2 \\ &= \dot{V}_1 + e_2[-\theta_1 x_2 - \theta_2 F_L + \theta_3 x_3 - \dot{x}_{2,d}] \\ &= -k_1 e_1^2 - k_2 e_2^2 < 0 \end{aligned}$$

The first and second equation of state space are stable.
The third step of backstepping. Set $i = 3$, define

$$e_3 = x_3 - x_{3,d}$$

The derivative of the above equation is

$$\begin{aligned} \dot{e}_3 &= \dot{x}_3 - \dot{x}_{3,d} \\ &= -\theta_4 x_2 - \theta_5 x_3 + \theta_6\sqrt{|p_s - \tanh(k_a x_4) x_3|}x_4 - \dot{x}_{3,d} \end{aligned}$$

Define the third Lyapunov function as

$$V_3 = V_2 + e_3^2/2$$

The derivative of above Lyapunov function is

$$\begin{aligned} \dot{V}_3 &= \dot{V}_2 + e_3\dot{e}_3 \\ &= \dot{V}_2 + e_3\left[-\theta_4 x_2 - \theta_5 x_3 + \theta_6\sqrt{|p_s - \mathrm{sgn}(x_4) x_3|}x_4 - \dot{x}_{3,d}\right] \end{aligned}$$

if

$$x_4 = x_{4,d} = \frac{-k_3 e_3 + \dot{x}_{3,d} + \theta_4 x_2 + \theta_5 x_3}{\theta_6\sqrt{|p_s - \mathrm{sgn}(x_4) x_3|}}$$

then

$$\begin{aligned} \dot{V}_3 &= \dot{V}_2 + e_3\dot{e}_3 \\ &= \dot{V}_2 + e_3\left[-\theta_4 x_2 - \theta_5 x_3 + \theta_6\sqrt{|p_s - \mathrm{sgn}(x_4)x_3|}x_4 - \dot{x}_{3,d}\right] \\ &= -k_1e_1^2 - k_2e_2^2 - k_3e_3^2 < 0 \end{aligned}$$

Then, the closed-loop control of the first three equations in the state space equation is stable.

In the fourth step, define

$$e_4 = x_4 - x_{4d}$$

The derivative of equation is

$$\begin{aligned} \dot{e}_4 &= \dot{x}_4 - \dot{x}_{4d} \\ &= -\theta_6 x_4 + \theta_7 u - \dot{x}_{4d} \end{aligned}$$

Define the forth Lyapunov function as

$$V_4 = V_3 + \frac{1}{2}e_4^2$$

The derivative equation of the above function is

$$\begin{aligned} \dot{V}_4 &= \dot{V}_3 + e_4\dot{e}_4 \\ &= \dot{V}_3 + e_4(-\theta_7 x_4 + \theta_8 u - \dot{x}_{4d}) \end{aligned}$$

If the input signal of the servo valve is defined as

$$u = \frac{-k_4e_4 + \dot{x}_{4d} + \theta_7 x_4}{\theta_8}$$

then

$$\begin{aligned} \dot{V}_4 &= \dot{V}_3 + e_4\dot{e}_4 \\ &= \dot{V}_3 + e_4(-\theta_7 x_4 + \theta_8 u - \dot{x}_{4d}) \\ &= -k_1e_1^2 - k_2e_2^2 - k_3e_3^2 - k_4e_4^2 < 0 \end{aligned}$$

The closed-loop state space equation of the whole system is stable.

5. Hardware-In-the-Loop Simulation Verification

In order to verify the correctness of the above online identification algorithm and the backstepping method, the corresponding program needs to be written and ported to the physical controller for verification. However, if the physical controller is directly connected to the entity of the controlled object, the direct control of the physical object is likely to cause unpredictable results, since the control algorithm is still in the debugging stage, which may result in damage to the controlled object or even cause personal injury in serious cases. The traditional solution is to simulate and verify the identification or control algorithm in a simulation environment and then, after this is successfully carried out, to transfer it to the controller to control the physical entity. The simulation environment is nonetheless different from the physical entity, and the algorithm, though successfully debugged in the simulation, may cause various unpredictable problems when transplanted to the physical controller. It would be very advantageous for control system engineers to be able directly debug the algorithm program of the physical controller in a relatively safe environment. HIL simulation technology is created to solve the above problems.

In this paper, the correctness of the online identification and the nonlinear control algorithm is verified using the HIL simulation technique. The HIL simulation technique replaces the physical entity of the controlled object with a mathematical model and operates together with a physical controller in a real-time environment. The physical controller is connected to the real-time solution of the controlled object through the I/O interface of the data acquisition card as if the controller were directly connected to the sensors of the physical system. Using the above method, it is not aware of the existence of the simulation model, as if it were running together with the physical entity of the controlled object in a real environment. More importantly, in this environment, the HIL simulation algorithm and the backstepping algorithm can be fully debugged because the controlled object is not a physical entity, and there is no need to worry about possible physical damage or personal injury caused by algorithm errors. At the same time, the successfully debugged physical controller can be directly connected to the physical entity of the controlled object, and the control of the physical entity of the controlled object can be basically and immediately successful based on HIL simulation in the early stage, thus improving the development efficiency and reducing the cost.

To improve the reliability of HIL simulation, two aspects need to be addressed.

1. The mathematical model of the physical entity that replaces the controlled object should be as accurate as possible.
2. The operating environment of the HIL simulation should be as real time as possible.

Point 1 guarantees the credibility of HIL simulation. If the mathematical model of the controlled object cannot accurately simulate the dynamic performance of the real physical entity, it deviates from the original intention of using HIL simulation. Point 2 ensures the reliability of HIL simulation, because only in a strongly real-time environment can the dynamic characteristics of the controlled physical entity be accurately reproduced, the real-time performance of the control algorithm guaranteed, and the reliability of controller verified.

Regarding point 1, the dynamic equation of the VCSC, as shown in (7), has been described in the previous section. Using this form, the requirements of the system dynamic characteristic equations for both online identification and backstepping simulation can be satisfied. The system parameters obtained from the identification can be directly adopted by the backstepping control algorithm, which ensures the coordination of the online identification algorithm and the backstepping control algorithm. Meanwhile, the equation of (7) takes into account the load spring stiffness, the internal leakage of the hydraulic cylinder, etc., and can reproduce all the dynamic characteristics of the controlled physical entity, and this characteristic equation is accurate.

Regarding point 2, considering the economy and feasibility, the solution adopted in this paper is to install the Preemp_RT patch to the kernel of Linux operating system to ensure the Linux is a hard real-time operating system. The most important feature of the hard real-time kernel is its preempt ability, where ready high-priority tasks are able to preempt low-priority tasks. This ensures the accuracy in timing of the real-time system. After actual testing, the timing accuracy of the Linux system with the Preempt_RT kernel installed can reach 1 ms and is very stable, thus fully meeting the requirements of this paper for system identification and backstepping control algorithm verification.

The validation of the HIL simulation consists of three main groups of experiments.

The first group of experiments verifies the performance of the HIL simulation environment, and the scheme used is the test comparison of 1 ms timer timing accuracy under a Windows environment and 1 ms timer timing accuracy after the Linux+Preempt_RT real-time kernel patch.

In the second group of experiments, the accuracy and reliability of system identification are verified under the HIL simulation environment.

In the third set of experiments, the accuracy and reliability of the backstepping algorithm are verified under the HIL simulation environment.

5.1. HIL Simulation Platform Hardware and Software

In order to verify the system identification and backstepping control method in a HIL simulation environment, this paper modifies the Ubuntu 18.04 system into a hard real-time by installing the Preempt_RT patch with the help of Ubuntu 18.04. A low-cost HIL simulation software and hardware platform is established through C language and development of the related software.

The physical hardware of the platform includes an industrial control computer with Ubuntu 18.04 system installed. There is one Advantech USB-5820 analog output data acquisition card and two USB-5817 input data cards. As the Linux operating system and its real-time kernel patch Preempt_RT are open source software, the procurement costs of Advantech data acquisition cards are relatively low under the premise of ensuring 16-bit acquisition accuracy, so the cost of entire HIL simulation platform is relatively low is comparison to commercial products.

The operation principle of the HIL simulation platform is that, with the help of the open source numerical algorithm library GSL [50], the system of differential equations described by the state space is numerically solved using the Runge–Kutta algorithm under the Linux+Preempt_RT real-time environment, and the results of solution operation are outputted by the USB-5820 data acquisition card for analog output (usually ±10 V or 4–20 mA signal). In this way, the physical object under control and its sensor output are simulated, providing a real-time input signal for the controller. At the same time, the HIL simulation platform, with the help of the USB-5817 analog input data acquisition card, reads the control signal of the physical controller (usually ±10 V or 4–20 mA signal) with a period of 1 ms, which is applied to the input of the state space equation, and the differential equation is solved and calculated based on the new control input to obtain the new state variable, which is then outputted with the help of USB-5820 data acquisition card. Outputting continues, so on and so forth, and the basic function of the HIL simulation platform is completed.

The hardware architecture of the HIL simulation platform developed in this paper is shown in Figure 2.

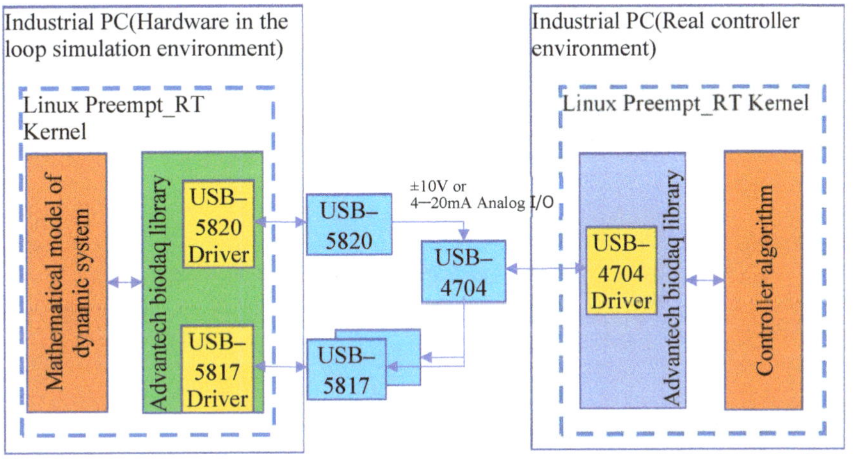

Figure 2. Hardware architecture schematic of the hardware-in-the-loop simulation platform.

The HIL simulation platform on which the experiments in this paper are conducted is shown in Figure 3.

Figure 3. Photo of the HIL simulation platform as a physical entity.

The development of software for this HIL simulation platform is realized using C language and calling library functions of Advantech data acquisition cards. Since Advantech data acquisition cards provide drivers for the Linux environment, it ensures the reading of high-precision control signals and the output of numerical calculation results in a real-time environment.

The software operation flow chart of the HIL simulation platform is shown in Figure 4.

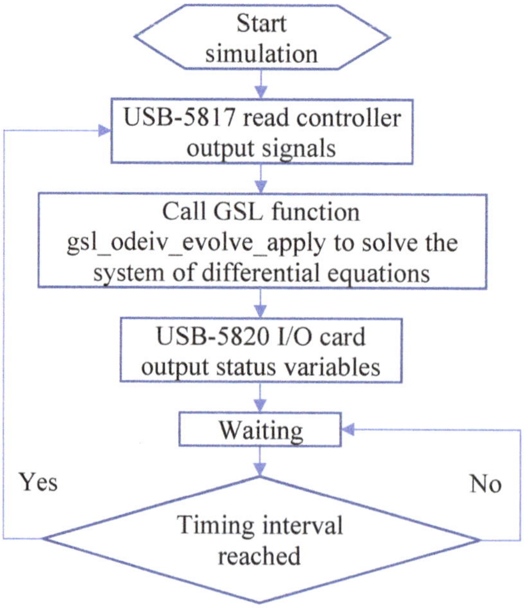

Figure 4. HIL simulation platform software operation flow chart.

To complete the HIL simulation experiment, cooperation of the physical controller is also required. In the latter, the cycle of data acquisition, operation, and output control signal is also executed with a timing cycle of 1 ms. The program flow chart is similar to Figure 4, except that the numerical solution part of the differential equation is replaced with either the system identification or the backstepping algorithm, depending on the content of control experiment. The former is used to estimate the key parameters of the system, and the latter is used to perform nonlinear control of the system. The execution process of the algorithm in the physical controller is divided into four main steps.

Step 1: Online identification, calculate the nonlinear coefficients of the electro-hydraulic cylinder system;

Step 2: Update the control parameters of the nonlinear backstepping algorithm;

Step 3: Calculate the next output $u(k+1)$ of the controller;

Step 4: Return to Step 2 and continue the loop.

The schematic diagram of the VCSC model injected by this hardware in the ring HIL simulation platform is shown in Figure 1. The key parameters of the whole model simulation are listed in Table 1.

Table 1. Key parameters of the electro-hydraulic servo system of the valve-controlled symmetrical cylinder.

Parameter	Value	Unit	Specification
k_c	1×10^5	N/m	Load spring stiffness
b_c	10	N/(m/s)	Viscous damping coefficient
D	0.032	m	Piston diameter
d	0.020	m	Rod diameter
g	9.81	m/s^2	Gravity of acceleration
m	100	kg	Load mass
F_L	1000	N	External load force
β_e	1.7×10^9	Pa	Elastic modulus of oil
C_d	0.62		Orifice flow coefficient
d_v	0.01	m	Servo valve spool diameter
w	$d_v\pi = 0.0314$	m	Valve area gradient
p_s	2.1×10^7	Pa	Supply pressure
k_a	1000		Gain of hyperbolic tangent function
ρ	890	kg/m^3	Oil density
τ_v	0.05	s	Servo valve time constant
k_v	0.015×10^{-3}	m/mA	Flow gain of valve
b	$d\pi = 0.0628$	m	Width of internal leakage gap
δ	1×10^{-4}	m	Height of internal leakage gap
μ	0.051	Pa s	Dynamic viscosity of oil
l	0.06	m	Length of internal leakage
C_l	$b\delta^3/(12\,\mu L)$		Internal leakage coefficient

5.2. Verification of the HIL Simulation Platform Performance in Real Time

In order to verify the real-time performance of the Ubuntu system after installing the Preempt_RT patch, a test is designed for comparing the performance between the Ubuntu18.04 and Windows 10 system in real time. The basic principle of the experiment is that in Ubuntu18.04 system with the Preempt_RT patch installed, the pulse voltage signal is sent out by the data acquisition card USB-5820 of Advantech at a time interval of 1 ms; under the same conditions in the Windows 10 system, using the high-precision multimedia timer and data acquisition card USB-5820 of Advantech, the pulse voltage signal with a period of 1 ms is also sent out. The periodic pulse voltage signals sent by two groups of software timers are collected by the logic analyzer, which is LA1010 produced by Kingst Company in Qingdao, China.

The logic analyzer sampling results for a 1 ms period pulse signal in the Windows 10 environment are shown in Figure 5.

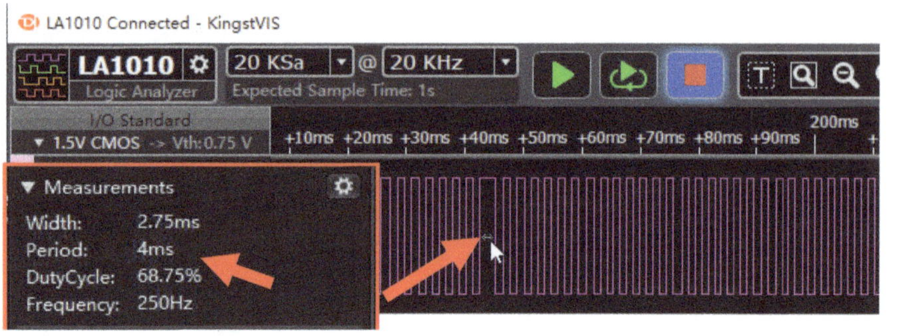

Figure 5. Timing accuracy test of high-precision timer under Windows 10 operating system.

It can be observed from the Figure 5 that when the pulse signal is sent at an interval of 1 ms, the period of the pulse signal should have been 2 ms. From the figure, it can be observed that although the vast majority of timing periods are close to 2 ms, there are individual cases of timing periods as long as 4 ms. That is to say, since Windows 10 is not a real-time system, it is not completely reliable to use it for high-precision timing.

In Ubuntu 18.04 with the Preempt_RT patch installed, the voltage signal is sent at 1 ms intervals, and the sampling results obtained with the help of the logic analyzer are shown in Figure 6.

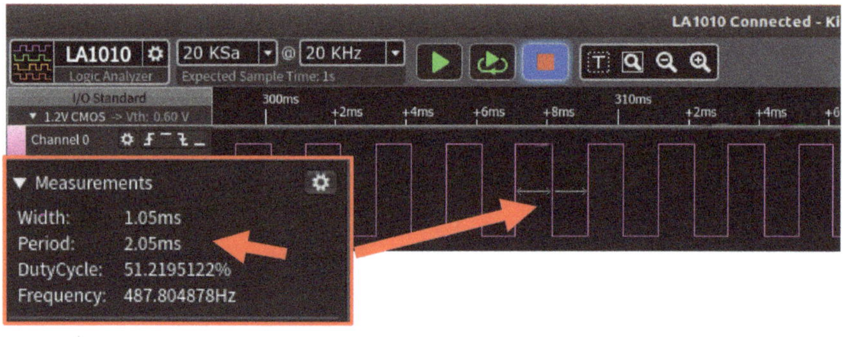

Figure 6. Timing accuracy test of high-precision timer under real-time Linux operating system.

From Figure 6, it can be observed that the periodic pulse signal issued under the real-time Linux system is uniform and stable. A timing period is selected for measurement at the maximum in the figure, and the obtained timing period is 2.05 ms, i.e., the maximum timing error is 50 µs.

Comparing Figures 5 and 6, it can be seen that the timing accuracy of the HIL simulation system based on the Linux real-time kernel patch developed in this paper can meet the requirements for real-time sampling and control. The results obtained from the HIL simulation experiments conducted using this platform are therefore credible.

The following paper will use this HIL simulation system to conduct identification and backstepping control experiments.

5.3. System Identification Experiments in HIL Simulation Environment

The system identification HIL simulation experiments were conducted on the HIL platform developed in the previous section. The process of system identification is as follows. The HIL simulation platform writes the differential equation of the identified system (valve-controlled symmetric cylinder) in C language on the HIL simulation platform, and the timer time interval is 1 ms. This solution is the output of the identified system, and the output signal is offered to the controller through the USB-5820 data acquisition card.

The controller is another industrial control computer that reads the output of the HIL simulation platform through the analog input function of USB-4704 and outputs a sinusoidal signal to the HIL simulation platform through the analog output function of USB-4704. The timing interval is also 1 ms, and so on and so forth. Between each timing cycle, the system identification algorithm, Equation (10), is invoked to generate the recursive identification results. The identification result is saved to provide the relevant parameters for the next backstepping control algorithm.

In this experiment, the input current signal is $u = 10\sin(2\pi t)\text{mA}$. The input and the displacement signal of the hydraulic cylinder output for the HIL simulation are shown in Figure 7, where the solid line indicates the input current signal and the dashed line indicates the hydraulic cylinder output displacement curve.

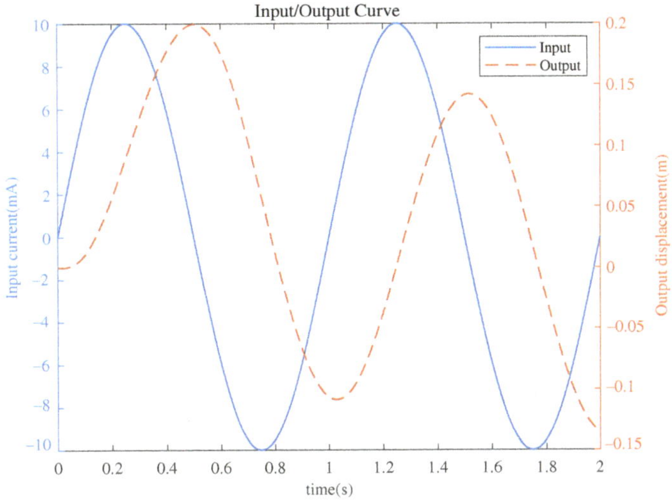

Figure 7. Input signal and output displacement curve in the system identification experiment.

Sampling plots of the main state variables and the derivatives of the acceleration and load pressure of the system during identification are shown in Figure 8. The three state variables, velocity, acceleration, and derivative of load pressure, correspond to y in Equation (13), and the values of each element of $\mathbf{\Psi}^T$ in the equation can be calculated from the state variables. Then, according to Equation (10), the value $\mathbf{\Theta}$ of each parameter to be identified can be calculated.

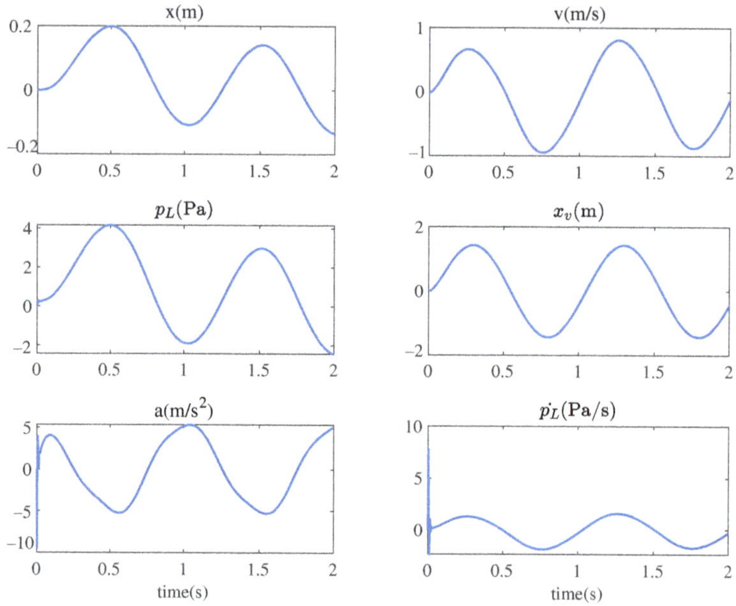

Figure 8. Curve diagram of key state variables in system identification experiment.

The iterative approximation curves of the system to identify each parameter are shown in Figure 9. Here, the dashed line is the real value of the identified parameter, which can be obtained according to the HIL simulation model equation. The solid line is the variation curve of the estimated value of the online identification parameter. It can be seen from Figure 9 that, with the accumulation of the number of online recognition iterations, the value of parameter rapidly approaches the true value.

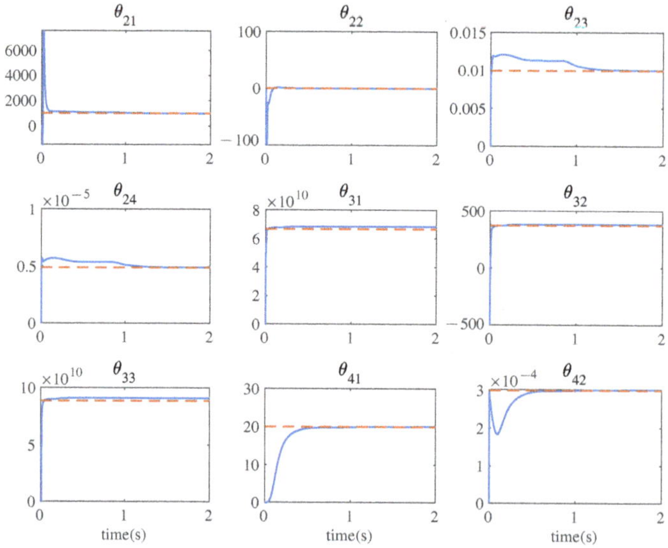

Figure 9. Iterative approximation curve of parameters in the system identification experiment.

To further test the correctness of system identification algorithm, a time-varying load is applied to the system with the help of the load simulation function of the platform, i.e., $F_L = [F_{L0} + 1000\sin(\pi t)]$. Under this time-varying load, the identification results of the system parameters are still close to the real values, as shown in Figure 10, and the identification results of the parameters are shown in column 5 of Table 2. Therefore, the system identification algorithm developed in this paper is demonstrated to be correct, and the effectiveness of using the semi-physical simulation platform to verify the online identification algorithm is also shown.

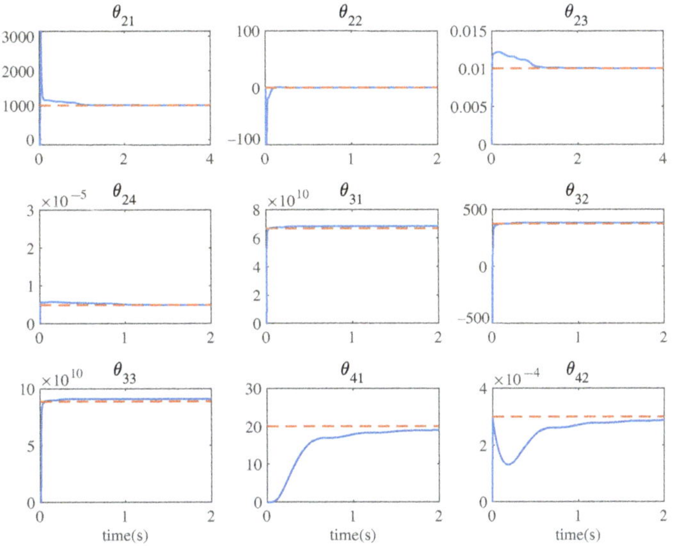

Figure 10. Iterative approximation curve of parameters in system identification experiment with time-varying load.

A comparison of the estimated and true values of the stability parameters obtained from system identification is shown in Table 2.

Table 2. Comparison between actual and estimated values in system identification of experimental parameters.

Parameters	Expression	Set Value of Hardware-in-the-Loop Simulation	Estimated Value	Estimated Value of Vary Load
θ_{21}	k_c/m	1000	1.0040×10^3	1.002×10^3
θ_{22}	b_c/m	0.1	0.0846	0.0801
θ_{23}	$1/m$	0.01	0.01	0.0100
θ_{24}	A_c/m	0.4901×10^{-5}	0.4921×10^{-5}	0.4902×10^{-4}
θ_{31}	$4\beta_e A_c/V_t$	6.6652×10^{10}	6.2156×10^{10}	6.8303×10^{10}
θ_{32}	$4\beta_e C_l/V_t$	372.3369	347.2963	381.5358
θ_{33}	$4\beta_e C_d w/(V_t\sqrt{\rho})$	8.8794×10^{10}	8.2822×10^{10}	9.0988×10^{10}
θ_{41}	$1/\tau_v$	20	20.0821	19.6383
θ_{42}	k_v/τ_v	0.3×10^{-3}	0.3014×10^{-3}	0.2953×10^{-3}

As can be seen from Table 2, the values of the parameters obtained from the system identification basically match the set values of the parameters in the HIL simulation environment. This demonstrates the correctness of the online identification algorithm and illustrates the effectiveness of its use in HIL simulation technology.

The parameter values obtained from the system identification are saved and prepared for the performance of backstepping control in the following.

5.4. Experiment of Backstepping Control Method in HIL Simulation Environment

Since all the parameters of the state space equation have been obtained based on system identification in the previous section, it is possible to develop the backstepping control program and perform HIL simulation verification of the backstepping control method for the valve-controlled symmetric cylinder system based on the basic principles of Section 3.

The difference from the previous section is that, instead of executing the system identification algorithm in each timing loop of the controller, the backstepping algorithm introduced in Section 3 is executed. The HIL simulation system, however, still uses 1 ms as the timing cycle to complete the fixed cycle of reading the data acquisition card signal, numerically solving the state equation of the VCSC system and outputting the solved state variables through the data acquisition card.

In order to illustrate the effectiveness of the nonlinear control algorithm, HIL simulation of the PID was also performed to verify the effectiveness of the PID control algorithm. The effect of the PID is compared with that of the nonlinear backstepping method.

The HIL simulation model of the VCSC system constructed with the data in Table 1 as key parameters is carried out for the PID and the backstepping control experiment as appropriate. The parameters of the conducted PID experiments are shown in Table 3.

Table 3. PID control parameters.

Control Parameter	Value
k_p	1000
k_i	0.01
k_d	10

The graph of simulation results for PID control is shown in Figure 11. The upper part of Figure 11 shows the graph for comparison between the command curve and the actual displacement curve, and the lower part shows the error values of the command curve and the actual output curve.

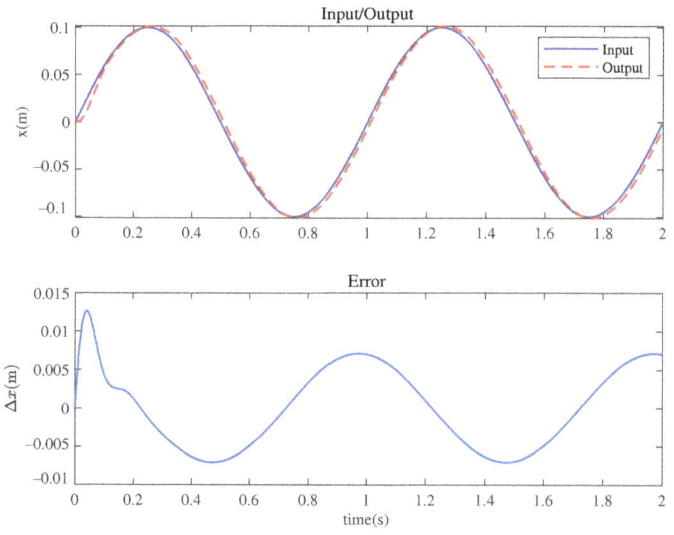

Figure 11. HIL simulation curve of PID control.

It can be observed from Figure 11 that the error of the system reaches a maximum of about 0.015 m with the current PID control parameters and then stabilizes at about 0.005 m.

The HIL simulation model of the valve-controlled symmetric cylinder system constructed with the key parameters in Tables 4 and 5 was subjected to two sets of nonlinear backstepping method control HIL simulation experiments. The control parameters of the first group of backstepping method are shown in Table 4.

Table 4. The first set of control parameters of the backstepping control method.

Control Parameter	Value
k_1	80
k_2	80
k_3	160
k_4	160

The first set of hardware-in-the-loop semi-physical simulation experiments for the backstepping method is shown in Figure 12. The command and output curves and the error between them are shown in this figure under the control of backstepping method. As can be seen in the lower part of Figure 12, the system error finally stabilizes within 0.02 m under the action of the control parameters in Table 4, which is approximately the same as the error under PID control.

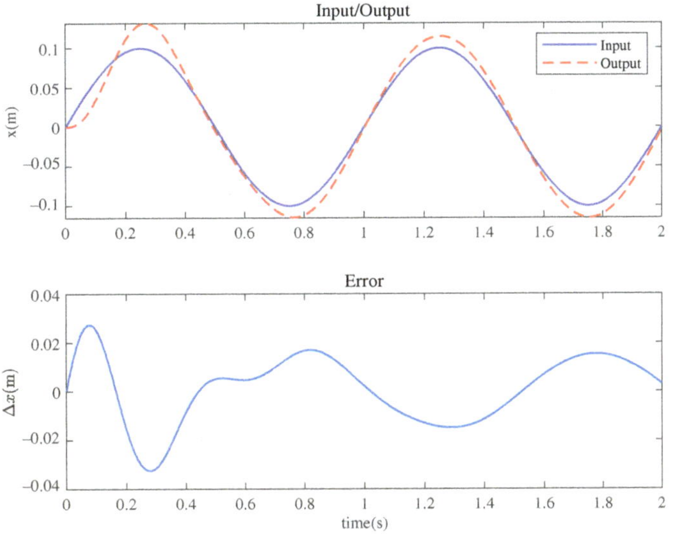

Figure 12. HIL simulation curve under the first set of control parameters of backstepping method.

The control parameters for the second set of backstepping methods are shown in Table 5 and have increased compared with the values in Table 4.

Table 5. The second set of control parameters of the backstepping control method.

Control Parameter	Value
k_1	160
k_2	160
k_3	320
k_4	320

The input–output and error curves of the nonlinear backstepping semi-physical simulation of the valve-controlled symmetric cylinder are shown in Figure 13.

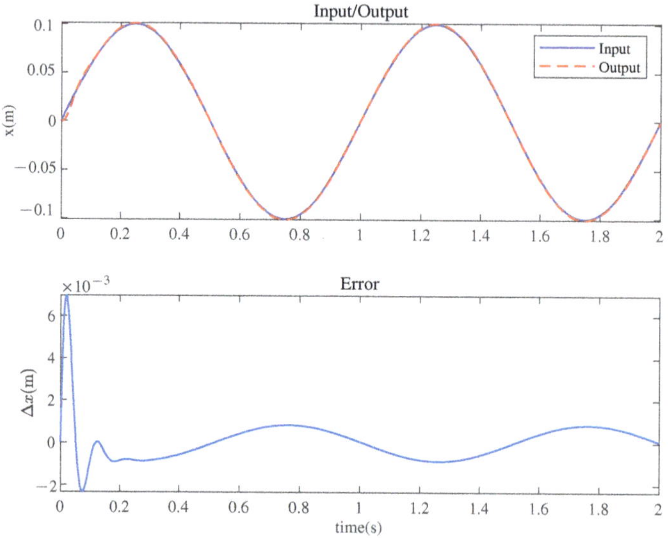

Figure 13. Hardware-in-the-loop simulation curve under the second set of control parameters of the backstepping method.

It can be seen from the Figure 13 that, when the values of the control parameters are appropriately increased, the maximum tracking error of the input and output signals is 0.007 m in the initial stage, and as the system operation tends to a steady state, the error rapidly decreases and stabilizes to less than 0.0005 m, which shows that the backstepping method has higher control accuracy compared with the PID control algorithm when the appropriate parameters are set.

6. Conclusions

A HIL simulation debugging method based on a Linux real-time kernel patch is proposed to address the problem of difficulty in debugging a EHS system online identification algorithm and nonlinear backstepping control method in practical applications. This developed HIL simulation experiment platform is developed based on Linux+Preempt_RT. Taking the VCSC EHS system as the control object, the online identification algorithm is developed, and the backstepping control algorithm is derived. Using the developed HIL simulation platform, the timer accuracy test is conducted, and the results show that the HIL simulation platform developed in this paper can achieve the required timing accuracy in hard real time. Furthermore, the HIL simulation platform was used to carry out the simulation of system identification and nonlinear backstepping method for VCSC. Using the HIL simulation for system identification, all the key parameters were accurately obtained and assigned to the simulation model of the VCSC in advance; at the same time, using the nonlinear backstepping control implemented by the parameters obtained from system identification, a better control effect was achieved than with the common PID control algorithm.

The above experiments demonstrate that the HIL simulation platform developed in this paper has reliable real-time performance, and the experimental results obtained from the system identification experiments and the nonlinear backstepping control algorithm conducted on this platform are credible. Since the HIL experiments used are simulations, the verification of physical controllers with system identification and control algorithms can be directly applied in engineering the physical entities of VCSCs.

Future research work will focus on building an experimental platform for the physical entity of the VCSC, transplanting the controller verified by the HIL simulation environment to the physical entity experimental platform and completing the system identification and nonlinear backstepping control experiments for the physical entity to further verify the effectiveness of the system identification and nonlinear backstepping control algorithm.

Author Contributions: Conceptualization, Q.L., F.L. and J.G.; methodology, Q.L.; software, K.W.; validation, K.W., H.Z. and Z.W.; formal analysis, D.S.; writing—original draft preparation, Q.L. and J.G.; writing—review and editing, D.S.; visualization, Q.L.; project administration, Q.L.; funding acquisition, D.S. All authors have read and agreed to the published version of the manuscript..

Funding: This research was funded by the National Nature Science Foundation of China, grant number 51775354.

Data Availability Statement: Not applicable.

Conflicts of Interest: The authors declare no conflict of interest.

Abbreviations

The following abbreviations are used in this manuscript:

EHS	electro-hydraulic servo
VCSC	valve-controlled symmetrical cylinder
HIL	hardware-in-the-loop

References

1. Jelali, M.; Kroll, A. *Hydraulic Servo-Systems: Modelling, Identification and Control*; Springer Science & Business Media: Berlin/Heidelberg, Germany , 2002.
2. Milecki, A.; Ortmann, J. Electrohydraulic linear actuator with two stepping motors controlled by overshoot-free algorithm. *Mech. Syst. Signal Process.* **2017**, *96*, 45–57. [CrossRef]
3. Maier, C.C.; Schröders, S.; Ebner, W.; Köster, M.; Fidlin, A.; Hametner, C. Modeling and nonlinear parameter identification for hydraulic servo-systems with switching properties. *Mechatronics* **2019**, *61*, 83–95. [CrossRef]
4. Jelali, M.; Schwarz, H. Nonlinear identification of hydraulic servo-drive Systems. *IEEE Control Syst. Mag.* **1995**, *15*, 17–22.
5. Wos, P.; Dindorf, R. Nonlinear modeling and parameter identification for electro-hydraulic servo system. In Proceedings of the 2019 20th International Carpathian Control Conference (ICCC), Krakow-Wieliczka, Poland, 26–29 May 2019; pp. 1–5.
6. Kwak, B.J.; Yagle, A.E.; Levitt, J.A. Nonlinear system identification of hydraulic actuator. Friction dynamics using a Hammerstein model. In Proceedings of the 1998 IEEE International Conference on Acoustics, Speech and Signal Processing, ICASSP'98 (Cat. No. 98CH36181), Seattle, WA, USA, 15 May 1998; Volume 4, pp. 1933–1936.
7. Ling, T.; Rahmat, M.; Husain, A. System identification and control of an Electro-Hydraulic Actuator system. In Proceedings of the 2012 IEEE 8th International Colloquium on Signal Processing and Its Applications, Malacca, Malaysia, 23–25 March 2012; pp. 85–88.
8. Papadopoulos, E.; Mu, B.; Frenette, R. On modeling, identification, and control of a heavy-duty electrohydraulic harvester manipulator. *IEEE/ASME Trans. Mechatron.* **2003**, *8*, 178–187. [CrossRef]
9. Kaddissi, C.; Kenne, J.P.; Saad, M. Identification and real-time control of an electrohydraulic servo system based on nonlinear backstepping. *IEEE/ASME Trans. Mechatron.* **2007**, *12*, 12–22. [CrossRef]
10. Sadeghieh, A.; Sazgar, H.; Goodarzi, K.; Lucas, C. Identification and real-time position control of a servo-hydraulic rotary actuator by means of a neurobiologically motivated algorithm. *ISA Trans.* **2012**, *51*, 208–219. [CrossRef]
11. Yuan, H.B.; Na, H.C.; Kim, Y.B. System identification and robust position control for electro-hydraulic servo system using hybrid model predictive control. *J. Vib. Control* **2018**, *24*, 4145–4159. [CrossRef]
12. Liang, X.W.; Ismail, Z.H. System identification and model predictive control using CVXGEN for electro-hydraulic actuator. *Int. J. Integr. Eng.* **2019**, *11*. [CrossRef]
13. Feng, H.; Yin, C.; Ma, W.; Yu, H.; Cao, D. Parameters identification and trajectory control for a hydraulic system. *ISA Trans.* **2019**, *92*, 228–240. [CrossRef]
14. Jinxing, S.; Hongxin, C.; Ke, F.; Hong, Z.; Huanliang, L. Parameter identification and control algorithm of electrohydraulic servo system for robotic excavator based on improved hammerstein model. *Math. Probl. Eng.* **2020**, *2020*, 9216019. [CrossRef]
15. Zhao, J.; Song, D.; Zhu, B.; Chen, Z.; Sun, Y. Nonlinear backstepping control of electro-hydraulic brake system based on bond graph model. *IEEE Access* **2020**, *8*, 19100–19112. [CrossRef]
16. Guo, Q.; Sun, P.; Yin, J.; Yu, T.; Jiang, D. Parametric adaptive estimation and backstepping control of electro-hydraulic actuator with decayed memory filter. *ISA Trans.* **2016**, *62*, 202–214. [CrossRef]
17. Ba, D.X.; Ahn, K.K.; Truong, D.Q.; Park, H.G. Integrated model-based backstepping control for an electro-hydraulic system. *Int. J. Precis. Eng. Manuf.* **2016**, *17*, 565–577. [CrossRef]
18. Yang, X.; Zheng, X.; Chen, Y. Position tracking control law for an electro-hydraulic servo system based on backstepping and extended differentiator. *IEEE/ASME Trans. Mechatron.* **2017**, *23*, 132–140. [CrossRef]

19. Tian, L.; Qian, L.; Chen, L.; Zhang, W. Sliding mode control based on backstepping method for electro-hydraulic single-rod actuator. In Proceedings of the 2015 IEEE International Conference on Information and Automation, Lijiang, China, 8–10 August 2015; pp. 2326–2329.
20. Kim, H.; Park, S.; Song, J.; Kim, J. Robust position control of electro-hydraulic actuator systems using the adaptive back-stepping control scheme. *Proc. Inst. Mech. Eng. Part I J. Syst. Control Eng.* **2010**, *224*, 737–746. [CrossRef]
21. Yin, X.; Lin, Y.; Li, W.; Gu, Y.; Lei, P.; Liu, H. Adaptive back-stepping pitch angle control for wind turbine based on a new electro-hydraulic pitch system. *Int. J. Control* **2015**, *88*, 2316–2326. [CrossRef]
22. Zang, W.; Zhang, Q.; Shen, G.; Fu, Y. Extended sliding mode observer based robust adaptive backstepping controller for electro-hydraulic servo system: Theory and experiment. *Mechatronics* **2022**, *85*, 102841. [CrossRef]
23. Ahn, K.K.; Nam, D.N.C.; Jin, M. Adaptive backstepping control of an electrohydraulic actuator. *IEEE/ASME Trans. Mechatron.* **2013**, *19*, 987–995. [CrossRef]
24. Li, X.; Zhu, Z.C.; Rui, G.C.; Cheng, D.; Shen, G.; Tang, Y. Force loading tracking control of an electro-hydraulic actuator based on a nonlinear adaptive fuzzy backstepping control scheme. *Symmetry* **2018**, *10*, 155. [CrossRef]
25. Feng, L.; Yan, H. Nonlinear adaptive robust control of the electro-hydraulic servo system. *Appl. Sci.* **2020**, *10*, 4494. [CrossRef]
26. Zibin, X.; Jianqing, M.; Ruan, J. Adaptive backstepping neural network control of electro-hydraulic position servo system. In Proceedings of the 2008 2nd International Symposium on Systems and Control in Aerospace and Astronautics, Shenzhen, China, 10–12 December 2008; pp. 1–4.
27. Lee, J.M.; Kim, H.M.; Park, S.H.; Kim, J.S. A position control of electro-hydraulic actuator systems using the adaptive control scheme. In Proceedings of the 2009 7th Asian Control Conference, Hong Kong, China, 27–29 August 2009; pp. 21–26.
28. Dashti, Z.A.S.; Jafari, M.; Gholami, M.; Shoorehdeli, M.A. Neural—Adaptive control based on backstepping and feedback linearization for electro hydraulic servo system. In Proceedings of the 2014 Iranian Conference on Intelligent Systems (ICIS), Bam, Iran, 4–6 February 2014; pp. 1–6.
29. Ziaei, K.; Sepehri, N. Modeling and identification of electrohydraulic servos. *Mechatronics* **2000**, *10*, 761–772. [CrossRef]
30. Vaezi, M.; Izadian, A. Piecewise affine system identification of a hydraulic wind power transfer system. *IEEE Trans. Control Syst. Technol.* **2015**, *23*, 2077–2086. [CrossRef]
31. Ishak, N.; Tajjudin, M.; Adnan, R.; Ismail, H. System identification and model validation of electro-hydraulic actuator for quarter car system. *ARPN J. Eng. Appl. Sci.* **2017**, *4*, 27–35.
32. Wos, P.; Dindorf, R. Practical parallel position-force controller for electro-hydraulic servo drive using on-line identification. *Int. J. Dyn. Control* **2016**, *4*, 52–58. [CrossRef]
33. Jin, L.; Wang, Q. Accurate model identification of the inertial mass dynamic of hydraulic cylinder with model uncertainty. *Proc. Inst. Mech. Eng. Part I J. Syst. Control Eng.* **2019**, *233*, 501–510. [CrossRef]
34. Ghazali, R.; Sam, Y.M.; Rahmat, M.F.; Zulfatman. On-line identification of an electro-hydraulic system using recursive least square. In Proceedings of the 2009 IEEE Student Conference on Research and Development (SCOReD), Serdang, Malaysia, 16–18 November 2009; pp. 471–474.
35. Ghazali, R.; Sam, Y.M.; Rahmat, M.F.; Zulfatman. Open-loop and closed-loop recursive identification of an electro-hydraulic actuator system. In Proceedings of the 2010 IEEE Conference on Robotics, Automation and Mechatronics, Singapore, 28–30 June 2010; pp. 285–290.
36. Ling, T.; Rahmat, M.; Husain, A.; Ghazali, R. System identification of electro-hydraulic actuator servo system. In Proceedings of the 2011 4th International Conference on Mechatronics (ICOM), Kuala Lumpur, Malaysia, 17–19 May 2011; pp. 1–7.
37. Ghazali, R.; Soon, C.; Jaafar, H.I.; Sam, Y.M.; Rahmat, M. System identification of electro-hydraulic actuator system with pressure and load effects. In Proceedings of the 2014 IEEE International Conference on Control System, Computing and Engineering (ICCSCE 2014), Penang, Malaysia, 28–30 November 2014; pp. 256–260.
38. Lee, K.H.; Baek, S.G.; Choi, J.Y.; Park, J.J.; Choi, H.R.; Moon, H.; Koo, J.C. System identification of hydraulic system using RLS algorithm for feedforward position control. In Proceedings of the 2015 12th International Conference on Ubiquitous Robots and Ambient Intelligence (URAI), Goyang, Republic of Korea, 28–30 October 2015; pp. 63–65.
39. Cologni, A.L.; Mazzoleni, M.; Previdi, F. Modeling and identification of an electro-hydraulic actuator. In Proceedings of the 2016 12th IEEE International Conference on Control and Automation (ICCA), Kathmandu, Nepal, 1–3 June 2016; pp. 335–340.
40. D'alfio, N.; Morgando, A.; Sorniotti, A. Electro-hydraulic brake systems: Design and test through hardware-in-the-loop simulation. *Veh. Syst. Dyn.* **2006**, *44*, 378–392. [CrossRef]
41. Yong, J.; Gao, F.; Ding, N.; He, Y. Design and validation of an electro-hydraulic brake system using hardware-in-the-loop real-time simulation. *Int. J. Automot. Technol.* **2017**, *18*, 603–612. [CrossRef]
42. Sorniotti, A.; Repici, G.M. Hardware in the loop with electro-hydraulic brake systems. In Proceedings of the 9th WSEAS International Conference on Systems, Athens, Greece, 11–13 July 2005.
43. Lee, Y.; Lee, W.S. Hardware-in-the-loop Simulation for Electro-mechanical Brake. In Proceedings of the 2006 SICE-ICASE International Joint Conference, Busan, Korea, 18–21 October 2006; pp. 1513–1516.
44. Heo, S.J.; Park, K.; Ahn, H.S. Design and implementation of hardware-in-the-loop simulator for EHB systems. In Proceedings of the SICE 2003 Annual Conference (IEEE Cat. No. 03TH8734), Fukui, Japan, 4–6 August 2003; Volume 1, pp. 612–616.
45. Velardocchia, M.; Sorniotti, A. *Hardware-in-the-Loop to Evaluate Active Braking Systems Performance*; Technical Report, SAE Technical Paper; SAE: Warrendale, PA, USA, 2005.

46. Montazeri-Gh, M.; Nasiri, M. Hardware-in-the-loop simulation for testing of electro-hydraulic fuel control unit in a jet engine application. *Simulation* **2013**, *89*, 225–233. [CrossRef]
47. Ferreira, J.A.; De Oliveira, J.E.; Costa, V.A. Modeling of hydraulic systems for hardware-in-the-loop simulation: A methodology proposal. In Proceedings of the ASME International Mechanical Engineering Congress and Exposition, Nashville, TN, USA, 14–19 November 1999; Volume 16479, pp. 33–39.
48. Merrit, H.E. *Hydraulic Control Systems*; John Whiley & Sons Inc.: New York, NY, USA, 1967.
49. Alleyne, A.; Liu, R. A simplified approach to force control for electro-hydraulic systems. *Control Eng. Pract.* **2000**, *8*, 1347–1356. [CrossRef]
50. Gough, B. *GNU Scientific Library Reference Manual*; Network Theory Ltd.: Godalming, UK, 2009.

energies

MDPI

Article

CVT for a Small Electric Vehicle Using Centrifugal Belt Pulley

Hyung-jin Do and Se-hoon Oh *

Department of Mechanical Engineering, Chung-Ang University, Seoul 06974, Republic of Korea
* Correspondence: osh@cau.ac.kr; Tel.: +82-2-820-5314

Abstract: In a conventional continuously variable transmission (CVT), the belt slides up and down along the slope of the pulley to shift gears. Efficiency is reduced because of slippage from the shifting of the belt. In this study, the centrifugal belt pulley for CVT were investigated. Instead of applying the belt slip method, a rail groove was machined on the side of the pulley, and a push piece pushed the belt up by centrifugal force along the rail groove. The CVT was designed and manufactured using this mechanism. A CVT tester was manufactured to validate the performance of the proposed transmission. The suitability of the CVT was verified through structural analysis.

Keywords: centrifugal belt pulley; efficiency tester; FEM analysis; plate; ratio

Citation: Do, H.-j.; Oh, S.-h. CVT for a Small Electric Vehicle Using Centrifugal Belt Pulley. *Energies* **2022**, *15*, 8800. https://doi.org/10.3390/en15238800

Academic Editors: Kan Liu and Wei Hu

Received: 19 September 2022
Accepted: 16 November 2022
Published: 22 November 2022

1. Introduction

A transmission is a device that modifies the speed and output of a vehicle by changing the number of revolutions of the power source [1,2]. They are usually classified into manual, automatic, and continuously variable transmissions (CVT) based on the operation method. A manual transmission is a device that allows the driver to shift gears based on the driving conditions to provide the necessary driving force and speed [3,4]. An automatic transmission is a device that automatically shifts gears in the transmission control unit (TCU) by determining the shift stage according to driving conditions [5–7]. Rubber belt CVT is used in a variety of off-road and on-road recreational vehicles [8,9]. Hybrid electric vehicles (HEVs) are equipped with CVTs [10,11]. In manual and automatic transmissions, shift shocks occur when the gears do not mesh during the shift process, resulting in a drop or sharp increase in output [12,13]. A CVT is a device that can shift to a different gear ratio without changing gears, unlike in a manual or automatic transmission. Because the CVT changes gears smoothly, there is no shift shock or power reduction during the shifting process [14–16].

A CVT with a belt pulley system is used in automobiles. Metal and rubber belts are usually used in light vehicles. In the wet method, oil is sprayed between the transmission and belt to resolve belt slippage and heat problems [17–19]. As described above, the CVT shifts gears by sliding the belt and the shift pulley, which causes power loss during the shifting process and generates severe heat in the belt and the shift pulley. Therefore, in contrast to the sliding transmission method of the belt and pulley during the shifting process, this study developed a CVT in which the outer diameter of the pulley was changed.

In a conventional CVT with a belt pulley, the pulley moves along the axial direction, and the belt slides up and down to change its speed. However, in the variable-diameter pulley of the CVT, rail grooves corresponding to the rotational direction were machined on both sides of the pulley, and a push piece generated centrifugal force along the rail grooves to change the outer diameter and speed of the pulley. Therefore, the shifting performance was determined by the angle and shape of the rail groove machined on the pulley plate.

In this study, Fortran and basic programs were used to analyze the movement of the push piece in the rail groove of the plate, and to design the rail groove. The shape and angle of the rail were selected to calculate the rail curve. A performance test of the CVT using the optimized centrifugal belt pulley method was conducted. In addition to the actual vehicle installation test, the CVT housing was fabricated such that the CVT could be installed without interfering with the differential gear and the motor in the actual vehicle. Finally, a driving tester was manufactured, and the durability was tested under uphill and unpaved road conditions to verify driving durability.

2. Principle of Shift Motion of a Centrifugal Belt Pulley

Figure 1 shows a manufactured pulley for ease of explanation and understanding.

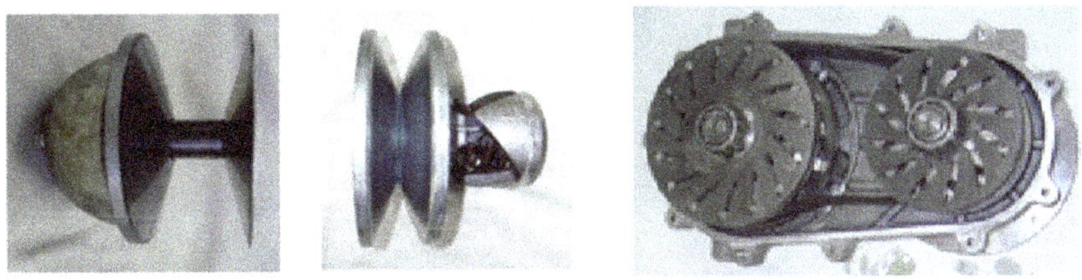

(a) (b)

Figure 1. (**a**) CVT structure; (**b**) Centrifugal belt pulley CVT structure.

The pulley in Figure 1a shows the CVT commonly used in micro-mini vehicles. This CVT is primarily used in vehicles with internal combustion engines. The belt slides up and down along the slope of the pulley to change the radius of the circle and speed.

The pulley in Figure 1b contains multiple parts. The belt is placed on the components. The speed is varied using centrifugal force to expand and contract these components.

Figure 1a shows that the belt slides along the slope of the pulley to shift gears in a conventional electric vehicle. Frictional heat is generated, and the efficiency decreases owing to slip shifting. In this study, a centrifugal belt-type CVT without belt slippage was designed and manufactured, as shown in Figure 1b.

2.1. Principle of the Centrifugal Belt-Type CVT

Figure 2 shows the slide plate coupled to the drive shaft rotates when the drive pulley rotates in the stationary state. As it rotates slowly, the push piece expands gradually, owing to the centrifugal force.

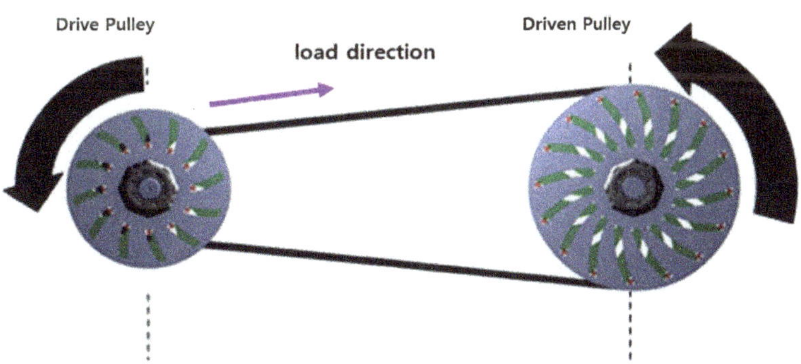

Figure 2. Principle of centrifugal belt CVT acceleration.

When there is a "load" on the wheel, the "load" generates a pulling force on the belt in a direction opposite to the direction of rotation of the drive pulley. The pressing piece was contracted (moved downward) in the direction of the slide plate's rail.

Figure 3 shows in the driven pulley, when the push piece rotates in the expanded state, the slide plate rotates along with the push piece. The slide plate was coupled to the driven shaft to reduce the output. Then, if a load is applied to the wheel and the driven shaft does not rotate in a direction opposite to the rotational direction, the slide plate expands according to the rail direction. The expansion range of the push piece expands to the same extent as that of the driven load.

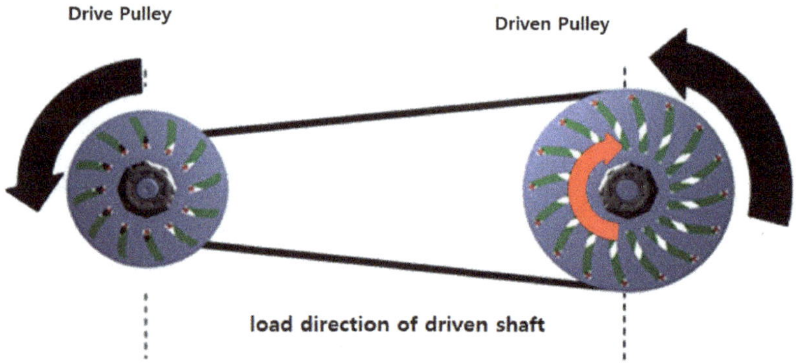

Figure 3. Deceleration principle of the centrifugal belt pulley CVT.

Design Condition of Centrifugal Belt-Type CVT

The centrifugal force at the maximum rotational speed of the drive pulley (maximum expansion state) must be greater than or equal to the magnitude of the centrifugal force and spring tension at the maximum rotational speed of the driven pulley (minimum contraction state):

$$\text{drive pulley F} \geqq \text{driven pulley F} + \text{spring tension}.$$

3. Centrifugal Belt Slide Plate Design Simulation

In a CVT with a centrifugal belt pulley, the gear ratio is selected based on the size of the outer diameter during the contraction and expansion of the pulley. The length of the rubber belt is selected based on the distance between the gear ratio and the shaft of the pulley. Similar to the principle of the sliding belt CVT, the shape of the rail plate is critical, because the speed change is based on the shape of the slide plate and guide plate. Therefore, two types of rails were designed and compared to determine the optimal deceleration efficiency according to the plate shape.

In this study, the movement of the push piece was calculated based on a 4 kW electric vehicle's CVT design. The driving force for the electric vehicle was supplied to the rear wheel via a 10:1 reducer and differential gear.

However, an excessive load on the battery and inverter initially occurred when driving uphill or at high speeds.

Therefore, to reduce the load on the motor during the initial operation, the belt length was calculated by setting the reduction ratio to 3:1 when the diameter of the drive pulley was extended from the minimum contracted state to the maximum state of the driven pulley.

The minimum diameter of the driven pulley was determined by calculating its belt length and the diameter of the drive pulley. The outer diameter of the driving plate was Ø122 mm, the diameter when contracted was Ø52 mm, the diameter when expanded was Ø109 mm, and the number of rail grooves was set at 12. The diameter of the driven plate was set at Ø167 mm, the diameter when expanded was Ø156 mm, and the diameter when contracted was Ø109 mm for contraction, with 16 rail grooves to form a smooth circle.

The final gear ratio was 3:1, and an initial driving force with a reduction ratio of 30:1 was provided.

Center distance: 175 mm

Belt length:

$$L_p = 2C + 1.57(D_p + d_p) + \frac{(D_p - d_p)^2}{4C} = 672\text{mm} \quad (1)$$

Spring constant: 0.498
Drive piece gross weight:

$$0.0314 \text{ kg} \times 12 = 0.3768 \text{ kg}$$

Driven piece gross weight:

$$0.0108 \text{ kg} \times 16 = 0.1728 \text{ kg}$$

The Excel sheet listed in Table 1 below contains the formulas given above as the primary design data. The initial design value was entered in the Excel sheet to calculate the remaining values. The design was carried out based on these values.

Table 1. Design datasheet.

Belt Length [mm]	Center Distance [mm]	[DR] Pulley Diameter [mm]	[DR] RPM	[DR] Centrifugal Force [N.m]	[DR] Pulley Diameter [mm]	[DR] Centrifugal Force [N.m]
692	175	52	1000	157.8	156.0	47.6
692	175	58	1100	212.9	151.8	73.7
692	175	64	1200	279.6	147.5	109.9
692	175	70	1300	359.0	143.	159.1
692	175	76	1400	452.0	138.3	224.9
692	175	82	1500	559.8	133.4	311.5
692	175	88	1600	683.6	128.4	424.4
692	175	94	1700	824.3	123.1	570.1
692	175	100	1800	983.1	117.6	757.3
692	175	109	1900	1193.9	108.8	1082.7

Two types of slide and guide plates were designed to compare the shift characteristics according to the shape of the plates.

The conventional design method is to design the rail groove of the guide plate as a straight line and the curves of the slide plate as a regular curve.

Another approach is to design the rail grooves of the slide plate and guide plate as a spline curve such that the part where the belt of the push piece is caught always faces upward.

Figure 4 shows that the spline curve is designed as a curve with normal "R" values.

When designed as a regular curve, a push piece moves along the rail and lies down. Therefore, the rail of the guide plate was designed as a straight line for multiple pieces to move simultaneously.

Figure 5 shows the slide plate and guide plate rail curves designed as spline curves. A spline curve is designed so that a piece moving along the rail can always move to the same position. The added benefit is that the guide plate and the slide plate can be processed simultaneously, even when processing the plate rail.

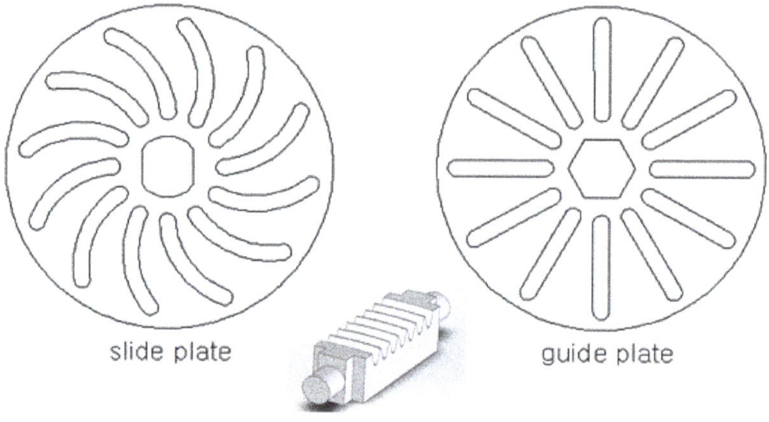

Figure 4. Slide plate designed with a normal curve and a guide plate designed with a straight line.

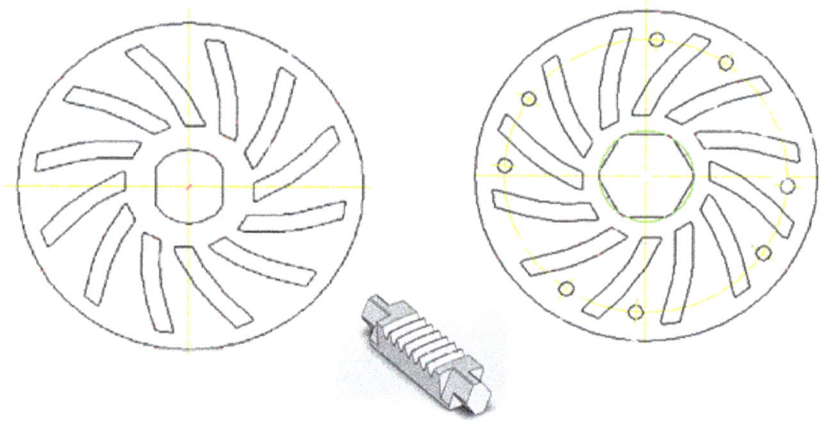

Figure 5. Slide plate and guide plate design with a spline curve.

Analysis of the Behavior of Push Piece

A Fortran program was written to perform the numerical motion analysis of the push piece. When input data is input as shown in Figure 6, the Fortran program sends the output data by investigating the stability and delivery efficiency of the push piece through dynamic analysis.

$$\sum Fx = 0 \tag{2}$$

$$\sum F1 - N2\cos x + \mu N2\sin x - N1\cos\theta 1 + \mu N1\sin\theta 1 = 0 \tag{3}$$

$$\sum Fy = 0 \tag{4}$$

$$N1\sin\theta 1 + \mu N1\cos\theta 1 - N2\cos x - \mu N2\sin x = 0 \tag{5}$$

Figure 7 shows the condition of the push piece does not rotate:

$$\mu N1 = \mu N2 \tag{6}$$

$$N1 = N2 \tag{7}$$

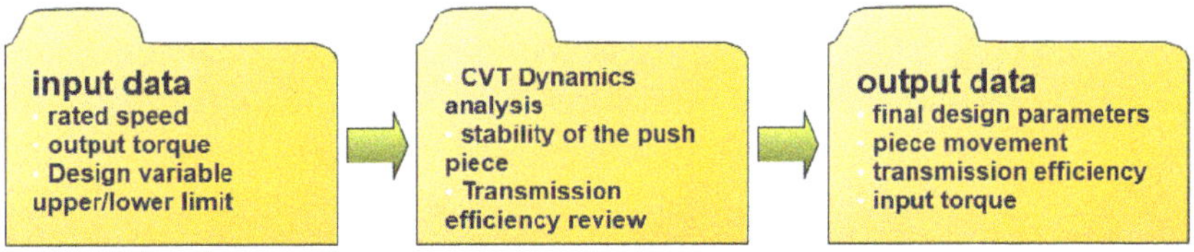

Figure 6. The principle of the design program for improving CVT performance.

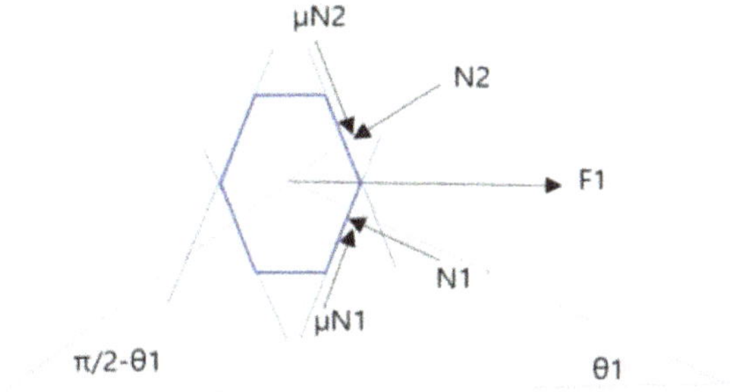

Figure 7. The force on the push piece.

A denotes the driving part, and the driven part is denoted by B.

Figure 8 shows O is the center of the rotary pulley and when centrifugal force is not applied, and the position of the spring is marked as R_A.

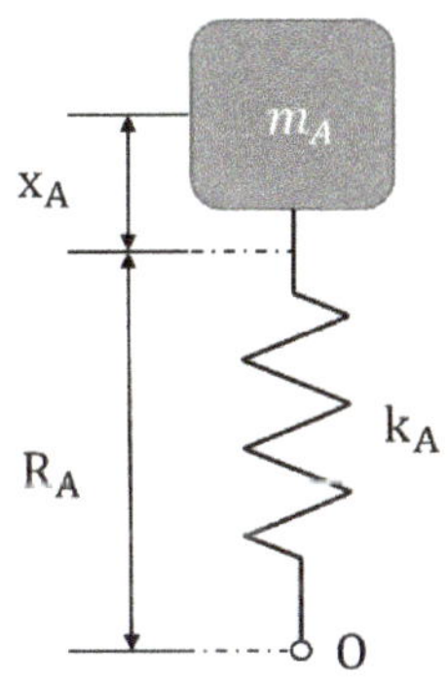

Figure 8. The force on the CVT for numerical analysis.

Subsequently, when it acts at angular velocity w_A, the centrifugal force is as follows:

$$m_A(x_A + R_A)\omega_A^2$$

Resilience in the spring will be as follows:

$$K_A x_A$$

At this time, if the tension caused by the belt is ignored, the driving unit will be $m_A(x_A + R_A)\omega_A^2 = K_A x_A$, and the displacement x_A will be:

$$x_A = \frac{m_A\omega_A^2}{K_A - m_A\omega_A^2}$$

If the same principle is applied to the driven part, then it is defined by the following equation:

$$x_B = \frac{m_B \omega_B^2}{K_B - m_B \omega_B^2}$$

The deceleration ratio is defined as follows (In this explanation of the principle, the effect on B is ignored):

$$i = \frac{x_A + R_A}{x_B + R_B}$$

If i is less than one, it slows down, and if i is more than one, it increases speed. Thus, increasing i means an increase in the velocity.

If the angular velocity is $w_A = 0$, the deceleration ratio is as follows:

$$i = \frac{R_A}{R_B}$$

The length of the belt is constant and can be simplified as follows. Of course, changes in the diameter ratio were omitted for a brief explanation:

$$\pi(x_A + x_B) + 2L_c = cosnt$$

If differentiated here, it can be written as follows:

$$\Delta x_A = -\Delta x_B$$

$i = \frac{x_A+R_A}{x_B+R_B}$ can be rewritten as:

$$i = \frac{\Delta x_A + R_A}{-\Delta x_A + R_B}$$

The velocity increases as the deceleration ratio increases.

The deceleration ratio has the smallest value because the angular velocity is zero at the start, which is a desirable characteristic in an electric vehicle. The centrifugal force, the length of the spring, and the angular speed of the driven part increase when the driving speed increases. When a load is applied to the driven part, the torque increases, and the driving part increases the engine torque, thus reducing the speed.

The spring and mass ratio at point B were ignored, but were considered in the actual design. To implement this principle, a spiral curve was added, and an auxiliary plate was used to extend it to a certain radius.

In this dynamical equation, the following variables were input and analyzed

According to the given spline angle α:

$$r_1 = f_1\ (m1, m2, N1, N2, \omega 1, T, K, \alpha, \mu 1, \mu 2) \tag{8}$$

$$r_2 = f_2\ (m1, m2, N1, N2, \omega 1, T, K, \alpha, \mu 1, \mu 2) \tag{9}$$

...

$$\eta = f_k\ (m1, m2, N1, N2, \omega 1, T, K, \alpha, \mu 1, \mu 2) \tag{10}$$

Here, the input variables are as follows:

r_1: Pin position radius of the input slot

r_2: Pin position radius of the output slot

η: Power-transmission efficiency

$m1$: Weight of input piece

$m2$: Weight of output piece

$N1$: Number of input pieces

$N2$: Number of output pieces.

$\omega 1$: Input rpm of motor

T: Output load torque

K: Tension spring factor of the output piece

$\mu 1$: Input plate slot, pin, and coefficient of friction

$\mu 2$: Output plate slot, pin, and coefficient of friction

The following functions are not explicit functions; therefore, they are difficult to solve mathematically. A trial-and-error approach using a computer dynamics program was used to generate a solution, which the provided graphs show below.

Figures 9–11 show a graph was drawn to compare the rail curvature angle and behavior of the push piece in each case of the normal curve and spline curve of the slide plate based on the ratio output value.

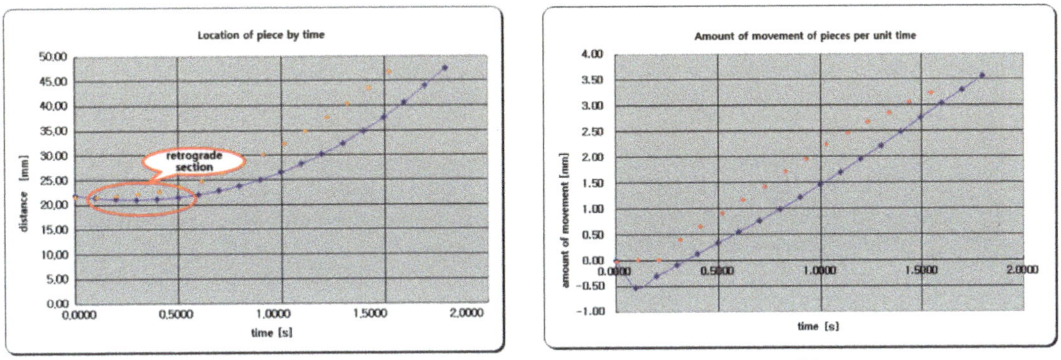

(a) (b)

Figure 9. When the normal (orange) and spline (blue) curve of the slide plate rail was 20 degrees at a constant speed: (**a**) Location of piece over time; (**b**) Amount of movement of pieces per unit time.

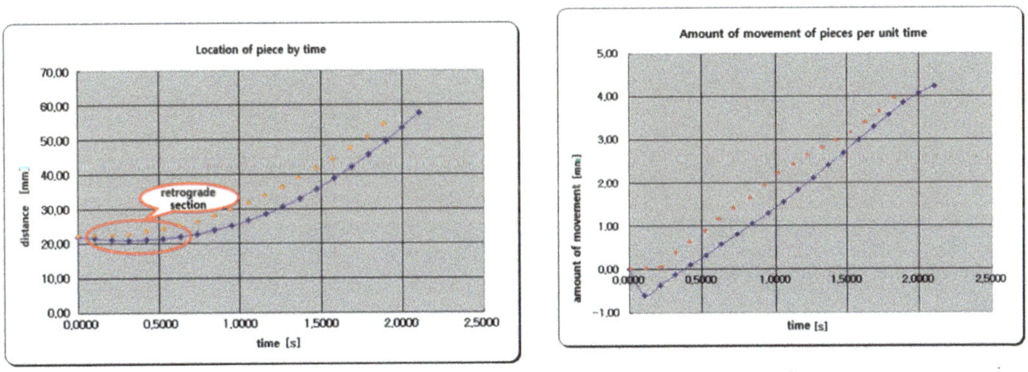

(a) (b)

Figure 10. When the normal (orange) and spline (blue) curve of the slide plate rail was 30 degrees at a constant speed: (**a**) Location of piece over time; (**b**) Amount of movement of pieces per unit time.

Figures 9 and 10 show the analysis was performed at an actual 1° interval at 20°, and there was no significant difference in the change in the spline curve between 20° and 30°.

The behavior of the push piece in the normal and spline curves at each angle was analyzed.

Figure 11a show Retrograde was observed at the beginning of the normal curve when analyz-ing the behavior of the fragments by angle.

Figure 11b there was a section where it did not ex-pand even after maximum expansion and was stagnant at 40°. It was established from the spline curve that the expansion time gradually increased as the angle increased.

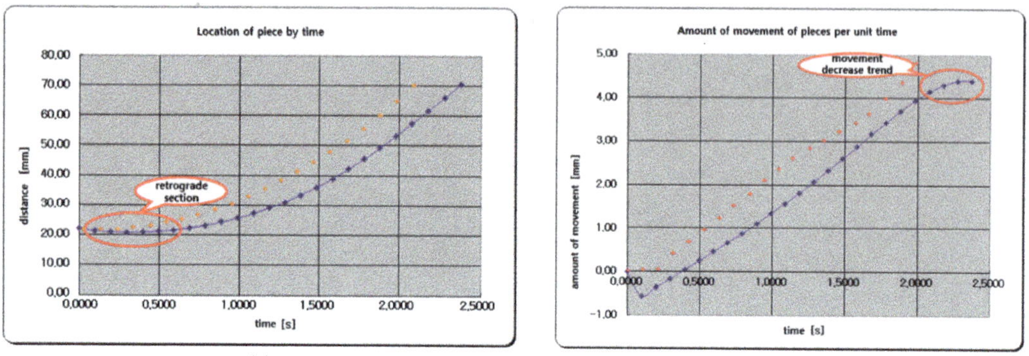

(a) (b)

Figure 11. When the normal (orange) and spline (blue) curve of the slide plate rail was 40 degrees at a constant speed; (**a**) Location of piece over time; (**b**) Amount of movement of pieces per unit time.

It stopped, due to interference, and then moved along the trajectory.

However, Figure 12 shows that the rail-grooved slide plate was designed using a spline curve. Therefore, the push piece moved immediately at the initial start without interference from the centrifugal force.

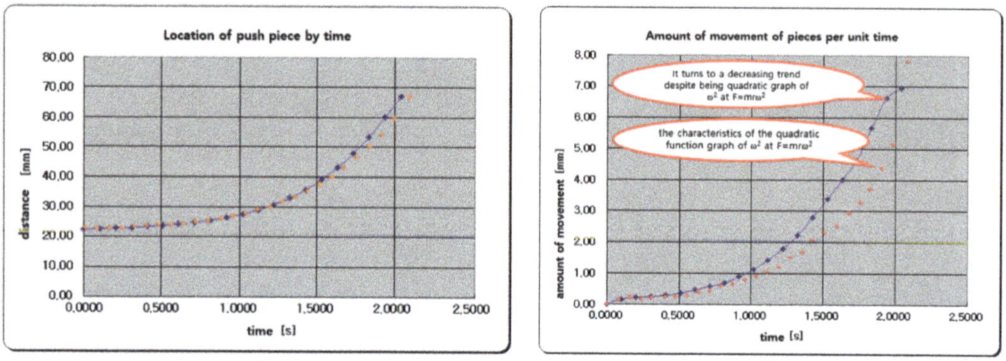

(a) (b)

Figure 12. Behavior analysis when the normal (orange) and spline (blue) of the plate rail was 28 degrees during acceleration; (**a**) Location of piece over time; (**b**) Amount of movement of pieces per unit time.

The movement amount was increased using a linear function method.

A guide plate can be designed and manufactured using splines when a rail groove is designed using a spline curve with centrifugal force F.

The slide-plate-like curve ensures the stability of the push piece during the design phase.

The plate structure was analyzed for its application to centrifugal pulley belt-type continuously variable transmission using the Abaqus program. Since the shaft was inserted in the plate hub, constraints were set on the plate hub, and carbon steel was used as material because SM45C was used in preparation.

As a result, it was found that the stress was the highest at the hub part where the load was generated when the shaft rotated, and the part where the distance between the hub part and the rail groove was the shortest, as shown in Figure 13. However, the maximum stress was 225.83 MPa, indicating that the yield strength of the SM45C selected as the plate material was about 490 MPa, and the tensile strength was above 686 MPa. Therefore, it was deemed that there would be no problem in using it.

Figure 13. Analysis of plate.

4. Centrifugal Belt Pulley

A CVT housing was designed for an electric vehicle with a 4-kW motor. It was manufactured so that the centrifugal belt pulley CVT could be attached to the speed reducer.

Although the power of existing electric vehicles is transferred from the motor to the speed reducer, the centrifugal belt pulley-type CVT was installed between the motor and the speed reducer by reinforcing the ribs inside the housing.

The motor was placed above the axle to reduce weight interference between the motor and vehicle components.

The distance between the axle of the driving pulley and driven pulley was set at 175 mm, and the belt length was set at 672 mm to calculate the contracted state of the driving pulley and the expanded state of the driven pulley.

The differential gear and L-shape were fixed, as shown in Figure 14b, so that the shaft could support the load of the motor and CVT.

Figure 14. Prototype production: (**a**) inside of housing; (**b**) assembly of reducer and housing.

5. Centrifugal Belt Pulley CVT Performance Tester

A performance tester was manufactured, as shown in Figure 15, to validate the performance of the centrifugal belt pulley CVT.

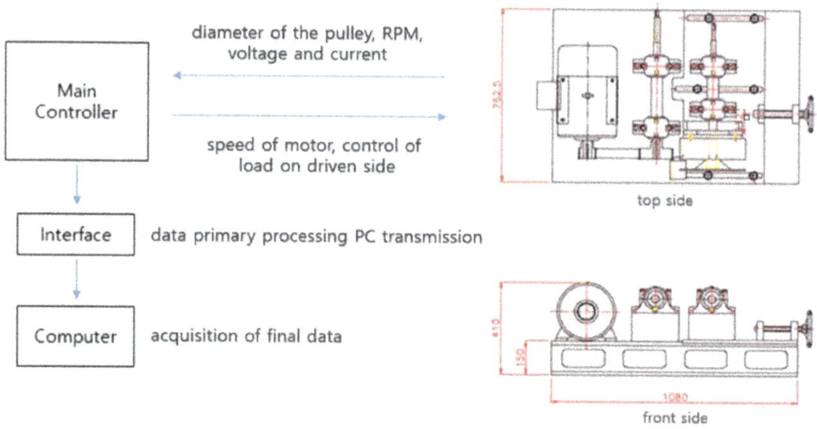

Figure 15. CVT tester.

The testbed of the performance tester was equipped with a motor that could rotate the transmission, and a brake that transmitted the load to the CVT.

In addition, a noncontact sensor was installed to measure the outer diameter of the pulley.

As shown in Figure 15, the main controller controlled the speed of the motor and the load on the driven side to determine the speed ratio of the pulley, the number of revolutions between the drive and driven plate, and the voltage and current. The measured values were processed at the interface and verified in the final calculation.

The software interface was configured such that the initial input values could be set and the final data could be generated and saved.

The driving speed was measured, and the driven plate, slip ratio, and transmission ratio are shown in Figure 17.

Figure 15 shows a performance test of the centrifugal belt pulley CVT was performed using a CVT tester. For the test method, Figure 16 shows the default value was entered on the main screen of the CVT tester, and the drive pulley was rotated by 2500 degrees. Figure 17 shows During the test, the change in the driven pulley and reduction ratio during acceleration were examined. In addition, after maximum acceleration, the change in the diameter and reduction ratio of the centrifugal belt pulley was confirmed when a load was applied to the driven shaft by braking.

As shown in Figure 18, when the drive pulley started to rotate, the driven pulley started to rotate as well. When the drive pulley was running at maximum speed, the rotational speed of the driven pulley was approximately equal to the rotational speed of the drive pulley. When a load was applied, at this point, the speed of the driven pulley dropped sharply and the reduction ratio rose to nearly 3:1; however, no significant change in the speed of the drive pulley was observed.

Driving durability tester

Figure 19 shows the driving tester manufactured to test the durability of the centrifugal belt.

The vehicle was equipped with a servomotor to control the accelerator. Additionally, a large roller that transmits rotation and load to the rear wheels, a power brake that delivers a load similar to that experienced when driving uphill, and a vibrator that simulates a road surface in poor conditions were configured to test durability. The vehicle was tied with a rope to prevent it from slipping off the rollers during testing.

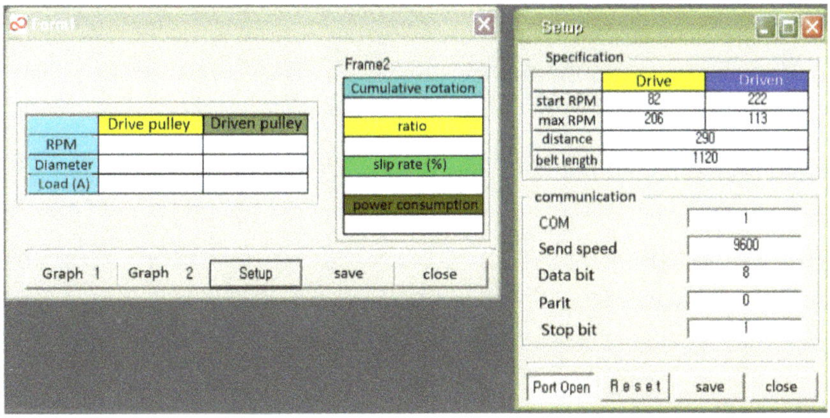

Figure 16. Main screen of computer output.

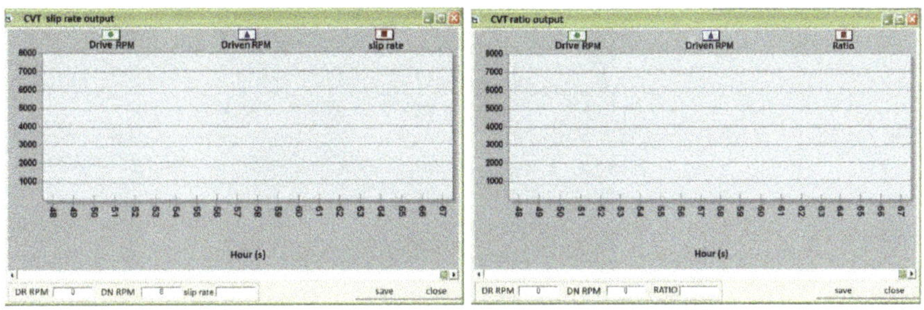

(a) (b)

Figure 17. Screenshot: (**a**) slip ratio; (**b**) transmission ratio.

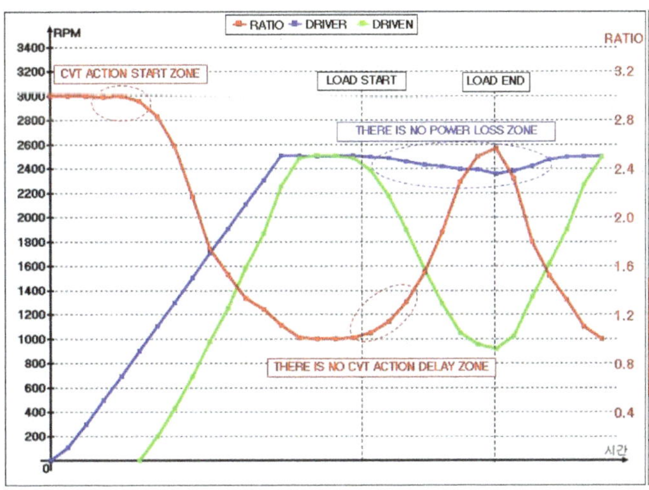

RPM		RATIO
DRIVER	DRIVEN	
300	100	3.00
496	166	2.99
696	232	3.00
903	305	2.96
1103	390	2.83
1297	500	2.59
1505	694	2.17
1708	980	1.74
1907	1250	1.53
2105	1587	1.33
2307	1867	1.24
2506	2256	1.11
2506	2489	1.01
2504	2510	1.00
2506	2509	1.00
2506	2489	1.01
2497	2386	1.05
2486	2174	1.14
2456	1894	1.30
2431	1568	1.55
2416	1287	1.88
2398	1048	2.29
2389	954	2.50
2356	917	2.57
2382	1025	2.32
2421	1347	1.80
2475	1623	1.52
2496	1897	1.32
2504	2268	1.10
2502	2495	1.00

Figure 18. CVT data.

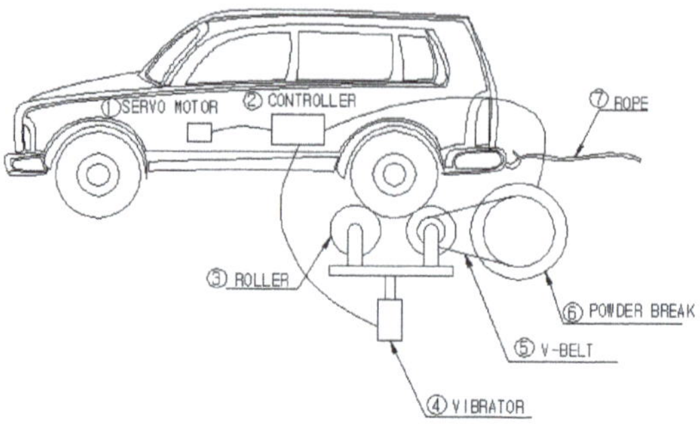

Figure 19. Driving tester concept.

As shown in Figure 20, a driving test and durability test were conducted, in which a centrifugal belt pulley was installed in a 4 kW electric vehicle with a CVT.

Figure 20. Actual vehicle installation.

6. Conclusions

In this study, experiments were conducted to demonstrate the acceleration, efficiency, and durability of a centrifugal belt-type CVT.

Section 1 is the introduction, and Section 2 described the differences between the existing CVT and the CVT developed in this study. The shifting principle of the new CVT was discussed as well. Section 3 detailed the design variables used to design the CVT and the reason for finding and designing the optimal rail curve. These curves were used to analyze the behavior of the push piece during shifting, and the strength of the CVT plate was examined using Abaqus software. Section 4 discussed photographs of the design and manufacturing of the transmission housing based on the above process. Section 5 detailed the tester that was manufactured for testing the CVT. In addition, the reasons and explanations for the manufactured parts were described.

In addition, a CVT performance tester was manufactured to test the performance of the proposed CVT. The housing for mounting the transmission in a vehicle was designed and manufactured. A driving tester was manufactured for the driving durability test, and driving and durability tests were performed. The findings of this study are summarized as follows:

(1) The reduction ratio was set at 3:1 to provide the initial driving force of a small electric vehicle equipped with a 10:1 reducer.
(2) Two types of plate rail grooves were designed so that multiple push pieces could rotate simultaneously. The behavior of the push part of a rail groove with a normal curve and a spline curve was analyzed using a Fortran program.
(3) It was verified that the rail groove using the spline curve showed excellent acceleration and responsiveness.
(4) The push piece's spline curve structure was designed for durability and safety.

(5) The shape and structure of the plate were designed to be compact and lightweight, considering the mileage of the vehicle. The housing, which protects against external shocks, was designed and manufactured for small CVTs with centrifugal belt pulleys, considering the distance between the axles.
(6) A performance tester equipped with a drive motor was manufactured, and a control circuit was configured in the main controller to measure the rotational speed and output values. The power-transmission efficiency of the final centrifugal belt pulley CVT was verified using this tester.
(7) Once the performance of the CVT was confirmed by the performance tester, the efficiency was confirmed by evaluating the rotation speed of the output versus the input and the corresponding gear ratio.
(8) The durability of the developed transmission was confirmed using a driving tester manufactured to simulate bad road conditions.
(9) A sudden belt slip occurred due to the regenerative braking operation, due to the characteristics of the traction motor in a downhill or elastic driving condition.
(10) If regenerative braking is required, then it should be possible to stop the motor without using a CVT.

Author Contributions: All authors contributed to this study. All authors designed the new method and conducted the experiments. S.-h.O. and H.-j.D. performed experiments and wrote the manuscript. All authors have read and agreed to the published version of the manuscript.

Funding: This work supported by the Technology Innovation Program (20001184, Development of ultrathin and short precision reducers and high torquetoweight ratio precision reducers for HRC robots) funded By the Ministry of Trade, Industry & Energy (MOTIE, Korea).

Conflicts of Interest: The authors declare no conflict of interest.

Abbreviations

CVT	Continuously variable transmission
TCU	Transmission control unit
HEV	Hybrid electric vehicle

References

1. Shin, Y.I.; Shin, S.H.; Oh, T.I.; Suh, J.S.; Song, C.K. Mass Reduction of Transmission Gears for Commercial Vehicles. *J. Korean Soc. Precis. Eng.* **2012**, *29*, 319–323. [CrossRef]
2. Youm, K.W.; Ham, S.H.; Oh, S.H. A Study on the Shift Motor Driving System Optimization of 4-WD Power Transformation Device. *J. Korean Soc. Precis. Eng.* **2013**, *30*, 1187–1192. [CrossRef]
3. Lee, I.K.; Moon, H.H.; Kim, J.J. Study for Examples of Fire Including Friction with Automotive Clutch, Manual Transmission and Tire System. *J. Korean Inst. Gas* **2015**, *19*, 49–53. [CrossRef]
4. Youm, K.W.; Ham, S.H. Study on Shift characteristic of Small reducer using Eccentric arm. *J. Korea Acad. Ind. Coop. Soc.* **2017**, *18*, 609–614. [CrossRef]
5. Park, K.H.; Jung, S.H.; Wee, H.; Lee, G.S.; Cho, S.H. A Study on the Fault Detection of Auto-transmission Using the Vibrational Characteristics of Roller Bearings. *J. Korean Soc. Noise Vib. Eng.* **2009**, *19*, 268–2736.
6. Kim, J.Y.; Yim, C.Y.; Lim, W.S. Development of Shift-By-Wire System for an Automatic Transmission Equipped Vehicle. *J. Korean Soc. Automot. Eng.* **2005**, *15*, 41–46.
7. Kwon, H.S. Assemblability Analysis of the Kinematic Configurations of Ravigneaux-Type Automatic Transmissions. *J. Korean Soc. Manuf. Process Eng.* **2019**, *18*, 49–58. [CrossRef]
8. Julió, G.; Plante, J.S. An experimentally-validated model of rubber-belt CVT mechanics. *Mech. Mach. Theory* **2011**, *46*, 1037–1053. [CrossRef]
9. Kim, S.M.; Zheng, C.; Lim, W.S.; Cha, S.W. Performance Analysis of the Rubber Belt type CVT System. *J. Korean Soc. Manuf. Technol. Eng.* **2011**, *20*, 376–381.
10. Kim, G.S.; Kwon, Y.W. The Development of Gradient Response CVT for a Small Size Electric Vehicle. *J. Korean Soc. Power Syst. Eng.* **2015**, *19*, 33–38. [CrossRef]
11. Yang, Y.; He, X.; Zhang, Y.; Qin, D. Regenerative Braking Compensatory Control Strategy Considering CVT Power Loss for Hybrid Electric Vehicles. *Energies* **2018**, *11*, 497. [CrossRef]

12. Luo, Y.; Baddour, N.; Liang, M. Dynamical Modeling of Gear Transmission Considering Gearbox Casing. *Conf. Mech. Vib. Noise* **2018**, *8*, 24–30. [CrossRef]
13. Karhula, H.; Nicolai, M.; Desmet, W. Exhaustive Synthesis and Analysis of Automotive 2-Stage Planetary Transmission Designs. *Int. Power Transm. Gearing Conf.* **2019**, *10*, 37–48. [CrossRef]
14. Manring, N.D.; Al-Ghrairi, T.S.; Vermillion, S.D. Designing a Hydraulic Continously Variable-Transmission (CVT) for Retrofitting a Rear-Wheel Drive Automobile. *J. Mech. Des.* **2013**, *135*, 3–13. [CrossRef]
15. Ryu, W.; Kim, H. Ratio Control of CVT by Considering the CVT System Efficiency. *J. Korean Soc. Automot. Eng.* **2006**, *14*, 157–163.
16. Zhu, H.; Zhang, Y.; Wang, Y. Study on the Slip Rate of CVT in Un-steady States. *IFAC-PapersOnLine* **2018**, *51*, 343–346. [CrossRef]
17. Dwivedula, R.V.; Pagilla, P.R. Effect of Compliance and Backlash on the Output Speed of a Mechanical Transmission System. *J. Dyn. Syst. Meas. Control* **2012**, *134*, 10–18. [CrossRef]
18. Xiang, C.L.; Huang, Y.; Liu, H. Response Sensitivity and the Assessment of Nonlinear Vibration Using a Nonlinear Lateral–Torsional Coupling Model of Vehicle Transmission System. *J. Vib. Acoust.* **2015**, *137*, 13–23. [CrossRef]
19. Moriwaki, I.; Nishiyama, M. Gear Noise Diagnosis System for Automobile Transmission Using Artificial Intelligence: Improvement of Judgement Performance. *Int. Power Transm. Gearing Conf.* **2000**, *6*, 695–700. [CrossRef]

energies

MDPI

Article

Mechanical Stress in Rotors of Permanent Magnet Machines—Comparison of Different Determination Methods

Christian Monissen [1], Mehmet Emin Arslan [1], Andreas Krings [2] and Jakob Andert [1,*]

1 Teaching and Research Area Mechatronics in Mobile Propulsion, Faculty of Mechanical Engineering, RWTH Aachen University, Forckenbeckstraße 4, 52074 Aachen, Germany
2 FEV Europe GmbH, Neuenhofstr. 181, 52048 Aachen, Germany
* Correspondence: andert@mmp.rwth-aachen.de

Abstract: In this work, different analytical methods for calculating the mechanical stresses in the rotors of permanent magnet machines are presented. The focus is on interior permanent magnet machines. First, an overview of eight different methods from the literature is given. Specific differences are pointed out, and a brief summary of the analytical approach for each method is provided. For reference purposes, a finite element model is created and simulated for each rotor geometry studied. A total of seven rotors rom representative automotive powertrains are considered in their specific speed range. The analytical methods are used to determine the maximum mechanical stress concentration factors for the seven rotor geometries, in which we are determined to find maximum mechanical stress as a final step of the analytical process. For each geometry and each respective operating speed range, the deviations from the finite element reference are determined. In addition, the error in the selected geometry variations is evaluated. A recommendation for the method with the lowest error considering all cases studied is given specifically for the stress in the airgap bridge and the central bridge.

Keywords: maximum mechanical stress; interior permanent magnet synchronous machine; finite element analysis; high-speed e-machine; stress concentration factor; analytical methods

Citation: Monissen, C.; Arslan, M.E.; Krings, A.; Andert, J. Mechanical Stress in Rotors of Permanent Magnet Machines—Comparison of Different Determination Methods. *Energies* **2022**, *15*, 9169. https://doi.org/10.3390/en15239169

Academic Editors: Kan Liu and Wei Hu

Received: 8 November 2022
Accepted: 30 November 2022
Published: 3 December 2022

1. Introduction

Mechanical stress analysis for rotor geometries is particularly important when considering the ongoing trend toward higher speeds in electric machines for traction applications. An overview of this is given in [1] for commercially available battery or hybrid electric vehicles.

The traction machines of the Toyota Prius across the generations can serve as an example. Here, it can be seen that with approximately the same rotor diameter range, the maximum speed has increased from 6000 rpm in the first Prius generation introduced in 1997 to 17,000 rpm in the fourth generation introduced in 2015 [2,3]. With increased speed and the same radial rotor dimensions, the resulting centrifugal forces increase accordingly. These centrifugal forces act on the rotor and cause high mechanical stresses in certain areas, depending on the rotor pole geometry [4,5].

The maximum mechanical stresses (MMS) in the rotor are limited by the yield strength of the materials. The commonly used electric steel materials have yield strength values between 320 and 500 MPa [6,7], and the electrical machines on the market for traction applications have MMS values mainly in that range. As an example, Figure 1 shows finite element analysis (FEA) results of the MMS of a rotor based on the electric machine of the Toyota Prius III at its maximum speed of 13,500 rpm.

The MMS estimation in a rotor configuration is typically carried out by static mechanical FEA [7–11]. However, the convergence of the solution strongly depends on the mesh size and quality. The time for re-meshing and solving after each iteration, as well as extra computational effort for MMS estimation, must be considered [12,13]. To reduce time and computational cost, several analytical methods using simplified equivalent models to determine the average mechanical stress in these rotors have been proposed in the literature.

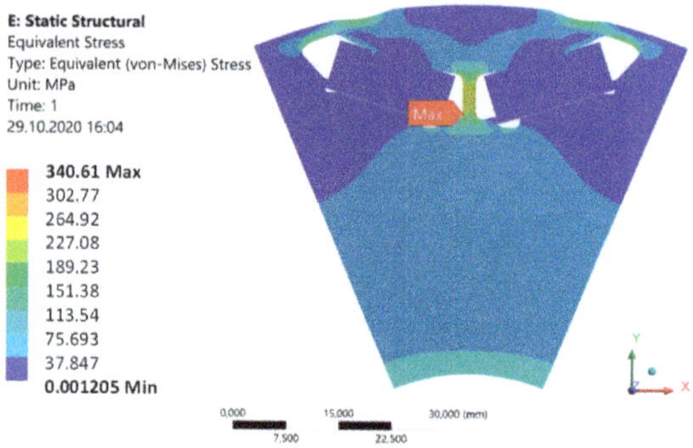

Figure 1. Distribution of von Mises stresses of Toyota Prius III rotor at 13,500 rpm.

To obtain a meaningful value for MMS based on average mechanical stress, a stress concentration factor (SCF) should be determined at the intersections where the geometric cross-section changes drastically [12–18]. Figure 2 shows a generic rotor pole of an interior permanent magnet synchronous machine (IPMSM). The areas where MMS mainly occurs are the intersection points of the air gap bridges and the central bridges (circled).

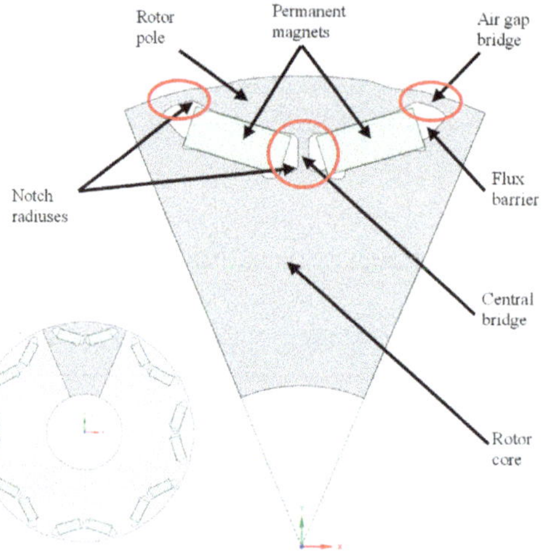

Figure 2. Definition of a single-pole rotor geometry definition marked areas where MMS can occur.

The approach of using equivalent mechanical models for the stress analysis in electrical machine rotor configurations is relatively new. Table 1 shows the methods studied with their publication date. The equivalent ring method (ERM), the centrifugal force method (CFM) and the beam theory method (BTM) are proposed to determine MMS on the air gap bridges using different approaches. The ERM method has evolved throughout the years with improvements and simplifications in the calculations. Therefore, there are five different ERM methods [12,13,15,19–21]. In this paper, the different ERM methods are chronologically referred to as ERM1 to ERM5. The central bridge method (CBM) and BTM method determine the MMS at the central bridges of V-shaped rotor geometries using different approaches [12,18,22].

Table 1. Analytical methods proposed in the literature for determining the maximum mechanical stress of an IPMSM.

Analytical Methods	Year of Publication	Author	Applied on
ERM1	2002–2006	Schätzer [19]–Binder et al. [20]	Air gap bridges
ERM2	2014	Li Yi et al. [21]	Air gap bridges
ERM3	2016	Chai et al. [12]	Air gap bridges
ERM4	2018	Chu et al. [15]	Air gap bridges
ERM5	2020	Chu et al. [13]	Air gap bridges
CFM	2020	Chu et al. [13]	Air gap bridges
BTM	2018	Kleilat et al. [18]	Air gap and central bridges
CBM	2016	Chai et al. [12]	Central bridges

Chu et al. [15] proposed an SCF for the air gap bridges on flat rotor geometries by using the polynomial surface fitting technique. Chai et al. [12] and Bremner et al. [17] proposed two different references: Peterson's stress concentration [16] and Roark's formulas for stress and strain [23] to determine SCF at points where MMS occur.

The listed analytical methods have been developed recently and are validated in the literature only for basic rotor geometries. Consequently, the comparison of accuracy and stability of these proposed methods needs to be further investigated. The evaluation of the most accurate and stable analytical methods is performed by applying the methods to current geometries of IPMSM traction drives available on the market. The results are compared to detailed FEA simulations in Ansys® Mechanical, Release 18.1. The most suitable method can finally be included in the pre-design processes and can significantly reduce the computational effort.

2. Materials and Methods

2.1. Analytical Methods to Determine the Maximum Mechanical Stress at the Air Gap Bridge and the Central Bridge

The process of each method and their main differences are shown in Figures 3 and 4.

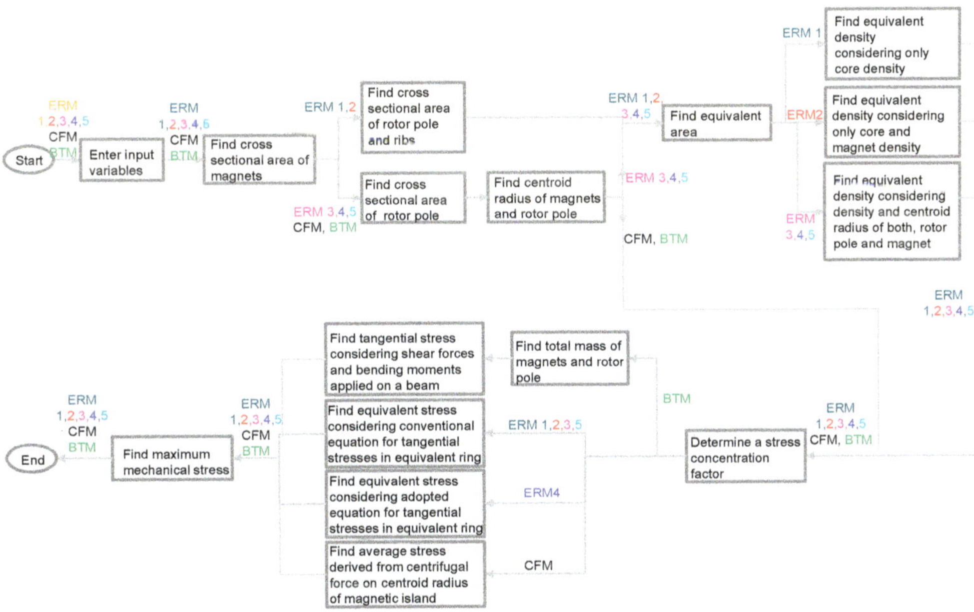

Figure 3. Methodology flow diagram of ERM1 to ERM5, CFM and BTM and general differences in the approaches.

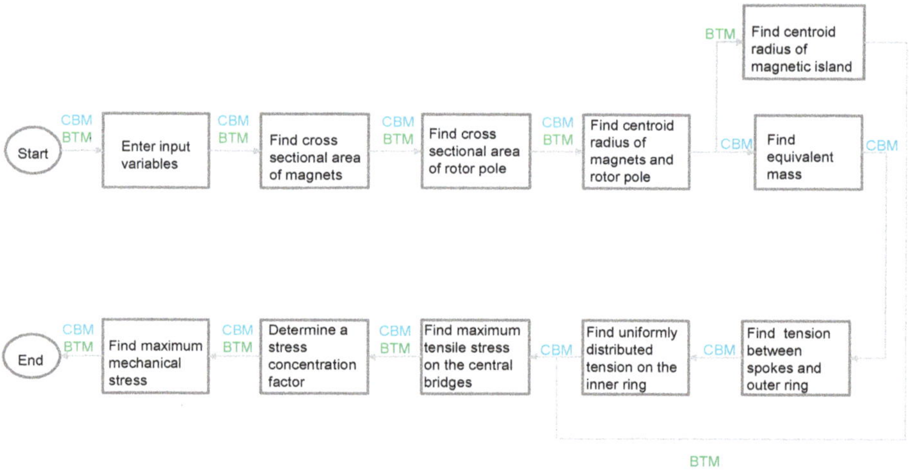

Figure 4. Methodology flow diagram of BTM and CBM and general differences in the approaches.

The most relevant parameters of an IPMSM are defined in Figure 5, the nomenclature of the regions is shown in Figure 1. The nomenclature and parameter definition for a V-shaped rotor geometry are analogous to the flat-shaped rotor geometries. R_0 is the outer radius of the rotor, d is the length of the airgap bridge, r_a is the outer radius, r_c is the inner radius of the magnet pocket, h is the magnet height, t is the length of central separation of magnet pockets, θ is the opening angle of the V-shape of the magnet, and α is the angle between the position of r_a to the center of the magnet pole.

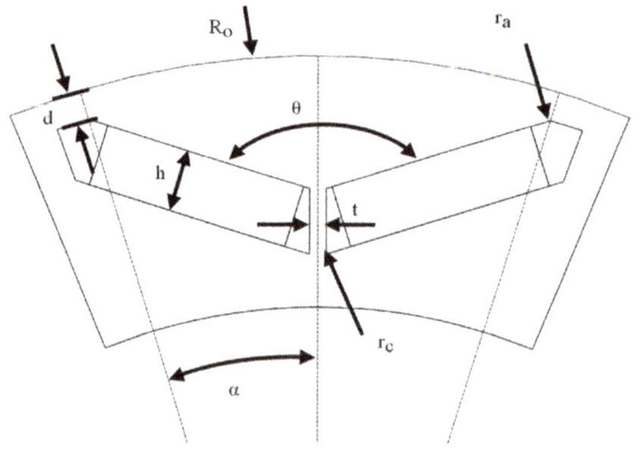

Figure 5. Parameter definition of a typical V-shaped rotor geometry (only the investigated parameters are shown).

Figure 6 shows the basic equivalent geometry approaches of each proposed method in the literature. In general, all methods define a simplified geometry, with an equivalent mass based on geometry and materials under investigation. In methods ERM2 to ERM5, the magnet-surrounding core material and the magnet are considered. For the ERM1 method only, the core material is considered. In ERM methods, only one ring is positioned at the airgap bridge representing the simplified geometry, as shown in Figure 6. With ERM1 and ERM2, no centroid radius of magnets and rotor poles are used, only the cross-sectional areas of rotor pole and ribs. With ERM3 to ERM5, these geometric quantities are considered, resulting in a more realistic representation of the model. All ERM methods use a sole ring at the outer radius of the rotor, which also limits the applicability of the stress estimation on the stress in the airgap bridge. This limitation must be considered when the maximal mechanical stress occurs in other regions of the rotor. Although a more detailed equivalent geometry is used in the CFM method, a similar limitation to the ERM method applies since no central bridge is represented. The stress in the central bridge can only be captured by the BTM and CBM methods mentioned above. Both methods also consider the central bridge by a beam or spoke, respectively, cf. Figure 6c,d. In the following, all mentioned methods are briefly recapitulated, and the equations of equivalent stress and maximum stress are given.

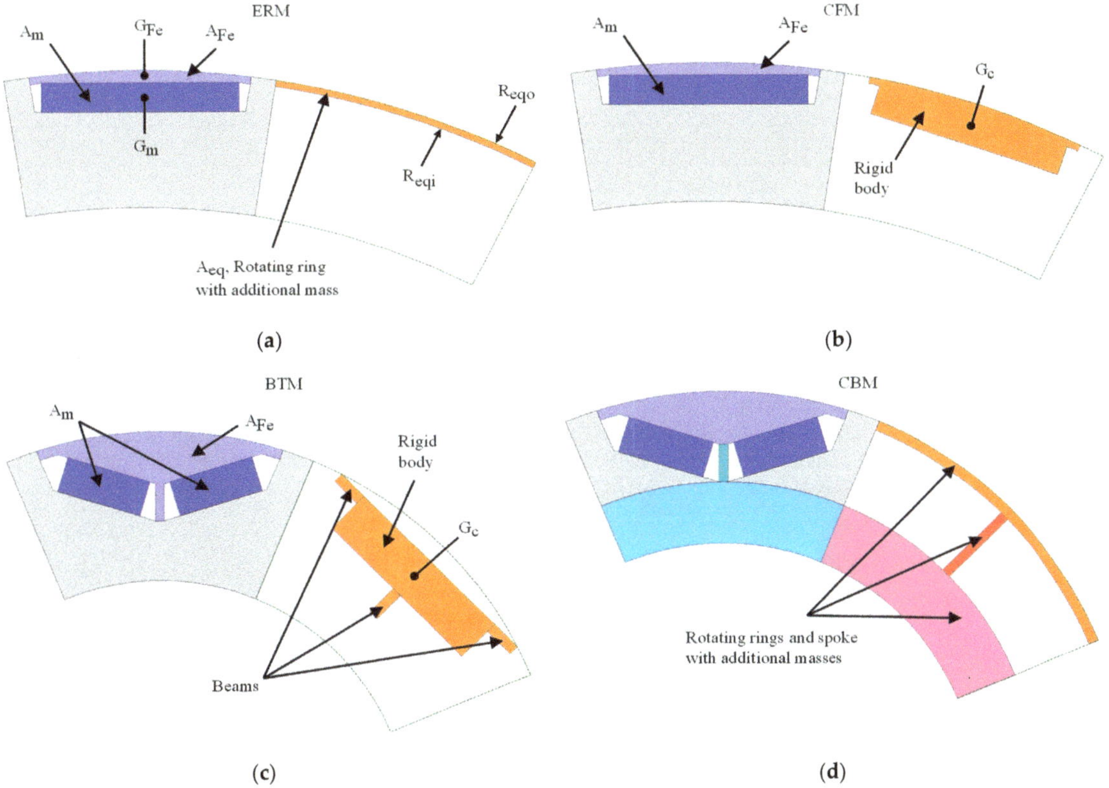

Figure 6. Real and equivalent models of analytical methods proposed in the literature: (**a**) ERM method, (**b**) CFM method, (**c**) BTM method, (**d**) CBM method.

First, the ERM1 method [19,20] is considered, which approximates the rotor pole and magnets as an equivalent ring model with additional mass density. The thickness of the ring is equal to the narrowest thickness of the air gap bridges. The authors introduced the equivalent density $\rho_{eq,erm1}$ as:

$$\rho_{eq,erm1} = \rho_{Fe} \cdot \frac{A_{Fe} + A_m}{A_{eq}}, \quad (1)$$

where ρ_{Fe} is the mass density of electric steel in kg/m^3. A_{Fe}, A_m and A_{eq} represent the cross-sectional area of the rotor pole, the permanent magnets and the equivalent ring, respectively, as shown in Figure 6a. It can be observed that only the density of the electric steel is assumed for the whole considered region.

The tangential stresses $\sigma_{eq,erm1}$ and MMS $\sigma_{eq,max1}$ inside the equivalent ring are calculated as:

$$\sigma_{eq,erm1} = \left(\frac{R_{eqo}+R_{eqi}}{2}\right)^2 \cdot \omega^2 \cdot \rho_{eq,erm1}, \quad (2)$$

$$\sigma_{eq,max1} = k_{air\ gap\ bridge} \cdot \sigma_{eq,erm1}, \quad (3)$$

Here, R_{eqo} and R_{eqi} represent the outer and inner radius of the equivalent ring, respectively, as shown in Figure 6a. ω is the rotational speed in rad/s, and $k_{air\ gap\ bridge}$ is the SCF at the air gap bridges, which is introduced in Section 2.2 Determination of Stress Concentration Factors.

Equation (1) is improved in the ERM2 method [21] by also considering the mass density of magnets:

$$\rho_{eq,erm2} = \frac{\rho_{Fe}A_{Fe}+\rho_m A_m}{A_{eq}}, \quad (4)$$

where ρ_m is the mass density of the magnet material.

ERM3 method uses Equation (4) and is further developed by taking the centroid radii of magnets and the rotor pole into account with:

$$\rho_{eq,erm3} = \frac{(R_{Fe}\rho_{Fe}A_{Fe}+R_m\rho_m A_m)\frac{1-\cos(\theta)}{2}}{R_0 A_{eq}}, \quad (5)$$

where R_{Fe} and R_m stand for the distance between the rotor center and the center of gravity of rotor pole G_{Fe} and center of gravity of the permanent magnets G_m, respectively. ρ_m is the mass density of the permanent magnet in kg/m^3. θ represents the angle between two magnets of a single pole in rad (see Figure 5). The centroid radius of the equivalent ring R_0 is calculated as:

$$R_0 = \left(\frac{R_{eqo}+R_{eqi}}{2}\right), \quad (6)$$

The calculations in Equation (5) are simplified in the ERM4 method [15], and Equation (2) is replaced by an adopted $\sigma_{eq,erm4}$ as:

$$\sigma_{eq,erm4} = \frac{\rho_{eq,erm4}\omega^2}{8}[(3+\nu)\left(R_{eqo}^2(1+R_{eqi})+R_{eqi}^2\right) - (1+3\nu)\,R_{eqi}^2], \quad (7)$$

where ν is the Poisson's ratio of electric steel, and $\rho_{eq,erm4}$ is the equival $\rho_{eq,erm4}$ is the equivalent density for the ERM4 method as:

$$\rho_{eq,erm4} = \frac{R_m\rho_m A_m + R_{Fe}\rho_{Fe}A_{Fe}}{R_0 A_{eq}}, \quad (8)$$

In the ERM5 method [14], the simplifications in the ERM4 method are used, but instead of using Equation (7) to determine the tangential stresses, the initial approach, Equation (2), is used again.

The CFM method considers rotor pole and magnets as a rigid body [13]. By performing a force analysis on that rigid body, average stresses at the air gap bridges σ_{cfm} are determined as:

$$\sigma_{cfm} = \frac{(\rho_m A_m+\rho_{Fe}A_{Fe})\omega^2 R_c}{2\sin(\frac{\alpha}{2})d} \quad (9)$$

where R_c is the distance between the rotor center and the center of gravity of magnets with rotor pole G_c as shown in Figure 6b. d is the air gap bridge thickness and α is half pole arc in rad as shown in Figure 5.

The BTM method simplifies the air gap bridges and central bridges to simple rectangular beams and considers the rotor pole with magnets as a rigid body, as can be seen in

Figure 6c [18,22]. Average mechanical stresses at the air gap bridges $\sigma_{btm,air\ gap\ bridge}$ and central bridges $\sigma_{btm,central\ bridge}$ are given as:

$$\sigma_{btm,air\ gap\ bridge} = \left(\frac{\sqrt{c2^2 + \left(6 \cdot \frac{c3}{d}\right)^2 \omega^2}}{l_{stack}} \right) \frac{m_{total} R_c}{d} \quad (10)$$

$$\sigma_{btm,central\ bridge} = \left(\frac{c1 \cdot \omega^2}{l_{stack}} \right) \frac{m_{total} R_c}{t} \quad (11)$$

where c1, c2 and c3 represent constant coefficients that quantify the relationship between centrifugal force, normal force and bending moment, respectively. The coefficients must be fitted to the FEA results. After the initial FEA calculation, these coefficients must be adjusted so that the FEA result matches the analytical result. l_{stack} and t are the stack length and the central bridge thickness, respectively. The total mass of the rotor pole with magnets m_{total} is expressed as:

$$m_{total} = (\rho_{Fe} A_{Fe} + \rho_m A_m) \cdot l_{stack} \quad (12)$$

In the CBM method, central bridges are equated with spokes, and the rotor pole with magnets and the rotor core are equated with an outer equivalent ring and an inner equivalent ring as in the conventional ERM methods [12]. The mechanical stresses at the central bridge σ_{cbm} are determined as:

$$\sigma_{cbm} = \frac{Y}{A_{sr}} \quad (13)$$

where A_{sr} is the radial cross-sectional area of the spokes, which is equal to the multiplication of the stack length and the thickness of the central bridge. The uniformly distributed tension on the inner ring Y is calculated as:

$$Y = X + \frac{q_s \omega^2}{2g} (R_2^2 - R_1^2) \quad (14)$$

where R_1 and R_2 are the inner and outer radii of the spokes. q_s is the spokes average weight per unit.

The tension between the spokes and the outer ring X is recalculated due to a probable mistake in the formula given in [13] as.

$$X = \frac{\frac{4\pi q_0 R_0^2}{EA_0} - \frac{nR_v q_s \omega^2 (R_2^2 - R_1^2)}{4\pi EA_v g} - \frac{(R_2 - R_1) q_s \omega^2}{2A_s gE} \cdot \left(R_2^2 - \frac{R_2^2 + R_2 R_1 + R_1^2}{3} \right)}{\frac{R_2 - R_1}{A_s E} + \lambda_0 + \frac{nR_v}{2\pi EA_v}} \quad (15)$$

where q_0 is the uniformly distributed inertial load, n is the number of spokes, E is the Young's modulus of electric steel, g is the gravitational acceleration, λ_0 is the flexibility coefficient of the outer ring, which express how the displacement along the centerline of the outer ring is influenced by the tension between the spokes and the outer ring, and R_v is the centerline radius of the inner ring. A_0, A_v, A_s represent the cross-sectional areas of the outer ring, the inner ring and the spokes. The equations for the defined symbols can be taken from [12].

2.2. Determination of Stress Concentration Factors

Chu et al. [13–15] proposed several methods to determine an accurate stress concentration factor (SCF) at the air gap bridges. The method proposed in [15] is considered in this study because it is applicable to a rotor geometry without considering a rotor pole number. The SCF at the air gap bridge $k_{air\ gap\ bridge}$ is determined as:

$$k_{air\ gap\ bridge} = 1.389-0.2374 \cdot \frac{d}{r_a} + 0.007181 \cdot \alpha + 0.03567 \cdot \left(\frac{d}{r_a}\right)^2$$
$$0.02148 \cdot \frac{d}{r_a} \cdot \alpha + 0.000745 \cdot \left(\frac{d}{r_a}\right)^3 - 0.001721 \cdot \left(\frac{d}{r_a}\right)^2 \cdot \alpha \quad (16)$$

where r_a represents the notch radius at the air gap bridge as shown in Figure 5. The case of a stepped flat bar under tensile stress from [16] proposed by [12] is taken as an equivalent case to determine SCF at the central bridge $k_{central\ bridge}$ as:

$$k_{central\ bridge} = 1.97-0.384 \cdot \left(\frac{2r_c}{t+2r_c}\right) - 1.018 \cdot \left(\frac{2r_c}{t+2r_c}\right)^2 + 0.43 \cdot \left(\frac{2r_c}{t+2r_c}\right)^3 \quad (17)$$

where r_c is the notch radius on the central bridge as shown in Figure 5, and the constant numbers are presented in Reference [16] to determine the SCF.

2.3. Initial Conditions and Assumptions

There are many factors such as thermal stress, rotor vibration, and unstable operation that affect the mechanical strength of a rotor. In this study, simplifications and initial conditions are assumed when evaluating the mechanical strength of IPMSMs in FEA. These assumptions can be considered feasible since the analytical approaches do not account for any of the simplified or neglected effects:

- Constant speed in steady state operation is assumed.
- Maximum deformation and stresses are mainly caused by centrifugal forces. The effect of electromagnetic forces between rotor and stator, and attraction forces between the permanent magnets and the rotor are considered negligible compared to centrifugal forces.
- Thermal effects are neglected.
- One pole in a two-dimensional model is used.
- The rotor core is assumed as an entity, and the lamination effects are considered negligible.
- Yield indicated by planar von Mises stress.
- The eccentricity of the rotor, the vibration, and the dynamic forces of the shaft are neglected.
- The analytical and numerical methods are based on the mono dimensional elastic theory. The inaccuracy of the methods proposed in this study increases with the mechanical stress above the yield point.

2.4. Investigated Machines

Seven rotor geometries with different rotor topologies, including those of the e-machines of the Nissan Leaf vehicle (rotor 3) and the Toyota Prius III vehicle (rotor 4), are used to investigate the stability of the analytical methods [1]. The main parameters of investigated geometries are presented in Table 2. To verify the general applicability of the analytical methods, different rotor topologies such as flat shape, V-shape, double V-shape and delta shape, different sizes, and different speed ranges are considered. One pole of each rotor topology is shown in Figure 7.

Table 2. Geometries of the investigated rotors.

Rotor Geometry	Geometry Definition/Unit				
	Outer Diameter R_o/mm	Max. Speed/rpm	Air Gap Bridge Thickness d/mm	Central Bridge Thickness t/mm	Angle between Magnets θ/degree
Rotor 1 flat shape	238	8000	1.26	-	-
Rotor 2 flat shape	206	11,200	1.8	-	-
Rotor 3 delta shape	130	10,400	0.7	-	-
Rotor 4 V-shape	160	13,500	1.94	1.8	146
Rotor 5 V-shape	206	11,200	2	2	170
Rotor 6 V-shape	146	18,100	2.3	3.4	137
Rotor 7 double V-shape	124	11,200	0.9	1.45	126

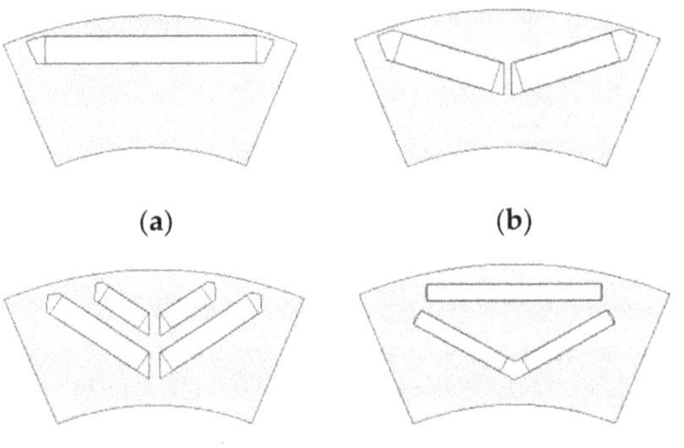

Figure 7. One pole view of the interior permanent magnet rotor geometries (**a**) flat shape, (**b**) V-shape, (**c**) double-V shape, (**d**) delta shape.

To determine the strength of the rotor geometries, FEA analyses were first performed to determine the location of the MMS and to visualize the stress distribution. Then, MMS at the air gap bridges for rotor 1, rotor 2 and rotor 3, and MMS at the central bridges for rotor 4, rotor 5, rotor 6 and rotor 7 were determined using the different analytical methods. To investigate the validity of these methods, the FEA results were taken as reference and the percentage difference between the FEA results and the analytical methods was determined.

The validity of the analytical methods was investigated over the speed range as well as through geometric design adjustments. To investigate the validity over the speed range, the speeds were changed from 1000 to 10,000 rpm. The design parameters shown in Figure 5 such as air gap bridge thickness (Figure 5d), magnet thickness (Figure 5h), notch radius on the air gap bridge r_a, notch radius on the central bridge r_c, and central bridge thickness were investigated by changing only the investigated parameter and keeping all other parameters constant. In addition, the entire rotor geometry was scaled to study the change in the stability of the analytical methods. The studied design parameters were changed in a reasonable range from half of the initial value to double the reference value.

The FEA analysis was performed using static structural analysis in ANSYS Workbench. Figure 8a shows the boundary conditions applied on a single rotor pole based on the assumptions and conditions given in reference papers [8,12,15,21]. A frictionless support was assumed on the symmetry axes to satisfy the symmetry condition. A cylindrical support on the inner surface of the rotor allowed the chosen nodes to move freely in radial direction, while preventing them to move on the tangential axis. Figure 8b shows the increased mesh density at the intersection points of air gap and central bridges where MMS

In order to investigate the behavior of the method over the entire speed range of the machines, Figures 9 and 10 show the MMS at the air gap bridge and the central bridge, respectively. The determined MMS in FEA and the selected analytical methods in the particular speed range are shown for rotors 1 to 3 and for rotors 4, 6 and 7. Additionally the error of each method is shown. It can be observed that for chosen methods, ERM 5 and CBM methods, the resulting error is stable over the whole speed range and is close to the given numeric value in Table 3.

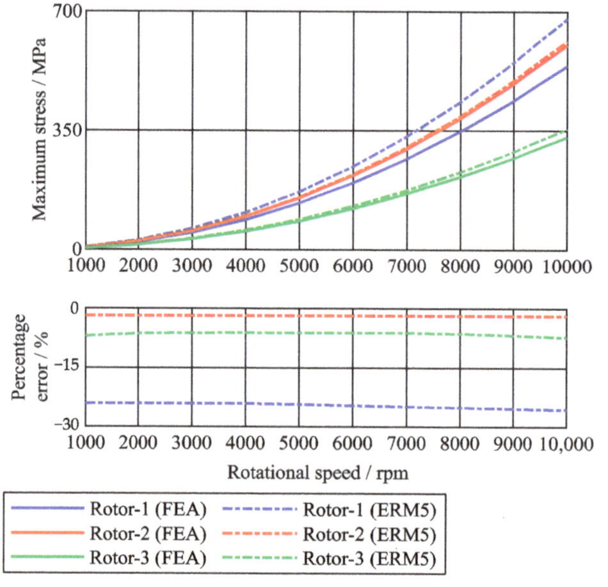

Figure 9. Stability analysis of the ERM5 method vs. rotor speed for different rotor geometries.

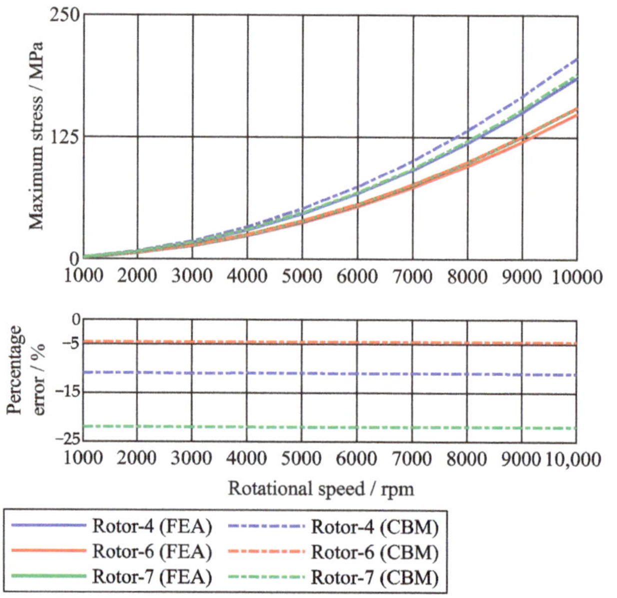

Figure 10. Stability analysis of the CBM method vs. rotor speed for different rotor geometries.

For development processes, the MMS could be investigated for geometry variations in a given rotor design. For this approach, a stable behavior of the method is required for specific geometry changes. The stability of selected rotor geometries under geometric design variations is presented in Figures 11 and 12. In this study, universal rotor geometry parameters are chosen, namely airgap bridge thickness, magnet thickness, notch radius and a general geometry scaling. The range of the parameters was chosen arbitrarily from 50% to 200% of the original value. This is a rather wide variation range especially for airgap and central bridge thickness, and the upper and lower limits might not be a typical rotor geometry, which could also affect the results of the analytical approaches and FEA simulations. The stability of the notch radius is only investigated for rotor 1 and rotor 4 geometries. Since the SCF is only dependent on the notch radius, but not dependent on the analytical methods, it is expected to have the same stability for different geometries. For the ERM5 method, magnet and core area will definitely be considered. However, the airgap bridge thickness will affect the actual stress in the rotor. This does not seem to be captured by the ERM5 method, as can be seen in Figure 11, where the error for an increased airgap bridge is increased for all rotor geometries (maximal by 25% and around 12% for rotor 2 and rotor 3). The geometry scaling seems to have the least effect on the resulting error; here, a slightly higher effect on the ERM5 is visible (below 10%) compared to the CBM (below 5%). In terms of overall error, geometry scaling can be performed accurately with ERM and CBM. For ERM5, a reduced magnet thickness compared to the original geometry shows a significant underestimation of the MMS, especially for rotor 3 (delta-shaped rotor). This again indicates that the more complex rotor geometry must be treated with caution using the simple ERM model. The variation of the central bridge thickness with regard to CBM is comparable to the variation of the airgap bridge thickness with ERM5. Examination of the CBM error with variation of the central bridge thickness shows the highest change in error (around 40% for rotor 6 and rotor 7), while variations in magnet thickness and geometry scaling are well below 10%.

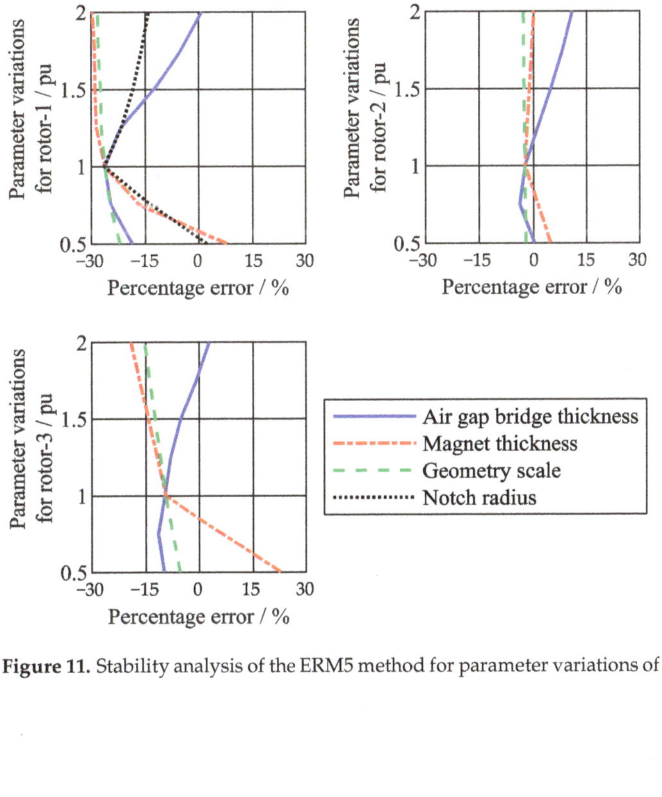

Figure 11. Stability analysis of the ERM5 method for parameter variations of various rotor geometries.

7. High-Strength Electrical Steel Strip: HS Grades for High Mechanical Stress [Internet]. Waelzholz. Available online: https://www.waelzholz.com/en/steel-materials/electrical-steel-strip/hs-grades.html (accessed on 19 September 2022).
8. Han, Z.; Yang, H.; Chen, Y. Investigation of the rotor mechanical stresses of various interior permanent magnet motors. In Proceedings of the 2009 International Conference on Electrical Machines and Systems, Tokyo, Japan, 15–18 November 2009; pp. 1–6.
9. Barcaro, M.; Meneghetti, G.; Bianchi, N. Structural Analysis of the Interior PM Rotor Considering Both Static and Fatigue Loading. *IEEE Trans. Ind. Appl.* **2014**, *50*, 253–260. [CrossRef]
10. Rao, J.; Qu, R.; Ma, J.; Xu, W. Investigate the influence of magnetic bridge design on mechanical strength and electromagnetic characteristics in high speed IPM machines. In Proceedings of the 2014 17th International Conference on Electrical Machines and Systems (ICEMS), Hangzhou, China, 22–25 October 2014; pp. 22–27.
11. Cha, K.-S.; Kim, D.-M.; Park, M.-R.; Yoon, M.-H.; Hong, J.-P. Multipolar High-Speed IPMSM Design for EV Traction Considering Mechanical Stress. In Proceedings of the 2016 IEEE 84th Vehicular Technology Conference (VTC-Fall), Montreal, QC, Canada, 18–21 September 2016; pp. 1–6.
12. Iacchetti, M.F.; Popescu, M.; Goss, J. Combined Analytic-Linear-FE Approach for Fast Stress Analysis of High-Speed Interior PM Rotors. In Proceedings of the 2019 IEEE International Electric Machines Drives Conference (IEMDC), San Diego, CA, USA, 12–15 May 2019; pp. 280–287.
13. Chai, F.; Li, Y.; Liang, P.; Pei, Y. Calculation of the Maximum Mechanical Stress on the Rotor of Interior Permanent-Magnet Synchronous Motors. *IEEE Trans. Ind. Electron.* **2016**, *63*, 3420–3432. [CrossRef]
14. Chu, G.; Dutta, R.; Rahman, M.F.; Lovatt, H.; Sarlioglu, B. Analytical Calculation of Maximum Mechanical Stress on the Rotor of Interior Permanent-Magnet Synchronous Machines. *IEEE Trans. Ind. Appl.* **2020**, *56*, 1321–1331. [CrossRef]
15. Chu, G.; Dutta, R.; Lovatt, H.; Sarlioglu, B.; Rahman, M.F. Analytical Calculation of Maximum Mechanical Stress on the Rotor of the Interior Permanent-Magnet Synchronous Machine. In Proceedings of the 2018 IEEE Energy Conversion Congress and Exposition (ECCE), Portland, OR, USA, 23–27 September 2018; pp. 255–261. [CrossRef]
16. Pilkey, W.D.; Pilkey, D.F.; Bi, Z. *Peterson's Stress Concentration Factors*; John Wiley & Sons: Hoboken, NJ, USA, 2020; 640p.
17. Bremner, R.D. Bridge stresses and design in IPM machines. In Proceedings of the IEEE EUROCON 2009, St. Petersburg, Russia, 18–23 May 2009; pp. 655–662.
18. Kleilat, I.; Langue, L.N.; Friedrich, G.; Vivier, S.; Kadri Benkara, K.E. Comparison between Assisted and Dual Phase synchronous reluctance machines for high speed applications. In Proceedings of the 2018 IEEE Energy Conversion Congress and Exposition (ECCE), Portland, OR, USA, 23–27 September 2018; pp. 1648–1655.
19. Schätzer, C. *Ein Verfahren zur Optimierung bei Elektrischen Maschinen mit Hilfe der Numerischen Feldberechnung*; Shaker: Aachen, Germany, 2002; 167p.
20. Binder, A.; Schneider, T.; Klohr, M. Fixation of buried and surface-mounted magnets in high-speed permanent-magnet synchronous machines. *IEEE Trans. Ind. Appl.* **2006**, *42*, 1031–1037. [CrossRef]
21. Yi, L.; Yulong, P.; Peixin, L.; Feng, C. Analysis of the rotor mechanical strength of interior permanent magnet synchronous in-wheel motor with high speed and large torque. In Proceedings of the 2014 IEEE Conference and Expo Transportation Electrification Asia-Pacific (ITEC Asia-Pacific), Beijing, China, 31 August–3 September 2014; pp. 1–5.
22. Kleilat, I.; Kadri Benkara, K.E.; Friedrich, G.; Vivier, S.; Moubayed, N.; Dib, R. Comparison of two Analytical Methods for Calculating the Maximum Mechanical Stress in the Rotor of High Speed Assisted Synchronous Reluctance Machines. In Proceedings of the 2019 IEEE Energy Conversion Congress and Exposition (ECCE), Baltimore, MD, USA, 29 September–3 October 2019; pp. 1669–1674.
23. Young, W.C.; Budynas, R.G.; Sadegh, A.M. *Roark's Formulas for Stress and Strain*, 8th ed.; McGraw Hill Professional: New York, NY, USA, 2011; 1073p.

machines

MDPI

Article

A Novel Suppression Method for Low-Order Harmonics Causing Resonance of Induction Motor

Pengyun Song [1,2], Yanghui Liu [1], Tao Liu [1,*], Huazhang Wang [1] and Liwei Wang [1]

[1] College of Electrical Engineering, Southwest Minzu University, Chengdu 610225, China
[2] Institute of National Productivity Development, Southwest Minzu University, Chengdu 610225, China
* Correspondence: liutao@stu.swun.edu.cn

Abstract: In the motor drive system of electric vehicles, there are some nonlinear factors, such as the deadtime and the conduction voltage drop of switching devices, which will generate low-order harmonics of the stator current and the torque ripple. The frequency of the harmonic may coincide with the natural frequency of the motor, so resonance may occur on the motor drive system. To reduce the noise caused by motor resonance, the characteristics of harmonic distortion caused by the deadtime, and the conduction voltage drop of the switching device, are analyzed firstly. Then, a motor vector control strategy with specific low order is proposed. The sixth-order harmonic resonance controller in d-q axis is introduced into the control loop, and the parameter designing principle of the controller is also presented. Without affecting the control performance of the current loop, the sixth-order harmonic of the stator current near the natural frequency can be significantly suppressed. Finally, the simulation and the experiment are carried out to certify the correctness and effectiveness of the proposed harmonic suppression method.

Keywords: induction motor (IM); harmonic suppression; motor resonance; controller parameter; deadtime; natural frequency

Citation: Song, P.; Liu, Y.; Liu, T.; Wang, H.; Wang, L. A Novel Suppression Method for Low-Order Harmonics Causing Resonance of Induction Motor. *Machines* **2022**, *10*, 1206. https://doi.org/10.3390/machines10121206

Academic Editor: Alejandro Gómez Yepes

Received: 3 November 2022
Accepted: 12 December 2022
Published: 13 December 2022

1. Introduction

The development of electric vehicles is an effective means to solve the problems of energy and environment [1], but serious vibration and noise exist in the motor drive system of the electric vehicle [2]. The output voltage of the inverter is a square wave instead of a sine wave [3], and contains abundant high order harmonics around the switching frequency and its integer multiples [4]. These higher-order harmonics will lead to serious electromagnetic vibration and noise, and the life and comfort of the electric vehicle are subsequently reduced [5,6]. To reduce the power of the harmonics near the switching frequency, some modulation strategies, such as the random carrier frequency [7] and periodic carrier frequency [8], are proposed. The above methods can disperse the main harmonics spectrum into a wider frequency band, a good suppression effect on the high-frequency noise of the motor caused by the harmonics of stator currents.

Ideally, there is no low-order harmonic in the inverter output voltage [9], but due to nonlinear factors such as the deadtime [10], the conduction voltage drop of switching devices and body diode [11], and the snubber capacitor [12], the lower order harmonics of the stator current will be generated. The natural frequency of the induction motor (IM) for electric vehicles is low, and the operating speed range of IM is wide, so the frequencies of these low-order harmonics may coincide with the natural frequencies of the motor, and the resonance may occur on the motor drive system of the electric vehicle [13]. In reference [14], a reasonable slot fit is selected to weaken the electromagnetic force wave related to electromagnetic vibration, and the vibration peak value of the IM at the natural frequency can be reduced. In reference [15], a Y-type capacitor is introduced at the DC terminal of the inverter; it is particularly effective to reduce the radiation emission of the vehicle in the above frequency range, and a simplified vehicle common-mode interference model is established. Compared to the above two methods for the structural improvements of the inverter or motor, it is more economical and practical to improve the modulation and control algorithm. Compensation based on pulse time is the most conventional and commonly used low-order harmonic suppression method. According to the polarity of the stator current, the voltage error caused by the deadtime and the conduction voltage drop of switching devices is converted into the duration of PWM pulse, and is subtracted in a switching period [16]. However, the compensation effect of this method is not very obvious. A modulation method without deadtime is proposed in [17]; according to the current polarity, only one switching tube of a phase bridge arm remains in the active state, and the other tube remains OFF state. The freewheeling is completed by the body diode, but the current harmonics caused by the conduction voltage drop of switching devices cannot be suppressed. In reference [18], a feedforward compensation of the average error on the command value is proposed, which can reduce low-order harmonic distortion, but the suppression effect is limited and depends on the sampling accuracy of the current polarity. Reference [19] proposes a scheme for suppressing the 5th and 7th harmonics based on a PI controller, this scheme needs to convert the 5th and 7th harmonic components of the stator current to d-q axis, and then four PI controllers are adopted to obtain the compensation amount of the 5th and 7th harmonics. Although the suppression effect is superior to other methods, the implementation process of this method is a little complicated. In order to suppress the low-order harmonics of the stator current, reference [16] optimizes the current PI controller parameters, but this scheme will reduce the dynamic response speed of the control system. Reference [13] proposes a current spectrum shaping control scheme in the natural frequency range based on a band-pass filter, but the optimal parameters design principle of the band-pass filter is not considered. In reference [20], a frequency adaptive selective harmonic control implemented in Z-plane is proposed for grid-connected inverters, and the fundamental frequency is fixed at 50 Hz. However, the speed range of the motor drive system is wide, the applicability of this method needs further verification. In reference [21], the resonance frequency is adjusted adaptively, however, whether the resonant controller with the same gain and different frequencies will affect the control performance of the current loop needs to be verified. Meanwhile, the calculation expression of the fundamental angle frequency contains IM parameters, rotor flux, stator current in q-axis and speed, so the motor parameters variation and the sampling accuracy will easily lead to the calculation error of the fundamental frequency.

In this paper, the characteristics of harmonic distortion caused by deadtime, and the conduction voltage drop of switching devices, are analyzed first. Then, the vector control scheme is improved, a sixth-order harmonic proportional resonance (PR) controller in d-q axis is introduced in the control loop, and the parameters of the PR controller are optimally tuned in theory. Finally, the simulation and experimental validations of the proposed method are carried out. The research steps are shown in Figure 1.

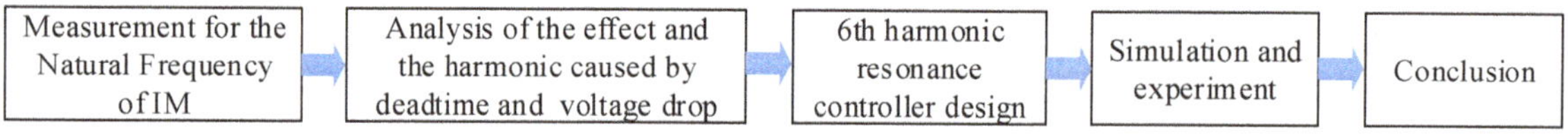

Figure 1. The chart of the research steps for this paper.

2. Simulation and Measurement for the Natural Frequency of IM

Firstly, the modal analysis of the IM is carried out in ANSYS/Workbench, and the IM parameters are shown in Table 1. One end of the IM is fixed, and a fixed constraint is applied to the other end, the natural frequencies of the IM are obtained by simulation and shown in Figure 2, the 1st to 4th natural frequencies are 850 Hz, 1152 Hz, 3125 Hz and 3893 Hz, respectively.

Table 1. The parameters of the IM.

Parameters	Value	Parameters	Value
Rated voltage	48 V	Rotor outer diameter	109 mm
Rated power	10 kW	Rotor inner diameter	36 mm
Maximum speed	6000 rpm	Rotor slot number	42
Pole pairs	2	Iron core length	180 mm
stator Outer diameter	188 mm	Stator resistance	0.047 Ω
Stator inner diameter	110 mm	Stator leakage inductance	81.5 μF
stator slot number	36	Rotor resistance	0.028 Ω
Rotor leakage inductance	81.3 μF	Excitation inductance	2.29 mH

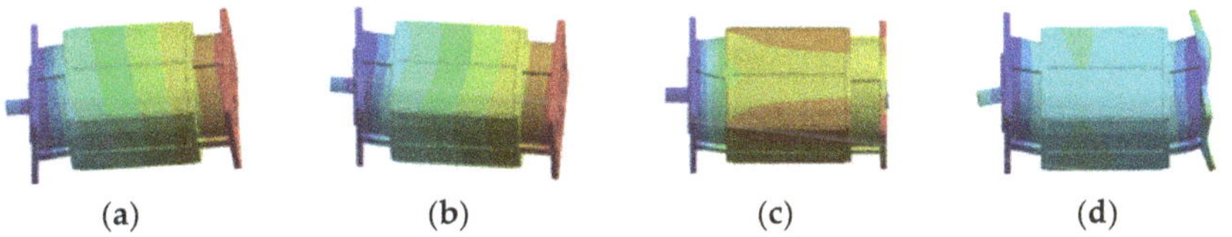

Figure 2. Simulation of the IM natural frequency. (**a**) 1st order modal; (**b**) 2nd order modal; (**c**) 3rd order modal; and (**d**) 4th order modal.

Then, the modal experiment is carried out, the hammering method is adopted to measure the IM natural frequencies, and the vibration intensity of the IM at different frequencies is shown in Figure 3. There are four vibration peak values, the corresponding frequencies are very close to the simulation results.

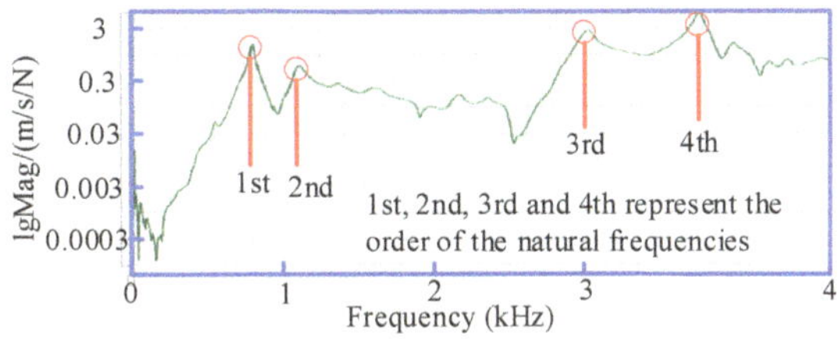

Figure 3. Experimental test of the IM natural frequency.

3. Analysis of Harmonic Characteristics Caused by Nonlinear Factors

Figure 4 shows the topology of the IM drive system fed by the three-phase voltage source inverter (VSI), where u_{dc} is the DC-link voltage, C_{dc} is the DC-link capacitor, S_1–S_6 is the power switching tube and i_a, i_b, i_c are the three-phase stator currents of the induction motor.

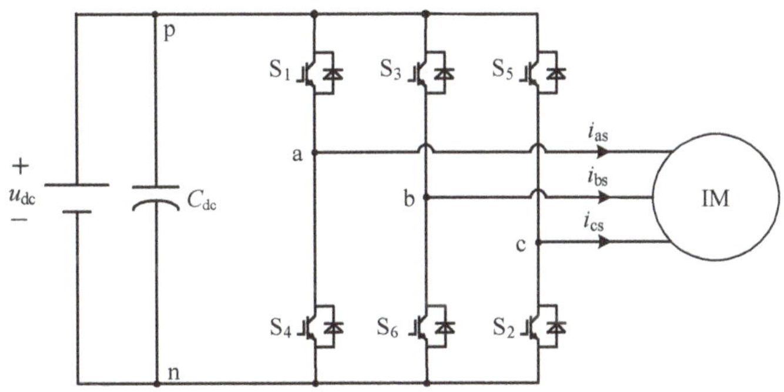

Figure 4. The topology of VSI-fed IM drive system.

Figure 5 shows the low-order harmonic distribution of the line-to-line stator voltage u_{ab} and the stator current i_{as} using SVPWM without deadtime and conduction voltage drop of switching devices. Except for the fundamental component, the harmonics of u_{ab} only include the high-order harmonic components around the switching frequency and its integer multiples. So only a few low-order harmonic components exist in the i_{as}.

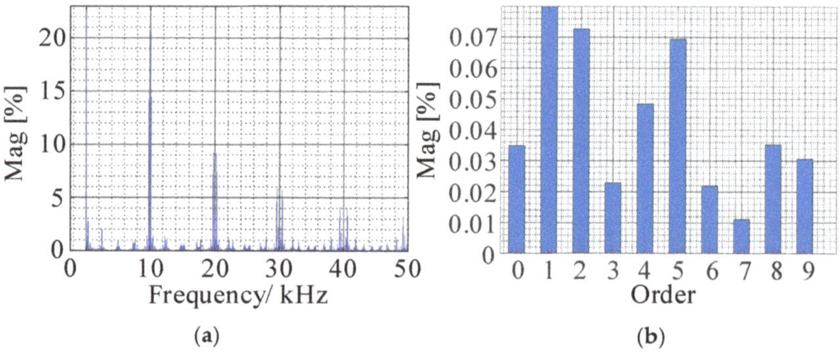

Figure 5. The harmonics distribution without deadtime and conduction voltage drop. (**a**) u_{ab}; (**b**) i_{as}.

If the switching on delay t_{on} and off delay t_{off} are considered, the dead time t_d should be added to prevent the short circuit of the DC battery, the actual switching signals of S_1 and S_4 are noted as p_1 and p_4, which are shown in Figure 6, an obvious timing error can be seen. At present, the low-order harmonic distribution of the stator current is shown in Figure 7a. Compared to Figure 5b, the 5th and 7th harmonics increase significantly. After setting the conduction voltage drops of the switching tube and body diode, noted as u_{vt} and u_{vd}, respectively, the 5th and 7th harmonics further increase, which is shown in Figure 7b. The output voltage u_{an} of phase a in a switching period T_s is shown in Figure 6, there is a nonlinear error Δu_{an} between u_{an} and the ideal voltage u^*_{an}, and Δu_{an} is affected by the polarity of i_{as}, which is expressed as:

$$\Delta u_{an} = \begin{cases} -[\tau u_{dc} + (d_a - \tau)u_{vt} + (1 - d_a + \tau)u_{vd}] & (i_a > 0) \\ \tau u_{dc} + (1 - d_a - \tau)u_{vt} + (d_a + \tau)u_{vd} & (i_a < 0) \end{cases} \quad (1)$$

where $\tau = (t_d + t_{on} - t_{off})/T_s$, d_a is the duty cycle. These low-order harmonics will increase the torque ripple. If the frequencies of these lower-order harmonics are close to the natural frequencies of the IM, the resonance will occur and the mechanical performance of the IM will deteriorate.

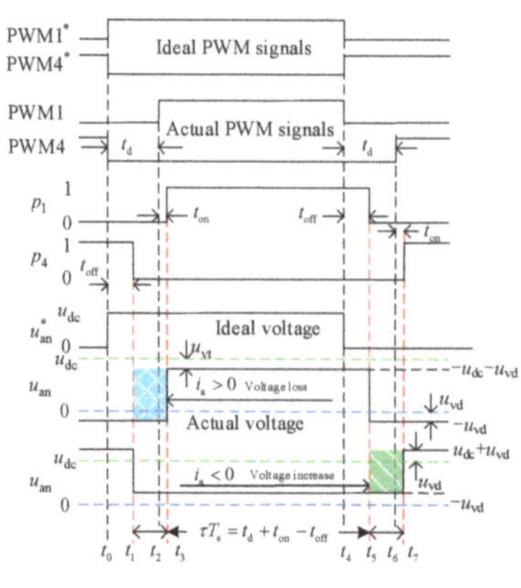

Figure 6. The effect diagram of the deadtime and voltage drops of the switching tube.

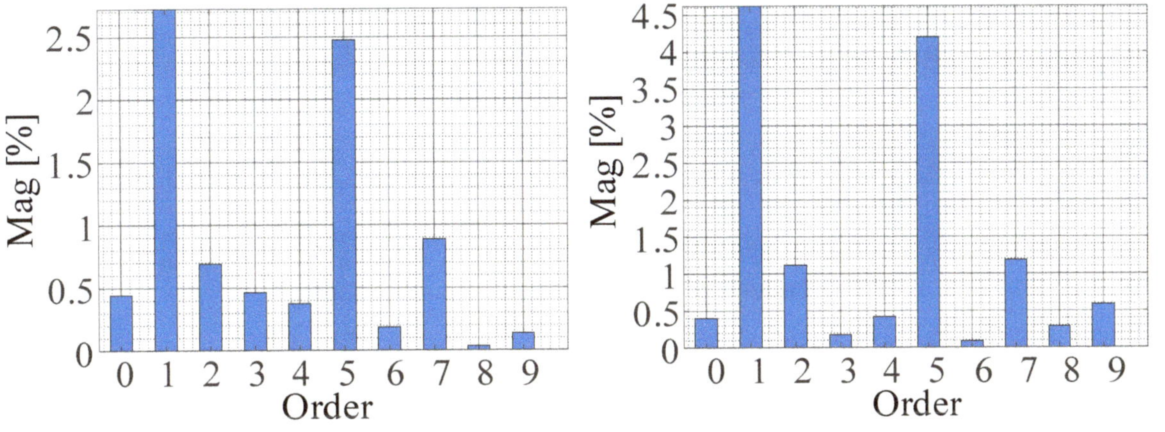

(a) (b)

Figure 7. The lower order harmonics distribution of i_{as}. (**a**) With deadtime; (**b**) with deadtime and conduction voltage drops.

The sequence of the three-phase 5th harmonic is negative, the expressions of the 5th harmonic component of the three-phase stator current are defined as follows:

$$\begin{cases} i_{a5} = I_5 \sin(5\omega t + \varphi_5) \\ i_{b5} = I_5 \sin(5\omega t + \varphi_5 + 120^\circ) \\ i_{c5} = I_5 \sin(5\omega t + \varphi_5 - 120^\circ) \end{cases} \tag{2}$$

where I_5 and φ_5 are the amplitude and phase of the 5th harmonic component of the stator current, ω is the fundamental angular frequency, respectively. After the coordinate transformation based on the rotor flux orientation, the expressions of the 5th harmonic component of the stator current in d-q axis are obtained as:

$$\begin{cases} i_{d5} = I_5 \cos(6\omega t + \varphi_5) \\ i_{q5} = I_5 \sin(6\omega t + \varphi_5) \end{cases} \tag{3}$$

The sequence of the three-phase 7th harmonic is positive, the expressions of the 7th harmonic component of the three-phase stator current are defined as follows:

$$\begin{cases} i_{a7} = I_7 \sin(5\omega t + \varphi_7) \\ i_{b7} = I_7 \sin(5\omega t + \varphi_7 - 120^\circ) \\ i_{c7} = I_7 \sin(5\omega t + \varphi_7 + 120^\circ) \end{cases} \tag{4}$$

where I_7 and φ_7 are the amplitude and phase of the 7th harmonic component of the stator current, respectively. The expressions of the 7th harmonic component of the stator current in d-q axis are expressed as:

$$\begin{cases} i_{d7} = I_7 \cos(6\omega t + \varphi_7) \\ i_{q7} = I_7 \sin(6\omega t + \varphi_7) \end{cases} \tag{5}$$

Figure 8 shows the low-order harmonics distribution of the stator current in d-q axis, except DC component, the 6th harmonic content is obviously higher than other harmonics in i_d and i_q. Therefore, suppression of the 6th harmonic component of i_d and i_q is equivalent to reducing the 5th and 7th harmonic components of the three-phase stator currents.

To further analyze the influence of k_r on stability, when k_r increases from 1 to 4, the pole distribution diagram of $G_{i\_cl}(s)$ is shown in Figure 12. As k_r increases, a pair of conjugate poles move to the right. When $k_r > 3.5$, $G_{i\_cl}(s)$ appears on the right half-plane pole, which leads to system instability.

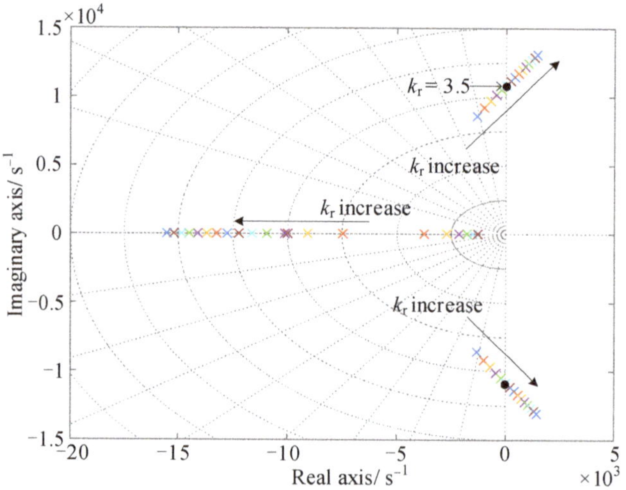

Figure 12. The pole distribution diagram of $G_{i\_cl}(s)$ under different k_r.

In order to theoretically prove the suppression effect of $G_R(s)$ on the 6th harmonic component in i_d, when k_r increases from 1 to 3, the bode plot of $G_{u\_i}(s)$ is shown in Figure 13. Due to the addition of $G_R(s)$, the amplitude around 1000 Hz is below 0 dB line. And with the increase of k_r, the suppression effect on the 6th harmonic component in i_d is more significant, and the band of the voltage harmonic frequency that can be suppressed by the $G_R(s)$ will also expand. Although another peak appears, the amplitude of the harmonic components of the inverter output voltage in d-q axis is very few, so it will not cause additional harmonic distortion. However, to prevent increasing the eleventh and thirteenth harmonics of stator current, the cut-off frequency of $G_{u\_i}(s)$ should be less than the twelfth harmonic frequency.

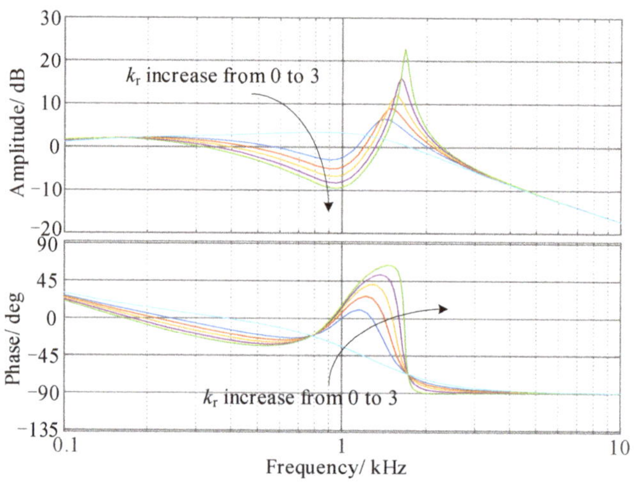

Figure 13. The bode plot of $G_{u\_i}(s)$ with resonance controller under different k_r.

4.3. Analysis of the Resonance Controller under Frequency Variation

According to the above analysis presented in Section 4.2, the suppression effect can be improved with the increase of k_r, however, the stability of the current loop is reduced, so k_r is set to 2.5 in this paper, and the transfer function of $G_R(s)$ is shown as:

$$G_R(s) = \frac{2505\pi s}{s^2 + 1002\pi s + (1002\pi)^2} \quad (11)$$

The bode plot of $G_{u\_i}(s)$ is shown in Figure 14. It can be seen that the resonance controller can suppress the stator voltage whose 6th harmonic frequency is between 0.35 kHz and 1.32 kHz. However, if the 6th harmonic frequency is between 0.065 kHz and 0.35 kHz, the corresponding fundamental frequency is between 10.8 Hz and 58.3 Hz, the resonance controller under k_r =2.5 and ω_n = 1002π rad/s is not able to suppress the harmonic distortion caused by the deadtime and the voltage drops of switching devices and diode.

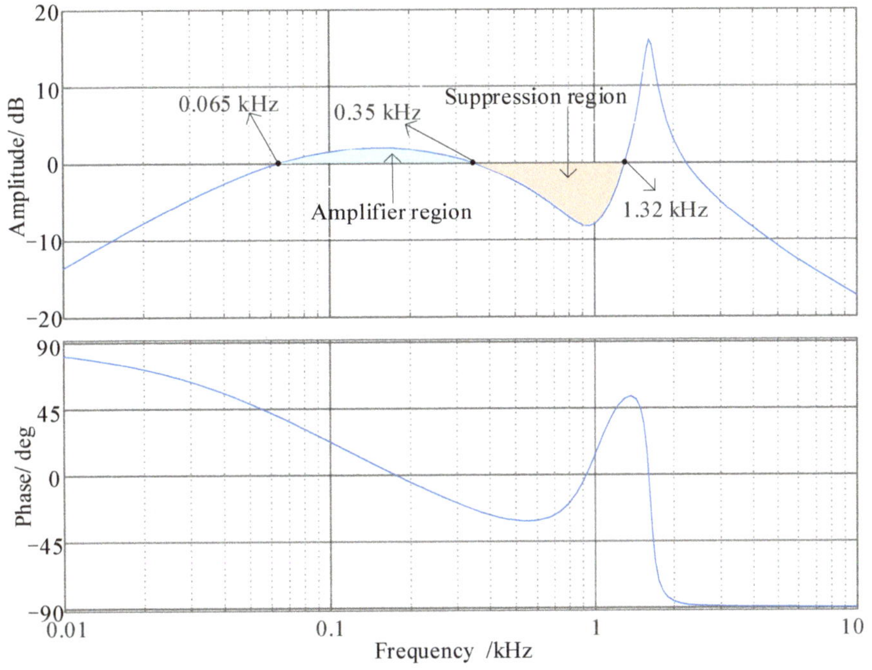

Figure 14. The bode plot of $G_{u\_i}(s)$ with resonance controller under k_r = 2.5 and ω_n = 1002π rad/s.

In the case where the fundamental frequency is less than 60 Hz. Another resonance controller is designed and adopted. In consideration of the stability and the harmonic suppression effect, according to the bode plot and the pole distribution diagram, the transfer function $G_R(s)$ is rewritten as:

$$G_R(s) = \frac{600\pi s}{s^2 + 100\pi s + 10000\pi^2} \quad (12)$$

At this time, the bode plots of $G_{i\_ol}(s)$, $G_{i\_cl}(s)$, and $G_{u\_i}(s)$ are shown in Figure 15. It can be seen that the cut-off frequency of $G_{i\_ol}(s)$ is 1 kHz, the phase margin of $G_{i\_ol}(s)$ is 45°, and $G_{i\_cl}(s)$ has a good DC following characteristics. Meanwhile, the resonance controller shown in Formula (10) can effectively suppress the disturbance from the stator voltage whose 6th harmonic frequency is less than 0.38 kHz.

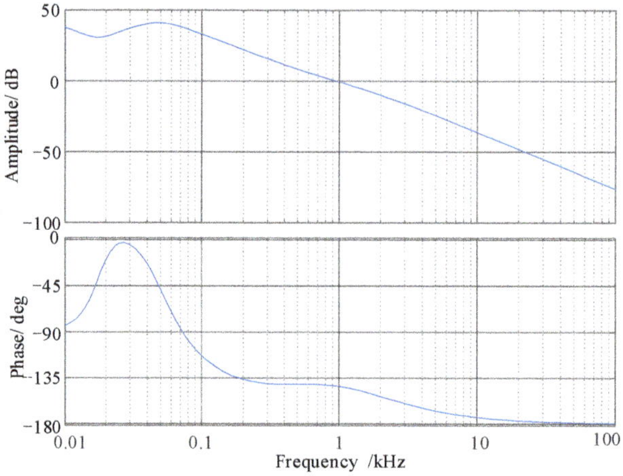

(**a**)

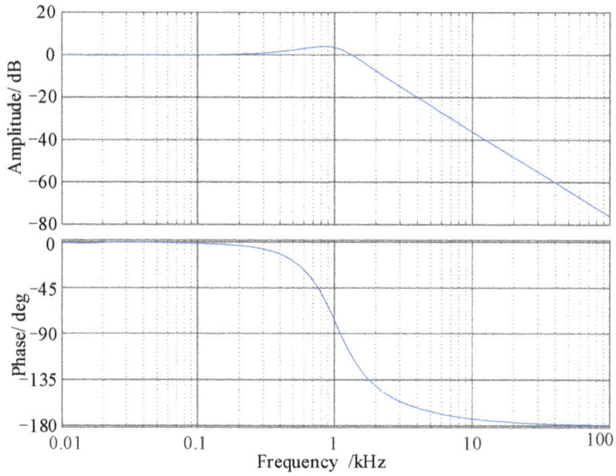

(**b**)

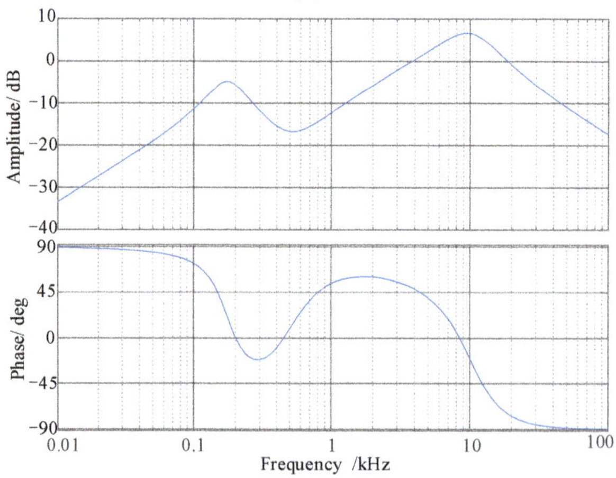

(**c**)

Figure 15. The bode plots the fundamental frequency is less than 60 Hz. (**a**) $G_{i\_ol}(s)$; (**b**) $G_{i\_cl}(s)$; and (**c**) $G_{u\_i}(s)$.

Since the two resonant controllers shown in Formulas (11) and (12) are applicable to different frequency ranges, the resonant controller needs to be selected according to the fundamental frequency. The fundamental angle frequency ω_s can be calculated by the follows:

$$\omega_s = n\omega_r + \frac{L_m R_r i_q}{L_r \psi_r} \quad (13)$$

where R_r is the rotor resistance, ψ_r is the rotor flux, ω_r is the speed, and n is the number of pole pairs. L_m and L_r are the mutual inductance and rotor inductance, respectively. The fundamental frequency $f_s = \omega_s/(2\pi)$, if $f_s > 60$ Hz, the $G_R(s)$ shown as (11) is adopted, otherwise, the $G_R(s)$ shown as (12) is adopted.

5. Simulation Results

In order to verify the suppression effect of the proposed method in this paper on the low-order harmonics in the stator current, the simulation model is established in the environment of Matlab/Simulink. The IM parameters are the same as in Table 1, other parameters are set in Table 2.

Table 2. Simulation parameters.

Parameters	Value	Parameters	Value
Switching cycle	10 kHz	Simulation step size	0.5 μs
Sampling cycle	10 kHz	Voltage drop of the switching devices	0.5 V
Deadtime	2 μs	Voltage drop of the diode	0.7 V

In order to unify with the digital control in the experiment, the controller composed PI and resonance is implemented by a Matlab function module in the environment of Matlab/Simulink, and the sampling time is set as 100 μs. According to the difference equation, the discrete control algorithm is written in the Matlab function module, which is shown as:

$$\begin{aligned} u_d^*(k) = \; & (3 - 2\xi\omega_n T_s)u_d^*(k-1) - (1 - 2\xi\omega_n T_s + \omega_n^2 T_s^2)u_d^*(k-2) \\ & + (k_i T_s + k_p)e(k) + 2k_r\xi\omega_n T_s e(k-1) - 2k_r\xi\omega_n e(k-2) \end{aligned} \quad (14)$$

where e is the current error. The fundamental frequency and the load are set as 167 Hz and N·m, respectively. Three different simulations are carried out under no compensation, pulse time compensation, and the 6th harmonic controller in d-q axis proposed in this paper, respectively. The simulation results under different conditions are shown in Figures 16–18, respectively. The THD, 5th harmonic, and 7th harmonic of the stator current, and the torque ripple, are summarized in Table 3; it can be seen that the low order harmonics suppression effect of the 6th harmonic controller in d-q axis is optimal.

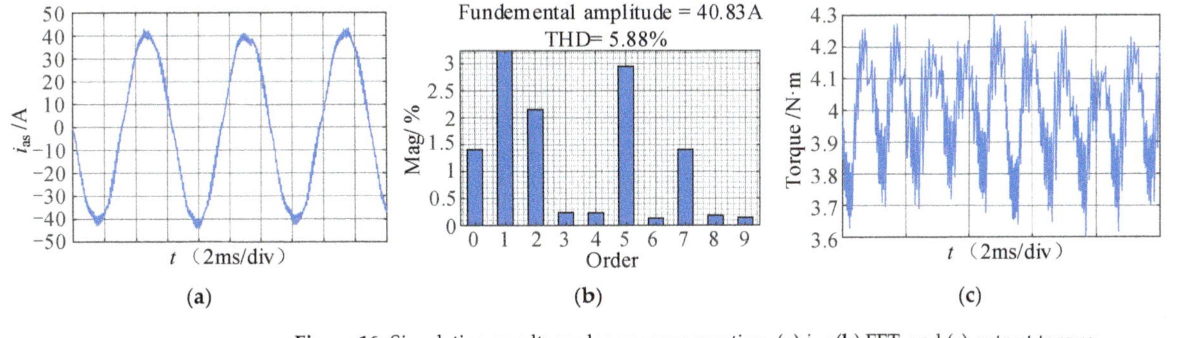

Figure 16. Simulation results under no compensation. (**a**) i_{as}; (**b**) FFT; and (**c**) output torque.

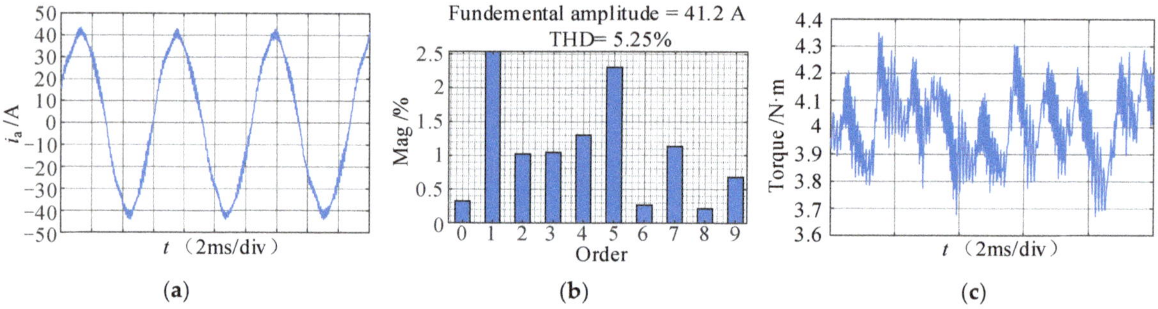

Figure 17. Simulation results under pulse time compensation. (**a**) i_{as}; (**b**) FFT; and (**c**) output torque.

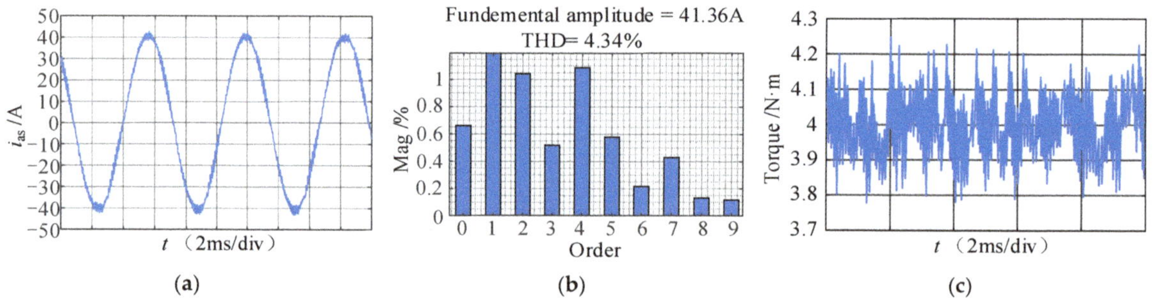

Figure 18. Simulation results under 6th harmonic controller. (**a**) i_{as}; (**b**) FFT; and (**c**) output torque.

Table 3. Comparison of the simulation results.

Compensation Method	THD	5th Harmonic	7th Harmonic	Torque Ripple (N·m)
No compensation	5.88%	2.9%	1.4%	3.6 to 4.3
Pulse time compensation	5.33%	2.3%	1.2%	3.7 to 4.3
6th harmonic controller	4.34%	0.58%	0.43%	3.8 to 4.2

To certify the effectiveness of the resonant controller shown in Formula (12) at low frequency, the fundamental frequency is set as 30 Hz, two simulations are carried out by the controller (11) and (12), respectively. The low-order harmonics distribution of the stator current is shown in Figure 19. It is obvious that the resonant controller of (11) cannot suppress the 5th and 7th harmonic components of the stator current. As opposed to this, the 5th and 7th harmonic components are significantly reduced when the resonant controller of (12) is adopted.

Next, the impact on the dynamic performance of the system from the resonant controller is verified. Firstly, the load torque is set to 4 N·m constantly, and the command value of speed changes from 3000 rpm to 4500 rpm, the dynamic simulation results with and without the resonant controller are shown in Figure 20. Then, the command value of speed is set to 4000 rpm, the load torque increases from 2 N·m to 4 N·m, and the dynamic simulation results with and without the resonant controller are shown in Figure 21. After the resonant controller is introduced, only a slight difference occurs in the dynamic waveforms of speed and torque, so the resonant controller has no impact on the dynamic performance.

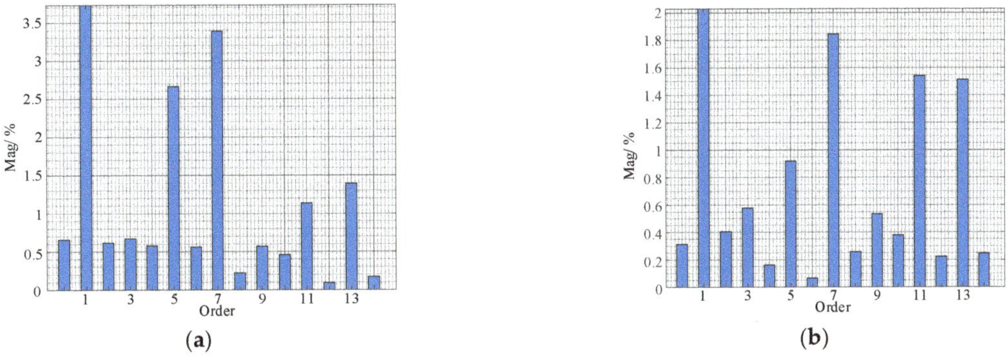

Figure 19. The FFT analysis results of the stator current under different resonant controllers. (**a**) $G_R(s)$ of (11); (**b**) $G_R(s)$ of (12).

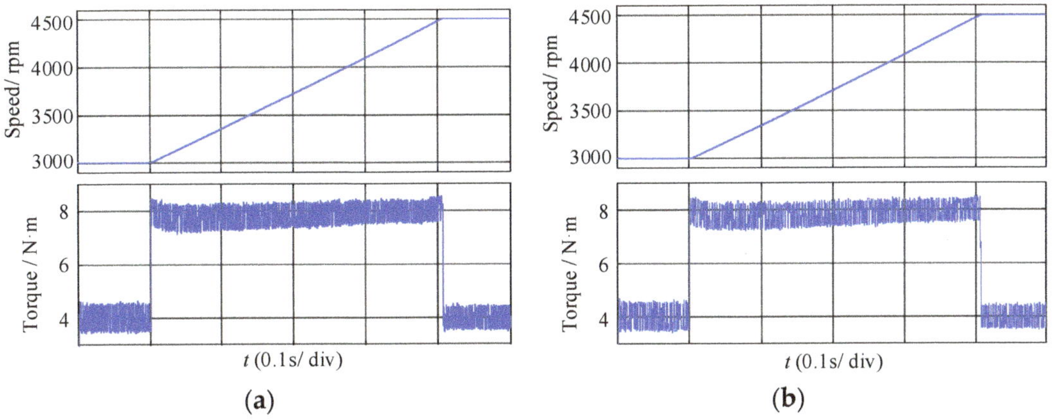

Figure 20. The dynamic simulation waveforms when the speed changes from 3000 rpm to 4500 rpm. (**a**) Without resonant controller; (**b**) with resonant controller.

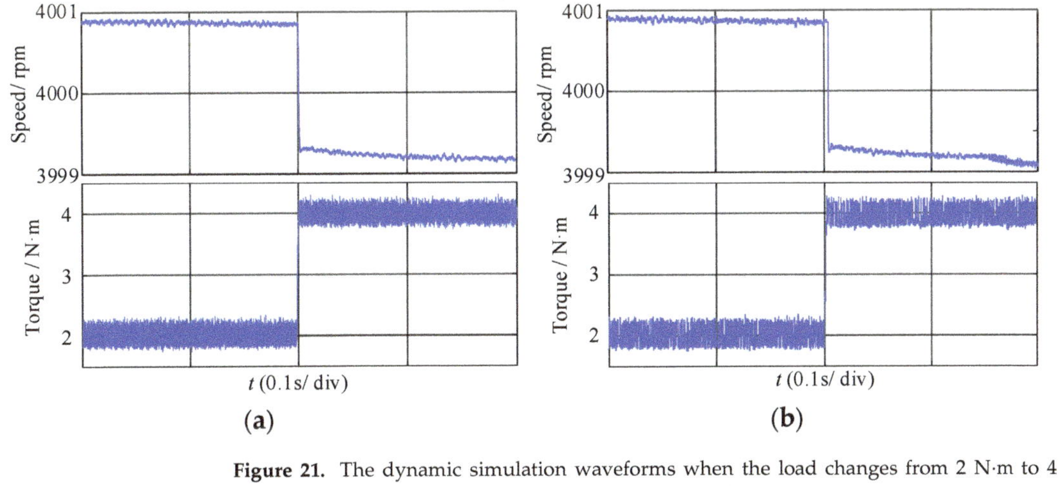

Figure 21. The dynamic simulation waveforms when the load changes from 2 N·m to 4 N·m. (**a**) Without resonant controller; (**b**) with resonant controller.

6. Experimental Results

The experimental prototype, shown in Figure 22, is established to verify the suppression effect of the proposed method on the low-order harmonics in the stator current. TMS320F28069 is used to execute the control algorithm, and three non-contact sensors MLX91205 is used to sample the stator current. The inverter bridge is composed by the MOSFET, the type is IPB042N10N, the DC-link voltage is 72 V, and is supplied by the battery pack. The parameters of the IM are the same as in Table 1, and the experimental parameters and operation are the same as in Table 2. The experimental results under no compensation, pulse time compensation, and the 6th harmonic controller in d-q axis are shown in Figures 23–25, respectively. Under the influence of deadtime and conduction voltage drop, without compensation, the stator current has serious harmonic distortion, the 5th harmonic content is 3.2%, the 7th harmonic content is 2.3%, and there is a little DC component. After the pulse time compensation, the content of the 5th and 7th harmonics and THD are slightly reduced, and the power spectral density at the natural frequency drops below −10 dB/Hz. By adopting the 6th harmonic controller in d-q axis proposed in this paper, the 5th harmonic content is only 1%, the 7th harmonic content 0.6%, and the THD decreases to 4.62%. At the same time, the DC component of the current is eliminated, and the power spectral density at all frequencies is less than −20 dB/Hz. The experimental results are consistent with the simulation results, which show that the proposed resonant controller in d-q axis can reduce the low-order harmonics of the stator current.

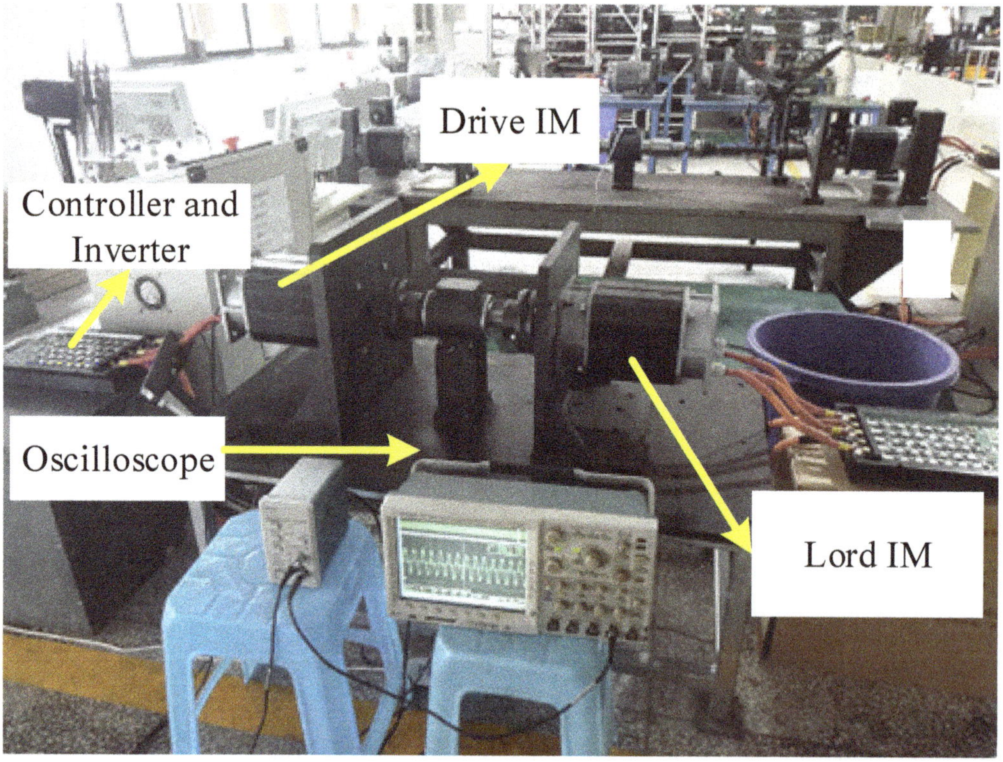

Figure 22. Experimental prototype.

To verify the effectiveness of the control algorithm in reducing vibration, a vibration measurement instrument is used to measure the vibration of the IM, and the measurement experiment platform is shown in Figure 26.

Figure 27 shows the power spectral density of the vibration within 1.8 kHz under no compensation, there are two peaks of vibration at the 1st and 2nd natural frequencies. Figure 28 shows the power spectral density of the vibration within 1.8 kHz under the 6th harmonic controller in d-q axis, the vibrations are significantly suppressed. The whole experimental results further prove that the 6th harmonic controller with its parameter design method in d-q axis has a good suppression effect on the 5th and 7th harmonics distortion of the stator current, which can not only reduce the torque ripple, but also weaken the vibration of the IM.

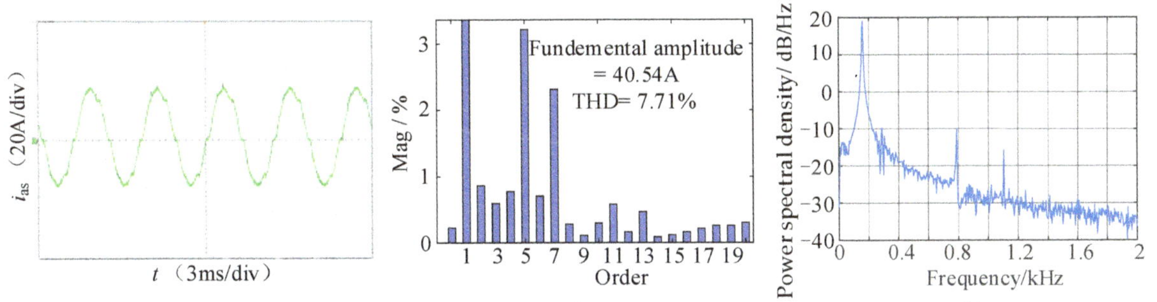

(**a**) (**b**) (**c**)

Figure 23. Experimental results under no compensation. (**a**) i_{as}; (**b**) FFT; and (**c**) power spectral density.

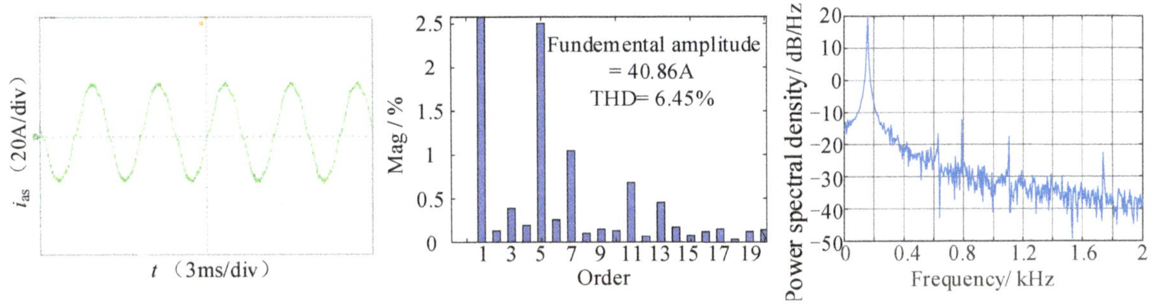

(**a**) (**b**) (**c**)

Figure 24. Experimental results under pulse time compensation. (**a**) i_{as}; (**b**) FFT; (**c**) Power spectral density.

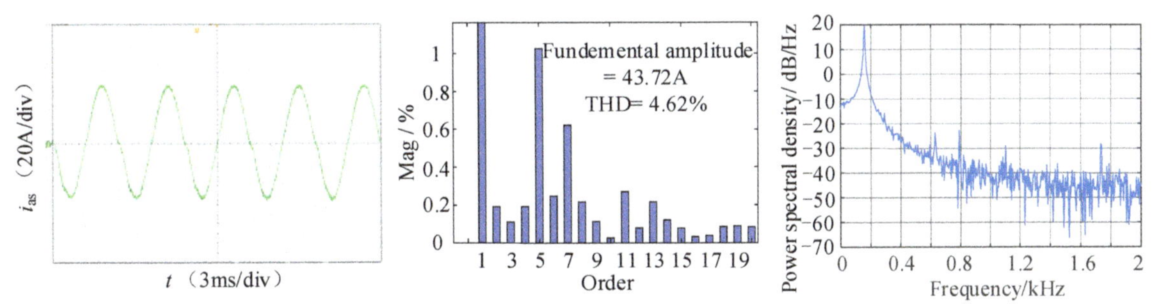

(**a**) (**b**) (**c**)

Figure 25. Experimental results under 6th harmonic controller. (**a**) i_{as}; (**b**) FFT; and (**c**) power spectral density.

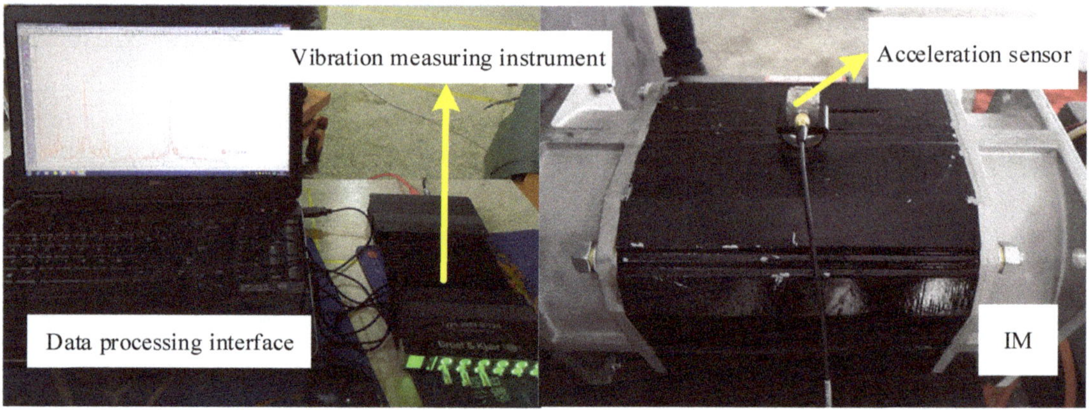

Figure 26. The vibration measurement platform.

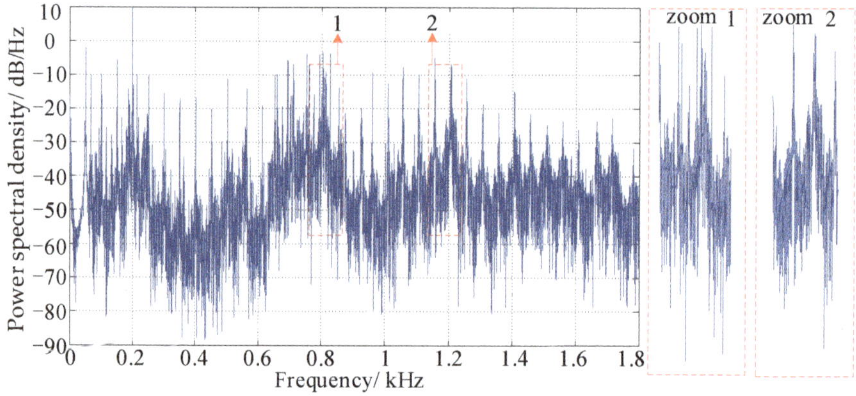

Figure 27. The power spectral density of vibration in IM without compensation.

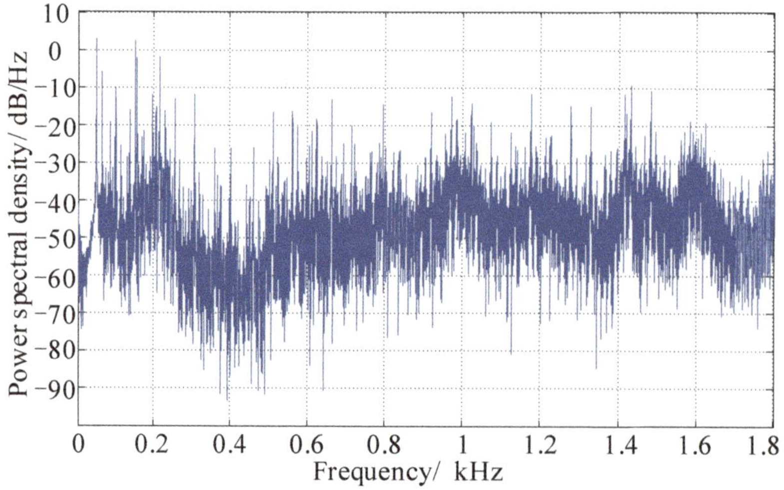

Figure 28. The power spectral density of vibration in IM 6th harmonic controller.

7. Conclusions

The simulation and experimental results prove that the proposed 6th harmonic controller with its parameter design method in d-q axis has good effects on reducing the 5th and 7th harmonic distortion of the stator current, weakening the low-frequency resonance of the motor, and improving the dynamic performance of the motor drive system. Finally, the following conclusions are formed:

(1) The dead time and the conduction voltage drop of the switching tube and body diode will produce nonlinear errors in the inverter output voltage. The current PI controller based on rotor flux orientation cannot suppress the 5th and 7th harmonic distortion of the stator current caused by the nonlinear voltage;
(2) Based on current loop stability and the harmonic suppression effect, according to the bode plot and the pole distribution diagram, the proposed 6th harmonic controller can effectively suppress the low-order harmonic distortion of the nonlinear voltage, without generating harmonic distortion of other frequencies, and does not affect the control performance of the current loop;
(3) Since another resonant controller for the low-speed range is introduced, low-order harmonic suppression at all operating frequencies can be achieved;
(4) Suppressing the 6th harmonic component of the stator current in d-q axis is an effective means to weaken the low-frequency resonance and vibration of the motor.

Author Contributions: Conceptualization and methodology, P.S.; simulation, Y.L.; validation and experiment, P.S. and T.L.; writing and supervision, L.W. and T.L.; review and editing, P.S.; funding acquisition, H.W. All authors have read and agreed to the published version of the manuscript.

Funding: This work was supported by "the Fundamental Research Funds for the Central Universities", Southwest Minzu University (Funder, Huazhang Wang; Funding number, 2021101).

Data Availability Statement: Not applicable.

Conflicts of Interest: The authors declare no conflict of interest.

References

1. Du, G.; Zou, Y.; Zhang, X.; Guo, L.; Guo, N. Heuristic Energy Management Strategy of Hybrid Electric Vehicle Based on Deep Reinforcement Learning with Accelerated Gradient Optimization. *IEEE Trans. Transp. Electrif.* **2021**, *7*, 2194–2208. [CrossRef]
2. Liang, J.; Howey, B.; Bilgin, B.; Emadi, A. Source of Acoustic Noise in a 12/16 External-Rotor Switched Reluctance Motor: Stator Tangential Vibration and Rotor Radial Vibration. *IEEE Open J. Ind. Appl.* **2020**, *1*, 63–73. [CrossRef]
3. Miao, Y.; Liu, H.; Zhang, W.; Liu, P. Optimal Control Strategy of Storage Inductor Current and Rotor Flux for Current Source Inverter Motor Drive System. *Proc. CSEE* **2019**, *39*, 2757–2767.
4. Song, P.; Liu, Y.; Liu, C. Research on Parameter Design and Control Method for Current Source Inverter–Fed IM Drive Systems. *Machines* **2022**, *10*, 922. [CrossRef]
5. Jiang, Z.; Ying, D.; Jian, S. Electromagnetic Noise Characteristics of Permanent Magnet Synchronous Motor Applied in Electric Vehicle. *Trans. China Electrotech. Soc.* **2016**, *31*, 53–59.
6. Dong, Z.; Liu, C.; Song, Z.; Liu, S. Suppression of Dual−Harmonic Components for Five−Phase Series−Winding PMSM. *IEEE Trans. Transp. Electrif.* **2022**, *8*, 121–134. [CrossRef]
7. Lai, Y.-S.; Chang, Y.-T.; Chen, B.-Y. Novel Random-Switching PWM Technique with Constant Sampling Frequency and Constant Inductor Average Current for Digitally Controlled Converter. *IEEE Trans. Ind. Electron.* **2013**, *60*, 3126–3135. [CrossRef]
8. Peng, H.; Heping, L.; Yiru, M.; You, X.; Guo, Q. Variable Switching Frequency Pulse Width Modulation for Induction Motors based on Current Ripple Peak Value. *Trans. China Electrotech. Soc.* **2020**, *35*, 4373–4383.
9. Liu, P.; Xu, J.; Sun, M.; Yuan, J.; Blaabjerg, F. New Discontinuous Space Vector Modulation Strategies for Impedance-Source Inverter with Superior Thermal and Harmonic Performance. *IEEE Trans. Ind. Electron.* **2022**, *69*, 13079–13089. [CrossRef]
10. Guo, Q.; Dong, Z.; Liu, H.; You, X. Nonlinear Characteristics Compensation of Inverter for Low-Voltage Delta-Connected Induction Motor. *Energies* **2020**, *13*, 590. [CrossRef]
11. Liu, P.; Xu, J.; Yang, Y.; Wang, H.; Blaabjerg, F. Impact of Modulation Strategies on the Reliability and Harmonics of Impedance-Source Inverters. *IEEE J. Emerg. Sel. Top. Power Electron.* **2020**, *8*, 3968–3981. [CrossRef]
12. Zhang, Z.; Xu, L. Dead-Time Compensation of Inverters Considering Snubber and Parasitic Capacitance. *IEEE Trans. Power Electron.* **2014**, *29*, 3179–3187. [CrossRef]
13. Liu, H.; Liu, Q.; Zhang, W.; Miao, Y.; Liu, P. Random PWM Technique for Acoustic Noise and Vibration Reduction in Induction Motors used by Electric Vehicles. *Trans. China Electrotech. Soc.* **2019**, *34*, 1488–1495.

14. Kobayashi, T.; Tajima, F.; Ito, M.; Shibukawa, S. Effects of Slot Combination on Acoustic Noise from Induction Motors. *IEEE Trans. Magn.* **1997**, *33*, 2101–2104. [CrossRef]
15. Gao, X.; Su, D. Suppression of a Certain Vehicle Electrical Field and Magnetic Field Radiation Resonance Point. *IEEE Trans. Veh. Technol.* **2018**, *67*, 226–234. [CrossRef]
16. Wu, Z.; Ding, K.; Yang, Z.; He, G. Analytical Prediction and Minimization of Deadtime-Related Harmonics in Permanent Magnet Synchronous Motor. *IEEE Trans. Ind. Electron.* **2021**, *68*, 7736–7746. [CrossRef]
17. Liu, H.; Lu, Y.; Wang, H.; Miao, Y. Dead-time Compensation Modulation Strategy of Subsection Integrated in Voltage Source Inverter. *Electr. Mach. Control* **2018**, *22*, 25–32.
18. Miao, Y.; Liu, H.; Wang, H.; Lu, Y. Average Current Feedforward Compensation Strategy of Overlap-Time Effect for Current Source Inverter. *Proc. CSEE* **2018**, *38*, 4183–4193.
19. Liao, Y.; Zhen, S.; Liu, R.; Yao, J. Torque Ripple Suppression of Permanent Magnet Synchronous Motor by the Harmonic Injection. *Proc. CSEE* **2011**, *21*, 119–127.
20. Yang, Y.; Zhou, K.; Wang, H.; Blaabjerg, F.; Wang, D.; Zhang, B. Frequency Adaptive Selective Harmonic Control for Grid-Connected Inverters. *IEEE Trans. Power Electron.* **2015**, *30*, 3912–3924. [CrossRef]
21. da Silveira, D.B.; Farias, J.V.M.; Pereira, H.A.; Luiz, A.-S.A.; Stopa, M.M.; Cupertino, A.F. Mitigation of torque oscillations caused by dc injection-based resistance estimation through resonant controllers. In Proceedings of the 2021 Brazilian Power Electronics Conference (COBEP), João Pessoa, Brazil, 7–10 November 2021; pp. 1–6.

machines

MDPI

Article

Reinforcement Learning Control of Hydraulic Servo System Based on TD3 Algorithm

Xiaoming Yuan [1,*], Yu Wang [1], Ruicong Zhang [1], Qiang Gao [2], Zhuangding Zhou [1], Rulin Zhou [3] and Fengyuan Yin [1]

1 Heavy Machinery Fluid Power Transmission and Control Laboratory of Hebei Province, Yanshan University, Qinhuangdao 066004, China
2 National Research Center of Pumps, Jiangsu University, Zhenjiang 212013, China
3 Beijing Tianma Intelligent Control Technology Co., Ltd., Beijing 100013, China
* Correspondence: yuanxiaoming@ysu.edu.cn; Tel.: +86-137-8056-0557

Abstract: This paper aims at the characteristics of nonlinear, time-varying and parameter coupling in a hydraulic servo system. An intelligent control method is designed that uses self-learning without a model or prior knowledge, in order to achieve certain control effects. The control quantity can be obtained at the current moment through the continuous iteration of a strategy–value network, and the online self-tuning of parameters can be realized. Taking the hydraulic servo system as the experimental object, a twin delayed deep deterministic (TD3) policy gradient was used to reinforce the learning of the system. Additionally, the parameter setting was compared using a deep deterministic policy gradient (DDPG) and a linear–quadratic–Gaussian (LQG) based on linear quadratic Gaussian objective function. To compile the reinforcement learning algorithm and deploy it to the test platform controller for testing, we used the Speedgoat prototype target machine as the controller to build the fast prototype control test platform. MATLAB/Coder and compute unified device architecture (CUDA) were used to generate an S-function. The results show that, compared with other parameter tuning methods, the proposed algorithm can effectively optimize the controller parameters and improve the dynamic response of the system when tracking signals.

Keywords: reinforcement learning; TD3; DDPG; hydraulic servo; Speedgoat

Citation: Yuan, X.; Wang, Y.; Zhang, R.; Gao, Q.; Zhou, Z.; Zhou, R.; Yin, F. Reinforcement Learning Control of Hydraulic Servo System Based on TD3 Algorithm. *Machines* **2022**, *10*, 1244. https://doi.org/10.3390/machines10121244

Academic Editors: Kan Liu and Wei Hu

Received: 28 November 2022
Accepted: 15 December 2022
Published: 19 December 2022

Publisher's Note: MDPI stays neutral with regard to jurisdictional claims in published maps and institutional affiliations.

1. Introduction

As a high-precision closed-loop control system, a hydraulic servo system can make the output (displacement, velocity or force) of a system follow the changes of the input quickly and accurately. This offers the unique advantages of fast response speed, large load stiffness and large control power and is widely used in the aerospace, heavy industry, robotics and national defense fields. However, as a high-order nonlinear system, the inherent nonlinearity, time-variability and parameter coupling of a hydraulic servo system pose challenges to the design of control methods [1]. The determination of controller parameters and the deployment of a control strategy are the main difficulties. Thus, the question of how to control a hydraulic servo system with high precision is of great significance to its popularization and application.

The use of reinforcement learning algorithms to solve control problems such as nonlinear engineering problems has been widely studied by many researchers. For instance, Wu, M. et al. [2] used a safe deep reinforcement learning (DRL) control method based on a safe reward shaping method and applied it to the constrained control for an electro-hydraulic servo system (EHSS). Chen, P. et al. [3] proposed a novel control strategy of speed servo systems based on deep reinforcement learning, which can achieve proportional–integral–derivative automatic tuning and effectively overcome the effects of inertia mutation and torque disturbance. Wyrwał, D. et al. [4] proposed a novel control strategy for a hydraulic cylinder based on deep reinforcement learning, which can automatically control the hydraulic system online so that the system can consequently maintain a consistently good control performance. Zhang, T. et al. [5] proposed a strategy based on deep reinforcement learning for the optimization of gain parameters of a cross-coupled controller, allowing the effective convergence of the gain parameters with the optimal intervals. The optimal gain parameters obtained by the proposed strategy can significantly improve the contour control accuracy in biaxial contour tracking tasks. Zamfirache, I.A. et al. [6] proposed a new control approach based on reinforcement learning (RL) that used policy iteration (PI) and a metaheuristic grey wolf optimizer (GWO) algorithm to train neural networks (NNs). In so doing they demonstrated good results for NN training and the solving of complex optimization problems. Shuprajhaa, T. et al. [7] developed a generic-data-driven modified proximal policy optimization (m-PPO) for an adaptive PID controller (RL-PID) based on reinforcement learning for the control of open-loop unstable processes, which eliminated the need for process modeling and pre-requisite knowledge on process dynamics and controller tuning.

Some studies have also integrated multiple comprehensive methods to achieve optimal control. Vaerenbergh, K.V. et al. [8] presented a practical application of a hybrid approach where reinforcement learning is the global layer to tune the controllers of every subsystem for the problem. It was shown that developing a centralized global controller for systems with many subsystems or complex interactions is usually very hard or even unfeasible. Lv, Y. et al. [9] used an RL-based approximate dynamic programming (ADP) structure to learn the optimal tracking control input of a servo mechanism, where unknown system dynamics were approximated with a three-layer NN identifier. Radac, M.B, and Lala, T. [10] proposed a Q-learning-like data-driven model-free (with unknown process dynamics) algorithm. They used neural networks as generic function approximators and validation on an active suspension system and it was shown to be easily amenable to artificial road profile disturbance generation for the optimal and robust control of a data-driven learning solution. Oh, T.H. et al. [11] proposed a DDPG-based deep RL method to simultaneously design and tune several notch filters and used a real industrial servo system with multiple resonances to demonstrate the proposed method effectively. They found the optimal parameters for several notch filters and successfully suppressed multiple resonances to provide the desired performances. Chen, W. et al. [12] proposed a novel adaptive law for the critical network in an RL framework to address the problem of nonlinear system control, which is driven by historical estimation errors but uses an auxiliary matrix instead of a historical data set, thus reducing the computational effort of the controller.

Reinforcement learning mainly learns and optimizes its own behavioral strategies through the idea and mechanism of trial and error. The behavior acts on the environment and obtains a response from the environment. The response is the evaluation index of the behavior. In the process of constantly interacting with the environment, the agent constantly changes their actions according to the rewards they get from the environment. With enough training, the agent can accumulate experience and then interact with the environment in a way that maximizes reward.

Although many studies have been undertaken in the practice of RL design and have laid a great theoretical foundation, there still exist some challenges and shortcomings. For instance, the classic reinforcement learning methods such as Q-learning do not have good generalizability and are usually only useful for specific tasks. Methods of control that have been optimized by neural network algorithms or genetic algorithms are usually effective only for specific cycle periods, lack on-line learning capabilities and have limited generalizability. Additionally, the experimental research on pure hydraulic servo systems has rarely been concerned with RL, and the verification of tests is relatively simple. As a result, most studies fail to set comparison tests or compile reinforcement learning algorithms for testing and verification.

In this study, the application of a reinforcement learning algorithm as the optimal control is presented Ref. [13]. This algorithm learns the optimal control strategy through direct interaction with objects. A deep neural network based on TD3 reinforcement learning training is introduced to realize this complex control [14–16]. Firstly, the feasibility and effectiveness of the controller are verified by simulation, and rapid prototype control is then carried out based on a Speedgoat prototype target machine. A GPU coder and LCT packaging tool are then used to deploy the reinforcement learning control program after the reinforcement learning training is completed. Finally, we verify the rationality of the control scheme and the effectiveness of the control algorithm by experiments.

2. Reinforcement Learning

Reinforcement learning [17] is a type of learning method in machine learning. It aims to construct and train agents to complete corresponding tasks in an unknown environment and its basic framework is shown in Figure 1 Since the actions in the learning process can affect the environment and the environment will then affect the subsequent actions, reinforcement learning can be regarded as, in essence, a closed-loop control, and the strategies of the agent allow it to complete the task in an optimal way through iterative updates, which is similar to the controller in the system [18,19].

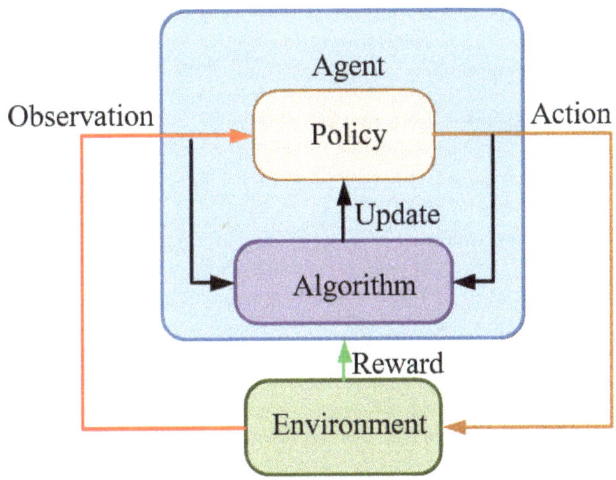

Figure 1. The basic framework of reinforcement learning.

Interactions between the agent and the environment in a period T are as follows:

(a) The *Agent* observes the system state after the previous action A;
(b) Under the current state of S and strategy P, agent undertakes action A by exploring the noise ϵ;
(c) The reward value R is obtained under the new environment state S, then the Agent updates the strategy P according to the reward value R;
(d) Repeat the above steps until the requirements are met.

Among these illustrated above, the reward value R is to evaluate the quality of the actions made by the agent in the environment. Since the reasonable setting of a reward function determines its convergence speed and stability, it is key for the agent to learn strategy effectively.

This paper takes a servo valve controlled asymmetric hydraulic cylinder system as the research object, builds a simulation environment based on a MATLAB/Simulink module, and carries out precise positional control of a hydraulic cylinder through a reinforcement learning algorithm. Additionally, the TD3 reinforcement learning algorithm was designed to interact with the environment so that ultimately the strategy of the control requirements can be satisfied. The servo valve controlled asymmetric hydraulic cylinder system is shown in Figure 2.

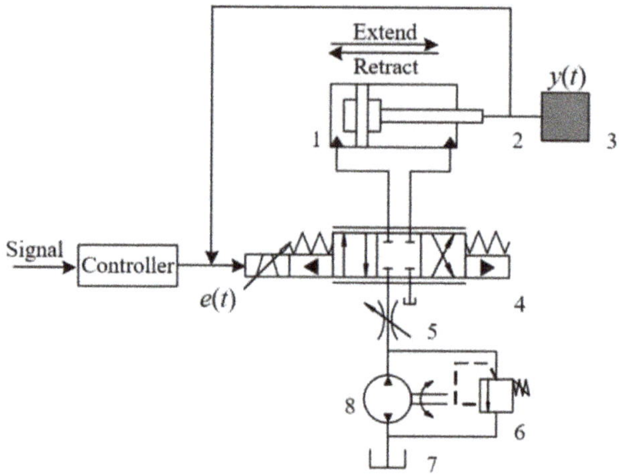

Figure 2. The schematic diagram of hydraulic servo system. 1. Asymmetric hydraulic cylinder 2. Sensor 3. Load 4. Servo hydraulic valve 5. Throttle hydraulic valve 6. Relief value 7. Hydraulic oil tank 8. Pump.

3. Design of the Algorithm

3.1. Actor–Critic Algorithm

The actor–critic [20] method combines the advantages of two methods: Actor and Critic. The method of Critic can be used to estimate a function of value that can lay a foundation to update the Actor function. As in Figure 3, there are two neural networks, one of these is the Actor network which determines the output in response to the environment, which can be either continuous or discrete. According to the combination of action and state on the output, the Critic network evaluates the value of the action of the Actor network. Meanwhile, the actual reward can be compared with the estimated value of the Critic to get the time difference error (TD-Error), which can judge how to adjust the parameters of the Critic network to obtain a more accurate estimate value. It can also judge the quality of the current action value [21,22], so as to achieve the purpose of updating the Actor parameters. In the process of the constant interaction between the agent and the environment, the weights of the Actor and Critic are updated and a relatively ideal control effect is achieved [23].

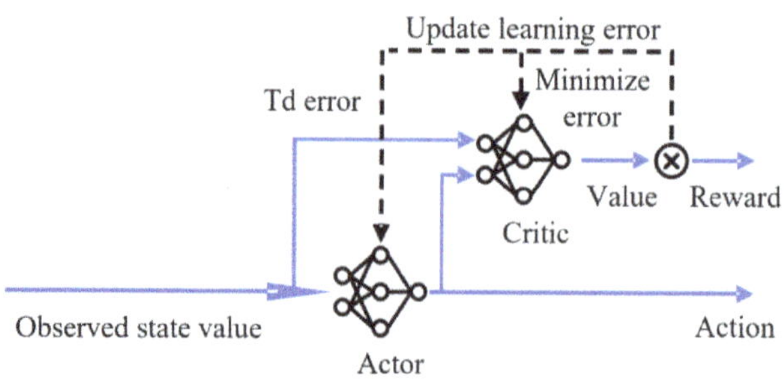

Figure 3. The Actor–Critic algorithm.

3.2. The Algorithm of DDPG and TD3

The depth deterministic policy gradient (DDPG) algorithm is a deep reinforcement learning algorithm based on Actor–Critic architecture for continuous action space [24]. The structure of a DDPG algorithm is shown in Figure 4.

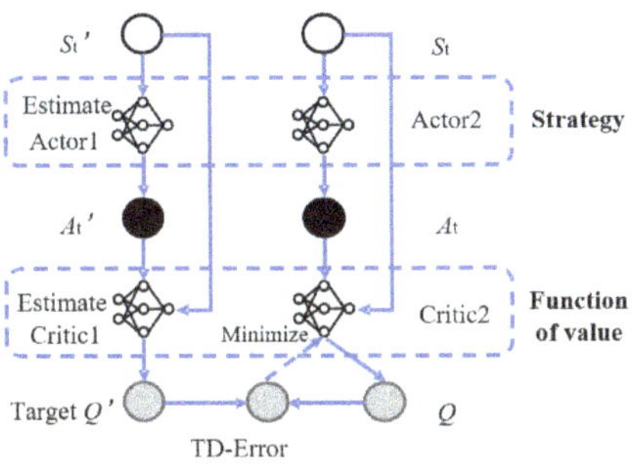

Figure 4. Structure of a DDPG algorithm.

The twin delayed deep deterministic policy gradient (TD3) algorithm (Figure 5) improves the network structure and update mode [25–27], which is based on DDPG algorithm.

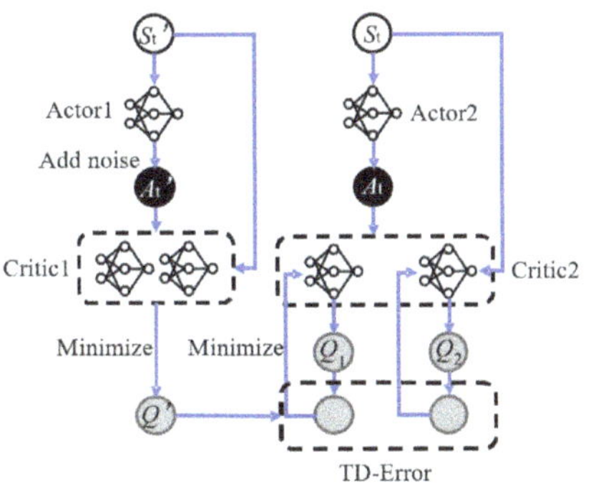

Figure 5. Structure of TD3 algorithm.

This uses the Actor–Critic framework and builds six neural networks to build the reinforcement learning body. The reward function is constructed based on the tracking error and error integral of the hydraulic position servo system, and the state set is composed of the output error of the position servo system. The TD3 algorithm has two advantages, which are as follows:

(i) A reduced overestimation of a dual Critic network.

Since noise θ appears in the sample value y, the target value E_θ of Q network in the real case of the learning process is:

$$y = r + \gamma \max_{a'} Q(s', a') \tag{1}$$

$$E_\theta[\max_{a'} Q(s', a') + \theta] \geq \max_{a'} Q(s', a') \tag{2}$$

After updating the Q function many times, errors will accumulate, leading to a number of bad states being assigned with higher value, and to a large deviation. A double Q-value network reduces estimation errors because it decouples the action and update operations [28]. The roles of the main and target networks and the update mode are shown in Table 1.

Table 1. Double Q-value network.

	Main Net	Target Net
Function	Approximation of action value function Q	Providing TD targets
Update	Updating the weights by minimum gradient method	Copy the Main weight

Using two independent Critics (sharing experience pool) in the TD3 algorithm, we can take the minimum value between the two Critics to eliminate the phenomenon of overestimation and to update the target value. However, although this may lead to an underestimated value for the strategy, this is preferable to overestimating it, as cumulative overestimation will make the strategy ineffective.

(ii) Smooth regularization of the target strategy.

Each step of the TD update produces a small error, which is more noticeable for approximate estimates. After many updates, a large number of errors will accumulate, eventually leading to an inaccurate Q value [29]. When the Actor and Critic are trained simultaneously, there may be a situation where training is unstable or divergent. In this paper, the target network is introduced by the delayed update of the policy. The method includes two aspects. First, a regularization of the parameters (add noise). Secondly, we update the target network every d times after updating Critic. The update objectives of Critic are calculated by the target network to improve the value function convergence, wherein the value function is updated at a higher frequency and the strategy is updated at a lower frequency.

3.3. Flow of Control

Reinforcement learning in this paper consists of PI parameters K_p and K_i to form an action set, and its learning process is shown in Figure 6 below.

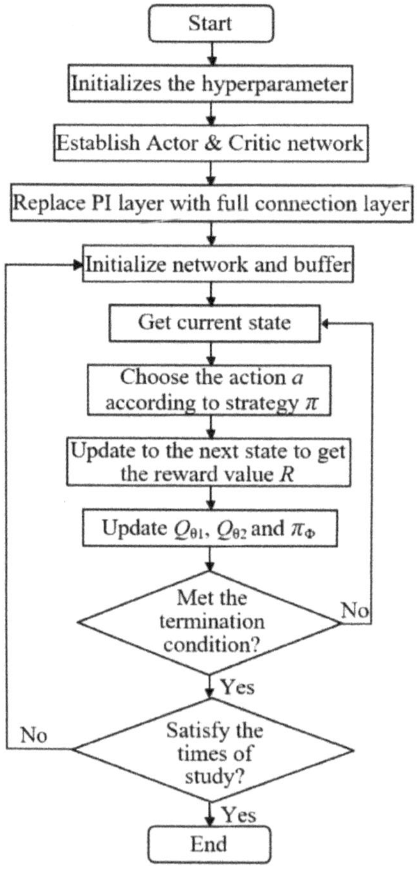

Figure 6. Flow chart of TD3 algorithm.

The TD3 servo control algorithm is shown in Figure 7. In the initial target network, we use a DQN mechanism and the initialization playback buffer B to eliminate the influence of catastrophic neural network forgetting [30,31]. By introducing a random noise with SARSA ideas ϵ (disturbance smooth value function in the action dimensions) to choose a actions and storage transition matrix to B, the memory pool can be formed. The process of updating the weights of Actor and Critic (π_φ, $Q_{\theta 1}$ and $Q_{\theta 2}$) is as follows:

$$\begin{cases} \widetilde{a} \leftarrow \pi_{\phi'}(s') + \epsilon \\ y \leftarrow r + \gamma \min\limits_{i=1,2} Q_{\theta'_i}(s', \widetilde{a}) \\ \epsilon \sim \text{clip}(\eta(0, \widetilde{\sigma}), -c, c) \end{cases} \tag{3}$$

Updating the Critic network with minimized loss function:

$$L_{\theta_i} = \min_{\theta_i} \frac{1}{N}(y - Q_{\theta_i}(s, a))^2 \tag{4}$$

Updating Actor networks using policy gradients:

$$\nabla_\phi J(\phi) \approx \frac{1}{N} \sum_i \nabla_a Q_{\theta_1}(s, a)\big|_{a=\pi_\phi(s)} \nabla_\theta \pi_\phi(s) \tag{5}$$

Reducing the cumulative error by updating the target network smoothly:

$$\begin{cases} \theta'_i \leftarrow \tau\theta_i + (1 - \tau)\theta'_i \\ \phi' \leftarrow \tau\phi + (1 - \tau)\phi' \end{cases} \tag{6}$$

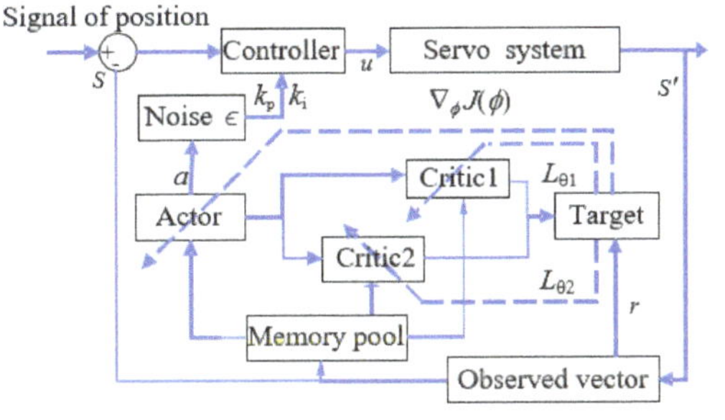

Figure 7. The servo control algorithm of TD3.

A neural network should be established before creating an Actor network, including an observation input and action output [32]. As a network, a *PI* controller can set error and error integral observation vectors on the fully connected layer.

$$u = K_p e(t) + K_i \int_0^t e(t)dt \tag{7}$$

$$w = [K_p, K_i] \tag{8}$$

Here, u is the output of the Actor network and w is the weight of the Actor network.

Since the gradient descent optimization may turn the weight negative, function Y should be executed to ensure that the weights are positive before setting the fully connected *PI* layer.

$$Y = \text{abs}(w) * [e(t), \int_0^t e(t)dt] \tag{9}$$

$$J = \lim_{T \to \infty} E(\frac{1}{T} \int_0^T [A(ref - y(t))^2 + Bu^2(t)]dt) \quad (10)$$

White noise ($E[n^2(t)] = 1$) is added to the model to simulate signal acquisition and external interference, so as to minimize the output control signal while ensuring minimal errors under the objective function standard.

4. The Establishment of the Environment

4.1. The Establishment of the Network

The reinforcement learning (RL) agent is used to replace the *PI* controller to keep the network structure of the two Critics in the same state (Figure 8), we then set the number of network neurons (Table 2).

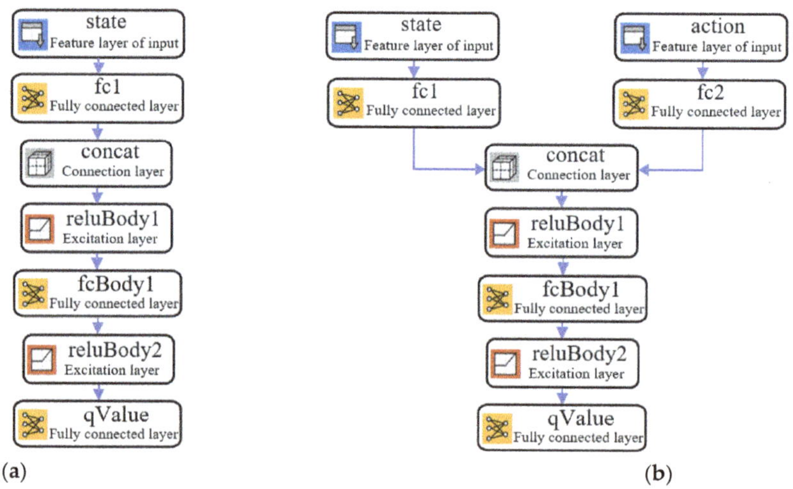

Figure 8. The neural network structure layer defined by TD3. (**a**) The network layer of the Actor and (**b**) the network layer of the Critic.

Table 2. Number of network neurons.

The Name of Layer	Critic Network	Actor Network
Layer of state	3 × 2	3 × 2
Layer of action	4 × 1	-
Fully connected layer 1	50 × 1	50 × 1
Fully connected layer 2	50 × 1	-
Connection layer	32 × 1	32 × 1
Layer of activation 1	25 × 1	25 × 1
Fully connected layer 3	50 × 1	50 × 1
Layer of activation 2	25 × 1	25 × 1
Fully connected layer 4	4 ×1	4 × 1

4.2. Establishment of the Environment

The environment receives operations from agents in reinforcement learning scenarios, outputs observations generated by the dynamic behavior of the environment model and generates rewards to measure the contribution of actions to the completion of tasks [33]. The observer structure is shown in Figure 9.

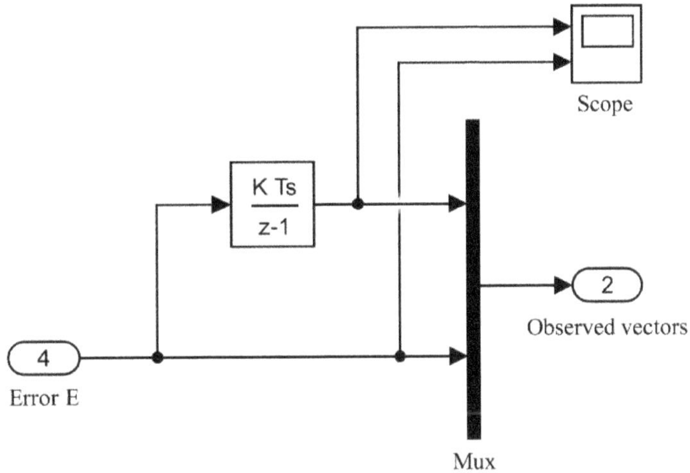

Figure 9. Observed vectors.

Mixed rewards are used for rewarding signals, which include continuous rewards and discrete rewards. Here, the discrete reward signals keep the system away from the bad state, while continuous reward signals improve the convergence by providing smooth rewards near the target state [34]. The following hybrid reward function R is designed for position control of the hydraulic servo system:

$$\begin{cases} r_1 = -10^a e_t^2 - u_t^2 \\ r_2 = -10^a (p \leq -125 || p \geq 125) \\ R = r_1 + r_2 \end{cases} \tag{11}$$

The larger a is, the better the network exploration is, but the longer the learning time and the higher the learning cost. When comprehensively considering the parameter $a \in (1,4)$, as the displacement is outside the range of activity, there will be a large reward and punishment ($P \in (125125)$). The program block diagram of the above Equation (11) is shown in Figure 10, and the Simulink reinforcement learning training program is shown in Figure 11 below.

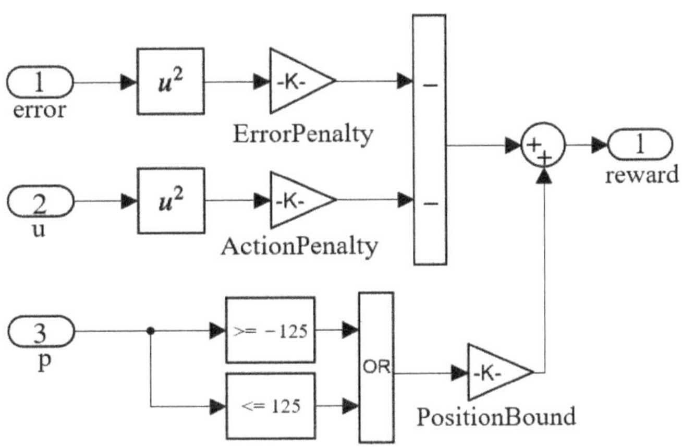

Figure 10. Calculation of the reward function.

References

1. Liu, Y.; Wang, A.; Wang, Z.; Liu, C. Research on modeling and model-free control of hydraulic servo system. *Mach. Des. Manuf.* **2019**, *5*, 10–13.
2. Wu, M.; Liu, L.; Yu, Z.; Li, W. Safe reward-based deep reinforcement learning control for an electro-hydraulic servo system. *Int. J. Robust Nonlinear Control.* **2022**, *32*, 7646–7662. [CrossRef]
3. Chen, P.; He, Z.; Chen, C.; Xu, J. Control Strategy of Speed Servo Systems Based on Deep Reinforcement Learning. *Algorithms* **2018**, *11*, 65. [CrossRef]
4. Wyrwał, D.; Lindner, T.; Nowak, P.; Białek, M. Control strategy of hydraulic cylinder based on Deep Reinforcement Learning. In Proceedings of the 2020 International Conference Mechatronic Systems and Materials (MSM), Bialystok, Poland, 1–3 July 2020; pp. 1–5.
5. Zhang, T.; Wu, C.; He, Y.; Zou, Y.; Liao, C. Gain parameters optimization strategy of cross-coupled controller based on deep reinforcement learning. *Eng. Optim.* **2021**, *54*, 727–742. [CrossRef]
6. Zamfirache, I.A.; Precup, R.E.; Roman, R.C.; Roman, R.C.; Petriu, E.M. Policy Iteration Reinforcement Learning-based control using a Grey Wolf Optimizer algorithm. *Inf. Sci. Int. J.* **2022**, *585*, 162–175. [CrossRef]
7. Shuprajhaa, T.; Sujit, S.K.; Srinivasan, K. Reinforcement learning based adaptive PID controller design for control of linear/nonlinear unstable processes. *Appl. Soft Comput.* **2022**, *128*, 109450. [CrossRef]
8. Vaerenbergh, K.V.; Vrancx, P.; Hauwere, Y.D.; Nowé, A.; Hostens, E.; Lauwerys, C. Tuning hydrostatic two-output drive-train controllers using reinforcement learning. *Mechatronics* **2014**, *24*, 975–985. [CrossRef]
9. Lv, Y.; Ren, X.; Zeng, T.; Li, L.; Na, J. Neural Network Tracking Control of Unknown Servo System with Approximate Dynamic Programming. In Proceedings of the 38th Chinese Control Conference, Guangzhou, China, 27–30 July 2019.
10. Radac, M.B.; Lala, T. Learning nonlinear robust control as a data-driven zero-sum two-player game for an active suspension system-ScienceDirect. *IFAC-PapersOnLine* **2020**, *53*, 8057–8062. [CrossRef]
11. Oh, T.H.; Han, J.S.; Kim, Y.S.; Yang, D.Y.; Cho, D.D. Deep RL Based Notch Filter Design Method for Complex Industrial Servo Systems. *Int. J. Control. Autom. Syst.* **2020**, *18*, 2983–2992. [CrossRef]
12. Chen, W.; Hu, J.; Xu, C.; Zhou, H.; Yao, J.; Nie, W. Optimal tracking control of mechatronic servo system using integral reinforcement learning. *Int. J. Control. Autom. Syst.* **2022**, *16*, 1–11. [CrossRef]
13. Ding, X.F.; Xu, X.L. Nonlinear Optimal Control of Hydraulic Servo System. *Chin. Hydraul. Pneum.* **2004**, *4*, 32–35.
14. Yuan, X.; Wang, W.; Zhu, X.; Zhang, L. Theoretical Model of Dynamic Bulk Modulus for Aerated Hydraulic Fluid. *Chin. J. Mech.* **2022**, *35*, 121. [CrossRef]
15. Wiens, T. Engine speed reduction for hydraulic machinery using predictive algorithms. *Int. J. Hydromechatronics* **2019**, *2*, 1. [CrossRef]
16. Rehab, I.; Tian, X.; Gu, F.; Ball, A.D. The influence of rolling bearing clearances on diagnostic signatures based on a numerical simulation and experimental evaluation. *Int. J. Hydromechatronics* **2018**, *1*, 16–46. [CrossRef]
17. Liu, Q.; Zhai, J.; Zhang, Z. A review of deep reinforcement learning. *Chin. J. Comput.* **2018**, *41*, 1–27.
18. Sung, O.; Jin, K. The design of a real-time simulator on the hydraulic servo system. *Int. J. Precis. Eng. Manuf.* **2003**, *4*, 9–14.
19. Yang, X.; He, H.; Wei, Q.; Luo, B. Reinforcement learning for robust adaptive control of partially unknown nonlinear systems subject to unmatched uncertainties. *Inf. Sci.* **2018**, *463–464*, 307–322. [CrossRef]
20. Chen, X.; Yang, Y. Adaptive PID control based on Actor-Critic learning. *Control. Theory Appl.* **2011**, *28*, 1187–1192.
21. Gao, Q.; Zhu, Y.; Liu, J. Dynamics modelling and control of a novel fuel metering valve actuated by two binary-coded digital valve arrays. *Machines* **2022**, *10*, 55. [CrossRef]
22. Chao, Q.; Xu, Z.; Tao, J.; Liu, C. Capped piston: A promising design to reduce compressibility effects, pressure ripple and cavitation for high-speed and high-pressure axial piston pumps. *Alex. Eng. J.* **2023**, *62*, 509–521. [CrossRef]
23. Li, J.; Ji, L.; Li, J. Optimal consensus control for unknown second-order multi-agent systems: Using model-free reinforcement learning method-ScienceDirect. *Appl. Math. Comput.* **2021**, *410*, 126451. [CrossRef]
24. Wei, X.; Zhang, Q.; Jiang, T.; Liang, L. Fuzzy adaptive deep reinforcement learning method for transient optimization of servo systems. *J. Xi'an Jiaotong Univ.* **2021**, *55*, 68–77.
25. Lee, D.; Lee, S.J.; Yim, S.C. Reinforcement learning-based adaptive PID controller for DPS. *Ocean. Eng.* **2020**, *216*, 108053. [CrossRef]
26. Zhen, H.; Zhai, H.; Ma, W.; Zhao, L.; Weng, Y.; Xu, Y.; Shi, J.; He, X. Design and tests of reinforcement-learning-based optimal power flow solution generator. *Energy Rep.* **2022**, *8*, 43–50. [CrossRef]
27. Chu, Z.; Sun, B.; Zhu, D.; Zhang, M.; Luo, C. Motion Control of Unmanned Underwater Vehicles Via Deep Imitation Reinforcement Learning Algorithm. *IET Intell. Transp. Systems* **2020**, *14*, 764–774. [CrossRef]
28. Zeng, R.; Zhou, J.; Liu, M. Transfer reinforcement learning algorithm for double Q network learning. *Appl. Res. Comput.* **2021**, *38*, 1699–1703.
29. Liu, P.; Bai, C.; Zhao, Y.; Bai, C.; Zhao, W.; Tang, X. Generating attentive goals for prioritized hindsight reinforcement learning. *Knowl.-Based Syst.* **2020**, *203*, 106–140. [CrossRef]
30. Yang, Y.; Li, J.; Peng, L. Multi-robot path planning based on a deep reinforcement learning DQN algorithm. *CAAI Trans. Intell. Technol.* **2020**, *5*, 177–183. [CrossRef]

31. Chen, J.; Yang, Z.; Liu, Q.; Wu, H.; Xu, Y.; Fu, Q. Heuristic Sarsa algorithm based on value function transfer. *J. Commun.* **2018**, *39*, 37–47.
32. Chen, J.; Zheng, M. A review of robot operating behavior based on deep reinforcement learning. *Robot* **2022**, *44*, 236–256.
33. Self, R.; Coleman, K.; He, B.; Kamalapurkar, R. Online Observer-Based Inverse Reinforcement Learning. In Proceedings of the 2021 American Control Conference (ACC), New Orleans, LA, USA, 25–28 May 2021; Volume 5, pp. 1959–1964.
34. Zhao, Y.; Qin, J.; Yuan, L. Intrinsic rewards that combine novelty and risk assessment. *Comput. Eng. Appl.* **2022**, 1–9.
35. Wei, L.; Luo, L.; Wang, Y.; Wang, G. Rapid Control Prototype Design of Electro-hydraulic Position Control Servo System Based on xPC Target. *Chin. Hydraul. Pneum.* **2018**, *7*, 24–28.
36. Maghareh, A.; Silva, C.E.; Dyke, S.J. Servo-hydraulic actuator in controllable canonical form: Identification and experimental validation. *Mech. Syst. Signal Process.* **2018**, *100*, 398–414. [CrossRef]
37. Lyu, L.; Chen, Z.; Yao, B. Advanced Valves and Pump Coordinated Hydraulic Control Design to Simultaneously Achieve High Accuracy and High Efficiency. *IEEE Trans. Control. Syst. Technol.* **2021**, *29*, 236–248. [CrossRef]

machines

Article

Crowning Method on Bearing Supporting Large Wind Turbine Spindle Considering the Flexibility of Structure of Shaft System

Xiangyang Liu [1], Rongjun Niu [1,*], Bin Wang [2], Shuai Zhang [3], Yongcun Cui [1] and Zhanli Zhang [1]

1 School of Mechatronics Engineering, Henan University of Science and Technology, Luoyang 471003, China
2 China State Shipbuilding Corporation Haizhuang Wind Power Co., Ltd., Chongqing 401123, China
3 School of Mechanical Engineering, Henan Institute of Technology, Xinxiang 453003, China
* Correspondence: niurj@haust.edu.cn

Abstract: To meet the precision design of bearings on large wind turbine spindles, a crowning method of bearing on wind turbine spindles considering the flexibility of the support structure is proposed. Firstly, a finite element (FE) model of the shaft system with a flexible structure is constructed by connecting the shaft and bearing through constraint equations (CE) and multi-point constraint (MPC) algorithms and replacing the bearing rollers with nonlinear spring elements and dampers. Then, the Newmark integration algorithm is used to solve the model and analyze the effect of the structure's rigidity on the load distribution of bearings. Then, perform convergence analysis of the sequences of the spring load distribution using a high-pass filter based on fast Fourier transform (FFT) and root mean square error (RMSE) to obtain a suitable number of replacement springs. Finally, a sub-model of the upwind bearing is constructed with structured mesh. With the maximum Von Mises stress of the roller profile as the design target, the optimal logarithmic crowning of the roller and its tolerance zone under the given working conditions are obtained. The results show that the FE model of the shaft system proposed has good convergence. The FE model of the shaft system considering the flexibility of the support structure can obtain more accurate load distributions of bearings and can make the accurate crowning design of the bearing rollers based on the actual working conditions. This provides support for the precision design of bearings in large shaft systems.

Keywords: tapered bearing; logarithmic crowning; FEM; nonlinear spring; damper; sub-model

Citation: Liu, X.; Niu, R.; Wang, B.; Zhang, S.; Cui, Y.; Zhang, Z. Crowning Method on Bearing Supporting Large Wind Turbine Spindle Considering the Flexibility of Structure of Shaft System. *Machines* **2023**, *11*, 28. https://doi.org/10.3390/machines11010028

Academic Editors: Kan Liu and Wei Hu

Received: 22 November 2022
Revised: 21 December 2022
Accepted: 22 December 2022
Published: 26 December 2022

1. Introduction

To solve the problem of energy shortage and environmental pollution, wind turbines are deployed in large numbers around the world. The bearings on the wind turbine spindle are the key rotary parts of wind power equipment, and their failure will cause great economic losses [1,2]. Because the bearings on wind turbine spindles are usually installed in pairs, the actual load of a single bearing is difficult to measure. At present, the traditional design theory of bearing is to use the design formula or experience to design a single bearing according to the given design load, but the structural flexibility of the shaft system cannot be considered in the design. In a shaft assembly, the contact stress distribution of bearing rollers designed by this method is not optimal and cannot meet the accuracy requirements. The finite element method (FEM) can consider the structural flexibility result by constructing the overall model of the shaft assembly, which can obtain a more accurate bearing load and stress distribution under the global load. It has higher calculation accuracy than the traditional design theory of bearing and can be used for the precision design of bearings in complex assembly. However, due to the large size, complex structure, and the large number of rollers of bearings on the wind turbine spindle, the FE analysis of it needs to include two sets of bearings and support structure at the same time, it is very difficult to carry out FE analysis. Therefore, the reliability design of large-size bearings, such as bearings on wind turbine spindles, has become increasingly prominent.

For the design and analysis of large-size bearings, the simplified FE method is now widely used by scholars. Smolnicki et al. [3] proposed a superelement-based FE model for large bearings, using superelements instead of the nonlinear contact actions between rollers and raceways, while the rings of the bearings are still modeled using solid elements. Chen et al. [4,5] used extension spring elements to replace the balls of a wind turbine pitch turntable bearing to analyze the effects of the rigidity of the support structure, the number of bolts, and the preload force on the bearing life. The effect of hub rigidity on the bearing load distribution was also analyzed. Plaza et al. [6] replaced the nonlinear contact actions between the balls and raceways of the pitch turntable bearing with extension spring elements and used superelements to model the rings of the bearing, thus significantly reducing the computational effort. Li et al. [7] proposed an equivalent FE modeling method to analyze the bearing and mounting structure by replacing the balls with nonlinear extension spring elements and using beam elements instead of solid mounting bolts. Aguirrebeitia et al. [8] developed a FE model of a four-point contact ball bearing based on nonlinear spring elements and calculated the static load-carrying capacity of the bearing, which was verified by comparing with the theoretical results, pointing out that the results of the FE model are more accurate than the theoretical results and the theoretical solution is more conservative. He et al. [9] used nonlinear springs instead of solid rollers to study the effect of raceway hardening on slewing bearing life. The use of nonlinear springs enabled the model to obtain results quickly. Śpiewak et al. [10] developed both a FE model and a theoretical analytical model of a double slewing bearing and verified them in comparison. It was suggested to replace each roller with one or more truss elements in the FE model of the bearing. To reduce the calculation scale of wire bearings in FEM, Martín et al. [11] proposed the method based on a custom matrix element and the replacement-element method for a roller based on a nonlinear spring element, among which the former has higher accuracy and the latter has a simpler operation process. The above studies have shown that the use of simplified elements instead of solid rollers in the analysis of bearings by the FEM can significantly reduce the calculation effort and has good accuracy, which is widely used in the analysis of large-size bearings.

The working load of bearings on large wind turbine spindles is extremely high. The profile of the roller or raceway has a direct impact on the stress distribution of the roller and the life of the bearing. The modification of the profile of the raceway or roller can optimize the load distribution. Kania [12] proposed a method for modification of the raceway profile of four-point contact rotary table bearings in a wind turbine, which improved the load-carrying capacity of bearings. However, compared with crowned rollers, this method increases the technical difficulty and production cost. Ju et al. [13] studied the effect of the power function profile and exponential profile on the stress concentration of bearing rollers and pointed out that the exponential profile can effectively reduce the stress concentration, but the unsmooth connection between the straight line of the roller and the exponential trim line is still prone to sudden stress changes. Fujiwara [14] and He [15] offered a method to study the optimal crowning amount of a logarithmic crowned roller, and the optimal crowning function was obtained by adjusting the control coefficient of load and control coefficient of contact. Based on the nonlinear equation of elastic contact, Li [16] established a numerical model of tapered roller bearings on wind turbine spindles and analyzed the load distribution of bearing with logarithmic crowned rollers, pointing out that the logarithmic crowning is widely accepted by bearing manufacturers. Zhang et al. [17] analyzed the effects of load, tilting angle, and the crowning curve of the roller on the stress distribution of rollers of three-row cylindrical roller turntable bearings in wind turbines, and proposed a crowning method using logarithmic curve and tangent arc for axial and radial rollers, respectively, based on the load characteristics of turntable bearings. Li et al. [18] constructed an overall model and a sub-model for a wind turbine with a large taper angle, respectively. The overall model analyzed the effect of axial bearing clearance on the load distribution of bearing, the sub-model analyzed the effect of different crowning shapes on the contact stress of the roller, and the use of simplified FEM greatly reduced the computational effort. The logarithmic crowning curve was found to enable the bearing to have a high load-carrying capacity under the studied working conditions. The above researches show that the logarithmic crowning makes the stress on the roller profile uniformly distributed under the non-extreme bias load condition, which is the optimal crowning method at present.

At present, the analysis of large-size bearings is widely focused on the analysis of individual bearings. For bearings on large wind turbine spindles, more accurate load distribution can be obtained by constructing a FE model of the shaft system, but the analysis type of the above traditional simplified method is still nonlinear, and the model is more difficult to converge after coupling the bearing further to the shaft system. The research of [17] considered the effect of mounting bolts on the load distribution of bearings but assumed that the design load of the bearing was known. The research of [4,5] established a FE model of a turntable bearing with a flexible support structure and investigated the effect of the rigidity of the support structure on the bearing life. However, the support structure is directly coupled to a single bearing and is not applicable to shaft systems of wind turbine spindles containing multiple sets of bearings. The research of [18] established a FE model of a flexible shaft system to analyze the bearings, but the coupling method of the FE model was not introduced in detail. To sum up, there is a lack of relevant literature on the design of spindle bearing considering the structure flexibility of complex shaft systems like large wind turbine spindles by using FEM.

In view of this, considering the flexibility of the support structure of the shaft system, a FE model of the shaft system of the wind turbine spindle is established based on CE, MPC algorithm, and nonlinear spring elements in ANSYS® 2020 R2 software. The critical damping technique is used to improve the convergence of the model. A FE modeling method with good convergence for the shaft system is proposed, which can obtain the accurate load distribution of the bearings by the external wind load of the shaft system and can optimize the bearing roller profile directly. Then the bearing sub-model with structured mesh is constructed. The design method of roller crowning's tolerance of large bearing is given with logarithmic crowning parameters as design variables and the maximum stress along the profile of the roller at the center of the contact area as the design target. The analysis flow of the work is shown in Figure 1.

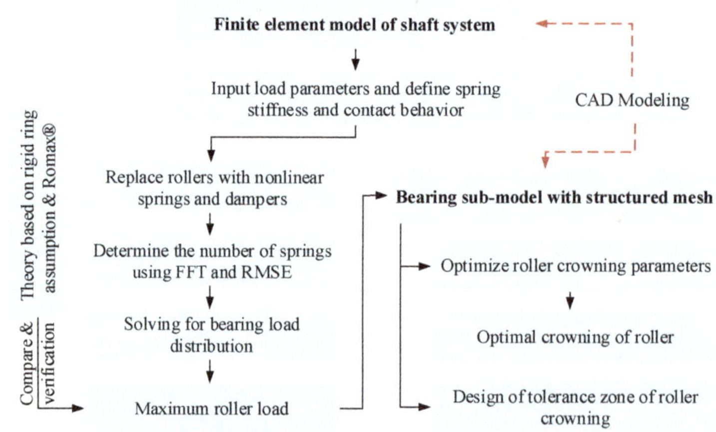

Figure 1. Design of logarithmic crowned roller for wind turbine spindle bearing.

2. FE Modeling Method of the Shaft Assembly

The main shaft assembly of a 6.25 MW wind turbine is used as the study object, and the installation structure consists of two sets of tapered roller bearings, a main shaft, and a bearing housing, as shown in Figure 2. The wind load is transmitted to the bearing through the hub and shaft, where L_1 = 2500 mm and L_2 = 2100 mm. Do not consider the effect of heat treatment and other effects on the material of the shaft, bearing, and housing in the analysis, using the same metal material parameters with a density of 7850 kg·m^{-3}, an elastic modulus of 2.1 × 10^{11} Pa, and a Poisson's ratio of 0.3. The structural parameters of the bearings are shown in Table 1. The wind load parameters are shown in Table 2.

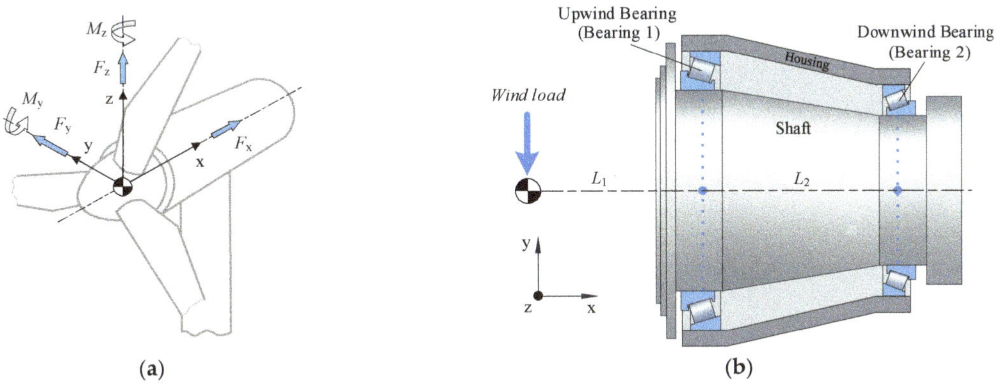

Figure 2. Structure and load of wind turbine spindle. (**a**) Load direction coordinate system; (**b**) Structure and load.

Table 1. Geometrical parameters of bearings.

Parameter	Value	
	Bearing 1	**Bearing 2**
Inner diameter d (mm)	1850	1600
Number of rollers Z	62	65
The contact angle of inner raceway α_i (°)	15.58	19.5
The contact angle of outer raceway α_e (°)	17.1	21.17
The contact angle of roller end–ring flange α_f (°)	83.52	69.67
Length of rollers l (mm)	121	98
Roller diameter in the middle D_w (mm)	95.085	73.26
Roller half taper angle β (°)	0.75	0.833
pitch diameter d_m (mm)	2043	1750

Table 2. Load parameters of the shaft system.

F_x/N	F_y/N	F_z/N	M_y/Nm	M_z/Nm
7.5 × 10^5	1.1 × 10^6	3 × 10^4	1700	−2700

2.1. Bearings Modeling

Due to the large size of bearings on the wind turbine spindle and the large number of rollers, it will generate massive nonlinear contact calculations if the solid elements are used to model the rollers. Meanwhile, it will introduce extremely strong nonlinearity to the FE model, which results in difficulty in solving the model.

In FE analysis of large-size bearings, modeling the rings with three-dimensional solid elements and the rollers with one-dimensional nonlinear elements can significantly reduce the computational effort. The simplification methods widely used are [4–6,19,20].

1. Nonlinear spring elements with special nonlinear stress-strain characteristics.
2. Superelements with the reduction matrix of structural stiffness.

Replacing rollers with nonlinear spring elements can significantly reduce the number of elements, and speed up the discrimination calculation of the contact algorithm. Compared with the superelement, it is more convenient to operate, so the nonlinear spring elements are adopted to replace the bearing rollers. The connection method is shown in Figure 3.

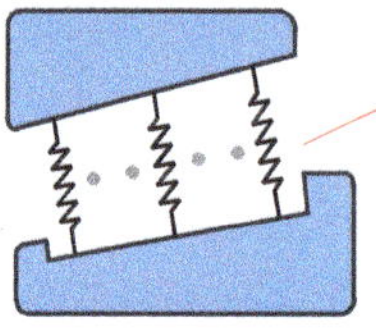

Figure 3. Replace rollers with nonlinear spring elements.

Nonlinear spring elements mainly consist of nonlinearity of state (Compressed Only) and nonlinearity of material (Load-deformation Curve). The nonlinearity of state allows the spring elements to better simulate the real physical contact process of rollers, and the nonlinearity of material allows the spring elements to accurately describe the stress-strain relationship of rollers. For ball bearings, each ball of the bearing can be replaced by one spring element, the length of which is equal to the diameter of the ball. For roller bearings, each roller should be replaced by two or more spring elements to simulate the bias load on the roller. The higher the number of spring elements replacing a roller, the closer the calculated result is to the actual one.

The larger the error of the FE model, the larger the fluctuation of the distribution curve of the spring reaction. To quantitatively determine the appropriate number of replacement springs, the root mean square error (RMSE) of the load distribution is used as an evaluation indicator to obtain the ideal number of replacement springs. The RMSE is the standard deviation, and the RMSE is very sensitive to the deviation of the data and can reflect the accuracy of the measured data. The smaller the value of the RMSE, the higher the accuracy of the measured value, so the RMSE can be used to evaluate the accuracy of the measured value.

For bearings on wind turbine spindles, the ideal load distribution has one peak value and is smooth. The sequence of spring reaction force distribution has the same low-frequency component, and the distribution of spring reaction force fluctuates irregularly along the ideal distribution due to the FE model error.

The sequence of the reaction force distribution X of a series of springs is

$$X = (x_1, x_2, \cdots, x_Z) \tag{1}$$

The high-frequency fluctuation characteristic sequence X_F of the data can be obtained by filtering sequence X with a high-pass filter based on FFT.

$$X_F = (x_{F1}, x_{F2}, \cdots, x_{FZ}) \tag{2}$$

Then the RMSE of sequence X is

$$\text{RMSE} = \sqrt{\frac{\sum_{i=1}^{Z}(X_i - \overline{X}_F)^2}{Z}} \tag{3}$$

where Z is the number of bearing rollers and $\overline{X}_F$ is the mean value of the sequence X_F.

Figure 4 shows the stress-strain characteristics of the nonlinear truss elements in the FE analysis, where $\varepsilon_{S(\Delta)}$ is the strain under the action of clearance S and clamp Δ. The truss element has a defined length d_r and cross-section, while the special material properties $\sigma(\varepsilon)$ can be defined, and such an element can be used as a nonlinear spring element [9,16]. The clearance of the bearing in the FE analysis can be achieved directly by translating the material property curve of the truss element as shown in Figure 4 along the transverse axis.

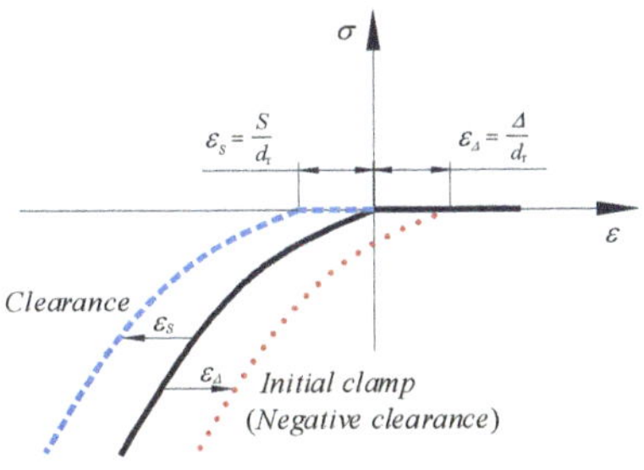

Figure 4. Stress-strain characteristics of the nonlinear spring element.

The mechanical properties of a roller can be simulated by assigning the stiffness curve of the roller to nonlinear spring elements. The contact type of tapered roller bearings is line contact. According to the Hertzian contact theory, the relationship between the contact load and the contact elastic deformation of the roller is [21]

$$Q = K_n \delta^{1.11} \tag{4}$$

where K_n is the total load-deformation constant between the roller and ring; δ is the total elastic deformation of the roller and two rings.

Palmgren gave a formula for calculating the elastic deformation of line contact [22].

$$Q = 0.71 \frac{l^{0.89}}{\eta} \delta^{1.11} \tag{5}$$

where l is the length of the roller, η is the combined elastic constant of two bodies in contact, and the formula is

$$\eta = \frac{1-\mu_1^2}{E_1} + \frac{1-\mu_2^2}{E_2} \tag{6}$$

where $\mu_{1(2)}$ is Poisson's ratio of the two bodies and $E_{1(2)}$ is the elastic modulus of the two bodies.

Comparing Equations (4) and (5), the load-deformation constant ($\mathrm{N/mm^{10/9}}$) between the roller and the rings is

$$K_\mathrm{n} = 0.71 \frac{l^{0.89}}{\eta} \tag{7}$$

The stiffness curve of the tapered roller can be approximated by Equations (4) and (7).

Assuming that each roller is replaced by n nonlinear spring elements, the load-deformation constant for each spring element is

$$K_\mathrm{spr} = \frac{K_\mathrm{n}}{n} \tag{8}$$

For the FE model of the whole shaft system, the nonlinear spring elements in the critical loading state perform rigid body movement, which may cause the singular stiffness matrix and make the FE model not converge. Such a simplified method is not very suitable for the FE analysis of large and complex structures. The use of critical damping technology based on nonlinear spring elements can realize the process of slow dynamic loading, i.e., a linear damper is established at the position of each nonlinear spring element, and the existence of the damper has no effect on the strain energy of the model and does not affect the solution accuracy. At the same time, the Newmark integration algorithm is used to solve the FE model by incremental integration method, and stable convergence results can be obtained quickly. The critical damping coefficient is determined by [23,24]

$$c = 2\sqrt{km} \tag{9}$$

where c is the critical damping constant, Ns/m; k is the stiffness of spring, N/m; m is the mass of roller, kg.

The method of replacing the rollers with nonlinear spring elements and linear dampers is shown in Figure 5.

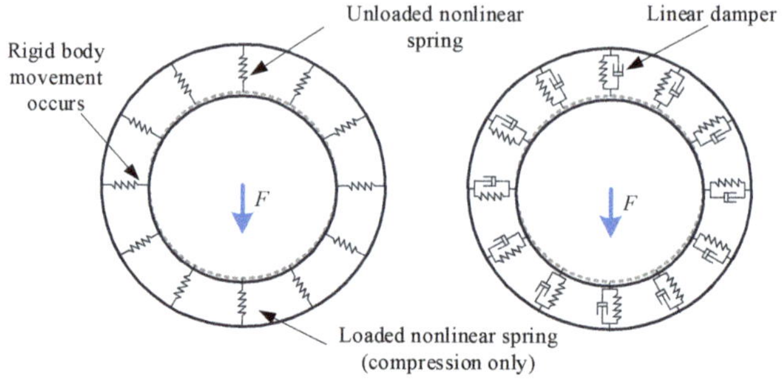

Figure 5. Replace rollers with nonlinear springs and dampers. (**a**) Replace rollers with springs; (**b**) Replace rollers with springs and dampers.

2.2. Main Shaft Modeling

The wind load is mainly distributed to the two support bearings through the main shaft, and the flexural deformation of the main shaft has a great influence on the load distribution of the bearings. Modeling of the main shaft with flexible beam elements. The bore surface of each bearing is coupled to a master node I at the center of the inner diameter surface through the CE. The MPC algorithm was used to constrain the degrees of freedom between the master node I and node J on the flexible beam, and the axial rotational degrees of freedom are simultaneously constrained to prevent shaft rotation. The wind load will be transferred to the bearing bore surface through the flexible beam. The shaft is connected to any of the bearings as shown in Figure 6, where q_I and q_J are the coordinate systems of degrees of freedom of the I and J nodes. The I and J nodes both have six degrees of freedom, and the I and J nodes spatially overlap.

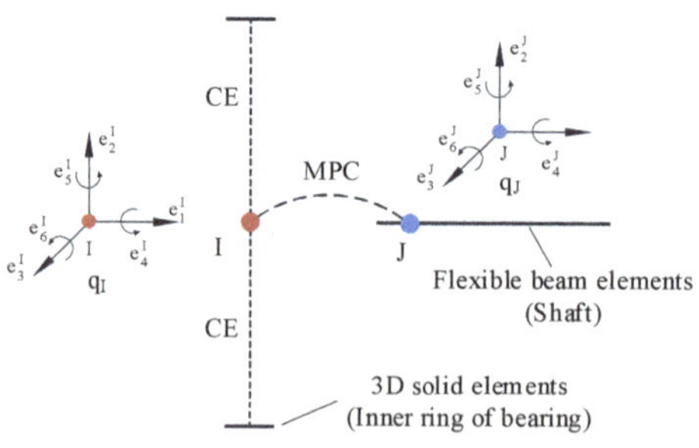

Figure 6. Shaft-bearing coupling diagram.

The degree of freedom constraint relationship of the two coupled nodes can be expressed as

$$C\left(q_I, q_J, t\right) = 0 \tag{10}$$

where q_I and q_J are the degrees of freedom corresponding to the coordinate system of degrees of freedom of the coupled nodes and t is the time.

2.3. Housing Modeling

In a wind turbine, the outer casing acts as the bearing housing for the bearings on the main shaft. The wind load is finally transferred to the outer casing through the shaft and bearings. The bearing housing holds the bearings on the spindle in place. Its elastic deformation under load affects the relative position of the outer rings of the bearings, and its flexibility directly affects the load distribution of the bearings just like that of the spindle. Ignoring its flexibility will result in too large spindle stiffness, which makes the calculation of the bearing load distribution inaccurate. Therefore, it is necessary to establish a flexible housing in the model. So the outer housing is modeled with 3D solid elements. To avoid the excessive introduction of nonlinearity in the system, the outer casing is coupled to the bearings' outer rings by sharing nodes.

The shaft, bearing rings, and rollers are assembled and the final FE connection method is shown in Figure 7.

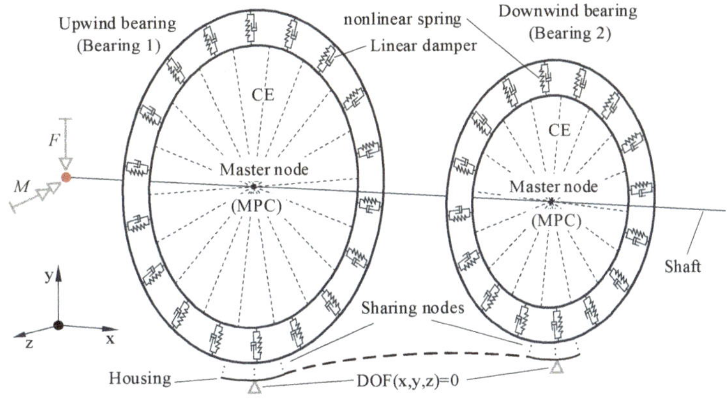

Figure 7. Coupling connection relationship of the shaft assembly.

For a roller, at least two springs are required to simulate tilt or offset load. To study the effect of the number of replacement springs on the load distribution of bearing rollers, each roller is replaced by 2, 3, and 5 nonlinear spring elements (denoted as 2 spr, 3 spr, and 5 spr), respectively, corresponding to the partial view of the model shown in Figure 8. It contains $n \times (Z_1 + Z_2)$ nonlinear spring elements, $n \times (Z_1 + Z_2)$ linear dampers, and 2 MPC elements.

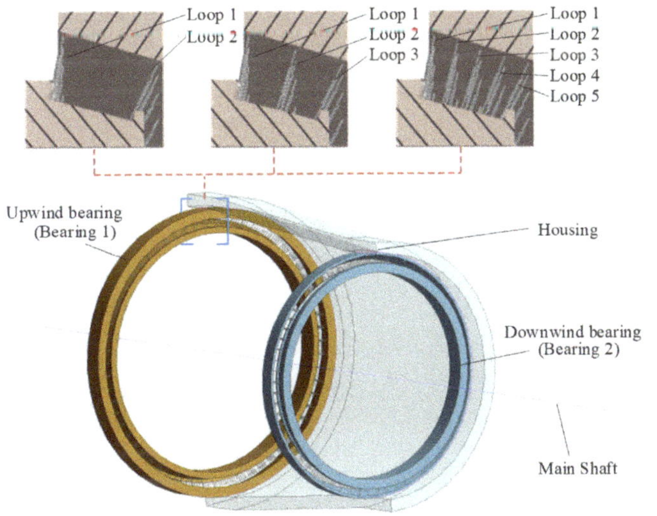

Figure 8. FE model of shaft system with different quantities of replacement springs.

3. Theoretical Calculation of Maximum Roller Load

Figure 9 shows the cross-section of bearings on the wind turbine spindle. Under the action of external load, the contact loads between the tapered roller and the inner raceway, outer raceway, and end-ring flange are Q_i, Q_e, and Q_f, respectively, and the contact angles are α_i, α_e, and α_f, respectively. The roller load satisfies the following equilibrium equation.

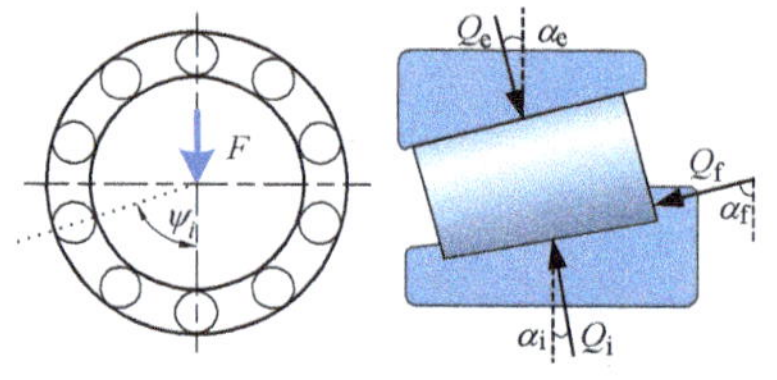

Figure 9. The positions and forces of the tapered rollers. (**a**) The angular positions of the rollers; (**b**) Forces of tapered rollers.

$$\begin{cases} Q_e \sin\alpha_e - Q_i \sin\alpha_i - Q_f \sin\alpha_f = 0 \\ Q_e \cos\alpha_e - Q_i \cos\alpha_i - Q_f \cos\alpha_f = 0 \end{cases} \tag{11}$$

It is further deduced that:

$$Q_i = Q_e \frac{\sin(\alpha_e + \alpha_f)}{\sin(\alpha_i + \alpha_f)} = c_i Q_e \tag{12}$$

$$Q_f = Q_e \frac{\sin(\alpha_e - \alpha_f)}{\sin(\alpha_i + \alpha_f)} = c_f Q_e \tag{13}$$

where

$$c_i = \frac{\sin(\alpha_e + \alpha_f)}{\sin(\alpha_i + \alpha_f)},\ c_f = \frac{\sin(\alpha_e - \alpha_f)}{\sin(\alpha_i + \alpha_f)}.$$

The tapered roller bearing installed on the wind turbine spindle is mainly affected by two forces, radial force F_r and axial force F_a, as shown in Figure 10. The radial displacement δ_r and axial displacement δ_a of the inner ring are generated relative to the outer ring under the action of the radial force F_r. Let the radial clearance of the bearing be zero, the radial displacement at the roller with position Angle ψ_i is

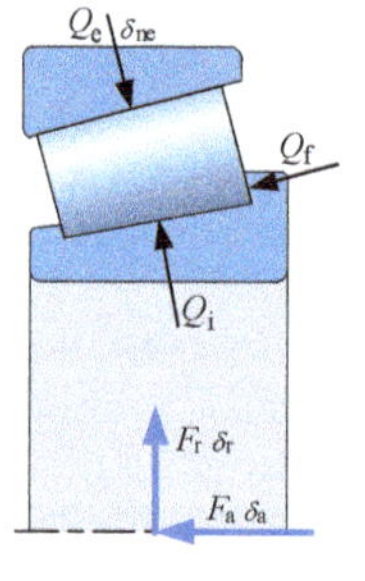

Figure 10. Tapered roller forces and bearing forces.

$$\delta_{ri} = \delta_r \cos\psi_i \tag{14}$$

The axial displacement components of all rollers are δ_a, then for the ith roller, the total displacement along the contact normal direction of the outer raceway is

$$\delta_{ni} = \delta_{ri} \cos\alpha_e + \delta_{ai} \sin\alpha_e \tag{15}$$

The load-deformation relation of the *i*th roller is [21,25]

$$Q_{ei} = K_{\mathrm{ne}} \delta_{\mathrm{n}i}^{1.11} \tag{16}$$

where K_{ne} is the total stiffness coefficient at the contact of the outer raceway, and the calculation formula is [25]

$$K_{\mathrm{ne}} = 8.06 \times 10^4 l^{0.89} \left[1 + c_i^{\ 0.89} \cos(\alpha_{\mathrm{e}} - \alpha_i)\right]^{-1.11} \tag{17}$$

where l is the length of the roller.

As shown in Figure 11, the components of Q_{ei} in the direction of F_{r} and F_{a} are, respectively,

$$Q_{\mathrm{r}i} = Q_{ei} \cos \alpha_{\mathrm{e}} \cos \psi_i \tag{18}$$

$$Q_{\mathrm{a}i} = Q_{ei} \sin \alpha_{\mathrm{e}} \tag{19}$$

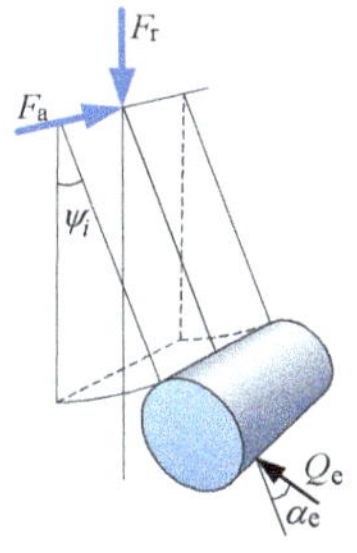

Figure 11. Spatial relationship between roller force and bearing forces.

Then the equilibrium equation of the bearing can be derived as

$$\begin{cases} F_{\mathrm{r}} - \sum\limits_{i=1}^{Z} Q_{\mathrm{r}i} = 0 \\ F_{\mathrm{a}} - \sum\limits_{i=1}^{Z} Q_{\mathrm{a}i} = 0 \end{cases} \tag{20}$$

Equation (20) is a nonlinear system of equations with δ_{r} and δ_{a} as unknown variables. The radial force F_{r} and axial force F_{a} of bearing 1 under wind load can be obtained by statics analysis and calculation of the shafting system. Combined with Equations (12) and (13), the theoretical values of the maximum load of the roller, such as Q_{emax}, Q_{imax}, and Q_{fmax} can be obtained by using iterative algorithms such as Newton's method. Since the general Q_{emax} is the largest, Q_{emax} is taken as the maximum roller load Q_{max}.

4. Sub-Model of Bearing

4.1. Logarithmic Crowning of Roller

The roller profile of bearings on the wind turbine spindle needs to be modified to reduce stress concentration and optimize load distribution. There is generally, power function crowning and logarithmic (exponential) crowning of the roller. Logarithmic crowning is not as good as power function crowning under extreme bias load conditions, but the stress distribution of the roller is more uniform under normal conditions [13,17], so the logarithmic crowning method is chosen for the study. The crowning zone of the tapered roller is shown in Figure 12.

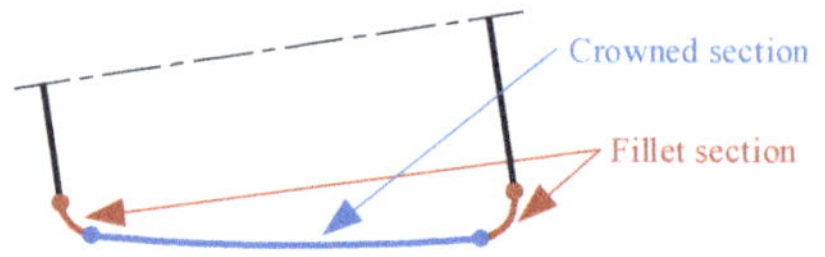

Figure 12. Crowned section and fillet section of tapered roller.

In practice, bearing rollers are often subjected to different and changing loads, yet the profile of the rollers in a set of bearing should remain consistent. Therefore, the maximum roller load is taken as the design load in the design of the roller's crowning.

In the FE model, n nonlinear spring elements are used to replace each roller, and the maximum value after superimposing the n loops spring reaction force is the maximum equivalent roller load $Q_{\max}$, and the expression is

$$Q_{\max} = (F_{\text{sum}i})_{\max} = \left(\sum_{j=1}^{n} F_{ji}\right)_{\max} \quad i = 1, 2, \ldots, Z \tag{21}$$

where $Q_{\max}$ is the maximum roller load, $F_{\text{sum}i}$ is the ith equivalent roller reaction force after superimposing n loops of spring reaction force, F_{ji} is the ith spring reaction force in loop j, and Z is the number of rollers.

The expression of the crowning function z of the tapered roller is

$$z = \frac{2Q}{\pi E' l_{we}} \times \ln\left[1 - \left(1 - 0.3033\frac{2b}{l_{we}}\right)(2x/l_{we})^2\right]^{-1} \tag{22}$$

where Q is the roller load, E' is the equivalent elastic modulus, b is the Hertzian contact half-width, and l_{we} is the crowning length.

The crowning profile of the roller is shown in Figure 13, where the maximum amount of crowning is denoted as $z_{\max}$.

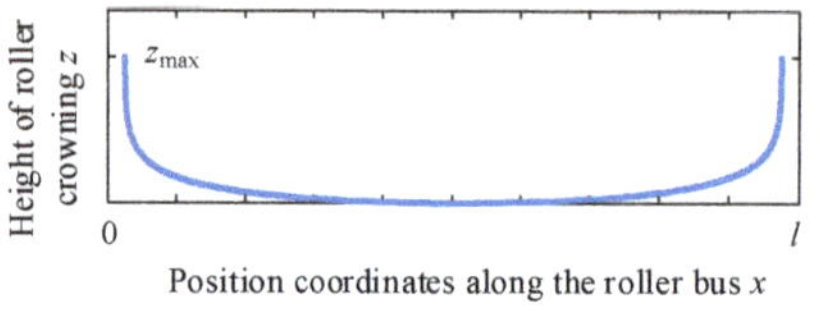

Figure 13. Logarithmic crowning profile of roller.

For line contact, according to Hertzian contact theory, the contact half-width b is

$$b = \left(\frac{4\eta Q}{\pi l \sum \rho}\right)^{1/2} \tag{23}$$

where $\Sigma\rho$ is the Curvature sum.

The formula for calculating $\Sigma\rho$ is as follows:

$$\sum \rho = \frac{2\cos\beta}{D_w} + \frac{2\gamma_i}{D_w(1-\gamma_i)} \tag{24}$$

where β is the tapered roller half-cone angle, D_{w} is the middle diameter of the roller, and $\gamma_i = D_w \cos\alpha_i / d_m$.

Defining $f_1 = 2Q/\pi E' l_{we}$ as the load parameter of logarithmic crowning and $f_2 = 1 - 0.3033 \times 2b/l_{we}$ as the width parameter of logarithmic crowning, the crowning function changes to

$$z = f_1 \times \ln\left[1 - f_2(2x/l_{we})^2\right]^{-1} \tag{25}$$

Equation (25) contains two crowning parameters f_1 and f_2. f_1 and f_2 are determined by the roller load Q and material parameters, and the roller load Q is the common influence variable. The research of [15] indicated that the effect of the parameter f_2 on the stress distribution of the roller is very small, while f_1 on the stress distribution of the roller is large. Only f_1 needs to be optimized and an adjustment-coefficient *cof* is added in front of Q in f_1 to make it variable, and the expression of the roller crowning function changes to

$$z = cof \times f_1 \times \ln\left[1 - f_2(2x/l_{we})^2\right]^{-1} \tag{26}$$

The reference values of the crowning parameters f_1, f_2, b, and z can be derived from theoretical calculations. Using the value of theoretically calculated z as a reference, the amount of crowning z can be adjusted by the adjustment-coefficient *cof*.

4.2. Mesh of Sub-Model

The profile of bearing roller crowning is an exact function curve and the sub-model needs to have a fine mesh to simulate the contact characteristics of the roller contact area. The mesh of the contact zone of the roller is refined, and the transition mesh is handled finely. The final mesh of the symmetric sub-model is shown in Figure 14, containing a total of 91,364 elements and 99,585 nodes.

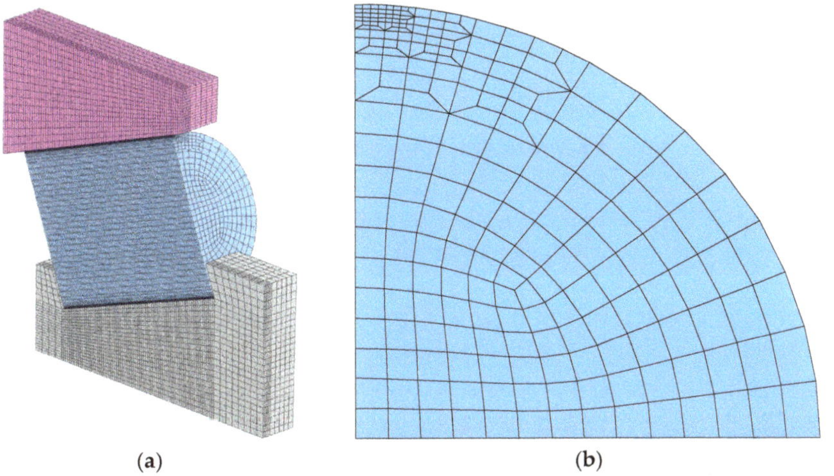

(a) **(b)**

Figure 14. Mesh of the symmetric sub-model. (**a**) The mesh of sub-model; (**b**) The mesh of roller cross-section.

For rollers without fillets, especially rollers with a straight profile, stress concentration is generated at the edge of the roller in the analysis with a very high calculated stress value, which neither does match the actual situation [13] nor is conducive to the stress assessment.

After meshing the geometry, the mesh of the roller edge needs to be shrunk in the radial direction by a certain amount, to meet the realistic fillet or chamfer features, the shrunk mesh of the roller edge is shown in Figure 15.

4.3. Boundary Conditions

Superimpose the n loops reaction forces of springs obtained from the FE model of the shaft system. The maximum value of the reaction force after superposition is used as the load boundary of the sub-model. Its direction is approximately the contact normal of the inner raceway. All degrees of freedom of the outer diameter surface of the outer ring of the bearing are fixed. The symmetry boundary is created on the symmetry surface of the rings and roller. Based on the direction of the force boundary vector, a coordinate system is established to constrain the other degrees of freedom of the inner ring bore surface except for the direction of the force. This is shown in Figure 16.

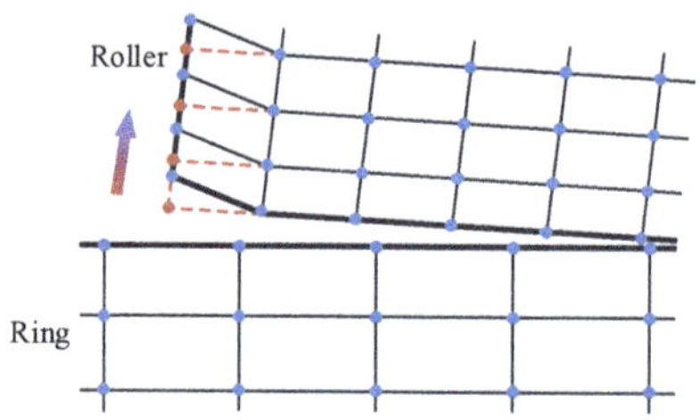

Figure 15. Radial shrinkage of mesh at roller edge.

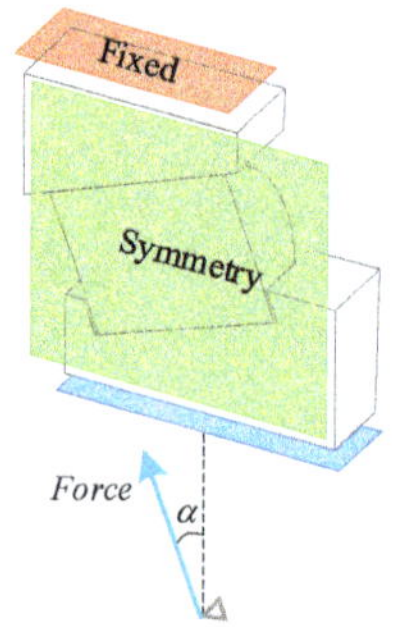

Figure 16. Boundary conditions of the sub-model.

On the symmetry plane, a path is created in the fillet and crowned section on the roller profile on the inner ring side, as shown in Figure 17, to extract the Von Mises stress of the nodes on the path.

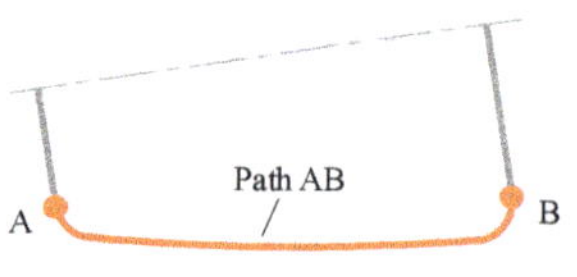

Figure 17. Path defined in post-processing.

5. Analysis and Discussion of Results

5.1. Effect of Damping on the Solution of the Shaft System

For the 5 spr case, the undamped model and the critically damped model of the FE model of the shaft system are solved, with the maximum displacement of the shaft system as the response index. The comparison of the response of the undamped and damped models is shown in Figure 18. It can be seen that the use of the critical damping technique can solve the FE model of the shaft system quickly and obtain a stable mechanical response.

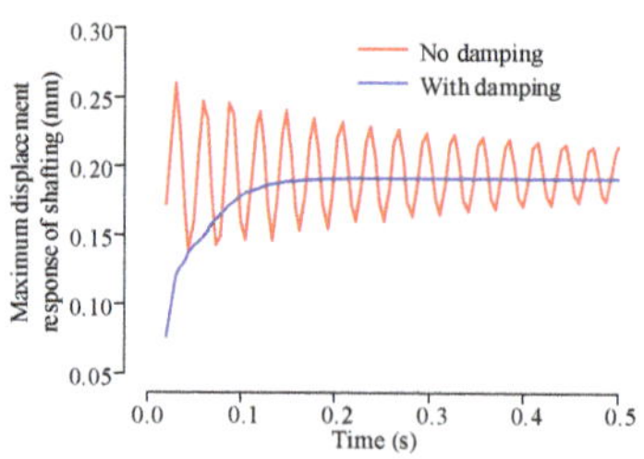

Figure 18. Mechanical response of FE model of the shaft system.

5.2. Influence of Structural Rigidity on the Shaft System

Figure 19 shows the deflection deformation of the shaft under load with a rigid/flexible structure (shaft and bearing housing). Figure 20 gives the relative position change of the profiles of the outer rings of two bearings.

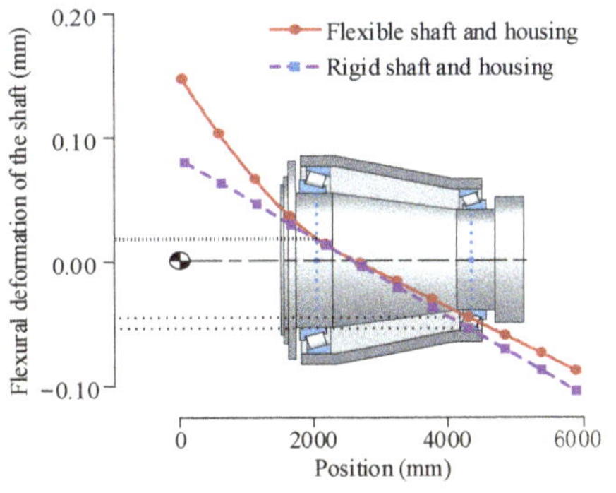

Figure 19. Comparison of shaft deformation under rigid and flexible structure.

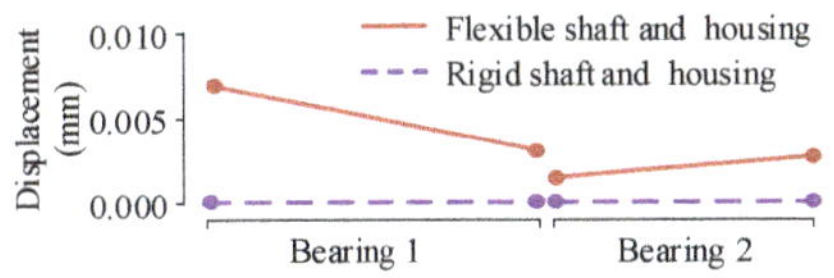

Figure 20. Relative position of the outer profiles of the outer rings of two bearings.

As can be seen in Figure 19, for a model with a rigid shaft and rigid housing, the shaft cannot deform flexibly and can only balance the load by deflecting an angle. In the model with a flexible shaft and a flexible bearing housing, the shaft can be bent and deformed flexibly and the deformation is greatest at the location of the loaded end of the shaft. This is also the location where the deflection is the largest compared to the rigid structure model.

As shown in Figure 20, the flexible housing allows the outer ring of the bearing to have a deflection deformation, which allows the bearing to better balance the load. The outer ring of bearing 1 has the largest deflection. However, combining the above factors, the displacement of the rigid shaft at the center position of bearing 1 has a smaller deviation compared to the flexible shaft, and the displacement of the rigid shaft at the center position of bearing 2 has a larger deviation compared to the flexible shaft. The large displacement of the rigid shaft will cause the load on the bearing to be larger, making the bearing life assessment biased, especially for bearing 2. Therefore, in the actual calculation for the bearing on the shaft system, the coupling effect of the flexibility of the structure on the bearing should be considered.

5.3. Effect of the Quantity of Replacement Springs on the Bearing Load Distribution

The n loops reaction forces of the springs for each case are extracted and shown in Figure 21. The equivalent roller load distribution is obtained after superimposing the n loops spring reaction force, and the equivalent roller load distribution under the three cases is shown in Figure 22. All the rollers of bearing 1 under the three cases are under load, and the trend was the same in all three cases, but the fluctuation of the equivalent roller load distribution is different in different cases.

The FE model considering the flexibility of the support structure of the shaft system can obtain a more accurate load distribution of bearing, and can accurately design the bearing profile based on actual working conditions. The FE modeling method proposed can be further used to construct the calculation and analysis system of the shaft assembly of a large wind turbine and to provide support for the precision design of bearings in large shaft systems.

Author Contributions: Conceptualization, R.N. and X.L.; methodology, R.N.; software, X.L.; validation, S.Z., Y.C. and Z.Z.; formal analysis, X.L.; investigation, X.L.; resources, B.W.; data curation, X.L.; writing—original draft preparation, X.L.; writing—review and editing, R.N.; visualization, X.L.; supervision, R.N.; project administration, R.N.; funding acquisition, R.N. All authors have read and agreed to the published version of the manuscript.

Funding: This research was funded by the National Natural Science Foundation of China, grant number 52005158, and the Key Research and Development Project of Shandong Province, grant number 2020CXGC011003.

Data Availability Statement: Not applicable.

Conflicts of Interest: The authors declare that they have no conflict of interest.

References

1. Doll, M.N.K.; Gary, L. Tribological advancements for reliable wind turbine performance. *Philos. Trans.* **2010**, *368*, 4829–4850.
2. Hart, E.; Clarke, B.; Nicholas, G.; Amiri, A.K.; Stirling, J.; Carroll, J.; Dwyer-Joyce, R.; McDonald, A.; Long, H. A review of wind turbine main bearings: Design, operation, modelling, damage mechanisms and fault detection. *Wind. Energy Sci.* **2020**, *5*, 105–124. [CrossRef]
3. Smolnicki, T.; Rusiński, E. Superelement-based modeling of load distribution in large-size slewing bearings. *J. Mech. Des.* **2007**, *129*, 459–463. [CrossRef]
4. Chen, G.; Wang, C.; Xiao, Z. Effects of supporting structure and bolt connection on the fatigue life and carrying capacity of a slewing bearing. *J. Eng. Tribol.* **2016**, *231*, 766–782. [CrossRef]
5. Chen, G.; Wen, J. Load Performance of Large-Scale Rolling Bearings With Supporting Structure in Wind Turbines. *J. Tribol.* **2012**, *134*, 041105. [CrossRef]
6. Plaza, J.; Abasolo, M.; Coria, I.; Aguirrebeitia, J.; Igor, F.B. A new finite element approach for the analysis of slewing bearings in wind turbine generators using superelement techniques. *Meccanica* **2015**, *50*, 1623–1633. [CrossRef]
7. Li, Y.; Jiang, D.; Song, L. Finite Element Analysis of Pitch Bearing in Wind Turbine Considering Influence of Installation Structures. *Acta Energ. Sol. Sin.* **2019**, *40*, 2021–2027. (In Chinese)
8. Aguirrebeitia, J.; Abasolo, M.; Avilés, R.; Igor, F.B. General static load-carrying capacity for the design and selection of four contact point slewing bearings: Finite element calculations and theoretical model validation. *Finite Elem. Anal. Des.* **2012**, *55*, 23–30. [CrossRef]
9. He, P.; Liu, R.; Hong, R.; Wang, H.; Yang, G.; Lu, C. Hardened raceway calculation analysis of a three-row roller slewing bearing. *Int. J. Mech. Sci.* **2018**, *137*, 133–144. [CrossRef]
10. Śpiewak, S. Methodology for calculating the complete static carrying capacity of twin slewing bearing. *Mech. Mach.* **2016**, *101*, 181–194. [CrossRef]
11. Martín, I.; Aguirrebeitia, J.; Heras, I.; Abasolo, M. Efficient Finite Element modelling of crossed roller wire race slewing bearings. *Tribol. Int.* **2021**, *161*, 107098. [CrossRef]
12. Kania, L.; Pytlarz, R.; Śpiewak, S. Modification of the raceway profile of a single-row ball slewing bearing. *Mech. Mach. Theory* **2018**, *128*, 1–15. [CrossRef]
13. Ju, S.H.; Horng, T.L.; Cha, K.C. Comparisons of contact pressures of crowned rollers. *J. Eng. Tribol.* **2000**, *214*, 147–156. [CrossRef]
14. Fujiwara, H. Logarithmic profile of rollers in roller bearing and optimization of the profile. *Trans. Jpn. Soc. Mech. Eng.* **2006**, *72*, 338–345. (In Japanese) [CrossRef]
15. He, Z.; Shao, M.; Ye, G. Optimum Design Method for Logarithmic Crowned Tapered Rollers. *Mech. Eng. Technol.* **2016**, *5*, 38–46. [CrossRef]
16. Li, Y.; Gao, Y. Internal load distribution of single-row tapered roller bearings doubly supporting main shaft of wind turbine. *Adv. Mech. Eng.* **2022**, *14*, 16878132221098895. [CrossRef]
17. Zhang, T.; Yang, X.; He, L.; Bu, Z.J.; Nie, W.; Li, M. Research on the Design Method of Roller Modification Curve for Wind Turbine. *J. Mech. Strength* **2021**, *43*, 183–190. (In Chinese)
18. Li, Y.; Xu, Z.; Liu, Q.; Wang, H. Analysis on Contact Behavior for Extra Large Size Tapered Roller Bearings in Wind Turbines Based on PERMAS. *Bearing* **2014**, *5*, 1–4. (In Chinese)
19. Liu, J.; Dong, H.; Wang, H.; Wang, Y.Y. Finite Element Analysis of Slewing Bearing Based on Rolling Element Solid Model and Substructure Technology. *Bearing* **2018**, *12*, 9–13. (In Chinese)

20. Kania, L. Modelling of rollers in slewing bearing calculations with the use of finite elements. *Mech. Mach. Theory* **2006**, *41*, 1359–1376. [CrossRef]
21. Harris, T.A.; Kotzalas, M.N. *Advanced Concepts of Bearing Technology: Rolling Bearing Analysis*, 5th ed.; CRC Press: Boca Raton, FL, USA, 2006; pp. 3–7.
22. Palmgren, A. *Ball and Roller Bearing Engineering*, 3rd ed.; SKF Industries Inc.: Philadelphia, PA, USA, 1959.
23. Wang, P.; Wang, Q.; Xu, X.; Chen, N. Fractional Critical Damping Theory and Its Application in Active Suspension Control. *Shock. Vib.* **2017**, *2017*, 2738976. [CrossRef]
24. Zeng, P.; Lei, L.; Fang, G. *Finite Element Analysis Guide: Modeling and Analysis of Structure*; China Machine Press: Beijing, China, 2011; pp. 428–430.
25. Lundberg, G.; Palmgren, A. Dynamics capacity of rolling bearings. *Acta Polytech.* **1947**, *1*, 7. [CrossRef]
26. Zhang, H.; Chen, S.; Dou, Y.; Fan, H.; Wang, Y. Mechanical model and contact properties of double row slewing ball bearing for wind turbine. *Rev. Adv. Mater. Sci.* **2021**, *60*, 112–126. [CrossRef]

applied sciences

MDPI

Article

Modified Design of Two-Switch Buck-Boost Converter to Improve Power Efficiency Using Fewer Conduction Components

Sunghwan Kim [1], Haiyoung Jung [2,*] and Seok-hyun Lee [1,*]

[1] Department of Electrical Engineering, Inha University, 100 Inha-ro, Michuhol-gu, Incheon 22212, Republic of Korea
[2] Department of Fire and Disaster Prevention, Semyung University, 65 Semyung-ro, Jecheon-si 27136, Republic of Korea
* Correspondence: hyjung@semyung.ac.kr (H.J.); plasma@inha.ac.kr (S.-h.L.)

Abstract: In this study, a modified design of a two-switch buck-boost (TSBB) converter is proposed to improve power efficiency using fewer conduction components, and the optimal power range is measured. The proposed TSBB converter operates in three topologies: buck, boost, and buck-boost, like the conventional TSBB converter. However, the proposed converter improves the power efficiency in the buck and buck-boost topologies by decreasing conduction loss using the diode in the switch-off section while maintaining the same number of semiconductors as that in the conventional TSBB converter. The power efficiency of the buck topology improves for the power range 10–80 W in the constant voltage (CV) and constant current (CC) modes; it increases on average by 0.75–1.36% and 0.83–2.27% in the CV and CC modes, respectively. The power efficiency of the buck-boost topology step-down improves for the 10–80 W in all modes. This increases the average by 0.73–0.99% and 3.33–4.75% in the CV and CC modes, respectively. The power efficiency of the buck-boost topology step-up increases on average by 1.65–2.00% for 10–80 W in the CV mode. In the CC mode, it increases by 2.17–2.77% on average for 10–50 W.

Keywords: converter; buck-boost converter; TSBB converter; conduction loss; switching loss; metal-oxide-semiconductor field effect transistor (MOSFET); switch; diode; voltage stress

Citation: Kim, S.; Jung, H.; Lee, S.-h. Modified Design of Two-Switch Buck-Boost Converter to Improve Power Efficiency Using Fewer Conduction Components. *Appl. Sci.* **2023**, *13*, 343. https://doi.org/10.3390/app13010343

Academic Editors: Kan Liu and Wei Hu

Received: 4 December 2022
Revised: 21 December 2022
Accepted: 22 December 2022
Published: 27 December 2022

1. Introduction

The DC–DC converter is a power conversion device that converts the received DC voltage to a DC voltage required by the system for transferring energy to the load [1]. It is used in several electronic devices for stabilizing the operation of systems; this can range from high-power applications such as solar photovoltaics, electric vehicles, and energy storage systems to low-power applications such as laptops, mobile phones, and portable batteries [2]. DC–DC converters are becoming miniaturized, and these parts are becoming denser because of the weight reduction and miniaturization of various application products [3]. DC–DC converters require high-power conversion efficiency for the reduction of energy consumption and long life [4]. Such high power-conversion efficiency is an important factor for a high-performance high-reliability DC–DC converter [5,6].

There are three topologies for non-isolated-type DC–DC converters: buck for step-down, boost for step-up, and buck-boost for both step-down and step-up. Figure 1 shows the circuit diagrams of each topology that consists of a switch, a diode, an inductor, and a capacitor. Step-down and step-up are determined by the duty ratio (d) and wiring of the semiconductors. Table 1 summarizes the current path and gain (G) of each topology based on switch operation; these topologies of the single switch type should be selected based on the input and output voltage specifications of the system; it is difficult to switch between topologies owing to the fixed element connection. Buck-boost topology with both step-down and step-up cannot be used in applications that require an output voltage of the same polarity as the input voltage because the polarity of the output voltage is opposite to that of the input voltage. Moreover, the power conversion efficiency decreases compared to that of other topologies because the voltage stress of the switch and diode equals the sum of the input voltage V_i and output voltage V_o. Therefore, the TSBB converter that can switch to a different topology based on system requirements is used frequently [7–10].

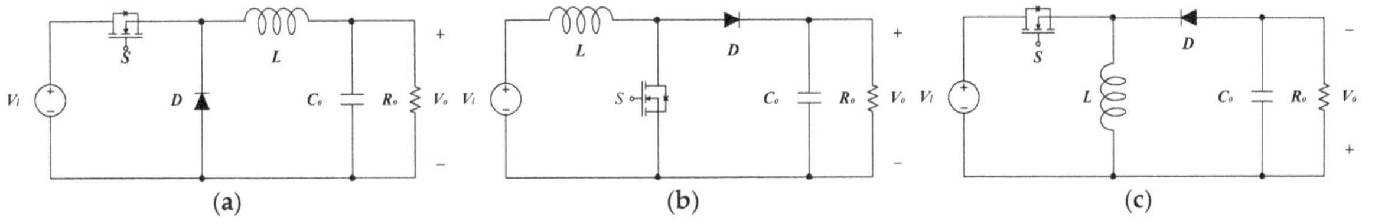

Figure 1. Circuit diagram: (**a**) buck; (**b**) boost; (**c**) buck-boost.

Table 1. Current path and gain in topologies.

Topology	Current Path		Gain (G)
	Switch ON	Switch OFF	
Buck	V_i-S-L-C_o	L-C_o-D	d
Boost	V_i-L-S	V_i-L-D-C_o	1/(1 − d)
Buck-Boost	V_i-S-L	L-C_o-D	d/(1 − d)

Figure 2 shows a circuit diagram of a conventional TSBB converter that comprises two switches, two diodes, an inductor, and a capacitor. This TSBB converter operates in the buck, boost, or buck-boost topology based on the operation of the switch [5–7]. Table 2 summarizes the switching operations of the conventional TSBB converter; it operates in the buck topology by the on/off switching of S_1 while S_2 is always off, and in boost topology by the on/off switching of S_2 while S_1 is always on. Further, the TSBB converter operates in the buck-boost topology by the on/off switching of both S_1 and S_2. It is easy to change the topology despite the increase in parts; the output voltage has the same polarity as the input in the buck-boost topology. Moreover, it is easy to select semiconductors because the voltage stress of the parts does not exceed the input voltage V_i and output voltage V_o [9–13].

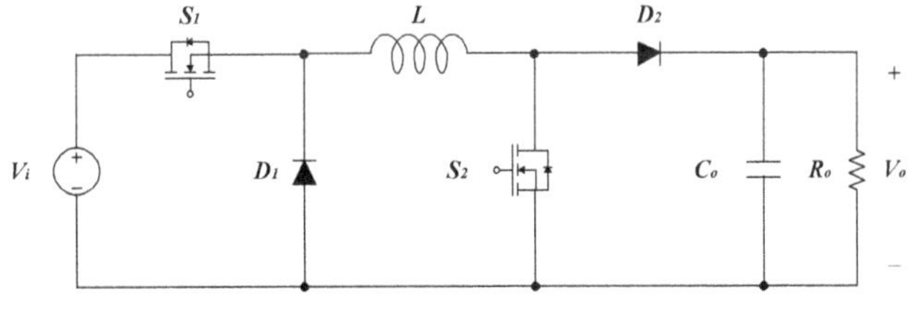

Figure 2. Circuit diagram of the conventional TSBB converter.

Table 3 presents the switching and conduction semiconductors in each topology of a single switch converter and a TSBB converter. The TSBB converter uses more parts than the single switch converter in each topology owing to the composition and wiring of the semiconductors; this increases the power loss and lowers the power efficiency. The loss of the TSBB converter increases because of the amount of the conduction loss of one diode than the single switch type in the buck topology and the amount of the conduction loss of one switch in the boost topology. Moreover, the power efficiency in the buck-boost topology decreases owing to the switching and conduction losses of one switch and one diode [5,6,10].

Table 2. Switch operation of the conventional TSBB converter.

Topology	Period	Component			
		S_1	S_2	D_1	D_2
Buck	Switch ON Switch OFF	ON OFF	Always OFF	OFF ON	Always ON
Boost	Switch ON Switch OFF	Always ON	ON OFF	Always OFF	OFF ON
Buck-Boost	Switch ON Switch OFF	ON OFF	ON OFF	OFF ON	OFF ON

Table 3. Switching and conduction semiconductors of single switch converters and the conventional TSBB converter.

Topology	Single Switch Type				Conventional TSBB			
	Switch ON		Switch OFF		Switch ON		Switch OFF	
	Switching	Conduction	Switching	Conduction	Switching	Conduction	Switching	Conduction
Buck	S	S	D	D	S_1	S_1, D_2	D_1	D_1, D_2
Boost	S	S	D	D	S_2	S_1, S_2	D_2	S_1, D_2
Buck-Boost	S	S	D	D	S_1, S_2	S_1, S_2	D_1, D_2	D_1, D_2

New circuits are suggested in [5,6] to prevent such a decrease in power efficiency. They demonstrated that power efficiency can be increased compared to those of the conventional TSBB converter by reducing the number of switching and conduction semiconductors in the current path. However, there is a trade-off between the number of semiconductors in the current path and the voltage stress. Therefore, in the new circuits of [5,6], the voltage stress of the semiconductors in the boost and buck-boost topologies increases, and it will lower power efficiency above a specific power range. The experimental results confirm the efficiency based on changes in the output current; however, it is difficult to find the optimal power range according to the changes in the input voltage and duty that influence the voltage stress of the semiconductors. To solve this problem, this study proposed a new type of TSBB converter for improving power efficiency and analyzing the optimal power range that can improve power efficiency. The proposed TSBB converter can increase power efficiency in buck and buck-boost topologies by reducing conduction loss caused by the diode in the switch-off section while using the same number of semiconductors as that of the conventional TSBB converter. Moreover, the optimal power range of each topology is analyzed by evaluating the effect of an increase in the voltage stress of semiconductors.

The contributions of this study are as follows:

- We investigated related research about the TSBB converter and proposed a modified design of the TSBB converter to improve power efficiency using fewer conduction components in the current path.
- We presented the optimal power range according to the buck, boost, and buck-boost topologies in CV and CC modes, and, in particular, divided into step-up/step-down sections in the buck-boost topology.
- We analyzed the power dissipation of the three topologies and explained why the CC mode of the buck-boost step-up is less efficient than conventional converters over a certain power range through analytic and experimental diode stress analysis.

The remainder of this paper is organized as follows. Section 2 describes the operation principle of the proposed TSBB converter. In Section 3, the power loss is compared between the proposed TSBB converter and the conventional TSBB converter in each topology by analyzing the switching and conduction losses. The experiment results are described in Section 4, and, finally, the conclusions are presented in Section 5.

2. Operation Principle

Figure 3 shows the circuit diagram of the proposed TSBB converter. Like the conventional TSBB converter, the proposed TSBB converter is composed of two switches, two diodes, an inductor, and a capacitor; further, it operates in three topologies based on the switching of S_1 and S_2, and it operates in the buck topology by the on/off switching of S_1, in the boost topology by the switching of S_2, and in the buck-boost topology by the simultaneous on/off switching of S_1 and S_2.

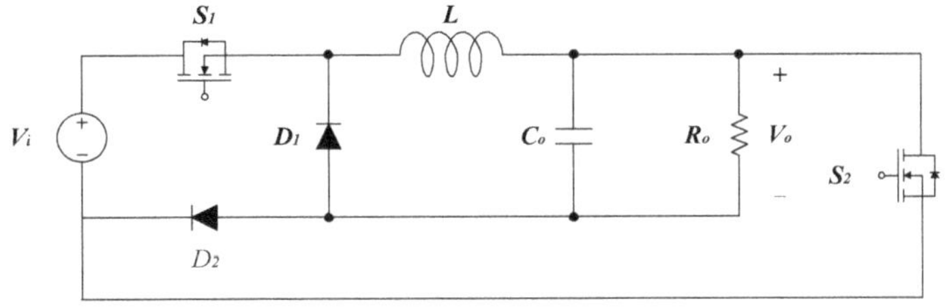

Figure 3. Circuit diagram of the proposed TSBB converter.

Table 4 summarizes the switching operations of the proposed TSBB converter. Table 4 indicates that the switching operations of S_1, S_2, and D_1 are the same as those of the conventional TSBB converter. However, D_2 performs on/off switching in the buck topology and is always off in the buck-boost topology.

Table 4. Switch operation of the proposed TSBB converter.

Topology	Period	Component			
		S_1	S_2	D_1	D_2
Buck	Switch ON	ON	Always	OFF	ON
	Switch OFF	OFF	OFF	ON	OFF
Boost	Switch ON	Always	ON	Always	OFF
	Switch OFF	ON	OFF	OFF	ON
Buck-Boost	Switch ON	ON	ON	OFF	Always
	Switch OFF	OFF	OFF	ON	OFF

Figure 4 presents the operation principle of the proposed TSBB converter in each topology. The switch-on/off sections are divided by the operation of the switch that transfers the energy to the inductor. The semiconductors located in the current path in each section are conduction semiconductors; the semiconductor that only operates in one of the switch-on/off sections is the switching semiconductor. Table 5 compares the switching and conduction semiconductors between the conventional and proposed TSBB converters. In comparison with the conventional TSBB converter, the proposed TSBB converter undergoes an increase in the switching loss of D_2 in the switch-on section and a decrease in the conduction loss of D_2 in the switch-off section. In the buck-boost topology, the switching and conduction losses of D_2 in the switch-off section decrease. Table 6 compares the stress between the conventional and proposed TSBB converters; they have the almost same voltage stress in buck and boost topologies, but in the buck-boost topology, the stress of D_1 increases to $V_i + V_o$.

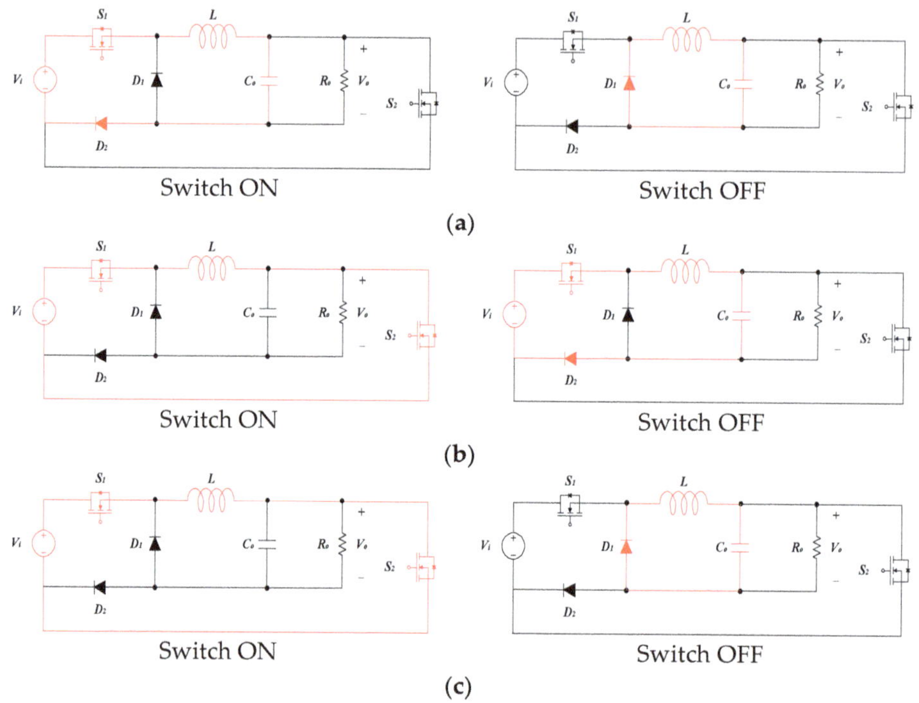

Figure 4. Operation principle of the proposed TSBB converter: (**a**) Buck; (**b**) Boost; (**c**) Buck-boost.

Table 5. Switching and conduction semiconductors of the conventional and proposed TSBB converters: (a) Switching semiconductors; (b) Conduction semiconductors.

Topology	Gain	Conventional		Ref. [5]		Proposed	
		Switch ON	Switch OFF	Switch ON	Switch OFF	Switch ON	Switch OFF
Buck	d	S_1	D_1	S_1, D_2	D_1	S_1, D_2	D_1
Boost	$1/(1-d)$	S_2	D_2	S_2	D_2	S_2	D_2
Buck-Boost	$d/(1-d)$	S_1, S_2	D_1, D_2	S_2	D_1	S_1, S_2	D_1

(a)

Topology	Gain	Conventional		Ref. [5]		Proposed	
		Switch ON	Switch OFF	Switch ON	Switch OFF	Switch ON	Switch OFF
Buck	d	S_1, D_2	D_1, D_2	S_1, D_2	D_1	S_1, D_2	D_1
Boost	$1/(1-d)$	S_1, S_2	S_1, D_2	S_2	S_1, D_2	S_1, S_2	S_1, D_2
Buck-Boost	$d/(1-d)$	S_1, S_2	D_1, D_2	S_2	D_1	S_1, S_2	D_1

(b)

Table 6. Comparison of stress in each TSBB converter: (a) Voltage stress; (b) Current stress.

Topology	Conventional			
	S_1	S_2	D_1	D_2
Buck	V_i	-	V_i	-
Boost	-	V_o	-	V_o
Buck-Boost	V_i	V_o	V_i	V_o

Table 6. *Cont.*

Topology	Ref. [5]			
	S_1	**S_2**	**D_1**	**D_2**
Buck	$\frac{C_{oss,D2}}{C_{oss,S1}+C_{oss,D2}}V_i$	-	V_i	$\frac{C_{oss,S1}}{C_{oss,S1}+C_{oss,D2}}V_i$
Boost	-	V_o	-	$\frac{C_{oss,S1}}{C_{oss,S1}+C_{oss,D2}}V_o$
Buck-Boost	$\frac{C_{oss,D2}}{C_{oss,S1}+C_{oss,D2}}V_i$ or $\frac{C_{oss,D2}}{C_{oss,S1}+C_{oss,D2}}V_o$	V_i+V_o	V_i+V_o	$\frac{C_{oss,S1}}{C_{oss,S1}+C_{oss,D2}}V_i$ or $\frac{C_{oss,S1}}{C_{oss,S1}+C_{oss,D2}}V_o$
Topology	**Proposed**			
	S_1	**S_2**	**D_1**	**D_2**
Buck	$\frac{C_{oss,D2}}{C_{oss,S1}+C_{oss,D2}}V_i(\cong V_i)$	-	V_i	$\frac{C_{oss,S1}}{C_{oss,S1}+C_{oss,D2}}V_i(\cong 0)$
Boost	-	V_o	-	V_o
Buck-Boost	$\frac{C_{oss,D2}}{C_{oss,S1}+C_{oss,D2}}V_i(\cong V_i)$	$\frac{C_{oss,D2}}{C_{oss,S2}+C_{oss,D2}}V_o(\cong V_o)$	V_i+V_o	-

(a)

Topology	Conventional/Proposed			
	S_1	**S_2**	**D_1**	**D_2**
Buck	$V_o\left(\frac{1}{R^1}+\frac{1-d}{2L\cdot f_{sw}}\right)$	-	$V_o\left(\frac{1}{R^1}+\frac{1-d}{2L\cdot f_{sw}}\right)$	$V_o\left(\frac{1}{R^1}+\frac{1-d}{2L\cdot f_{sw}}\right)$
Boost	-	$V_i\left[\frac{1}{(1-d)^2\cdot R^1}+\frac{d}{2L\cdot f_{sw}}\right]$	-	$V_i\left[\frac{1}{(1-d)^2\cdot R^1}+\frac{d}{2L\cdot f_{sw}}\right]$
Buck-Boost	$V_i\left[\frac{d}{(1-d)^2\cdot R^1}+\frac{d}{2L\cdot f_{sw}}\right]$	$V_i\left[\frac{d}{(1-d)^2\cdot R^1}+\frac{d}{2L\cdot f_{sw}}\right]$	$V_i\left[\frac{d}{(1-d)^2\cdot R^1}+\frac{d}{2L\cdot f_{sw}}\right]$	-

(b)

[1] Output resistance.

3. Analysis of Semiconductor Power Loss

The power loss of a TSBB converter includes the losses of the switch, diode, inductor, and capacitor when various parasitic components in the circuit are ignored. The power loss of the two converters is determined by the power loss of the semiconductors assuming that the losses of the inductor and capacitor are the same between the conventional and proposed TSBB converters. Therefore, the increase or decrease in power efficiency is determined by the operation of the switch and diode.

3.1. Switch and Diode Current

Figure 5 presents the switch and diode current of the proposed TSBB converter in each topology. In the buck topology, the inductor current flows through S_1 and D_2 during the switch-on section and through D_1 during the switch-off section. The inductor current I_L is the output current I_o and $\Delta I_L = \frac{V_o(1-d)}{L\cdot f_{sw}}$ by the volt-sec balance law at switching frequency f_{sw}, the inductor current I_{max}, I_{min} flowing through switch S_1 and diode D_1/D_2 can be expressed as:

$$I_{max} = I_o + \frac{\Delta I_L}{2} = V_o\left(\frac{1}{R} + \frac{1-d}{2L\cdot f_{sw}}\right) \quad (1)$$

$$I_{min} = I_o - \frac{\Delta I_L}{2} = V_o\left(\frac{1}{R} - \frac{1-d}{2L\cdot f_{sw}}\right) \quad (2)$$

The inductor current flows through S_1 and D_2 during the switch-on section and through D_1 during the switch-off section in the boost topology. Since the inductor current

I_L is $\frac{I_o}{1-d}$ and $\Delta I_L = \frac{d \cdot V_i}{L \cdot f_{sw}}$, the inductor current I_{max}, I_{min} flowing through switch S_1/S_2 and diode D_2 can be expressed as:

$$I_{max} = I_o + \frac{\Delta I_L}{2} = V_i \left[\frac{1}{(1-d)^2 \cdot R} + \frac{d}{2L \cdot f_{sw}} \right] \tag{3}$$

$$I_{min} = I_o - \frac{\Delta I_L}{2} = V_i \left[\frac{1}{(1-d)^2 \cdot R} - \frac{d}{2L \cdot f_{sw}} \right] \tag{4}$$

In buck-boost topology, the inductor current flows through S_1 and D_2 during the switch-on section and through D_1 during the switch-off section. Since the inductor current $I_L = \frac{I_o}{1-d}$ and $\Delta I_L = \frac{d \cdot V_i}{L \cdot f_{sw}}$ same as boost topology, the inductor current I_{max}, I_{min} flowing through switch S_1/S_2 and diode D_2 can be expressed as:

$$I_{max} = I_o + \frac{\Delta I_L}{2} = V_i \left[\frac{d}{(1-d)^2 \cdot R^1} + \frac{d}{2L \cdot f_{sw}} \right] \tag{5}$$

$$I_{min} = I_o - \frac{\Delta I_L}{2} = V_i \left[\frac{d}{(1-d)^2 \cdot R^1} + \frac{d}{2L \cdot f_{sw}} \right] \tag{6}$$

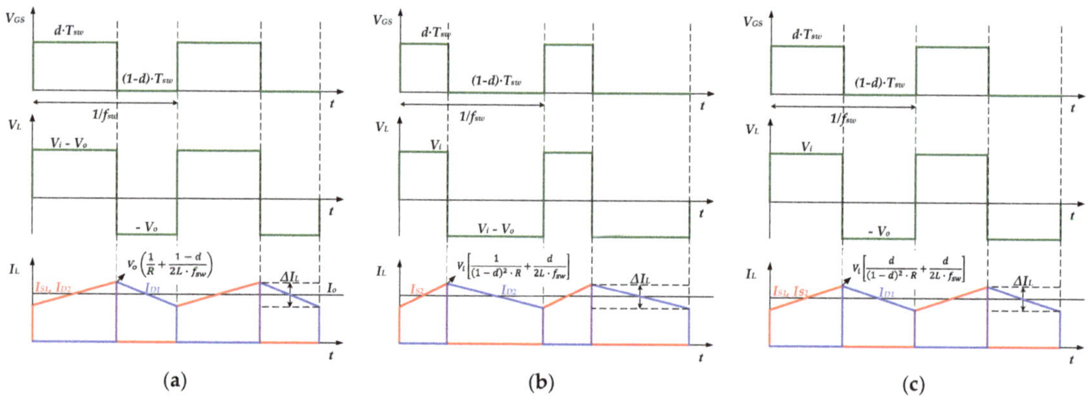

Figure 5. Switch and diode current of the proposed TSBB converter: (**a**) Buck; (**b**) Boost; (**c**) Buck-boost.

The power loss of a TSBB converter includes the losses of the switch, diode, inductor, and capacitor when various parasitic components in the circuit are ignored. The power loss of the two converters is determined by the power loss of the semiconductors assuming that the losses of the inductor and capacitor are the same between the conventional and proposed TSBB converters. Therefore, the increase or decrease in the power efficiency is determined by the operation of the switch and diode.

The losses of semiconductors are divided into switching and conduction losses. A switching loss occurs during the transient time of the switching operation. Although the switching loss is 0 in the ideal condition, it is caused by the time delay attributed to the parasitic resistance and parasitic capacitance at the time of turn-on or turn-off [14,15]. The conduction loss is caused by the current that flows by the turn-on of the semiconductor and on the resistance of the semiconductor [16–18].

3.2. Switching Loss

The switching loss of MOSFET, $P_{S,SW}$ is divided into switch turn-on loss, $P_{S,SW,ON}$ switch turn-off loss $P_{S,SW,OFF}$, and output capacitance loss $P_{S,SW,Coss}$ as [5,15,17]:

$$P_{S,SW} = P_{S,SW,ON} + P_{S,SW,OFF} + P_{S,SW,Coss} \quad (7)$$

The switch turn-on and turn-off losses are difficult to calculate because of the nonlinear characteristics of the drain-source voltage v_{DS} and drain current i_D. Therefore, they can be determined by applying linear approximation in the rising and falling sections of v_{DS} and i_D. The output capacitance loss of MOSFET can be determined by calculating the stored energy of the capacitor because the energy is charged in the output capacitance when the MOSFET turns off, and is discharged when the MOSFET turns on. The MOSFET switching loss can be represented by using the turn-on time t_{on}, turn-off time t_{off}, switching frequency f_{sw}, and output capacitance C_{oss} of the MOSFET as [5,17,18]:

$$\begin{aligned} P_{S,SW} &= \tfrac{1}{2}\cdot v_{DS}\cdot i_D\cdot t_{on}\cdot f_{sw} + \tfrac{1}{2}\cdot v_{DS}\cdot i_D\cdot t_{off}\cdot f_{sw} + \tfrac{1}{2}\cdot C_{oss}\cdot {v_{DS}}^2\cdot f_{sw} \\ &= \tfrac{1}{2}\cdot v_{DS}\cdot i_D\cdot (t_{on} + t_{off})\cdot f_{sw} + \tfrac{1}{2}\cdot C_{oss}\cdot {v_{DS}}^2\cdot f_{sw} \end{aligned} \quad (8)$$

The output capacitance of the MOSFET is several tens to hundreds of picofarads. Therefore, it is negligible compared to the switch turn-on and turn-off losses and can be represented as [5,6,17,18]:

$$P_{S,SW} = \frac{1}{2}\cdot v_{DS}\cdot i_D\cdot (t_{on} + t_{off})\cdot f_{sw} \quad (9)$$

The switching loss $P_{D,SW}$ of the diode can be divided into the switch turn-on loss $P_{D,SW,ON}$ and the switch turn-off loss $P_{D,SW,OFF}$ it is represented as:

$$P_{D,SW} = P_{D,SW,ON} + P_{D,SW,OFF} \cong P_{D,SW,OFF} \quad (10)$$

The switching loss of the diode can be approximated as the switch turn-off loss because the loss of the diode caused by the reverse recovery in the turn-off section is considerably larger than the turn-on loss [19,20].

Figure 6 shows the reverse recovery characteristic of the diode [6]. In an ideal diode, the current flows when the voltage is applied in the forward direction, and no current flows when the voltage is applied in the reverse direction. However, in an actual operation, the reverse current flows for a certain time before it reaches zero when the diode is turned off after a forward current flow. Here, the time during which the reverse current flows is referred to as the reverse recovery time (t_{rr}); the maximum value of the reverse current that flows in the diode is referred to as the repetitive peak reverse current (I_{RRM}). The turn-off loss $P_{D,SW,OFF}$ of the diode when the reverse voltage applied to the diode is V_R and the switching frequency is f_{sw} can be expressed as [21,22]:

$$P_{D,SW} = \frac{1}{2}\cdot V_R\cdot I_{RRM}\cdot t_{rr}\cdot f_{sw} \quad (11)$$

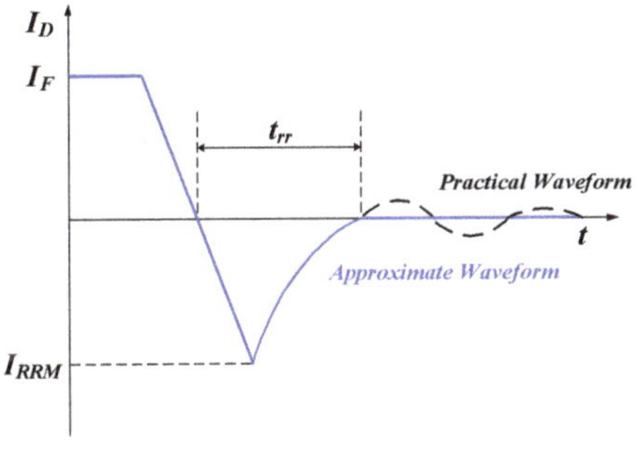

Figure 6. Diode reverse recovery characteristics.

3.3. Conduction Loss

The instantaneous value of the conduction loss of the MOSFET, $P_{S,CD}(t)$ can be expressed by $v_{DS}(t)$, $i_D(t)$, and the drain-source resistance $R_{DS(ON)}$ in the complete switch turn-off section as [6,15,16,19]:

$$P_{S,CD}(t) = v_{DS}(t) \cdot i_D(t) = i_D{}^2(t) \cdot R_{DS(ON)} \tag{12}$$

The average value can be determined by integrating Equation (6) over the switching period T_{sw} as:

$$P_{S,CD}(t) = \frac{1}{T_{sw}} \int_0^{T_{sw}} i_D{}^2(t) \cdot R_{DS(ON)}\, dt \tag{13}$$

$$P_{S,CD} = I_{D(rms)}{}^2 \cdot R_{DS(ON)} \tag{14}$$

The instantaneous value of the conduction loss of the diode $P_{D,CD}(t)$ can be expressed by the forward voltage drop $v_F(t)$, forward current $i_F(t)$, and diode resistance R_D as [5,6]:

$$P_{D,CD}(t) = v_F(t) \cdot i_F(t) + R_D \cdot i_F{}^2(t) \tag{15}$$

The average value can be determined by integrating Equation (8) over the switching period T_{sw} as [5,12,15]:

$$P_{D,CD}(t) = \frac{1}{T_{sw}} \int_0^{T_{SW}} \left\{ v_F(t) \cdot i_F(t) + R_D \cdot i_F{}^2(t) \right\} dt \tag{16}$$

$$P_{D,CD} = V_F \cdot I_{F(AVG)} + R_D \cdot I_{F(RMS)}{}^2 \tag{17}$$

4. Analysis of Power Loss in Topologies

Table 5 summarizes the switching and conduction semiconductors of the conventional and proposed TSBB converters in each topology. Table 6 presents the voltage and current stresses of the conventional and proposed TSBB converters in each topology. In each topology, the switching loss that has reflected the voltage stress and conduction loss based on the switch on/off time can be determined and compared.

4.1. Buck Topology

In Table 5, the switching semiconductors of the conventional TSBB converter are S_1/D_1, and the conducting semiconductors are $S_1/D_1/D_2$. Thus, the switching loss $P_{CON,SW}$ and conduction loss $P_{CON,CD}$ can be expressed, respectively, as:

$$P_{CON,SW} = P_{SW,S1} + P_{SW,D1} \tag{18}$$

$$P_{CON,CD} = P_{CD,S1,ON} + P_{CD,D2,ON} + P_{CD,D1,OFF} + P_{CD,D2,OFF} \tag{19}$$

The switching semiconductors of the proposed TSBB converter are $S_1/D_1/D_2$, and those of the conducting semiconductor are $S_1/D_1/D_2$. Therefore, the switching loss $P_{PRO,SW}$ and conduction loss $P_{PRO,CD}$ can be expressed as:

$$P_{PRO,SW} = P_{SW,S1}\left(\frac{C_{oss,D2}}{C_{oss,S1} + C_{oss,D2}} V_i\right) + P_{SW,D1}(V_i) + P_{SW,D2}\left(\frac{C_{oss,S1}}{C_{oss,S1} + C_{oss,D2}} V_i\right) \tag{20}$$

$$P_{PRO,CD} = P_{CD,S1,ON} + P_{CD,D2,ON} + P_{CD,D1,OFF} \tag{21}$$

The switching loss varies by the voltage stress; however, the internal voltage of semiconductors is V_i for both the conventional and proposed TSBB converters. Consequently, the switching losses of the semiconductors are the same. Therefore, the difference in power loss between the two converters can be expressed as:

$$P_{CON} - P_{PRO} = P_{CD,D2,OFF} - P_{SW,D2}\left(\frac{C_{oss,S1}}{C_{oss,S1} + C_{oss,D2}}V_i\right) \quad (22)$$

This difference can be determined by the conduction and switching losses in the switch-off section of D_2. Since the switch-off section is $(1-d)\cdot T_{SW}$, the lower the duty ratio, the higher the efficiency of the proposed TSBB converter.

4.2. Boost Topology

In the boost topology, the switching semiconductors are S_2/D_2, and the conducting semiconductors are $S_1/S_2/D_2$. Thus, the switching loss $P_{CON,SW}$ $(= P_{PRO,SW})$ and the conduction loss $P_{CON,CD}$ $(= P_{PRO,CD})$ can be expressed as:

$$P_{CON,SW} = P_{PRO,SW} = P_{SW,S2} + P_{SW,D2} \quad (23)$$

$$P_{CON,CD} = P_{PRO,CD} = P_{CD,S1,ON} + P_{CD,S2,ON} + P_{CD,S1,OFF} + P_{CD,D2,OFF} \quad (24)$$

There is no difference in the efficiency between the conventional and proposed TSBB converters because there is no change in the operation of semiconductors.

4.3. Buck-Boost Topology

In Table 5, both the switching and conducting semiconductors of the conventional TSBB converter are $S_1/S_2/D_1/D_2$. Therefore, the switching loss $P_{CON,SW}$ and conduction loss $P_{CON,CD}$ can be expressed as:

$$P_{CON,SW} = P_{SW,S1}(V_i) + P_{SW,S2}\,(V_o) + P_{SW,D1}(V_i) + P_{SW,D2}(V_o) \quad (25)$$

$$P_{CON,CD} = P_{CD,S1,ON} + P_{CD,S2,ON} + P_{CD,D1,OFF} + P_{CD,D2,OFF} \quad (26)$$

Both the switching and conducting semiconductors of the proposed TSBB converter are $S_1/S_2/D_1$. Therefore, the switching loss $P_{PRO,SW}$ and conduction loss $P_{PRO,CD}$ can be, respectively, expressed as:

$$P_{PRO,SW} = P_{SW,S1}\left(\frac{C_{oss,D2}}{C_{oss,S1}+C_{oss,D2}}V_i\right) + P_{SW,S2}\left(\frac{C_{oss,D2}}{C_{oss,S2}+C_{oss,D2}}V_o\right) + P_{SW,D1}(V_i + V_o) \quad (27)$$

$$P_{PRO,CD} = P_{CD,S1,ON} + P_{CD,S2,ON} + P_{CD,D1,OFF} \quad (28)$$

The voltage stress of each semiconductor is indicated in parentheses since the switching loss varies by the voltage stress. The loss difference between the conventional and proposed TSBB converters is determined by the switching loss of D_1/D_2 and the conduction loss of D_2 in the switch-off section, and it can be expressed as:

$$P_{CON} - P_{PRO} = P_{SW,D1}(V_i) + P_{SW,D2}(V_o) + P_{CD,D2,OFF} - P_{SW,D1}(V_i + V_o) \quad (29)$$

There are no switching and conduction losses by D_2 because the proposed TSBB converter D_2 does not operate. However, the voltage stress of D_1 increases to $V_i + V_o$, the voltage stress and reverse current of the diode in a linear section are $P_{SW,D1}(V_i) + P_{SW,D2}(V_o) \cong P_{SW,D1}(V_i + V_o)$. Therefore, the loss is determined by the conduction loss of the switch-off section of D_2. The power efficiency of the proposed TSBB converter is higher than that of the conventional TSBB converter. In contrast, the switching loss by D_1, $P_{SW,D1}(V_i + V_o)$, increases in the section where the reverse current of the diode increases sharply owing to the voltage stress. Therefore, the power efficiency of the proposed TSBB converter is lower than that of the conventional TSBB converter.

5. Experimental Results

Figure 7 shows a prototype of a 100 W TSBB converter fabricated to verify the improved power efficiency. The Arduino controller generates a 5 V pulse width modulation (PWM) to control the MOSFET driver IC input; upon receiving this signal, the MOSFET driver IC converts it to an 18 V drive signal and transfers it to the gate of the MOSFET. The specifications of the components are 20% or more larger than the calculated maximum stress considering the various input/output conditions of the experiment. Table 7 summarizes the detailed specifications of the components; Table 8 presents the maximum values of the voltage stress measured in the CV/CC modes.

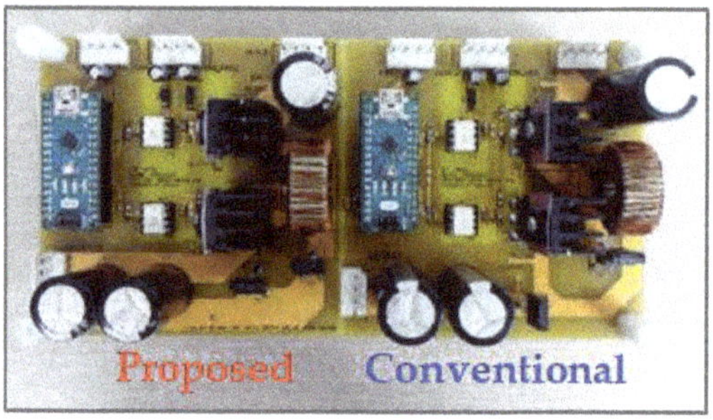

Figure 7. 100 W TSBB prototype.

Table 7. Components specifications.

Component	Part Name	Specification
PWM Generator	Arduino Nano	5 V, 16 MHz
MOSFET Driver	HCPL-J312	Output Peak Current = 2.5 A, Input Current = 7–16 mA Supply Voltage = 15–30 V, Input Capacitance = 60 pF Rise Time = 0.1 us, Fall Time = 0.1 us
MOSFET	RCX510N25	Drain-Source Voltage V_{DSS} = 250 V Gate-Source Voltage V_{GSS} = ±30 V Drain Current I_D = 51 A Static Drain-Source ON-State Resistance = 48 mΩ Output Capacitance = 350 pF Rise Time = 300 ns, Fall Time = 210 ns
Diode	RF2001T3D	Reverse voltage (DC) V_R = 300 V Forward voltage V_F = 1.3 V (at I_F = 10 A) Average Rectified Forward Current I_F = 20 A Reverse recovery time t_{rr} = 25 ns (at I_F = 0.5 A, I_R = 1 A, I_{rr} = 0.25 × I_R)
Inductor	CH270125	Cross Section = 0.654 cm^2, Path Length = 6.35 cm Window Area = 1.56 cm^2, Volume = 4.154 cm^3 AL Value = 157 nH/Turn^2, Permeability μ = 125 Inductance = 250 uH
Electrolytic Capacitor	100YXG820MEFC18 × 40	Rated Voltage (V_{dc}) = 100 V Rated ripple current = 2330 mA (at 100 kHz) Leakage Current = 3 μA, Impedance = 20 °C, 100 kHz Dissipation Factor(MAX) tanδ = 0.08 Capacitance = 820 uF

The experiments are conducted in the CV/CC modes of the three topologies. The experiment for the buck-boost topology is conducted separately for the step-down and step-up sections. In the CV mode, the power efficiency is measured for the output power range of 10–80 W at the switching frequency f_{sw} = 100 kHz of the output current. The power efficiency was measured at three duty ratios to analyze the power efficiency based on the change in the duty ratio. If the input/output voltage variation ratio V_{var} is defined as the ratio of the difference between the input and output voltages to the input voltage, it can be expressed using the input voltage V_i and the output voltage V_o as:

$$V_{var}\ (\%) = \frac{|V_i - V_o|}{V_i} \times 100 \tag{30}$$

In each topology, the power efficiency was measured at the duty ratios where V_{var} was 25%, 33%, and 50%. Table 9 shows the duty ratio based on V_{var} in each topology. Figure 8a–d shows PWM waveforms of the conventional and the proposed TSBB converter in each topology implemented using Arduino Nano. Figure 8e–h presents waveforms of the inductor current, the gate-source voltage, and the output voltage of the conventional and proposed TSBB converters measured in CC mode at V_{var} = 33% (V_o ≅40 V).

Table 8. Measured maximum voltage stresses on components in CV and CC modes.

V_{var}	Device	CV								CC							
		Conventional (V_{max})				Proposed (V_{max})				Conventional (V_{max})				Proposed (V_{max})			
		Buck	Boost	Buck-Boost		Buck	Boost	Buck-Boost		Buck	Boost	Buck-Boost		Buck	Boost	Buck-Boost	
				Step-Down	Step-Up			Step-Down	Step-Up			Step-Down	Step-Up			Step-Down	Step-Up
25%	S_1	53.3	-	53.3	32.0	53.3	-	53.3	32.0	80.0	-	80.0	80.0	80.0	-	80.0	80.0
	S_2	-	41.3	46.0	45.5	-	41.6	47.5	46.8	-	104.5	67.2	114.8	-	105.2	68.9	116.9
	D_1	53.3	-	53.3	32.0	53.3	-	100.8	78.8	80.0	-	80.0	80.0	80.0	-	148.9	196.9
	D_2	-	41.3	46.0	45.5	-	41.6	-	-	-	104.5	67.2	114.8	-	105.2	-	-
33%	S_1	60.0	-	60.0	30.0	60.0	-	60.0	30.0	80.0	-	80.0	80.0	80.0	-	80.0	80.0
	S_2	-	41.3	45.9	44.9	-	41.7	47.5	46.3	-	111.9	59.0	121.0	-	112.0	60.5	123.0
	D_1	60.0	-	60.0	30.0	60.0	-	107.5	76.3	80.0	-	80.0	80.0	80.0	-	140.5	203.0
	D_2	-	41.3	45.9	44.9	-	41.7	-	-	-	111.9	59.0	121.0	-	112.0	-	-
50%	S_1	80.0	-	80.0	26.7	80.0	-	80.0	26.7	80.0	-	80.0	80.0	80.0	-	80.0	80.0
	S_2	-	41.8	46.3	45.5	-	41.7	47.9	46.9	-	126.5	43.2	138.5	-	127.3	44.6	141.2
	D_1	80.0	-	80.0	26.7	80.0	-	127.9	73.6	80.0	-	80.0	80.0	80.0	-	124.6	221.2
	D_2	-	41.8	46.3	45.5	-	41.7	-	-	-	126.5	43.2	138.5	-	127.3	-	-

Table 9. Duty ratio according to V_{var} in each topology.

V_{var}	Buck	Boost	Buck-Boost	
			Step-Down	Step-Up
25%	0.750	0.200	0.430	0.556
33%	0.667	0.250	0.400	0.571
50%	0.500	0.330	0.330	0.600

5.1. Buck Topology

Figure 9 shows the power efficiency of the buck topology in the CV and CC modes. If the power efficiency difference P_{Diff} is positive (+), it means efficiency improvement; if it is negative (-), it indicates an efficiency decline. The power efficiency improves in every condition from 10–80 W because of the removal of D_2 in the switch-off section in the CV and CC modes. P_{Diff} increases as the switch-off section becomes longer, i.e., when there is an increase in the duty ratio. In the same duty ratio condition, P_{Diff} decreases with an increase in the input voltage and output current. The improvement effect is large in the low-power region. The proposed converter can achieve a greater efficiency improvement effect in a region where the voltage conversion is large, and the output current is low.

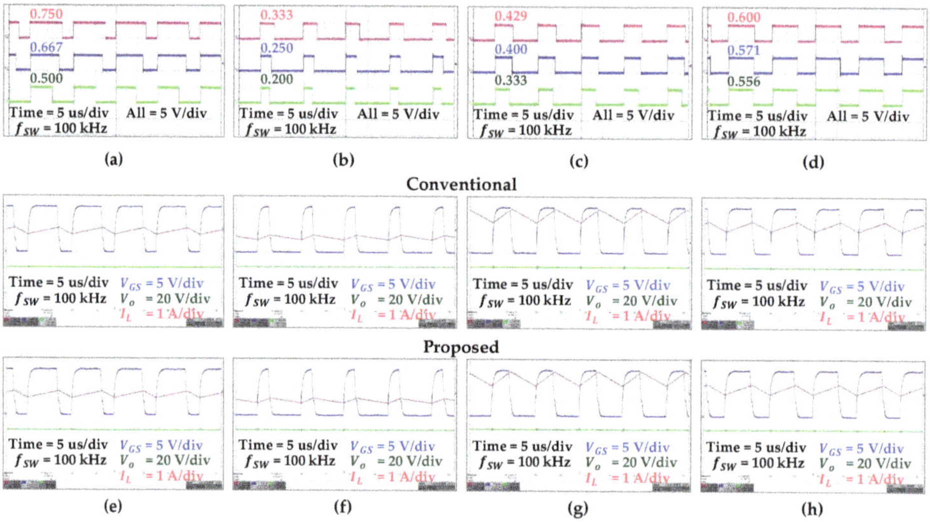

Figure 8. Waveforms of PWM, inductor current I_L, gate-source voltage V_{GS}, and output voltage V_o: PWM of (**a**) buck; (**b**) boost; (**c**) buck-boost step-down; (**d**) buck-boost step-up; I_L, V_{GS}, V_o of (**e**) buck (at d = 0.667; I_L = 1.5 A); (**f**) boost (at d = 0.250, I_L = 0.75 A); (**g**) buck-boost step-down (at d = 0.400, I_L = 1.5 A); and (**h**) buck-boost step-up (at d = 0.571, I_L = 0.75 A).

Duty	V_I (V)	$I_{O\_100\%}$ (A)	V_{var}
0.750	53.33	2.00	25%

Duty	V_I (V)	$I_{O\_100\%}$ (A)	V_{var}
0.667	60.00	2.00	33%

Duty	V_I (V)	$I_{O\_100\%}$ (A)	V_{var}
0.500	80.00	2.00	50%

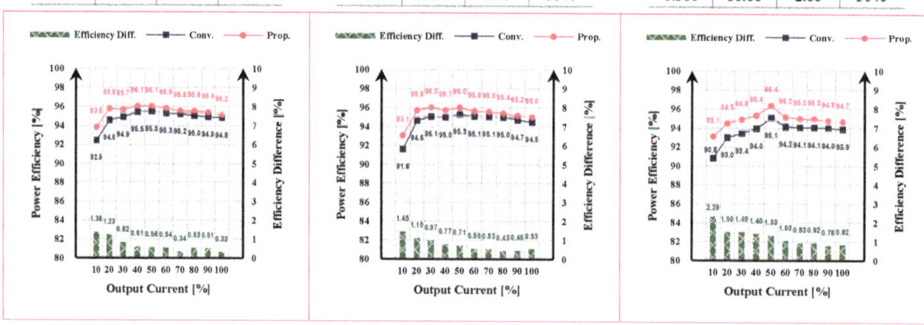

(a)

Duty	I_O (A)	V_{var}
0.750	1.33	25%

Duty	I_O (A)	V_{var}
0.667	1.50	33%

Duty	I_O (A)	V_{var}
0.500	2.00	50%

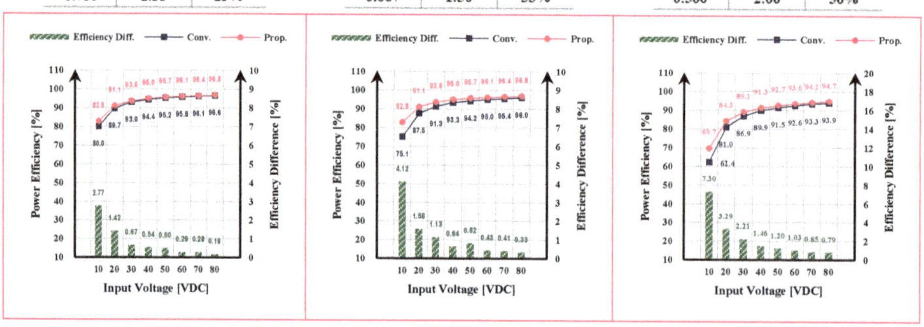

(b)

Figure 9. Power efficiency comparison of the buck topology: (**a**) CV mode; (**b**) CC mode.

5.2. Boost Topology

Figure 10 shows the power efficiency of the boost topology. The proposed converter shows the same power efficiency in the measured error range because it has the same number of semiconductors and the same stress as the conventional converter.

Duty	V_I (V)	$I_{O\_100\%}$ (A)	V_{var}
0.200	32.00	2.00	25%

Duty	V_I (V)	$I_{O\_100\%}$ (A)	V_{var}
0.250	30.00	2.00	33%

Duty	V_I (V)	$I_{O\_100\%}$ (A)	V_{var}
0.333	26.67	2.00	50%

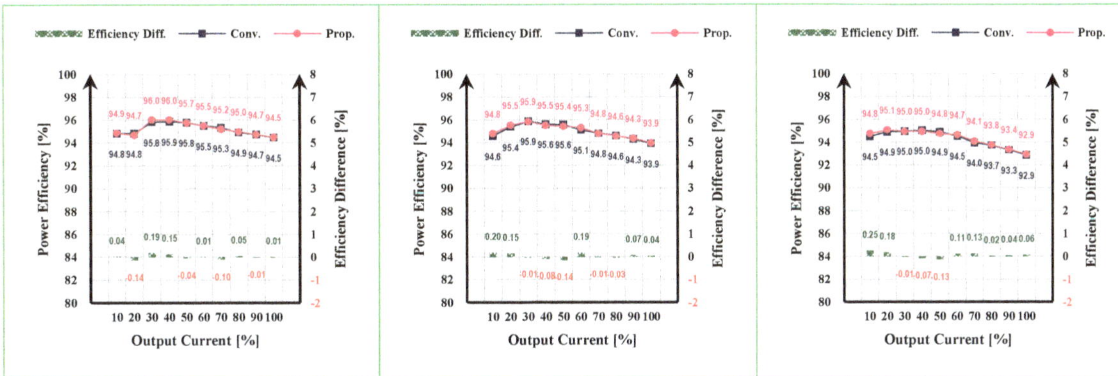

(a)

Duty	I_O (A)	V_{var}
0.200	0.80	25%

Duty	I_O (A)	V_{var}
0.250	0.75	33%

Duty	I_O (A)	V_{var}
0.333	0.67	50%

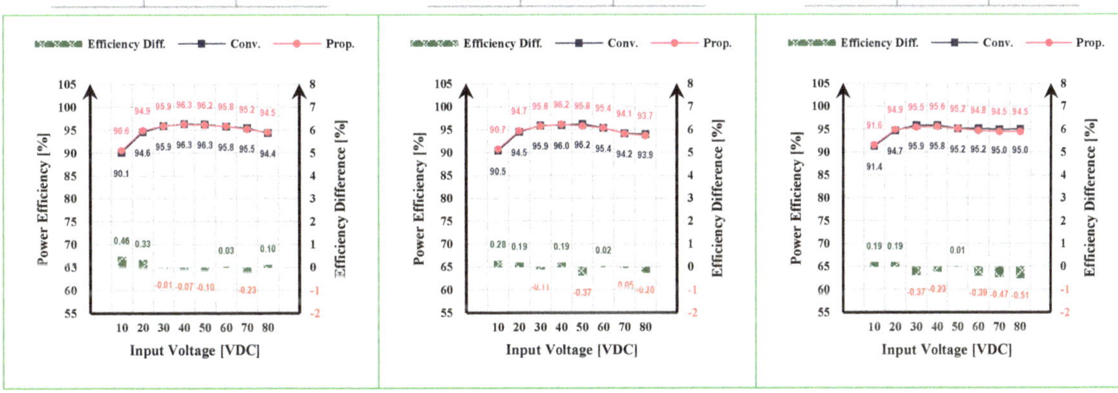

(b)

Figure 10. Power efficiency comparison of the boost topology: (**a**) CV mode; (**b**) CC mode.

5.3. Buck-Boost Topology

5.3.1. Step-Down

Figure 11 shows the power efficiency in the step-down section of the buck-boost topology in the CV and CC modes. The power efficiency is improved in every condition because of the removal of D_2 in the switch-off section. In the CV mode, the efficiency improvement becomes larger with an increase in the output current; P_{Diff} increases in the high-power region. In the CC mode, the power efficiency improves with a decrease in the input voltage, and P_{Diff} increases in the low-power region. In the same duty condition, the voltage stress of the diode D_1 increases with the input voltage; P_{Diff} decreases with an increase in loss.

Duty	V_I (V)	$I_{O\_100\%}$ (A)	V_{var}
0.429	53.33	2.00	25%

Duty	V_I (V)	$I_{O\_100\%}$ (A)	V_{var}
0.400	60.00	2.00	33%

Duty	V_I (V)	$I_{O\_100\%}$ (A)	V_{var}
0.333	80.00	2.00	50%

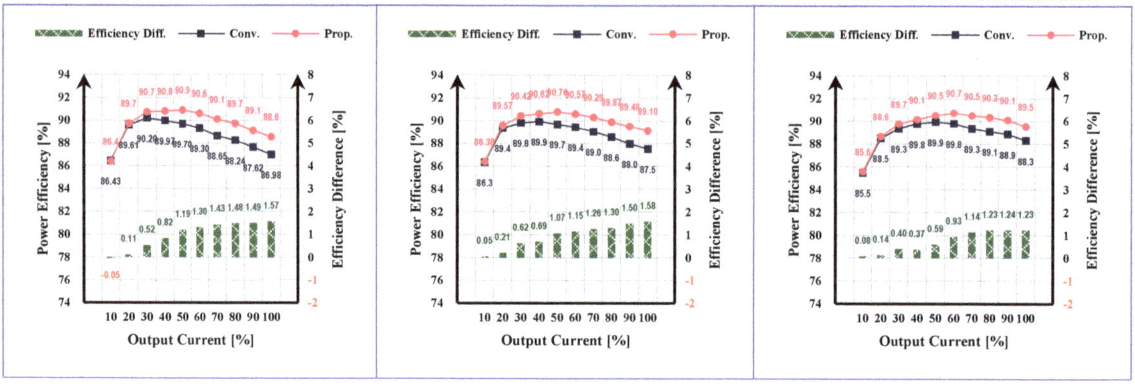

(a)

Duty	I_O (A)	V_{var}
0.429	1.33	25%

Duty	I_O (A)	V_{var}
0.400	1.50	33%

Duty	I_O (A)	V_{var}
0.333	2.00	50%

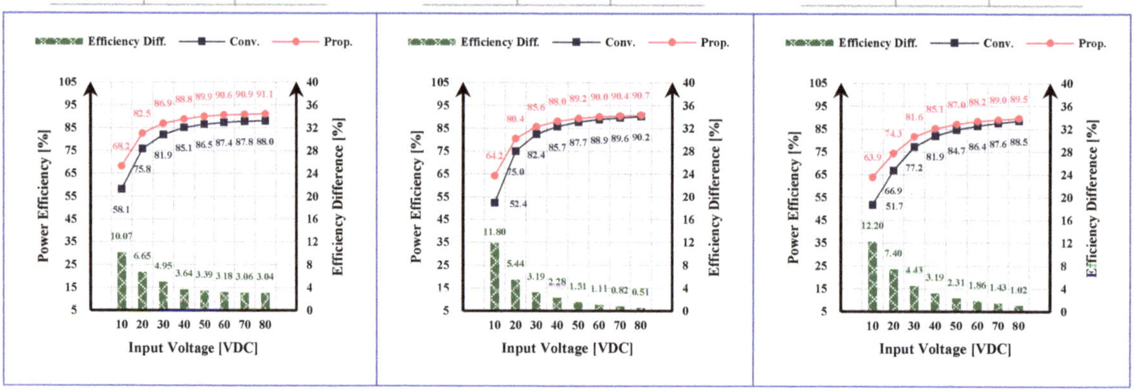

(b)

Figure 11. Power efficiency comparison of the buck-boost topology (step-down): (**a**) CV mode; (**b**) CC mode.

5.3.2. Step-Up

Figure 12 shows the power efficiency of the step-up buck-boost topology in the CV and CC modes. In the CV mode, the efficiency improves in general to 10–80 W and P_{Diff} increases in the high-power region where the output current increases. In the CC mode, P_{Diff} decreases with an increase in V_{var} and the input voltage. The power efficiency decreases more than the conventional converter at a power of 50 W or higher. In the step-up section, the output voltage increases with the input voltage, and the voltage stress of diode D_1 also increases. The efficiency decreases more than the conventional converter because the resulting diode loss exceeds the increase in the efficiency obtained by the removal of D_2 in the switch-off section. Therefore, the proposed converter is suitable for applications below 50 W in the step-up section of the buck-boost topology.

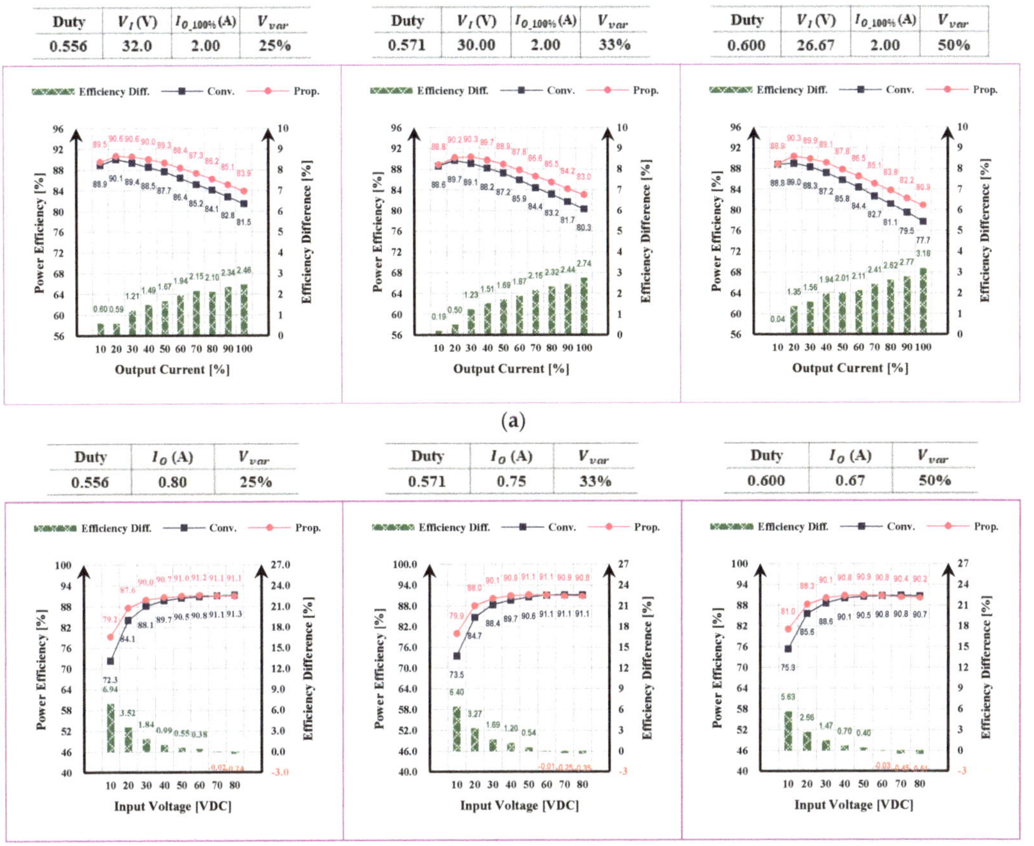

Figure 12. Power efficiency comparison of the buck-boost topology (step-up): (**a**) CV mode; (**b**) CC mode.

5.3.3. Diode Stress (Buck-Boost Step-Up, CC Mode)

Figure 13 shows the analytic and experimental diode power loss in the CC mode in the step-up section of the buck-boost topology. A positive power loss difference L_{Diff} implies that the diode loss of the proposed TSBB converter is large, whereas if it is negative, it implies a large diode loss of the conventional TSBB converter.

To calculate the diode loss, information on several parameters was referenced from the datasheet and all data are typical values at 25 degrees. Parameters not provided in the datasheet were used to calculate the diode loss with some assumptions.

- Forward voltage V_F: obtained from the V_F-I_F characteristic curve.
- Equivalent resistance R_D: obtained from the V_F-I_F characteristic curve by linear approximation under the current conditions used in the experiment.
- Reverse recovery time T_{rr} and peak reverse recovery current I_{rrm}: estimated and calculated using value 16 nsec at the forward current I_F = 1.0 A and the reverse current I_R = 0.5 A.
- Peak reverse recovery voltage V_{rrm}: assumed $\frac{dI_F}{dI_R} = 0.5$ at $V_R\left(1+\frac{dI_F}{dI_R}/\frac{dI_F}{dI_R}\right)$ and corrected coefficient by temperature with V_R characteristic curve.
- Ignored any other parasitic factors like internal inductance, capacitance, and so on.

Due to several assumptions and uncertain parameter values, the two results are slightly different. This is expected to be due to the reverse recovery characteristics that change exponentially with the increase in temperature and voltage stress, and also to be affected by parasitic components on the PCB and errors in measurement equipment.

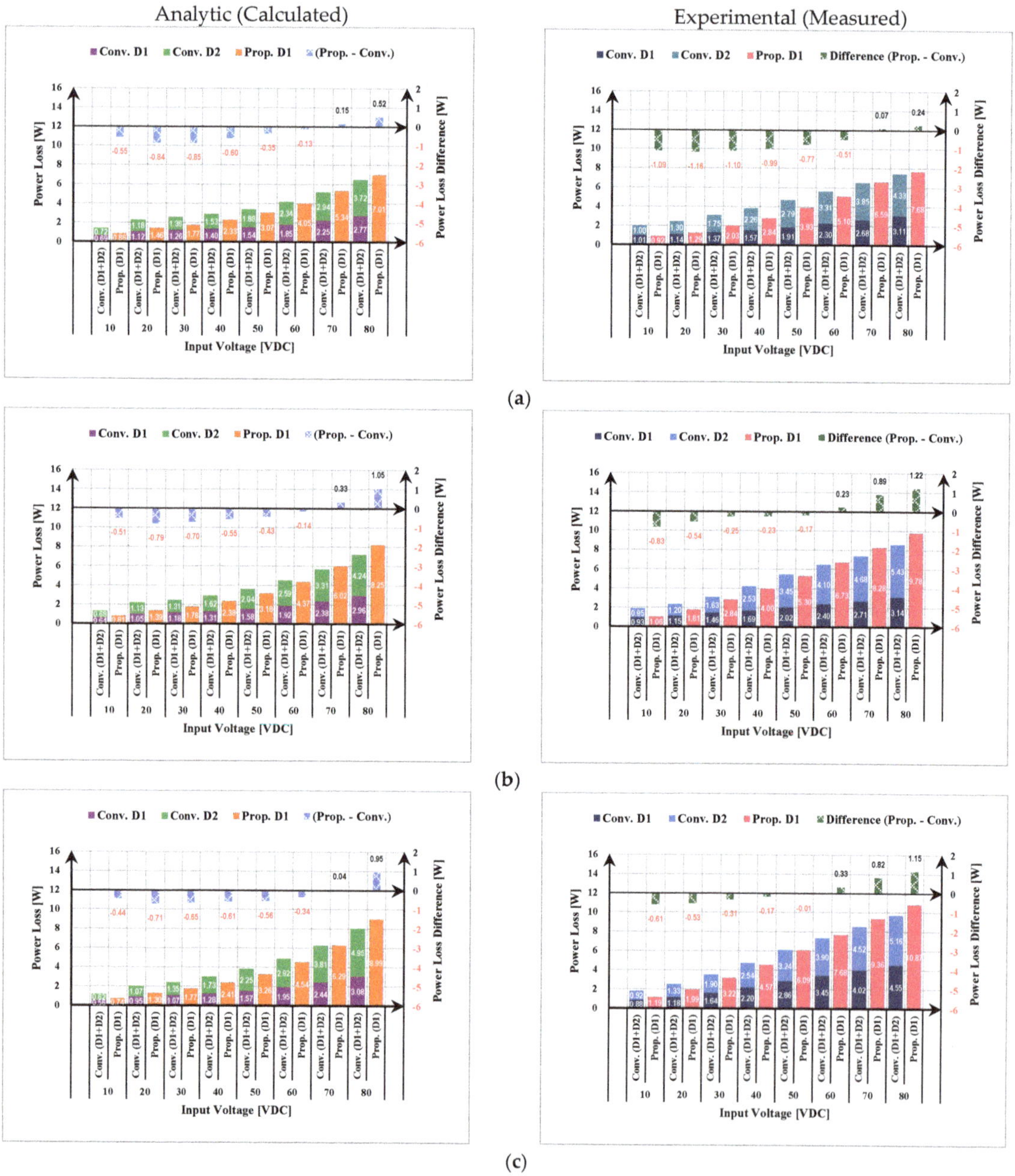

Figure 13. Analytic and experimental diode power loss of the buck-boost topology step-up period in the CC mode: (**a**) Duty = 0.556, I_O = 0.8 A; (**b**) Duty = 0.571, I_O = 0.75 A; and (**c**) Duty = 0.600, I_O = 0.667 A.

Despite these differences, both results show a similar trend in which diode losses rapidly increase with voltage stress over a certain region, thereby diode power loss of the proposed TSBB converter is bigger than the conventional converter over a certain input voltage. The increase in the D_1 loss of the proposed TSBB converter becomes larger than the power efficiency improvement by the removal of D_2. The power efficiency is reversed at the point where the difference in diode loss changes from negative to positive (+). For the proposed TSBB converter, the power efficiency increases at 60 W or lower at the duty ratio of 0.556; the power efficiency increases at 50 W or lower at the duty ratios of 0.571 and 0.600.

6. Conclusions

This study proposed a modified design of the TSBB converter to improve power efficiency using fewer conduction components and measured the optimal output power range. The proposed TSBB converter improved power efficiency in buck and buck-boost topologies by reducing the conduction loss caused by the diode in the switch-off section power efficiency. A 100 W prototype was designed and fabricated to verify the improvement. Experiments were conducted in the CV/CC modes of three topologies, and the power efficiency was measured for 10–80 W. In the buck topology, power efficiency improved in the entire power range of 10–80 W; it increased on average by 0.75–1.36% and 0.83–2.27% in the CV and CC modes, respectively. In the buck-boost topology step-down, the power efficiency improved in the entire power range of 10–80 W; it improved on average by 0.73–0.99% and 3.33–4.75% in the CV and CC modes, respectively. In the buck-boost topology step-up, the power efficiency increased on average by 1.65–2.00% in the entire power range of 10–80 W in the CV mode, and by 2.17–2.77% in the power range of the 10–50 W in the CC mode.

In future research, we will study how to reduce the conduction and switching losses by reducing the voltage and current stress of semiconductors for efficiency improvement. In addition, we will analyze the effect on output ripple under various conditions by using inductance, capacitance, and switching frequency as design variables, and study how these parameters affect converter efficiency.

Author Contributions: Conceptualization, S.K. and H.J.; methodology, S.K. and H.J.; software, S.K. and H.J.; validation, S.K. and H.J.; formal analysis, S.K. and H.J.; investigation, S.K. and H.J.; data curation, S.K. and H.J.; writing—original draft preparation, S.K. and H.J.; writing—review and editing, S.K. and H.J.; supervision, H.J. and S.-h.L.; project administration, H.J. and S.-h.L.; funding acquisition, H.J. All authors have read and agreed to the published version of the manuscript.

Funding: This work is supported by the Korea Agency for Infrastructure Technology Advancement (KAIA) grant funded by the Ministry of Land, Infrastructure and Transport (Grant RS-2021-KA164174).

Informed Consent Statement: Informed consent was obtained from all subjects involved in the study.

Data Availability Statement: Data sharing not applicable.

Conflicts of Interest: The authors declare no conflict of interest.

References

1. Zhou, X.; Sheng, B.; Liu, W.; Chen, Y.; Wang, L.; Liu, Y.F.; Sen, P.C. A high-efficiency high-power-density on-board low-voltage dc–dc converter for electric vehicles application. *IEEE Trans. Power Electron.* **2021**, *36*, 12781–12794. [CrossRef]
2. Jin, F.; Nabih, A.; Chen, C.; Chen, X.; Li, Q.; Lee, F.C. A high efficiency high density dc/dc converter for battery charger applications. In Proceedings of the 2021 IEEE Applied Power Electronics Conference and Exposition (APEC), Phoenix, AZ, USA, 14–17 June 2021.
3. Yang, C.; Xie, S.; Mao, L.; Zhang, Z. Efficiency improvement on two-switch buck-boost converter with coupled inductor for high-voltage applications. *IET Power Electron.* **2014**, *7*, 2846–2856. [CrossRef]
4. Sahu, B.; Rincón-Mora, G.A. A low voltage, dynamic, noninverting, synchronous buck-boost converter for portable applications. *IEEE Trans. Power Electron.* **2004**, *19*, 443–452. [CrossRef]
5. Moon, B.; Jung, H.Y.; Kim, S.H.; Lee, S.H. A modified topology of two-switch buck-boost converter. *IEEE Access* **2017**, *5*, 17772–17780. [CrossRef]

6. Jung, H.Y.; Kim, S.H.; Moon, B.; Lee, S.H. A new circuit design of two-switch buck-boost converter. *IEEE Access* **2018**, *6*, 47415–47423. [CrossRef]
7. Chandran, C.; Chandran, L.R. Two switch buck boost converter for power factor correction. In Proceedings of the 2015 International Conference on Technological Advancements in Power and Energy (TAP Energy), Kollam, India, 24–26 June 2015.
8. Veerachary, M.; Khuntia, M.R. Design and analysis of two-switch-based enhanced gain buck–boost converters. *IEEE Trans. Ind. Electron.* **2021**, *69*, 3577–3587. [CrossRef]
9. Ahmad, A.A.; Abrishamifar, A. A simple current mode controller for two switches buck-boost converter for fuel cells. In Proceedings of the 2007 IEEE Canada Electrical Power Conference, Montreal, QC, Canada, 25–26 October 2007.
10. Son, H.S.; Kim, J.K.; Lee, J.B.; Moon, S.S.; Park, J.H.; Lee, S.H. A new buck–boost converter with low-voltage stress and reduced conducting components. *IEEE Trans. Ind. Electron.* **2017**, *64*, 7030–7038. [CrossRef]
11. Fan, H. Design tips for an efficient non-inverting buck-boost converter. *Analog Appl. J.* **2014**, 20–25.
12. Chen, J.; Maksimovic, D.; Erickson, R.W. Analysis and design of a low-stress buck-boost converter in universal-input PFC applications. *IEEE Trans. Power Electron.* **2006**, *21*, 320–329. [CrossRef]
13. MOSFET Power Losses and How They Affect Power-Supply Efficiency. Available online: https://www.ti.com/lit/an/slyt664/slyt664.pdf (accessed on 5 November 2022).
14. Infineon Application Note. Available online: https://application-notes.digchip.com/070/70-41484.pdf (accessed on 5 November 2022).
15. Xiong, Y.; Sun, S.; Jia, H.; Shea, P.; Shen, Z.J. New physical insights on power MOSFET switching losses. *IEEE Trans. Power Electron.* **2009**, *24*, 525–531. [CrossRef]
16. Lirio, L.E.A.; Bellar, M.D.; Neto, J.A.M.; Dos Reis, M.S.; Aredes, M. Switching losses analysis in sic power mosfet. In Proceedings of the 2015 IEEE 13th Brazilian Power Electronics Conference and 1st Southern Power Electronics Conference (COBEP/SPEC), Fortaleza, Brazil, 29 February 2016.
17. Ren, Y.; Xu, M.; Zhou, J.; Lee, F.C. Analytical loss model of power MOSFET. *IEEE Trans. Power Electron.* **2006**, *21*, 310–319.
18. Power MOSFET Basics: Understanding MOSFET Characteristics Associated with the Figure of Merit. Available online: https://www.vishay.com/docs/71933/71933.pdf (accessed on 5 November 2022).
19. Shen, Z.J.; Xiong, Y.; Cheng, X.; Fu, Y.; Kumar, P. Power MOSFET switching loss analysis: A new insight. In Proceedings of the 2006 IEEE Industry Applications Conference and Forty-First IAS Annual Meeting, Tampa, FL, USA, 8–12 October 2006.
20. Rodríguez, M.; Rodríguez, A.; Miaja, P.F.; Lamar, D.G.; Zúniga, J.S. An insight into the switching process of power MOSFETs: An improved analytical losses model. *IEEE Trans. Power Electron.* **2010**, *25*, 1626–1640. [CrossRef]
21. Li, X.; Jiang, J.; Huang, A.Q.; Guo, S.; Deng, X.; Zhang, B.; She, X. A SiC power MOSFET loss model suitable for high-frequency applications. *IEEE Trans. Ind. Electron.* **2017**, *64*, 8268–8276. [CrossRef]
22. Ivanovic, Z.; Blanusa, B.; Knezic, M. Power loss model for efficiency improvement of boost converter. In Proceedings of the 2011 XXIII International Symposium on Information, Communication and Automation Technologies, Sarajevo, Bosnia and Herzegovina, 27–29 October 2011.

applied sciences

MDPI

Article

Adaptive Fractional-Order Anti-Saturation Synchronous Control for Dual-Motor Systems

Yongbin Zhong, Jian Gao * and Lanyu Zhang

State Key Laboratory of Precision Electronic Manufacturing Technology and Equipment, Guangdong University of Technology, Guangzhou 510006, China
* Correspondence: gaojian@gdut.edu.cn; Tel.: +86-1356-0125-827

Featured Application: High-speed synchronous motion system for precision electronic packaging equipment.

Abstract: The synchronization error of a dual-motor system will seriously affect the motion profile accuracy. To solve this problem, an adaptive fractional-order anti-saturation synchronous control method based on fractional-order frequency-domain control theory is proposed in this paper. On the one hand, the proposed method performs a compensation on the closed-loop feedback control loop to unify the frequency-domain characteristics for a dual-motor system. With the frequency-domain characteristics' unification module, the dual-motor system will have the same response performance regarding the input signal. On the other hand, considering that the nonlinear problem of control voltage saturation will also cause the asynchronization problem of the dual-motor system, the proposed method involves an adaptive fractional-order anti-saturation module to prevent voltage saturation and eliminate the nonlinear effects. The experimental results verify that the proposed method can accurately avoid the saturation effect and effectively reduce the synchronization error of the dual-motor system, with a root-mean-square synchronization error reduction of 80.974%. Hence, the proposed method provides an effective solution for the high-precision synchronous motion of a dual-motor system.

Keywords: dual-motor system; synchronous motion; fractional-order frequency-domain control theory; saturation effect

Citation: Zhong, Y.; Gao, J.; Zhang, L. Adaptive Fractional-Order Anti-Saturation Synchronous Control for Dual-Motor Systems. *Appl. Sci.* **2023**, *13*, 2307. https://doi.org/10.3390/app13042307

Academic Editors: Kan Liu and Wei Hu

Received: 4 January 2023
Revised: 6 February 2023
Accepted: 8 February 2023
Published: 10 February 2023

1. Introduction

With the development of power electronics technology and high-performance permanent magnet materials, the permanent magnet linear synchronous motor (PMLSM) has gained significant attention. Because its structure has no mechanical friction problem, a large thrust force, and small volume, the PMLSM is widely used in transportation systems, machine tools, precision positioning platforms, and other fields [1,2]. In addition, the market demand for chip semiconductors has increased in recent years, leading to the rapid development of the semiconductor manufacturing industry; therefore, the high-speed and high-acceleration motion platform built by the PMLSM has gradually become an essential part of the equipment [3]. However, the performance requirements are becoming increasingly stringent, and the single-motor driving method can no longer meet the high-acceleration requirement. To overcome the thrust force limit of the PMLSM and further improve the efficiency of semiconductor manufacturing equipment, a dual-motor driving method has been designed and widely used [4,5]. A redundant direct-drive gantry platform is a common application of the dual-motor driving method. The Y-degree-of-motion freedom is driven by two PMLSMs simultaneously, and the rigid connection between the PMLSMs ensures that the thrust force can act on the beam together, realizing high speed, high acceleration, and a large stroke motion [6].

A dual-actuator system can usually achieve a higher acceleration motion, but it also introduces the asynchronous motion problem, which will cause tension in the connection structure and decrease the service life or even seriously damage the mechanical structure [7]. Scholars have proposed a variety of control methods for the synchronous motion of a dual-motor system, including the series synchronous control method [8], parallel synchro-nous control method [9], and cross-coupling synchronous control method [10]. Because the cross-coupling synchronous control method can adjust the control voltage of each motor in real time according to the synchronization error, it has certain advantages in ensuring the synchronous motion of the dual-motor system [11,12]. The control structure of the cross-coupling synchronous control method is simple; therefore, the accuracy of the control parameter selection is the key factor that determines the synchronous motion performance of the dual-motor system. Some scholars improved the synchronous control method by considering cross-coupling, such as the variable-gain cross-coupling synchronous control method [13], fuzzy neural network cross-coupling synchronous control method [14], and self-tuning cross-coupling synchro-nous control method [15]. These methods can improve the real-time adjustment ability of the parameters of the cross-coupling controller, thereby ensuring the adaptability of the control parameters under different synchronization errors. However, accurately establishing the parameter adjustment law of the cross-coupling synchronous control method requires long-term pre-learning, and the control parameter adjustment algorithm has a high degree of complexity, making it difficult for practical applications. Considering that the characteristics of the PMLSM system can be obtained through system identification, several methods were studied, for example, a model-based feed-back-feedforward decoupling control method [16], model-based adaptive synchronous control method [17], robust immersion and invariance adaptive coordinated control method [18], adaptive thrust-allocation based synchronous control method [19], and other related methods [20–22]. The introduction of model information can reduce the design complexity of the parameter adjustment method and provide a basis for control parameter selection, thus effectively suppressing the dual-motor system's synchronization error. However, these synchronous control methods adjust the control voltage of the PMLSM through position compensation or force compensation when synchronization error is observed. Since there is a time delay, the methods will have difficulty eliminating the synchronization error completely.

Through investigation, we found that the synchronization error is caused by the characteristic difference in the dynamic response of the dual-motor system, and, because the fractional-order deferential operator can accurately describe the control frequency-domain characteristics required by the controlled object, there are several control methods based on fractional-order control theory [23–25]. In literature [26], the authors developed a feedforward control method for PMLSM based on the fractional-order control theory. Through the accurate description of the system's control frequency-domain characteristics, the fractional-order feedforward control method improved the PMLSM tracking accuracy effectively for a single-motor system. However, for a dual-motor system, besides the tracking accuracy of each motor, the synchronization accuracy of the two motors is the key issue and needs to be further studied. Therefore, considering the advantage of the fractional-order control theory, this paper proposes an adaptive fractional-order anti-saturation synchronous control method to improve the synchronization accuracy of the dual-motor system. Through compensation, we unified the frequency-domain characteristics of the motors in a dual-motor system to ensure they have the same dynamic response characteristics. With the design of the frequency-domain characteristics' unification module (FDC-UM), the dynamic response characteristics of the two motors were adjusted to be consistent. Then, the adaptive fractional-order anti-saturation module (AFOAM) was designed to avoid the saturation effect of the control voltage. By calculating the theoretical peak voltage, the parameters of the AFOAM were determined by the relationship between the parameter selection and the FDC adjustment of the fractional-order lead-lag controller (FOLLC). Through the AFOAM, the dual-motor system can prevent the control voltage from reaching the voltage limitation and thus guarantee the suppression of the FDC-UM for the synchronization error. Therefore, the proposed adaptive fractional-order anti-saturation synchronous method can actively suppress the generation of the synchronization error and avoid the saturation effect, ensuring a dual-motor system with high-precision synchronous motion performance.

The remainder of this paper is organized as follows. Section 2 introduces the kinetic model of the dual-motor system and the proposed adaptive fractional-order anti-saturation synchronous control method in detail. In Section 3, based on the established relationship between the FDC and the dynamic response characteristics of the motor systems, the FDC-UM and the AFOAM are designed and finalized with the required parameters. In Section 4, the experimental setup is described, and experimental work is performed to verify the effectiveness and superiority of the proposed method. Section 5 presents the discussion, and Section 6 presents the conclusions.

2. The Proposed Synchronous Control Method for the Dual-Motor System

2.1. Kinetic Model of the Dual-Motor System

A schematic of the redundant direct-drive gantry platform is shown in Figure 1. A cross beam was installed on two linear guides arranged in parallel and jointly driven by two PMLSMs. Motors Y_1 and Y_2 were rigidly connected to the beam, and the stators were installed on the base where the two linear guides were located. For the sake of simplicity, this study only focused on the synchronous control of the *Y*-axis, and it is assumed that the motor X is rigidly connected to the cross-beam. Therefore, from Figure 1 and the dynamic analysis in literature [27], the kinetic model of the dual-motor system can be obtained as follows:

$$\begin{cases} M_1\ddot{y}_1 + c_1\dot{y}_1 + k_1\xi_1 = F_{m1} - F_{r1} - F_{c1} \\ M_2\ddot{y}_2 + c_2\dot{y}_2 + k_2\xi_2 = F_{m2} - F_{r2} + F_{c2} \end{cases}, \tag{1}$$

where $F_{ci} = f\,(y_1, y_2)$ is the coupling force, which is related to the displacements of motors Y_1 and Y_2, and i = 1, 2. F_{mi} is the thrust force, F_{ri} is the friction, M_i is the mass, c_i is the viscosity coefficient, k_i is the elasticity coefficient, y_i is the displacement of the motor, and ξ_i is the deformation between the mover and slider. The control method typically adopted by a dual-motor system is also shown in Figure 1. Each motor has an independent feedback controller (FBC); by designing the reasonable friction compensator (FC) and the synchronous control method, the synchronous motion performance of the dual-motor system is guaranteed.

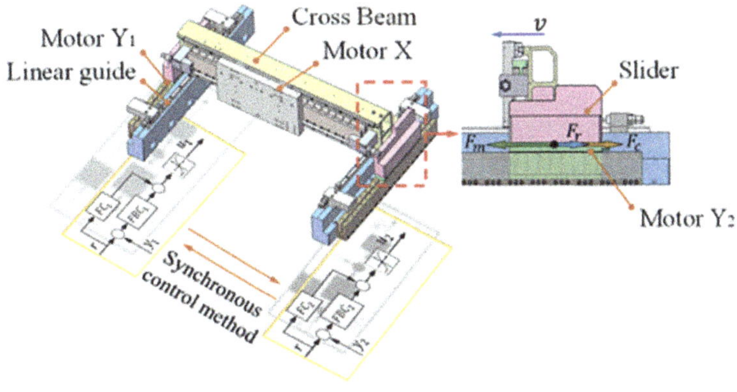

Figure 1. Schematic of the structure and control strategy of redundant direct-drive gantry platform.

The relationship between the motor thrust force and control system characteristics can be obtained as follows:

$$\begin{cases} F_{m1} = K_{u1}u_1 = K_{u1}sat\{u_{FC1} + u_{FBC1}\} \\ F_{m2} = K_{u2}u_2 = K_{u2}sat\{u_{FC2} + u_{FBC2}\} \end{cases}, \tag{2}$$

where K_{ui} is the thrust force constant of the motor, *sat* {} is the saturation function, and u_{FCi} and u_{FBCi} are the control voltages of FC and FBC, respectively. Because the coupling force is related to the displacements of motors Y_1 and Y_2, it appears with the generation of the synchronization error. When the synchronization error is eliminated, the influence of the coupling force on the motor movement will also be negligible, so the kinetic model can be re-expressed as follows:

$$\begin{cases} M_1\ddot{y}_1 + c_1\dot{y}_1 + k_1\xi_1 = F_{m1} \\ M_2\ddot{y}_2 + c_2\dot{y}_2 + k_2\xi_2 = F_{m2} \end{cases}. \tag{3}$$

We can establish the following transfer function of each motor system according to Equation (3):

$$g_{yi}(s) = \frac{b_{0i}s^{m_i} + b_{1i}s^{(m-1)_i} + \cdots + b_{mi}}{s^{n_i} + a_{1i}s^{(n-1)_i} + \cdots + a_{ni}}, \tag{4}$$

where a_{ni} and b_{mi} are the parameters of the denominator and numerator polynomials, respectively, and n_i and m_i are the orders of the denominator and numerator polynomials, respectively. By designing a proper feedback controller $c_i(s)$, the transfer function between the input signal r and output signal y_i is as follows:

$$G_i(s) = \frac{y_i(s)}{r(s)} = \frac{c_i(s)g_{yi}(s)}{1 + c_i(s)g_{yi}(s)}. \tag{5}$$

From Equation (5), we can calculate the amplitude and phase characteristics of each closed-loop feedback control loop in the dual-motor system, and these characteristics determine the dynamic response characteristics of the motor system. Therefore, we can know when the dual-motor system is in a synchronous-state, and the motion of the motor system is mainly affected by its electric characteristic, structure, and control system, leading to the differences in the dynamic response characteristics between the motor systems.

2.2. Description of the Proposed Method

To make the dual-motor system have consistent dynamic response characteristics and realize the two motors' synchronous motion, we proposed a frequency-domain characteristic compensation-based adaptive fractional-order anti-saturation synchronous control method, which contains two modules: the FDC unification module (FDC-UM), which can adjust the dynamic response characteristics of the two motors to be consistent, and the adaptive fractional-order anti-saturation module (AFOAM), which can avoid the saturation effect of the control voltage adaptively. Figure 2 illustrates the working principle of the proposed synchronous control method. The FDC-UM performs the FDC unification through compensation in the feedback control loops. With the FDC-UM, the FDC of the dual-motor system can be unified. We have

$$\begin{cases} A_1 = A_2 = A_f \\ \varphi_1 = \varphi_2 = \varphi_f \end{cases}, \tag{6}$$

where A_1, A_2 are the amplitude characteristics of the Equation (5), and φ_1, φ_2 are the phase characteristics. A_f, φ_f are the amplitude and phase characteristics unified by the FDC-UM. In addition to the FDC-UM, the AFOAM is designed to prevent the peak voltage of the motion from reaching the voltage limitation. With these modules, the amplitude and phase characteristics are determined to be A_a and φ_a according to the input signal spectrum and theoretical calculation of the peak voltage. With the FDC unification and the elimination of the nonlinear saturation effect, the dual-motor system will possess consistent dynamic response characteristics for the two motors and thus ensure their synchronous motion.

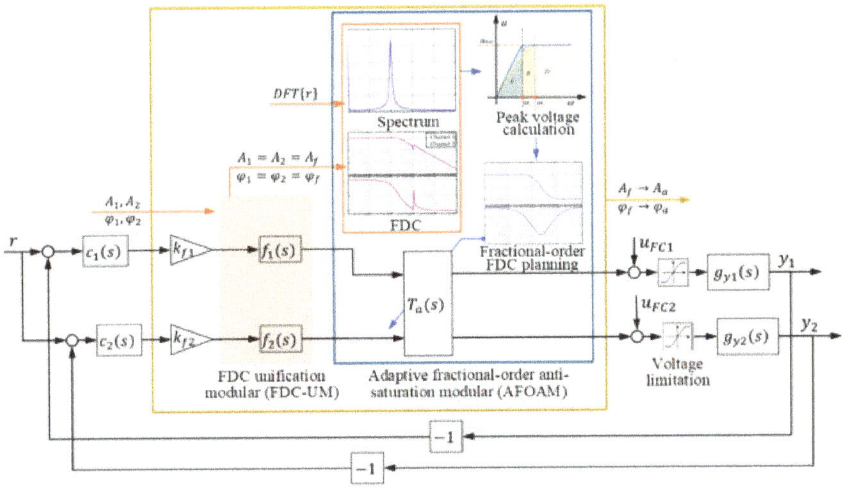

Figure 2. The transfer block diagram of the proposed adaptive fractional-order anti-saturation synchronous control method.

3. Adaptive Fractional-Order Anti-Saturation Synchronous Control Method

3.1. FDC Unification Module (FDC-UM)

According to the frequency-domain control theory, the dynamic response characteristics of the LTI system are determined by the FDC of the control system. In literature [28], the relationship between the input and output signals of the control system was analyzed in detail, and the output signal corresponding to the input signal $r = \sin(\omega_r t)$ was obtained as follows:

$$y = K\sin(\omega_r(t - \Delta t)). \tag{7}$$

The signals have the same frequency ($f = \omega_r/2\pi$), but there is a certain time delay Δt_i and an amplitude ratio K_i between them. Therefore, the synchronization error expression of the dual-motor system can be defined as follows:

$$e_{syn} = K_1 \sin(\omega_r(t - \Delta t_1)) - K_2 \sin(\omega_r(t - \Delta t_2)), \tag{8}$$

$$\Delta t_i = \varphi_i/\omega_r, \tag{9}$$

$$K_i = 10^{A_i/20}, \tag{10}$$

where $\varphi_i(\omega) = \arctan(\mathrm{Im}\,[G_i(\omega)]/\mathrm{Re}\,[G_i(\omega)])$ and $A_i(\omega) = 20\log_{10}|G_i(\omega)|$ are the phase and amplitude characteristics of the transfer function shown in Equation (5), respectively. According to Equations (8)–(10), to achieve high-synchronous motion performance of the dual-motor system, both feedback control systems must have consistent FDC. That is, the dual-motor system must satisfy the following condition:

$$C(s)F(s)G_Y(s)K_F = 0, \tag{11}$$

where $C(s) = [c_1(s), c_2(s)]$, $G_Y(s) = \mathrm{diag}(g_{y1}(s), g_{y2}(s))$. $F(s) = \mathrm{diag}(f_1(s), f_2(s))$, in which $f_i(s)$ is the transfer function of the FDC-UM. $K_F = [k_{f1}, -k_{f2}]^T$ is the gain matrix used to adjust the control performance of the system. According to Equations (5) and (11), we can derive the following when $\varphi_1 \geq \varphi_2$ in the main frequency domain.

$$\begin{cases} f_1(s) = \left(k_{f1}c_1(s)g_{y1}(s)\right)^{-1} k_{f2}c_2(s)g_{y2}(s) \\ f_2(s) = 1 \end{cases}, \tag{12}$$

When $\varphi_1 < \varphi_2$ in the main frequency domain,

$$\begin{cases} f_1(s) = 1 \\ f_2(s) = \left(k_{f2}c_2(s)g_{y2}(s)\right)^{-1} k_{f1}c_1(s)g_{y1}(s) \end{cases}. \tag{13}$$

After the compensation, the transfer function of the feedback control system is as follows:

$$G_{fi}(s) = \frac{y_i(s)}{r(s)} = \frac{k_{fi}f_i(s)c_i(s)g_{yi}(s)}{1 + k_{fi}f_i(s)c_i(s)g_{yi}(s)}. \tag{14}$$

Based on the analysis performed above, we know that the dynamic response characteristics of the control system are related to the frequency of the control signal. Through compensation, the dual-motor system meets the requirements of $A_{f1}(\omega) = A_{f2}(\omega)$ and $\varphi_{f1}(\omega) = \varphi_{f2}(\omega)$, so the two motor systems possess a unified dynamic response characteristic and thus can achieve a synchronous motion. To verify the effectiveness of the FDC-UM, we used the chirp signal for the experiments. Figure 3 shows the working principle of the FDC-UM and the synchronization error comparison by the control methods of the PID feedback control method (Method 1 in Figure 3a), the force compensation cross-coupling synchronous control method (Method 2 in Figure 3b), and the FDC-UM based synchronous control method (Method 3 in Figure 3c). Figure 3d shows that both Method 2 and Method 3 can ensure excellent synchronous-motion performance of the dual-motor system, and the FDC-UM-based synchronous control method achieves the best performance, which can avoid the effects of control signal variation in frequency.

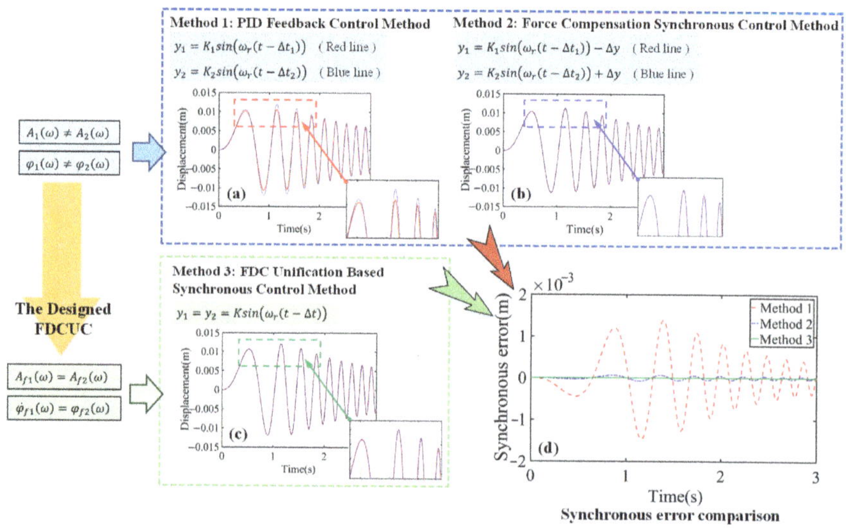

Figure 3. Working principle of the FDC-UM and comparison of synchronization errors with three different control methods: PID feedback method, force compensation method, and the proposed FDC-UM method.

Since the proposed FDC-UM is used for the LTI system, when the control voltage overpasses the voltage limitation, the control voltage will induce a nonlinear behavior and affect the dynamic response characteristics of the dual-motor system, as shown in Equation (15). When the total control voltage of the feedback control system u_{FBC} and the friction compensator u_{FC} are beyond the voltage limitation, the control voltage u is limited to the highest (lowest) voltage u_{max} ($-u_{max}$).

$$u = \begin{cases} -u_{\max}, \ for \ (u_{FBC} + u_{FC}) \le -u_{\max}, \\ u_{FBC} + u_{FC}, \\ u_{\max}, \ for \ (u_{FBC} + u_{FC}) \ge u_{\max}. \end{cases}, \quad (15)$$

Figure 4 shows the influence of the saturation effect on the synchronization error of the dual-motor system. In Figure 4a, the synchronization errors of the dual-motor system are compared with the PID feedback control method and the FDC-UM-based synchronous control method. Figure 4b shows the control voltage of the motor system in the range of $[-u_{max}, u_{max}]$. We can see that the saturation effect influences the synchronization errors of the two control methods and deteriorates the performance of the FDC-UM. Therefore, to ensure the synchronous motion performance of the dual-motor system, the saturation effect needs to be eliminated.

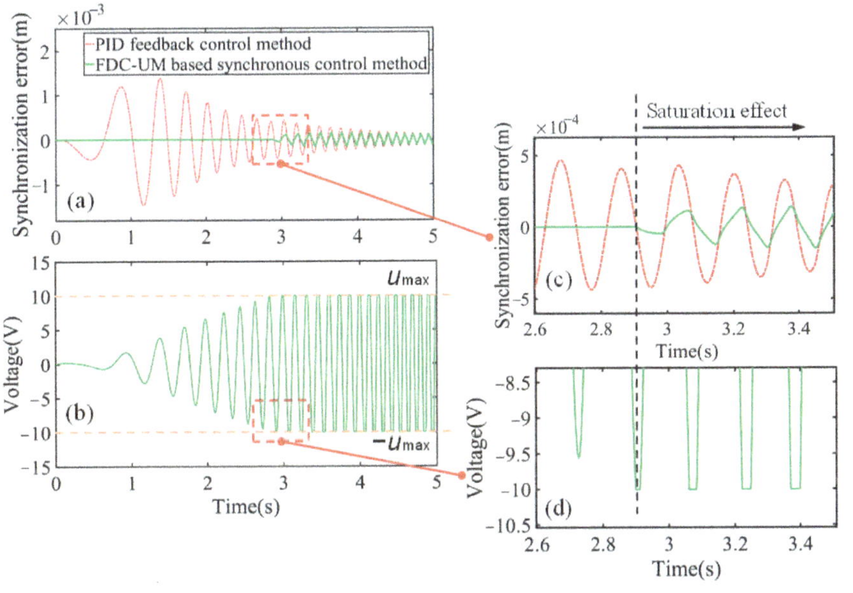

Figure 4. The saturation effect of the control voltage on the synchronization error of the dual-motor system: (**a**) under different control methods; (**b**) control voltage of the motor system in the range of $[-u_{max}, u_{max}]$; (**c**) partial enlarged view of (**a**); (**d**) partial enlarged view of (**b**).

3.2. Adaptive Fractional-Order Anti-Saturation Module (AFOAM)

3.2.1. Peak Voltage Calculation and Amplitude Characteristic Adjustment Value Determination

To avoid the nonlinear problem caused by the saturation effect, it is necessary to recognize the determinants of the control voltage in the feedback control loop and calculate the required suppression value of the control voltage. According to the dynamic response characteristics of the control system and Equations (4) and (14), the transfer function from the input signal r to the feedback control voltage u_{FBCi} is as follows:

$$G_{ui}(s) = \frac{u_{FBCi}(s)}{r(s)} = \frac{k_{fi}f_i(s)c_i(s)}{1 + k_{fi}f_i(s)c_i(s)g_{yi}(s)}. \tag{16}$$

In the frequency domain, $s = j\omega$, so the amplitude characteristic $A_{ui}(\omega)$ of Equation (16) is $20\log_{10}(|G_{ui}(\omega)|)$. Because the input signal amplitude in the frequency domain can be obtained by spectrum analysis, the calculation expression of the theoretical peak voltage of the control system can be expressed as

$$u_i^{\max} = |s_p(\omega_k)|10^{A_{ui}(\omega_k)/20} + u_{FCi}, \tag{17}$$

where $|s_p(\omega_k)|$ is the amplitude spectrum of the input signal when $\omega = \omega_k$, and u_{FCi} is the friction compensation voltage. Therefore, the amount by which the theoretical peak voltage exceeds the voltage limitation can be calculated as follows:

$$\widetilde{u}_i = \mathrm{sgn}_i(u)|u_i^{\max} - u_{\max}|, \tag{18}$$

in which the symbolic function for determining whether the theoretical peak voltage exceeds the voltage limitation is defined as follows:

$$\mathrm{sgn}_i(u) = \begin{cases} 1, \ for \ u_i^{\max} > u_{\max}, \\ 0, \ for \ u_i^{\max} \leq u_{\max}. \end{cases} \tag{19}$$

According to Equation (17), we know that the control voltage can be changed by adjusting the spectrum of the input signal or the amplitude characteristic of the control system. Since the main work of this paper is the design of the synchronous control system, the adjustment of the input signal will directly affect the trajectory of the dual-motor system. Therefore, in order to avoid the saturation effect, the amplitude characteristic $A_{ui}(\omega)$ at ω_k must be adjusted to be the value as shown in Equation (20).

$$\overline{A}_{ui}(\omega_k) = 20\log_{10}\left|10^{A_{ui}(\omega_k)/20} - \widetilde{u}_i/|s_p(\omega_k)|\right|. \tag{20}$$

The amplitude characteristic adjustment value of each motor system is

$$\Delta A_i = \overline{A}_{ui}(\omega_k) - A_{ui}(\omega_k). \tag{21}$$

To ensure that the adjustment of the amplitude characteristic does not affect the unification of the dynamic response characteristics of the dual-motor system, it is necessary to unify the adjustment value to a smaller value ΔA, so that the anti-saturation module can satisfy the adjustment requirements of both feedback control loops simultaneously. The amplitude characteristic adjustment value can be determined as follows:

$$\Delta A = -MAX\{|\Delta A_1|, |\Delta A_2|\}. \tag{22}$$

3.2.2. AFOAM Design and Implementation

A. AFOAM design

Combined with the above theoretical peak voltage calculation, we can obtain the saturation situation of the control voltage by analyzing the input signal when the frequency domain characteristics of the dual-motor system are determined. On this basis, we developed an AFOAM based on fractional-order frequency-domain control theory to realize the specific amplitude characteristic adjustment in which the saturation status of the control voltage is determined by analyzing the input signal when the frequency-domain characteristics of the dual-motor system are determined. The implementation block diagram of AFOAM is shown in Figure 5. When the input signal is determined, the controller calculates the peak voltage according to Equation (17) to adjust the control system to avoid the saturation effect, i.e., not adjust the control parameters in real time during motion.

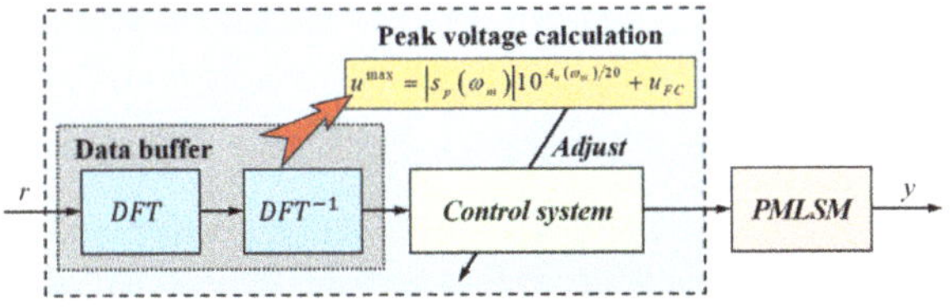

Figure 5. Implementation block diagram of AFOAM.

As shown in Equation (20), we only need to adjust the amplitude characteristic at ω_k. Since the adjustment at a specific frequency will affect the amplitude and phase characteristics of the surrounding frequency and will further influence the dynamic response performance of the feedback control system, we propose the AFOAM in this paper to adjust the specific FDC accurately and flexibly by introducing the fractional-order lead-lag controller (FOLLC) as the fractional-order FDC planning target. The expression for the FOLLC is described as follows:

$$C_f(s) = K_C\left(\frac{1+\lambda x s^{\alpha}}{1+\lambda s^{\alpha}}\right), \tag{23}$$

where K_C, λ, and x are coefficients. The order α can be any real number in the range (0, 2), which enables the FOLLC to accurately describe the specified FDC at the specified frequency through accurate parameter selection. The transfer function from the input signal r to the feedback control voltage u_{FBCi} with the AFOAM is as follows:

$$\overline{G}_{ui}(s) = \frac{k_{fi}f_i(s)c_i(s)T_a(s)}{1+k_{fi}f_i(s)c_i(s)g_{yi}(s)T_a(s)}, \tag{24}$$

where $T_a(s)$ represents the transfer function of the AFOAM. With the AFOAM, the voltage amplitude characteristic of the feedback control system is changed as

$$\overline{A}_{ui}(\omega) = A_{ui}(\omega) + A_f(\omega), \tag{25}$$

where $A_f(\omega)$ is the amplitude characteristic of the FOLLC. According to the transfer block diagram of the proposed adaptive fractional-order anti-saturation synchronous control method, the transfer function of AFOAM can be expressed as follows:

$$T_a(s) = \frac{K_C(1+\lambda x s^{\alpha})}{(1+\lambda s^{\alpha} - K_C(1+\lambda x s^{\alpha}))k_{fi}f_i(s)c_i(s)g_{yi}(s) + \lambda s^{\alpha} + 1}. \tag{26}$$

B. Parameter determination

It can be seen from Equation (26) that the parameters of the AFOAM can be determined by selecting the parameters of the FOLLC. Therefore, we first need to establish the relationship between the parameters of the FOLLC and the FDC adjustment. Equation (23) can be rewritten as

$$C_f(j\omega) = K_C\left(\frac{1+x\lambda(j\omega)^{\alpha}}{1+\lambda(j\omega)^{\alpha}}\right). \tag{27}$$

The amplitude and phase characteristics of the FOLLC can be calculated as follows:

$$A_f(\omega) = 20\log_{10}\left(K_C\sqrt{\frac{1+x^2\lambda^2\omega^{2\alpha}+2x\lambda\omega^{\alpha}\cos(\alpha\pi/2)}{1+\lambda^2\omega^{2\alpha}+2\lambda\omega^{\alpha}\cos(\alpha\pi/2)}}\right), \tag{28}$$

$$\varphi_f(\omega) = \tan^{-1}\left[\frac{(x-1)\lambda\omega^{\alpha}\sin(\alpha\pi/2)}{1+(x+1)x\lambda^2\omega^{2\alpha}\cos(\alpha\pi/2)}\right], \tag{29}$$

Let

$$g(\omega) = K_C\sqrt{\frac{1+x^2\lambda^2\omega^{2\alpha}+2x\lambda\omega^{\alpha}\cos(\alpha\pi/2)}{1+\lambda^2\omega^{2\alpha}+2\lambda\omega^{\alpha}\cos(\alpha\pi/2)}}. \tag{30}$$

The logarithmic function is monotonic, and thus, $g(\omega)$ represents the trend of $A_f(\omega)$. Because the FOLLC has equal-order properties, that is, the orders of the differential operator in the numerator and denominator polynomials are the same, we can deduce that the FOLLC has the following properties: (1) $A_f(\omega)|_{\omega\to 0} = 20\log_{10}(K_C)$ and $A_f(\omega)|_{\omega\to+\infty} = 20\log10(K_C x)$; (2) $\varphi_f(\omega)|_{\omega\to 0} = \varphi_f(\omega)|_{\omega\to+\infty} = 0$; (3) when $\omega = \omega_m$, the FOLLC reaches its extreme phase characteristic value. By deriving Equation (29), we obtain

$$\omega_m = \left(\lambda\sqrt{x}\right)^{-1/\alpha}, \tag{31}$$

The parameter λ can be obtained by Equation (32).

$$\lambda = \left(\sqrt{x}\omega_m^{\alpha}\right)^{-1}, \tag{32}$$

Because the middle- and low-frequency domains are the main control-frequency domains of the motor system, the FDC adjustment should avoid excessive influence on the amplitude and phase characteristics in these areas. However, owing to the continuity of the FDC adjustment, this influence cannot be completely avoided. To minimize the influence of the FDC adjustment caused by the AFOAM on the outer area of the specified angular frequency, especially the middle- and low-frequency domains, we need to let $K_C = 1$. According to the above analysis, to minimize the influence range of the FDC adjustment, we determine $A_f(\omega)|_{\omega\to+\infty} = \Delta A$ and $|A_f(\omega_k) + \Delta A| \leq \varepsilon_1$. Because the logarithmic function is a monotonic function, $A_f(\omega_k)$ satisfies $|\dot{g}(\omega_k)| \leq \varepsilon_2$ simultaneously, where ε_1 and ε_2 represent the minimum values. Thus, parameter x can be calculated by Equation (33).

$$x = 10^{\Delta A/20}, \tag{33}$$

To simplify the calculation, we let $A_f(\omega_k) \approx \Delta A$ and substitute it into Equation (28), and then we can obtain the relationship between ω_j and α by the following expression:

$$\omega_j = \omega_k\left(-2\sqrt{x}\cos(\alpha\pi/2)/(x+1)\right)^{2/\alpha}, \tag{34}$$

where ω_j satisfies the condition $\omega_m = (\omega_k\omega_j)^{1/2}$, and $(\omega_k-\omega_j)$ is the range of the FDC adjustment. Since the variables x, ω_k, and ω_j are not less than zero, the order α should be in the range of [1, 2). We can see that when the order $\alpha = 1$, $\omega_j = 0$, which means that the FDC adjustment of the AFOAM will influence the amplitude and phase characteristics at all frequencies. Under this condition, because $\omega_m = (\omega_k\omega_j)^{1/2}$, ω_m will be zero by calculation, so the parameters of the AFOAM cannot be determined uniquely and adaptively. Therefore, to reduce the influence of the FDC adjustment, the order must be in the range of (1, 2).

Through the derivation of Equation (30), we obtain

$$\dot{g}(\omega_k) = \frac{\left(1+\frac{\omega_k^{\alpha}}{\omega_j^{\alpha}}+2\frac{\omega_k^{\alpha/2}}{\sqrt{x}\omega_j^{\alpha/2}}\cos(\alpha\pi/2)\right)^{1/2}\left((x-1)\alpha\frac{\omega_k^{(\alpha/2-1)}}{\sqrt{x}\omega_j^{\alpha/2}}\cos(\alpha\pi/2)+\frac{\omega_k^{\alpha/2}}{\sqrt{x}\omega_j^{\alpha/2}}+\sqrt{x}\frac{\omega_k^{\alpha/2}}{\omega_j^{\alpha/2}}+\frac{\omega_k^{\alpha}}{\omega_j^{\alpha}}\cos(\alpha\pi/2)\right)}{\left(1+x\frac{\omega_k^{\alpha}}{\omega_j^{\alpha}}+2\sqrt{x}\frac{\omega_k^{\alpha/2}}{\omega_j^{\alpha/2}}\cos(\alpha\pi/2)\right)^{1/2}\left(1+\frac{\omega_k^{\alpha}}{x\omega_j^{\alpha}}+2\frac{\omega_k^{\alpha/2}}{\sqrt{x}\omega_j^{\alpha/2}}\cos(\alpha\pi/2)\right)^{2}}, \tag{35}$$

For Equation (35), $|\dot{g}(\omega_k)| \leq \varepsilon_2$, so there is a minimum value ε, making $|\dot{g}(\omega_k) - \varepsilon| = 0$. Combining Equations (34) and (35), and the above conditions, the optimal model for parameter α can be deduced as follows:

$$\begin{cases} \min Z(\alpha) = \omega_k - \omega_k\left(-2\sqrt{x}\cos(\alpha\pi/2)/(x+1)\right)^{2/\alpha} \\ \text{s.t.} \quad 1 < \alpha < 2. \\ \qquad\quad \left|A_f(\omega_k) + \Delta A\right| \leq \varepsilon_1. \\ \qquad\quad \left|\dot{g}(\omega_k) - \varepsilon\right| = 0. \end{cases}, \tag{36}$$

Through Equation (36), parameter α can be solved and determined. Then, parameters λ and ω_m can be calculated by combining Equations (31)–(33), and $\omega_m = (\omega_k\omega_j)^{1/2}$. Therefore, through the establishment of the relationship between the FOLLC and FDC adjustment, the parameters K, x, λ, and α of the AFOAM can be determined properly.

4. Experiment

4.1. Experimental System

Based on the above theoretical analysis and derivation, we implemented the proposed adaptive anti-saturation synchronous control system using a dSPACE prototyping system and conducted experiments on the redundant direct-drive gantry platform to verify the improvement in the synchronous motion performance of the proposed method. Figure 6 shows the experimental setup. The Y-direction motion of the redundant direct-drive gantry platform was driven by AUM3 PMLSMs from Akribis. Each motor was equipped with RGS20-S grating from Renishaw, with a resolution of 0.1 μm.

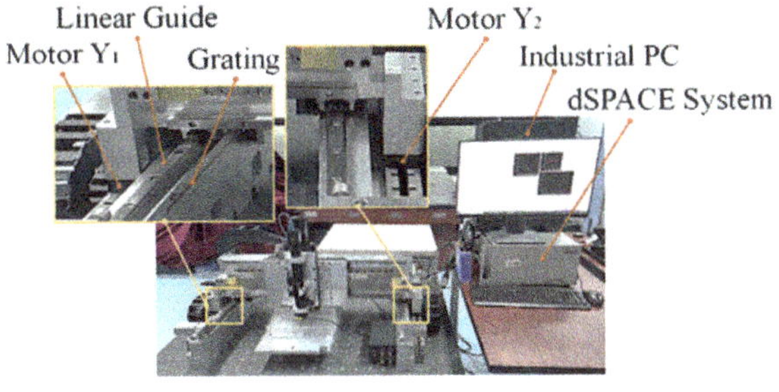

Figure 6. Experimental system.

The beam had a length of 836 mm, and its weight exceeded 25 kg, which caused the synchronous motion performance of the dual-motor system to be significantly affected by the synchronous control method. During the experiments, the dSPACE system controlled the PMLSMs in real time, and the feedback signals were provided by the gratings. By establishing the mathematical model of the motor system and using a frequency sweep experiment to identify the model parameters, the transfer functions of the motor systems were calculated as follows:

$$\begin{cases} g_{y1}(s) = \dfrac{0.9183s^3 + 5.994s^2 + 8959s + 8499}{s^5 + 16.92s^4 + 1.061e04s^3 + 7.916e04s^2 + 5.448e04s} \\ g_{y2}(s) = \dfrac{0.8818s^3 + 8.583s^2 + 8918s + 2.223e04}{s^5 + 12.51s^4 + 1.104e04s^3 + 3.237e04s^2 + 4.215e04s} \end{cases}, \tag{37}$$

The dual-motor system adopted a proportional–integral–derivative (PID) feedback control method. According to the controlled object models shown in Equation (37), the Ziegler–Nichols tuning rules were used to obtain a set of PID parameters that were more consistent with the synchronous-motion requirements of the dual-motor system. The PID controller and its parameters are as follows:

$$\begin{cases} c_1(s) = \left(27.41s^2 + 388.52s + 1.05\right)/s \\ c_2(s) = \left(28.12s^2 + 275.44s + 1.12\right)/s \end{cases}, \tag{38}$$

Figure 10 shows the calculated amplitude spectrum of the motion planning with different accelerations under the same acceleration time, and Figure 11 shows the calculated amplitude spectrum of the motion planning with different velocities under the same acceleration. It can be seen from Figure 10 that, as the acceleration increases, the amplitude spectrum at the non-zero frequency increases, but all accelerations are at a similar frequency. The results in Figure 11 show that, under the same acceleration, with an increase in velocity, the amplitude spectrum at the non-zero frequency increases, but the frequencies are all retained within a small range.

Based on the spectrum analysis of the input signal, the voltage amplitude characteristics of the motor system at the corresponding angular frequency were calculated. Table 1 shows the comparison results of the theoretical and experimental peak voltages under different motion planning conditions. The relative error δ in the table is calculated as follows:

$$\delta = |u_{cal} - u| / u \times 100\%, \tag{44}$$

where u is the experimental peak voltage, and u_{cal} is the calculated theoretical peak voltage. In the experimental results shown in Table 1, the maximum relative error between the experimental and the calculated theoretical peak voltages was 3.86%, which proves the accuracy of the calculation of the theoretical peak voltage; thus, Equation (17) can guarantee the accurate fractional-order FDC planning for the adaptive fractional-order anti-saturation synchronous control method.

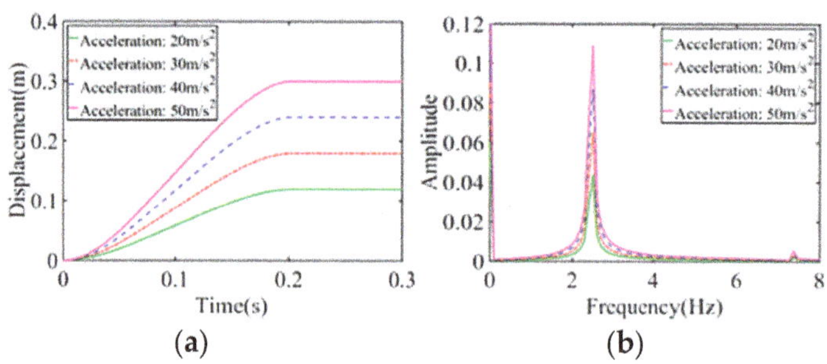

Figure 10. Calculated amplitude spectrum of different accelerations under the same acceleration time (80 ms): (**a**) displacement comparison; (**b**) amplitude spectrum comparison.

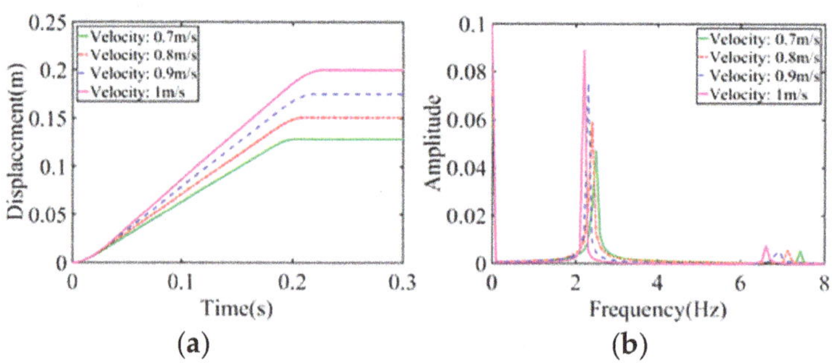

Figure 11. Calculated amplitude spectrum of different velocities under the same acceleration (50 m/s^2): (**a**) displacement comparison; (**b**) amplitude spectrum comparison.

Table 1. Comparison of calculated theoretical and experimental peak voltages.

Motion Planning		Amplitude Spectrum		Voltage Amplitude Characteristic of the Motor Y_1 (dB)	Peak Voltage (V)		δ
Acceleration (m/s^2)	Velocity (m/s)	Frequency (Hz)	Peak Value		Experiment	Calculation	
20	0.8	2.5	0.0435	50.3265	14.8562	14.2828	3.86%
30	1.2	2.5	0.0653	50.3265	22.2843	21.4406	3.79%
40	1.6	2.5	0.0870	50.3265	29.7124	28.5655	3.86%
50	2.0	2.5	0.1088	50.3265	37.1405	35.7233	3.82%

4.2.2. Synchronization Accuracy Verification

To further verify the correctness of the calculation for the proposed adaptive fractional-order anti-saturation synchronous control method and its effectiveness in improving the synchronization accuracy of the dual-motor system, the cubic curve shown in Equation (43) was selected to perform the high-speed high-acceleration motion experiments. According to the above analysis and experiments of the FDC-UM, the parameters of the PID feedback controller selected in the experiments are shown in Equation (38), and those of the friction compensator are shown in Equations (40) and (41). When the dual-motor system was controlled by the proposed synchronous control method, the AFOAM analyzed the spectrum of the input signal and the FDC of the control system and then used Equation (36) to calculate the order of the FOLLC as 1.3190. According to Equations (31)–(33), $\omega_j = 7.1740$, $K_C = 1$, $x = 0.8055$, and $\lambda = 0.1381$. Therefore, using Equation (26), the transfer function of the AFOAM was calculated as Equation (45). According to Equation (45), we could draw the adjustment curve of the voltage amplitude characteristic realized by the AFO-AM, as shown in Figure 12a. An FDC comparison of the motor system with and without the designed AFO-AM is shown in Figure 12b.

The anti-saturation module can avoid synchronization errors owing to the voltage limitation; therefore, compared with the force compensation cross-coupling synchronous control method, the proposed synchronous control method can avoid the artificial secondary parameter adjustment of the control system to achieve high-synchronization accuracy under different motion plannings. Figure 13 shows the experimental results of the synchronous motion of the dual-motor system. Under force compensation cross-coupling synchronous control, both motor systems were affected by long-term saturation, which led to a large synchronization error. However, there was no control voltage saturation under the proposed synchronous control, as shown in Figure 13b, so the synchronization error was reduced obviously, as shown in Figure 13c. Table 2 shows the comparison results of experimental data under three different motion plannings, in which we use the commonly used indicators, such as absolute maximum value (|MAX|), mean absolute error (MAE), and root mean square error (RMSE) to compare the synchronization error under different control methods. Under the proposed adaptive fractional-order anti-saturation synchronous control, the root-mean-square synchronization error of the dual-motor system can be reduced by more than 77.580% compared with that under the force compensation synchronous control. The comparison results of the three sets of experimental data verify that the proposed adaptive fractional-order anti-saturation synchronous control method has obvious advantages in improving the synchronization accuracy of the dual-motor system.

$$T_a(s) = \frac{1 + 0.1112s^{1.3190}}{0.0269s^{1.3190}\left(\frac{25.17s^5 + 521.1s^4 + 2.479e05s^3 + 3.714e06s^2 + 3.311e06s + 8924}{s^6 + 16.92s^5 + 10610s^4 + 79160s^3 + 54480s^2}\right) + 0.1381s^{1.3190} + 1}. \quad (45)$$

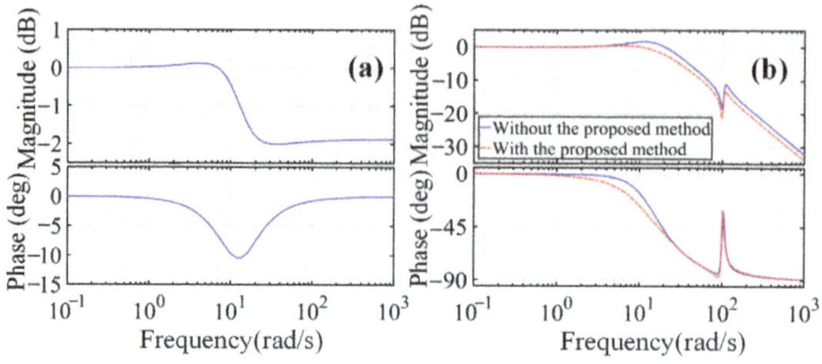

Figure 12. FDC adjustment of the proposed adaptive fractional-order anti-saturation method: (**a**) fractional-order FDC planning target for the voltage amplitude characteristic; (**b**) FDC comparison of the two motor systems with and without the proposed method.

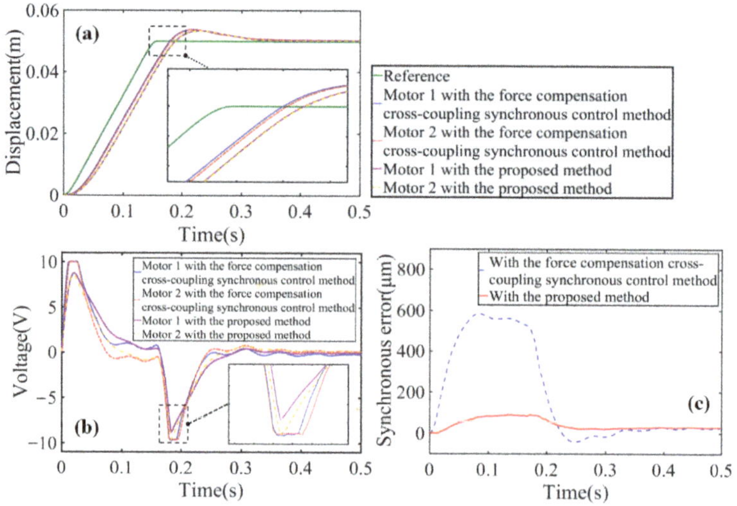

Figure 13. Synchronous motion experimental results of the dual-motor system under different synchronous control methods: (**a**) displacement comparison; (**b**) control voltage comparison ($|u_{max}|$ = 10V); (**c**) synchronization error comparison.

Table 2. Experimental data comparison results of the synchronous motion experiments.

Experiment	Motion Planning			Control Method	Synchronization Error (μm)			RMSE Reduction (%)
	Displacement (m)	Velocity (m/s)	Acceleration (m/s^2)		|MAX|	MAE	RMSE	
1	0.05	0.35	30	Method 2 Proposed method	584.500 89.800	117.259 36.987	221.443 42.132	- 80.974
2	0.05	0.5	30	Method 2 Proposed method	561.600 52.900	80.388 31.436	161.193 32.338	- 79.938
3	0.05	0.5	50	Method 2 Proposed method	769.000 141.400	106.681 40.019	220.616 49.454	- 77.580

Method 2: Force compensation cross-coupling synchronous control method.

5. Discussion

Synchronization error of a dual-motor system will usually lead to degradation of equipment performance. Several researchers focused on the cross-coupling synchronous control method, by reducing the synchronization error from the aspects of the compensation method and the control parameter tunning method, to improve the synchronous motion performance of a dual-motor system [19,30]. However, the cross-coupling synchronous control method is usually used to compensate the control voltage after the synchronization error occurred, so there is a time delay between the generation and the compensation of the synchronization error. As shown in Figure 3, under the control of the chirp signal, although the cross-coupling synchronous control method could greatly reduce the synchronization error of the dual-motor system, it was difficult to eliminate the synchronization error completely due to the constant changes in the amplitude and frequency of the input signal and the inconsistent response characteristics of the dual-motor system. The experimental results showed the relationship between the dynamic response characteristics and the FDC of the motor systems. With the proposed adaptive fractional-order anti-saturation synchronous control method, the FDC of the dual-motor system was adjusted to be consistent to guarantee the same dynamic response output of the dual-motor system. Compared with the cross-coupling synchronous control method, our method can suppress the generation of the synchronization errors of the dual-motor system and eliminate the influence of the saturation effect and thus can achieve a better synchronous-motion control performance, as shown in Table 2. It can be seen from the experimental results in Figure 13 that our method can effectively avoid the non-linear phenomenon of control voltage saturation and ensure the high-precision synchronization accuracy of the dual-motor system. Considering the problem of the hysteresis nonlinearity and gap nonlinearity of the dual-motor system, the synchronous control performance of the proposed method may be affected. Next, we will tackle this problem to further improve the synchronous control method for the dual-motor system.

6. Conclusions

In this study, we proposed an adaptive fractional-order anti-saturation synchronous control method to deal with the asynchronous problem of the dual-motor system. With the elimination of the differences in FDC by the designed FDC-UM, both motor systems possessed consistent dynamic response characteristics. Based on the theoretical peak voltage calculation, we further developed an adaptive fractional-order anti-saturation module (AFOAM) to eliminate the nonlinear saturation effect caused by the control voltage saturation and thus effectively improved the synchronization accuracy of the dual-motor system. The experimental results showed that the maximum relative error between the calculated and experimental peak voltage was only 3.86%, which confirmed the correctness of the theoretical calculation of the peak voltage in this paper. On this basis, the proposed adaptive fractional-order anti-saturation synchronous control method could prevent the control voltage from reaching the limitation and reduce the root-mean-square synchronization error from 221.443 μm to 42.132 μm, with a reduction of 80.974%. The experimental results demonstrated that the proposed method can effectively suppress the inconsistent characteristics of the dual-motor system and avoid the nonlinear saturation effect. Therefore, the proposed method can effectively improve the synchronization accuracy of the dual-motor system.

Author Contributions: Writing–original draft, Y.Z.; Writing—review & editing, J.G. and L.Z. All authors have read and agreed to the published version of the manuscript.

Funding: This work was supported by the National Natural Science Foundation of China under Grant No. 52075106, No. 51905108, and No. U20A6004.

Conflicts of Interest: The authors declare no conflict of interest.

References

1. Tan, L.; Gao, J.; Luo, Y. Super-twisting sliding mode control with defined boundary layer for chattering reduction of permanent magnet linear synchronous motor. *J. Mech. Sci. Technol.* **2021**, *35*, 1829–1840. [CrossRef]
2. Yu, W.; Yang, G.; Li, Z. A tubular linear motor structure suitable for large thrust. *J. Mech. Sci. Technol.* **2021**, *35*, 4987–4995. [CrossRef]
3. Liu, W.; Shu, F.; Xu, Y. Iterative learning based neural network sliding mode control for repetitive tasks: With application to a PMLSM with uncertainties and external disturbances. *Mech. Syst. Signal Process.* **2022**, *172*, 108950. [CrossRef]
4. Chung, S.-U.; Kim, J.-W.; Woo, B.-C. Design and experimental validation of doubly salient permanent magnet linear synchronous motor for precision position control. *Mechatronics* **2013**, *23*, 172–181. [CrossRef]
5. Chen, S.-Y.; Chen, C.-S.; Yang, Z.-W. Self-tuning cross-coupled two degree-of-freedom PID control for position synchronization of dual linear motors. *Appl. Math. Model.* **2018**, *64*, 214–234. [CrossRef]
6. Chen, R.; Yan, L.; Jiao, Z. Dynamic modeling and analysis of flexible H-type gantry stage. *J. Sound Vib.* **2019**, *439*, 144–155. [CrossRef]
7. Nguyen, V.H.; Kim, W.-J. Error analysis to minimize cross-axis couplings in 6-DOF motion systems with a single moving part. *Precis. Eng.* **2020**, *63*, 49–61. [CrossRef]
8. Sarachik, P.; Ragazzini, J.R. A 2-dimensional feedback control system. In *Transactions of the American Institute of Electrical Engineers, Part II: Applications and Industry*; IEEE: New York, NY, USA, 1957; Volume 76, pp. 55–61.
9. Koren, Y. Cross-Coupled Biaxial Computer Control for Manufacturing Systems. *J. Dyn. Syst. Meas. Control* **1980**, *102*, 265–272. [CrossRef]
10. Huo, F.; Poo, A.-N. improving contouring accuracy by using generalized cross-coupled control. *Int. J. Mach. Tools Manuf.* **2012**, *63*, 49–57. [CrossRef]
11. Chen, W.; Wang, D.; Geng, Q. Robust adaptive cross-coupling position control of biaxial motion system. *Sci. China Technol. Sci.* **2016**, *59*, 680–688. [CrossRef]
12. Zhu, X.; Yao, B.; Tao, G.; Wang, Q.; Cao, J. Adaptive robust synchronous control of a individual metering dual-cylinder pneumatic system with composite parallel method. In Proceedings of the 2014 IEEE/ASME International Conference on Advanced Intelligent Mechatronics (AIM), Besançon, France, 8–11 July 2014; pp. 304–309.
13. Koren, Y.; Lo, C.C. Variable-Gain Cross-Coupling Controller for Contouring. *CIRP Ann.* **1991**, *40*, 371–374. [CrossRef]
14. Chou, P.-H.; Chen, C.-S.; Lin, F.-J. DSP-based synchronous control of dual linear motors via Sugeno type fuzzy neural network compensator. *J. Frankl. Inst.* **2012**, *349*, 792–812. [CrossRef]
15. Huang, Z.; Song, G.; Li, Y. Synchronous control of two counter-rotating eccentric rotors in nonlinear coupling vibration system. *Mech. Syst. Signal Process.* **2019**, *114*, 68–83. [CrossRef]
16. García-Herreros, I.; Kestelyn, X.; Gomand, J. Model-based decoupling control method for dual-drive gantry stages: A case study with experimental validations. *Control Eng. Pract.* **2013**, *21*, 298–307. [CrossRef]
17. Teo, C.S.; Tan, K.K.; Lim, S.Y. Dynamic modeling and adaptive control of a H-type gantry stage. *Mechatronics* **2007**, *17*, 361–367. [CrossRef]
18. Wang, K.; Liu, X.; Su, H. Robust immersion and invariance adaptive coordinated control for a class of biaxial gantry systems. *J. Control Decis.* **2021**, *9*, 289–300. [CrossRef]
19. Li, C.; Li, C.; Chen, Z. Adaptive thrust allocation based synchronization control of a dual drive gantry stage. *Mechatronics* **2018**, *54*, 68–77. [CrossRef]
20. Gordon, D.J.; Erkorkmaz, K. Precision control of a T-type gantry using sensor/actuator averaging and active vibration damping. *Precis. Eng.* **2012**, *36*, 299–314. [CrossRef]
21. Wang, Z.; Hu, C.; Zhu, Y. Trajectory modification method based on frequency domain analysis for precision contouring motion control systems. *Mech. Syst. Signal Process.* **2021**, *158*, 107646. [CrossRef]
22. Zou, M.; Fang, P.; Peng, H. Study on synchronization characteristics for self-synchronous vibration system with dual-frequency and dual-motor excitation. *J. Mech. Sci. Technol.* **2019**, *33*, 1065–1078. [CrossRef]
23. Nirmala, R.J.; Balachandran, K.; Trujillo, J.J. Null controllability of fractional dynamical systems with constrained control. *Fract. Calc. Appl. Anal.* **2017**, *20*, 553–565. [CrossRef]
24. Yaghooti, B.; Hosseinzadeh, M.; Sinopoli, B. Constrained control of semilinear fractional-order systems: Application in drug delivery systems. In Proceedings of the 2020 IEEE Conference on Control Technology and Applications (CCTA), Montreal, QC, Canada, 24–26 August 2020; pp. 833–838.
25. Rhouma, A.; Bouani, F.; Bouzouita, B. Model predictive control of fractional order systems. *J. Comput. Nonlinear Dyn.* **2014**, *9*, 031011. [CrossRef] [PubMed]
26. Zhong, Y.; Gao, J.; Zhang, L. Fractional-order feedforward control method for permanent magnet linear synchronous motor based on frequency-domain adjustment theory. *Mech. Syst. Signal Process.* **2023**, *190*, 110115. [CrossRef]
27. Dai, L.; Li, X.; Zhu, Y. The Generation Mechanism of Tracking Error During Acceleration or Deceleration Phase in Ultraprecision Motion Systems. *IEEE Trans. Ind. Electron.* **2019**, *66*, 7109–7119. [CrossRef]
28. Tao, J.; Mercan, O. A study on a benchmark control problem for real-time hybrid simulation with a tracking error-based adaptive compensator combined with a supplementary proportional-integral-derivative controller. *Mech. Syst. Signal Process.* **2019**, *134*, 106346. [CrossRef]

29. Yoon, J.Y.; Trumper, D.L. Friction microdynamics in the time and frequency domains: Tutorial on frictional hysteresis and resonance in precision motion systems. *Precis. Eng.* **2019**, *55*, 101–109. [CrossRef]
30. Deng, Z.; Shang, J.; Nian, X. Synchronization controller design of two coupling permanent magnet synchronous motors system with nonlinear constraints. *ISA Trans.* **2015**, *59*, 243–255. [CrossRef]

applied sciences

MDPI

Article

Study on Stator-Rotor Misalignment in Modular Permanent Magnet Synchronous Machines with Different Slot/Pole Combinations

Danilo Riquelme [1,*], Carlos Madariaga [2], Werner Jara [1], Gerd Bramerdorfer [3], Juan A. Tapia [2] and Javier Riedemann [4]

[1] School of Electrical Engineering, Pontificia Universidad Católica de Valparaíso, Valparaíso 2362804, Chile
[2] Department of Electrical Engineering, University of Concepcion, Concepcion 4070386, Chile
[3] Department of Electrical Drives and Power Electronics, Johannes Kepler University Linz, 4040 Linz, Austria
[4] Department of Electronic and Electrical Engineering, University of Sheffield, Sheffield S1 3JD, UK
* Correspondence: danilo.riquelme.s@mail.pucv.cl

Featured Application: The results of this work can be included in the design stage of modular permanent magnet synchronous machines for fault-tolerant applications.

Abstract: Addressing stator-rotor misalignment, usually called eccentricity, is critical in permanent magnet (PM) machines since significantly high radial forces can be developed on the bearings, which can trigger a major fault and compromise the structural integrity of the machine. In this regard, this paper aims to provide insight into the unaddressed identification and analysis of the impact of eccentric tolerances on relevant performance indices of permanent magnet synchronous machines (PMSMs) with modular stator core. Static and dynamic eccentricity are assessed for different slot/pole combinations through the finite element method (FEM), and the results are compared with those of PMSMs with a conventional stator core. The unbalanced magnetic forces (UMF), cogging torque, back-emf, and mean torque variations are described and related to the eccentricity magnitude and classification. The main findings indicate that severe radial forces and significant additional cogging torque harmonics are generated because of eccentricity. Additionally, it is found that the main differences between modular PMSMs and conventional PMSMs rely on the value of slots per pole per phase.

Keywords: cogging torque; dynamic eccentricity; finite element analysis; modular stator; permanent magnet; static eccentricity; stator-rotor misalignment; tolerance analysis

Citation: Riquelme, D.; Madariaga, C.; Jara, W.; Bramerdorfer, G.; Tapia, J.A.; Riedemann, J. Study on Stator-Rotor Misalignment in Modular Permanent Magnet Synchronous Machines with Different Slot/Pole Combinations. *Appl. Sci.* **2023**, *13*, 2777. https://doi.org/10.3390/app13052777

Academic Editors: Kan Liu and Wei Hu

Received: 17 January 2023
Revised: 13 February 2023
Accepted: 17 February 2023
Published: 21 February 2023

1. Introduction

In safe-critical applications, the uninterrupted and reliable operation of the electrical machine is requisite, which is described by their fault-tolerance capability. This is the case, for instance, of aeronautical, aerospace, and electromobility applications [1–3]. As a concept, fault-tolerant machines should provide acceptable levels of electric, magnetic, physical, and thermal isolation of phase windings [3], and can also consider multiphase windings [4,5] or limiting short-circuit current capability [6,7].

In addition to the fault-tolerance capability, electrical machines must provide enough power density, efficiency, and electromagnetic performance to meet the application's requirements. In this regard, electrical machines based on permanent magnets (PMs) are well known as the most promising technology in terms of power density and efficiency [8–11], but they lack inherent fault-tolerance capability [12]. It is desirable that PMSMs resist winding open-circuit and short-circuit faults without compromising the integrity of magnets. Nevertheless, since their electromagnetic performance is unmatchable by other conventional topologies so far, considerable efforts have been carried out in the last few years to enhance the fault-tolerant capability of PM machines [2–5,7,12–14]. In the technical literature, adopting a modular stator core, as illustrated in Figure 1b, has been shown to be a feasible solution to enhance the fault-tolerant capability of PM machines [15–23]. This obeys mechanical and electromagnetic reasons, since PM synchronous machines with modular stator core (MPMSMs) offer several advantages in high-power density and fault-tolerance applications [15]: they can provide physical, thermal, and electromagnetic insulation between phases, lower material requirements (laminations) than conventional topologies and a high slot fill factor [16]. Notwithstanding, the cost of MPMSMs developing these features is the presence of manufacturing and assembly challenges, decrease in mean torque, and increase in torque ripple [15,18,19]. As can be seen from Figure 1b, the stator structure in MPMSM is segmented and, therefore, the attachment of the modules to the stator body can provide the design with additional manufacturing and assemblies tolerances (uncertainties) when compared with conventional PMSMs (Figure 1a).

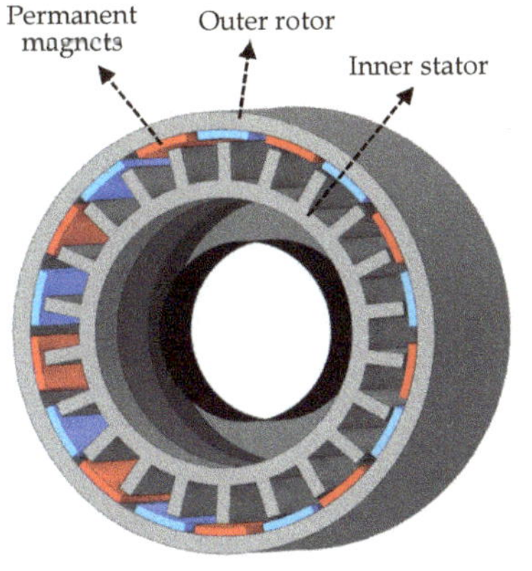

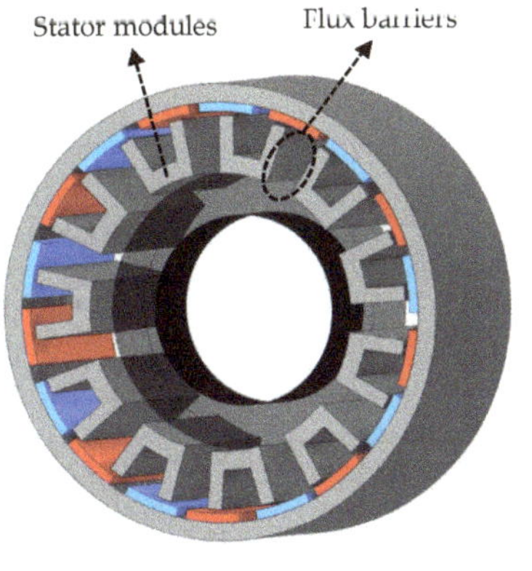

Figure 1. Three-dimensional representation of stator/rotor structure of a 24-slot, 20-pole PMSM: (**a**) conventional PMSM with non-segmented stator (monolithic), (**b**) PMSM with modular stator core (MPMSM).

Recent studies have approached the analysis of some of these uncertainties, giving insight into the effect of design parameters and dimensional tolerances on the machine performance [15,17,18]. Although these studies are relevant for describing the impact of dimensional tolerances on the performance of MPMSMs and provide guidelines for devising more robust designs, there is a critical uncovered aspect that compromises the reliability and continuous operation of these machines: stator-rotor misalignment.

Stator-rotor misalignment, which can be classified into static eccentricity or dynamic eccentricity depending on the dynamics of the misalignment [24,25], wears down the bearings since it generates vibration and pulsing forces on their structure [26,27] and can truncate the fault-tolerance capability of electrical machines. In addition, eccentricity affects the performance of several topologies in terms of the generation of radial forces in the rotor structure [27–29], cogging and ripple torque increase [24,30], input current harmonic distortion when the machine is fed with inverters [31], and back-emf unbalance [30].

Studies on eccentricity effects are particularly relevant in PM machines, as the magnets on the rotor can generate significantly high radial forces on the stator structure and the bearings [24], which can trigger a major fault. The impact of eccentricity on PMSMs with non-segmented stators has been covered in [25–27,29–31], focusing on quantifying and analyzing the effects on the electromagnetic performance of the machine. Particularly, [29] disclosed a crucial design tendency that can be included in the early stages of the machine design: the electromagnetic performance penalization due to eccentricity strongly depends on the machine slot-pole combination, becoming especially relevant as the slot count and pole count get closer.

In turn, the impact of eccentricity on MPMSMs has been recently addressed in [24] for axial-even eccentricity in a single 24-slot, 28-pole machine: relevant radial forces appear as expected, but the cogging torque, back-emf, and electromagnetic torque seem to be insensitive. The authors concluded that the addressed slot/pole combination was mostly insensitive to eccentricity even when the slot count and pole count differed by four, but the study required to be expanded to more slot/pole combinations. The comparison between the results of [24,29] suggests that the effect of slot/pole combinations on PMSMs with eccentricity may be different from their modular counterpart. This could also be inferred from the analysis of [18], which states that, in presence of manufacturing tolerances, the periodicity of the stator core changes when adopting a modular structure, affecting the cogging torque main period and its sensitiveness to assembly tolerances for the same slot/pole combination.

All things considered, if the fault-tolerance capability of a MPMSM is to be assured, its electromagnetic performance and structure integrity should not be greatly affected by mild degrees of eccentricity. Nevertheless, and despite of its relevance, the effects of eccentricity on the performance of MPMSMs for different slot/pole combinations and their inclusion in the early design stages of the machine remains uncovered in the technical literature so far.

The aim of this work is, therefore, to identify and analyze the impact of eccentric tolerances on relevant performance indices of MPMSMs for different slot/pole combinations, comparing their response to that of PMSMs with conventional stators. Both static and dynamic eccentricity are analyzed by means of the finite element method (FEM) for five slot/pole combinations, evaluating the radial forces that the rotor structure is exposed to cogging torque, back-EMF, and rated torque. The relevance of this paper lies with providing a quick yet reliable comparison of the MPMSM performance between slot/pole combinations and the reasons behind the observed phenomena, critical in the design stage of a MPMSM, as well as disclosing relevant similarities and differences with respect to PMSMs with conventional stator core. This paper is organized as follows: in Section 2, the selected topology and its main data are presented; in Section 3, the methodology and details regarding the considered eccentricity types are described; in Section 4, the results of key performance indices when the machine has eccentricity tolerances are presented and discussed. The conclusions are drawn at the end of the paper.

2. Selected Topology and Slot/Pole Combinations

In order to compare the effect of eccentricity on different slot/pole combinations of both modular and non-modular PMSMs, tooth coil windings (TCW) were adopted since they allow a higher power/weight ratio due to the shorter end winding turns, lower cogging torque, and compactness when compared with the distributed windings [32]. The modular machine is considered to have U-Shaped stator segments and surface-mounted magnets as depicted in Figure 1b as they have been of interest in the last years for fault-tolerance applications [15,21]. The main data of the machines and their design parameters are presented in Table 1 and schematized in Figure 2. The machines assessed in this work aim to provide maximum mean torque within dimensional constraints, following the optimization criteria established in [17,21]. These machines represent low-power scaled protoypes with a rated power of 5 [kW], considering a current density of 10 A/mm^2 and a speed range of 0 to 6000 [RPM].

In turn, the slot/pole combinations were selected with the aim of covering different values of slots per phase per pole (q) given by:

$$q = \frac{Q_s}{2pm}, \tag{1}$$

where Q_s is the slot number, $2p$ is the pole number, and m is the number of phases. For the case of TCW-PMSM, it is recommended to adopt a value of q between 0.25 and 0.5 [32], which is the reason why the selected slot/pole combinations are within this range. The combinations evaluated in this work are summarized in Table 2, which also shows their winding layouts. For all considered designs, the coil pitch is equal to 1 slot, as documented in [32].

Table 1. Main data of the machines (MPMSM and PMSM).

Symbol	Quantity	Value
d	Effective core length	70.0 mm
r_{se}	Stator outer radius	133.0 mm
δ_g	Airgap length	2.0 mm
b_s	Slot width	24.8 mm *
h_s	Slot height	27.5 mm
b_t	Tooth width	10.0 mm *
h_{ys}	Height of the stator yoke	10.5 mm
h_{yr}	Height of the rotor yoke	10.0 mm
h_{pm}	PM height	5.0 mm
B_r	PM remanence	1.1 T
μ_r	Relative recoil permeability	1.04

* Reference tooth width and slot width values were considered for the 28-slot, 20-pole machines, as suggested in [17,18], and they were adjusted for another slot/pole combinations to develop similar saturation levels, based on [17,21].

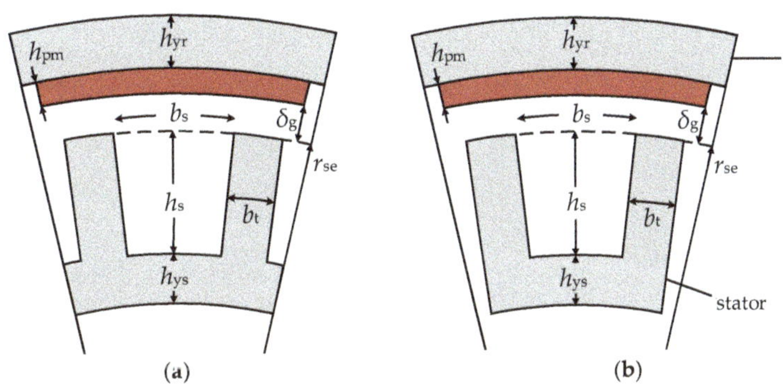

Figure 2. Schematics of selected PMSMs: (**a**) conventional PMSM with non-segmented stator, (**b**) PMSM with modular stator core.

Table 2. Selected slot/pole combinations, including their winding layout considering TCW.

<table>
<tr><th>Slot Number</th><th>Pole Number</th><th>q</th><th>$\Delta(Q_s, 2p)$</th><th>Winding Layout</th></tr>
<tr><td>18</td><td>12</td><td>1/2</td><td>6</td><td>… | A A′ | B B′ | C C′ | …</td></tr>
<tr><td>18</td><td>20</td><td>3/10</td><td>2</td><td>… | C′ A | A′ A′ | A A | A′ B | B′ B′ | B B | B′ C | C′ C′ | C C | …</td></tr>
<tr><td>24</td><td>20</td><td>2/5</td><td>4</td><td>… | C′ A | A′ A′ | A B′ | B B | B′ C | C′ C′ | C A′ | A A | A′ B | B′ B′ | B C′ | C C | …</td></tr>
<tr><td>24</td><td>22</td><td>4/11</td><td>2</td><td>…A | A′ A′ | A…A | A′ A′ | A B′ | B B | B′… B′ | B B | B′ C | C′ C′ | C … C | C′ C′ | C…
$Q_s/6$ $Q_s/6$ $Q_s/6$</td></tr>
<tr><td>24</td><td>28</td><td>2/7</td><td>4</td><td>… | C′ A | A′ A′ | A B′ | B B | B′ C | C′ C′ | C A′ | A A | A′ B | B′ B′ | B C′ | C C | …</td></tr>
</table>

In addition, machines with dissimilar differences between the slot number (Q_s) and the pole number ($2p$) are considered in this work from 2 to 6, as addressed in [33].

3. Types of Eccentricities Evaluated and Assessed Performance Indicators

In this research, both static eccentricity (SE) and dynamic eccentricity (DE) are assessed by means of FEM. When SE is present, and as depicted in Figure 3a, the rotor and stator structures are not coaxial and a non-uniform airgap distribution that does not vary with the rotation of the rotor is generated. On the other hand, when DE is present, the rotor and stator are not coaxial and, additionally, the rotational axis and the rotor geometry axis do not match. As shown in Figure 3b, this creates a non-uniform airgap distribution that circulates with the rotor spinning.

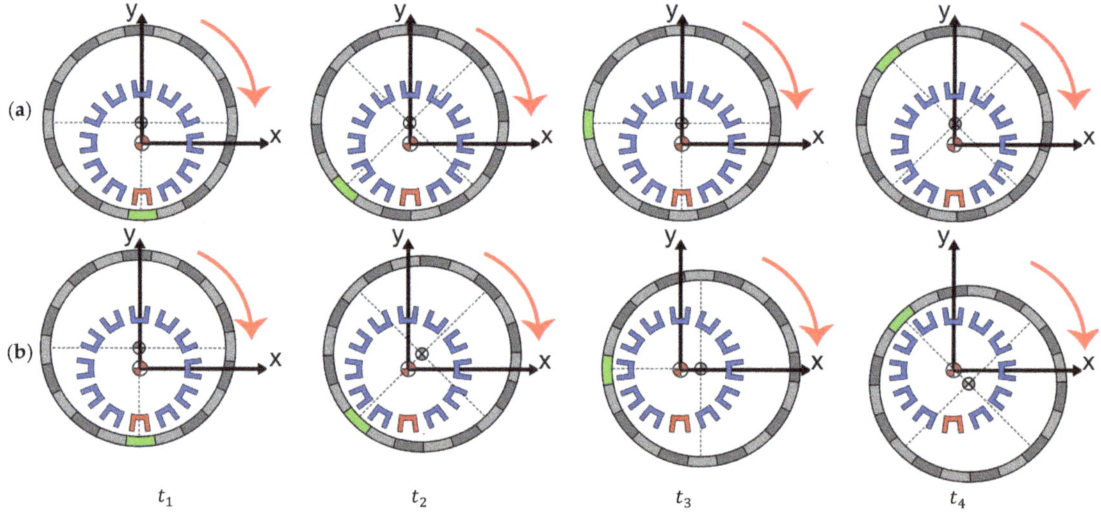

Figure 3. Representation of eccentricity types depending on its dynamics: (**a**) SE, in which the position of the minimum airgap is fixed, (**b**) DE, in which the position of the minimum airgap changes as the rotor structure rotates. Four arbitrary time instants are represented in the figure.

This work considers the difference between the nominal airgap length and the minimum airgap length generated by the stator/rotor misalignment as the magnitude of eccentricity as suggested in [24,30,33]. In addition, potential axis deflections and axial variations of the eccentricity magnitude are neglected. As a result, five eccentricity magnitudes are studied to cover from very low eccentricity (12.5% of the nominal airgap length) to high eccentricity (half the nominal airgap length), as presented in Table 3.

The effects of eccentricity on the performance of both PMSMs and MPMSMs are studied by means of: (i) the analysis of the generated radial forces, also called unbalanced magnetic force (UMF), (ii) cogging torque, (iii) back-emf, and (iv) mean torque, as these indicators have proven to be affected in conventional topologies for at least one slot/pole combination.

Table 3. Eccentricity magnitudes analyzed in the selected PMSMs.

Eccentricity Magnitude (% of Nominal Airgap Length)	Eccentricity Magnitude (mm)	Severity
12.5	0.25	Very low
25.0	0.50	Low
37.5	0.75	Medium
50.0	1.00	High

The following sections present the results organized so as to cover the impact of eccentricity on each performance indicator separately. In each case, the results are firstly presented in detail for one slot/pole combinations to show graphics of relevant curves and their harmonic spectrum. From these graphics, the main numerical indicators are extracted and summarized in tables for all the addressed slot/pole combinations, in order to allow an organized comparison and analysis of results. All results are obtained by means of 2D FEM simulations carried out with the commercial package Ansys Electronics.

4. Results and Discussion: Unbalanced Magnetic Forces

One of the most critical consequences of eccentricity is the generation of unbalanced magnetic forces between the rotor and stator structures. In the case of radial-flux topologies, this may translate into radial forces of considerable magnitude that affect the bearings of the machine and whose dynamics depend on the eccentricity type. In this section, net forces acting on the rotor are assessed considering that in $t = 0$ the minimum airgap position is in $\theta = \frac{3}{2}\pi$ rad, as indicated in time instant t_1 of Figure 3.

4.1. Evaluation on a 24-Slot, 22-Pole PMSM and MPMSM

As an example, in Figure 4 the outcomes of the FEM evaluation of a 24-slot, 22-pole PMSM and a 24-slot, 22-pole MPMSM are presented for both SE and DE. The results were extracted considering the rotor structure completes one full turn. In Figure 4a,b the x-axis and y-axis forces acting on the rotor structure are shown when DE is present for the modular and the conventional machine, respectively. The X represents the average radial forces generated in a full rotation, which is zero for the case of DE. It may be noted that a rotating force vector is generated in both cases, with an almost constant magnitude and low ripple (visualized as saw teeth in the force circumference periphery).

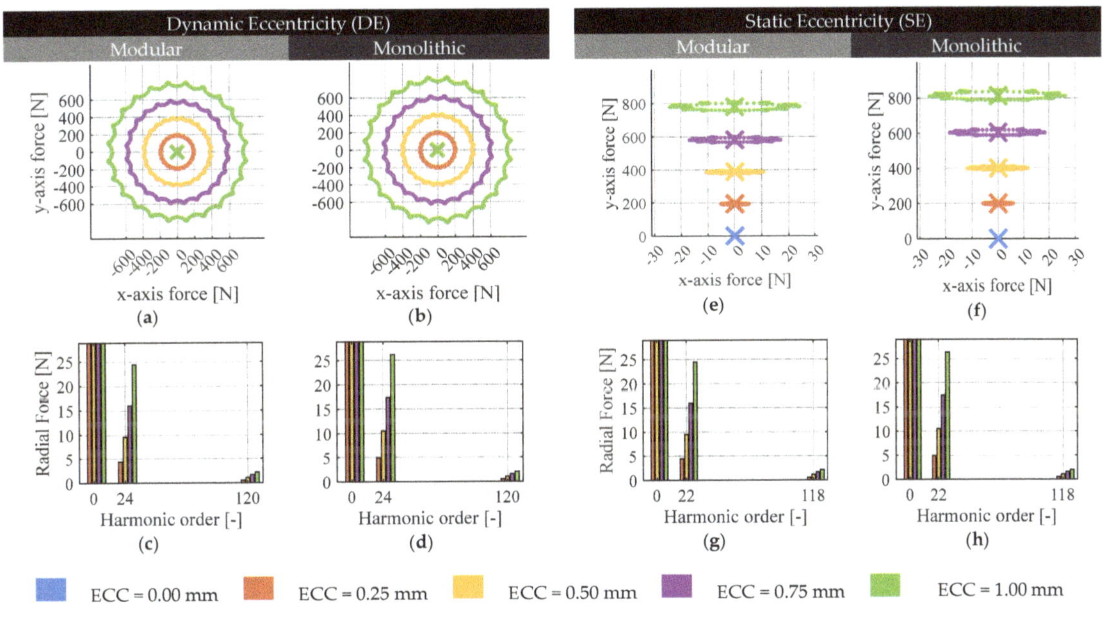

Figure 4. FE evaluation of radial forces acting on the rotor structure in the presence of DE (left column) and SE (right column) for a 24-slot, 22-pole PMSM and MPMSM: (**a**) x-axis and y-axis forces of DE on MPMSM, (**b**) x-axis and y-axis forces of DE on PMSM, (**c**) radial force spatial harmonic content of DE on MPMSM, (**d**) radial force spatial harmonic content of DE on PMSM, (**e**) x-axis and y-axis forces of SE on MPMSM, (**f**) x-axis and y-axis forces of SE on PMSM, (**g**) radial force spatial harmonic content of SE on MPMSM, (**h**) radial force spatial harmonic content of SE on PMSM.

From Figure 4c,d, which show the spatial harmonic content of the radial forces, it can be seen that the spatial frequency of the radial force ripple matches the slot pitch of the machine (HO = 24). This can be explained since for DE the relative position between the magnet symmetry axis and teeth symmetry axis is periodical and repeats each $360/Q_s$ mechanical degrees. Similar findings were described for the cogging torque in conventional PMSMs in [34]. In turn, in Figure 4e,f, the x-axis and y-axis forces acting on the rotor structure are shown when SE is acting in the modular and the conventional machine, respectively. The X represents the average radial forces generated in a full rotation, which points from the minimum airgap position towards the center of rotation for the case of SE. It may be noted that the x-axis forces are not zero and configure the radial force ripple. From Figure 4g,h, which show the spatial harmonic content of the radial forces, it can be seen that the spatial frequency of the radial force ripple matches the pole pitch of the machine (HO = 22). This is because in SE the relative position between the magnet symmetry axis and teeth symmetry axis is periodical and repeats each $360/2p$ mechanical degrees, as described in [24].

In summary, relevant features of the radial force results can be summarized by the following indicators, useful for evaluating and comparing the different slot/pole combinations in Table 2.

- Mean value of the radial force acting on the rotor structure, measured in N.
- Peak-to-peak value of the radial force, representing the radial force ripple and measured in N.

4.2. Comparison of Slot/Pole Combinations for PMSMs and MPMSMs

In Table 4, the average forces generated by different DE magnitudes are presented for both the modular and monolithic machines considering different slot/pole combinations. In the first place, it may be noted that not all selected slot/pole combinations exhibit unbalanced magnetic forces in the absence of eccentricity. Secondly, and according to the tendencies of Figure 4a,b, it can be seen that the generated radial forces are directly proportional to the eccentricity magnitude. This means that the radial force created by eccentricity can be quickly predicted from a single ECC value.

Table 4. Average UMF of different PMSM and MPMSM slot/pole combinations in the presence of DE.

Q/2p \ ECC	0 mm (Faultless)		0.25 mm		0.50 mm		0.75 mm		1.00 mm	
	Mod [N]	Mon [N]	Mod [N]	Mon [N]	Mod [N]	Mon [N]	Mod [N]	Mon [N]	Mod [N]	Mon [N]
18S 12P	0	0	92	116	184	231	277	348	370	466
18S 20P	0	0	148	151	296	304	446	458	599	614
24S 20P	0	0	181	194	362	390	546	588	733	790
24S 22P	0	0	192	200	385	401	580	604	780	813
24S 28P	0	0	203	209	408	419	617	634	833	856

From Table 4, it can be also noted that the unbalanced magnetic force increases as the slot number and pole number increase. The effect of the slot number can be noted from the comparison of the 18-slot, 20-pole machine to the 24-slot, 20-pole machine, which develops a consistent 20% force increase for all the evaluated ECC values. In consequence, it can be stated that the slot count has a medium impact on the radial forces generated by eccentricity and should not be neglected. In turn, the pole count effect can be deduced from the comparison of the 24-slot, 20-pole and the 24-slot, 28-pole machine, the latter having 12% more radial force. On the other hand, it is observed that in some slot/pole combinations, the radial force generated in the monolithic machine is significantly different from the modular machine. This can be explained by the flux density penalization of modular machines with a high value of q (See Equation (1)). To this end, Figure 5 shows the flux density lines for two slot/pole combinations that have different q values. Figure 5a,b show the flux lines when q is high, in which case the pole number is much smaller than the slot number. In this case, the flux lines require to travel through several teeth, crossing the stator yoke which is absent between the modules of MPMSMs. Therefore, in the modular machine, the flux lines will cross the gap between modules, circulating across a high-reluctance path that penalizes the airgap flux density and, subsequently, the radial forces. Instead, Figure 5c,d show the flux lines when q is lower, in which case the pole number is closer to the slot number. In this situation, flux lines need to use fewer teeth to close the path, which is the reason why in modular machines the lines can circulate using a single module and not penalize the airgap flux density to a high extend.

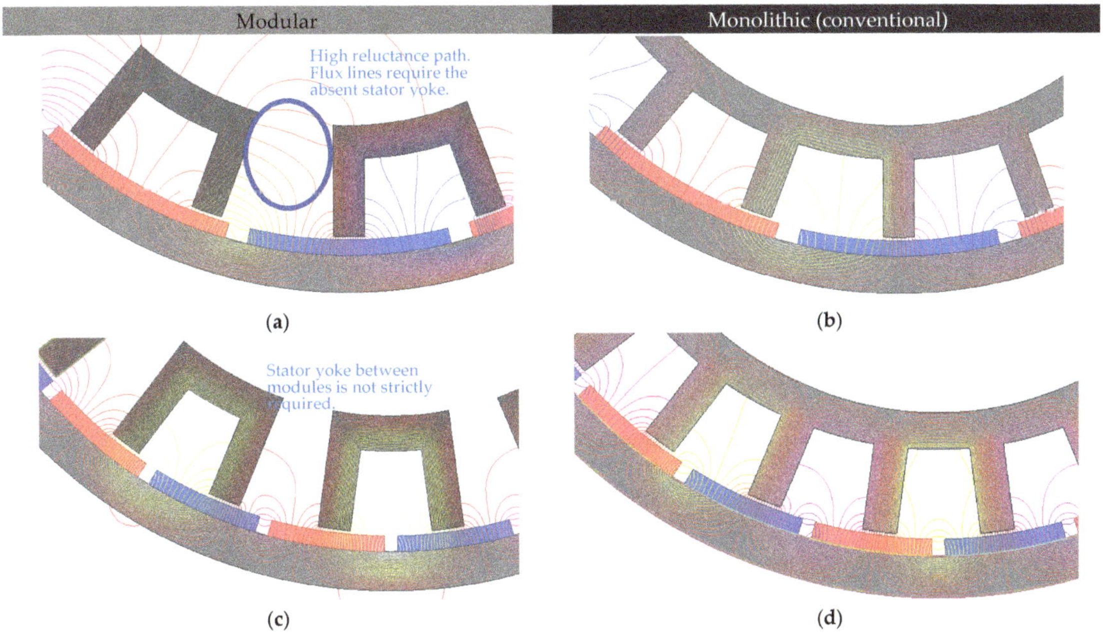

Figure 5. Close-up to flux density lines for different values of q: (**a**) Modular 18-slot, 12-pole, $q = 0.5$, (**b**) monolithic, 18-slot, 12-pole, $q = 0.5$, (**c**) modular, 24-slot, 22-pole, $q = 0.36$ (**d**) monolithic, 24-slot, 22-pole, $q = 0.36$.

In Table 5, the maximum radial force (ECC = 1 mm) for each slot/pole combination is presented and compared between modular and monolithic machines. It can be noted that, effectively, the difference between the radial forces generated on the modular machine vs. on the monolithic machine is higher as the value of q is greater.

Table 5. Difference between the radial force on modular and monolithic machines in terms of the number of slots per pole per phase.

Q/2p	q	F_{max} [N] Mod	F_{max} [N] Mon	ΔF_{max}
18S 12P	1/2	370	466	25.9%
18S 20P	3/10	599	614	2.5%
24S 20P	2/5	733	790	7.8%
24S 22P	4/11	780	813	4.2%
24S 28P	2/7	833	856	2.7%

In Table 6, the radial force ripple generated by different magnitudes of DE is presented for each slot/pole combination. It is worth noting that slot/pole combinations with a lower difference between the slot count and the pole count exhibit the highest force ripple. This is particularly relevant since it translates into higher-frequency vibration and force pulses that can reach the bearings. For instance, for the 18-slot, 20-pole machines, the force ripple is around 40% of the average radial force, which is already significant in magnitude, and it generates a force that pulses 18 times for each machine rotation. For all slot/pole combinations, the main harmonic order (HO) of the torque ripple matches the slot number, and there are no significant differences between the modular and the monolithic machines in this regard.

Table 6. Radial force ripple for different slot/pole combinations of PMSM and MPMSM in the presence of DE.

Q/2p \ ECC	0 mm (Faultless)		0.25 mm		0.50 mm		0.75 mm		1.00 mm		Main HO	
	Mod [N]	Mon [N]	Mod [N]	Mon [N]	Mod [N]	Mon [N]	Mod [N]	Mon [N]	Mod [N]	Mon [N]	Mod [-]	Mon [-]
18S 12P	0	0	4	2	12	4	23	7	38	9	18	18
18S 20P	0	0	81	83	160	164	235	242	304	314	18	18
24S 20P	0	0	3	1	6	3	10	7	22	13	24	24
24S 22P	0	0	10	11	22	24	37	39	55	58	24	24
24S 28P	0	0	1	1	2	1	3	3	7	5	24	24

Table 7 summarizes the average value of the radial force generated by different SE magnitudes for both the modular and monolithic machines. In concordance with the tendencies of Figure 4e,f and similarly to DE, it can be seen that the generated radial forces are directly proportional to the eccentricity magnitude. Nevertheless, and as occurred in the case of DE, modular machines have significantly lower radial forces when the value of q is high, for the reasons explained in Figure 5. From Table 7, it can also be seen that the unbalanced magnetic force increases as the slot number and pole number increase. Finally, when comparing Table 4 with Table 7, it may be appreciated that the force magnitude generated by DE is similar to that of SE.

Table 7. Average UMF for different slot/pole combinations of PMSM and MPMSM in the presence of SE.

$Q_s/2p$ \ ECC	q	0 mm (Faultless)		0.25 mm		0.50 mm		0.75 mm		1.00 mm	
		Mod [N]	Mon [N]	Mod [N]	Mon [N]	Mod [N]	Mon [N]	Mod [N]	Mon [N]	Mod [N]	Mon [N]
18S 12P	1/2	0	0	92	115	183	231	276	348	370	466
18S 20P	3/10	0	0	148	151	296	303	446	457	599	614
24S 20P	2/5	0	0	180	194	362	389	546	587	733	790
24S 22P	4/11	0	0	192	200	384	400	580	604	779	812
24S 28P	2/7	0	0	203	208	408	419	617	634	833	855

In Table 8, the radial force ripple generated by different magnitudes of SE is presented for each slot/pole combination. Similar to DE, slot/pole combinations with a lower difference between the slot count and the pole count exhibit the highest force ripple. Again, the 18-slot, 20-pole machines provide the highest ripple of around 40% of the average radial forces. Different to what was observed for DE, in the case of SE, the main HO of the torque ripple matches the pole number, and there are no considerable differences between the modular and the monolithic machines.

Table 8. Radial force ripple for different slot/pole combinations of PMSM and MPMSM in the presence of SE.

ECC / $Q_s/2p$	0 mm (Faultless)		0.25 mm		0.50 mm		0.75 mm		1.00 mm		Main HO	
	Mod [N]	Mon [N]	Mod [N]	Mon [N]	Mod [N]	Mon [N]	Mod [N]	Mon [N]	Mod [N]	Mon [N]	Mod [-]	Mon [-]
18S 12P	0	0	4	2	12	4	23	6	36	8	12	12
18S 20P	0	0	81	83	160	165	236	244	307	318	20	20
24S 20P	0	0	2	1	5	3	10	6	20	12	20	20
24S 22P	0	0	11	12	22	24	37	40	55	60	22	22
24S 28P	0	0	1	0	2	1	3	2	7	5	28	28

4.3. Summary: DE and SE on the UMF of PMSMs and MPMSMs

In summary, it was found that radial forces on machines with SE and DE scale with the slot number and pole number, and that the severity of these forces is lower in the case of modular machines depending on the value of q. Moreover, a force ripple of significant magnitude appears in machines in which the slot number and pole number are close to each other.

5. Results and Discussion: Cogging Torque

5.1. DE: Evaluation of 24-Slot, 22-Pole PMSM and MPMSM

In Figure 6, the cogging torque outcomes of the FE evaluation of a 24-slot, 22-pole PMSM and a 24-slot, 22-pole MPMSM are presented for DE. The results were extracted considering the rotor structure completes 360 mechanical degrees to correctly obtain the harmonic content of the cogging torque signal. However, only the main period (when eccentricity is present) is shown in the figure in order to provide a clearer visualization. In Figure 6a,c, the cogging torque waveforms are shown for a 24-slot, 22-pole MPMSM and PMSM respectively. In turn, Figure 6b,d present the harmonic spectrum of the cogging torque waveform for both machines, modular and conventional, when dynamic eccentricity (DE) is present.

The main period of the cogging torque was found to be 1.36 mechanical degrees when no eccentricity is applied in both cases. This changes when DE is present: the main period changes to 15 degrees, which corresponds to the slot pitch. This can be explained since in conventional PMSMs, the cogging torque main period is defined by the periodicities of the stator and rotor structures when no eccentricity is present (native harmonic content) and is accounted by [34]:

$$T_{T\ \mathrm{NHC}} = \frac{360}{\mathrm{LCM}(Q_s, 2p)}. \quad (2)$$

This results in 1.36° for the 24-slot, 22-pole machine. However, when dimensional tolerances are affecting the machine, the stator/rotor periodicity breaks. For the case of DE, the minimum airgap is rotating with the rotor and the modules are stationary, and hence, the relative position between magnet symmetry axis and teeth symmetry axis is periodical and repeats each $360/Q_s$ mechanical degrees. Therefore, the main cogging torque period should change to the additional harmonic components generated, given by:

$$T_{T\ \mathrm{AHC\ DE}} = \frac{360}{Q_s}, \quad (3)$$

which happens to occur for the 24-slot, 22-pole machine and generates a significant increase in the peak-to-peak value of the cogging torque. The cogging torque after eccentricity is therefore comprised of the native harmonic components (faultless) and the additional harmonic components (AHC) as per [34]:

$$T_{\mathrm{cogg}}(\alpha) = T_{\mathrm{NHC}}(\alpha) + T_{\mathrm{AHC}}(\alpha) \quad (4)$$

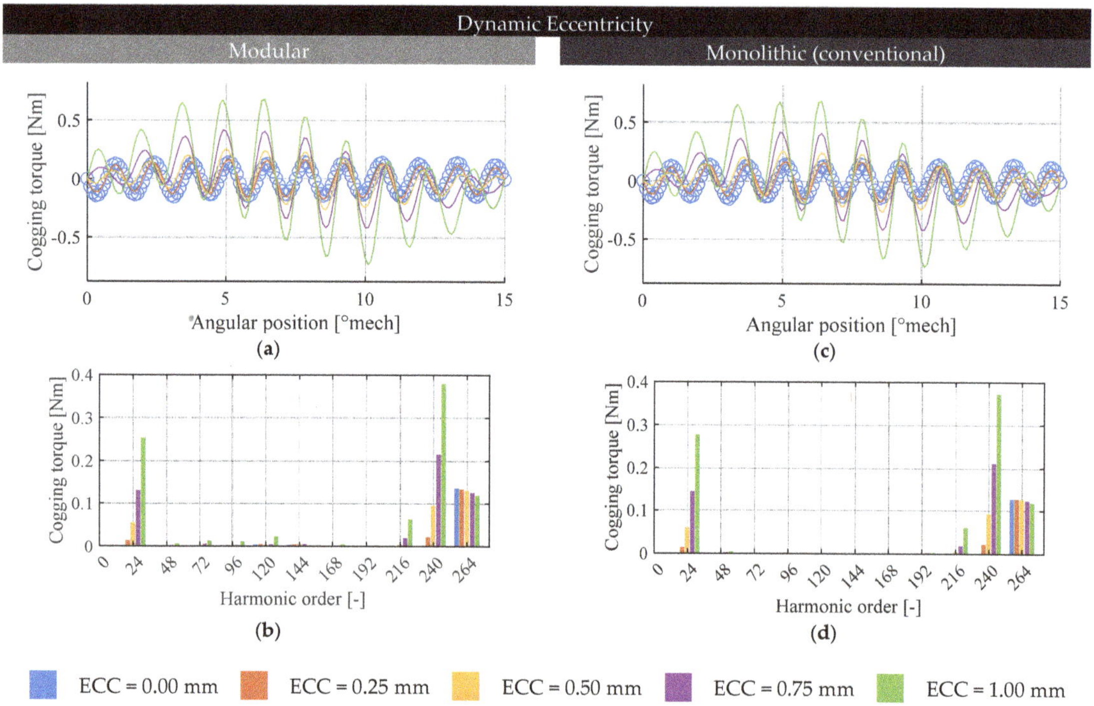

Figure 6. FE evaluation of cogging torque in the presence of DE for the 24-slot, 22-pole PMSM and MPMSM: (**a**) cogging torque waveform for the modular machine, (**b**) cogging torque spectrum for the modular machine, (**c**) cogging torque waveform for the monolithic machine, (**d**) cogging torque spectrum for the monolithic machine.

From Figure 6, it can be noted that significant cogging torque components with HO = 24 and their multiples are generated with DE, which results in a peak-to-peak value increase of up to 400%. Nevertheless, this does not hold true for all evaluated slot/pole combinations.

In summary, relevant features of the cogging torque results can be summarized by the following indicators, useful for evaluating and comparing the different slot/pole combinations of Table 2.

- NHC of the cogging torque, measured in Nm;
- AHC of the cogging torque, measured in Nm;
- AHC of the cogging torque, measured in percentage of the NHC.

5.2. DE: Comparison of Slot/Pole Combinations for PMSMs and MPMSMs

In Table 9, the natural harmonic component of cogging torque for different DE magnitudes is presented. Both the modular and monolithic machines are evaluated considering different slot/pole combinations. From Table 9, it may be noted that the cogging torque magnitude is very different depending on the slot/pole combination and it does not have a clear scalation with respect to the slot count or pole count separately. In this regard, it is commonly accepted in faultless machines to consider the cogging torque HO (HO_{NHC}) as a comparative indicator for cogging torque magnitude ([35], see (2)). The higher the HO, the lower the main period of the cogging torque and the lower its magnitude should be. As can be seen from Table 9, the NHC of the cogging torque is barely affected by DE, and HO_{NHC} is able to compare slot/pole combinations regardless of the eccentricity magnitude. It may draw the attention that a significant difference in the NHC magnitude can be observed for the 24-slot, 20-pole and the 24-slot, 28-pole machines when comparing the modular and the monolithic machine. This is related to a stator/rotor periodicity break generated by the modular structure of U-shape MPMSMs. In those cases, HO_{NHC} should be corrected to include the periodicity provided by the modular stator core. As an example, in the case of the 24-slot, 28-pole machine, the main period of the cogging torque is 84 for the modular machine, which explains the difference regarding the monolithic machine.

Table 9. Cogging torque NHC of different PMSM and MPMSM-slot/pole combinations in the presence of DE.

ECC / $Q_s/2p$	HO_{NHC}	0 mm (Faultless)		0.25 mm		0.50 mm		0.75 mm		1.00 mm	
		Mod [Nm]	Mon [Nm]	Mod [Nm]	Mon [Nm]	Mod [Nm]	Mon [Nm]	Mod [Nm]	Mon [Nm]	Mod [Nm]	Mon [Nm]
18S 12P	72	19.11	22.42	19.13	22.44	19.17	22.46	19.24	22.49	19.34	22.54
18S 20P	180	0.63	0.66	0.62	0.65	0.59	0.62	0.54	0.56	0.47	0.48
24S 20P	120	3.92	2.70	3.91	2.71	3.88	2.77	3.84	2.86	3.83	2.99
24S 22P	264	0.27	0.25	0.27	0.25	0.26	0.25	0.25	0.25	0.24	0.24
24S 28P	168	2.28	0.72	2.28	0.73	2.30	0.74	2.34	0.77	2.40	0.81

In Table 10, the peak-to-peak value of the additional harmonic components of cogging torque for different DE magnitudes is summarized. Both the modular and monolithic machines are evaluated considering different slot/pole combinations. From Table 10, it is clear that the generated AHC of cogging torque is very different from a one slot/pole combination to another. Nevertheless, and contrary to what was observed for NHC, the AHC does not scale with HO_{NHC}. This can be explained by analyzing airgap flux density. As developed in [33], a simple analytical model of the additional airgap flux density radial and circumferential components accounting for the eccentricity can be expressed by:

$$B_{r_{ecc}} = B_{sr}\lambda_{ecc} \tag{5}$$

$$B_{\alpha\_ecc} = B_{s\alpha}\lambda_{ecc} \tag{6}$$

where B_{sr} and $B_{s\alpha}$ are the radial and circumferential air flux densities (without the presence of eccentricity) with harmonic order n, and λ_{ecc} is the real component of the equivalent complex permeance that represent the influence of eccentricity given by:

$$\lambda_{ecc} = 1 + \sum_{v} \lambda_v \cos(v\alpha - \chi v\omega t) \tag{7}$$

where λ_v is the magnitude of the v-th order component of equivalent permeance representing the eccentricity and χ allows to select between SE ($\chi = 0$) and DE ($\chi = 1$). Since the stator slotting effect generates field harmonics of order $n = mp \pm \mu Q_s$, then the additional field harmonics due to eccentricity are $n \pm v$. By virtue of this, there are two scenarios in which eccentricity have an impact on the cogging torque:

- If both the additional flux density due to eccentricity and the flux density without eccentricity share harmonic spatial orders, eccentricity has an influence on the cogging torque;

- If the additional flux density due to eccentricity has a component of spatial order l resulting from different values of n and v, then eccentricity can also contribute to the cogging torque (for instance, if $l = n_1 \pm v_1 = n_2 \pm v_2$);
- In the case of machines having $2p = Q_s \pm 2$ (18-slot, 20-pole and 22-slot, 24-pole), the interaction between spatial harmonics from n and $n \pm 2$ generate the AHC; in the case of machines that have $2p = Q_s \pm 4$ (24-slot, 20-pole and 28-slot, 24-pole), the interaction between spatial harmonics from n and $n \pm 4$ generate the AHC; in the case of machines that have $2p = Q_s \pm 6$ (18-slot, 12-pole), the interaction between spatial harmonics from n and $n \pm 6$ generate the AHC. It can be noted that several field harmonic components interact in machines with $2p = Q_s \pm 2$, and interacting harmonic components decrease as the difference between the slot number and the pole number is higher. The result of this analysis is that slot/pole combinations with the slot number being close to the pole number ($2p = Q_s \pm 2$) should have a more significant effect of eccentricity on cogging torque that machines with it slot number very different from its pole number ($2p = Q_s \pm 4$ and $2p = Q_s \pm 6$). This is verified from the results presented in Table 10.

Table 10. Peak-to-peak value of cogging torque AHC of different PMSM and MPMSM slot/pole combinations in the presence of DE.

ECC / $Q_s/2p$	HO_{NHC}	$\Delta(Q_s, 2p)$	**0.25 mm**		**0.50 mm**		**0.75 mm**		**1.00 mm**	
			Mod [Nm]	Mon [Nm]	Mod [Nm]	Mon [Nm]	Mod [Nm]	Mon [Nm]	Mod [Nm]	Mon [Nm]
18S 12P	72	6	0.06	0.06	0.12	0.15	0.26	0.29	0.45	0.48
18S 20P	180	2	0.20	0.20	0.74	0.78	1.59	1.70	2.80	2.95
24S 20P	120	4	0.04	0.06	0.13	0.09	0.32	0.22	0.63	0.50
24S 22P	264	2	0.08	0.09	0.30	0.31	0.72	0.73	1.39	1.36
24S 28P	168	4	0.02	0.02	0.05	0.04	0.13	0.09	0.29	0.21

The results of the 18-slot, 12-pole machines versus that of the 24-slot, 28-pole machines may draw attention since they do not exactly follow this tendency: the 18-slot, 12-pole machine has higher absolute AHC than the 24-slot, 28-pole machine, but it has a higher difference between the slot count and the pole count. However, the analysis must consider the proportion between the AHC generated by eccentricity and the NHC. From Table 9, it can be recalled that the 18-slot, 12-pole machine has a significantly higher cogging torque NHC than the 24-slot, 28-pole machine, which translates into AHC being proportionally lower than that of the 24-slot, 28-pole machine.

Table 11 is created to account for this proportion and presents the relative cogging torque increase of the different slot/pole combinations in the presence of DE. It may be appreciated that the relative cogging torque increase effectively depends on the difference between the slot count and the pole count, as indicated in the spatial field harmonic analysis. Cogging torque of modular and monolithic machines have a similar response to eccentricity, although some slot/pole combinations have mild differences due to the break of stator/rotor periodicity, as indicated in [18].

Table 11. Relative cogging torque increase of different PMSM and MPMSM slot/pole combinations in the presence of DE.

ECC / $Q_s/2p$	$\Delta(Q_s, 2p)$	0.25 mm		0.50 mm		0.75 mm		1.00 mm	
		Mod [%]	Mon [%]	Mod [%]	Mon [%]	Mod [%]	Mon [%]	Mod [%]	Mon [%]
18S 12P	6	0.19	0.22	0.53	0.65	1.15	1.30	2.07	2.18
18S 20P	2	27.23	26.21	102.72	105.80	238.69	236.47	422.04	424.21
24S 20P	4	0.29	0.69	1.33	3.05	4.42	7.62	11.00	17.45
24S 22P	2	12.58	16.57	72.73	89.18	180.61	215.49	374.00	435.19
24S 28P	4	0.41	0.11	1.62	2.12	4.64	8.61	10.33	26.39

5.3. SE: Evaluation on 24-Slot, 22-Pole PMSM and MPMSM

In Figure 7, the cogging torque outcomes of the FE evaluation of a 24-slot, 22-pole PMSM and a 24-slot, 22-pole MPMSM are presented for SE. Similar to DE results, SE results were extracted considering the rotor structure completes one full turn to correctly obtain the harmonic content of the cogging torque signal. However, only the main period (when eccentricity is present) is presented in order to provide a clearer visualization. In Figure 7a,c, the cogging torque waveform is shown for a 24-slot, 22-pole MPMSM and PMSM, respectively. In turn, Figure 7b,d present the harmonic spectrum of the cogging torque waveform for both machines, modular and conventional, when SE is present.

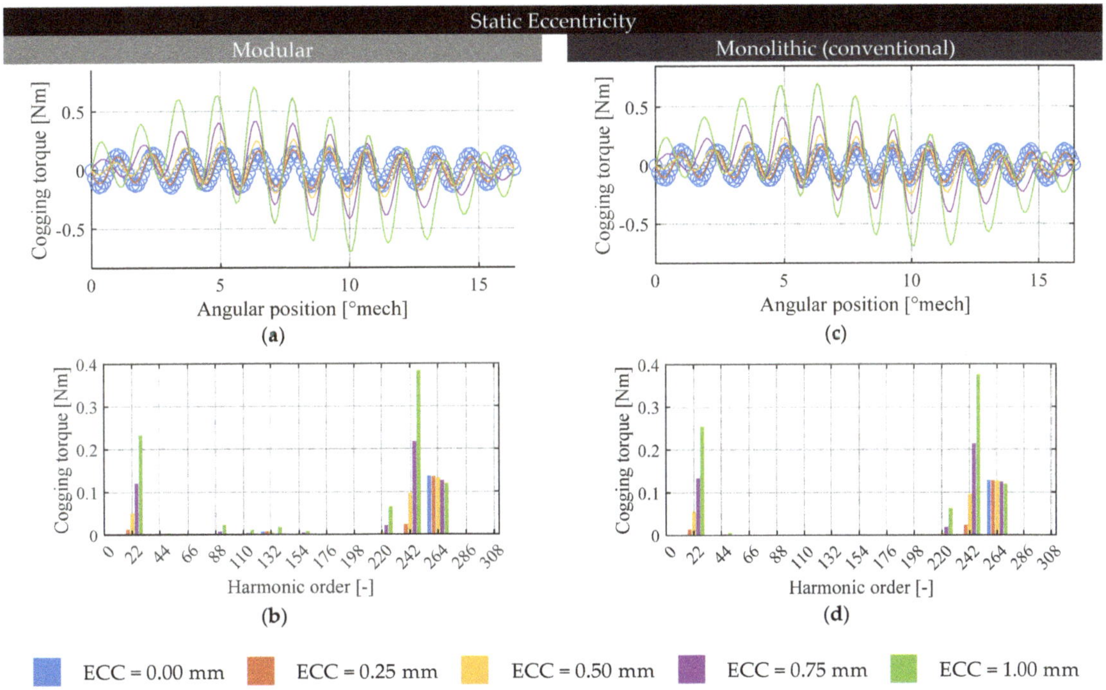

Figure 7. FE evaluation of cogging torque in the presence of SE for 24-slot, 22-pole PMSM and MPMSM: (**a**) cogging torque waveform for the modular machine, (**b**) cogging torque spectrum for the modular machine, (**c**) cogging torque waveform for the monolithic machine, (**d**) cogging torque spectrum for the monolithic machine.

When SE is present, the main period of the cogging torque changes from 1.36 to 16.36 mechanical degrees, which corresponds to the pole pitch. From (2), it may be noted that the cogging torque main period is defined by the periodicities of the stator and rotor structures when no eccentricity is present, but eccentricity breaks the stator/rotor periodicity. For the case of SE, the minimum airgap is stationary, and the relative position between the magnet symmetry axis and teeth symmetry axis depends exclusively on the movement of the poles, which is periodical and repeats each $360/2p$ mechanical degrees. Therefore, the main cogging torque period should change to the additional harmonic components generated, given by:

$$T_{T\ \mathrm{AHC\ SE}} = \frac{360}{2p}, \tag{8}$$

From Figure 7, it can be noted that significant cogging torque components with HO = 22 and its multiples are generated with SE, which results in a peak-to-peak value increase of up to 400%. Nevertheless, this does not hold true for all evaluated slot/pole combinations.

5.4. SE: Comparison of Slot/Pole Combinations for PMSMs and MPMSMs

In Table 12, the natural harmonic component of cogging torque for different SE magnitudes is presented. Both the modular and monolithic machines are evaluated considering different slot/pole combinations. From Table 12, it can be seen that the NHC of the cogging torque is barely affected by SE and, therefore, the natural cogging torque HO ($T_{T\ NHC}$) can be used as a comparative indicator for cogging torque magnitude. Similar to the results obtained for SE, from Table 12, it can be noted that HO_{NHC} is able to compare slot/pole combinations regardless of the eccentricity magnitude.

Table 12. Cogging torque NHC of different PMSM and MPMSM slot/pole combinations in the presence of SE.

ECC / $Q_s/2p$	HO_{NHC}	0 mm (Faultless)		0.25 mm		0.50 mm		0.75 mm		1.00 mm	
		Mod [Nm]	Mon [Nm]	Mod [Nm]	Mon [Nm]	Mod [Nm]	Mon [Nm]	Mod [Nm]	Mon [Nm]	Mod [Nm]	Mon [Nm]
18S 12P	72	19.10	22.42	19.13	22.44	19.16	22.46	19.20	22.49	19.24	22.54
18S 20P	180	0.63	0.66	0.62	0.66	0.60	0.63	0.57	0.59	0.52	0.54
24S 20P	120	3.90	2.66	3.90	2.68	3.87	2.74	3.83	2.82	3.82	2.96
24S 22P	264	0.27	0.25	0.27	0.25	0.26	0.25	0.25	0.25	0.24	0.24
24S 28P	168	2.28	0.72	2.28	0.73	2.30	0.75	2.34	0.78	2.40	0.81

In Table 13, the relative cogging increase of the cogging torque for different SE magnitudes is summarized, which is calculated as the ratio between the peak-to-peak value of the AHC and the NHC. Both the modular and monolithic machines are evaluated considering different slot/pole combinations. From Table 13, it is clear that the generated AHC of cogging torque is very different from a slot/pole combination to another. It may be appreciated that the relative cogging torque increase depends on $\Delta(Q_s, 2p)$ in a similar manner to what was observed for DE (see Table 11). This tendency is defined by the spatial field harmonics that interact to generate cogging torque for each machine as per (5) to (7). Furthermore, the cogging torque of modular and monolithic machines have a similar response to eccentricity.

Table 13. Relative cogging torque increase of different PMSM and MPMSM slot/pole combinations in the presence of SE.

ECC / $Q_s/2p$	HO_{NHC}	$\Delta(Q_s, 2p)$	0.25 mm		0.50 mm		0.75 mm		1.00 mm	
			Mod [%]	Mon [%]	Mod [%]	Mon [%]	Mod [%]	Mon [%]	Mod [%]	Mon [%]
18S 12P	72	6	0.19	0.22	0.54	0.66	1.17	1.31	2.05	2.19
18S 20P	180	2	28.92	24.87	117.38	107.98	260.77	244.69	462.91	439.94
24S 20P	120	4	0.41	0.83	1.52	2.89	4.63	7.72	11.22	17.86
24S 22P	264	2	12.31	16.10	67.51	87.25	178.98	210.45	373.15	417.45
24S 28P	168	4	0.55	0.59	1.46	2.72	4.37	9.48	10.14	25.80

5.5. Summary: SE and DE on Cogging Torque of PMSMs and MPMSMs

In summary, it was found that the cogging torque can be separated into two components: NHC, which are originally present in the faultless machine, and AHC, which appears due to eccentricity. In this regard:

- NHC is not affected by eccentricity, and its magnitude can be related to the slot/pole combination by means of the main harmonic order of the cogging torque, which can be determined as the least common multiple of slot and pole number.
- AHC is affected by eccentricity and can translate into severe peak-to-peak increases of the cogging torque. Machines that have a slot number close to their pole number are very sensitive to eccentricity and develop a higher cogging torque increase. In terms of magnitude, there are no significant differences between DE and SE. However, the order of the AHC generated by DE are multiples of the slot number, and the order of the AHC generated by SE are multiples of the pole number.
- Modular machines may have different native harmonic components when compared to their monolithic counterparts due to the stator/rotor periodicity break when adopting a modular stator core. Nevertheless, AHC is generated to a similar extent in both modular and monolithic machines.

These findings can be used for diagnosis ends: in machines with their slot number close to their pole number, anomalous cogging torque increases with certain harmonic orders can indicate the presence of SE or DE.

6. Results and Discussion: Back-Emf

6.1. DE: Evaluation on a 24-Slot, 22-Pole PMSM and MPMSM

In Figure 8, the back-emf outcomes of the FE evaluation of a 24-slot, 22-pole PMSM and a 24-slot, 22-pole MPMSM are presented for DE. The results were extracted considering the rotor structure completes one full turn to correctly obtain the harmonic content of the back-emf signal, which is depicted in the figures. In Figure 8a,c, the three-phase back-emf waveforms are presented for a 24-slot, 22-pole MPMSM and PMSM, respectively. In turn, Figure 8b,d show the harmonic spectrum of the back-emf for both machines, modular and conventional, when DE is present.

From Figure 8, it can be noted that the back-emf waveform is composed of a high magnitude fundamental component and a third-order harmonic component of low magnitude. In addition, it is evident that eccentricity does not have a significant impact on the back-emf magnitude, and no significant unbalance between phases is observed either, contrary to what was suggested in [30].

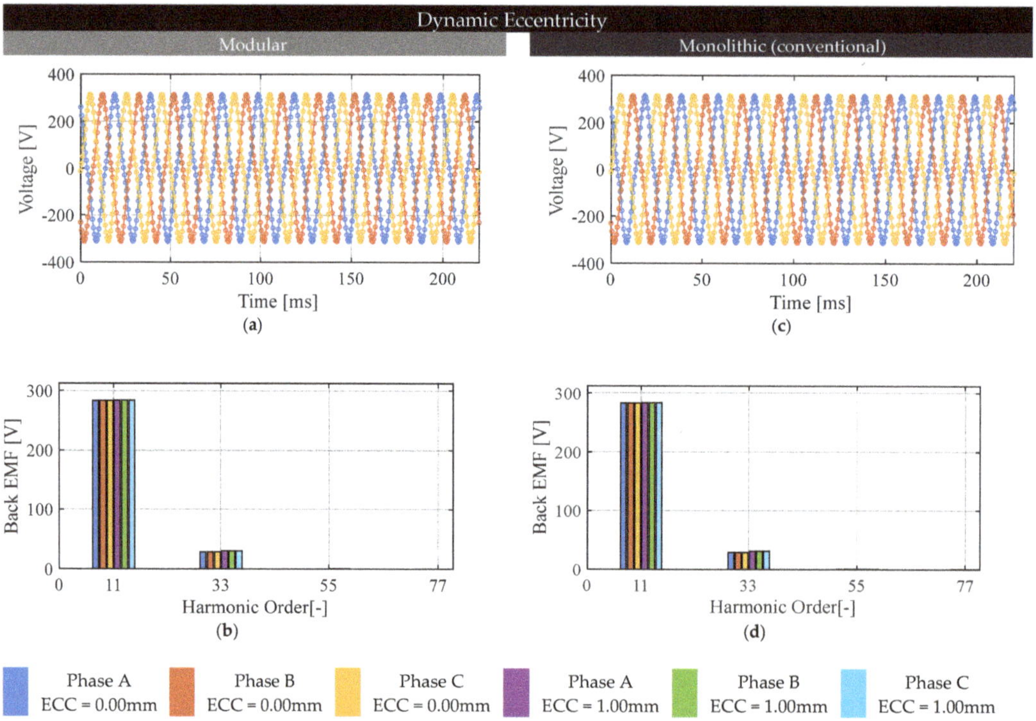

Figure 8. FE evaluation of the back-emf in the presence of DE for a 24-slot, 22-pole PMSM and MPMSM: (**a**) back-emf waveform for the modular machine, (**b**) back-emf spectrum for the modular machine, (**c**) back-emf waveform for the monolithic machine, (**d**) back-emf spectrum for the monolithic machine.

A significant back-emf unbalance was observed in [30] when evaluating eccentricity on a 9-slot, 8-pole PMSM with conventional stator core. This can be explained by analyzing the spatial distribution of phases of that 9-slot, 8-pole machine when compared to an 8-slot, 12-pole PMSM, which are represented in Figure 9a,b, respectively.

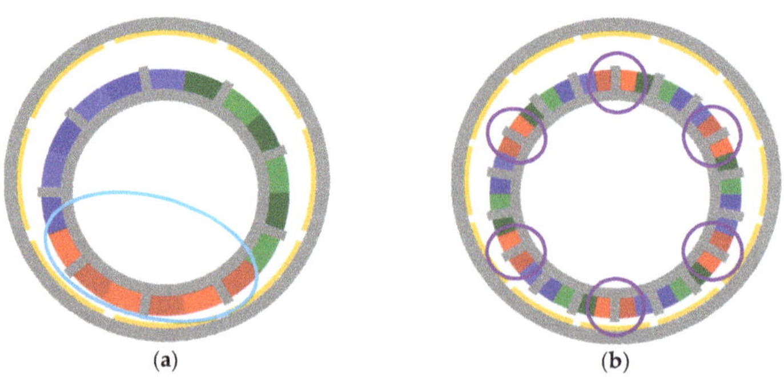

Figure 9. Schematics of (**a**) a 9-slot, 8-pole PMSM with conventional stator core subject to eccentricity. In this case, the phase windings form a single group of conductors (light blue oval) distributed in a specific part of the stator circumference and (**b**) an 18-slot, 12-pole PMSM with conventional stator core subject to eccentricity. In this case, the phase windings form six group of conductors (light blue oval) evenly distributed in the stator circumference, damping the potential back-emf unbalance.

For the case of the 9-slot, 8-pole PMSM, when eccentricity is applied, the minimum airgap position can increase or decrease the back-emf magnitude of a phase depending on how close it is to the phase windings, since it can strengthen or weaken the linked flux. As can be seen from the schematics of Figure 9a, phase A is strengthened as the airgap in front of phase A is smaller, and the back-emf of phase B and C should be lower since the equivalent airgap is larger than the original. In that case, the phase windings are not evenly distributed in the stator circumference, which translates into an unbalanced back-emf waveform when eccentricity is present. By the contrary, in the case of the 18-slot, 12-pole machine shown in Figure 9b, the phase windings are forming six groups of conductors evenly distributed in the stator circumference. Even when the rotor and stator are misaligned, the flux linked by a phase is strengthened on one side of the machine but weakened on the other side of the machine, which results into a low-to-null variation of the back-emf. The number of evenly distributed phase groups will be denoted by N_{pg} from now on.

In summary, relevant features of the back-emf results can be summarized by the following indicators, useful to evaluate and compare the different slot/pole combinations of Table 2.

- Fundamental component of the back-emf, measured in V;
- Third-order harmonic of the back-emf, measured in V;
- Maximum unbalance between phases, measured in V.

6.2. DE: Comparison of Slot/Pole Combinations for PMSMs and MPMSMs

In Table 14, the fundamental component of back-emf in the presence of the maximum DE magnitude (1.00 mm) is presented for both the modular and monolithic machines considering different slot/pole combinations. As in the case of the 24-slot, 22-pole machine, no significant effect is observed in the back-emf fundamental and third-order harmonics caused by eccentricity.

Table 14. Fundamental component and third-order harmonic of back-emf for different PMSM and MPMSM slot/pole combinations in the presence of DE.

ECC / $Q_s/2p$		0 mm (Faultless)						1.00 mm					
		Modular [V]			Monolithic [V]			Modular [V]			Monolithic [V]		
	HO	A	B	C	A	B	C	A	B	C	A	B	C
18S 12P	1st	247.5	247.5	247.5	263.7	263.8	263.7	247.6	247.7	247.6	263.7	263.9	263.7
	3rd	21.9	21.9	21.9	24.6	24.0	24.6	22.0	22.0	22.0	24.8	24.4	24.9
18S 20P	1st	281.0	281.1	281.0	275.4	275.5	275.4	281.7	281.7	281.6	276.2	276.2	276.2
	3rd	31.6	31.6	31.6	35.9	36.0	36.0	33.3	33.3	33.3	37.8	37.8	37.8
24S 20P	1st	276.7	276.7	276.6	275.9	275.9	275.8	277.5	277.5	277.5	276.7	276.6	276.6
	3rd	37.9	37.9	37.9	37.8	37.8	37.9	40.5	40.6	40.6	39.7	39.7	39.7
24S 22P	1st	283.4	283.4	283.3	283.2	283.2	283.2	284.1	284.1	284.1	283.9	283.9	283.9
	3rd	28.7	28.7	28.7	29.0	28.9	29.0	30.7	30.7	30.7	30.8	30.8	30.9
24S 28P	1st	300.9	300.9	300.9	292.5	292.5	292.5	303.4	303.4	303.4	294.8	294.8	294.8
	3rd	9.5	9.5	9.7	18.9	18.9	19.1	9.6	9.6	9.8	19.8	19.8	19.8

In Table 15, the maximum unbalance between phases in the presence of the maximum DE magnitude (1.00 mm) is presented for both the modular and monolithic machines considering different slot/pole combinations. It can be noted that the number of phase groups evenly distributed in the stator circumference is a strong indicator of the magnitude of back-emf unbalance for a given slot/pole combination. No significant differences were found by comparing the results of modular and monolithic machines.

Table 15. Maximum back-emf unbalance for different PMSM and MPMSM slot/pole combinations in the presence of DE.

ECC	Modular [V]	Monolithic [V]	N_{pg}
18S 12P	0.526	0.657	6
18S 20P	10.386	9.091	2
24S 20P	1.331	1.312	4
24S 22P	6.660	6.286	2
24S 28P	1.065	1.301	4

6.3. SE: Evaluation on a 24-Slot, 22-Pole PMSM and MPMSM

In Figure 10, the back-emf outcomes of the FE evaluation of a 24-slot, 22-pole PMSM and a 24-slot, 22-pole MPMSM are presented for SE. The results were extracted considering the rotor structure completes one full turn to correctly obtain the harmonic content of the back-emf signal, which is depicted in the figures. In Figure 10a,c, the three-phase back-emf waveforms are presented for a 24-slot, 22-pole MPMSM and PMSM, respectively. In turn, Figure 10b,d show the harmonic spectrum of the back-emf for both machines, modular and conventional, when SE is present.

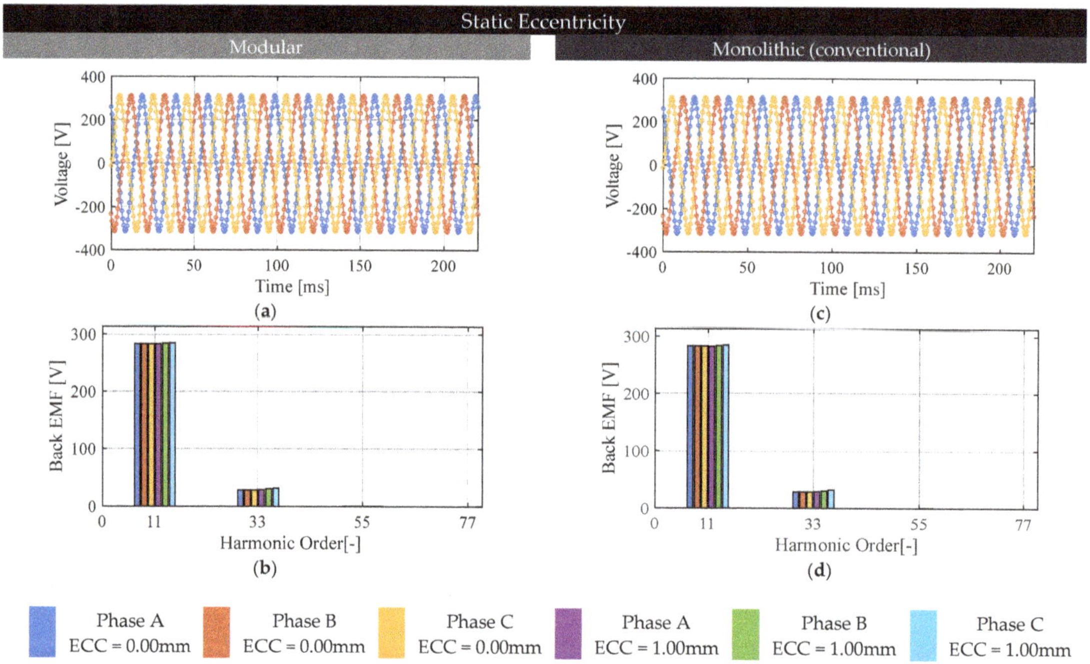

Figure 10. FE evaluation of the back-emf in the presence of SE for a 24-slot, 22-pole PMSM and MPMSM: (**a**) back-emf waveform for the modular machine, (**b**) back-emf spectrum for the modular machine, (**c**) back-emf waveform for the monolithic machine, (**d**) back-emf spectrum for the monolithic machine.

From Figure 10, and similar to what was observed in Figure 8, eccentricity does not have a significant impact on the back-emf magnitude, and no significant unbalance between phases is observed either.

6.4. SE: Comparison of Slot/Pole Combinations for PMSMs and MPMSMs

In Table 16, the fundamental component of back-emf in the presence of the maximum SE magnitude (1.00 mm) is presented for both the modular and monolithic machines considering different slot/pole combinations. As in the case of the 24-slot, 22-pole machine, no significant effect is observed in the back-emf fundamental and third-order harmonics caused by eccentricity.

Table 16. Fundamental component and third-order harmonic of back-emf for different PMSM and MPMSM slot/pole combinations in the presence of SE.

ECC / $Q_s/2p$		0 mm (Faultless)						1.00 mm					
		Modular [V]			Monolithic [V]			Modular [V]			Monolithic [V]		
	HO	A	B	C	A	B	C	A	B	C	A	B	C
18S 12P	1st	247.2	247.3	247.2	263.5	263.7	263.5	247.5	247.6	247.4	263.7	263.8	263.7
	3rd	21.9	21.9	21.9	24.5	24.1	24.5	21.9	22.0	22.1	24.6	24.6	24.6
18S 20P	1st	280.7	280.8	280.7	275.1	275.2	275.1	281.2	280.6	283.0	275.3	275.4	277.5
	3rd	31.7	31.7	31.7	36.0	36.1	36.1	30.5	32.9	36.97	36.2	36.2	41.3
24S 20P	1st	276.7	276.8	276.7	281.7	281.7	281.7	277.4	277.4	277.3	282.3	282.3	282.2
	3rd	37.9	37.8	37.9	38.6	38.6	38.6	40.3	40.4	40.4	40.4	40.4	40.5
24S 22P	1st	283.4	283.4	283.3	283.2	283.2	283.2	283.1	284.0	285.0	282.9	283.9	284.9
	3rd	28.7	28.7	28.7	29.0	29.0	29.0	29.3	30.8	32.2	29.6	31.0	32.2
24S 28P	1st	300.9	301.0	300.9	292.5	292.5	292.5	303.2	303.2	303.1	294.6	294.6	294.5
	3rd	9.5	9.5	9.6	19.0	19.0	19.1	9.7	9.7	9.7	19.8	19.8	19.8

In Table 17, the maximum unbalance between phases in the presence of the maximum SE magnitude (1.00 mm) is presented for both the modular and monolithic machines considering different slot/pole combinations. As in the case of DE, it is clear that the number of phase groups evenly distributed in the stator circumference is a strong indicator of the magnitude of back-emf unbalance for a given slot/pole combination. However, in the case of SE, this unbalance is static, which depends on the position of the minimum airgap, and it does not change from one phase to another. Notwithstanding, no significant differences were found by comparing the results of modular and monolithic machines.

Table 17. Maximum back-emf unbalance for different PMSM and MPMSM slot/pole combinations in the presence of SE.

ECC	Modular [V]	Monolithic [V]	N_{pg}
18S 12P	0.453	0.527	6
18S 20P	7.134	7.284	2
24S 20P	0.391	0.367	4
24S 22P	4.012	3.934	2
24S 28P	0.397	0.539	4

6.5. Summary: SE and DE on Back-Emf of PMSMs and MPMSMs

In summary, it was found that eccentricity has a low impact on the back-emf of the evaluated slot/pole combinations. In this regard:

- Back-emf magnitude is not affected by eccentricity, which was observed for all the evaluated slot/pole combinations and can translate into a low impact of eccentricity on the mean torque;
- Slot/pole combinations having a high number of evenly distributed phase groups (see Figure 9) are less likely to develop back-emf unbalance. That is the case for the 18- slot, 12-pole, 24-slot, 20-pole, and 24-slot, 28-pole machines.

7. Results and Discussion: Mean Torque

7.1. DE: Comparison of Slot/Pole Combinations for PMSMs and MPMSMs

In Table 18, the mean torque for different DE magnitudes is presented. Both the modular and monolithic machines are evaluated considering different slot/pole combinations. From Table 18, it can be seen that the mean torque is not affected by DE regardless of the slot/pole combination, which was expected from the back-emf results: since the flux linkage is not considerably penalized by eccentricity, the mean torque should not be decreased either.

Table 18. Mean torque of different PMSM and MPMSM slot/pole combinations in the presence of DE.

ECC / $Q_s/2p$	0 mm (Faultless)		0.25 mm		0.50 mm		0.75 mm		1.00 mm	
	Mod [Nm]	Mon [Nm]	Mod [Nm]	Mon [Nm]	Mod [Nm]	Mon [Nm]	Mod [Nm]	Mon [Nm]	Mod [Nm]	Mon [Nm]
18S 12P	147.8	171.8	148.0	172.0	147.8	171.8	147.9	171.8	148.0	171.9
18S 20P	218.9	226.1	218.7	225.8	218.9	226.0	218.9	226.0	219.0	226.2
24S 20P	208.4	215.4	207.6	214.6	207.6	214.6	208.1	215.1	208.3	215.3
24S 22P	216.2	218.4	216.2	218.4	216.3	218.5	216.4	218.6	216.6	218.8
24S 28P	207.4	232.4	207.2	232.2	207.4	232.4	207.8	232.7	208.5	233.4

It could be noted that the 18-slot, 12-pole modular machine develops significantly lower mean torque than the monolithic counterpart. This can be explained by the same analysis carried out for the radial forces in Section 4.2 (see Figure 5): the absence of ferromagnetic material between stator modules penalizes the flux linkage in machines having a high value of q.

7.2. SE: Comparison of Slot/Pole Combinations for PMSMs and MPMSMs

In Table 19, the mean torque developed by the machines for different SE magnitudes is summarized, considering different slot/pole combinations. From Table 19, and complementing the findings of Table 18, it is apparent that the mean torque is not affected by DE either. The mean torque difference between machines with eccentricity and faultless machines is always lower than 0.5%.

Table 19. Mean torque of different PMSM and MPMSM slot/pole combinations in the presence of SE.

ECC / $Q_s/2p$	0 mm (Faultless)		0.25 mm		0.50 mm		0.75 mm		1.00 mm	
	Mod [Nm]	Mon [Nm]	Mod [Nm]	Mon [Nm]	Mod [Nm]	Mon [Nm]	Mod [Nm]	Mon [Nm]	Mod [Nm]	Mon [Nm]
18S 12P	147.7	171.7	147.7	171.8	147.7	171.7	147.7	171.8	147.8	171.8
18S 20P	218.2	225.3	218.4	225.5	218.4	225.5	218.6	225.7	218.7	225.9
24S 20P	207.9	214.9	208.0	215.0	208.0	215.0	208.3	215.3	208.4	215.3
24S 22P	216.2	218.4	216.2	218.4	216.3	218.5	216.4	218.6	216.6	218.8
24S 28P	207.3	232.2	207.2	232.2	207.4	232.4	207.8	232.7	208.1	232.9

7.3. Summary: SE and DE on the Mean Torque of PMSMs and MPMSMs

In summary, it was found that eccentricity has a negligible impact on the mean torque of the evaluated slot/pole combinations. Modular machines may develop a lower mean torque than monolithic machines depending on the number of slots per pole per phase.

8. Conclusions

In this paper, the impact of eccentric tolerances on relevant performance indices of MPMSMs was identified and described for different slot/pole combinations, as well as compared with equivalent machines with conventional monolithic stator. Both static and dynamic eccentricity were assessed for five slot/pole combinations, and the radial forces, cogging torque, back-EMF, and mean torque were analyzed. It was found that:

- Radial forces on machines with SE and DE scale with the slot number and pole number, and that the severity of these forces is lower in the case of modular machines with a high value of slots per pole per phase. In addition, a high-frequency force ripple of significant magnitude is present in machines with a slot number and pole number close to each other.
- When assessing the cogging torque, additional harmonic components (AHC) of severe magnitude are generated due to eccentricity. In this regard, machines which slot number is close to its pole number are very sensitive to eccentricity and develop a substantial cogging torque increase of up to 400%. There are no significant differences between DE and SE in terms of the cogging torque increase magnitude. Nevertheless, the harmonic order of the AHC generated by DE corresponds to the slot number and multiples, while the order of the AHC generated by SE is equivalent to pole number and its multiples.
- Back-emf magnitude and mean torque are not affected by eccentricity, indistinctly for DE or SE. This was observed for all the evaluated slot/pole combinations. However, slot/pole combinations with a low number of evenly distributed phase groups can develop a mild back-emf unbalance.

The findings of this analysis can be used in the design stage of a MPMSM and aim to enable the inclusion of the sensitiveness of each performance indicator to quickly compare the performance between slot/pole combinations. Furthermore, the observed results for the cogging torque and radial forces can be used for diagnosis ends: in MPMSMs with a slot number close to their pole number, anomalous cogging torque increases can indicate the presence of SE or DE, which can be distinguished by the harmonic order of the cogging torque signal.

Author Contributions: Conceptualization, D.R., C.M., and W.J.; methodology, D.R., W.J., and G.B.; software, D.R. and C.M.; validation, D.R. and C.M.; formal analysis, D.R. and W.J.; investigation, D.R.; resources, W.J., J.R., J.A.T., and G.B.; data curation, D.R. and C.M.; writing—original draft preparation, D.R., C.M., W.J., G.B., and J.R.; writing—review and editing, D.R., W.J., G.B., and J.A.T.; visualization, D.R.; supervision, W.J., J.A.T., and G.B.; project administration, W.J.; funding acquisition, W.J., J.R., and G.B. All authors have read and agreed to the published version of the manuscript.

Funding: This work is supported in part by the Agencia Nacional de Investigación y Desarrollo (ANID), Chile through grant ANID-PFCHA/Doctorado Nacional/2020-21200350, project FONDECYT REGULAR #1201667, and project FONDEF ID21I10099.

Institutional Review Board Statement: Not applicable.

Informed Consent Statement: Not applicable.

Data Availability Statement: Data are contained within the article.

Conflicts of Interest: The authors declare no conflict of interest.

References

1. Ullah, S.; McDonald, S.P.; Martin, R.; Benarous, M.; Atkinson, G.J. A Permanent Magnet Assist, Segmented Rotor, Switched Reluctance Drive for Fault Tolerant Aerospace Applications. *IEEE Trans. Ind. Appl.* **2019**, *55*, 298–305. [CrossRef]
2. Kang, H.; Zhou, L.; Wang, J.; Peng, X. Investigation of multiphase modular stator SPM with fault tolerance. In Proceedings of the 2014 17th International Conference on Electrical Machines and Systems (ICEMS), Hangzhou, China, 22–25 October 2014; pp. 1572–1575. [CrossRef]
3. Zhao, W.; Xu, L.; Liu, G. Overview of permanent-magnet fault-tolerant machines: Topology and design. *CES Trans. Electr. Mach. Syst.* **2018**, *2*, 51–64. [CrossRef]

4. Zheng, P.; Sui, Y.; Zhao, J.; Tong, C.; Lipo, T.A.; Wang, A. Investigation of a Novel Five-Phase Modular Permanent-Magnet In-Wheel Motor. *IEEE Trans. Magn.* **2011**, *47*, 4084–4087. [CrossRef]
5. Liu, G.; Chen, M.; Zhao, W.; Chen, Q.; Zhao, W. Design and Analysis of Five-Phase Fault-Tolerant Interior Permanent-Magnet Vernier Machine. *IEEE Trans. Appl. Supercond.* **2016**, *26*, 1–5. [CrossRef]
6. Dusek, J.; Arumugam, P.; Brunson, C.; Amankwah, E.K.; Hamiti, T.; Gerada, C. Impact of Slot/Pole Combination on Inter-Turn Short-Circuit Current in Fault-Tolerant Permanent Magnet Machines. *IEEE Trans. Magn.* **2016**, *52*, 1–9. [CrossRef]
7. Jiang, X.; Xu, D.; Gu, L.; Li, Q.; Xu, B.; Li, Y. Short-Circuit Fault-Tolerant Operation of Dual-Winding Permanent-Magnet Motor Under the Four-Quadrant Condition. *IEEE Trans. Ind. Electron.* **2019**, *66*, 6789–6798. [CrossRef]
8. Bianchini, C.; Torreggiani, A.; David, D.; Davoli, M.; Bellini, A. Fault Tolerance Analysis of a Ironless PM Machine for Energy Storage. In Proceedings of the 2020 IEEE Energy Conversion Congress and Exposition (ECCE), Detroit, MI, USA, 11–15 October 2020; pp. 4499–4504. [CrossRef]
9. Deng, W.; Zuo, S. Electromagnetic Vibration and Noise of the Permanent-Magnet Synchronous Motors for Electric Vehicles: An Overview. *IEEE Trans. Transp. Electrif.* **2019**, *5*, 59–70. [CrossRef]
10. Tahanian, H.; Aliahmadi, M.; Faiz, J. Ferrite Permanent Magnets in Electrical Machines: Opportunities and Challenges of a Non-Rare-Earth Alternative. *IEEE Trans. Magn.* **2020**, *56*, 1–20. [CrossRef]
11. Bird, J.Z. A Review of Electric Aircraft Drivetrain Motor Technology. *IEEE Trans. Magn.* **2022**, *58*, 1–8. [CrossRef]
12. Gan, C.; Chen, Y.; Qu, R.; Yu, Z.; Kong, W.; Hu, Y. An Overview of Fault-Diagnosis and Fault-Tolerance Techniques for Switched Reluctance Machine Systems. *IEEE Access* **2019**, *7*, 174822–174838. [CrossRef]
13. Sarigiannidis, A.G.; Beniakar, M.E.; Kakosimos, P.E.; Kladas, A.G.; Papini, L.; Gerada, C. Fault Tolerant Design of Fractional Slot Winding Permanent Magnet Aerospace Actuator. *IEEE Trans. Transp. Electrif.* **2016**, *2*, 380–390. [CrossRef]
14. Chen, H.; Liu, X.; Zhao, J.; Demerdash, N.A.O. Magnetic-Coupling Characteristics Investigation of a Dual-Rotor Fault-Tolerant PMSM. *IEEE Trans. Energy Convers.* **2018**, *33*, 362–372. [CrossRef]
15. Petrov, I.; Di, C.; Lindh, P.; Niemelä, M.; Repo, A.; Pyrhönen, J. Fault-tolerant modular stator concentrated winding permanent magnet machine. *IEEE Access* **2020**, *8*, 7806–7816. [CrossRef]
16. Jara, W.; Petrov, I.; Tapia, J.; Lindh, P.; Pyrhönen, J. Analytical model of tooth-coil winding permanent magnet synchronous machines with modular U-Shape stator. In Proceedings of the 2016 XXII International Conference on Electrical Machines (ICEM), Lausanne, Switzerland, 4–7 September 2016; pp. 145–151.
17. Madariaga, C.; Jara, W.; Tapia, J.; Riedemann, J.; Bramerdorfer, G.; Castro, P.; Sarlioglu, B. Analytical model and sensitivity analysis of tooth-coil-winding permanent magnet synchronous machine with modular U-shape stator. In Proceedings of the 2019 IEEE Energy Conversion Congress and Exposition (ECCE), Baltimore, MD, USA, 29 September–3 October 2019; pp. 1761–1768.
18. Madariaga, C.; Jara, W.; Riquelme, D.; Bramerdorfer, G.; Tapia, J.; Riedemann, J. Impact of tolerances on the cogging torque of tooth-coil-winding PMSMs with modular stator core by means of efficient superposition technique. *Electronics* **2020**, *9*, 1594. [CrossRef]
19. Li, G.J.; Zhu, Z.Q.; Chu, W.Q.; Foster, M.P.; Stone, D.A. Influence of flux gaps on electromagnetic performance of novel modular PM machines. *IEEE Trans. Energy Convers.* **2014**, *29*, 716–726. [CrossRef]
20. Wang, H.; Zhu, H.; Ding, S.; Dai, Y.; He, C.; Wang, X. A New Partitioned Stator Field Modulation Machine With H-Shape Permanent Magnet Excitation. *IEEE Trans. Magn.* **2022**, *58*, 1–6. [CrossRef]
21. Perez, E.; Jara, W.; Madariaga, C.; Tapia, J.; Bramerdorfer, G.; Riedemann, J.; Pyrhönen, J. Comparison of Optimized Fault-Tolerant Modular Stator Machines with U-shape and H-shape Core Structure. In Proceedings of the Comparison of Optimized Fault-Tolerant Modular Stator Machines with U-shape and H-shape Core Structure, Vancouver, BC, Canada, 10–14 October 2021; pp. 4254–4259.
22. Khan, S.; Pasund, A.; Ahmad, N.; Ahmed, S.; Khan, H.A.; Cheema, K.M.; Milyani, A.H. Performance Investigation and Cogging Torque Reduction in a Novel Modular Stator PM Flux Reversal Machine. *Energies* **2022**, *15*, 2261. [CrossRef]
23. Wang, A.; Li, S. Investigation of E-Core Modular Permanent Magnet Wind Turbine. *Energies* **2020**, *13*, 1751. [CrossRef]
24. Riquelme, D.; Jara, W.; Madariaga, C.; Tapia, J.; Bramerdorfer, G.; Riedemann, J. Impact of Static and Dynamic Eccentricity on the Performance of Permanent Magnet Synchronous Machines with Modular Stator Core. In Proceedings of the 2021 IEEE Energy Conversion Congress and Exposition (ECCE), Vancouver, BC, Canada, 10–14 October 2021; pp. 3775–3780. [CrossRef]
25. Ebrahimi, B.M.; Faiz, J.; Roshtkhari, M.J. Static-, Dynamic-, and Mixed-Eccentricity Fault Diagnoses in Permanent-Magnet Synchronous Motors. *IEEE Trans. Ind. Electron.* **2009**, *56*, 4727–4739. [CrossRef]
26. Wang, Y.; Liu, K.; Hua, W.; Zhang, C.; Wu, Z.; Zhang, H. Analysis and Detection of Rotor Eccentricity in Permanent Magnet Synchronous Machines Based on Linear Hall Sensors. *IEEE Trans. Power Electron.* **2022**, *37*, 4719–4729. [CrossRef]
27. Zhang, A.; Bai, Y.; Yang, B.; Li, H. Analysis of Nonlinear Vibration in Permanent Magnet Synchronous Motors under Unbalanced Magnetic Pull. *Appl. Sci.* **2018**, *8*, 113. [CrossRef]
28. Bessous, N.; Chemsa, A.; Sbaa, S. New Vision about the Mixed Eccentricity Fault Causes in Induction Motors and its relationship with the Rolling Element Bearing Faults: Analytical model dedicated to the REB faults. In Proceedings of the 2018 International Conference on Communications and Electrical Engineering (ICCEE), El Oued, Algeria, 17–18 December 2018; pp. 1–11. [CrossRef]
29. Zhu, Z.Q.; Jamil, M.L.M.; Wu, L.J. Influence of slot and pole number combinations on unbalanced magnetic force in permanent magnet machines. In Proceedings of the 2011 IEEE Energy Conversion Congress and Exposition, Phoenix, AZ, USA, 17–22 September 2011; pp. 3291–3298. [CrossRef]

30. Park, J.-C.; Park, S.-H.; Kim, J.-H.; Lee, S.-G.; Lee, G.-H.; Lim, M.-S. Diagnosis and Robust Design Optimization of SPMSM Considering Back EMF and Cogging Torque due to Static Eccentricity. *Energies* **2021**, *14*, 2900. [CrossRef]
31. Lee, H.K.; Bang, T.K.; Woo, J.H.; Shin, H.S.; Choi, J.Y. Electromagnetic Characteristic Analysis of Permanent Magnet Synchronous Machine Considering Current Waveform According to Static Rotor Eccentricity. *Appl. Sci.* **2020**, *10*, 8453. [CrossRef]
32. Madariaga, C.; Jara, W.; Tapia, J.A.; Pyrhönen, J.; Lindh, P.; Riedemann, J.A. Closed-Form Solution for the Slot Leakage Inductance of Tooth-Coil-Winding Permanent Magnet Machines. *IEEE Trans. Energy Convers.* **2019**, *34*, 1572–1580. [CrossRef]
33. Zhu, Z.Q.; Wu, L.J.; Jamil, M.L.M. Influence of Pole and Slot Number Combinations on Cogging Torque in Permanent-Magnet Machines With Static and Rotating Eccentricities. *IEEE Trans. Ind. Appl.* **2014**, *50*, 3265–3277. [CrossRef]
34. Gašparin, L.; Fišer, R. Intensity of the native and additional harmonic components in cogging torque due to design parameters of permanent-magnet motors. In Proceedings of the 2009 International Conference on Power Electronics and Drive Systems (PEDS), Taipei, Taiwan, 2–5 November 2009; pp. 1062–1067. [CrossRef]
35. Guo, L.; Wang, H. Research on Stator Slot and Rotor Pole Combination and Pole Arc Coefficient in a Surface-Mounted Permanent Magnet Machine by the Finite Element Method. *World Electr. Veh. J.* **2021**, *12*, 26. [CrossRef]

electronics

MDPI

Article

Transmission Performance of Halbach Array Cylindrical Permanent Magnet Governor

Yonglong Zhu [1,2,†], Hai Wang [1,2,*,†], Henian Li [1,2], Chunlai Yang [1,2] and Jingsong Gui [3]

1 School of Mechanical Engineering, Anhui Polytechnic University, Wuhu 241000, China
2 Anhui Key Laboratory of Advanced Numerical Control & Servo Technology, Wuhu 241000, China
3 Wuhu Ruilong Robot Technology Co., Ltd., Wuhu 241000, China
* Correspondence: wanghai@ahpu.edu.cn; Tel.: +86-0135-1553-3761
† These authors contributed equally to this work.

Abstract: A novel cylinder permanent magnet governor (CPMG) with Halbach segmentation is proposed in this paper. In order to improve the transmission performance of the CPMG, different permanent magnet (PM) arrangement methods are adopted. To achieve a fair comparison result, all the PMs are of the same size. The main magnetic fluxes are considered to obtain a comprehensive equivalent magnetic circuit model of the CPMG with Halbach array and analytical output torque that is calculated. The analytical method of transmitted torque for CPMG is then presented. Additionally, the effect of the average output torque of CPMG under parameters of the thickness of the copper rings, the slip rate and the effective coupling of the copper rings are investigated. Finally, the prototype platform is ready for testing on the field. The results were consistent with the results of the simulation, and the error was kept within the range of 5%. This research can provide a theoretical and practical reference for the optimal design of the transmission characteristic of CPMG.

Keywords: permanent magnet governor; Halbach array; magnetic field characteristics; output torque

Citation: Zhu, Y.; Wang, H.; Li, H.; Yang, C.; Gui, J. Transmission Performance of Halbach Array Cylindrical Permanent Magnet Governor. *Electronics* **2023**, *12*, 1161. https://doi.org/10.3390/electronics12051161

Academic Editor: Ahmed Abu-Siada

Received: 4 January 2023
Revised: 24 February 2023
Accepted: 24 February 2023
Published: 27 February 2023

1. Introduction

As a new type of transmission equipment, cylindrical permanent magnet governor (CPMG) has attracted much attention in recent years and is widely used in fan and pump systems [1,2]. Compared to traditional mechanical connections, cylindrical permanent magnet governor (CPMG) uses magnetic field coupling between a motor and a load shaft to transfer torque in the transmission system without any direct mechanical connections, which can produce less mechanical friction with over-torque protection ability as well as soft start through adjusting the coupling ratio between the conductor and magnet [3–5]. Unlike an axial permanent magnet governor, CPMG is a simple structure with fewer axial forces during operation. The permanent magnet governor operates in noncontact mode, which avoids vibration operation.

In recent years, many researchers have carried out a series of studies on permanent magnet governors. A large number of researchers are dedicated to the analysis and design optimization of the PMG with 2D and 3D FEA [6–11]. These works provide ideas on how to select meshing for FEA models, reduce numerical errors, selection of design parameters and constraints for the optimization process. Taqavi Omolbanin has calculated flux densities in different parts of the axial-flux permanent magnet machines combining the solution of Maxwell's equations and the magnetic equivalent circuit (MEC), taking into account the effects of permanent magnet (PM) reluctance, leakage flux between PM-to-PM, PM-to-rotor leakage flux at different PM edges, and the fringing flux effect. By considering these leakage fluxes as a factor called variation function in the design procedure of the machine, the analysis time is dramatically reduced compared to numerical methods while maintaining high reliability [12,13]. These works provide ideas on how to analyze the modeling of CPMG. Similarly, CPMG can also be called transmitted torque calculation considering theoretical and FEA models [12] and detailed theoretical and experimental analysis of CPMG under steady-state and transient operation [14]. In-cheol Kim designed a magnetic coupling that improves the water tightness of a marine current turbine (MCT). The results show that the new design is watertight and has lower mechanical losses [15]. Tian Mengmeng and others designed a novel flux-adjustable permanent magnet coupler (PMC) with a double-layer permanent magnet rotor (PMR) and used a two-layer arrangement of permanent magnets N-S to achieve optimum control of the speed of the torque [15–17]. Lu designed a general 3D analytical method for calculating the magnetic field distributions and eddy currents [18]. These researches are mainly focused on improving output torque through structural design for high torque applications. Furthermore, by focusing on the recent trends in electromagnetic devices, the design of magnetic gear could be mentioned here. Magnetic gears also have a similar structure to that of the CPMG, and the only difference is the inclusion of an extra component called modulation pieces between the two rotors [18–24].

From the research mentioned above, the methods are large-costly and time-consuming. This paper aims to improve the output Torque performance of the CPMG drives for different operating conditions. A comparative analysis of different magnets arrangement to ensure a torque production capability higher than traditional arrangement and the analysis is performed considering the same number of magnets. The traditional design of permanent magnet governor is mainly focused on magnetic fields and transmission properties with typical N- and S-arrangement magnetization. Halbach magnet array can generate the strongest magnetic field with the least amount of magnets, which is proposed by Klaus Halbach [24]. Halbach's permanent magnet array has a unidirectional magnetic field, which improves the air gap magnetic flux density, increases the sinusoidal magnetic field of the air gap and enhances transmission stability [20–27]. A cylindrical permanent magnet governor with a Halbach magnet array is proposed in this paper. The main magnetic fluxes are considered to obtain a comprehensive magnetic circuit model. The analytical method of transmitted torque for CPMG is then presented. Based on the comparative analysis, the CPMG model with the best performance is prototyped and tested on the inductor motor dynamometer testbed. The following sections provide a detailed description of the proposed work.

2. Materials and Methods

2.1. Cylindrical Permanent Magnet Governor with Halbach Magnet Array

CPMG is composed of the conductor rotor, which is mainly composed of a conductor barrel, a copper ring and the permanent magnet rotor. The permanent magnet rotor contains a permanent magnet tray and PMs, where the PMs are embedded in the permanent magnet tray. As shown in Figure 1, the conductor rotor is connected to the motor shaft, and the output axis is connected to the permanent magnet rotor. There is no contact between the conductor and the permanent magnet rotor, which allows a certain alignment error during installation.

When the motor is in operation, the conductor barrel connected to the motor shaft is rotating, which cuts the magnetic field generated by the permanent magnet array. An induced current is generated in the copper ring. A Lorentz force is applied to the copper ring, which generates a torque to drive the permanent magnet array. There is always a rotation speed difference between the conductor and permanent magnet array in order to ensure the presence of output torque. The rotation speed of the output shaft and output torque can be adjusted by regulation of the effective coupling area between the conductor and the permanent magnet array.

The schematic diagram of the specific speed regulation principle is shown in Figure 2. The output torque and rotation speed can be tuned through the adjustment of the effective coupling ratio (the ratio of the actual coincidence area between the conductor and the permanent magnet to the full coincidence area). The maximum output torque can be realized when the permanent rotor is fully coupled to the rotor of conductors (100% effective coupling ratio).

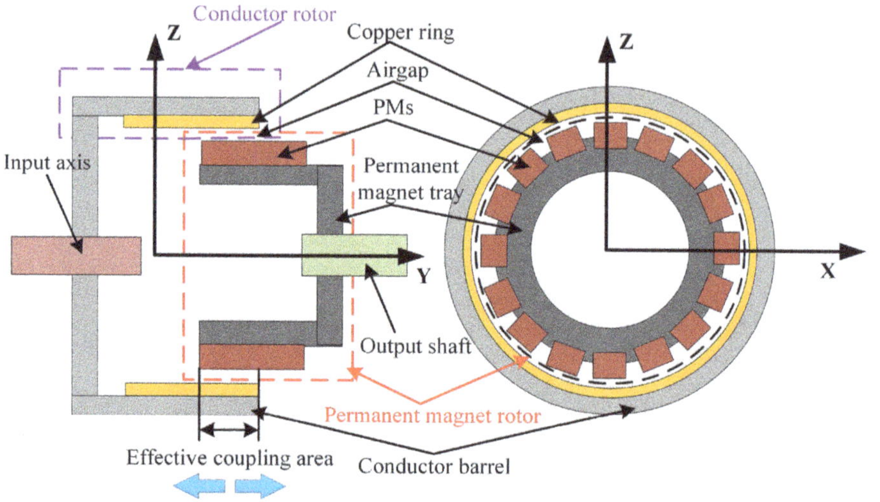

Figure 1. Structure of CPMG.

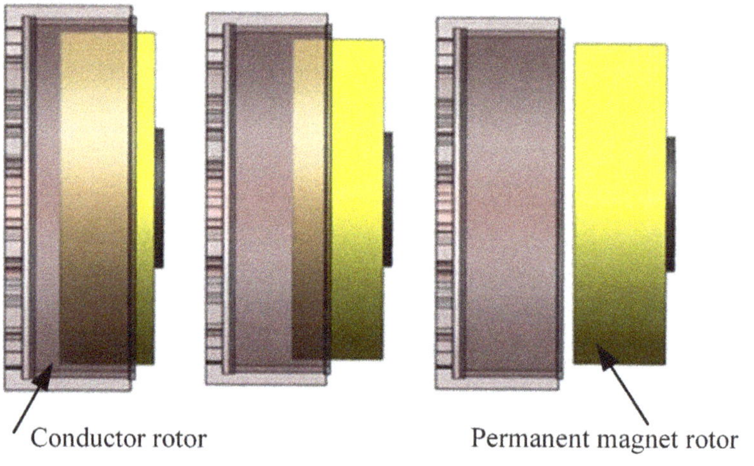

Figure 2. Speed regulation principle of CPMG.

2.2. *Equivalent Magnetic Circuit of Halbach Permanent Magnet Cell and Simulation Model*

In order to simplify the process for analysis, the main components of CPMG were unfolded radially into a 2D model. The magnetic field distribution was assumed to be linear in CPMG without consideration of leakage. The equivalent model of a Halbach cell is shown in Figure 3. The permanent magnet's geometry was considered to be rectangular shaped, which can be equalized to the magnetomotive. The main flux paths (solid line) were where it passes through PMs, airgap, copper ring, conductor barrel and permanent magnet tray. The leakage flux paths (dashed line) were mainly generated between adjacent permanent magnets.

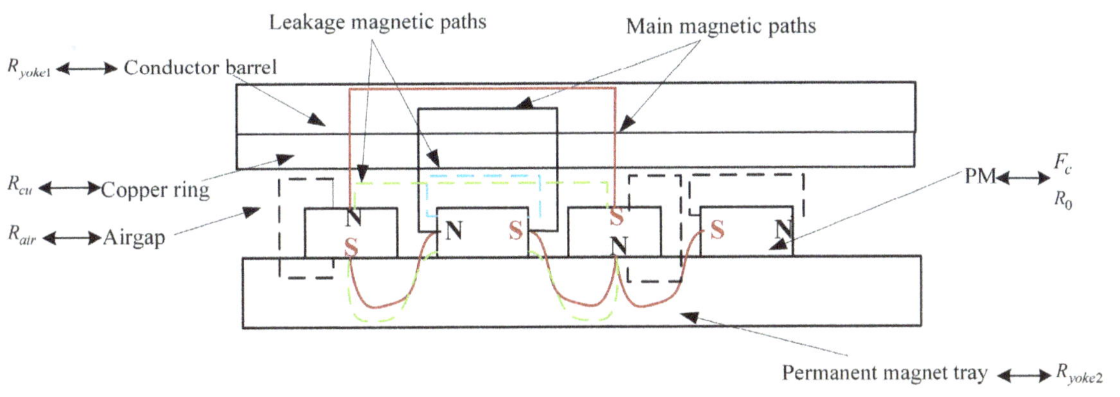

Figure 3. Equivalent magnetic model of a Halbach permanent magnet cell.

On the basic, before-mentioned Equivalent magnetic model of a Halbach permanent magnet cell analysis, the permanent magnet's geometry is considered to be rectangular shaped, which is equalized to be magnetomotive. The Conductor barrel is regarded as R_{yoke2}, the permanent magnet tray is regarded as R_{yoke1}, the copper ring is regarded as R_{cu}, and the space in between the adjacent permanent magnets is regarded as R_L. Therefore, the magnetic reluctance circuit of the Halbach permanent magnet cell is presented in Figure 4, where only a pair of Halbach array magnets is considered on account of the symmetry of the magnetic circuit.

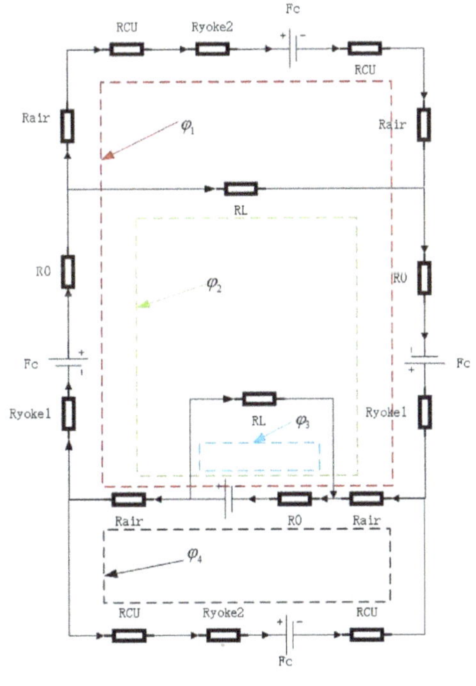

Figure 4. Magnetic reluctance circuit of Halbach permanent magnet cell.

According to the principles of equal magnetic circuits, a magnetic circuit was studied under a pair of magnetic poles, and a permanent magnet was regarded as the magnetomotive force F_c, in series with the internal reluctance r_0 of the permanent magnet, to form the magnetomotive force source.

$$F_c = H_c h_{mp} \tag{1}$$

where H_c is the coercive force of the permanent magnet, and h_{mp} is the magnetization direction length of the permanent magnet. When the permanent magnet governor is operating, the induced electromotive force in the copper ring during the cutting of the magnetic field generated by the permanent magnet array is created by the eddy current, which increases the armature magnetomotive force F_a.

The magnetic reluctance in the circuit is calculated below:

$$R = \frac{L}{\mu S} \tag{2}$$

where L represents the length and S represents the cross-section of the magnet circuit, and μ represents the magnetic permeability.

According to the current KCL law, the magnetic circuit equations of the Halbach magnet cell are built as below:

$$\begin{cases} \left(3R_0 + 2R_{Yoke1} + 4R_{air} + 2R_{Cu} + R_{yoke2}\right)\varphi_1 + (3R_0 + 2R_{Yoke1} + 2R_{air})\varphi_2 \\ +R_0\varphi_3 + (2R_{air} + R_0)\varphi_4 = 3F_m - F_a \\ (3R_0 + 2R_{Yoke1} + 2R_{air})\varphi_1 + (2R_0 + 2R_{Yoke1} + 2R_{air} + R_L)\varphi_2 + R_0\varphi_3 \\ +(2R_{air} + R_0)\varphi_4 = 3F_m \\ R_0\varphi_1 + R_0\varphi_2 + (R_0 + R_L)\varphi_3 + R_0\varphi_4 = F_m \\ (2R_{air} + R_0)\varphi_1 + (2R_{air} + R_0)\varphi_2 + R_0\varphi_3 + (2R_{air} + 2R_{Cu} + R_{Yoke2} + R_0) \\ = F_m - F_a \end{cases} \tag{3}$$

In Equation (3), φ_1, φ_2, φ_3 and φ_4 represent the flux of air between the four magnets. Magnetic induction strength in the air gap

$$B = \frac{\phi}{A_0} \tag{4}$$

where ϕ is the average magnetic flux of the pair of polar poles and A_0 is a cross-sectional area that corresponds with the air gap.

The average electromotive force is calculated as follows:

$$\overline{F} = \frac{\pi}{2\sqrt{2}} B\omega_s N_p A_0 \tag{5}$$

The induced current inhibits the penetration of magnetic fields in the ring of copper, and it only allows magnetic fields to penetrate at a specific depth, which is called the skin effect. Due to the skin effect, the inducing currents are concentrated at the surface of the copper ring. The current decays rapidly beyond the depth of penetration. The penetration is decreased with the rotation frequency. The penetration depth expression is as follows

$$\delta = \sqrt{\frac{2}{\omega\sigma\mu}} \tag{6}$$

$$\omega = \frac{\pi\omega_s N_P}{30} \tag{7}$$

In Equation (6), ω is a radial frequency, σ is copper conductivity, and μ represents the relative permeability of copper rings. In Equation (7), ω_s is the speed difference and N_P is the number of pole pairs.

The induced eddy current region is generated by a single permanent magnet on the copper ring, which is equivalent to a circular region with a diameter of d. Therefore, the reluctance of the ring of the eddy current is shown below:

$$dR = \rho \frac{2\pi r}{\delta dr} \tag{8}$$

The area of the eddy current region is as follows:

$$A_m = \pi \left(\frac{d}{2} \right)^2 \tag{9}$$

The instantaneous eddy current obtained from the above equation is:

$$i = \int_0^{\frac{d}{2}} di = \int_0^{\frac{d}{2}} \frac{\varepsilon}{dR} \tag{10}$$

The eddy current loss generated by the permanent magnet corresponding to the copper ring is

$$P_{Cu} = \frac{n \cdot A_m^2 \cdot B^2 \cdot \omega^2 \cdot \delta}{16 \cdot \pi \cdot \rho \cdot \sin(\omega^2 t)} \tag{11}$$

where n is the number of permanent magnet blocks.

Since the permanent magnet governor is connected to a constant torque load, according to the energy conservation law, which can be obtained:

$$T_1\omega_1 = T_2\omega_2 + P_{Cu} \tag{12}$$

Simulation Model

The simulation model of the CPMG is built by using Maxwell Software, as shown in Figure 5. It is mainly composed of a conductor rotor and a permanent magnet rotor, and a band motion domain. The conductor rotor contains a conductor barrel and a copper ring, where the copper ring is welded to the inner wall of the conductor barrel. The permanent magnet rotor contains a permanent magnet tray and PMs (NdFeB), where the permanent magnets are embedded in the permanent magnet array tray by means of a Halbach array. The number of PMs is 36. In addition, The band motion domain is a finite element mesh restriction and not an actual structure.

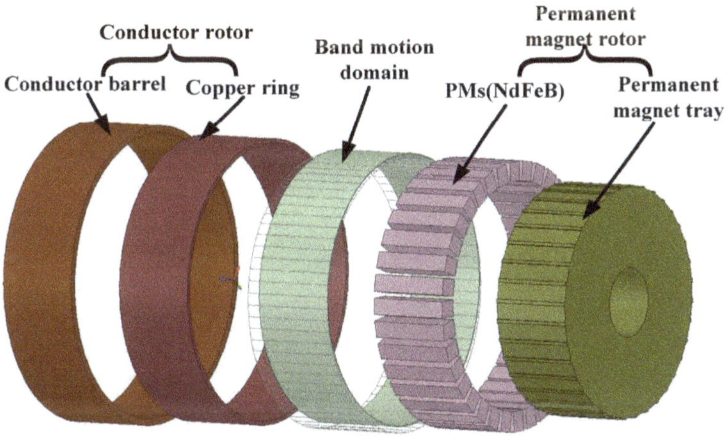

Figure 5. Simulation model of CPMG.

The material of the CPMG parts is given, of which the material of the conductor rotor and the permanent magnet rotor is Steel_1008, and the permanent magnet is selected N48H.

Due to the components of the copper ring, the induced currents in the copper ring are crucial for the transfer of the output torque to the CPMG. Therefore, it is particularly important to pay attention to the meshing subdivision of permanent magnets and the copper ring in the 3D FEM, as shown in Figure 6. In order for the higher precision results to be obtained, the quality of meshing of CPMG should be above 0.3. Moreover, it should be noted that 3D FEM results with stabilization and convergence are taken as valid data, the residual error of which is less than the set value ($\varepsilon < 0.003$).

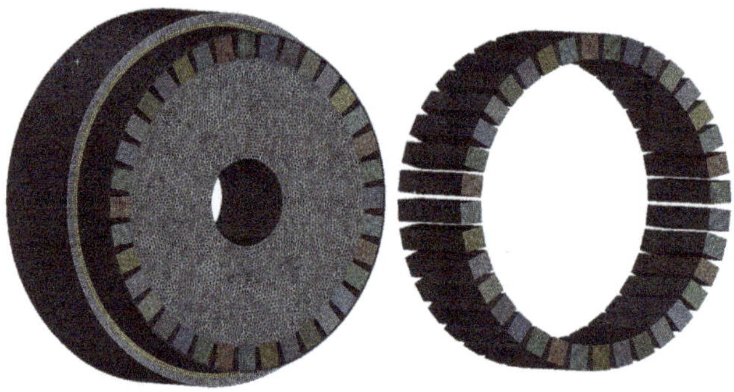

Figure 6. Meshing of CPMG.

3. Analysis of Results

3.1. Magnetic Field Distribution

The output torque of CPMG is determined by the magnet field in the air gap between the conductor and the permanent magnet. As shown in Figure 7, there are three types of permanent magnet arrangements in Figure 7.

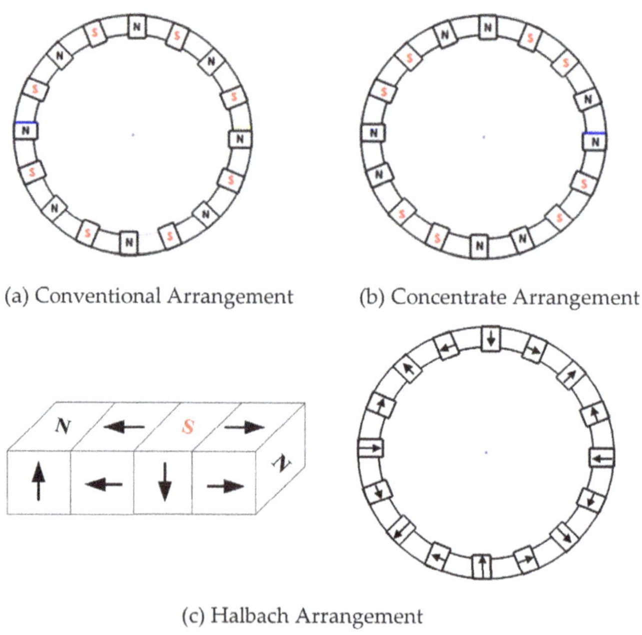

Figure 7. Permanent magnet arrangement. (The direction of the arrow represents the direction of the magnetic poles).

The air gap flux in the above three arrangements is compared and analyzed. The distribution of magnetic field lines is studied under static magnetic field conditions, and the boundary condition is 0 vector boundary. In Figure 8, the magnetic flux density distributions are shown, and the magnetic field is more dense in the Halbach arrangement CPMG than in the conventional and concentrated arrangement. However, a small part of the magnetic field lines directly returns to the adjacent S pole through the air gap, which causes a few flux leakage. Most of the magnetic flux lines are concentrated in the air gap and the conductor side, which realizes the effect of unilateral magnetic concentration.

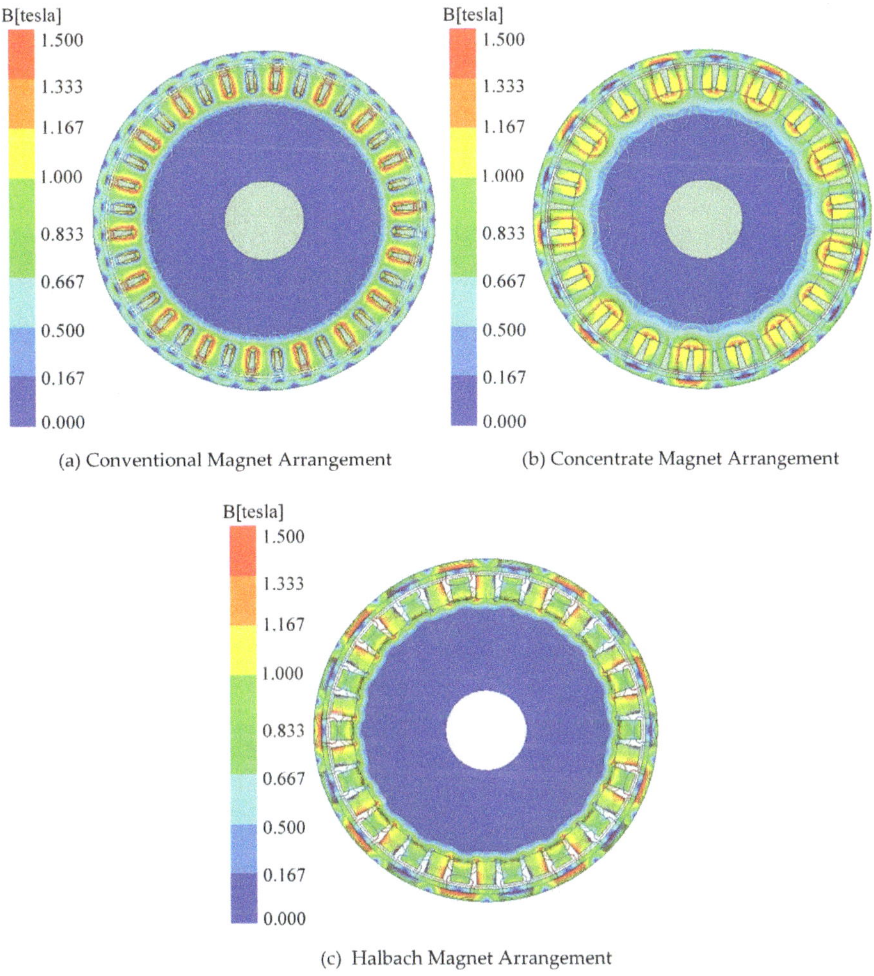

Figure 8. Magnetic flux density distribution in CPMG.

The magnetic flux density of CPMG under the different PMs arrangements was quantitively compared. According to Figure 9, the density of magnetic flux density in the air gap is different under different magnet arrangements. It is clear that the flux density in the airgap of the Halbach arrangement is more than that of the conventional arrangement and concentrate arrangement under the same number of magnets. Because of the symmetries of a permanent magnet, the magnetic flux density curve shows the sine-variations under the Halbach arrangement, which can improve transmission stability. The magnetic characteristics in the air gap with different arrangements are listed in Table 1.

The average magnet density in air gaps with the Halbach arrangement is 13% and 6.2% higher than the other two arrangements, respectively.

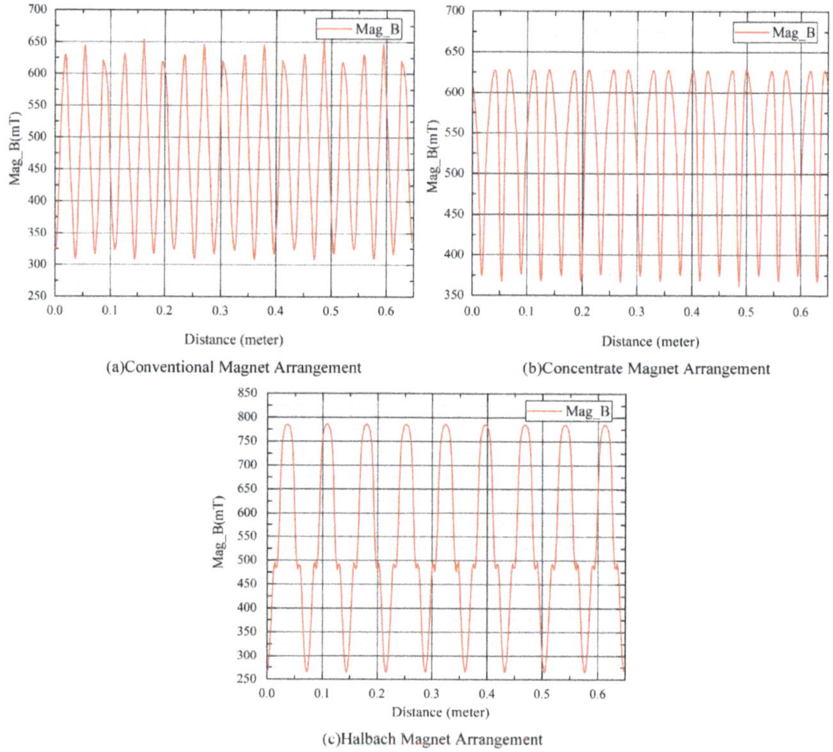

Figure 9. Magnetic flux density in the air gap along the tangential direction.

Table 1. Comparison of the performance between different magnet arrangements.

Type of Permanent Magnet Arrangement	Conventional Arrangement	Concentrate Arrangement	Halbach Arrangement
peak magnetic induction line (Wb/m)	0.012	0.013	0.015
average magnetic flux density (mT)	497.98	511.30	545.15
Peak magnetic flux density (mT)	653.59	610.65	785.88
Average output torque (N·m)	269.40	287.26	320.30

At the beginning of the simulation, the input rotation speed was maintained at 1000 rpm, and the output rotation speed was maintained at 900 rpm. The transient output torque of CPMG under different magnets arrangements are shown in Figure 10. It is clear that the Halbach arrangement has the highest output torque. The detailed comparison is listed in Table 1. As shown in Table 1, The average output torque of the Halbach arrangement is increased by 15.89% and 10.31%, which is compared to the conventional arrangement and concentrate arrangement.

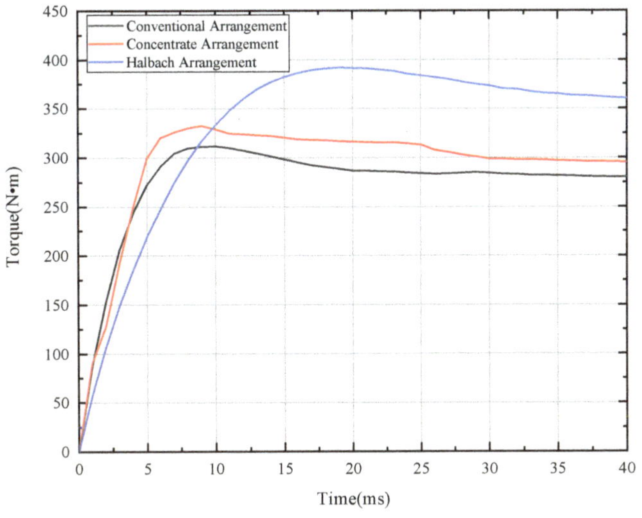

Figure 10. Output torque of CPMG under different magnets arrangements.

3.2. Eddy Current Density Distribution in the Copper Rings

When the slip rate is 0.01, and the effective coupling ratio between the conductor and permanent magnet is 100%, the distribution of the eddy current density in the copper ring is illustrated in Figure 11. As we can see in Figure 11, the eddy current density of the permanent magnet in the copper ring is the largest area mapped to the area of the copper ring. At a maximum density of 2.52×10^7 Am^{-2}, the eddy current density between adjacent permanent magnet mapping areas dropped sharply and reduced to 3.20×10^6 Am^{-2}. The center of the loop formed by the induced eddy current in the copper ring is located in the center between the two mapping regions of permanent magnets. The induced magnetic field generated by the circular eddy current interacts with the permanent magnetic field, thereby providing torque for CPMG.

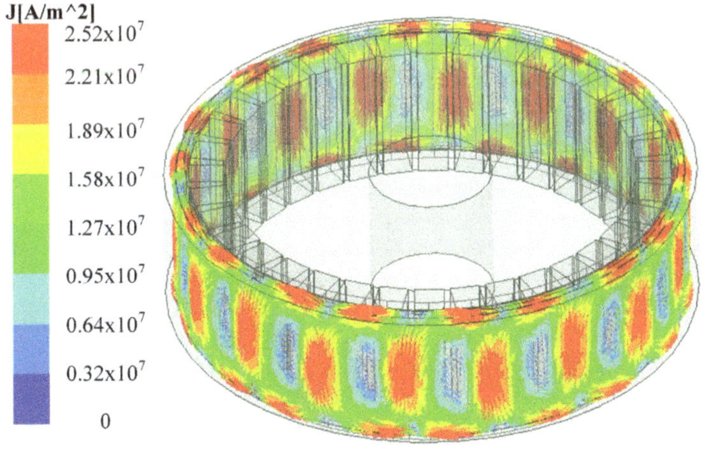

Figure 11. Eddy current distribution across copper rings.

3.3. Ohmic Loss in the Copper Ring

The induced current generated in the copper coil of the permanent magnet regulator produces Ohmic loss due to the existence of the coil reluctance. The Ohmic loss distribution of the copper ring under 0.1 slip rate and 100% effective coupling ratio is shown in Figure 12. The maximum Ohmic loss is concentrated in the center of the copper ring. Moreover, the Ohmic loss is increased with the effective coupling ratio between the conductor and permanent magnet.

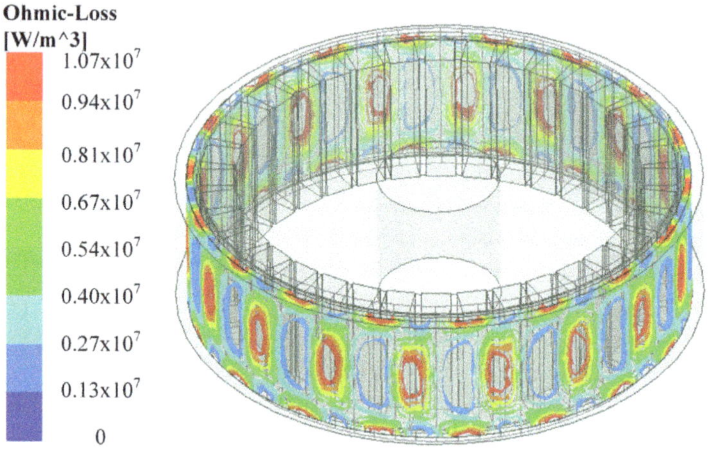

Figure 12. Ohmic loss distribution in the copper rings.

3.4. Output Torque of CPMG

During the simulation, the input rotation speed is maintained at 1000 rpm, and the output rotation speed is adjusted from 600 to 950 rpm, which means the slip rate range from 0.05 to 0.4. The thickness of the copper ring is 6 mm. The output torque of CPMG under the different slip rates and effective coupling ratio is listed in Table 2 and shown in Figure 13. The output torque is increased with the effective coupling ratio at a constant slip rate. The output torque is increased first, then decrease the slip with a constant effective coupling ratio, whose peak value is 74.56 N·m at 230 rpm output rotation speed (slip rate = 0.23) with a 25% effective coupling rate. The peak output torque is 185.20 N·m, 316.01 N·m and 390.28 N·m at 180 rpm (slip rate = 0.18) with 50%, 75% and 100% effective coupling rates, respectively.

Table 2. Average output torque at different coupling ratio(N·m).

Slip Rate	25%	50%	75%	100%
0.05	30.10	89.83	149.51	179.49
0.1	53.75	153.95	258.58	320.30
0.15	67.42	179.15	307.52	384.38
0.16	67.84	184.23	314.96	388.46
0.17	68.12	185.04	315.95	390.26
0.18	69.48	185.20	316.01	390.28
0.19	70.12	184.74	313.28	388.82
0.2	72.64	180.70	307.63	386.12
0.25	73.13	172.62	289.09	364.51
0.3	71.31	162.94	267.45	340.86
0.35	68.54	152.96	250.59	317.23
0.4	65.08	141.98	230.19	291.81

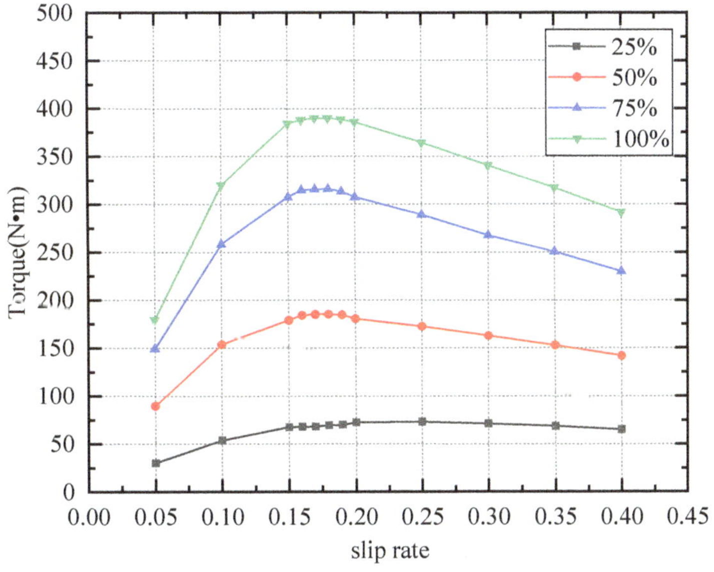

Figure 13. Average output torque at different effective coupling ratio.

3.5. Influence of Copper Ring Thickness on Transmission Performance

In the transmission process of the CPMG, the main function of the copper ring is to be cut to magnetic wire induced by the permanent magnets, generating eddy currents. However, the changes in the thickness of the copper ring cause a change in resistance to copper; thus, the thickness of the copper ring has an important effect on transmission performance.

At the beginning of the simulation, the input rotation speed was maintained at 1000 rpm, and the output rotation speed was maintained at 900 rpm. The transient output torque of CPMG 25%, 50%, 75% and 100% effective coupling ratio under difference copper ring thickness is shown in Figure 14.

As shown in Figure 14, the transient output torque of CPMG has an overshoot at the beginning due to a small damping ratio of the system. The output torque reached a stable value after dozens of ms. The transient output torque of CPMG with different copper ring thicknesses is shown in Figure 13. The transient output torque increases with the copper ring thickness.

The stable output torque for a different copper ring thickness under a constant effective coupling ratio and slip rate is listed in Table 3. The table shows the output torque peak value at 6 mm thickness and 100% effective coupling ratio, which is 340.05 N·m. The output torque increases with the copper ring thickness. However, the minimum value of the air gap must be greater than 5 mm taking into account the assembly problems and costs in the actual production process, which is limited to the copper ring thickness. Here, we take the maximum copper ring thickness as 6 mm.

Table 3. Output torque at different copper ring thicknesses.

Copper Ring Thickness (mm)	Effective Coupling Ratio (%)			
	25	50	75	100
6	57.75	163.33	275.57	340.05
5	50.40	147.72	247.39	304.18
4	42.49	128.97	214.60	261.73
3	33.95	104.95	175.63	212.46

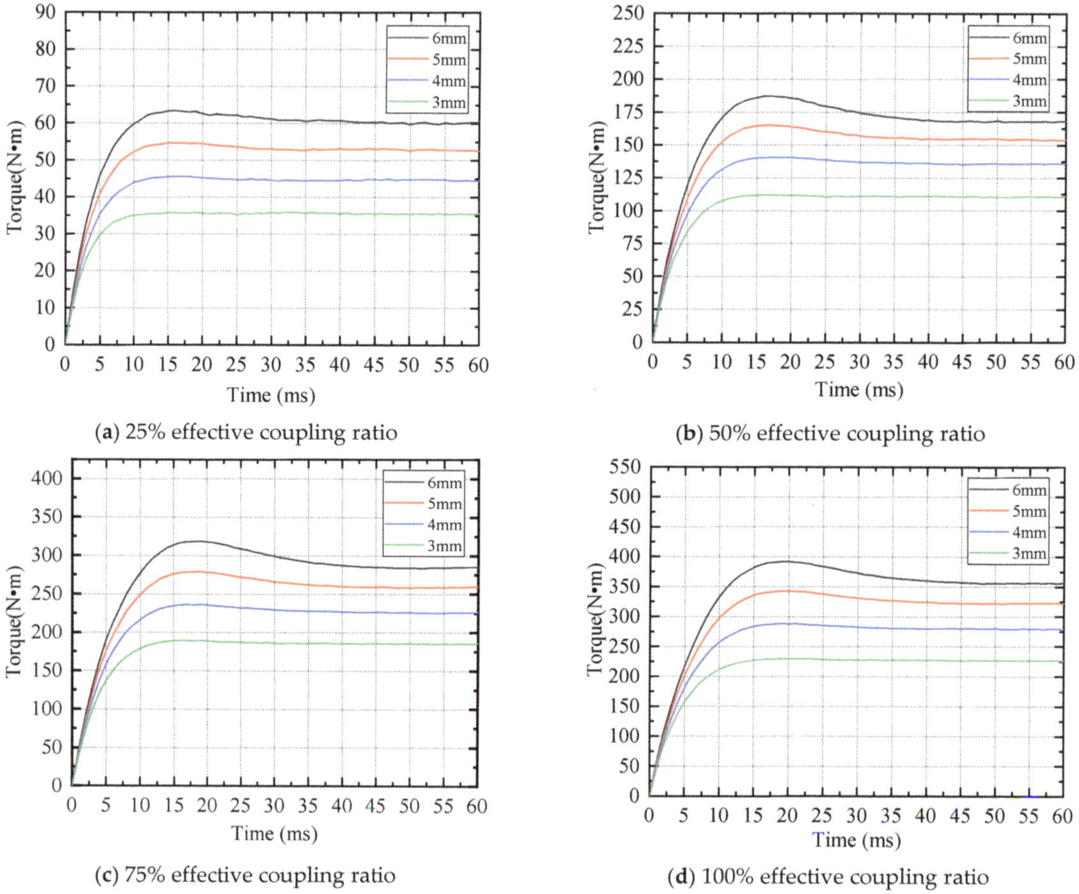

Figure 14. Output torque of CPMG under different copper ring thicknesses.

4. Experimental Analysis

4.1. Setup of Experimental Platform

To verify the correctness of the transmission characteristic model and the accuracy of the calculation results. An experimental platform was constructed, as shown in Figure 15, including an eddy current dynamometer providing the required output load, which can change the load by adjusting its control current. A torque sensor measures the output torque; the CPMG can be tested; and the YE2-200L-4 three-phase asynchronous motor is an electromotor providing the driving torque, where the rated output power of the motor is 30 kW, the rated rotation speed is 1470 rpm, and the rated current is 56.8 A. Two frequency converters control the rotating speed of the electromotor and eddy current dynamometer, and the testing instrument displays the output torque. An IKTM1000 electric actuator is connected to the rotor of the conductor to realize the adjustment of the effective coupling ratio. All these components were installed on the test platform, as shown in Figure 16.

The thickness of the copper ring is measured in advance of the test, and the copper ring is assembled in the inner wall of the conductor barrel. The Gaussian meter was used to measure the air gap magnetic field after assembly. Finally, the CPMG was started, and the torque data were sampled by the test equipment.

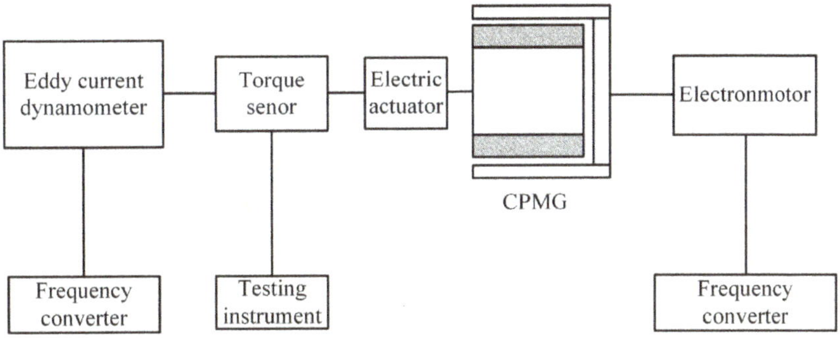

Figure 15. Basic structure diagram of test platform.

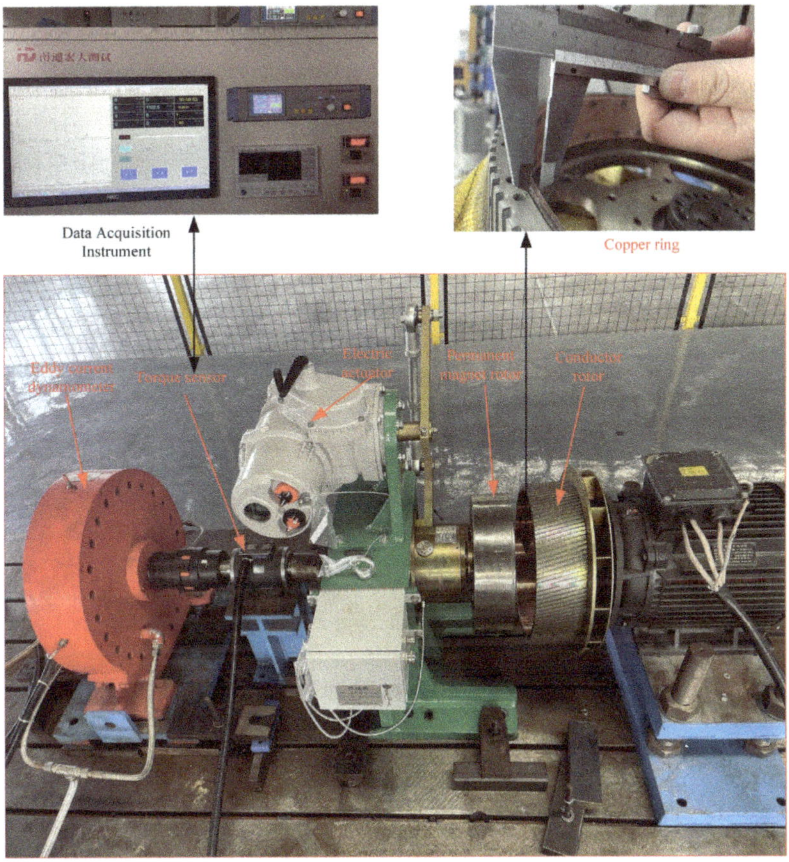

Figure 16. Test prototype construction.

4.2. Output Torque Test

The CPMG 100% effective coupling ratio, 6 mm thickness of the copper ring for testing, and start motor were selected. The CPMG input speed (speed of the conductor motor) was maintained at 1000 rpm; the output torque of the Permanent Magnet Rotor was tested at 950 rpm, 900 rpm, 850 rpm, 800 rpm,750 rpm, 700 rpm, 650 rpm, 600 rpm (slip rate range from 0.05 to 0.4), respectively; and the average air gap flux measured by the Gauss meter was substituted into the torque equation derived in Section 2.2 through MATlab software to be calculated, which is the analytical torque. The output torque test was carried out on the experimental platform. The experimental data, which are T_{test1}, were obtained with the slip rate range from 0.05 to 0.4 and compared with the theoretical calculation data and simulation results, as shown in Figure 17. The experimental results show that the variation trend of the output torque is consistent with numerical simulation. The maximum output torque is reached when the slip rate is 0.2. The error of experimental results of output torque compared with analytical and numerical counterparts are less than 5%, under 100% effective coupling ratio and 6 mm copper ring thickness. The output torque of CPMG does not always increase with the increased slip rate. This is due to the adverse effects of the inductive magnetic field of permanent magnets. In addition, the slip rate increases the skin effect, which reduces the equivalent permeability in the copper ring, thereby reducing the intensity of the induced current in the copper ring, leading to the transfer of torque that is gradually reduced.

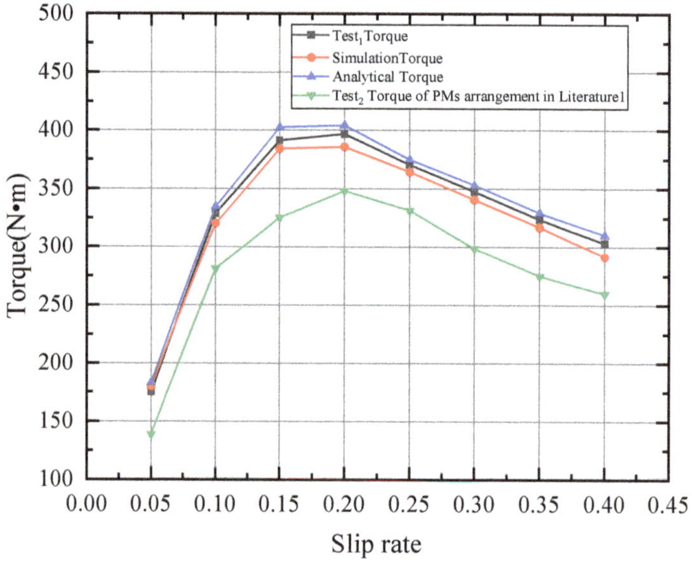

Figure 17. Comparison of average output torque under different slip rates.

According to the magnet arrangement in Ref. [1], the same size and quantity of PMs were installed in the CPMG for testing. Additionally, the CPMG was operated, and the torque data were sampled by the test equipment. The detailed comparison is listed in Table 4. During the test, the output torque of the Halbach array was higher than the magnets' arrangement in Ref. [1]. Furthermore, the peak torque of the Halbach array was increased by 12.21% when compared to the magnet arrangement in Ref. [1].

Table 4. Comparison of output torque under different speed differences.

Slip Rate	T_{test1}	T_{test2}	T_{sim}	T_{ana}	err_{sim}	err_{ana}
0.05	174.89	138.75	179.49	183.62	2.63%	4.99%
0.1	328.62	281.41	320.30	334.82	2.53%	1.89%
0.15	391.25	325.48	384.38	402.77	1.75%	2.94%
0.2	396.73	348.25	386.12	404.42	2.67%	1.94%
0.25	370.43	331.38	364.51	375	1.74%	1.23%
0.3	347.52	298.82	340.86	353.13	1.92%	1.61%
0.35	323.56	274.89	317.23	329.35	1.96%	1.79%
0.4	303.21	259.63	291.81	310.23	3.75%	2.31%

In Table 4, T_{test1} represents the test measurements of the Halbach array, T_{sim} is simulated stable torque, T_{test2} represents the test measurements of PMs arrangement in Ref. [1]. T_{ana} is the theoretical torque of the theoretical analysis of the principle of the equivalent magnetic circuit. err_{sim} is the error between the simulation and the experimentation. err_{ana} is the error between the theoretical analysis and the experimentation.

In order to test the effect of the thickness of the copper ring on the output torque of the CPMG, a series of copper ring thicknesses of 6, 5, 4 and 3 mm, respectively, were installed in the inner walls of the conductor barrel and the tested output torque in the different effects of the coupling. During the test procedure, the speed of the input CPMG was maintained at 1000 rpm, and then the speed of the permanent magnet rotor was adjusted at 900 rpm (the slip rate was 0.1) to test the output torque of the CPMG. The test results were obtained and compared with the numerical simulation results, as shown in Figure 18. The torque measured by the sensor and obtained by a finite element is shown below in Table 5. When the effectiveness of the coupling is constant, the torque of the output increases with the increasing thickness of the ring of copper. The errors of experimental results of output torque compared with the numerical counterpart are less than 5%.

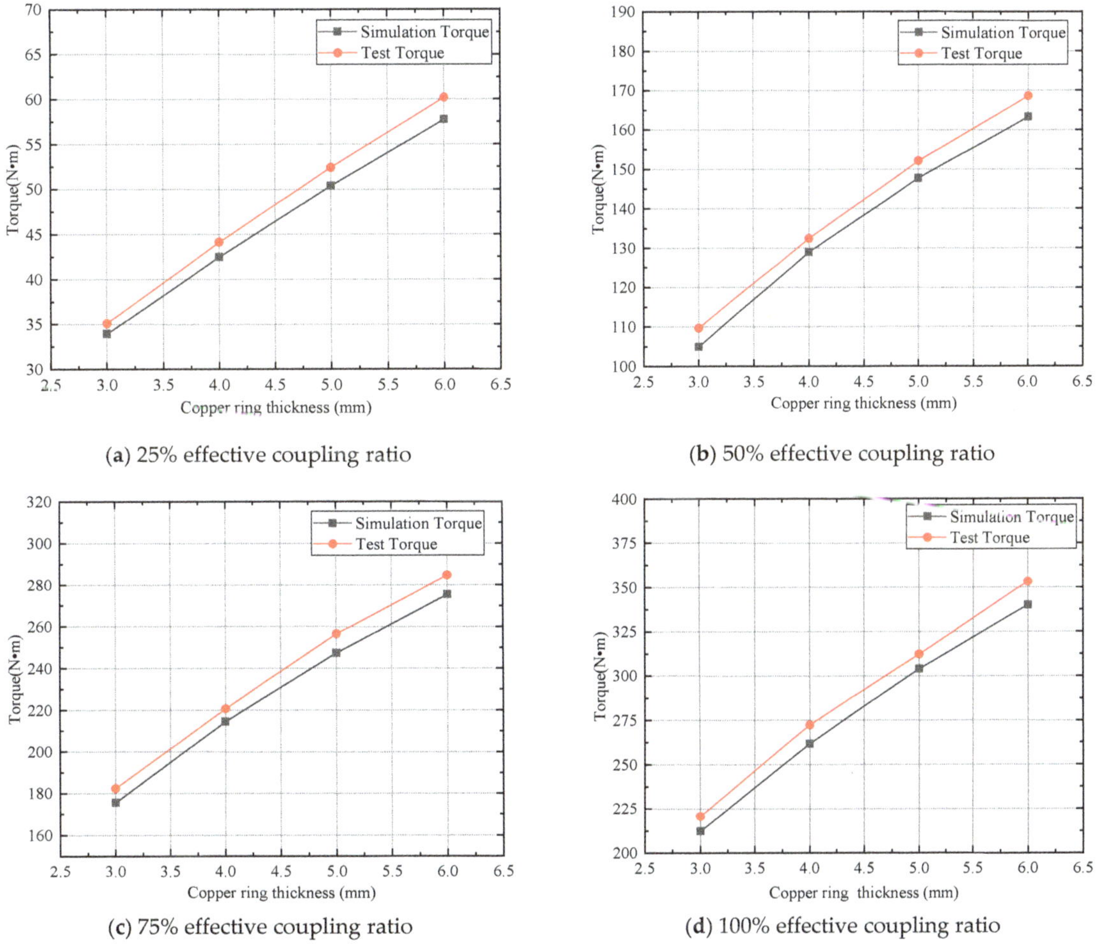

(**a**) 25% effective coupling ratio (**b**) 50% effective coupling ratio

(**c**) 75% effective coupling ratio (**d**) 100% effective coupling ratio

Figure 18. Output torque simulation and test comparison under different thicknesses.

Table 5. Comparison of experimental and simulation results of output torque under different copper ring thicknesses.

Effective Coupling Ratio (%)	h_{copper} (mm)	T_{sim} (N·m)	T_{test} (N·m)	Error (%)
25	3	33.95	35.14	3.86
	4	42.49	44.18	3.82
	5	50.40	52.47	3.94
	6	57.75	60.21	4.08
50	3	104.95	109.72	4.34
	4	128.97	132.51	2.67
	5	147.72	152.18	2.93
	6	163.33	168.69	3.18
75	3	175.63	182.56	3.79
	4	214.60	220.87	2.84
	5	247.39	256.64	3.6
	6	275.57	284.73	3.21
100	3	212.46	220.94	3.84
	4	261.73	272.42	3.93
	5	304.18	312.46	2.65
	6	340.05	353.48	3.78

5. Conclusions

This paper shows a comparative analysis of the performance of the CPMG under the three different types of PM arrangements. Considering the same number of PMs and the cost, the Halbach array is suitable for high-torque appliances, such as high-power fans. The main magnetic fluxes are considered to obtain a comprehensive equivalent magnetic circuit model of the CPMG with Halbach array, and analytical output torque is calculated. Moreover, the transmission properties of the CPMG under the Halbach array were studied, taking into account the effects of slip rate, effective coupling ratio and thickness of the copper ring on the torque output of the CPMG under the Halbach array. By comparing the simulation results with the field-test data of the CPMG, it was found that the simulation results match well with the measured data, which validates the effectiveness of the proposed model and calculation method. The main conclusion of this paper is summarized as follows:

1. The air gap flux density under the Halbach arrangement is 13% and 6.2% higher than the other two arrangements, respectively. The average output torque of the Halbach arrangement is increased by 15.89% and 10.31%, which is compared with the conventional arrangement and concentrate arrangement.
2. When the effective coupling ratio and the thickness of the copper ring are constant, the output torque increases first and then drops with the slip rate increasing, which reaches the maximum torque of around 0.2. The results were consistent with the results of the simulation, and the error was kept within the range of 5%. The output torque of the Halbach array is higher than the magnets' arrangement in Ref [1]. The peak torque of the Halbach array is 12.21% higher when compared to the magnets' arrangement in Ref. [1].
3. When the slip rate is constant, the output torque increases with the increase in the effective coupling ratio (between conductor and magnet), and the thickness of the copper ring increases.

Author Contributions: Conceptualization, H.W. and C.Y.; methodology, H.W., Y.Z. and C.Y.; software, J.G.; validation, C.Y.; formal analysis, C.Y. and Y.Z.; investigation, H.W.; resources, C.Y.; data curation, H.W. and H.L.; writing—original draft preparation, H.W.; writing—review and editing,

Y.Z.; visualization, H.W. and C.Y.; supervision, C.Y.; project administration, H.W. and C.Y.; funding acquisition, J.G. All authors have read and agreed to the published version of the manuscript.

Funding: This research was funded by "The overseas study visit and training program for outstanding young backbone talents of Anhui Province, grant number gxgwfx2021035", "Graduate Student Innovation Project of Anhui Province, grant number 2022xscx097 ", "The Innovation team of Anhui Polytechnic University, Special display and Imaging Technology Anhui Technology Innovation Center open project, grant number 2020AJ06001 " and "Anhui Polytechnic University-Jiujiang District Industrial Collaborative Innovation Special Fund Project, grant number 2022cyxtb4".

Conflicts of Interest: The authors declare no conflict of interest.

Sample Availability: Samples of the compounds are available from the authors.

Abbreviations

The following abbreviations are used in this manuscript:

DPMG	disk permanent magnet governor;
CPMG	cylindrical permanent magnet governor;
MEC	magnetic equivalent circuit;
MCT	marine current turbine;
PM	permanent magnet;
PMR	permanent magnet rotor;
KCL	Kirchhoff current law.

References

1. Cheng, X.; Liu, W.; Zhang, Y.; Liu, S.; Luo, W. A Concise Transmitted Torque Calculation Method for Pre-design of Axial Permanent Magnetic Coupler. *IEEE Trans. Energy Convers.* **2020**, *35*, 938–947.
2. Seo, S.W.; Kim, Y.H.; Lee, J.H.; Choi, J.Y. Analytical torque calculation and experimental verification of synchronous permanent magnet couplings with Halbach arrays. *AIP Adv.* **2020**, *8*, 056609. [CrossRef]
3. Wang, L.; Jia, Z.Y.; Zhang, L. Investigation on the accurate calculation of the temperature field of permanent magnet governor and the optimization method of heat conduction. *Case Stud. Therm. Eng.* **2019**, *13*, 100360. [CrossRef]
4. Guo, B.; Li, D.; Shi, J.; Gao, Z. A Performance Prediction Model for Permanent Magnet Eddy-Current Couplings Based on the Air-Gap Magnetic Field Distribution. *IEEE Trans. Magn.* **2022**, *58*, 1–9. [CrossRef]
5. Dolisy, B.; Mezani, S.; Lubin, T.; Levrque, J. A new analytical torque formula for axial field permanent magnets coupling. *IEEE Trans. Energy Convers.* **2015**, *30*, 892–899. [CrossRef]
6. Yang, X.; Liu, Y.; Wang, L. An Improved Analytical Model of Permanent Magnet Eddy Current Magnetic Coupler Based on Electromagnetic-Thermal Coupling. *IEEE Access* **2020**, *8*, 95235–95250. [CrossRef]
7. Shi, J.; Suo, S.; Meng, G. The theoretical calculation model of torque transmission in permanent-magnet couplers. *AIP Adv.* **2021**, *11*, 025303. [CrossRef]
8. Orlova, S.; Konuhova, M.; Kamolins, E.; Otankis, R. Design of magnetic couplings for bioreactors: Analytical treatment and optimization. In Proceedings of the 2018 20th European Conference on Power Electronics and Applications (EPE'18 ECCE Europe), Riga, Latvia, 17–21 September 2018.
9. Sun, K.; Shi, J.; Cui, W.; Meng, G. Theoretical Computational Model for Cylindrical Permanent Magnet Coupling. *Electronics* **2021**, *10*, 2026. [CrossRef]
10. Deshan, K.; Dazhi, W.; Wenhui, L.; Sihan, W.; Zhong, H. Analysis of a novel flux adjustable axial flux permanent magnet eddy current coupler. *IET Electr. Power Appl.* **2023**, *17*, 181–194. [CrossRef]
11. Park, J.; Paul, S.; Chang, J.; Hwang, T.; Yoon, J. Design and comparative survey of high torque coaxial permanent magnet coupling for tidal current generator. *Electr. Power Energy Syst.* **2020**, *120*, 105966. [CrossRef]
12. Taqavi, O.; Taghavi, N. Development of a Mixed Solution of Maxwell's Equations and Magnetic Equivalent Circuit for Double-Sided Axial-Flux Permanent Magnet Machines. *IEEE Trans. Magn.* **2021**, *57*, 1–11. [CrossRef]
13. Yang, C.; Peng, Z.; Tai, J.; Zhu, L.; Telezing, B.J.K. Torque characteristics analysis of slotted-type eddy-current couplings using a new magnetic equivalent circuit model. *IEEE Trans. Magn.* **2020**, *56*, 1–8. [CrossRef]
14. Lubin, T.; Mezani, S.; Rezzoug, A. Experiment and analytical analyses of axial magnetic coupling under steady-state and transient operations. *IEEE Trans. Ind. Electron.* **2014**, *61*, 4356–4365. [CrossRef]
15. Kim, I.C.; Wata, J.; Tongphong, W.; Yoon, J.S.; Lee, Y.H. Magnetic Coupling for a 10 kW Tidal Current Turbine: Design and Small Scale Experiments. *Energies* **2020**, *13*, 5725. [CrossRef]
16. Lukočius, R.; Vilkauskas, A.; Marčiulionis, P.; Grigaliūnas, V.; Nakutis, Ž.; Deltuva, R. An Analysis of Axial Magnetic Coupling Force and Torque Dependencies on Its Structure Parameters Using a 3D FEM. *Appl. Sci.* **2022**, *12*, 6546. [CrossRef]

17. Tian, M.; Zhao, W.; Wang, X.; Wang, D.; Yang, Y.; Diao, J.; Ma, X. Analysis on a Novel Flux Adjustable Permanent Magnet Coupler With a Double-Layer Permanent Magnet Rotor. *IEEE Trans. Magn.* **2018**, *54*, 1–5. [CrossRef]
18. Lu, X.; He, X.; Jin, P.; Huang, Q.; Yang, Y.; Diao, J.; Ma, X. General 3D Analytical Method for Eddy-Current Coupling with Halbach Magnet Arrays Based on Magnetic Scalar Potential and H-Functions. *IEEE Trans Magn.* **2021**, *14*, 8458. [CrossRef]
19. Lee, H.J.; Joung, H.K.; Kim, C.H. Optimal Design of the Halbach Array of the Halbach Array of Magnetic Coupling. In Proceedings of the 2022 25th International Conference on Electrical Machines and Systems(ICEMS), Shanghai, China, 29 November–2 December 2022; pp. 1–5.
20. Li, Y.; Hu, Y.; Guo, Y.; Song, B.; Mao, Z. Analytical Modeling and Design of Novel Conical Halbach Permanent Magnet Couplings for Underwater Propulsion. *J. Mar. Sci. Eng.* **2021**, *9*, 290. [CrossRef]
21. Li, Z.; Zhang, L.; Qu, B.; Wang, D. Evaluation and analysis of novel flux-adjustable permanent magnet eddy current couplings with multiple rotors. *IET Electr. Power Appl.* **2021**, *15*, 754–768. [CrossRef]
22. McGilton, B.; Crozier, R.; McDonald, A.; Mueller, M. Review of magnetic gear technologies and their applicants in marine energy. *IET Reneaable Power Gener.* **2018**, *12*, 174–181. [CrossRef]
23. Mateev, V.; Marinova, I. Loss estimation of magnetic gears. *Electr. Eng.* **2020**, *102*, 387–399. [CrossRef]
24. Zhao, X.; Niu, S. Design and optimization of a new magnetic-geared pole-changing hybrid excitation machine. *IEEE Trans. Ind. Electron.* **2017**, *12*, 9943–9952. [CrossRef]
25. Huang, J.; Fu, W.; Niu, S.; Zhao, X. Comparative Analysis of Different Permanent Magnet Arrangements in a Novel Flux Modulated Electric Machine. *IEEE Access* **2021**, *1*, 14437–14445. [CrossRef]
26. Yang, K.; Zhao, F.; Wang, Y. Analysis of Double Layer Permanent Magnet Flux Reversal Machines with Different Permanent Magnet Arrangements in Stator. *IEEE Trans. Magn.* **2021**, *57*, 1–5. [CrossRef]
27. Li, B.; Yang, B.; Xiang, F.; Guo, J. Optimal Design of a New Rotating Magnetic Beacon Structure Based on Halbach Array. *Appl. Sci.* **2022**, *12*, 10506. [CrossRef]

machines

Article

Design and Verification of Adaptive Adjustable Output Control on Micro Spray Gun †

Jiun-Hung Lin [1], Chih-Hong Chen [2] and Shih-Tsang Tang [3,*]

[1] College of Electrical Engineering and Computer Science, National Kaohsiung University of Science and Technology, Kaohsiung 824005, Taiwan
[2] Department of Electronic Engineering, Kun Shan University, Tainan 710303, Taiwan
[3] Department of Biomedical Engineering, Ming Chuan University, Taoyuan 333321, Taiwan
* Correspondence: sttang@mail.mcu.edu.tw
† This paper is an extended version of our paper published in 2022 International Conference on Fuzzy Theory and Its Applications (iFUZZY), Kaohsiung, Taiwan, 3–5 November 2022; Available online: https://doi.org/10.1109/iFUZZY55320.2022.9985228.

Abstract: The general spray gun is used for industrial large-area spraying, and there is less demand for different pressures and the accuracy of spraying pressure, so mechanical pressure regulators are mostly used. However, as the demand for artistic innovation continues to grow, it promotes the advent of the micro spray gun. The micro spray gun is currently commonly known as an airbrush. The micro spray gun is mainly used for fine drawing, so it must provide different pressures with high precision pressures, but the existing mechanical regulators cannot meet this requirement. For these unmet requirements, this study proposed a solution for PID (proportional-integral-derivative) control micro spray gun system. The results showed that the PID control could effectively provide various stable output pressures of the micro spray gun. The pressure-varying range of 30 kPa could rapidly return to the target value in 10 s (the usual spraying time). The proposed solution then presents better spraying effects.

Keywords: micro air pump; micro spray gun; pressure control; pulse width modulation

Citation: Lin, J.-H.; Chen, C.-H.; Tang, S.-T. Design and Verification of Adaptive Adjustable Output Control on Micro Spray Gun. *Machines* **2023**, *11*, 354. https://doi.org/10.3390/machines11030354

Academic Editors: Kan Liu and Wei Hu

Received: 5 February 2023
Revised: 3 March 2023
Accepted: 3 March 2023
Published: 4 March 2023

1. Introduction

A compressor is a mechanical device used for increasing distinct compressible fluid or air pressure, whereas an air compressor is the commonest one using air output technology. An air compressor is an inevitable and important piece of equipment in various industries converting the mechanical energy of a motor into air pressure energy. Compressed air is used as the power source for industrial pneumatic tools, paint-spraying machines, and sand-blast equipment. Many air compressors are applied to paint spraying with spray guns [1,2]. Such a traditional sprayer is generally equipped with a mechanical air pressure regulator to adjust the pressure output with manual mode, and a pneumatic throttle valve is used for tuning the pressure for paint output. An air compressor running with an electric motor, rather than a mechanical principle, is called an air pump. The principle of the air pump is to exhaust from an enclosed space or to generate air from an enclosed space. When it is used for small-size vacuum pumping, it is also called a micro air pump, or a micro air compressor.

With either mechanical or electric air compressors, the paint going through the spray gun might be interfered by air pressure or the external environment. It would result in the change of air current in the spray gun to cause unstable working of the spray gun or affect the spraying efficiency and effectiveness [3]. The control valve of current spray gun systems generally balances the air manually, without the function to regulate signal feedback. It is rather difficult to regulate air output along with experience or condition during the work, and air pressure might not be stabilized. Due to the rapid development of microcontrollers in past years, there are many studies on the application of control technology to medium sprayers, in which variable control technique is added to largely reduce unstable output, e.g., intermittent and continuous variable rate spraying based on pulse width modulation [4–7]. Pulse width modulation (PWM) changes the winding driving module through impulse signals to achieve the variable spray coating through the duty ratio of pulse-width signals. Researchers also control the spraying amount by flow control with pressure monitoring and pulse width modulation [8–11]. Such variable pressure control is limited the regulation range and appears unstable atomization characteristics, but the principle is simple and the cost of constructing control circuits is lower. The technology of adaptive air compressors, therefore, is a long-term development direction.

A pressure-monitoring output control technology is proposed for micro spray guns in art drawing, body painting, and cosmetic fields [12–14]. Such micro air pumps cannot use mechanical air pressure modulators and air receivers due to the cost and size limit that the pressure output is modulated with simple power control. A proportional-integral-derivative (PID) control algorithm with pressure feedback is therefore developed to provide stable air output for the micro air compressor in the sprayer as well as modulate the air output. The motor valve in the air pump is modulated by pulse width modulation, according to the algorithm calculation result, so that the spray gun presents pressure feedback and adaptability to achieve the variable spraying effect. We implemented an adaptive micro spray gun simulation platform in 2022 [15]. The simulation results show that the PID algorithm is feasible in the control of the micro-spray gun device. Therefore, in this study, we applied the method on the real device, and further determined the ranges of the algorithm parameters and convergence conditions. We applied the pulse-like driving waves to observe the influences of the algorithm parameters. At the same time, the performances of algorithm convergence are analyzed. All considerations are for determining the operation optimization.

2. Materials and Methods

The basic components for this system contain a micro air pump, DC motor drive module, and PID control unit.

2.1. DC Motor Drive Module

The micro air pump used in this study is a kind of electric diaphragm pump with an intake and exhaust structure, Figure 1. During the compression, the air intake is closed, while the exhaust port is open to form the positive pressure; during the stretch, the exhaust port is closed, and the air intake is open to form negative pressure. The pressure difference appears on the air intake of the air pump and the ambient pressure, and under the pressure difference, the air is sucked into the pump chamber. It is the entire exhaust motion of the micro air pump. Unlike a large air pump requiring lubricating oil and vacuum pump oil, a micro air pump would not pollute the working medium and shows the advantages of small size, low noise, maintenance-free, and continuous running [16].

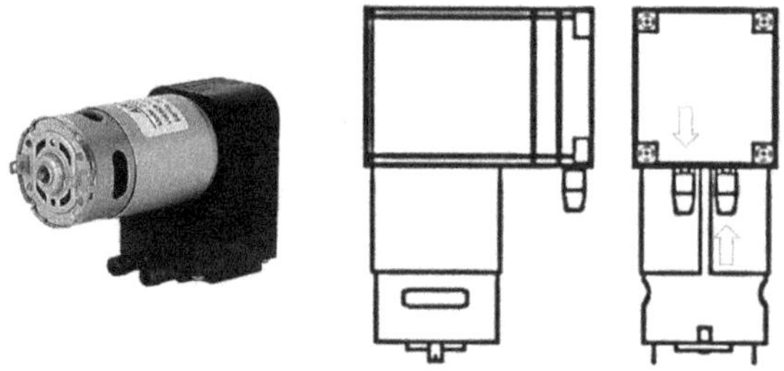

Figure 1. Picture and side shot of a micro electric air pump.

The DC motor revolution basically presents analog modulation and pulse-width modulation. Analog modulation linearly modulates the input voltage or current of a motor; since the signals show linear changes, they are more easily interfered with by ambient electric noise and affected stability. Furthermore, control signals would easily drift with time and appear large power dissipation and worse energy efficiency. Moreover, the motor torque depends on the current going through the motor winding; when the voltage on the motor winding is changed, the current going through the motor winding would be unstable and the motor torque is also unstable. Unlike the instability of analog modulation, pulse width modulation is often used for DC motor drives as well as power conversion of DC/DC converters and AC/DC converters [17]. PWM, in principle, outputs the required power through switching impulse, and pulse amplitude and cycle are constant. The output power is controlled with conducting pulse width. The voltage output, therefore, is the average of the ratio of the corresponding conducting and pulse cycle time, called duty ratio or duty cycle which the unit is a percentage (%). Since the motor torque depends on the current going through motor winding, the increase in duty ratio also increases power output, and the motor speed would be relatively faster. With PWM control, the impulse voltage is comparatively constant, and the current stability is influenced that the motor torque output being relatively stable.

2.2. PID Control Unit

Stable air output is required in this study and the adaptive control presents is important. In industrial control, PID control is a mature control system with broad applications [18–20]. It is often used for control feedback to stabilize output signals. PID control technology is suitable for a system and a controlled object not being completed understood or the system parameters not being acquired through effective measurement. A PID controller is composed of a proportional unit (*P*), integral unit (*I*), and derivative unit (*D*), whose parameters of proportional gain, integral gain, and differential gain could be used for modulating the system output characteristics. Figure 2 shows the basic architecture of the PID control module.

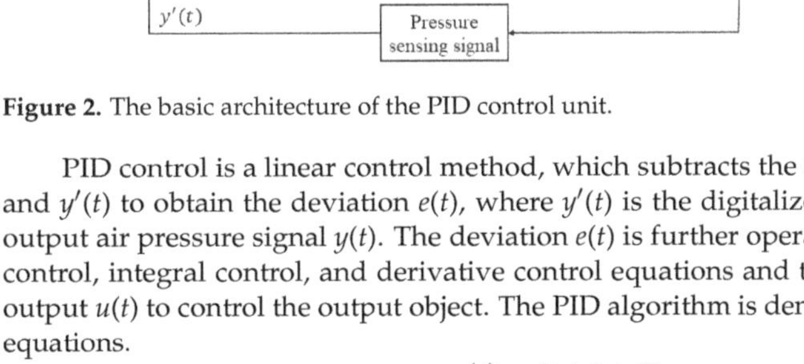

Figure 2. The basic architecture of the PID control unit.

PID control is a linear control method, which subtracts the desired target value $n(t)$ and $y'(t)$ to obtain the deviation $e(t)$, where $y'(t)$ is the digitalized version of the actual output air pressure signal $y(t)$. The deviation $e(t)$ is further operated in the proportional control, integral control, and derivative control equations and then combined with the output $u(t)$ to control the output object. The PID algorithm is denoted with the following equations.

$$u(t) = P + I + D \tag{1}$$

$$P = K_p \times e(t) \tag{2}$$

$$I = K_i \times \int_0^t e(\tau)d\tau \tag{3}$$

$$D = K_d \times \frac{d}{dt}e(t) \quad (4)$$

where

K_p: proportional gain adjustable parameter,
K_i: integral gain adjustable parameter,
K_d: differential gain adjustable parameter,
t: current time,
τ: variable of integration, with the number from 0 to current time t.

When the system is continuously operated, the PID control unit must constantly modulate proportional gain, integral gain, and differential gain parameters according to the feedback of the barometric pressure sensing signal to have the system achieve optimal control, i.e., achieving system stability without divergence or excessive oscillation of output signals. When PID is applied to the micro spray gun (airbrush) system that can switch the spray on and off, its robustness can make the spray gun maintain a constant output speed and allow the spray gun to maintain the required output speed under different working conditions. Through the feedback control, the motor can reduce the jitter when starting and stopping, reducing the wear or damage of the motor.

2.3. Experimental Design

2.3.1. System Architecture

Figure 3 depicts the PID control process, mainly containing two dotted square boxes of software function and hardware function. The software box includes a PID control algorithm and outputs pulse width modulation through programming. NI LabVIEW (National Instruments Co., Austin, TX, USA), the graphic high-level programming language, is used for software development. The hardware box covers a multifunction I/O card (NI USB-6211, National Instruments Co., Austin, TX, USA), a DC motor drive module, and a pressure sensor module. The $u(t)$ in the PID control block is eventually driven for the output unit, which contains a pulse width modulation converting module and DC motor drive module to drive the micro air pump. In order to acquire the feedback signal output for the parameter modulation in the PID control module, a barometric pressure sensor is selected for capturing the pressure change in the pipe, Figure 3.

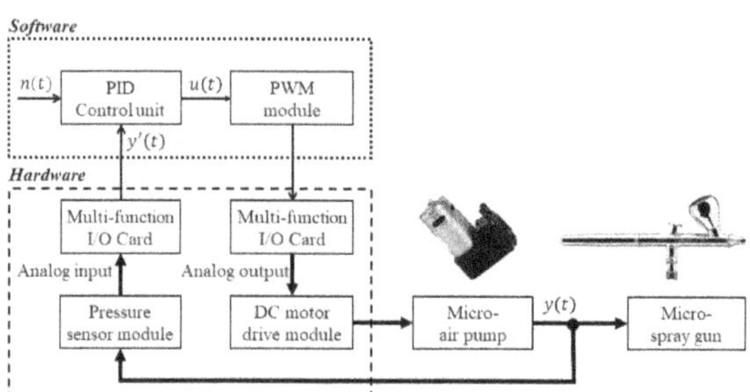

Figure 3. PID control test architecture of micro air pump.

NI USB-6211 isolated multifunction I/O card, with digital and analog output/input function and the highest digital-analog resolution and sampling rate of up to 16 bits and 250 k samples per second, is utilized as the digital-analog interface card. The hardware interface card reads analog signals from the pressure sensor module as real-time feedback data for the PID control module, and the analog output refresh rate is 250 k samples per second, of the interface card, which could generate high-frequency impulse shape to modulate accurate duty ratio for controlling the micro air pump.

The small air pump, Airpon D2028 (Ningbo Forever Electronic Appliance Co., Ltd., Ningbo, Zhejiang, China) with a working pressure from −70 to +250 kPa, is used as the micro air pump in this study. DC motor drive module with dull-channel full-bridge motor driver chip produced by STMicroelectronics company (Geneva, Switzerland) is used for driving the micro air pump. The driver module with the features of high operating voltage, large output current, strong driving capability, low calorific power, and strong anti-jamming capability is often used for driving relay, solenoid, solenoid valve, DC motor, and stepping motor. To reduce shock at the beginning of the air pump start, the micro air pump is directly fixed on the plastic mat with plastic ties, Figure 4a. Two control signal wires of the air pump are connected to the DC motor drive module so that the exhaust would first pass a unidirectional valve to prevent airflow from backflow and affecting motor operation. The exhaust pipe of the micro air pump, through a three-way pipe, as the dotted square in Figure 4a, is connected to the micro spray gun and a pressure sensor module.

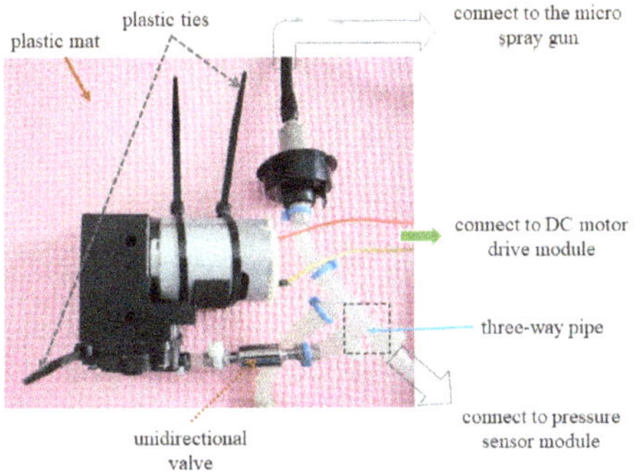

(**a**)

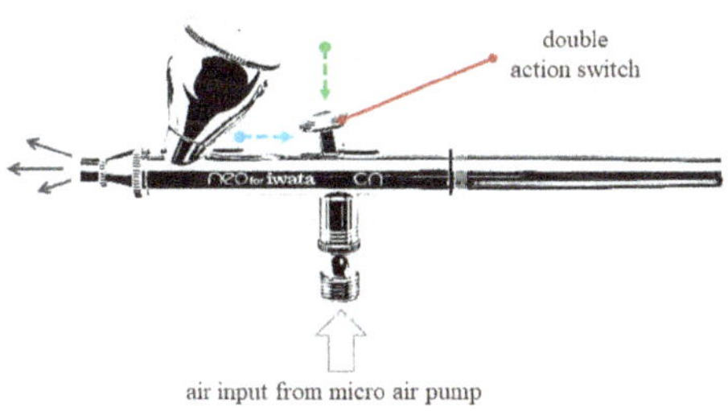

(**b**)

Figure 4. Micro air pump and micro spray gun configuration. (**a**) Picture of micro air pump fixed on the plastic mat material and the I/O contact layout. (**b**) Picture of micro spray gun and the schematic diagram of exhaust switch start. The green arrow stands for pressing down the switch, and the blue arrow reveals pressing the switch to the back of the spray gun.

To test the control effect of the sprayer of the air pump, an airbrush is added to the back of the micro air pump. The system is called a micro spray gun. The professional spray gun (Iwata NEO N4500 Airbrush, Anest Iwata Co., Nishio-shi, Aichi, Japan) is adopted; it is the spray gun suitable for dealing with details or large areas. Figure 4b shows the picture of the spray gun. The micro spray gun sprays paint through air pressure, and the spray gun is generally equipped with an air exhaust switch. The spray gun applies dual action function, as green and blue arrows in Figure 4b. The exhaust volume is controlled by pressing the button or pressing the back of the spray gun. The air input source comes from the micro air pump shown in Figure 4a. The connection point can be seen from the plastic hose above the picture.

To observe the pressure change in the system operating process, another side of the three-way pipe is connected to a pressure sensor module, which integrates a Wheatstone bridge sensor and signals processing chip and covers circuits of a differential amplifier, automated calibration, and temperature compensation. The pressure sensor module could transmit the pressure in the pipe, through the analog-digital conversion of a multifunction I/O card, to the software box for algorithm operation.

2.3.2. Characteristics of the Micro Air Pump

To test the air pressure changes after the multifunction I/O card driving the micro air pump, the PID control algorithm is not utilized, but simply using the multifunction I/O card to output the PWM signals to the DC motor drive module for the micro air pump exhaust. A program is planned with software to change the duty ratio of PWM signal output from 0% to 100% in 20 s, meaning that the duty ratio would increase by 1% per 200 ms. Finally, the multifunction I/O card would read and record the pressure in the pipe.

After the 20 s experimental recording, Figure 5a shows the air pressure change without pressing the micro spray gun switch, as a black solid line, and the air pressure change when pressing the micro spray gun switch, as a gray solid line. The left vertical axis is the pressure change. From the data in Figure 5a,b, the micro air pump appears slow pressure rise on PWM being 55% in both experiments, i.e., the motor smoothly runs at about 11 s, and the initial running of the motor appears on the PWM duty ratio about 55–65%. As shown in the red square in Figure 5b, there is a comparatively large pressure fluctuation. The pressure fluctuation is considered as the micro air pump exhaust not being stable yet, and the pressure signal change would also affect the algorithm in successive PID control experiments. In this case, the duty ratio of the PWM output is controlled at >70% in the successive PID experiment, as to set the micro air pump pressure above 40 kPa, to stabilize the system. It is also discovered in Figure 5b that a large amount of exhaust appears when the micro spray gun switch is pressed, and the maximal pressure difference, under the same PWM output control, is about 30 kPa; it could be understood from the green area in the figure. From Figure 5b, it can be found that when the PWM output is in the range of 70% to 100%, the pressure fluctuation does not exceed 10 kPa. Therefore, assumed that the $e(t)$ of pressure should be controlled below 10 kPa when the algorithm could stabilize the convergence.

The relationship between the driving motor speed and the current of the micro air compressor is quite complicated, and it will vary due to factors such as structure, power supply, and load. This study used a 12 V micro air compressor. When the PWM duty ratio reaches the maximum percentage rate (100%), the motor will also reach the highest speed. When the motor starts from the static, the speed of the motor is not a linear relationship with the driving current, and it must be known according to the actual experiment. Therefore, to understand the relationship between the speed and current of the driving motor, the current corresponding to different PWM duty ratios must be measured. The correlation characteristic curve of the two is shown in Figure 5c. There are five slope lines in Figure 5b that indicate the pressure delta to PWM duty ratio. The slope is in order p3 > p2 > p5 > p4 > p1, and the unit is kPa/%. The driving current of the corresponding PWM duty ratio is shown in Figure 5c. Compared with Figure 5b,c, which would present the driving currents at each stage. The Area-A area of Figure 5c are the currents when the PWM duty ratio is below 55%. Due to insufficient driving current, the motor is almost un-rotated. Area-B was just starting, so the pressure shown in Figure 5b has not yet stabilized. When the micro spray gun is not started, it can be seen from Figure 5b,c that the current value of the Area-C is large, the current delta is relatively large, and the pressure slope is largest. Currently, the PWM duty ratio is about 69–77%. The current in Area-D is relatively stable. So, when the PWM duty ratio is more than 77%, it can reach stability. When the micro spray gun starts to spray, the p4 slope line in Figure 5b corresponds to the Area-C, D, and E in Figure 5c, the PWM duty ratio is more than 69%, and the current is relatively stable. This should be because the gas starts to discharge, lifting the previous closed chamber effect, so that pump can operate stably; improve efficiency.

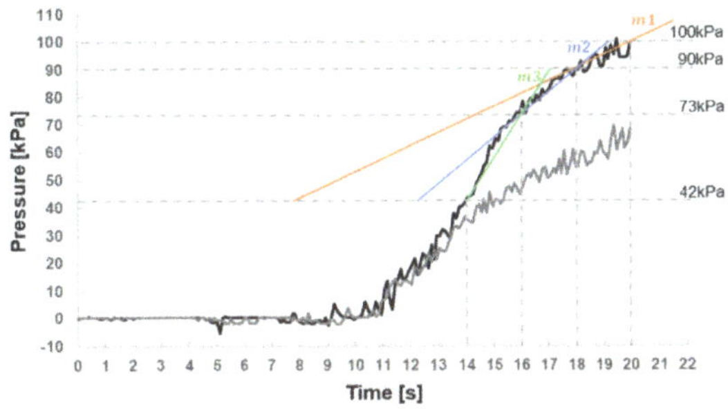

(a)

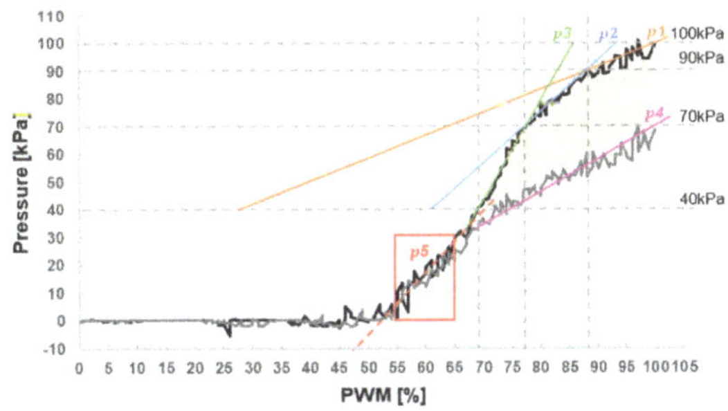

(b)

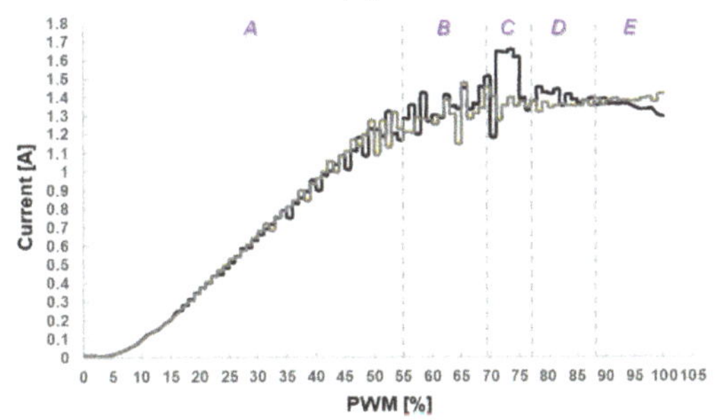

(c)

Figure 5. Numerical change of pressure sensor in the micro air pump operating process without PID control algorithm. (**a**) The cross plot of time and pressure sensor value, PWM duty ratio, (**b**) The cross plot of PWM duty ratio and pressure sensor value. (**c**) The cross plot of PWM duty ratio and current value.

2.3.3. Convergence Condition

The algorithm convergence condition is further defined in the section. In the PID control experiment, the $e(t)$ is generally observed by modulating the parameters to analyze the most suitable parameters for the expected target value n(t). m_1, m_2, and m_3 in Figure 5a are used for denoting the pressure above 40 kPa changing with time. For instance, m_3 shows a 15.5 kPa rise per second, m_2 reveals a 8.5 kPa rise per second, and m_1 appears at a 5 kPa rise per second. It also reveals that the PWM output could have the pressure rise 10 kPa within 2 s when the PID parameter is modulated to a stable stage. From another point of view, when the parameter could have the algorithm converge, the $e(t)$ in this system could control it under 10 kPa within 2 s.

Aiming at the monitoring time interval of the $e(t)$ getting into the convergence state at least higher than 2 s, the monitoring time interval is set to 5 s in our experiment. That is, when the $e(t)$ could be inhibited in the minimal convergence error amount of 10 kPa in 5 s intervals, the experiment achieves the convergence condition. It would be easier to comprehend from the curve in Figure 6. When the pressure of $e(t)$ starts to drop below the convergence critical threshold (10 kPa), the change maintaining below 10 kPa in 5 s is observed. When it conforms to the condition within 5 s, the time point dropping down to 10 kPa is called the convergence time (T_c), also the convergence starting point.

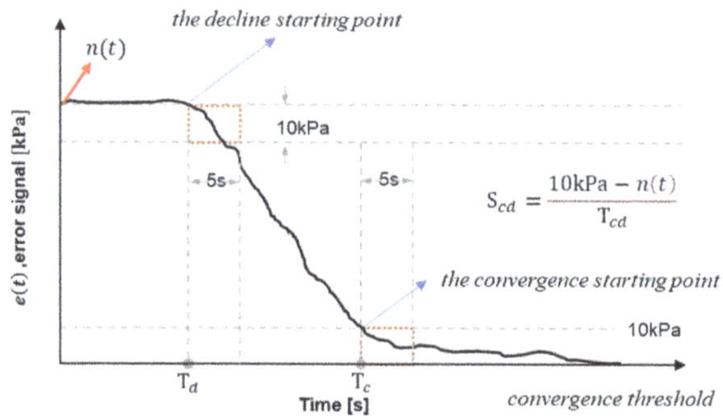

Figure 6. Definition points of convergence condition aiming at PID control algorithm observing error amount.

When the PID algorithm is executed, it would appear distinct fall trend due to different target values $n(t)$. In this case, the convergence condition in the previous paragraph is defined; that is, when the $e(t)$ starts to be lower than the $n(t)$, the observation would be continued for 5 s. When $e(t)$ is lower than $n(t)$ in the 5 s, it is called the decline time (T_d), i.e., the decline starting point. To compare the convergence condition under different parameters, the convergence and decline intervals (T_{cd}) of $e(t)$ is defined as the time difference between T_c and T_d. The convergence and decline slope (S_{cd}) is subtracting the target value $n(t)$ from the minimal convergence error amount (10 kPa) and then divided by T_{cd}, as the equation in Figure 6. Incidentally, the $e(t)$ convergence change curve presented in Figure 6 is the data collected continuously for 0.1 s, and then reproduced with the data of 1 s after averaging 10 points.

3. Results

3.1. *Effects under Different Target Values*

To verify the effect of the PID control algorithm on the micro air pump, the experimental design and software planning are preceded according to the system architecture in Figure 3. A micro spray gun is added on the back of the micro air pump but exhausted by not pressing the switch of the micro-spray gun. According to the data explanation in Section 2.3.2, the micro spray gun is suitable for the application to exhaust volume above 40 kPa. $n(t)$ is therefore set to 40, 50, 60, 70, 80, 90, and 100 kPa as the target value for the algorithm.

The PID controller could generally acquire the close parameters through trial and error. First, according to the experience, merely K_p is added to the system. K_p is set as the minimum to observe the convergent tendency of $e(t)$. If the system response is slow, increase the K_p value, otherwise decrease the K_p value to obtain the range suitable for the experimental parameters. So, with these rules of thumb, we finally set the differential gain adjustable parameter K_d to 0, the integral gain adjustable parameter K_i is set to 0.1, and the proportional gain adjustable parameter K_p is modulated to 0.1, 0.05, 0.01, and 0.005 to observe the convergence condition.

Figure 7 shows the condition of $e(t)$ being continuously monitored for 100 s when the target value $n(t)$ is set to 60 kPa, and the PID parameters are set with four K_p values. The smaller $e(t)$ reveals the closer $y'(t)$ from the pressure sensor to the target value $n(t)$.

According to the convergence condition described in Section 2.3.3, the PID algorithm monitors the stability when $e(t)$ gets in the minimal convergence error amount (10 kPa). From Figure 7, it is discovered that the decline point T_d could be achieved in about 2–3 s when the K_p is set to 0.1. However, the $e(t)$ is hard to achieve the convergence condition in a short period; it takes about 96 s to drop down to the convergence critical threshold (10 kPa). Strictly speaking, the parameter has the system approach the divergence. The other three K_p values reach the convergence condition; when K_p is 0.05, 0.01, and 0.005, the T_c appears at 9 s, 27 s, and 52 s, respectively. Such a result reveals that the fastest speed to achieve the convergence critical threshold appears when K_p is set to 0.05.

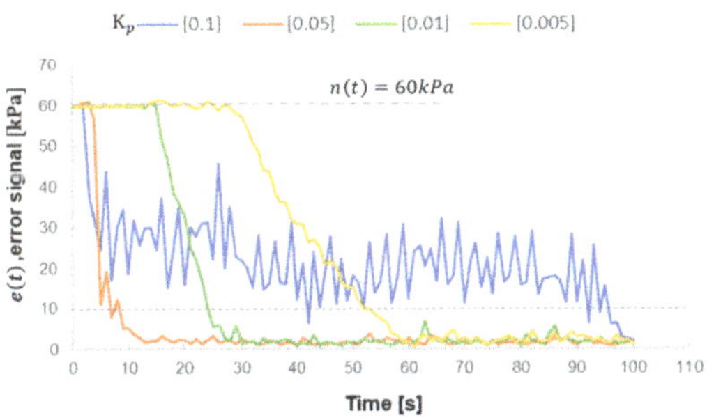

Figure 7. The changing trend of each error amount $e(t)$, when target value $n(t)$ is set to 60 kPa, and set PID parameter K_p = 0.1, 0.05, 0.01, and 0.005.

The experimental data are organized in Figure 8. The x-coordinate in Figure 8a is the target value $n(t)$ with the PID algorithm. In this case, the selection of K_p indeed presents distinct effects on T_c, under the same target value. K_p being 0.005, compared with 0.05 and 0.01, appears longer convergence time; when K_p is 0.1, signal oscillation appears at 60 kPa and 70 kPa, without being stably adapted. Furthermore, the convergence time T_c, relative to the target pressure, is out of proportion, and, even when $n(t)$ is 60 kPa, T_c is longer than K_p being set to 0.005. In this case, when K_p is set to 0.1, $n(t)$ should be above 80 kPa to really conform to the convergence condition; it could be viewed from the blue round in Figure 8a. In Figure 8b, convergence decline slope S_{cd} is applied to present the difference under a different target value. It is discovered that the S_{cd} change to the convergence situation of each target value is not big, when K_p is 0.005 and 0.01, maximally dropping the $e(t)$ amount about 7.8 kPa per second. When K_p is 0.05, as shown in Figure 8b, the slope of S_{cd} increases with the increase of $n(t)$, which means that the algorithm can achieve a convergence effect at a similar time when $n(t)$ is between 40 and 100 kPa. As shown in the orange triangle in Figure 8a, the overall T_c are very close.

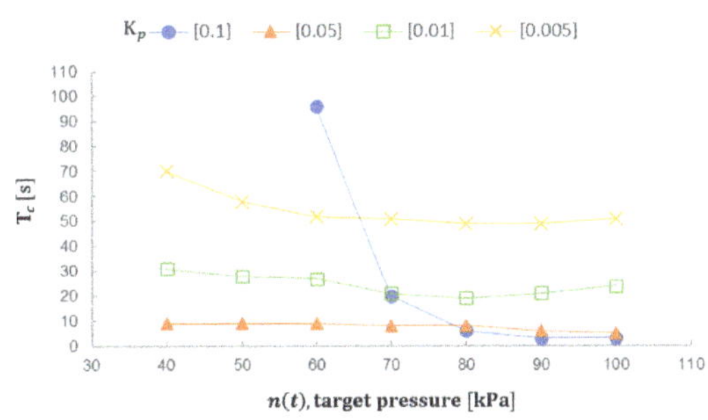

(a)

Figure 8. *Cont.*

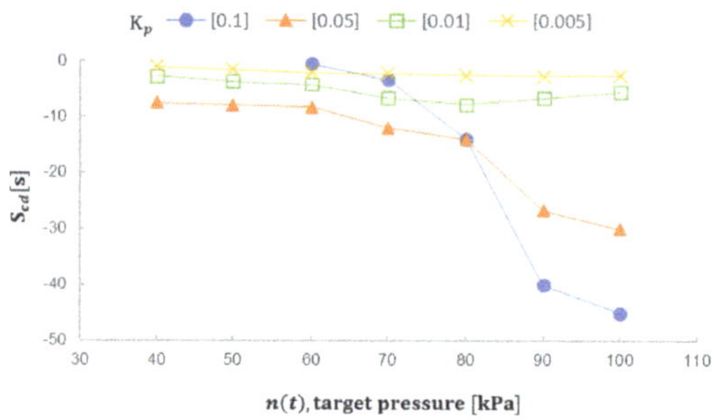

(b)

Figure 8. When K_p = 0.1, 0.05, 0.01, and 0.005, (**a**) The different $n(t)$ values and the corresponding convergence time T_c and (**b**) The different $n(t)$ values and the corresponding convergence decline slope S_{cd}.

3.2. Effects under Pulse-like Variables

When using a micro spray gun, pressing down the switch for spraying paint results in dropping pressure in the pipe, with a change of about 30 kPa. In this case, a pulse-like sequence variable, target value $n(t)$, is regarded as the target to understand the adaption of PID parameters to different pressure changes. It simulates two target amounts, 65 kPa, and 95 kPa, in the time interval as the alternate output. The sequence data, according to the software planning, would change the $n(t)$ of the system at 20, 40, and 60 s, as the switch between 65 kPa and 95 kPa of the grey curve in Figure 9.

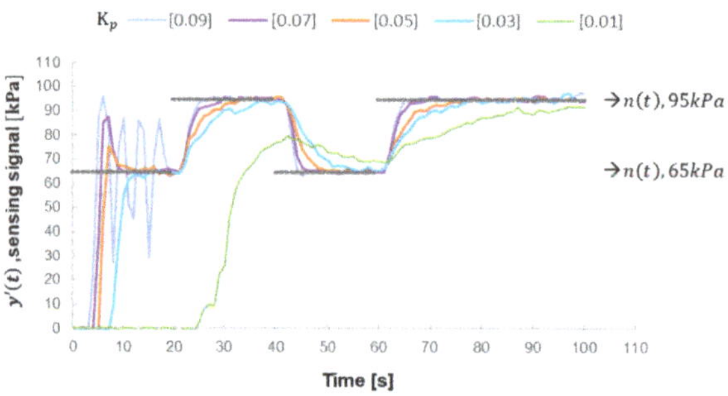

Figure 9. The grey line indicates a pulse-like curve simulating the target value $n(t)$ in 65 kPa and 95 kPa, and converged at 20, 40, and 60 s. Output pressure $y'(t)$ under PID parameter K_p = 0.1, 0.05, 0.01, and 0.005.

Since better convergence appears on K_p being 0.05, the intervals ±0.02 and ±0.04 are selected for observing the effect of such K_p on pulse-like sequence data, while K_i remains 0.1 and K_d remains 0. From Figure 9, five curves show the results of tracing the target value with the PID control technology, when K_p is 0.01, 0.03, 0.05, 0.07, and 0.09.

In the first 20 s, $n(t)$ as 65 kPa is the target of PID when the micro air pump is just started; it is discovered that the system appears to overshoot when K_p is 0.09. The result is similar to Figure 8a in that oscillation is more easily appear when K_p is 0.1 and $n(t)$ is under 70, it is hard to get into the region of convergence. When K_p is 0.01, it is difficult to catch the target value of 65 kPa in the first 20 s, due to the relatively smaller value. The convergence effect could be achieved at about 10 s with the other three parameters of K_p being 0.03, 0.05, and 0.07; that is, the rising speed and the effect of overshoot, when K_p appears 0.05 ± 0.02, are more acceptable. Overall, the overshoot and time for rising to the target value under K_p being 0.05 show better adaptability on 20 s after the system startup.

After 20 s, the target value $n(t)$ is programmed to 95 kPa and the standard of convergence is soon achieved when K_p is 0.09, as it is close to 0.1. According to the observation of Figure 8a, it would get into the region of convergence soon when K_p is 0.1 and $n(t)$ is above 90. The experimental result with the parameters is therefore reasonable. When K_p is 0.09 and the system is modulated to a steady state, it would not appear unstable error variation as in the starting, even when the target value is changed to 65 kPa after 40 s. When K_p is 0.01, although it takes a long time to keep up with the target value $n(t)$, it has begun to catch up gradually, but $e(t)$ is still much larger than other parameters. With the rest three parameters, e.g., K_p = 0.03, 0.05, and 0.07, the change is stable along with the target value and the convergence is achieved within a short period. Furthermore, after 20 s and the overshoot phenomenon as in the system starting does not appear when K_p are 0.05 and 0.07.

3.3. *Effects of PID Convergence*

As the system architecture in Figure 3, when pressing down the switch on the spray gun, the spray nozzle starts to spray air and immediately reduces the air pressure in the pipe. To have the air maintain certain output, the system would enhance the exhaust volume of the micro air pump through the PID algorithm, i.e., enhancing the percentage of PWM.

Figure 10a shows the pressure change $y'(t)$ before/after pressing the switch of the micro spray gun. The experimental parameters selected K_p to be 0.05, K_i remained at 0.1, and K_d is 0. The grey line in Figure 10a shows the system locking the target value $n(t)$ on 65 kPa. In this case, the system, after starting, would catch up with the set target value $n(t)$ according to the parameters set by PID. The result reveals to catch up to the target value at about 10 s. When the time achieves 20 s, the switch of the micro spray gun, as the green arrow indicated in Figure 4b, is pressed. It is discovered that the pressure, after pressing the switch, immediately drops about 25 kPa, and PID would control output after the system detects the pressure change. As a result, when the pressure is reduced, PID would increase the percentage of PWM to enhance the exhaust output volume to maintain the pressure at the target value of 65 kPa. Figure 10a shows that it takes about 10 s, after pressing down the switch, to have the pressure return the target value. The process could be viewed through the change in PWM modulation. Figure 10b displays the PWM change after pressing down the switch of the micro spray gun. When the system achieves stability, the PWM modulation value does not appear to large changes, e.g., PWM stably maintains 70% after starting for 10–20 s, as the red dotted double arrows in Figure 10b. When the spray gun switch is pressed at 20 s, PWM starts to increase the output percentage; after 30 s, PWM almost remains at 95%, without much modulation, as the black dotted arrow in Figure 10b.

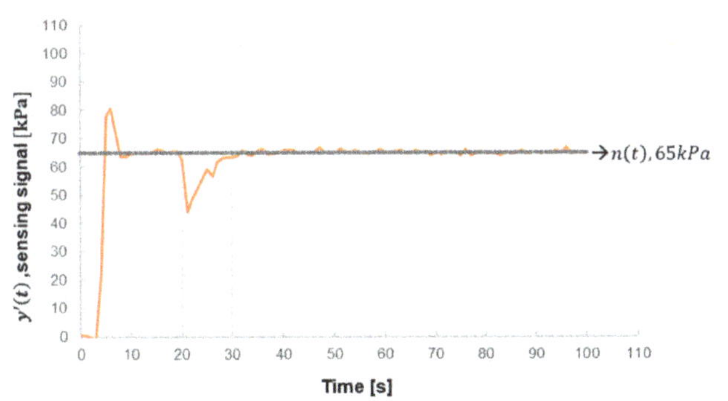

(a)

Figure 10. *Cont.*

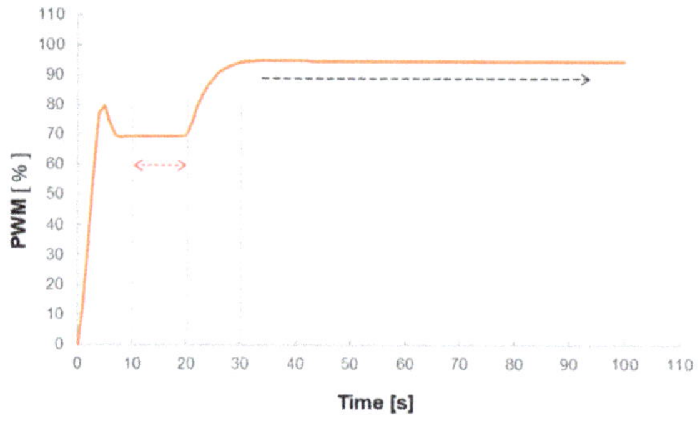

(b)

Figure 10. When the PID parameter is set $K_p = 0.05$, $K_i = 0.1$, $K_d = 0$, (**a**) The sensed pressure change, and (**b**) The change in PID modulating PWM output, before and after pressing down the micro spray gun.

4. Discussions

Aiming at the experiment results, three points are discussed.

1. The experimental data reveal that the internal structure of the micro air pump presents certain effects on the control response, rather than simply a linear state. Consequently, the experiment results shown in Figure 5a could verify the better pressure output range of the micro air pump above 40 kPa. The exhaust, when the spray gun is pressed, results in dropping pressure in the pipe to influence the overall pressure output. In this case, the maximal pressure sensed is about 70 kPa when PWM is modulated to the maximal 100%, with the difference of about 30 kPa from the spray gun not being pressed. In the experiment in Section 3.3, the preset target value of 65 kPa would drop down to 40 kPa after pressing the spray gun. The low pressure during paint spray would affect the spray quality and result in the irregular operation of the motor, due to the unstable air pump, damaging the life. Automatic control technology could effectively change the exhaust output volume and stabilize the operation of the air pump. This study verifies the driving module and algorithm control technology of the automatic system. The experiment results after pressing the spray gun proves that proper parameter modulation allows the system design to return the target pressure within a specified period.
2. The proportional gain adjustable parameter K_p determines the ratio of the output response to the error signal. In general, increasing K_p will increase the responsiveness of the control system, but if the proportional gain is too large, the process variable will start to oscillate. If K_p is increased further, the vibration will become larger and larger, and the system will become unstable, and even uncontrollable vibration may occur. As the result in Figure 7, $K_p = 0.1$ is obviously a larger value in this experiment; under the same target value $n(t)$, the error is comparatively difficulty to be controlled in the convergence range. Perhaps, K_p appears faster convergence time and convergence speed when $n(t)$ is above 80 kPa with several larger target values $n(t)$, as in Figure 8a. Nevertheless, the higher instability results in $n(t)$ being 60 and 70 kPa, the convergence time being largely prolonged, and the convergence time, with $n(t)$ being 60 kPa, is even longer than it when K_p is 0.005. Particularly, it could not get into the convergence condition when $n(t)$ is 40 kPa and 50 kPa. Such applications would cause several unpredictable errors and even non-practical ones. When smaller K_p is selected, the error signal is less sensitive and the response to the target signal being suddenly interfered with or appearing larger variable would be too slow and the output control signal is not large enough to catch up with the variable quantity. As in Figure 9, the target value $n(t)$ in the moment of conversion at 65 kPa and 95 kPa, when K_p is 0.01, does not cause system instability or divergence, but the speed to catch up with the

target value is indeed slower. In real applications, it is difficult to predict or limit users to stably press the spray gun so that the pressure variable would happen at any time. The pulse-like target variable experiment planned in Section 3.2 is used for coping with such pressure variables with proper parameter settings so that the micro spray gun system could be normally operated.

3. After pressing the micro spray gun, the pressure of the feedback signal immediately drops to have the PID system make larger output modulation. It reveals the same operation idea of pulse-like variable target value; but the pressure disturbance in the pipe, after pressing the spray gun, cannot be predicted. Figure 10a shows that $y'(t)$, after returning the target value, still appears slight fluctuation. However, the PID output control could have the PWM stably increase, without overshoot of the system, Figure 10b. It is therefore considered that K_p being set to 0.05 could effectively catch up with the target within a short period, and selecting proper K_i could accelerate the system approaching the target and eliminate the steady state error on K_p.

5. Conclusions

The micro spray gun (airbrush) is primarily used for fine art drawing, and the quality of the spraying depends on the accuracy and stability of the air pressure. However, traditional mechanical regulation may not always meet the necessary requirements. Our study introduces the PID method for the first time to improve the spray quality, even without an air receiver. The experiment results show that, under a proper selection of K_p and K_i, the target pressure changes within 30 kPa could achieve the convergence condition in 10 s. The results as well show that the instant pressure difference of 25 kPa after spraying could be increased to the target pressure in 10 s.

Consequently, the entire system architecture design could realize the adaptive micro sprayer, effectively reduces wear, and enhances the service life. The system architecture designed in this study would provide the reference for the realization of miniaturized microcontroller in the future to achieve the goal of adaptive portable micro spray gun equipment.

Author Contributions: Conceptualization, J.-H.L. and C.-H.C.; methodology, C.-H.C.; software, J.-H.L.; validation, S.-T.T.; formal analysis, S.-T.T.; investigation, C.-H.C.; resources, J.-H.L.; data curation, S.-T.T.; writing—original draft preparation, C.-H.C.; writing—review and editing, J.-H.L. and S.-T.T.; visualization, J.-H.L.; supervision, S.-T.T.; project administration, S.-T.T.; funding acquisition, J.-H.L. All authors have read and agreed to the published version of the manuscript.

Funding: This research received no external funding.

Conflicts of Interest: The authors declare no conflict of interest.

References

1. Sudduth, K.A.; Borgelt, S.C.; Hou, J. Performance of a chemical injection sprayer system. *Appl. Eng. Agric.* **1995**, *11*, 343–348. [CrossRef]
2. Erkan, U.; Huseyin, G.; First, K. A review of electrostatic spraying for agricultural applications. *J. Agric. Mach. Sci.* **2016**, *12*, 229–233.
3. Liu, Z.B. High-Precision Spray Gun Spraying Control System Based on Linear Regulation. China Patent CN113019732A, 2021.
4. Giles, D.K.; Bensalem, E. Spray droplet velocity and energy in intermittent flow from hydraulic nozzles. *J. Agric. Eng. Res.* **1992**, *51*, 101–112. [CrossRef]
5. Deng, W.; Zhao, C.; Chen, L.; Wang, X. Comparison of spray characteristics for three types of variable spray. *J. Agric. Eng. Res.* **2016**, *XLVII 512*, 148–156. [CrossRef]
6. Grella, M.; Gioelli, F.; Marucco, P.; Zwertvaegher, I.; Mozzanini, E.; Mylonas, N.; Nuyttens, D.; Balsari, P. Field assessment of a pulse width modulation (PWM) spray system applying different spray volumes: Duty cycle and forward speed effects on vines spray coverage. *Precis. Agric.* **2022**, *23*, 219–252. [CrossRef]
7. Orti, E.; Cuenca, A.; Perez, M.; Torregrosa, A.; Ortiz, C.; Rovira-Mas, F. Preliminary evaluation of a blast sprayer controlled by pulse-width-modulated nozzles. *Sensors* **2022**, *22*, 4924. [CrossRef] [PubMed]
8. Maski, D.; Durairaj, D. Effects of electrode voltage, liquid flow rate, and liquid properties on spray chargeability of an air-assisted electrostatic-induction spray-charging system. *J. Electrost.* **2010**, *68*, 152–158. [CrossRef]

9. Chen, S.; Yin, D.; Wei, X.; Pei, W. Design and simulation of variable weed spraying controller based on adaptive neural fuzzy inference system. *J. Drain. Irrig. Mach. Eng.* **2011**, *29*, 272–276.
10. Patel, M.K.; Ghanshyam, C.; Kapur, P. Characterization of electrode material for electrostatic spray charging: Theoretical and engineering practices. *J. Electrost.* **2013**, *71*, 55–60. [CrossRef]
11. Sasaki, R.S.; Teixeira, M.M.; Fernandes, H.C.; Paulo, M.; Denílson, E.R.; Cleyton, B. Parameters of electrostatic spraying and its influence on the application efficiency. *Rev. Ceres* **2013**, *60*, 474–479. [CrossRef]
12. Shilkrot, R.; Maes, P.; Paradiso, J.A.; Zoran, A. Augmented airbrush for computer aided painting. *ACM Trans. Graph.* **2015**, *34*, 1–11. [CrossRef]
13. Mohamed, A.A.S. An Investigation into the History of the Airbrush and the Impact of the Conservation Treatment of Airbrushed Canvas Paintings. Ph.D. Thesis, Northumbria University, Newcastle upon Tyne, UK, 2016.
14. Chen, Y.C. Study on the Airbrush Makeup Applied to the Whole Fashion Design Combining with the Concept of Multi-Media. Master's Thesis, Department of Cosmetics Science, Vanung University, Taoyuan, Taiwan, 2022.
15. Lin, J.H.; Tang, S.T.; Chen, C.H.; Cheng, Y.T.; Tai, C.C.; Lin, M.H. The Prototype Design and Verification of Intelligent Variable Output Control Apply on Micro-spray Gun Simulation Platform. In Proceedings of the 2022 International Conference on Fuzzy Theory and Its Applications (iFUZZY), Garden Villa, Kaohsiung, Taiwan, 3–5 November 2022.
16. Yan, H. Diaphragm Air Pump. China Patent CN201714643U, 2011.
17. Jalnekar, R.M.; Jog, K.S. Pulse-width-modulation techniques: A review. *IETE J. Res.* **2020**, *46*, 175–183. [CrossRef]
18. Li, Y.; Ang, K.H.; Chong, G.C.Y. Patents, software and hardware for PID control—An overview and analysis of the current art. *IEEE Control Syst. Mag.* **2006**, *26*, 42–54.
19. Lopez-Gomez, J.; Vargas-Trevino, M.A.D.; Vergara-Limon, S.; Vargas-Trevino, M.; Gutierrez-Gutierrez, J.; Palomino-Merino, A.D.; Martinez-Solis, F.; Felix-Beltran, O. Influence of PWM torque control frequency in DC motors by means of an optimum design method. *IEEE Access* **2020**, *8*, 80691–80706. [CrossRef]
20. Zhang, R.; Gao, L.X. Research on motor control and simulation based on PID and internet of things system. *Microprocess. Microsyst.* **2021**, *80*, 103602. [CrossRef]

applied sciences

MDPI

Article

Analysis of Explicit Model Predictive Control for Track-Following Servo Control of Lunar Gravity Compensation Facility

Yonggui Zheng †, Meng Liu †, Hao Wu * and Jun Wang

School of Aeronautic Science and Engineering, Beihang University, Beijing 100191, China; zhengyonggui@buaa.edu.cn (Y.Z.)
* Correspondence: haowu@buaa.edu.cn
† These authors contributed equally to this work.

Abstract: The Lunar Gravity Compensation Facility (LGCF) is a critical component in ground tests for a crewed lunar roving vehicle (CLRV). The track-following servo subsystem's performance is of critical importance in the LGCF, as it needs to achieve high-precision tracking of the CLRV's fast, wide range of motion in the horizontal direction. The subsystem must also operate within various constraints, including those related to speed, acceleration, and position. These requirements introduce new challenges to both the design and control of the subsystem. To tackle these challenges, this paper employs a Permanent-Magnet Synchronous Motor (PMSM) vector control method based on Space Vector Pulse Width Modulation (SVPWM) to achieve accurate speed tracking. Additionally, this paper presents an Explicit Model Predictive Control (EMPC) strategy for precise position servo control of the track-following system under multi-parameter constraints. The simulation model of the track-following servo subsystem is established based on the above methods. The simulation results demonstrate that the position tracking error of the gravity compensation system, constructed using the above method combined with EMPC control, is less than 0.2 m. The control performance of the EMPC is significantly better than those of the PI and LQI controllers. The influence of errors on the drawbar pull is within 12.5%, and its effect on the compensation force is negligible. These results provide theoretical support for the design of a track-following servo subsystem.

Keywords: lunar gravity compensation facility; track-following servo subsystem; PMSM; explicit MPC

Citation: Zheng, Y.; Liu, M.; Wu, H.; Wang, J. Analysis of Explicit Model Predictive Control for Track-Following Servo Control of Lunar Gravity Compensation Facility. *Appl. Sci.* **2023**, *13*, 4411. https://doi.org/10.3390/app13074411

Academic Editors: Kan Liu and Wei Hu

Received: 11 March 2023
Revised: 29 March 2023
Accepted: 29 March 2023
Published: 30 March 2023

1. Introduction

The crewed lunar roving vehicle (CLRV) is essential in Chinese plans to build a lunar base. The CLRV is an essential and indispensable exploration tool that not only transports astronauts to great distances from the lunar module but also ensures their safety as they move on the lunar surface [1]. To mitigate the potential issues of the CLRV, it is imperative to conduct a comprehensive ground test of the full rover before launch. Furthermore, the differences in gravity between the Moon and Earth give rise to distinct steering properties and driving sensations for the astronauts using the CLRV. Therefore, it is essential to construct a ground test facility on Earth that accurately simulates lunar gravity to facilitate astronaut training and evaluate the CLRV's performance.

A problem common to the ground test facility is simulating a low-gravity environment similar to the spacecraft's natural working environment [2]. Simulating a low-gravity field is an essential issue in the motion performance experiment of the planetary rover. The simulation method of offsetting part of the lunar rover's gravity by an external force is called gravity compensation. The primary ways to achieve artificial micro-gravity include free-fall testing [3], the air-bearing table [4], neutral buoyance [5,6], and the suspension system [7,8]. Among the available methods, the suspension system [9,10] is widely adopted because of its relatively simple structure, easy construction, and 3-D simulation with unlimited time.

2. Methodology

In this section, we present the mechanical design of the track-following servo subsystem. The subsystem comprises a moving crane system, a moving trolley system, and multiple motors that enable the track-following servo in both the X and Y directions, as illustrated in Figure 2. The crane is driven by four motors, and the trolley is driven by two motors. The motors power the steel wheels of the crane and trolley to move on the rail. The servo motors transmit power to the system, which is then transmitted to the low-speed shaft end of the reducer via the coupling. The reducer's low-speed shaft end outputs to the driving wheel shaft through another coupling, and the driving wheels propel the crane and trolley to follow the CLRV with high precision.

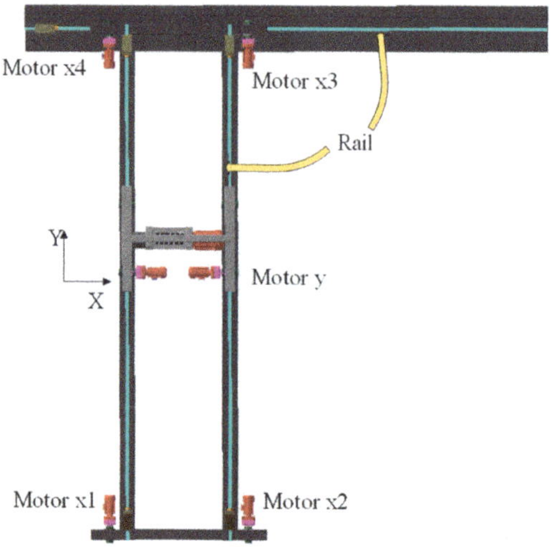

Figure 2. Schematic diagram of the subsystem movement.

Then, based on the mechanical design of the subsystem, suitable equipment is selected, and a simulation model of the track-following servo subsystem is established. In this paper, the servo motors used are PMSMs. The track-following servo subsystem comprises a position servo closed-loop and closed-loop control of the motor speed and current; the crane control model is illustrated in Figure 3. The position control loop is situated on the outer layer of the double closed-loop, which includes the speed and current loops of the servo motor. The position control loop receives a reference signal and subsystem state variables and calculates the expected velocity increment of the crane or trolley in a given unit of time. This value is then passed to the motor's speed loop controller. The speed loop controller calculates the expected current output by comparing the expected speed with the actual speed. The motor's current loop controller then calculates the expected voltage input based on the expected current and the actual current. Together, these three loops form the track-following servo subsystem.

To achieve high-precision control of the crane and trolley, the PMSM vector control method based on SVPWM is adopted for fast response in speed control. The PI control method, with its advantages of being a simple algorithm, having good stability, and having high reliability in the motor speed control system, is widely used for controlling motor speed and current. Therefore, the PI method is applied to the PMSM vector control system, as depicted in Figure 3.

The performance of the outermost position servo loop plays a crucial role in the track-following servo subsystem's tracking performance. However, the position loop's control is limited by the electromechanical system's kinematics and dynamics. To overcome this multi-parameter constraint, the present paper proposes utilizing the EMPC controller within the position servo closed-loop.

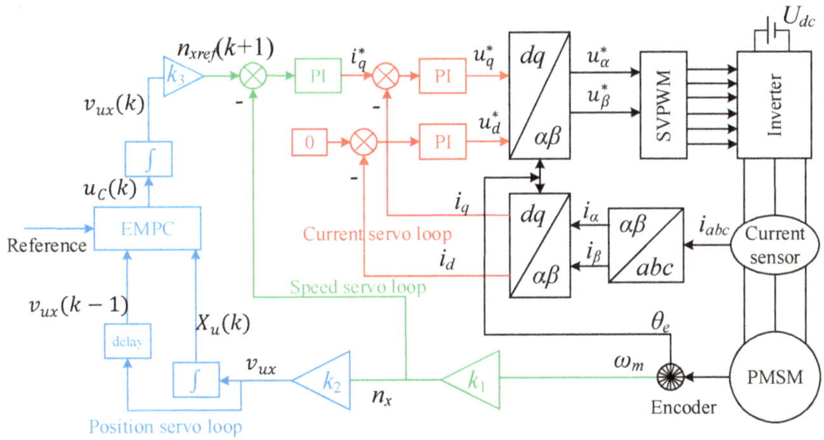

Figure 3. Three-loop control of the crane in track-following servo subsystem.

This section mainly describes the establishment of the mathematical model for the three-loop control system, including the mathematical model and model parameters of the PMSM, as well as the mathematical model of the position loop.

2.1. Mathematical Model of PMSM

As shown in Figure 3, this paper adopts vector control technology based on SVPWM to realize the decoupling control of torque and excitation components. Following the stator flux linkage orientation control rules, the d-axis of the reference coordinate is aligned with the motor flux direction, where $i_d = 0$. The stator flux linkage component on the d-axis is represented as $\psi_f = \psi_d$. The voltage equation of the PMSM is formulated in the two-phase synchronous rotating reference frame.

$$\begin{cases} \frac{di_d}{dt} = \frac{1}{L_d}(-Ri_d + P_n\omega_m L_q i_q + u_d) \\ \frac{di_q}{dt} = \frac{1}{L_q}\left(-Ri_q - P_n\omega_m L_d i_d + u_q - P_n\omega_m\psi_f\right) \\ \frac{d\omega_m}{dt} = \frac{1}{J}(T_e - T_L - B_u\omega_m) \end{cases} \quad (1)$$

The torque reference (electromagnetic torque) T_e in the following equation is used:

$$T_e = 1.5P_n\left[\psi_f i_q + (L_d - L_q)i_d i_q\right] \quad (2)$$

where R is the stator resistance (Ω); L_d and L_q are the inductances (H) of the stator on the d and q axes, respectively; i_d and i_q are the current (A) of the stator on the d and q axes, respectively; u_d and u_q are the voltage (V) of the stator on the d and q axes, respectively; P_n is the number of pole pairs of the rotor; J is the inertia; ω_m is the angular velocity measured from the motor; T_L is a constant load torque; B_u is the mechanical damping constant.

Suppose that the load torque of the four motors in the X direction is equal, the output torque and speed are equal, the load torque of the two motors in the Y direction is equal, and the output torque and speed are equal. In order to build the mathematical model of the motor, it is necessary to select the specific motor first according to the required motor torque and speed.

The motor torque T_{out} is divided into load torque T_u for overcoming rolling friction and accelerating torque T_a for maximum acceleration.

$$T_{out} = q_{motor}(T_u + T_a) = q_{motor}(\mu m g r_w + m a r_w) \quad (3)$$

where the gross load hauled by the crane is about $m_x = 12,000$ kg, and the gross load hauled by the trolley is about $m_y = 2000$ kg. The number of motors for the moving crane system and the moving trolley system are $q^X_{motor} = 4$ and $q^Y_{motor} = 2$, respectively. The coefficient of rolling friction with railroad steel wheels on steel rails is $\mu = 0.05$, and the radius of the steel wheel is $r_w = 0.2$ m.

The design speed of the crane is larger than the maximum speed (v_{rmax}) of the CLRV, so $v^{max}_{ux} = 5$ m/s is set. Additionally, the VLRV moves in a small range in the Y direction, so $v^{max}_{uy} = 3.5$ m/s is set. The v_{ux} and v_{ux} are the actual speeds of the crane and trolley, respectively.

According to Equation (4), the max speeds of motors n^{max}_x and n^{max}_y can be calculated:

$$n = \frac{60r_g}{2\pi r_w} v_u \tag{4}$$

where the gear ratio of the crane motors is $r^X_g = 10$ and the gear ratio of the trolley motors is $r^Y_g = 10$.

According to the calculation, the motors are selected. The parameters of the motors are shown in the following Table 1.

Table 1. Permanent-magnet synchronous motors parameters table.

Parameters	Motors x (1, 2, 3, 4)	Motors y (1, 2)
bus voltage U_{dc}	380 V	380 V
stator rated current I_N	63 A	20 A
stator inductance $L_d = L_d$	0.0083 H	0.0024 H
stator resistance R	0.376 Ω	0.369 Ω
permanent magnet flux ψ_f	0.9720 Wb	0.8815 Wb
the rotor moment of inertia J	0.2026 kg·m^2	0.0182 kg·m^2
the number of pole pairs P_n	1	1
viscous damping B_u	0.00464 N·m·(s·rad^{-1})	0.003 N·m·(s·rad^{-1})
the rated speed n	2500 rpm	2500 rpm
The rated torque T_e	123 N·m	37.4 N·m

In the system, the motors' speeds are n_x and n_y, and the actual angular velocities of the motors are ω_{mx} and ω_{my}, respectively.

$$v_{ux} = \frac{2\pi r_w n_x}{60 r^X_g} = \frac{r_w \omega^X_m}{r^X_g} \tag{5}$$

$$v_{uy} = \frac{2\pi r_w n_y}{60 r^Y_g} = \frac{r_w \omega^Y_m}{r^Y_g} \tag{6}$$

In this paper, the PI control is adopted in the motor's current loop and speed loop control. The primary constraints during the controller design are the voltage and current constraints on the quadrature axis and direct axis. The constraints are as follows:

$$\begin{array}{l} \sqrt{\left(i_d^2 + i_q^2\right)} \le \sqrt{2} I_N \\ \sqrt{\left(u_d^2 + u_q^2\right)} \le \frac{U_{dc}}{\sqrt{3}} \end{array} \tag{7}$$

As this is a real physical system, constraints on the states and inputs have to be considered during the controller design given. In the wheel/rail transmission, the premise for obtaining different traction forces is not to destroy the adhesion moment between the wheel/rail. In this paper, the calculation formula of the train wheel/rail adhesion coefficient

u_j is used to calculate the maximum traction torque. When the train speed $v < 50$ km/h, the u_j is as follows [26]:

$$\mu_j = 0.24 + \frac{12}{100 + 8v_{max}} \tag{8}$$

Then, the maximum traction torque is

$$T_j = \frac{\mu_j m g r_w}{r_g q_{motor}} \tag{9}$$

The maximum accelerations a_{ux}^{max} and a_{uy}^{max} can be calculated from the maximum output torque T_{out}^{max}:

$$T_{out}^{max} = min\{\text{rated torque } T_e,\ \text{maximum traction torque } T_j\} \tag{10}$$

The maximum accelerations of the crane and trolley are $a_{ux}^{max} = 1.558\ \text{m/s}^2$ and $a_{uy}^{max} = 2.608\ \text{m/s}^2$, respectively.

Additionally, this system adopts 17-bit incremental rotary encoders to measure the angular velocity and rotor-position of the motors. The measurement error of the motor rotor-position is $\pm 0.001°$, and the measurement error of the motor angular velocity is ± 0.2 rad/s. Rotor-position measurement noise adds Gaussian white noise to the θ_e, the amplitude is 0.001°, and the sampling period is 1/10,000 s. Angular velocity measurement noise adds Gaussian white noise to the angular velocity feedback ω_m, the amplitude is 0.2 rad/s, and the sampling period is 1/10,000 s. The current measurement noise also adds Gaussian white noise to the feedback loop of i_d and i_q, the noise amplitude is 0.2 A, and the sampling period is 1/10,000 s. The discrete signals θ_e, i_d, and i_q are delayed by one sampling period. Then, the signals are passed to the PI controllers or $dq/\alpha\beta$ converters.

2.2. Position Servo Loop Model

In this paper, the position servo loop receives the lifting points' information from the position and orientation measurement system. The EMPC controller outputs the desired movement increment Δv_{ux} of the crane in the X direction and the desired movement increment Δv_{uy} of the trolley in the Y direction. In this section, the subscript C denotes the crane and the subscript T denotes the trolley.

Under the fixed coordinates OXY on the ground, the kinematic equation of the trolley and crane in X and Y directions is as follows:

$$u(k) = \begin{pmatrix} \dot{X}_u(k) \\ \dot{Y}_u(k) \end{pmatrix} = \begin{pmatrix} v_{ux}(k) \\ v_{uy}(k) \end{pmatrix} \tag{11}$$

where X_u and Y_u are the position of the crane and trolley in the coordinate system OXY, respectively.

In Figure 3, n_{xref} (rpm) is the reference speed of the motor that drives the crane to move in the X direction, and n_{yref} (rpm) is the reference speed of the motor that drives the trolley to move in the Y direction. The relationship between n_{ref} and $u(k)$ is as follows:

$$n_{ref}(k+1) = \begin{pmatrix} n_{xref}(k+1) \\ n_{yref}(k+1) \end{pmatrix} = \begin{pmatrix} \frac{60r_{gx}}{2\pi r_w} v_{ux}(k) \\ \frac{60r_{gy}}{2\pi r_w} v_{uy}(k) \end{pmatrix} \tag{12}$$

When designing the position loop servo controller, the position servos of the X and Y directions are designed independently. In engineering, the system control generally adopts the incremental control method, and the incremental control is to output the increment Δv of the control variable v every period:

$$v_{ux}(k) = v_{ux}(k-1) + \Delta v_{ux}(k) = v_{ux}(k-1) + u_C(k)T_s \tag{13}$$

$$v_{uy}(k) = v_{uy}(k-1) + \Delta v_{uy}(k) = v_{ux}(k-1) + u_T(k)T_s \tag{14}$$

The reference signal can be regarded as a state variable:

$$X_u(k+1) = X_u(k) + v_{ux}(k)T_s \tag{15}$$

$$Y_u(k+1) = Y_u(k) + v_{uy}(k)T_s \tag{16}$$

The state-space models for the moving crane system and the moving trolley system are as follows:

$$\begin{aligned} x_C(k+1) &= A_C x_C(k) + B_C u_C(k) \\ y_C(k) &= C x_C(k) \end{aligned} \tag{17}$$

$$\begin{aligned} x_T(k+1) &= A x_T(k) + B u_T(k) \\ y_T(k) &= C_T x_T(k) \end{aligned} \tag{18}$$

where

$$\begin{gathered} \begin{bmatrix} -40 \\ -5 \end{bmatrix} \leq x_C(k) = \begin{bmatrix} X_u(k) \\ v_{ux}(k-1) \end{bmatrix} \leq \begin{bmatrix} 40 \\ 5 \end{bmatrix} \\ \begin{bmatrix} -10 \\ -3.5 \end{bmatrix} \leq x_T(k) = \begin{bmatrix} Y_u(k) \\ v_{uy}(k-1) \end{bmatrix} \leq \begin{bmatrix} 10 \\ 3.5 \end{bmatrix} \\ \begin{bmatrix} -1.558 \\ -2.608 \end{bmatrix} \leq \begin{pmatrix} u_C(k) \\ u_T(k) \end{pmatrix} \leq \begin{bmatrix} 1.558 \\ 2.608 \end{bmatrix} \\ A_C = A_T = \begin{bmatrix} 1 & T_s \\ 0 & 1 \end{bmatrix}, \; B_C = B_T = \begin{bmatrix} {T_s}^2 \\ T_s \end{bmatrix}, \\ C_C = C_T = \begin{bmatrix} 1 & 0 \end{bmatrix}. \end{gathered}$$

The deviation in the crane and trolley's position in the X and Y directions from the desired trajectory converges to zero through the proposed control scheme. To implement Equations (17) and (18) in an explicit MPC scheme, a zero-order hold method with a sampling time of 0.005 s is utilized in this paper.

3. Explicit Model Predictive Control

To ensure robustness and meet complex constraints, this study employs MPC [27,28] as a crucial optimal control scheme. At each sampling time k, the optimal control law of the inner loop is solved by formulating and solving a control problem that is then transformed into an online quadratic programming (QP) problem. The cost function used in this control problem is defined as follows:

$$
\begin{gathered}
\min_{\Delta u}\left\{J(U,x(k)) = x_{k+N_p|k}^{T} P x_{k+N_p|k} + \sum_{i=0}^{N_p-1}\left[(y_{k+i|k} - y_{k+i|k}^{ref})^{T} Q_y (y_{k+i|k} - y_{k+i|k}^{ref}) + x_{k+i|k}^{T} Q x_{k+i|k} + u_{k+i|k}^{T} R u_{k+i|k}\right]\right\} \\
s.t.\ x_{k|k} = x(k) \\
x_{k+i+1|k} = A_d x_{k+i|k} + B_d u_{k+i|k} \\
y_{k+i|k} = C_d x_{k+i|k} \\
u_{min} \le u_{k+i|k} \le u_{max}, i = 0,1,\cdots,N_c \\
x_{min} \le x_{k+i|k} \le x_{max}, i = 0,1,\ldots,N_p \\
u_{k+i|k} = K x_{k+i|k},\ N_c < i \le N_p
\end{gathered} \tag{19}
$$

where A_d, B_d, and C_d are the discrete-time versions of the system, input, and output matrices, respectively. N_p is the prediction horizon and N_c is the control horizon. The notation $x_{k+i|k}$ represents the predicted value of x at i steps ahead of k. Here, Q, R, and P are the weighting matrices for the state, input, and terminal state, respectively. Q_y is the output weighting matrix used to measure tracking error.

In addition, the output can be obtained with corresponding dimensions K and P by solving the discrete-time algebraic Riccati equation [29], given by

$$
\begin{gathered}
P = A_d^T P A_d - (A_d^T P B_d)\left(R + B_d^T P B_d\right)^{-1}(B_d^T P A_d) + Q \\
K = \left(R + B_d^T P B_d\right)^{-1} B_d^T P A_d
\end{gathered} \tag{20}
$$

Using these weighting matrices, the MPC controller computes a sequence of optimal vectors U that minimize the cost function J.

Although MPC offers several advantages, such as optimal control and handling of constraints, the online optimization process can lead to a significant computational burden, which is a major drawback. To address this issue, Alberto proposed a new MPC scheme that can reduce the computational load [30].

The equations for predicting the state vector $X(k)$ can be obtained through the following derivation:

$$
X(k) = \Theta x(k) + \Lambda U(k) \tag{21}
$$

where

$$
X(k) = \begin{bmatrix} x(k+1|k) \\ \vdots \\ x(k+N_p|k) \end{bmatrix}, U(k) = \begin{bmatrix} u(k|k) \\ \vdots \\ u(k+N_p-1|k) \end{bmatrix},
$$

$$
\Theta = \begin{bmatrix} A_d \\ A_d{}^2 \\ \vdots \\ A_d{}^{N_p} \end{bmatrix}, \Lambda = \begin{bmatrix} B_d & 0 & 0 & \cdots & 0 \\ A_d B_d & B_d & 0 & \cdots & 0 \\ A_d{}^2 B_d & A_d B_d & B_d & \cdots & 0 \\ \vdots & \vdots & \vdots & & \vdots \\ A_d{}^{N_p-1} B_d & A_d{}^{N_p-2} B & A_d{}^{N_p-3} B_d & \cdots & B_d \end{bmatrix}.
$$

Equations (19) and (21) can be reformulated as follows:

$$
\begin{gathered}
V(x(k)) = U(k)^T\left(\Lambda^T Q_{N_p} \Lambda + R_{N_c}\right)U(k) + 2x(k)^T \Theta^T Q_{N_p} \Lambda U(k) \\
+ x(k)^T\left(\Theta^T Q_{N_p} \Theta + Q\right)x(k)
\end{gathered} \tag{22}
$$

where

$$\hat{Q} = \begin{bmatrix} Q & & & \\ & \ddots & & \\ & & Q & \\ & & & P \end{bmatrix}, \hat{R} = \begin{bmatrix} R & & \\ & \ddots & \\ & & R \end{bmatrix}.$$

By defining $H = (\Lambda^T \hat{Q} \Lambda + \hat{R})$, $F = \Theta^T \hat{Q} \Lambda$, and $G = \Theta^T Q_{N_p} \Theta + Q$, Equation (22) can be formulated as the following equivalent form:

$$\begin{array}{c} V(x(k)) = \frac{1}{2} U(k)^T H U(k) + x(k)^T F U(k) \\ s.t.\ \hat{A} U(k) \leq W + E x(k) \end{array} \tag{23}$$

where

$$\hat{A} = \begin{bmatrix} \Lambda_i \\ -\Lambda_i \end{bmatrix}, i = 1, \cdots, N_p,\ W = \begin{bmatrix} x_{max} \\ -x_{min} \end{bmatrix},\ E = \begin{bmatrix} -A_d^i \\ A_d^i \end{bmatrix}, i = 1, \cdots, N_p.$$

In Equation (23), Λ_i indicates the ith row of Λ.

By defining $z \triangleq U(k) + H^{-1} F^T x(k)$, the reformulated optimization problem presented in Equation (23) can be expressed as a mixed-integer quadratic programming (mp-QP) problem, as shown below [24]:

$$\begin{array}{c} V_z(x) = \min\limits_z \frac{1}{2} z^T H z \\ s.t.\ \hat{A} z \leq W + S x(k) \end{array} \tag{24}$$

where $S \triangleq E + \hat{A} H^{-1} F^T$ and $V_z(x) = V(x(k)) - \frac{1}{2} x(k)^T (G - F H^{-1} F^T) x(k)$.

The Karush–Kuhn–Tucker (KKT) optimization conditions are used for the above problem [30]:

$$\begin{array}{c} Hz + \hat{A}^T \lambda = 0, \lambda \in \mathbb{R}^q \\ \lambda^i (\hat{A}^i z - W^i - S^i x) = 0, i = 1, \ldots, q \\ \lambda \geq 0 \\ \hat{A} z \leq W + Sx \end{array} \tag{25}$$

Solving the above equations:

$$\lambda_a = -\left(\hat{A}_a H^{-1} \hat{A}_a^T \right)^{-1} (W_a + S_a x) \tag{26}$$

$$z = H^{-1} \hat{A}_a^T \left(\hat{A}_a H^{-1} \hat{A}_a^T \right)^{-1} (W_a + S_a x) \tag{27}$$

Based on the above equation, it can be observed that Equations (26) and (27) are linear affine functions of the state $x(k)$. Furthermore, it is apparent that the control variable U is also a linear affine function of the state $x(k)$ when considered in conjunction with equation $z = U + H^{-1} F^T x(0)$. The subscript a indicates the active constraint, and the subscript i indicates the inactive constraint.

Based on the KKT condition, it is evident that the validity of the above equation is contingent on satisfying the inequality constraints:

$$-\left(\hat{A}_a H^{-1} \hat{A}_a^T \right)^{-1} (W_a + S_a x(0)) \geq 0 \tag{28}$$

$$\hat{A} H^{-1} \hat{A}_a^T \left(\hat{A}_a H^{-1} \hat{A}_a^T \right)^{-1} (W_a + S_a x(0)) \leq W + Sx(0) \tag{29}$$

The polyhedral set (critical region) CR_0 can be formed:

$$CR_0 = \left\{ x \in \mathbb{R}^n \middle| \lambda_a \geq 0,\ \hat{A}_i z \leq W_i + S_i x \right\} \tag{30}$$

By applying different constraints, more critical regions are formed for different groups of states. Therefore, a map of states to optimal control inputs is eventually created.

4. Controller Design

4.1. Desired Path

First, the reference signal needs to be determined. In this system, the input signal should be given according to the driving characteristics of the CLRV. The performance test of the LVR should include the ability to accelerate, decelerate, drive at a uniform speed, move forward, reverse, turn, and climb over slopes of about 20°. Suppose the maximum speed of the pre-researched CLRV is $v_{rmax} = 4$ m/s with $a_{rmax} = 1$ m/s^2.

Additionally, the signal-to-noise ratio of a_{rx} and a_{ry} is 20 dB. After integrating the accelerations a_{rx} and a_{ry}, the speeds v_{rx} and v_{ry} can be obtained. Then, integrating the speeds v_{rx} and v_{ry}, the position information X_r and Y_r of the CLRV can be obtained, respectively. Figure 4 illustrates the position X_r as well as the speed v_{rx} of path one of the CLRV in the X direction. When the CLRV is going uphill or downhill, its speed in the X direction quickly decreases before stabilizing as it ascends, and quickly increases before stabilizing as it descends.

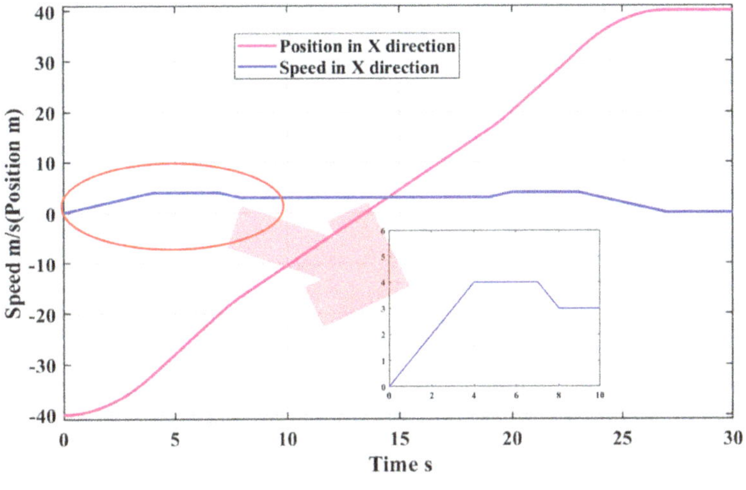

Figure 4. Desired path one for track-following servo control is plotted in this figure. The initial position of the CLRV is (−40, 0). Path one is the trajectory of the CLRV moving in the X-direction. The trajectory consists of uniform acceleration, uniform deceleration, uniform motion, and changes in X-directional speed due to uphill and downhill slopes.

Path two is the trajectory of the CLRV moving forward and backward in the Y-direction. The speed v_{ry} and position Y_r changes in the Y direction of path two are shown in Figure 5.

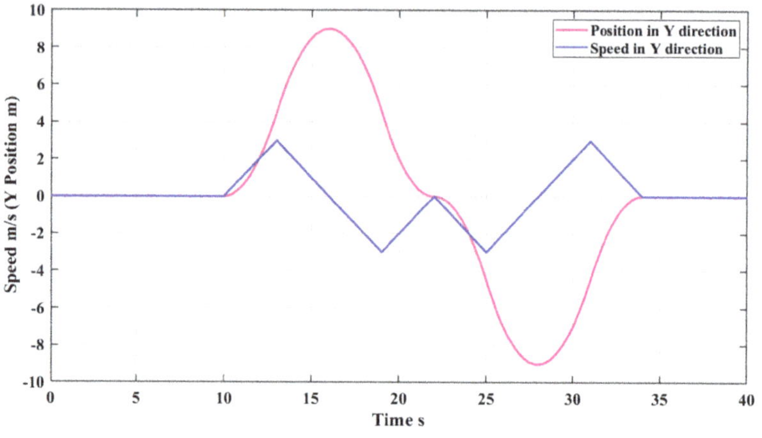

Figure 5. Desired path two for track-following servo control is plotted in this figure. The initial position of the CLRV is (−40, 0).

Path three represents the trajectory of the CLRV as it makes a turn with a radius of 10 m and a velocity of 3 m/s. Path four is the trajectory of the CLRV navigating a turn while avoiding obstacles. The speeds of the CLRV in the *X* and *Y* directions for paths three and four are also displayed in Figures 6 and 7, respectively.

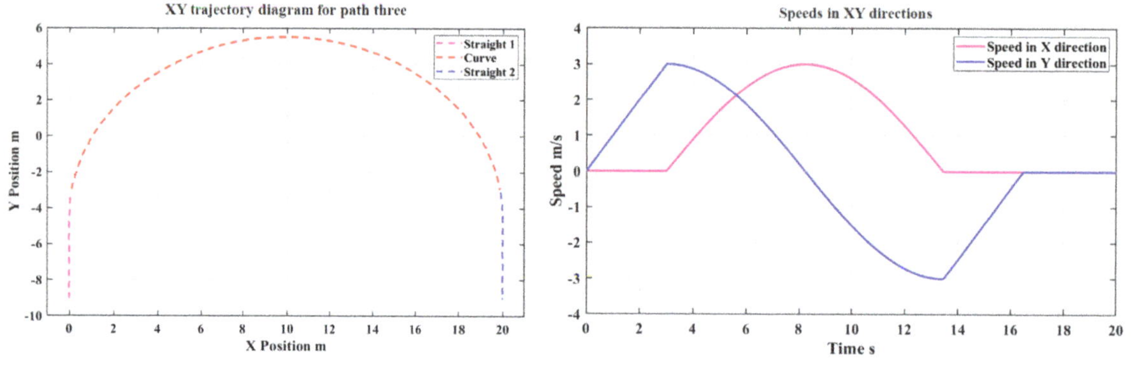

Figure 6. CLRV's motion trajectory and speeds in the *XY* directions for desired path three. The initial position of the CLRV is (0, −9). This path mainly consists of three parts: a section of straight motion with uniform acceleration, a turning curve with a radius of 10 m, and a section of straight motion with uniform deceleration.

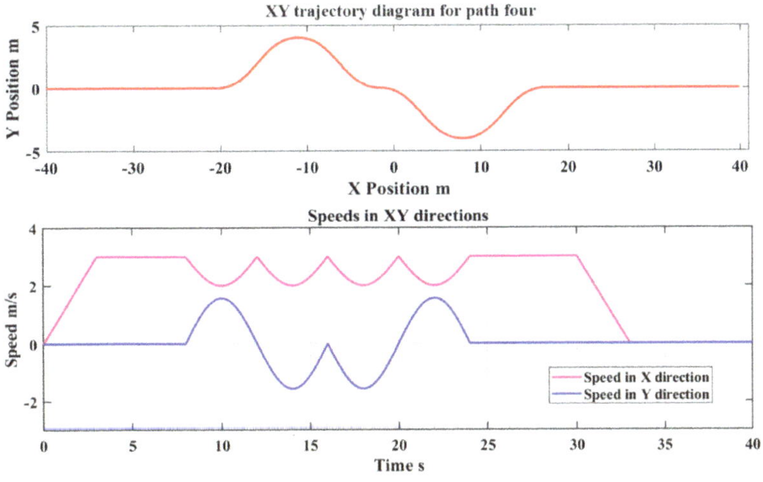

Figure 7. CLRV's motion trajectory and speeds in the XY directions for desired path four. The initial position of the CLRV is (−40, 0). This path mainly comprises five parts: a uniformly accelerated straight motion part, two turning curves, uniform straight motions, and a uniformly decelerated straight motion part.

In order to accurately control the crane, the controller must handle digital values obtained from sensors rather than continuous values of the states. Therefore, it is necessary to convert the desired path signal from continuous time to discrete time for more accurate representation. This system uses a monocular vision system to measure the position and orientation of lifting points, with a range of $-40 \leq X \leq 40$ m (or $-10 \leq Y \leq 10$ m) and positioning accuracy of ± 10 mm. The processing frequency of the visual system is 20 Hz. The present positions of the crane and the trolley in the X and Y directions are measured by the high-precision magnetic scale, the response time is 1 ms, and the measurement error is better than ± 10 μm.

The position measurement noise is added to the output of the desired path with Gaussian white noise. The Gaussian noise amplitude is 0.01 m and the sampling time is 0.05 s. The sampling period of the desired path is 0.05 s, the discrete signal is delayed by one sampling period, and then the signal is passed to the position servo loop controller.

4.2. Design of Explicit MPC Controllers

This section elaborates on the design of the explicit model predictive controller, including the controller structure and the process of determining the weighting matrices Q, R, and Q_y, are shown in the following Table 2. In this section and the next section, the superscript C refers to the crane and the superscript T refers to the trolley.

Table 2. EMPC Controller design parameters.

Parameters	Value
Predicted horizon N_p	$N_p^C = 12, N_p^T = 6$
Control horizon N_c	$N_c^C = N_c^T$=2
State weighting matrix Q	$Q^C = Q^T = \begin{bmatrix} 0.00001 & 0 \\ 0 & 0.00001 \end{bmatrix}$
Output weighting matrix Q_y	$Q_y^C = Q_y^C = 10,000$
Input weighting matrix R	$R^C = R^T = 0.08$
Sampling time T_s	5 ms

In this paper, the EMPC controller, designed using the MPT toolbox 3.0 [31], generates an optimal input while satisfying the system's constraints. The performance of the EMPC scheme can be tuned by appropriately adjusting the weighting matrices Q, R, and Q_y, and by a suitable choice of the prediction horizon N_p and N_c. The EMPC controller for the crane moving system has 993 critical regions, whereas the EMPC controller for the trolley moving system has 685 critical regions. Figure 8 shows the feasible critical regions of the trolley moving system's controller.

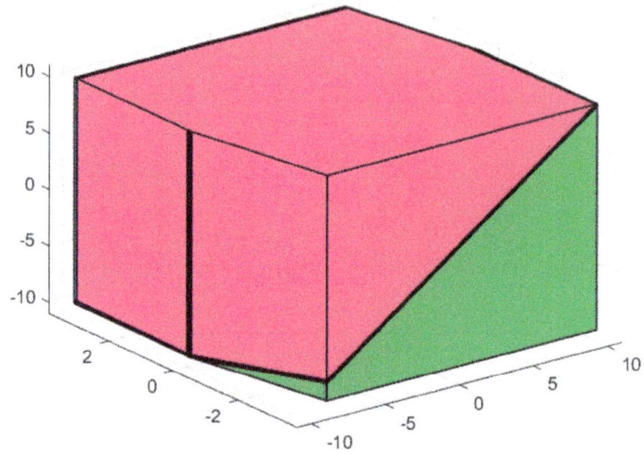

Figure 8. Critical regions. The figure illustrates the critical regions generated by the EMPC controller for the moving trolley system.

4.3. Linear-Quadratic-Integral Controller Design

In this paper, the LQI controller [32] is also employed for the position-loop servo control and is compared with the performance of the EMPC controller. Taking crane control as an example, in the design of the LQI controller, the ultimate goal is to track the position reference of the lunar rover. Therefore, in practice, the optimal control scheme should be applied as follows:

$$u_c^{LQI} = -K_{LQI}^C \begin{bmatrix} x \\ x_i \end{bmatrix} = -K_{LQI}^C \begin{bmatrix} e_X(k) \\ v_{ux}(k-1) \\ \int e_X(k) \end{bmatrix} \tag{31}$$

where $e_X(k) = X_r(k) - X_u(k)$ is the tracking error. Assuming that the reference signal is constant in the future, the state equation for the reference signal is:

$$X_r(k+1) = X_r(k) \tag{32}$$

Combining Equations (31), (32), and (17) results in the following equation:

$$\begin{bmatrix} e_X(k+1) \\ v_{ux}(k) \\ \int e_X(k+1) \end{bmatrix} = \begin{bmatrix} 1 & -T_s & 0 \\ 0 & 1 & 0 \\ 1 & -T_s & 1 \end{bmatrix} \begin{bmatrix} e_X(k) \\ v_{ux}(k-1) \\ \int e_X(k) \end{bmatrix} + \begin{bmatrix} -T_s{}^2 \\ T_s \\ -T_s{}^2 \end{bmatrix} u_C(k) \tag{33}$$

Similarly, we can obtain the spatial state equation for position tracking control of the trolley by following the same procedure as we did for the crane moving system. In both the crane and trolley moving systems for position servo control, the weighting matrices used in the LQI controllers are as follows:

$$Q_{LQI}^C = \begin{bmatrix} 40,000 & & \\ & 0.00001 & \\ & & 1 \end{bmatrix}, \; Q_{LQI}^T = \begin{bmatrix} 10,000 & & \\ & 0.00001 & \\ & & 1 \end{bmatrix}, \\ R_{LQI}^C = R_{LQI}^T = 0.1, \tag{34}$$

During the parameter selection process of the weight matrix, special consideration is given to prioritize accurate tracking. As a result, the error state is given a high penalty. Additionally, the integral error state is also assigned a high penalty to minimize the deviation from the steady state as much as possible. Furthermore, the plant input weight is chosen relatively high to prevent any significant changes in the system. Then, the state feedback gain K_{LQI} for the LQI controllers can be determined by solving the algebraic Riccati equation, as illustrated in Equation (20).

5. Results and Discussion

In this system, the impact of the tracking error of the track-following system mainly includes the impact on the drawbar pull force F_{DP} [33] and gravity compensation force F_C of the CLRV.

Supposing that the upper lifting point is positioned 10 m above the CLRV, the drawbar pull force error in the horizontal direction caused by the tracking error denoted as ΔF_h can be expressed as:

$$\frac{\Delta F_h}{\frac{5}{6}G_c} = \frac{\delta P}{10} \tag{35}$$

where G_c is the weight of the CLRV and δP is the tracking error.

The gravity compensation force error ΔF_v in the vertical direction caused by the position tracking error is:

$$\Delta F_v = F_C\left[1 - cos\left(arctan\frac{\delta P}{10}\right)\right] \tag{36}$$

Supposing that the drawbar pull factor of the CLRV's wheel with a radius of 0.4 m on loose soil is $u_{DP} = 0.8$, the maximum drawbar pull force and gravity compensation force can be roughly expressed as:

$$F_{DP} = u_{DP}W_N \tag{37}$$

$$F_C = \frac{5G_c}{6} \tag{38}$$

where W_N is the wheel vertical load. Hence, the maximum drawbar pull of the entire rover can be estimated by setting W_N as $1/6G_c$.

Combining Equations (35)–(38) gives:

$$\varepsilon_h = \frac{\Delta F_h}{F_{DP}} \tag{39}$$

$$\varepsilon_v = \frac{\Delta F_v}{F_C} \tag{40}$$

where ε_h and ε_v are the error impact factors to measure the influence of tracking errors.

This section presents and discusses the simulation results of the track-following servo control for paths one, two, three, and four using the PI, LQI, and EMPC controllers.

Figure 9 presents a comparison of simulation results for track-following servo control along path one, using three different controllers: PI, LQI, and EMPC. The "Reference" curve shows the real-time position of the CLRV. The figure displays the tracking results, tracking errors, and the crane speed of the track-following servo for each controller.

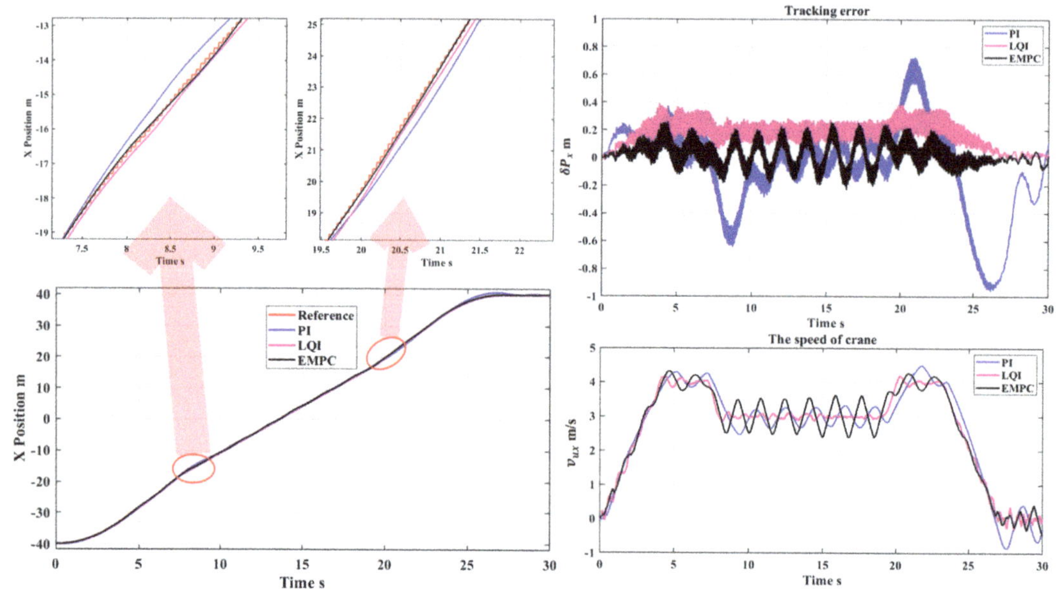

Figure 9. The simulation results' comparison of tracking path one used the PI, LQI, and EMPC controllers. The figures present the following results: the results of the crane tracking desired path one in the X direction; tracking errors δP_x between the crane's position and the desired position along path one in the X direction; the crane speed v_{ux} of tracking path one.

It can be observed from Figure 9 that the EMPC controller has the best performance and the smallest tracking error against the disturbance caused by the rapid change in the X-direction speed of the CLRV, followed by the LQI and PI controllers. Moreover, by examining the motion speed diagram of the crane, it can be inferred that the tracking stability of LQI is highest.

As shown in Figure 9, the positional deviation resulting from the track-following servo of the PI controller is significantly greater than those of the EMPC and LQI controllers. This is primarily due to the high inertia of the crane, which poses challenges in tracking the desired path. To overcome this, a simple PI controller with higher proportional coefficients is employed, leading to improved response speed but degraded steady-state performance. Interestingly, LQI has a lower settling time than PI and EMPC, and its stability is best. This can be attributed to its very high weight in the state of the tracking error, which helps it reach the steady state earlier. However, LQI suffers from delay errors between the desired path and the crane position, resulting from its reliance on the output value of the error integrator.

In contrast, the EMPC controller enhances control performance by generating optimal inputs while satisfying constraints. The model predictive control algorithms used in EMPC are more complex, taking into account various parameters and predicting the most optimal path for possible trajectories. As a result, the EMPC controller performs better at minimizing errors compared to controllers that use a simple cost function or a set of gains to correct deviations.

Figure 9 shows the tracking error of tracking desired path one using the EMPC controller. P_x refers to the real-time error of the crane and lunar rover in the X direction, while P_y refers to the real-time error of the trolley and lunar rover in the Y direction. In the tracking servo of path one, there is only an error in the X direction. The tracking error $\delta P = \delta P_x$ and the tracking error of the EMPC controller for path one is within 0.2 m.

Figure 10 depicts the simulation results of the trolley tracking path two, which involves frequent acceleration, deceleration, and forward and reverse movements in the Y direction.

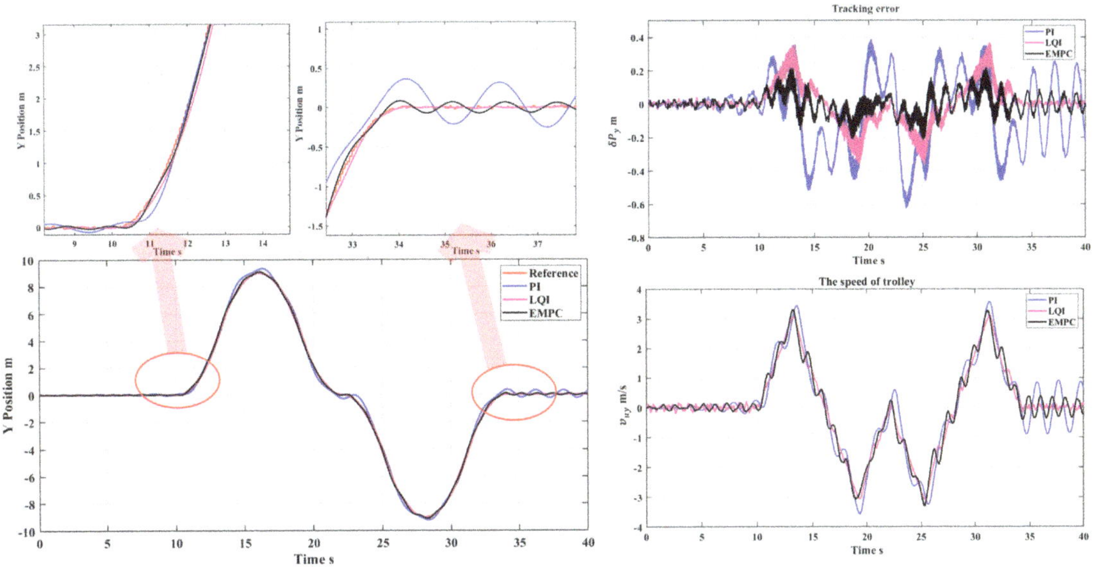

Figure 10. Results of the trolley tracking desired path two. The figures present the following results: the results of trolley tracking desired path one in the Y direction; tracking errors δP_y between the trolley's position and the desired position along path two in the Y direction; the trolley speed v_{uy} of tracking path two.

Figure 10 shows the simulation results of the trolley tracking path two in the Y direction. The tracking error is mainly caused by the trolley's inertia. Comparing it with Figure 9, it can be observed that as the crane's inertia is much greater than that of the trolley, the error will be even greater when the crane tracks the expected path with rapid changes in the X-directional speed. The results shown in Figure 10 show the trolley's speed v_{uy} of tracking path two. From the tracking error plot and trolley speed plot, it can be seen that large errors usually occur during the time periods when the trolley undergoes acceleration and deceleration switching. To minimize the impact of the crane's large inertia on tracking performance, it is recommended to avoid frequent acceleration and deceleration of the lunar rover in the X direction. Instead, it is advisable to conduct forward and reverse movement performance testing solely in the Y direction to achieve optimal results.

In the tracking servo of path two, there is only an error in the Y direction. The tracking error $\delta P = \delta P_y$ and the tracking error of the EMPC controller for path two is within 0.2 m.

Figure 11 depicts the simulation results of the crane and trolley tracking desired path three, which follows a turning motion trajectory of the CLRV with a radius of 10 m, and operates at a speed of 3 m/s. On the other hand, Figure 12 illustrates the crane and trolley tracking path four, which represents the driving trajectory of the CLRV when encountering continuous turns during obstacle avoidance. The CLRV operates at a speed of around 3 m/s while following this path.

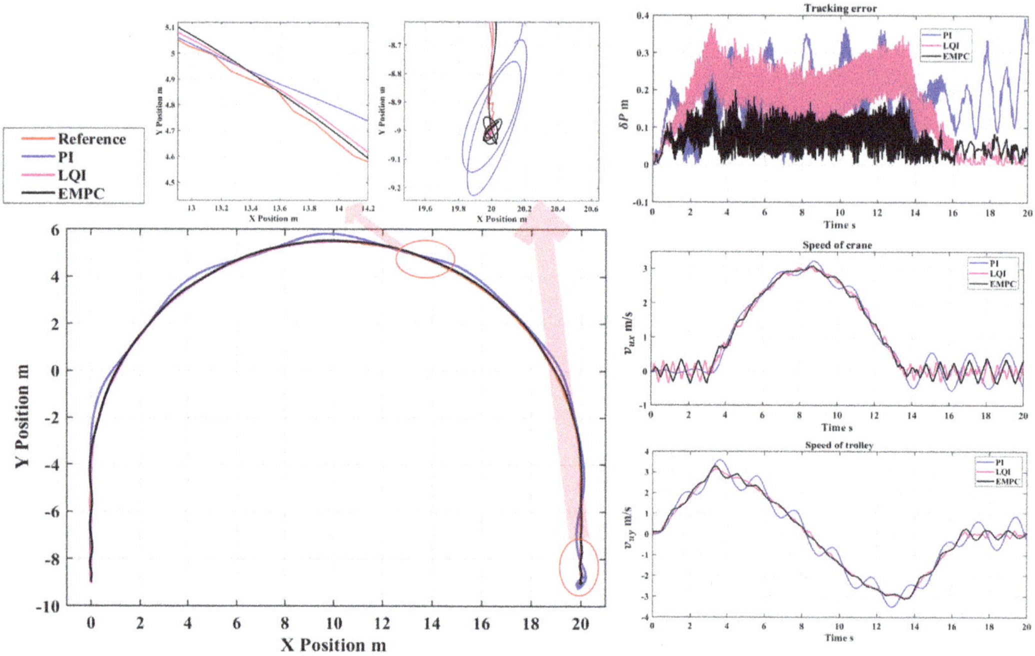

Figure 11. The results of tracking desired path three by the crane and trolley are presented. The figures illustrate the following outcomes: a comparison between the trajectory of the upper lifting point and the expected trajectory under three different controllers, the tracking error $\delta P = \sqrt{\delta P_x^2 + \delta P_y^2}$ of the track-following servo subsystem, and the movement speed of both the crane and trolley.

Figures 11 and 12 illustrate that the EMPC controller results in the smallest tracking error, followed by LQI and PI controllers. Furthermore, the results suggest that the LQI controller exhibits excellent stability, as the movement speed of the crane and trolley remains relatively stable in response to dynamic disturbances. On the other hand, when using a PI controller for a large inertia system, the overshoot can be relatively large, and it can be challenging to balance the response speed and stability.

Figures 11 and 12 displays the simulation results of crane and trolley tracking for desired paths three and four, utilizing the EMPC controller. The tracking error $\delta P = \sqrt{\delta P_x^2 + \delta P_y^2}$ is basically less than 0.2 m.

The simulation results show that the maximum tracking errors δP of the track-following servo subsystem using EMPC controllers are almost less than 0.2 m. According to Equations (39) and (40), it can obtain the influence factor of the tracking error on the CLRV motion in the horizontal direction:

$$\varepsilon_h < 0.125 \tag{41}$$

In addition, the error impact factor in the vertical direction is as follows:

$$\varepsilon_v < 0.00045 \tag{42}$$

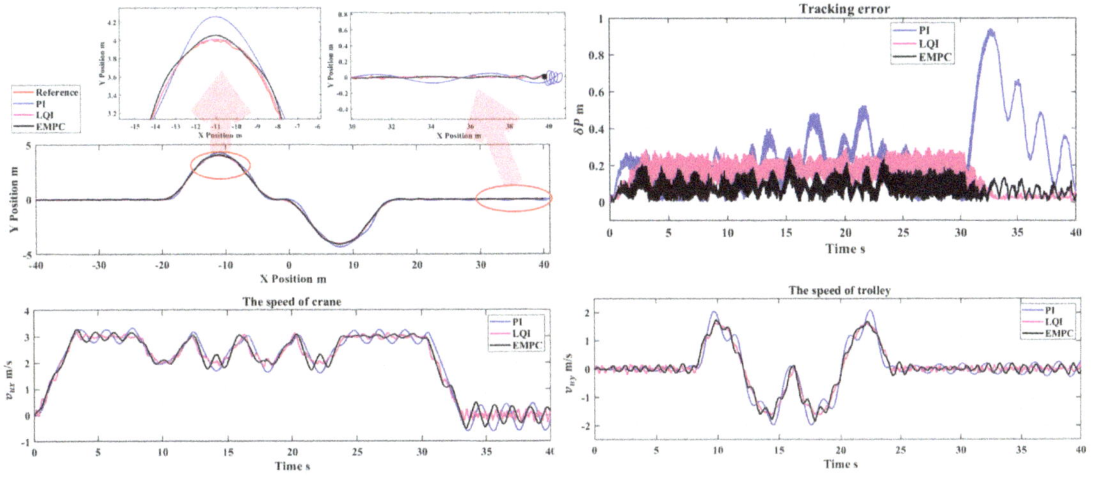

Figure 12. The results of tracking desired path four by the crane and trolley are presented. The figures illustrate the following outcomes: a comparison between the trajectory of the upper lifting point and the expected trajectory under three different controllers, the tracking error $\delta P = \sqrt{\delta P_x^2 + \delta P_y^2}$ of the track-following servo subsystem, and the movement speed of both the crane and trolley.

The simulations yield some key findings: the EMPC demonstrates a notably impressive performance for the system, with explicit MPC delivering the objectively best results. While the PI and LQI controllers show comparatively inferior results, they remain reliable options with their own strengths. The PI controller is easy to apply, while the LQI controller exhibits excellent robustness and stability. However, their application is limited for multi-constraint control systems. Although the EMPC shows good performance under different expected trajectories, it also has some limitations in terms of stability, especially in dealing with large disturbances in the X direction (i.e., when the CLRV speed changes rapidly). In such cases, the crane may lose control. To address this issue, future work will focus on conducting relevant methods to further analyze the stability of the proposed EMPC controller.

In conclusion, EMPC is a controller that can provide the smallest possible dynamic tracking error and steady-state error under the condition of dealing with multiple constraints, and it has better control performance. Moreover, frequent acceleration and deceleration should be avoided in the X direction when driving the CLRV, to reduce the impact of the large inertia of the crane on the controller's tracking performance.

Based on the simulation results, we compare the tracking performance of the track-following servo subsystem of the suspension gravity compensation system to that of the system designed by Liu et al. [2]. Although the maximum tracking error of Liu et al.'s system is approximately 0.1 m, its crane employs open-loop control, making it only suitable for tracking a slow-moving unmanned lunar rover. In contrast, the research presented in this paper provides a solution to the theoretical research gap of the track-following servo subsystem suitable for CLRV experiments.

6. Conclusions

This paper proposes a design and control scheme for a track-following servo subsystem of a LGCS suitable for CLRV ground testing:

1. The track-following servo subsystem consists of a crane, trolley, and servo motors. The crane moves along the bridge, tracking the real-time position of the CLRV, while the trolley moves along the crane girder to follow the real-time position of the CLRV. Three-loop control models are established for both the crane and trolley, including the position loop, motor speed loop, and current loop.

2. The driving force of the track-following servo subsystem is provided by a PMSM. Specific parameters for the motor and transmission mechanism are determined based on the subsystem's motion capability. PMSM vector control based on the SVPWM is adopted to achieve fast response and precise control of the motor speed. PI controllers are employed in both the current and speed loops of the motors.
3. In the position loop control of the track-following servo subsystem, explicit MPC control is introduced for multi-parameter constraint control. This paper presents an EMPC controller suitable for the track-following servo subsystem, including the cost function design and offline calculation process. The weighting matrices, prediction horizon, and control horizon are also determined.

Especially, the effectiveness of the proposed EMPC controller is demonstrated through a comparison with PI and LQI controllers. The simulation results show that the maximum tracking error δP of the track-following servo subsystem is consistently below 0.2 m. The impact of the errors on the drawbar pull is within 12.5%, and their effect on the compensation force is negligible. The simulation results also suggest that, due to the large inertia of the crane, the CLRV should avoid frequent forward, backward, acceleration, and deceleration in the X direction to reduce the influence of the crane's large inertia on tracking performance. These simulation results provide valuable theoretical support for designing a track-following servo subsystem that is suitable for CLRV ground testing.

In our future work, our objective is to develop a track-following servo subsystem that is suitable for ground testing of the CLRV. The primary contribution of this paper is the development of an EMPC controller for the position control loop, along with PMSM vector control, which has demonstrated excellent tracking performance for the subsystem. However, implementing the EMPC controller in engineering settings may pose significant challenges. In our upcoming research, we plan to refine our work further, including addressing potential deviations in motor load torque, exploring cooperative control of multiple motors, and enhancing the robustness of the EMPC controller. These are essential considerations that must be addressed to fully realize the potential of the proposed approach in practical applications.

Author Contributions: Conceptualization, M.L. and J.W.; methodology, Y.Z.; software, Y.Z.; validation, Y.Z., H.W. and M.L.; formal analysis, H.W.; investigation, H.W.; resources, Y.Z.; data curation, Y.Z.; writing—original draft preparation, Y.Z.; writing—review and editing, Y.Z.; visualization, Y.Z.; supervision, H.W.; project administration, H.W. All authors have read and agreed to the published version of the manuscript.

Funding: This research received no external funding.

Institutional Review Board Statement: Not applicable.

Informed Consent Statement: Not applicable.

Data Availability Statement: The data presented in this study are available on request from the corresponding author. All data are calculated by the author and have been included in this paper.

Acknowledgments: In addition, we would like to express our appreciation to all the individuals who provided administrative and technical support throughout this research.

Conflicts of Interest: We declare that there is no conflict of interest regarding the submission of this manuscript, and that all authors have provided their approval for publication. We confirm that this work represents original research that has not been previously published, nor is it currently being considered for publication elsewhere, in its entirety or in part. We can also confirm that all listed authors have reviewed and approved the final version of the manuscript.

References

1. Liang, Z.; Gao, H.; Ding, L.; Deng, Z. Approach to imitate maneuvering of Manned lunar roving vehicle under lunar gravity using a terrestrial vehicle. *Mechatronics* **2015**, *30*, 383–398. [CrossRef]
2. Liu, Z.; Gao, H.B.; Deng, Z.Q. Design and Implementation of a Large-Scale Gravity Compensation System for Lunar Rover. *Appl. Mech. Mater.* **2013**, *385–386*, 759–767. [CrossRef]

3. Callens, N.; Ventura-Traveset, J.; De Lophem, T.-L.; De Echazarreta, C.L.; Pletser, V.; Van Loon, J.J.W.A. ESA Parabolic Flights, Drop Tower and Centrifuge Opportunities for University Students. *Microgravity Sci. Technol.* **2010**, *23*, 181–189. [CrossRef]
4. Kwok-Choon, S.; Buchala, K.; Blackwell, B.; Lopresti, S.; Wilde, M.; Go, T. Design, fabrication, and preliminary testing of air-bearing test vehicles for the study of autonomous satellite maneuvers. In Proceedings of the 31st Florida Conference on Recent Advances in Robotics, Orlando, FL, USA, 10–11 May 2018.
5. Alves, F.; Park, J.; McCarty, L.; Rabelo, R.; Karunasiri, G. MEMS Underwater Directional Acoustic Sensor in Near Neutral Buoyancy Configuration. *Sensors* **2022**, *22*, 1337. [CrossRef]
6. Davis, K.; Meginnis, I. Testing of the NASA Exploration Extravehicular Mobility Unit Demonstration (xEMU Demo) Architecture at the Neutral Buoyancy Laboratory. In Proceedings of the 49th International Conference on Environmental Systems, Boston, MA, USA, 7–11 July 2019.
7. Zhao, Z.; Fu, K.; Li, M.; Li, J.; Xiao, Y. Gravity compensation system of mesh antennas for in-orbit prediction of deployment dynamics. *Acta Astronaut.* **2020**, *167*, 1–13. [CrossRef]
8. Jia, J.; Jia, Y.; Sun, S. Preliminary design and development of an active suspension gravity compensation system for ground verification. *Mech. Mach. Theory* **2018**, *128*, 492–507. [CrossRef]
9. Jin, Y.; Wu, R.; Liu, W.; Tang, X. Visual servo for gravity compensation system. *Neurocomputing* **2017**, *269*, 256–260. [CrossRef]
10. Liu, Z.; Niu, F.; Gao, H.; Yu, H.; Ding, L.; Li, N.; Deng, Z. Design, analysis, and experimental validation of an active constant-force system based on a low-stiffness mechanism. *Mech. Mach. Theory* **2018**, *130*, 1–26. [CrossRef]
11. Hičár, M.; Ritók, J. Robust crane control. *Acta Polytech. Hung.* **2006**, *3*, 91–101.
12. Cunningham, T. *System Requirements Document for the Active Response Gravity Offload System (Argos)*; NASA Engineering Directorate document AR&SD-08007; NASA: Washington, DC, USA, 2010.
13. Kemurdjian, A.; Khakhanov, U.A. Development of Simulation Means for a Gravity Forces. In *Robotics 2000*; American Society of Civil Engineers: Reston, VA, USA, 2000. [CrossRef]
14. Preliminary Design Study of a Lunar Gravity Simulator Interim Report [EB/OL]. Available online: https://ntrs.nasa.gov/search.jsp?R=19670022528 (accessed on 1 January 2023).
15. Zwerger, T.; Mercorelli, P. Using a Bivariate Polynomial in an EKF for State and Inductance Estimations in the Presence of Saturation Effects to Adaptively Control a PMSM. *IEEE Access* **2022**, *10*, 111545–111553. [CrossRef]
16. Ullah, K.; Guzinski, J.; Mirza, A.F. Critical Review on Robust Speed Control Techniques for Permanent Magnet Synchronous Motor (PMSM) Speed Regulation. *Energies* **2022**, *15*, 1235. [CrossRef]
17. Bida, V.M.; Samokhvalov, D.V.; Al-Mahturi, F.S. PMSM vector control techniques—A survey. In Proceedings of the 2018 IEEE Conference of Russian Young Researchers in Electrical and Electronic Engineering (EIConRus), Moscow, Russia, 29 January–1 February 2018; IEEE: Piscataway, NJ, USA, 2018.
18. Kaiqi, Z. The study of improved PI method for PMSM vector control system based on SVPWM. In Proceedings of the 2011 IEEE Industry Applications Society Annual Meeting, Orlando, FL, USA, 9–13 October 2011; IEEE: Piscataway, NJ, USA, 2011. [CrossRef]
19. Xu, Z.; Tao, N.; Du, M.; Liang, T.; Xia, X. Damage Prediction for the Starter Motor of the Idling Start-Stop System Based on the Thermal Field. *SAE Int. J. Commer. Veh.* **2017**, *10*, 443–453. [CrossRef]
20. Barata, B.; Lundgre, K. Methods for Servo Position Control. Master's Thesis, Chalmers University of Technology, Gothenburg, Sweden, 2019.
21. Brandão, A.S.M.; Lima, D.M.; Filho, M.V.A.D.C.; Normey-Rico, J.E. A Comparative Study on Embedded MPC for Industrial Processes. In Proceedings of the Congresso Brasileiro de Automatica—CBA, João Pessoa, Brazil, 8–12 September 2018; Volume 1. No. 1.
22. Wang, H.; Liu, B.; Ping, X.; An, Q. Path Tracking Control for Autonomous Vehicles Based on an Improved MPC. *IEEE Access* **2019**, *7*, 161064–161073. [CrossRef]
23. Nubert, J.; Koehler, J.; Berenz, V.; Allgower, F.; Trimpe, S. Safe and Fast Tracking on a Robot Manipulator: Robust MPC and Neural Network Control. *IEEE Robot. Autom. Lett.* **2020**, *5*, 3050–3057. [CrossRef]
24. Bemporad, A.; Morari, M.; Dua, V.; Pistikopoulos, E.N. The explicit linear quadratic regulator for constrained systems. *Automatica* **2002**, *38*, 3–20. [CrossRef]
25. Oravec, J.; Klaučo, M. Real-time tunable approximated explicit MPC. *Automatica* **2022**, *142*, 110315. [CrossRef]
26. Zhang, W.; Chen, J.; Wu, X.; Jin, X. Wheel/rail adhesion and analysis by using full scale roller rig. *Wear* **2002**, *253*, 82–88. [CrossRef]
27. Hewing, L.; Wabersich, K.P.; Menner, M.; Zeilinger, M.N. Learning-Based Model Predictive Control: Toward Safe Learning in Control. *Annu. Rev. Control Robot. Auton. Syst.* **2020**, *3*, 269–296. [CrossRef]
28. Berberich, J.; Kohler, J.; Muller, M.A.; Allgower, F. Data-Driven Model Predictive Control With Stability and Robustness Guarantees. *IEEE Trans. Autom. Control* **2021**, *66*, 1702–1717. [CrossRef]
29. Wang, L. Numerical Algorithms of the Discrete Coupled Algebraic Riccati Equation Arising in Optimal Control Systems. *Math. Probl. Eng.* **2020**, *2020*, 1841582. [CrossRef]
30. Bemporad, A. Explicit Model-Predictive Control. In *Encyclopedia of Systems and Control*; Springer: Berlin, Germany, 2014.
31. Herceg, M.; Kvasnica, M.; Jones, C.N.; Morari, M. Multi-parametric toolbox 3.0. In Proceedings of the 2013 European Control Conference (ECC), Zurich, Switzerland, 17–19 July 2013; IEEE: Piscataway, NJ, USA, 2013.

32. Ahmadi, A.; Mohammadi-Ivatloo, B.; Anvari-Moghaddam, A.; Marzband, M. Optimal Robust LQI Controller Design for Z-Source Inverters. *Appl. Sci.* **2020**, *10*, 7260. [CrossRef]
33. Shibly, H.; Iagnemma, K.; Dubowsky, S. An equivalent soil mechanics formulation for rigid wheels in deformable terrain, with application to planetary exploration rovers. *J. Terramech.* **2005**, *42*, 1–13. [CrossRef]

energies

MDPI

Article

Super-Twisting Sliding Mode Control to Improve Performances and Robustness of a Switched Reluctance Machine for an Electric Vehicle Drivetrain Application †

Rabia Sehab [1,*], Ahmad Akrad [1] and Yakoub Saadi [2]

[1] ESTACA'Lab—ESTACA, Ecole Supérieure des Techniques Aéronautiques et de Construction Automobile, 53000 Laval, France
[2] ICube, CNRS (UMR 7357) INSA Strasbourg, University of Strasbourg, 67000 Strasbourg, France
* Correspondence: rabia.sehab@estaca.fr

† This paper is an extended version of our paper published in 2017 IEEE Vehicle Power and Propulsion Conference (VPPC), Belfort, France, 11–14 December 2017; Performance Comparison between Conventional and Robust Control for the Powertrain of an Electric Vehicle Propelled by a Switched Reluctance Machine.

Abstract: In electric vehicles, performances of electric vehicle drivetrains depend on the electric machine and the control. Switched Reluctance Machines (SRMs) are today an alternative to rare earth magnets machines such as Permanent Magnet Synchronous Machine (PMSM), which is used in the vehicle drivetrain. Because of its high nonlinear behavior, the classical control designed for SRMs is not sufficient to obtain good performances. The objective of this paper is to make performance and robustness comparisons of the designed robust controllers considering the high nonlinear behavior of SRMs. Sliding Mode Control (SMC) and Super-Twisting Sliding Mode Control (STSMC) are developed and validated by simulation for the velocity control loop and the current control loops of the control strategy. However, an evaluation of their performances compared to classical control based on PI controllers is carried out. For a robustness comparison, a variation of SRM parameters is carried out by simulation using the three controllers. Finally, an experimental validation on a developed test bench using the three controllers is conducted to show that Super-Twisting Sliding Mode Control (STSMC) is the best in terms of performances and robustness for an electric vehicle application.

Keywords: electric vehicle; switched reluctance machine; PI control; sliding mode control; super-twisting sliding mode control; performance; robustness

Citation: Sehab, R.; Akrad, A.; Saadi, Y. Super-Twisting Sliding Mode Control to Improve Performances and Robustness of a Switched Reluctance Machine for an Electric Vehicle Drivetrain Application. *Energies* **2023**, *16*, 3212. https://doi.org/10.3390/en16073212

Academic Editors: Kan Liu and Wei Hu

Received: 21 February 2023
Revised: 17 March 2023
Accepted: 21 March 2023
Published: 2 April 2023

1. Introduction

Nowadays, electric vehicles (EVs) are gaining increased attention due to environmental and energy concerns. EVs can limit environmental impacts and reduce greenhouse gases. In the vehicle drivetrain, the main component is the electric machine. In such an application, high power density, high torque density, wide speed range, and efficiency are of primary importance [1]. To meet these demands, Rare Earth Magnetic Material (REMM) has been widely used in different types of electric machines dedicated to the EV applications such as the sintered Neodymium Iron Boron and Samarium Cobalt [2]. However, the cost of REMM-based machines has increased over several years. Moreover, due to the limited resources, the use of REMM-based machines in EVs applications is now being challenged. In fact, many researchers and industrials are working on other machines that can be competitive in terms of size, efficiency, and torque density. This is the case of the Wound-Rotor Synchronous Machine (WRSM), which is designed and manufactured by Renault for the first urban electric vehicle; Renault Zoe and the Switched Reluctance Machine (SRM) is used today by Land Rover in the electric vehicle Land Rover'110 defender.

Considering the requirements, Switched Reluctance Machines (SRMs) represent an alternative. They not only feature a salient pole stator with concentrated coils, which provides earlier winding and shorter end turn than other types of electric machines, but also feature a salient pole rotor, which has neither conductors nor magnets. Simplicity in its construction makes the SRM manufacturing and its maintenance inexpensive. Its high reliability and performance at high-speed range with a constant power, fault-tolerant operation capability, and the simplicity of the power converter [3,4] make it a very interesting candidate for electric vehicles propulsion. However, the SRM exhibits high torque ripples and acoustic noises [5]. However, these drawbacks can be significantly reduced with an optimal SRM mechanical design [6] and a good control strategy [7–10]. In [10], for example, the dynamic performance of direct torque control (DTC) method is improved by replacing the PI controller of the SRM velocity loop by a sliding mode controller associated with a disturbance observer.

Unlike most of the other types of electric machines, the SRM is highly nonlinear and operates in saturation to maximize the output torque. Moreover, the torque, the current, and the position are strongly coupled through nonlinear characteristics. Consequently, the design of the controllers is tedious [11]. Because of the high requirements of drivability for electric vehicles, the control of speed and torque is an important issue [12].

In the literature, there is a huge number of control techniques for speed and/or current control of SRM. Each one has its own advantages and disadvantages. For example, we can mention direct torque control [10], the variable gains proportional-integral (PI) controller [13], LPV adaptive controllers [14], fuzzy logic, and artificial neural network controllers [15,16]. The Variable Structure Control (VSC) with a sliding mode control has also been applied in several works to the speed control of SRM [17,18]. Despite its good performances, the chattering is a serious drawback for EV because it excites non-modeled dynamics that may cause unexpected problems and instability [19]. A robust controller based on H-infinity approaches has been proposed in [20]. Moreover, Pulse Width Modulation (PWM), hysteresis current regulation [21], and model predictive controllers (MPC) [22] are commonly used to drive the SRM. Unfortunately, they are not suited to eliminate the chattering phenomenon of SRM. In this paper, PI control, Sliding Mode Control, and higher-order sliding mode control based on Super-Twisting Algorithm are simultaneously proposed for currents and velocity loops of the suggested control strategy. Even they are robust, Super-Twisting Sliding Mode Control remains the best way to eliminate the phenomenon of chattering, particularly for electric vehicle applications. Indeed, with the improvement of vehicle drivetrain performances using SRM and STSM control, other applications can be improved since they depend on the dynamic behavior of the vehicle where the model of vehicle drivetrain is considered. In vision-aided intelligent vehicle sideslip, angle estimation is based on a dynamic model [23]. Similarly, automated vehicle sideslip angle estimation considers signal measurement characteristics [24].

The main contributions of this paper are focused on the following:

- Performance comparison of the designed Sliding Mode Control (SMC) and the Super-Twisting Sliding Mode Control (STSMC) for the current and the velocity control loops of SRM control strategy with the designed classical control. These controllers are developed and validated by simulation in [12];
- Robustness Comparison of the designed Sliding Mode Control (SMC) and Super-Twisting Sliding Mode Control (STSMC) for the current and the velocity control loops of SRM control strategy with the classical control by varying SRM physical parameters;
- Implementation and validation of the three designed controllers [12] on a developed test bench using a multicore dSpace 1005 with a SRM and a DC machine to create a load torque;
- Finally, performances comparison is carried with the collected experimental data showing that the Super-Twisting Sliding Mode Control is the best algorithm to select for the improvement of electric vehicle drivetrain performance.

This paper is organized as follows: The model of the SRM and SRM power stage converter are presented in Sections 2 and 3, respectively. A short description of PI, SMC, and STSMC control design is proposed in Section 4. Section 5 is devoted to the validation of the designed controllers by simulation and the comparison with the PI controllers' results. Section 6 gives the description of the developed test bench, and the experimental validation of the designed controllers. Conclusion and perspectives are drawn in Section 7.

2. SRM Modeling

The SRM has a simple construction, but the solution of its mathematical model is relatively difficult due to its dominant nonlinear behavior [25] related to the flux linkage and the torque, which both depend on the current phase and the rotor position.

The electromagnetic model of the equivalent circuit of one phase is [3] the following:

$$V_j = R_j.I_j + \frac{\partial \psi_j(\theta, I)}{\partial t} \quad (1)$$

with $j = 1, 2, 3, 4$.

The flux linkage ψ_j of Equation (1) is a nonlinear characteristic depending on two variables: the phase current I (0 to 80A) and the rotor position θ (0 to 60°). In this study, Figure 1 shows the flux linkage characteristics of the used SRM with the topology 8S/6R given in the Appendix A.

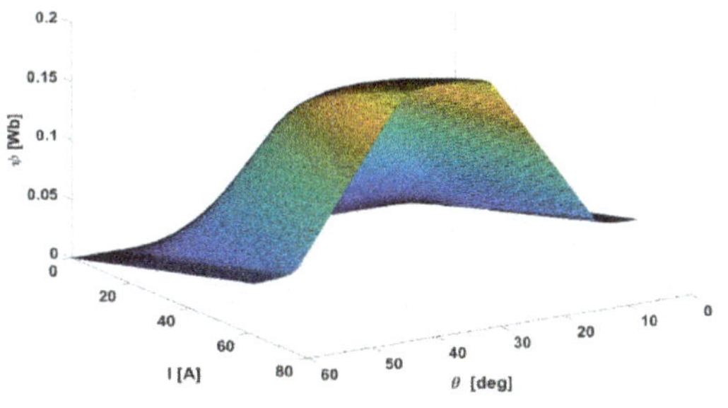

Figure 1. Flux in terms of position and phase current.

The electromechanical model of the SRM associated with a load and viscous friction torques can be expressed as follows [3]:

$$\frac{d\Omega}{dt} = \frac{1}{J}(T_e(\theta, I) - f_r\Omega - T_L) \quad (2)$$

The torque T_e of Equation (2) is, according to Figure 2, a nonlinear characteristic which depends on the phase current I (0 to 80A) and the rotor position θ (0 to 60°).

The torque T_e (Equation (3)) given by Figure 2 is computed from the sum of the instantaneous torque (Equation (4)) developed by phase j according to [3]:

$$T_e = \sum_{1}^{j=4} T_{phase_j} \quad (3)$$

where:

$$T_{phase_j} = \frac{1}{2}\frac{dL(\theta)}{d\theta}I_j^2 \quad (4)$$

All the variables of the model are given by the following: θ: Rotor position, Ω: Angular velocity of rotor, J: Moment of inertia (rotor), T_e: Total electromagnetic torque, f_r: Friction Coefficient, T_L: Load torque, I_j: Current in the jth phase, ψ_j: Flux linkages in jth phase, V_j: Voltages of jth phase, R_j: Resistance of the jth phase, L: Instantaneous inductance.

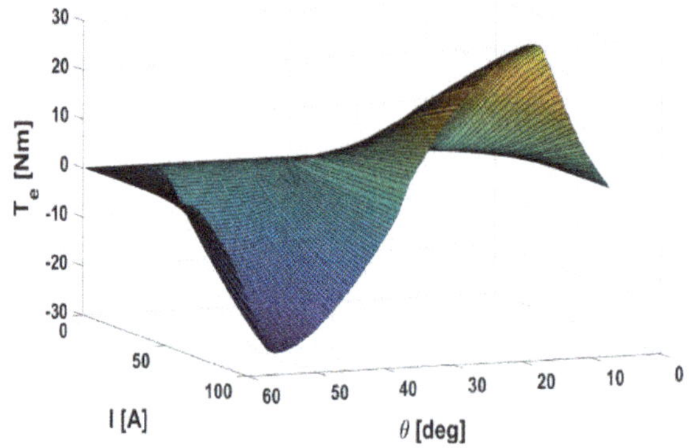

Figure 2. Torque in terms of position and phase current.

3. SRM Power Stage Converter

As the SRM is four phases, four asymmetric H-bridges are selected to supply the stator windings because of the simple design, control, and low cost. Indeed, different types of converters are listed in the literature to supply the SRM [26]. Figure 3 gives a view of the selected converter supplying each phase of the SRM. The DC link voltage Vdc to supply the converter is fixed according to the SRM specification at 250 V/61 A. The supply of each phase is carried out by each asymmetric H-bridge. On the other hand, PWM signals for supplying the drivers of the transistors are provided by the current loops of the implemented control strategy, which is given in Section 4. Using MATLAB/Simscape, the power stage to supply the SRM is modelized by four asymmetric H-bridges.

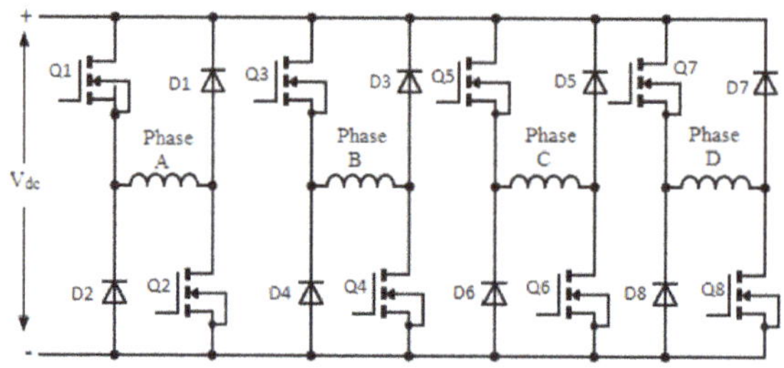

Figure 3. SRM four asymmetric H-bridges.

4. SRM Control Strategy

For velocity or torque control, different control strategies are suggested [12]. Direct and indirect torque control are among the control strategies. The control strategy selected for the velocity control is based on two cascade control loops: velocity and current loops [12]. Figure 4 shows the block diagram of this control strategy. The velocity control loop provides, according to the velocity point (*) the total torque setpoint (*) of SRM to a torque sharing function in charge of generating torque setpoint (*) for each phase. From each phase torque setpoint (*), a phase current setpoint (*) is deduced using the reverse of the nonlinear torque characteristics given by Figure 2. The deduced current sept points (*) are finally provided to the four current control loops to generate PWM signals of the converter.

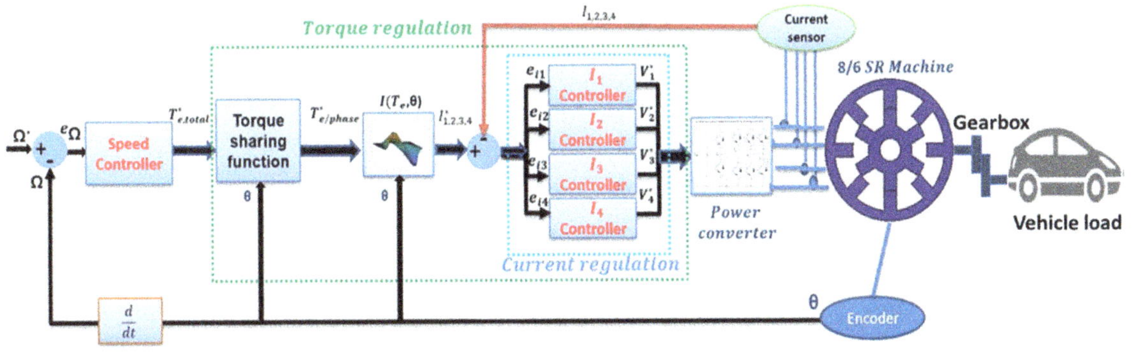

Figure 4. Block diagram of the control strategy.

Based on the control strategy given by Figure 4, PI, SMC, and STSMC controllers are designed for velocity and current control loops. SMC and STSM are developed separately for current and velocity control loops in [12] using the nonlinear model of the SRM given in Section 2.

4.1. PI Controllers

PI controllers are widely used in industries due to their simplicity and low cost. Moreover, their implementation in analog or digital hardware is simple and easy. Under limited operating conditions, they perform well and their steady state performance is good. For any PI controller, the control design is defined by the following:

$$u(t) = K_p e(t) + K_i \int_0^t e(\Gamma) d\Gamma \tag{5}$$

where $e(t)$ is the error defined as the difference between the desired set point and the measured variable. K_p and K_i are the proportional and integral gains.

For the velocity control loop, $e(t)$ is defined by the following:

$$e_\Omega(t) = \Omega^*(t) - \Omega(t) \tag{6}$$

Ω^* and Ω are the velocity set point and the measured velocity, respectively.

For the current loop, $e(t)$ is given by the following:

$$e_i(t) = i^*(t) - i(t) \tag{7}$$

i^* and i are the current set point and the measured current, respectively.

The PI parameters K_p and K_i are designed for current and velocity loops using the classical methods of control.

4.2. Sliding Mode Control (SMC)

The design of SMC can be achieved in two steps. In the first step, we define the sliding mode, which is a surface (switching function) that is invariant to the controlled dynamics, where the controlled dynamics are exponentially stable and where the system tracks the desired set point. In the second step, the control law is designed in such a way that it should guarantee steering the system trajectories towards the sliding surface [27].

4.2.1. SMC Controller for Velocity Loop

The switching function is defined as follows:

$$S_1(t) = e_\Omega(t) + \lambda_1 \int_{-\infty}^t e_\Omega(\Gamma) d\Gamma \tag{8}$$

where λ_1 is a positive constant.

The aim is that the error will converge to zero exponentially (if $S_1(t) = 0$ then $e_\Omega = 0$). The state space variables are defined as the following:

$$\begin{cases} x_1(t) = \int_0^t e_\Omega(\Gamma)d\Gamma \\ x_2(t) = e_\Omega(t) \end{cases} \dot{x_1}(t) = x_2(t) \tag{9}$$

Using (8) and (9), we can write the following:

$$S_1(t) = x_2(t) + \lambda_1 x_1(t) \tag{10}$$

According to (9), the dynamical model of SRM, (2) can be rewritten as follows:

$$\dot{x_2} = \frac{1}{J}T_e - \frac{1}{J}f_r\Omega^* - \frac{1}{J}f_r x_2 - \Omega^* - \frac{1}{J}T_L \tag{11}$$

In a state space representation, (11) becomes the following:

$$\begin{cases} \dot{x_1} = x_2 \\ \dot{x_2} = f(t) + g.u + d \end{cases} \tag{12}$$

with $f(t) = -\frac{1}{J}f_r\Omega^*(t) - \frac{1}{J}f_r e_\Omega(t) - \Omega^*(t), g = \frac{1}{J}, d = -\frac{1}{J}T_L$, and $u = T_e$.

The equivalent SMC law (u_{eq}) that ensures the asymptotic convergence of the velocity error towards zero verifies the following equality [27]:

$$S_1(t) = \dot{S_1}(t) = 0 \tag{13}$$

From (10) and (12), we deduce the equivalent control input as follows:

$$u_{eq} = g^{-1}(-f(x) - d - \lambda_1 x_2) \tag{14}$$

u_{eq} represents the total reference torque ($T^*_{e,total}$).

Additionally, the discrete SMC law (u_d) that guarantees the reachability of SMC in finite time is warranted if the following is the case [27]:

$$S_1\dot{S_1} < 0 \tag{15}$$

From (10), (15) can be rewritten as the following:

$$S_1\dot{S_1} = S_1(f(x) + g.u + d + \lambda_1 x_2) \tag{16}$$

if we set the following:

$$u_d' = g.u = g.u_d \tag{17}$$

and

$$u_d' = -(f(x) + d + \lambda_1 x_2) - C_1\, sign(S_1) \tag{18}$$

where C_1 is a positive constant. Therefore, the SMC law that ensures the asymptotic convergence of the velocity towards zero in finite time is given by the following:

$$u = u_{eq} + u_d \tag{19}$$

4.2.2. SMC Controller for Current Loop

For the current SMC controller, the sliding surface is defined as the following:

$$S_2(t) = K_1 e_i(t) + K_2 \int_{-\infty}^{t} e_i(\Gamma)d\Gamma \tag{20}$$

where K_1 and K_2 are two positive coefficients.

The reason for this choice is obvious. If $S(t) = 0$, then $\dot{e}=-\frac{K_2}{K_1}e$, so the current error will converge to zero exponentially.

Equation (1) can be written as follows:

$$V_j = R_j.I_j + \frac{\partial\psi_j(\theta, I_j)}{\partial\theta}\frac{d\theta}{dt}+\frac{\partial\psi_j(\theta, I_j)}{\partial I_j}\frac{dI_j}{dt} \tag{21}$$

in which $\frac{d\theta}{dt} = \Omega$, where $\frac{\partial\psi_j(\theta,I_j)}{\partial I_j}$ represents the self-inductance of the phase and $\frac{\partial\psi_j(\theta,I_j)}{\partial\theta}\Omega$ is the back counter-electromotive force (EMF) produced in the jth phase. From (21), we can write the following:

$$\frac{dI_j}{dt} =(V_j - R_j.I_j).\frac{\partial I_j}{\partial\psi_j(\theta, I_j)} - \frac{\partial I_j}{\partial\theta}\Omega \tag{22}$$

The equivalent voltage (V_{eq}) obtained by the following:

$$S_2(t) = \dot{S_2}(t) = 0 \tag{23}$$

From (22) and (23), the equivalent voltage (V_{eq}) for each current regulation is written as the following:

$$V_{eq}= R_jI_j{}^*+\frac{\partial\psi_j(\theta, I_j)}{\partial\theta}\Omega+\frac{\partial\psi_j(\theta, I_j)}{\partial I}\frac{dI_j{}^*}{dt}+(R_j - \frac{\partial\psi_j(\theta, I_j)}{\partial I}\frac{K_2}{K_1})e_i(t) \tag{24}$$

Finally, the current SMC controller law is defined as the following:

$$V_{SMC} = V_{eq} - C_2 sign(S_2) \tag{25}$$

with C_2 is a positive constant.

4.3. *Super Twisting Sliding Mode Control (STSMC)*

The implementation of sliding mode control presents an undesirable phenomenon of oscillation, which is known as "chattering". The Super-Twisting Sliding Mode control (STSMC) has been developed to avoid chattering in variable structure control (VSC) for the case of systems that have a relative degree equal to one [27].

Consider sliding variable dynamics given by a system with a relative degree of two, the following is the case:

$$\dot{y_1}(t) = \varphi_{ST}(y_1, t)+Y_{ST}(y_1, t)u_{ST}(t) \tag{26}$$

where $y_1(t)$ is the sliding function S, in which φ_{ST} and Y_{ST} are uncertain functions with the upper and lower bounds of (27) and (28), respectively, and $u_{ST}(t)$ is the scalar control input.

$$|\varphi_{ST}(y_1, t)| \leq \phi_{ST} \tag{27}$$

$$0< Y_{mST} \leq Y_{ST}(y_1, t) \leq Y_{MST} \tag{28}$$

The control signal $u_{ST}(t)$ can be given as the sum of two terms [27]:

$$u = u_1(t) + u_2(t) \tag{29}$$

$$\dot{u_1}(t)= \begin{cases} u(t)_{ST} if |u(t)_{ST}| > U \\ -W sign(y_1(t)) else \end{cases} \tag{30}$$

$$u_2(t) = \begin{cases} -\lambda|S_0|^{\rho} sign(y_1(t)) if |y_1(t)| > S_0 \\ -\lambda|y_1(t)|^{\rho} sign(y_1(t)) else \end{cases} \tag{31}$$

where U is the control value boundary and S_0 is a boundary layer around the sliding surface S. The sufficient condition of limited time convergence is the following:

$$\begin{cases} W > \frac{\phi_{ST}}{Y_{mST}} \\ \lambda^2 > \frac{4\phi_{ST}}{Y_{mST}^2} \frac{Y_{MST}(W+\phi_{ST})}{Y_{mST}(W-\phi_{ST})} \\ 0 < \rho \leq 0.5 \end{cases} \tag{32}$$

4.3.1. STSMC Controller for Velocity Loop

To compare the SMC and STSMC, we use the same switching function expression used in SMC controller design:

$$S_3(t) = y_1(t) = e_{\Omega}(t) + c\int_{-\infty}^{t} e_{\Omega}(\Gamma)d\Gamma, c > 0 \tag{33}$$

The aim of this mathematical development is to write the systems having a relative degree equal to one and to compare it with (26) to find the sufficient conditions and to use the control law defined in (29).

From (2), (33) can be rewritten as the following:

$$\dot{y_1} = \dot{\Omega}^* + \frac{1}{J}T_L + \frac{1}{J}f_r\Omega^* + \left(-\frac{f_r}{J} + c\right)e_{\Omega} - \frac{1}{J}T_e \tag{34}$$

Defining the variable as the following:

$$\varphi_{ST} = \dot{\Omega}^* + \frac{1}{J}T_L + \frac{1}{J}f_r\Omega^* + \left(-\frac{f_r}{J} + c\right)e_{\Omega} \tag{35}$$

$$Y_{ST} = 1 \tag{36}$$

$$u_{ST} = -\frac{1}{J}T_e \tag{37}$$

Equation (34) becomes the following:

$$\dot{y_1} = \varphi_{ST} + Y_{ST}u_{ST} \tag{38}$$

If we set the following:

$$\begin{cases} 0 < Y_{mST} = 0.5 \leq Y_{ST} = 1 \leq Y_{MST} = 2 \\ u_{ST} = -\frac{1}{J}T_e < U = \frac{1}{J}T_{Max} \\ \phi_{ST} = \left|\dot{\Omega}^* + \frac{T_L}{J} + \frac{f_r\Omega^*}{J} + \left(-\frac{f_r}{J} + c\right)e_{\Omega}\right| + \left|\frac{T_e}{J}\right| \end{cases} \tag{39}$$

where T_{Max} is the maximum torque of the SRM. Conditions in (27) and (28) are satisfied.

To design the control law defined in (29), it is necessary to choose the controller parameters that verify (32).

4.3.2. STSMC Controller for Current Loop

In the same way as in the previous paragraph, for the STSMC current controller, the switching surface is defined as the following:

$$S_4(t) = y_2(t) = K_3e_i(t) + K_4\int_{-\infty}^{t} e_i(\Gamma)d\Gamma \tag{40}$$

Putting $e_i(t)$ with its expression (21) in (38), we can write the following:

$$\dot{y_2} = K_3 e_i{}^* + K_4 e_i + K_3 R_j.I_j \frac{\partial I_j}{\partial \psi_j(\theta, I_j)} - K_3 \frac{\partial I_j}{\partial \theta}\Omega + - \frac{\partial I_j}{\partial \psi_j(\theta, I_j)} V_j \tag{41}$$

where K_3 and K_4 are two positive numbers.

Defining the variable as the following:

$$\varphi_{ST}{}' = K_3 e_i{}^* + K_4 e_i + K_3 R_j.I_j \frac{\partial I_j}{\partial \psi_j(\theta, I_j)} - K_3 \frac{\partial I_j}{\partial \theta}\Omega \tag{42}$$

$$Y_{ST}{}' = \frac{\partial I_j}{\partial \psi_j(\theta, I_j)} \tag{43}$$

$$u_{ST}{\prime} = -V_j \tag{44}$$

Equation (41) is expressed as the following:

$$\dot{y_2} = \varphi_{ST}{}' + Y_{ST}{}' u_{ST}{}' \tag{45}$$

where

$$\begin{cases} 0 < Y'_{mST} = 0.5 \frac{\partial I_j}{\partial \psi_j(\theta, I_j)} \leq Y'_{ST} = \frac{\partial I_j}{\partial \psi_j(\theta, I_j)} \leq Y'_{MST} = 2 \frac{\partial I_j}{\partial \psi_j(\theta, I_j)} \\ u'_{ST} = \left|-V_j\right| < U' = V \\ \phi'_{ST} = \left| K_3 e_i{}^* + K_4 e_i + K_3 R_j.I_j \frac{\partial I_j}{\partial \psi_j(\theta, I_j)} \right| + \left| K_3 \frac{\partial I_j}{\partial \theta}\Omega \right| \end{cases} \tag{46}$$

It satisfies the conditions in (27) and (28).

5. Performances of the Designed Controllers

For the velocity and current loops, controllers are designed for both using PI Controllers, Sliding Mode Controller (SMC), and Super-Twisting Sliding Mode Controller (STSMC). Simulations are performed at a reduced scale for a vehicle drivetrain (Figure 4) using Matlab/Simulink to evaluate the proposed controllers [12]. The objectives are to minimize the velocity error and to ensure a smooth total torque waveform of the SRM. A velocity profile from zero to SRM maximum velocity (10,000 rpm) is chosen (Figure 5) with a load torque $T_L = 8$ Nm applied, in steady state, at t = 1.5 s, for a duration of 2 s. In addition, SRM torque viscous friction defined by fr*Ω where fr is the viscous friction coefficient (see Appendix A) is considered for all the simulation. The velocity profile tracking of the PI, SMC, and STSMC controllers and torques responses are presented in Figures 5 and 6, respectively. Indeed, Figure 5 shows that the velocity profile is well followed using the three controllers except the PI controller. Indeed, with the zooms displayed in Figure 5, STSM control is the best in terms of tracking. In Figure 6, the torque responses confirm on the one hand that the STSM control is the best, in terms of response time, to follow the torque profile, and on the other hand, as shown in the zoom of Figure 6, the torque ripple is considerably reduced in steady state compared to the SMC. In addition, the motor torque (current) at standstill is not zero because of the applied load torque of 8Nm for a duration of 3 s. However, viscous friction torque is zero because it depends on the motor velocity.

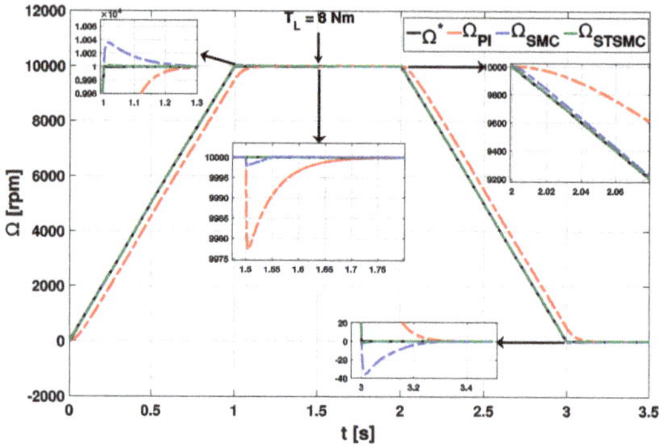

Figure 5. Velocity responses for the designed controllers.

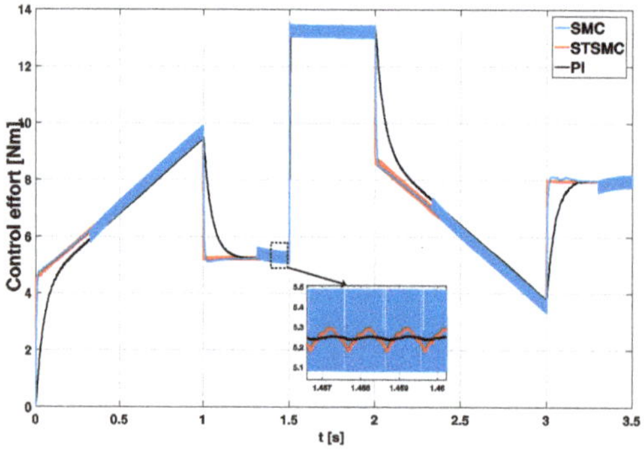

Figure 6. Torque responses for the designed controllers.

Based on these simulation results, Table 1 summarizes performances of the three controllers, namely rise time, maximum steady state error, maximum overshoot of velocity responses given by Figure 5, and torque ripples computed from torques responses given by Figure 6. Finally, performances comparison is carried out to show that Super-Twisting Sliding Mode Control is the best algorithm to provide the best performances for an electric vehicle drivetrain using a Switched Reluctance Machine.

Table 1. Performances comparison of the designed controllers.

Performances	PI	SMC	STSMC
Rise time	0.105 s	0.08 s	0.01 s
Max of steady state error	6%	0.3%	0.1%
Max of overshoot	27‰	23‰	8‰
Torque ripple at maximum velocity	14.5%	13.9%	12%

6. Robustness of the Designed Controllers

In the context of an electric vehicle application, the study of the robustness of the proposed controllers is necessary to check the limits of performances degradation with respect to the variations of the SRM physical parameters. The parameters to be varied are the resistance of the stator windings of a phase, the moment of inertia, the viscous friction coefficient, and the vehicle load torque whose variations are due to driving conditions (acceleration/deceleration), road slope, and state of the road (speed bumps, presence of obstacles, etc.). Based on the designed controllers where performances are evaluated by simulation in [12], a second simulation creating the variation of physical parameters of the vehicle drivetrain is carried with the designed controllers PI, SMC, and STSMC.

6.1. Stator Resistance Windings Variation

As the stator resistance windings varies because of temperature and ageing of the SRM, two values are chosen with −50% and +100% of the nominal value. Using the same simulator with the three controllers, namely PI, SMC and STSMC, response velocities and stator currents responses are collected. Figure 7a (with PI control), Figure 7b (with SMC), and Figure 7c (with STSMC) show the effect of these variations. Indeed, if the stator resistance of a phase decreases, an overshoot occurs; if the stator resistance of a phase increases, dynamic responses are slow. In addition, the zooms associated with these figures for the current and velocity responses show the interest to use the STSMC for the resistance varyies.

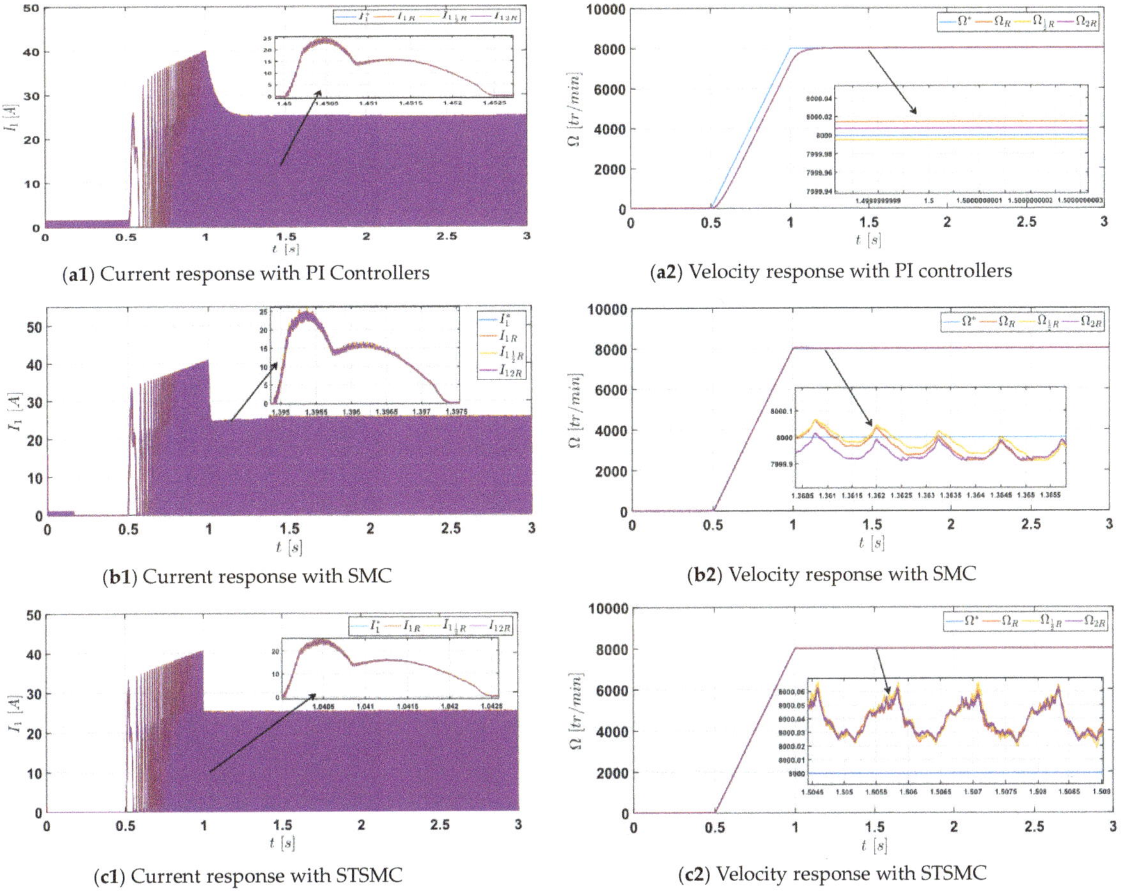

(**a1**) Current response with PI Controllers

(**a2**) Velocity response with PI controllers

(**b1**) Current response with SMC

(**b2**) Velocity response with SMC

(**c1**) Current response with STSMC

(**c2**) Velocity response with STSMC

Figure 7. Velocity and current responses with the three controllers.

6.2. Load Inertia Variation

As the vehicle inertia varies because of the vehicle load inertia, two values are chosen with −50% and +100% from the SRM inertia nominal value. Velocity responses show the effect of these variation on the velocity responses with the PI controller (Figure 8a), SMC (Figure 8b), and STSM (Figure 8c).

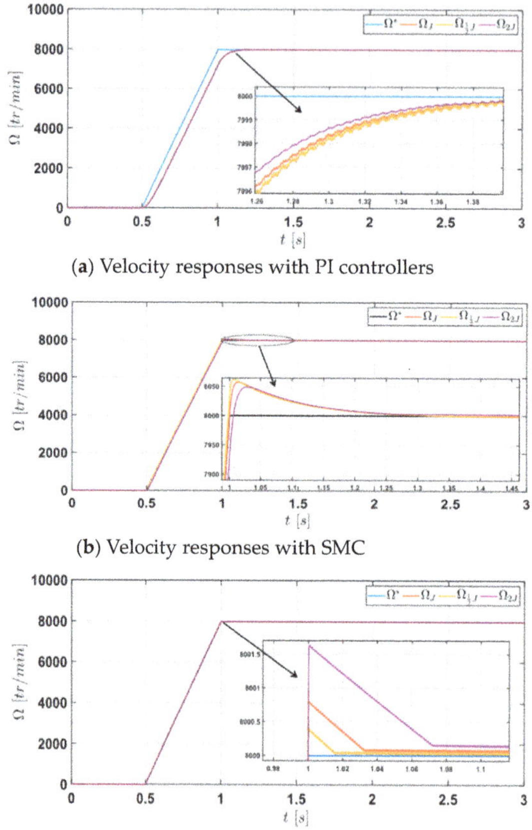

(a) Velocity responses with PI controllers

(b) Velocity responses with SMC

(c) Velocity responses with STSMC

Figure 8. Velocity responses with the three controllers.

6.3. Viscous Friction Coefficient Variation

For the viscous friction coefficient, a variation of 1/2 f_r and 2 f_r of the friction coefficient is carried out by simulation using the three controllers. Figure 9a–c show the velocity and torque responses. Indeed, STSMC is more robust compared to the other controllers. The zooms show that the errors are neglected when the viscous friction coefficient varies.

6.4. Load Torque Variation

For load torque variation, a first step load torque of 12 Nm is applied at t = 1.5 s and a second step of 16 Nm is applied at t = 2 s. Figure 10a–c show torque and velocity responses using the three controllers. Indeed, the load torque change is well compensated by the motor torque. In addition, as shown in the zoom of velocity responses, the overshoot error is neglected.

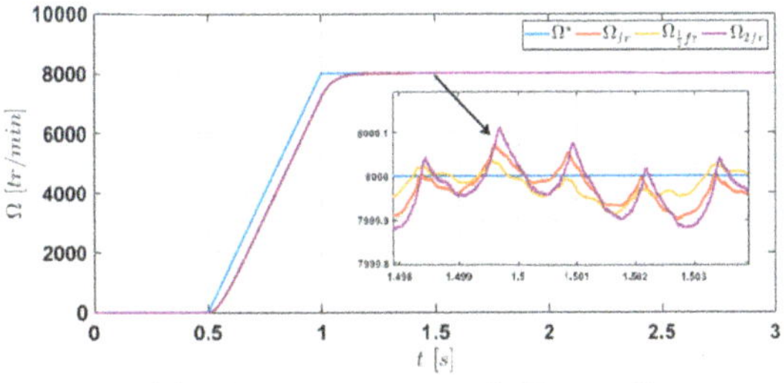

(**a**) Velocity responses with PI controllers

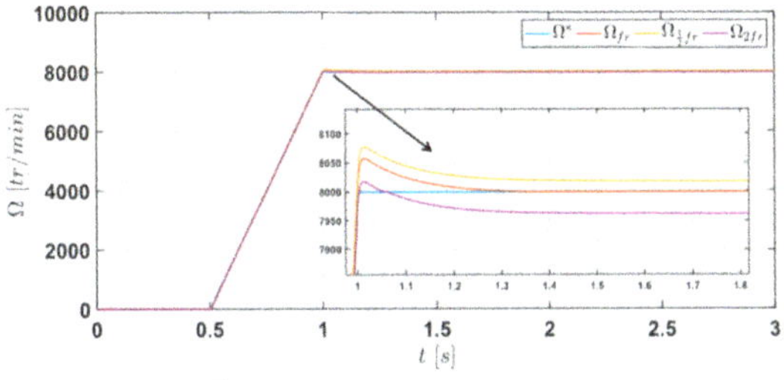

(**b**) Velocity responses with SMC

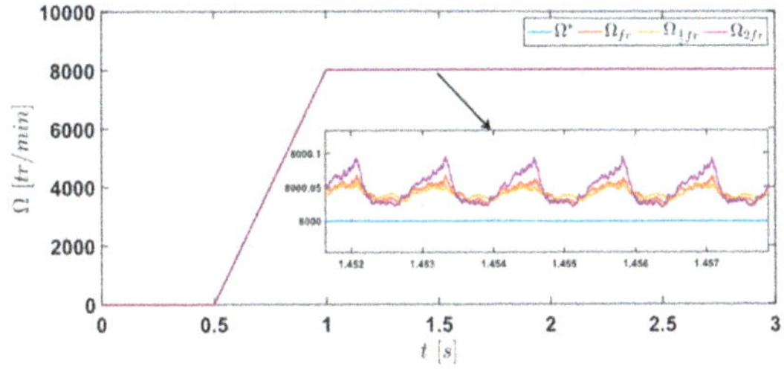

(**c**) Velocity responses with STSMC

Figure 9. Velocity responses with the three controllers.

Based on the responses of Figures 7–10, a quantification of the maximum error in steady state is carried out for current and velocity responses, while from torque responses, torque error is quantified in % using the following formula:

$$\Delta T_e(\%) = \frac{T_{emax} - T_{emin}}{T_{eaverage}}$$

Table 2 summarizes the computed values of the maximum errors at the maximum velocity using PI control, SMC, and STSMC.

Table 2. Evaluation of the robustness of the three controllers.

Physical Parameter Variation	PI	SMC	STSMC
Resistance R	Velocity emax = 2 rpm	Velocity emax = 0.8 rpm	Velocity emax = 0.5 rpm
	Current emax = 1.3 A	Current emax = 0.97 A	Current emax = 0.3 A
Moment of inertia J	Velocity emax = 5 rpm	Velocity emax = 5 rpm	Velocity emax = 1.5 rpm
Friction Coefficient f_r	Velocity emax = 3 rpm	Velocity emax = 40 rpm	Velocity emax = 0.5 rpm
load torque T_L	Velocity emax = 48 rpm	Velocity emax = 8 rpm	Velocity emax = 5 rpm
	Torque emax in % = 0.6	Torque emax in % = 0.1	Torque emax in % = 0.06

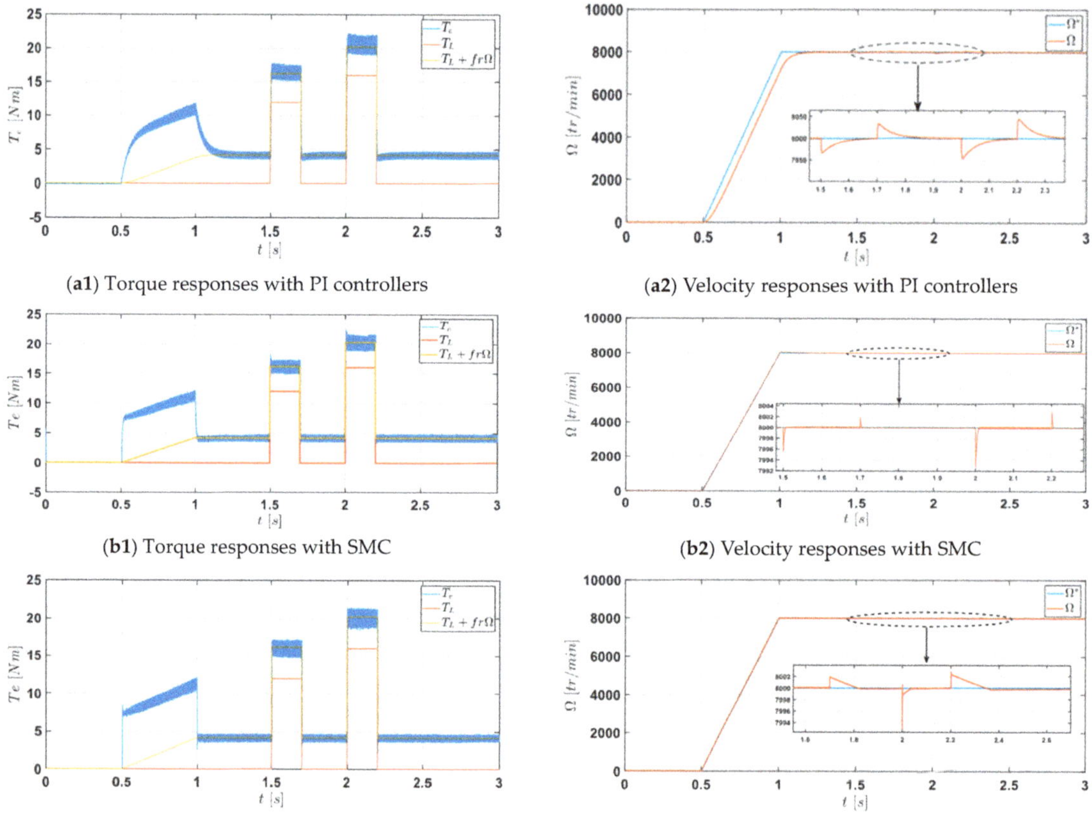

(**a1**) Torque responses with PI controllers

(**a2**) Velocity responses with PI controllers

(**b1**) Torque responses with SMC

(**b2**) Velocity responses with SMC

(**c1**) Torque responses with STSMC

(**c2**) Velocity responses with STSMC

Figure 10. Velocity and torque responses with the three controllers.

In comparing the maximum errors of the three controllers, STSMC also shows its added value in robustness. Indeed, for all physical parameter variations, the maximum errors are the smallest ones for velocity, current, and torque responses. However, STSMC remains the best controller in terms of performance and robustness for electric vehicle applications. Finally, in Table 3, performance and robustness of the three controllers shows that Super-Twisting Sliding Mode Control is the best comparing to the PI and sliding mode controllers.

Table 3. Performances and robustness evaluation of the three controllers.

		Controller		
		PI	**SMC**	**STSMC**
Performances	Tracking	-	++	++
	Rapidity	+	++	++
	Precision	+	++	++
	Torque ripple	+	-	++
Robustness	R	+	+	++
	f_r	+	+	++
	J	+	++	++
	T_L	+	++	++

++: Very good, +: Good, -: Poor.

7. Experimental Validation

For the experimental validation, an emulation of the electric vehicle drivetrain is carried out at a reduced scale. As shown in Figure 11, a test bench is set up using a Switched Reluctance Machine of 8.3 kW (1) and a DC machine (2) with a resistive load (3) to emulate the vehicle load torque. The power supply of SRM is provided by four asymmetric half-bridges (4) supplied by a DC link of 250 V (7). In addition, the PWM signals to provide to the drivers of the IGBTs of each half-bridge converter are computed from the control strategies implemented on a multicore dSpace 1005 platform (5). The provided PWM signals are amplified from 5 V to 15 V using an electronic stage (8). For all the implemented control strategies, four current sensors (6) are used to provide, in real time, the stator currents through A/D inputs of the dSpace panel.

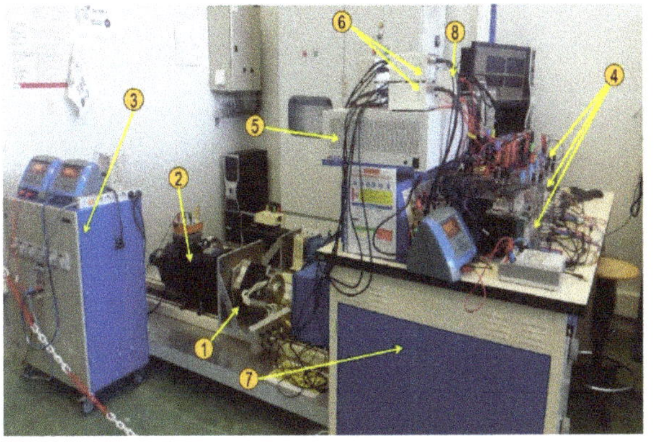

Figure 11. Test bench SRM-DC machine.

The real-time interface RTI of the implemented control strategies is mainly composed of four blocks. Figure 12 gives the general block scheme of the implemented control strategies, namely PI, SMC, and STSMC with a sample time of 10^{-5} s and a PWM frequency of 18 kHz.

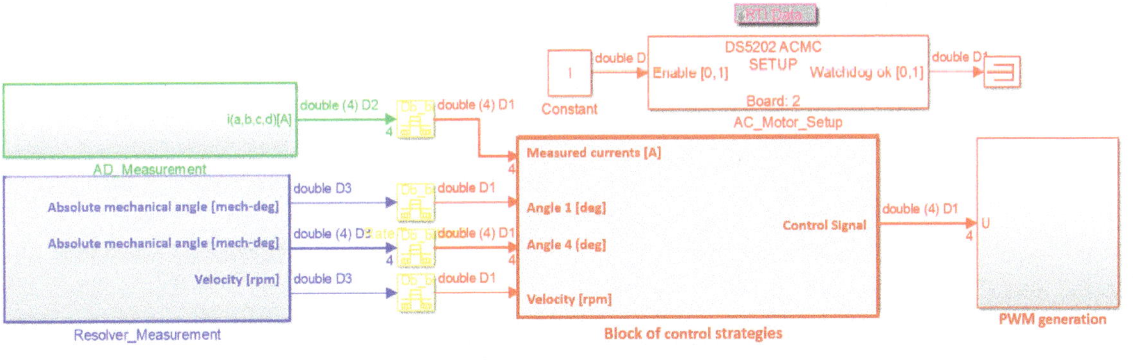

Figure 12. RTI block scheme of the implemented control strategies. 1—block of the four measured currents i_a, i_b, i_c, and i_d. 2—block of the measured position and velocity with a resolver mounted on the shaft of the SRM. 3—block of the control strategies with the designed robust controllers; PI, SMC, and STSM implemented separately. 4—block of PWM signal generation.

For the experimental validation, a velocity profile (*) of 3000 rpm max is chosen with acceleration, steady state, and deceleration phases for a duration of 180 s. Based on the velocity responses of Figure 13, responses errors are plotted in Figure 14a–c showing, respectively, a zoom of the reached velocity in steady state errors with PI, SMC, and STSMC.

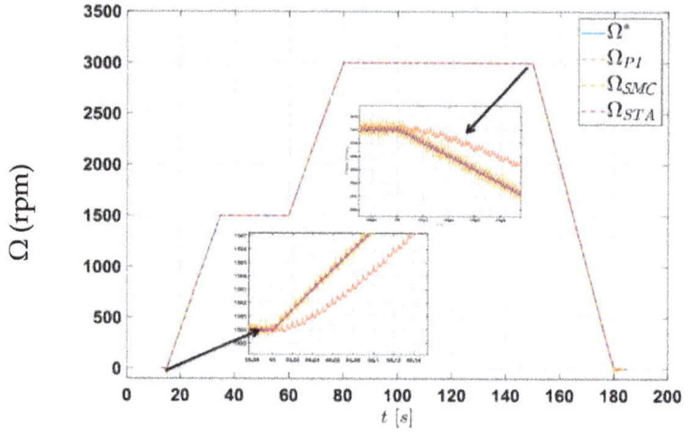

Figure 13. Velocity responses of the implemented robust controllers.

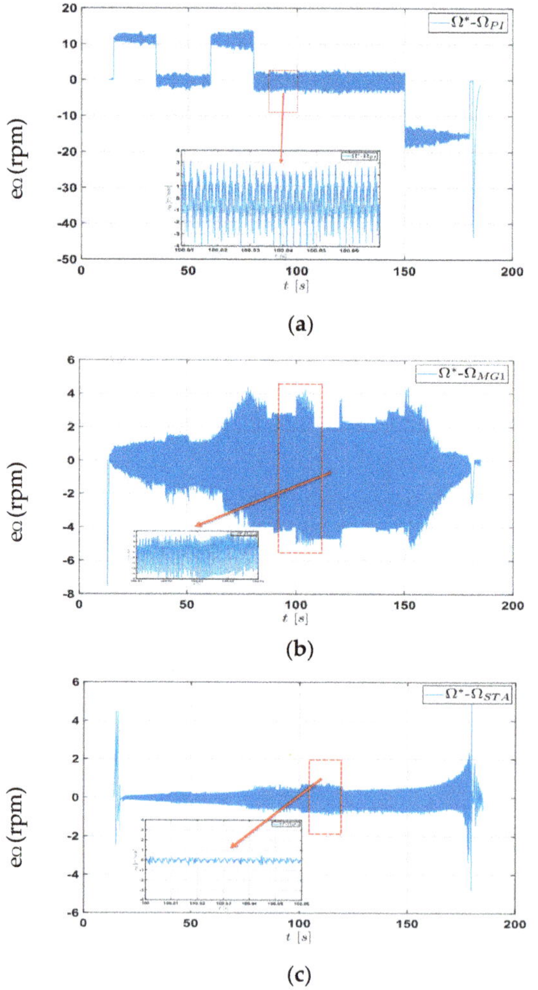

Figure 14. Zoom on the velocity responses in steady state.

Indeed, the best robust control given the minimum of oscillations around the steady state value is STSMC and, consequently, the minimum error in a steady state compared to other controllers. Moreover, when comparing the response of SMC and STA, the velocity response ripples, as shown in the zooms of Figure 13, are significantly improved due to the elimination of the chattering phenomenon.

Finally, simulation and experimental results confirm that, according to the performance comparison, STSMC is the best algorithm to be selected for an electric vehicle drivetrain with a Switched Reluctance Machine.

For the same controllers, current responses for a current set point (*) are collected from the test bench and plotted to make a performance comparison. Figures 15–17 show the current responses of phase 1 using PI, SMC, and STSMC, while Figures 18–20 show the corresponding current error responses.

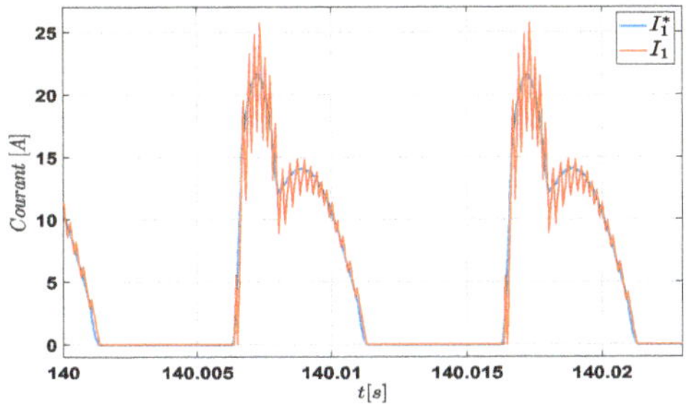

Figure 15. Current response of phase 1 using PI control.

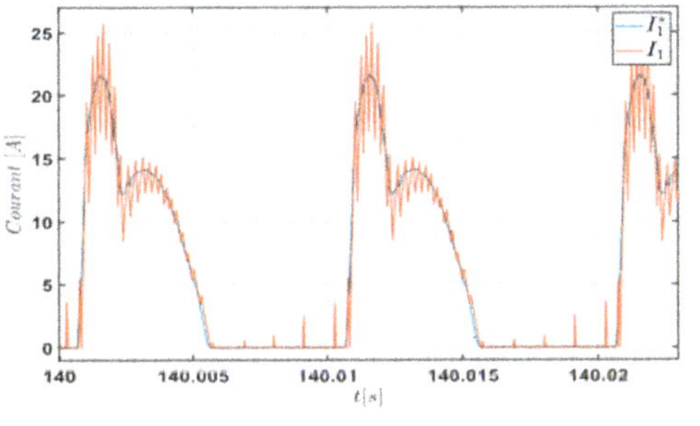

Figure 16. Current response of phase 1 using SMC.

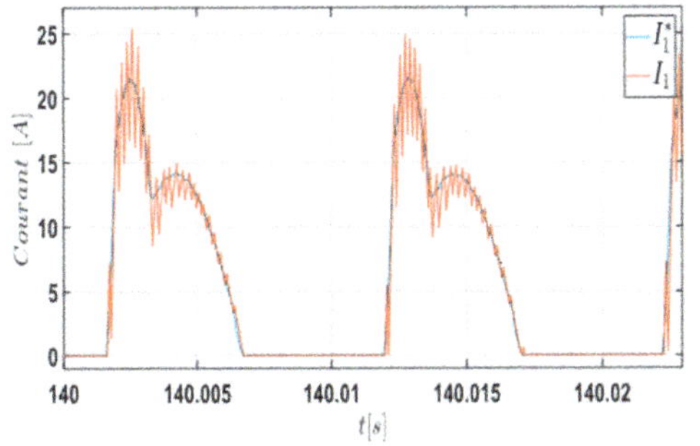

Figure 17. Current response of phase 1 using STSMC.

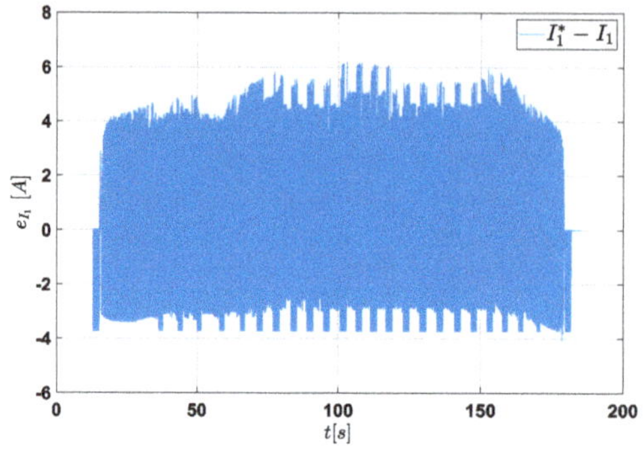

Figure 18. Current error response of phase 1 using PI control.

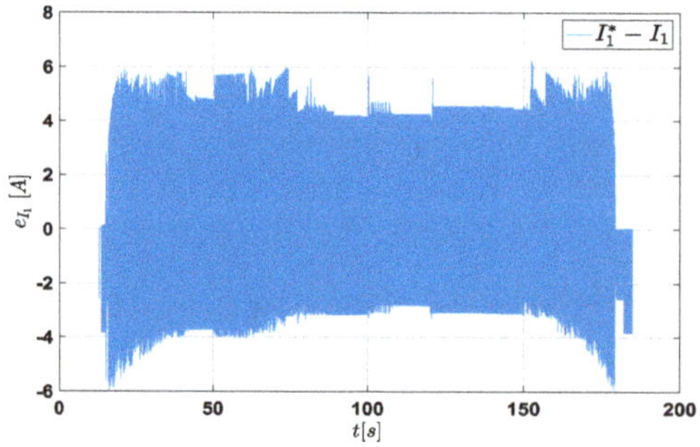

Figure 19. Current error response of phase 1 using SM.

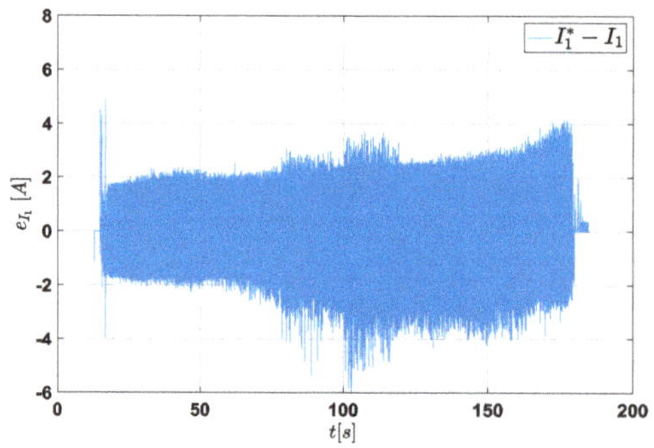

Figure 20. Current error response of phase 1 using STSMC.

According to the experimental current responses, it is shown that STSMC implemented in the current loop is the best algorithm to obtain minimum oscillations around the steady state value and, consequently, minimum torque ripple. Finally, the three robust controllers implemented simultaneously in the velocity loop and in the four current loops of the SRM control strategy confirm that STSMC performances are better compared to the other robust controls, namely PI and SM controllers.

8. Conclusions

In this paper, STSMC, SMC, and PI controllers are successfully validated and implemented to deal simultaneously with the velocity and current tracking problem of a Switched Reluctance Machine for electric vehicle applications. A cascade control strategy is adopted to control separately the SRM velocity and each phase current.

- The simulation and the experimental results show and confirm the benefits of using STSMC in terms of tracking performances and torque ripple minimization, thanks to the significant reduction in the chattering.
- The performances of electric vehicle drivetrain are improved using a Switched Reluctance Machine and STSMC. Both could be a new solution for future electric vehicles drivetrains: SRM as a fault-tolerant design component without rare earth material and STSMC as a robust control to obtain better performances and comfort in the electric vehicle.
- Improvement of performances and robustness of electric vehicle drivetrain using switched reluctance machine can improve performances of other functions of safety in electric vehicle applications and/or in autonomous vehicles where the dynamic behavior of the vehicle on the road depends on the dynamic behavior of the vehicle drivetrain in closed loop.
- This work can be extended to deal with electric and mechanical faults of the electric vehicle drivetrain to show the added value of SRM as a fault tolerant design machine.
- Finally, this study can be extended to Four-Wheel Independent Control Electric Vehicles (FWIC-EV) using Switched Reluctance Machines and STSMC.

Author Contributions: Validation, Y.S.; Writing—original draft, R.S.; Writing—review & editing, A.A. All authors have read and agreed to the published version of the manuscript.

Funding: This research received no external funding.

Data Availability Statement: Not applicable.

Conflicts of Interest: The authors declare no conflict of interest.

Appendix A

Parameter	Value
Topology	8 S/6 R
Phase number	4
Power supply (DC)	250 V
Maximum current	61 A
Nominal Power	8 kW
Maximum torque T_{Max}	20 Nm
Maximum speed	10,000 rpm
Phase resistance R	0.0404 Ohm
Moment of inertia J	0.0043 Kg/m^2
Viscous friction coefficient fr	0.005 Nm/s

References

1. Husain, I. *Electric and Hybrid Vehicles: Design Fundamentals*; CRC Press: Boca Raton, FL, USA, 2003.
2. Widmer, J.D.; Martin, R.; Kimiabeigi, M. Electric vehicle traction motors without rare earth magnets. *Sustain. Mater. Technol.* **2015**, *3*, 7–13. [CrossRef]
3. Krishnan, R. *Switched Reluctance Motor Drives: Modeling, Simulation, Analysis, Design, and Applications*; CRC Press: Boca Raton, FL, USA, 2001.
4. Biczyski, M.; Sehab, R.; Whidborne, J.; Krebs, G.; Luk, P. Fault-tolerant Switched Reluctance Motor Propulsion System for eVTOLs. In Proceedings of the 12th EASN International Conference on "Innovation in Aviation & Space for opening New Horizons, Barcelona, Spain, 18–21 October 2022.
5. Colby, R.S.; Mottier, F.M.; Miller, T.J.E. Vibration modes and acoustic noise in a four-phase switched reluctance motor. *IEEE Trans. Ind. Appl.* **1996**, *32*, 1357–1364. [CrossRef]
6. Chiba, A.; Kiyota, K.; Hoshi, N.; Takemoto, M.; Ogasawara, S. Development of a Rare-Earth-Free SR Motor With High Torque Density for Hybrid Vehicles. *IEEE Trans. Energy Convers.* **2015**, *30*, 175–182. [CrossRef]

7. Takiguchi, M.; Sugimoto, H.; Kurihara, N. Acoustic Noise and Vibration Reduction of SRM by Elimination of Third Harmonic Component in Sum of Radial Forces. *IEEE Trans. Energy Convers.* **2015**, *30*, 883–891. [CrossRef]
8. Rana, A.K.; Raviteja, A. A mathematical torque ripple minimization technique based on nonlinear modulating factor for switched reluctance motor drives. *IEEE Trans. Ind. Electron.* **2022**, *69*, 1356–1366. [CrossRef]
9. Ding, W.; Liu, G.; Li, P. A hybrid control strategy of hybridexcitation switched reluctance motor for torque ripple reduction and constant power extension. *IEEE Trans. Ind. Electron.* **2020**, *67*, 38–48. [CrossRef]
10. Sun, X.; Wu, J.; Lei, G.; Guo, Y.; Zhu, J. Torque ripple reduction of srm drive using improved direct torque control with sliding mode controller and observer. *IEEE Trans. Ind. Electron.* **2021**, *68*, 9334–9345. [CrossRef]
11. Kelly, L.; Cossar, C.; Miller, T. *Electronic Control of Switched Reluctance Machines*; Newnes Power Engineering Series; Newnes: Oxford, UK, 2001.
12. Saadi, Y.; Sehab, R.; Chaibet, A.; Boukhnifer, M.; Diallo, D. Performance comparison between conventional and robust control for the powertrain of an Electric Vehicle propelled by a Switched Reluctance Machine. In Proceedings of the IEEE Vehicle Power and Propulsion Conference (VPPC), Arlington, TX, USA, 9–12 September 2007; pp. 1–6.
13. Hannoun, H.; Hilairet, M.; Marchand, C. Gain-scheduling pi current controller for a switched reluctance motor. In Proceedings of the IEEE International Symposium on Industrial Electronics, Vigo, Spain, 4–7 June 2007; pp. 1177–1182.
14. Ouddah, N.; Boukhnifer, M.; Chaïbet, A.; Monmasson, E. Robust LPV current control of switched reluctance motor. In Proceedings of the 22nd Mediterranean Conference on Control and Automation, Palermo, Italy, 16–19 June 2014.
15. Ghani, M.; Frah, N.; Tamjis, M. Vector Control of Switched Reluctance Motor Using Fuzzy Logic and Artificial Neutral Network. In Proceedings of the International Conference on Electrical, Electronics, and Optimization Techniques (ICEEOT), Chennai, India, 3–5 March 2016; pp. 4412–4417.
16. Ling, F.; Ma, M.; Yang, Q.; Li, F. Torque Ripple Reduction of Switched Reluctance Motor by Segmented Harmonic Currents Injection Based on Adaptive Fuzzy Logic Control. In Proceedings of the 14th IEEE Conference on Industrial Electronics and Applications (ICIEA), Xi'an, China, 19–21 June 2019.
17. John, G.; Eastham, A. Speed control of switched reluctance motor using sliding mode control strategy. In Proceedings of the IAS '95. Conference Record of the 1995 IEEE Industry Applications Conference Thirtieth IAS Annual Meeting, Orlando, FL, USA, 8–12 October 1995; Volume 1, pp. 263–270.
18. Ruiwei, Z.; Xisen, Q.; Liping, J.; Yingch, Z.; Jintong, N. An adaptive sliding mode current control for switched reluctance motor. In Proceedings of the IEEE Conference and Expo Transportation Electrification Asia-Pacific (ITEC Asia-Pacific), Beijing, China, 31 August–3 September 2014.
19. Canale, M.; Fagiano, L.; Ferrara, A.; Vecchio, C. Comparing Internal Model Control and Sliding-Mode Approaches for Vehicle Yaw Control. *IEEE Trans. Intell. Transp. Syst.* **2009**, *10*, 31–41. [CrossRef]
20. Ouddah, N.; Boukhnifer, M.; Chaibet, A.; Monmasson, E. Robust controller designs of switched reluctance motor for electrical vehicle. In Proceedings of the 22nd Mediterranean Conference on Control and Automation, Palermo, Italy, 16–19 June 2014.
21. Peng, F.; Emadi, A. A digital PWM current controller for switched reluctance motor drives. In Proceedings of the IEEE Transportation Electrification Conference and Expo (ITEC), Dearborn, MI, USA, 15–18 June 2014; pp. 1–6.
22. Li, X.; Shamsi, P. Model Predictive Current Control of Switched Reluctance Motors with Inductance Auto-Calibration. *IEEE Trans. Ind. Electron.* **2016**, *63*, 3934–3941. [CrossRef]
23. Liu, W.; Xiong, L.; Xia, X.; Lu, Y.; Gao, L.; Song, S. Vision-aided intelligent vehicle sideslip angle estimation based on a dynamic model. *IET Intell. Transp. Syst.* **2020**, *14*, 1183–1189. [CrossRef]
24. Liu, W.; Xiong, L.; Xia, X.; Yu, Z. Intelligent vehicle sideslip angle estimation considering measurement signals delay. In Proceedings of the IEEE Intelligent Vehicles Symposium (IV), Changshu, China, 26–30 June 2018; pp. 1584–1589.
25. Husain, I.; Hossain, S. Modeling, Simulation, and Control of Switched Reluctance Motor Drives. *IEEE Trans. Ind. Electron.* **2005**, *52*, 1625–1634. [CrossRef]
26. Pollock, C.; Williams, B. A unipolar converter for a Switched Reluctance Motor. *IEEE Trans. Ind. Appl.* **1990**, *26*, 222–228. [CrossRef]
27. Bartolini, G.; Ferrara, A.; Usai, E. Chattering avoidance by second order sliding mode control. *IEEE Trans. Autom. Control* **1998**, *43*, 241–246. [CrossRef]

applied sciences

MDPI

Article

Structure Optimization and Control Design of Electronic Oxygen Regulator

Dongsheng Jiang *, Yue Liu, Haowen Yang, Xingxing Fang, Binbin Qian and Hui Li

School of Mechanical and Electrical Engineering, Anhui Jianzhu University, Heifei 230009, China; liuyue566@stu.ahjzu.edu.cn (Y.L.)
* Correspondence: jiangds922@163.com

Abstract: The oxygen regulator is the core component of the aircraft life support system, which adjusts the flow and pressure of the breathing gas according to the pilot's breathing needs. In response to the problem that structural parameters are difficult to adjust and prone to jitter when the indirect oxygen regulator system is stable, a direct oxygen regulator is designed using a stepper motor to drive a lung-type flapper, replacing the diaphragm lever-type structure of the indirect oxygen regulator. Due to the nonlinearity and time-varying nature of the dynamic characteristics of oxygen regulators, a single-neuron PID control strategy based on online identification of RBF neural networks is proposed to improve the PID control performance. The RBF neural network is used to identify the Jacobian information of the controlled object, and the single-neuron PID controller completes the online adjustment of the controller parameters to realize the intelligent control of the system. Simulation experimental studies are conducted to verify the performance of the direct oxygen regulator. The result analysis verifies the excellence of the single-neuron PID control strategy based on online recognition of the RBF neural network to improve the system performance.

Keywords: electronic oxygen regulator; structural optimization; RBF neural network; single neuron; life support system

Citation: Jiang, D.; Liu, Y.; Yang, H.; Fang, X.; Qian, B.; Li, H. Structure Optimization and Control Design of Electronic Oxygen Regulator. *Appl. Sci.* **2023**, *13*, 5431. https://doi.org/10.3390/app13095431

Academic Editors: Kan Liu and Wei Hu

Received: 31 March 2023
Revised: 21 April 2023
Accepted: 23 April 2023
Published: 27 April 2023

1. Introduction

With the development of onboard electronics, electronic oxygen supply regulators using electronic technology and sensor technology for control began to appear [1,2]. It is mainly through the pressure sensor that one can feel the small pressure difference, with the use of the solenoid valve to control the opening of the valve, or using the micro-motor-driven diaphragm to indirectly drive the opening of the oxygen supply valve to achieve oxygen supply regulation [3]. The performance of the electronic oxygen supply regulator has greatly improved, and it can adapt to the low outlet pressure of molecular sieve oxygen system and has the advantages of small size, light weight, high flow rate, low suction resistance, and fast dynamic response, which is the future development direction of airborne oxygen supply regulators [4–6].

At present, the actuator of the electronic oxygen regulator is mainly mechanical, and the control part adopts the electronic auxiliary regulation [7]. The system is dominated by mechanical actuator regulation and is supplemented by electronic regulation control to improve the comprehensive performance of the system, where the oxygen supply regulator and anti-charge regulator are integrated into a system, which compared to the mechanical control system performance makes it a great improvement [8]. This indirect oxygen regulator indirectly controls the opening and closing of the pulmonary oxygen supply valve by controlling the residual pressure valve with a motor. However, there are still disadvantages in that the mechanical oxygen regulator mechanism parameters are difficult to adjust, and the pressure is easy to fluctuate.

In the research on oxygen regulator structures, Wan et al. studied the calculation of the structural parameters of an indirect oxygen regulator [9]. The performance parameters of the indirect oxygen regulator were analyzed by Zou et al. [10]. Concerning the electronic control of oxygen regulators, Yu et al. used switching control [11] and Sun et al. applied fuzzy control to design a control law for an electronic oxygen regulator [12]. Li et al. used expert control to automatically adjust the PID parameters, which solved the deficiency of fixed control parameters [5]. Jiang et al. pointed out that disturbance suppression is a key issue in the control of oxygen regulators. Generalized predictive control (GPC) [13], automatic disturbance rejection control (ADRC) [14], and the adaptability has been applied to controllers to improve performance [15].

In the control application of an oxygen regulator, there are unknown perturbations acting on the system such as gas source pressure changes and suction flow changes in the oxygen regulator, and it is difficult to achieve the desired control effect with a fixed parameter PID regulator. Therefore, the control method of a single-neuron PID is adopted, and an RBF neural network is introduced for online identification of system output to provide system dynamic information for the learning process of single-neuron PID.

Through the study and analysis of an indirect oxygen regulator, it is found that the parameters of the indirect oxygen regulator are not easily adjustable, and they jointly affect the overall performance. It is necessary to optimize the structure of the indirect oxygen regulator by changing the parameters and proposing a direct oxygen regulator to solve the jitter problem caused by the diaphragm lever mechanism. An intelligent control method is applied to improve the breathing resistance and pressure fluctuation of the oxygen regulator. This paper designs a direct oxygen regulator structure based on the indirect oxygen regulator by merging the lung type valve mechanism and the electronically controlled residual pressure valve mechanism of the indirect electronic oxygen regulator, and optimizing it to a direct motor-driven valve mechanism to close or open the valve for oxygen supply and establish a mathematical model of respiratory pressure control. In this control system structure, the RBF neural network is used to identify the Jacobian information of the controlled object and the online adjustment of the controller parameters is completed by the single-neuron PID controller to realize the intelligent control of the system. The control performance of the structure's design and control method on pressure fluctuation and breathing resistance is verified through a simulation and experiment.

2. Oxygen Regulator Principle Analysis

2.1. Indirect Oxygen Regulator Principle

Combined with the structure principle of indirect oxygen regulator in Figure 1, when the pilot inhales, the P2 pressure measurement point detects the negative pressure generated by the pilot's inhalation, and the electronic control device controls the electronically controlled residual pressure valve to move forward to establish the residual pressure chamber pressure (P2 pressure measurement point); the residual pressure chamber pressure acts on the main lung valve diaphragm, which deforms the main lung valve diaphragm to open the main valve to supply air to the pilot. When the pilot exhales, P1 detects the exhalation action, and the electronic control device controls the electronically controlled residual pressure valve to move backward and remove the residual pressure chamber pressure, and the main lung valve diaphragm returns to its initial state to close the main valve and stop supplying air to the pilot. When the height rises, the electronic control device controls the electronically controlled residual pressure valve to move forward and establish the safety residual pressure (P2 pressure measurement point). The breathing chamber pressure is controlled by the residual pressure chamber pressure through the main lung type valve diaphragm 8 and the main valve to establish the safety residual pressure.

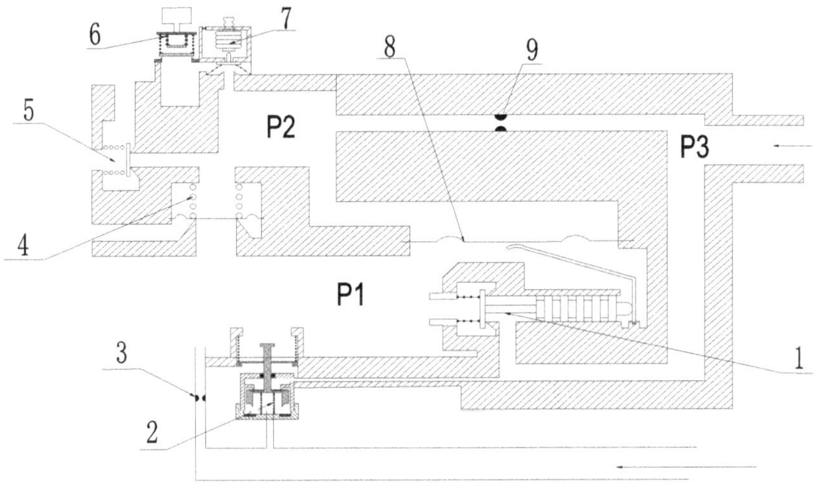

Figure 1. Indirect oxygen regulator structure schematic. P1: breathing chamber pressure measuring point; P2: residual pressure chamber pressure measuring point; P3: inlet pressure measuring point; 1: pulmonary valve; 2: air valve; 3: limiting hole; 4: ventilation valve; 5: safety valve; 6: residual pressure valve; 7: high residual pressure valve; 8: membrane; 9: sizing hole.

2.2. Optimized Design

The literature [15] proposes that the phenomenon of positive pressure fluctuations is caused by a combination of inlet pressure fluctuations and diaphragm deformation. At low lung flux, the effects of both causes are similar. At high lung flux, diaphragm deformation becomes the main cause of significant positive pressure fluctuations.

To address this one drawback of the indirect oxygen regulator, a structural solution design was focused on the oxygen supply assembly. The lung-type flapper mechanism and the electronically controlled residual pressure flapper mechanism of the indirect electronic oxygen supply anti-charge regulator are combined and optimized to close or open the flapper for oxygen supply by direct motor drive. The design of removing the diaphragm can fundamentally improve the system performance of the oxygen regulator.

The stepper motor is driven to open or close the oxygen supply door of the lung mechanism according to the parameters of cockpit height, acceleration, and product input and output pressure collected by the sensor set. The opening and closing of the oxygen supply door of the lung mechanism is realized by the electronic control device. The schematic sketch of the pulmonary mechanism is shown in Figure 2. The flow restrictor provides a continuous flow of oxygen supply. When the control or power failure occurs in the electronically controlled oxygen supply, and normal oxygen supply cannot be carried out, oxygen regulation must be implemented to downgrade the oxygen supply. When a power failure occurs or the electronic control device monitors the abnormal output of the subsystem, the electronic control device drives the normally open solenoid valve down to turn on the continuous flow and implement the downgraded oxygen supply. The oxygen supply assembly mainly consists of a shell assembly, a linear stepping motor, a normally open solenoid valve, a lung mechanism, a flow limiter, a time delay mechanism, a safety valve mechanism, an anti-asphyxiation mechanism, and a pressure limiting mechanism. The structure's schematic diagrams are shown in Figures 2 and 3, respectively.

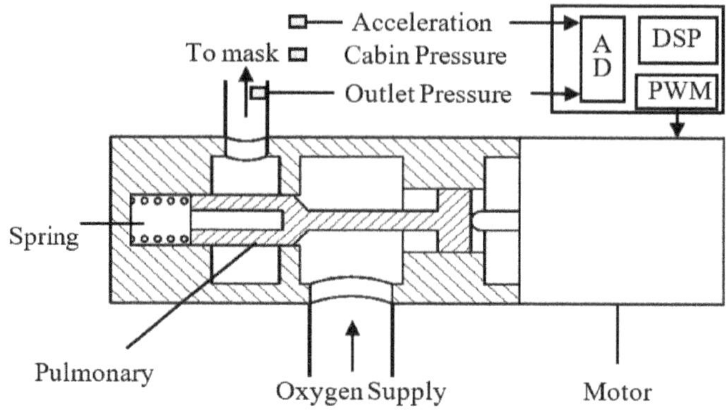

Figure 2. Electronic oxygen regulator working principle diagram.

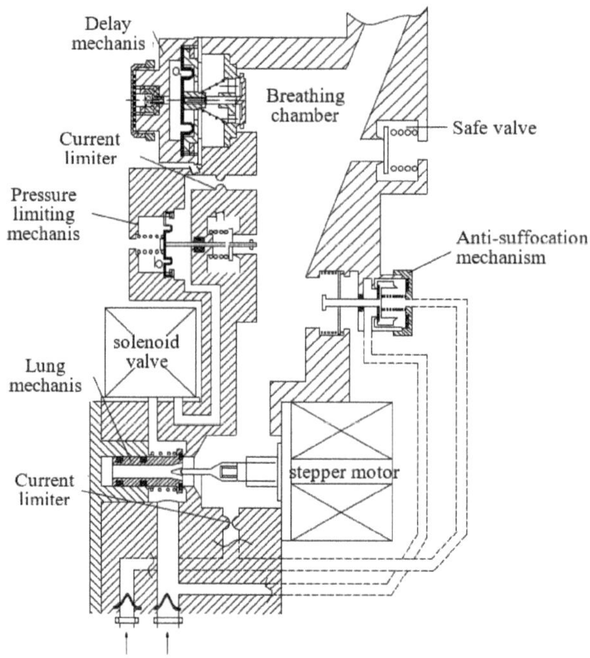

Figure 3. Direct oxygen regulator structure schematic.

3. Control Method Design

3.1. Mathematical Modeling

The mathematical model of the oxygen regulator consists of two parts: the pulmonary valve and the stepper motor. Due to the complex structure of the entire oxygen regulator, model identification using system identification techniques is very difficult. Combining the advantages of identification techniques and mechanistic analysis, both the lung valve and stepper motor models are based on mechanistic analysis.

Before building the mathematical model, it can be assumed that the gas is regarded as an ideal gas and satisfies the ideal gas equation [16]. The process of gas flow through the positive pressure chamber and the valve is very short, so the heat exchange can be neglected and the gas flow state is regarded as an isentropic process [15]. Considering the gas in the suction chamber as a control body, the dynamic equations of the gas pressure and flow in the suction chamber are as follows:

$$\frac{dp}{dt} = \frac{\mathrm{K} \cdot R_g \cdot T \cdot \delta_m}{V} \quad (1)$$

$$\delta_m = G_{in} - G_{out} \quad (2)$$

where p is the gas pressure in the suction chamber, T is the gas temperature, K is the gas adiabatic index, R_g is the gas constant, V is the volume of the suction chamber, δ_m is the mass change, G_{in} is the gas mass flow rate into the suction chamber, and G_{out} is the gas mass flow rate out of the suction chamber.

The gas flow area of the valve is calculated as follows:

$$A = \pi \cdot (d - x \sin\theta \cos\theta) x \cdot \sin\theta \quad (3)$$

where d is the diameter of the valve hole, θ is the valve half taper, and x is the valve opening.

The gas flow in the suction chamber can be considered an isentropic flow [16]. When expressing the gas mass flow equation, there are two cases.

In the first case, when $p/p_{in} > \varepsilon$

$$G_{in} = \frac{C_d \cdot A \cdot p_{in}}{\sqrt{R_g \cdot T_{in}}} \sqrt{\frac{2\kappa}{\mathrm{K} - 1} \left(\frac{p}{p_{in}}\right)^{\frac{2}{\kappa}} \left(1 - \left(\frac{p}{p_{in}}\right)^{\frac{\kappa}{\kappa-1}}\right)} \quad (4)$$

In the second case, when $p/p_{in} \leq \varepsilon$

$$G_{in} = \frac{C_d \cdot A \cdot p_{in} \cdot B}{\sqrt{R_g \cdot T_{in}}} \sqrt{\frac{2\kappa}{\kappa + 1} \left(\frac{2}{\kappa + 1}\right)^{\frac{2}{\kappa-1}}} \quad (5)$$

where C_d is the flow coefficient of the flapper, A is the flow area of the flapper, $\varepsilon = \left(\frac{2}{\kappa+1}\right)^{\frac{\kappa}{\kappa-1}}$ is the critical pressure ratio, p_{in} is the inlet gas pressure, and T_{in} is the inlet gas temperature.

The following force balance equation is satisfied during the motion of the stepper motor:

$$m\frac{d^2x}{dt} = F_m - c\frac{dx}{dt} - kx + F_d \quad (6)$$

$$F_m = k_m \cdot i \quad (7)$$

where m is the mass of the moving part, F_m is the electromagnetic force, c is the viscous damping factor, k is the spring stiffness, and F_d is the disturbance force, i is the coil current, and k_m is the electromagnetic force coefficient.

The mathematical model can be synthesized as follows:

$$\begin{cases} \dot{p} = f_1(p, w_1) + b_1 x \\ \ddot{x} = f_2(x, \dot{x}, w_2) + b_2 u \\ y = p \end{cases} \quad (8)$$

where w_1 and w_2 are unknown perturbations.

$$f_1(p, w_1) = \frac{\kappa \cdot R_g \cdot T}{V}(k_1 p - G_{out}) \quad (9)$$

$$f_2(x, \dot{x}, w_2) = -\left(\frac{c}{m} + \frac{k_m k_E}{mR}\right)\dot{x} - \frac{k}{m}x + \frac{F_d}{m} \quad (10)$$

$$b_1 = \frac{\kappa \cdot R_g \cdot T}{V} k_2 \tag{11}$$

$$b_2 = \frac{k_m}{mR} \tag{12}$$

In the simulation calculation, G_{out} is calculated by the human lung capacity model. The variation of human lung capacity can be regarded as a sinusoidal law variation.

3.2. Controller Design

According to the oxygen regulation principle of the direct oxygen regulator, the electronic control device drives the stepper motor to open or close the oxygen supply valve of the pulmonary mechanism according to parameters such as cockpit height, acceleration, and product input and output pressure collected by the sensor group. The goal of the controller design is, on the one hand, to regulate u to keep the output pressure y close to a given reference input; on the other hand, to regulate u to counteract the effects of various disturbances w. In both cases, the smaller the control error $e = r - y$, the smaller the suction resistance and pressure fluctuations. During the operation of the oxygen regulator, the output gas pressure was affected by various disturbances such as pilot suction flow, change in gas source pressure, a disturbance force acting on the flapper, change in motor parameters, etc. The structural parameters of the whole control system are time-varying and non-linear, and the PID regulator with fixed parameters cannot get the ideal control quality. A single-neuron PID control method is used, and an RBF neural network is introduced for online identification of the system output to provide system dynamic information for the learning process of the single-neuron PID [17]. The control system structure is shown in Figure 4. In Figure 4, SNPID is the single-neuron PID controller and RBFNNI is the RBF neural network discriminator.

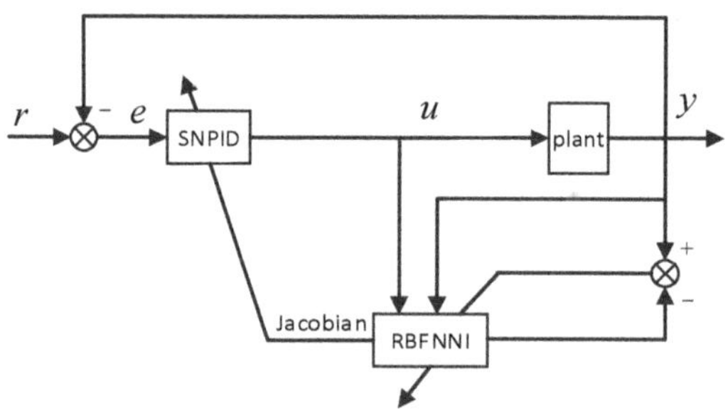

Figure 4. Structure of a single-neuron PID control system based on RBF identification.

An adaptive single-neuron PID control method based on dynamic RBF neural network identification is proposed in the reference paper [18], where a dynamic neural network identifier identifies the system model online, obtains online tuning information of the PID parameters, and the single-neuron controller realizes the self-tuning of the controller parameters. The system has the advantages of high adaptability and robustness compared with the traditional RBF neural network-based PID control method. The single-neuron PID controller based on RBF neural network recognition consists of a single-neuron PID controller and a dynamic RBF neural network recognizer. The dynamic RBF neural network is used to model the system, obtain a predictive mathematical model of the system, replace the actual output of the object with the predicted output of the model, and then design the single-neuron PID controller. The network recognition structure adopts a series-parallel structure to ensure the stability of the system.

The single-neuron PID controller is a controller that combines PID control laws and neural networks [19], and the implicit layer contains three units that are proportional (P), integral (I), and differential (D), and the structure of this controller is shown in Figure 5. Adaptivity is achieved through the adjustment of single-neuron weighting coefficients and the commonly used learning rules are the unsupervised Hebb learning rule, supervised Delta learning rule, and supervised Hebb learning rule [19]. In this paper, the controller performs online rectification of the weights based on the gradient descent method and the Jacobian information provided by the RBFNNI. The control and learning algorithms are as follows:

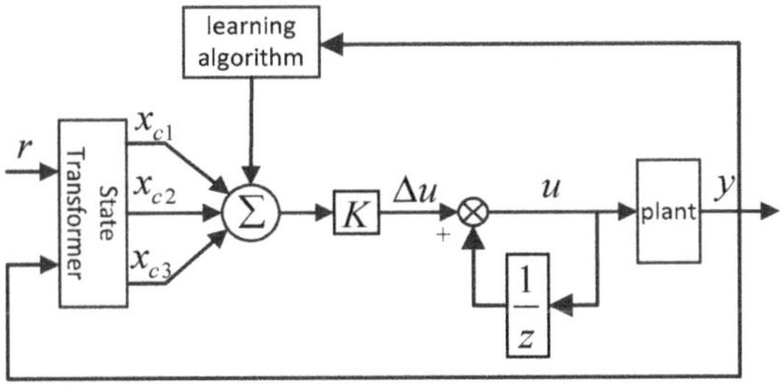

Figure 5. Single-neuron controller structure.

Let the system control error be $e(k) = r(k) - y(k)$. Take the neuron input as:

$$\begin{cases} x_{c1}(k) = e(k) - e(k-1) \\ x_{c2}(k) = e(k) \\ x_{c3}(k) = e(k) - 2e(k-1) + e(k-2) \end{cases} \tag{13}$$

Neuronal outputs are:

$$u(k) = u(k-1) + K\sum_{i=1}^{3} v_i(k)x_{ci}(k) \tag{14}$$

where $v_i(k)$ is the weight coefficient of a single-neuron network, and $K > 0$ is the scale factor of single neuron. The performance index function of the controller was taken as:

$$J_c = \frac{1}{2}e^2(k) \tag{15}$$

According to the gradient descent method, the adjustment of the single-neuron weighting coefficient is:

$$\Delta v_i(k+1) = -\eta_i \frac{\partial J_c}{\partial v_i(k)} = -\eta_i \frac{\partial J_c}{\partial y}\frac{\partial y}{\partial u}\frac{\partial u}{\partial v_i(k)} = \eta_i e(k)\frac{\partial y}{\partial u}Kx_{ci}(k) \tag{16}$$

where n is the learning rate and y can be replaced by the RBFNN discrimination information u. From this, the adjusted weighting coefficient was obtained as:

$$v_i(k+1) = v_i(k) + v_i(k+1) \tag{17}$$

To ensure the convergence and robustness of the control algorithm Equations (16) and (17), the learning algorithm was normalized [20] to obtain:

$$\begin{cases} u(k) = u(k-1) + K\sum\limits_{i=1}^{3} v_i(k)x_{ci}(k) \\ v_i'(k) = v_i(k) / \sum\limits_{i=1}^{3} |v_i(k)| \end{cases} \tag{18}$$

3.3. Simulation Results

For the structural characteristics of the direct oxygen regulator, the performance of the direct oxygen regulator is analyzed by a simulation focusing on the pulmonary mechanism and pressure limiting mechanism, and the simulation structure is represented in Figure 6. The respiratory flow rate of the user varies approximately according to the sinusoidal law. A joint simulation of the model was implemented using amsim and simulink. The sampling time is 0.001 s. The parameters used in the simulation and their values are listed in Table 1.

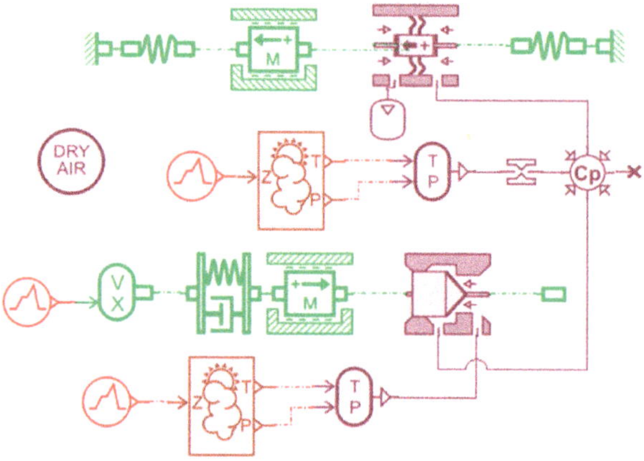

Figure 6. Oxygen regulator simulation structure.

Table 1. Structural parameters.

Parameter	Value
Spring stiffness (N/m)	470
Volume of chamber (cm^3)	13
Diameter of hole (mm)	10
Diameter of poppet (mm)	15
Orifice area (mm^2)	0.1

Oxygen supply is mainly achieved by stepper motors opening or closing the oxygen supply valve of the pulmonary mechanism. The simulation structure includes only the motor and its driven lung mechanism, the pressure limiting mechanism, and the breathing chamber. The principle of the pulmonary mechanism is mainly a motor pushing the pulmonary valve, and the simplified simulation mechanism is represented in Figure 6. The pressure profile of the output respiratory chamber fluctuates after a given respiratory pressure profile.

It can be seen in Figure 7 that the pressure in the breathing chamber is essentially consistent with the reference pressure and that it follows this well.

Combined with the single-neuron PID control strategy based on online identification of the RBF neural network, the sampling time in the simulation is 0.001 s and the main parameters are shown in Table 2. The network structure of the RBFNNI is 3-6-1, and the network input is $x = [u(k)\ y(k)\ y(k-1)]$, $\eta = 0.2$, $\alpha = 0.05$. SNPID in taking $K = 6$, $\eta_1 = 0.2$, $\eta_2 = 0.001$, and $\eta_3 = 0.1$.

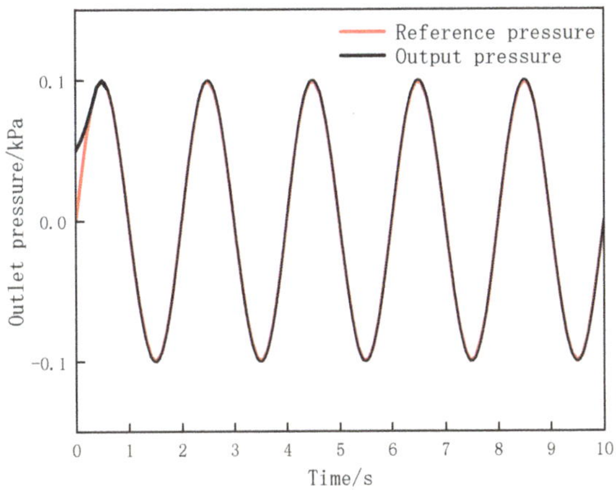

Figure 7. Respiratory chamber outlet pressure curve.

Table 2. Main parameters in the simulation.

Parameter (Unit)	Value
m (kg)	0.037
K (N/m)	700
d (m)	0.07
Θ (°)	11
R (Ω)	15
K_m (N/A)	23
K_E (V/(m/s))	23
V (m^3)	0.0005
κ	1.4
R_g ($m^2/(s^2 \cdot K)$)	260
Cd	0.8

At a lung ventilation rate of 44 L/min and inlet gas pressures of 40 kPa, 70 kPa, 150 kPa, and 300 kPa, the oxygen regulator outlet pressure curve is shown in Figure 8.

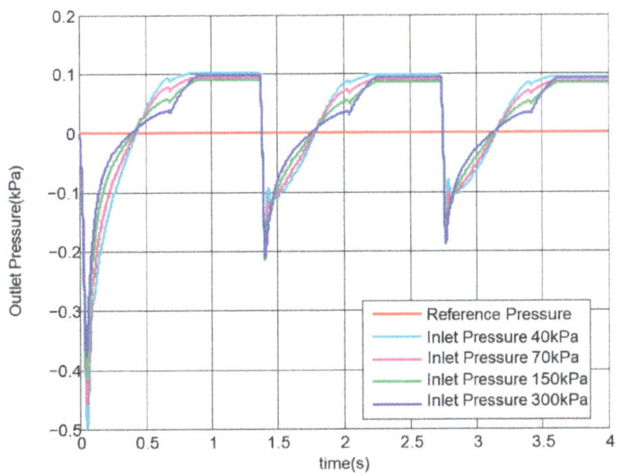

Figure 8. Outlet pressure curve of the regulator with 44L/min ventilation.

The reference pressure is used as both the regulation target value and the dividing line of respiratory resistance. An outlet pressure value above the reference pressure indicates expiratory resistance and an outlet pressure value below the reference pressure indicates inspiratory resistance. After entering the respiratory cycle, the desired outlet pressure should converge to the reference pressure. The peak pressure fluctuations in Figure 8 are between 0.1 kPa and −0.2 kPa. Near the positive and negative peak pressures, the outlet pressure is essentially the same; at the respiratory pressure transition, with an inlet pressure of 300 kPa, the regulator output pressure converges more closely to the reference pressure. Changes in the regulator's inlet gas pressure have little effect on the output pressure, and the oxygen regulator is able to output a stable pressure. When the inlet gas pressure is higher, the regulator output pressure is closer to the reference pressure.

4. Experimental Analysis

The electronic oxygen regulator designed in this paper was tested under ground test chamber conditions using mechanical lungs for pressurization, with industrial oxygen as the gas source and a pressurized oxygen supply mask for simulation experiments.

4.1. Experimental Projects

The indirect oxygen regulator was compared with the designed direct oxygen regulator for the experiment.

1. Breathing resistance test

The flow characteristics of the oxygen regulator are mainly ensured by the pulmonary mechanism, and the suction resistance is the main characteristic of this mechanism. To verify the flow characteristics of the direct oxygen regulator, a breathing resistance experiment was designed. Respiratory resistance tests were performed on mechanical lungs with lung ventilation volumes of 20 L/min, 30 L/min, and 44 L/min at an inlet gas pressure setting of 150 kPa, and the pressure fluctuations in the mask chamber were measured. As shown in Figure 9, piece 4 should be connected to piece 5; piece 12 should be opened, the breathing gas resistance test mode should be selected; switch 1 should be turned on, the pressure reducer 2 should be turned on; the pressure reducer 2 should be turned on according to the pressure gauge 3 to establish the pressure of 150 kPa; the dummy lung 9 should open in order to establish a 20 L/min, a 30 L/min, and a 44 L/min lung gas exchange volume; the value of the pressure gauge 11 should be recorded.

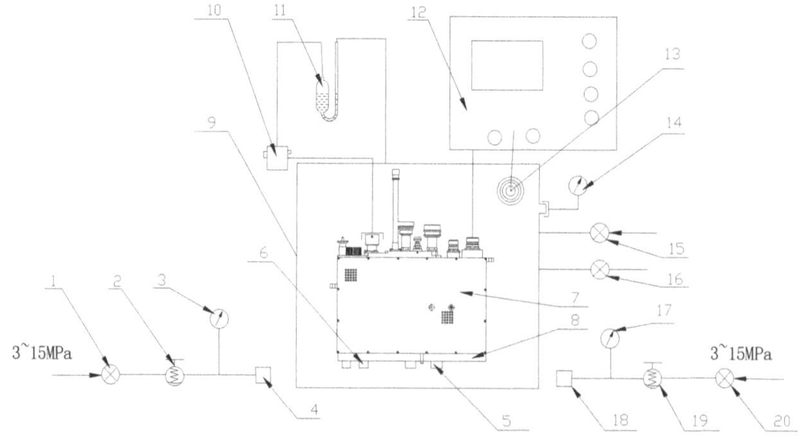

Figure 9. Experimental equipment connection diagram. 1: high pressure switch; 2: reducer adjustable; 3: pressure gauge; 4: connector; 5: anti-load inlet; 6: main oxygen inlet; 7: regulator; 8: tooling for simulating gas source distributor; 9: vacuum box; 10: mask model; 11: oil pressure gauge; 12: exciter; 13: height sensor; 14: altimeter; 15: vacuum switch; 16: deflation switch; 17: pressure gauge; 18: connector; 19: reducer adjustable; 20: high pressure switch.

2. Pressurization test

In high-altitude and overload situations, a certain amount of positive pressure needs to be established in the oxygen supply mask through active control based on the altitude and overload values to achieve pressurized respiratory protection and to improve the pilot's tolerance in high-altitude and overload environments. Calculate the target pressure according to the cockpit height and control and adjust the target pressure of the oxygen supply. The cockpit height is expressed by atmospheric pressure, and the relationship between the oxygen supply target pressure and the cockpit height is shown in Figure 10 [21]. When the cockpit altitude $PH < 16.5$ kPa, the target pressure value of the oxygen supply is shown in the equation.

$$PB = -0.756PH + 14.14 \quad (19)$$

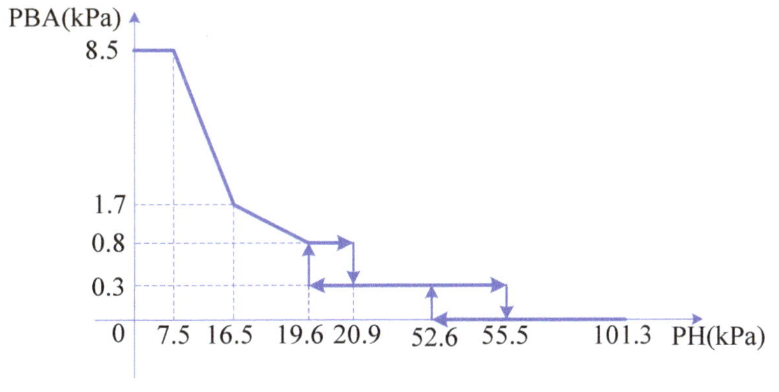

Figure 10. Relationship between cockpit height and oxygen supply target pressure curve.

When the oxygen supply regulator works in pressurized breathing regulation, it needs to pressurize the mask chamber at 2.0 kPa, 5.0 kPa, 7.5 kPa, and 8.5 kPa, as well as measure the pressure value of the mask chamber. It should connect according to Figure 10; connect piece 4 to piece 6, turn on switch 1, turn on pressure reducer 2 to establish a pressure of 2 kPa according to the pressure gauge 3; turn on switch 1, turn on pressure reducer 2 to establish a pressure of 0.2 MPa according to the pressure gauge 3; turn on switch 15 and slowly rise to 16 kPa according to the pressure gauge 14; record the residual pressure establishment height and residual pressure value.

4.2. Analysis of Experimental Results

The pressure profiles of the mask chamber at different lung fluxes at an inlet gas pressure of 150 kPa are shown in Figure 11. The positive peak of the waveform indicates the maximum expiratory pressure, and the negative peak indicates the inspiratory pressure [17]. At a lung ventilation volume of 20 L/min, the maximum inspiratory resistance was 0.25 kPa and the maximum expiratory resistance was 0.1 kPa. At a lung ventilation rate of 30 L/min, the maximum inspiratory resistance was 0.27 kPa and the maximum expiratory resistance was 0.13 kPa. At a lung ventilation rate of 44 L/min, the maximum inspiratory resistance was 0.4 kPa and the maximum expiratory resistance was 0.25 kPa. Comparing the appeal data with the ideal boundary values of respiratory resistance in Table 3, the peak resistance was all within the ideal boundary and much smaller than the boundary of respiratory resistance [22].

Table 3. Ideal boundary of respiratory resistance.

Ventilation Volume (L/min)	Expiratory Resistance (kPa)	Inspiratory Resistance (kPa)
20	≤0.29	≤0.49
30	≤0.49	≤0.59
44	≤1.08	≤0.88

The experimental data on the respiratory resistance of the indirect electronic oxygen supply anti-charge regulator are shown in Table 4. In comparison with the indirect oxygen regulator, it can be seen that the output of the direct oxygen regulator has less breathing resistance, which greatly enhances the comfort of the pilot when in use.

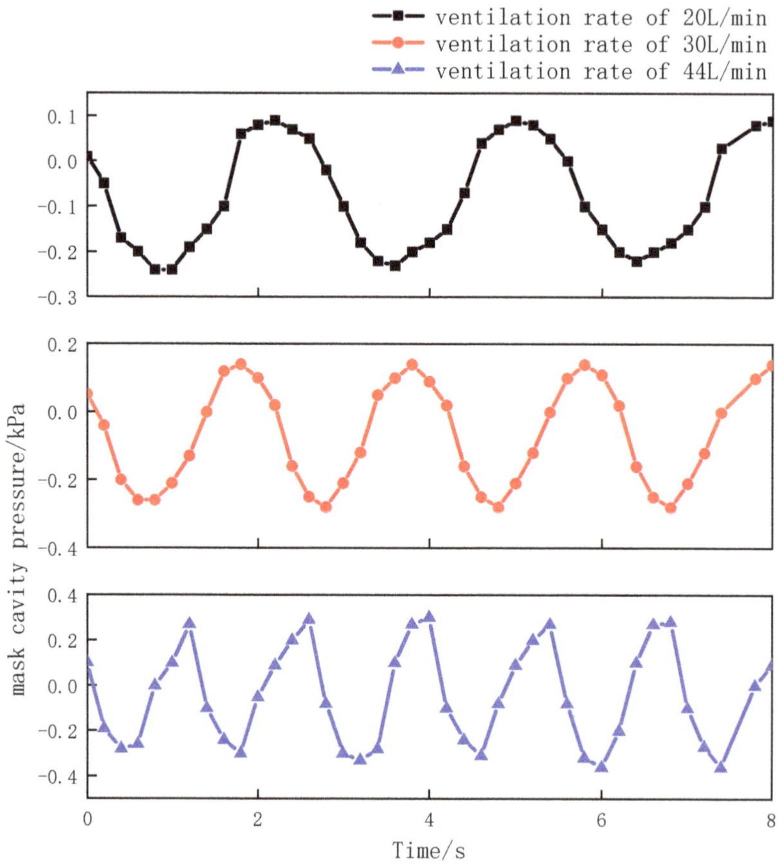

Figure 11. Mask pressure curves under different lung fluxes.

Table 4. Suction resistance of indirect oxygen regulator.

Lung Ventilation (L/min)	Peak Inspiratory Resistance (kPa)
20	0.49
30	0.59
44	0.88

When the oxygen supply regulator was used for pressurized breathing, the target pressure values of the pressurization test were 2.0 kPa, 5.0 kPa, 7.5 kPa, and 8.5 kPa, respectively. The pressure curves of the mask cavity are shown in Figure 12a,b.

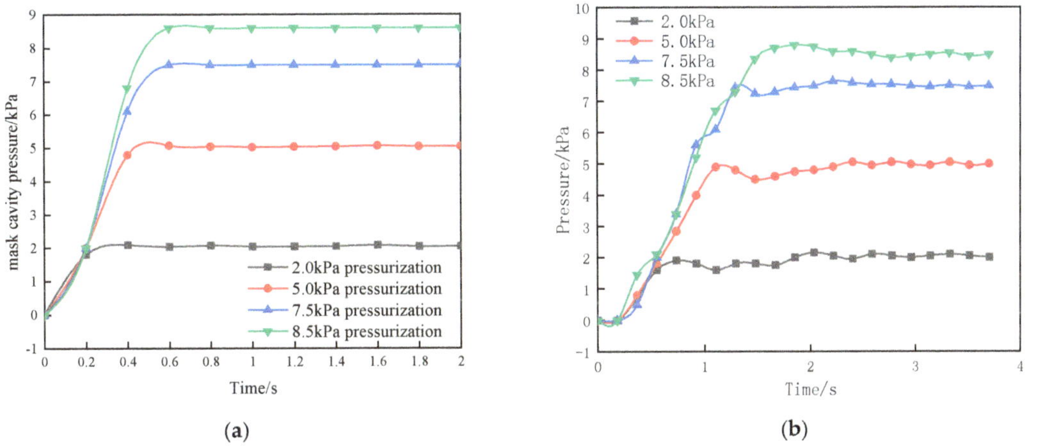

Figure 12. (**a**) Pressure curve of the mask chamber during pressurization; (**b**) indirect oxygen regulator mask chamber pressure.

During pressurization at 2.0 kPa, the pressure values can reached the target values at 0.4 s. When the pressurization value was at 8.5 kPa, the pressure value reached the stable target value at 0.6 s. The pressurization test process of the electronic oxygen regulator builds pressure quickly at different pressure targets. The speed of pressure building was within 0.8 s for all the different pressure tests. The smaller the target pressure of the pressurization test, the faster the pressure-building speed. For example, during the pressurization process at 2.0 kPa, the pressure value can reach the target value in 0.4 s. After the indirect oxygen regulator reaches the regulated pressure, there is a significant pressure fluctuation. However, indirect oxygen regulators have a longer pressure build-up time and exhibit significant pressure fluctuations.

The direct oxygen regulator solves the problem of pressure fluctuation of the indirect oxygen regulator, and the single-neuron PID control strategy based on online recognition of the RBF neural network improves the system performance, resulting in less breathing resistance and faster pressure building during pressurized breathing with the direct oxygen regulator.

5. Conclusions

In this paper, a direct oxygen regulator is designed for the problems of jittering output pressure and the difficult adjustment of structural parameters of an indirect oxygen regulator, and a stepper motor was used to drive the lung type valve, replacing the diaphragm lever-type structure of traditional indirect oxygen regulator.

A direct oxygen regulator with a single-neuron PID control strategy based on online recognition of the RBF neural network is designed by combining the current advanced intelligent control theory. The simulation results show that the control strategy makes the oxygen regulator have better control accuracy and system response speed.

The control performance of the oxygen regulator was verified experimentally. This control method reduces respiratory resistance by 50% and improves respiratory comfort compared to indirect oxygen regulators at different lung fluxes. In pressurized breathing, the output pressure fluctuation does not exceed 0.1kPa and the pressure build-up time is within 0.6 s, both of which show better performance than the indirect oxygenator regulator.

Author Contributions: Conceptualization, D.J. and Y.L.; methodology, D.J.; software, Y.L. and H.Y.; validation, Y.L., X.F. and B.Q.; investigation, H.L.; resources, D.J.; writing—original draft preparation, Y.L.; writing—review and editing, D.J.; All authors have read and agreed to the published version of the manuscript.

Funding: This research was funded by the Anhui Provincial University Natural Science Research Project (K2210033) and the Provincial Natural Science Research Projects of Anhui Provincial Universities (KJ2021A0635).

Institutional Review Board Statement: Not applicable.

Informed Consent Statement: Not applicable.

Data Availability Statement: The data is available from the corresponding author upon reasonable request.

Conflicts of Interest: The authors declare no conflict of interest.

References

1. Fan, Y.; Sun, Q.; Dong, F.; Chen, Q. An adaptive oxygen regulator control system based on particle swarm optimization and back propagation neural network. *Control Theory Appl.* **2020**, *3*, 687–695.
2. Siska, W.D., Jr.; Robert, C. Electromechanical Oxygen Valve and Regulator. U.S. Patent No. 7,677,529, 16 March 2021.
3. Yan, S.H.I.; Yixuan, W.A.N.G.; Maolin, C.A.I.; Zhang, B.; Jian, Z.H.U. An aviation oxygen supply system based on a mechanical ventilation model. *Chin. J. Aeronaut.* **2018**, *31*, 197–204.
4. Jiang, D.; Bu, X.; Lin, G.; Sun, B.; Huang, J.; Fang, L.; Zhao, H. Control design and experimental verification of three-bed airborne oxygen generation system. *J. Beijing Univ. Aeronaut. Astronaut.* **2018**, *44*, 6.
5. Li, X.; Lin, G.; Zeng, Y.; Wu, F. Design of electronic oxygen regulator PID control system based on LabVIEW. *Comput. Meas. Control* **2016**, *3*, 80–83.
6. Jiang, Y.; Sun, Q.; Tan, P.; Chen, Z. Modeling and Simulation of an Electronic Oxygen Regulator Based on All-Coefficient Adaptive Control. *J. Dyn. Syst. Meas. Control* **2016**, *138*, 081010. [CrossRef]
7. Yuxin, J.; Qinglin, S.; Zengqiang, C.; Sanpeng, D. Modeling and simulation of an electronic oxygen regulator based on generalized predictive control algorithm. In Proceedings of the 2015 34th Chinese Control Conference (CCC), Hangzhou, China, 28–30July 2015; IEEE: Piscataway, NJ, USA, 2015; pp. 4067–4072.
8. Jiang, D.; Jin, H.; Sun, B.; Lin, G.; Bu, X.; Zhao, H. Design and experiment research of pressure control cavity on electronic oxygen regulator. In Proceedings of the 2017 8th International Conference on Mechanical and Aerospace Engineering (ICMAE), Prague, Czech Republic, 22–25 July 2017; IEEE: Piscataway, NJ, USA, 2017; pp. 656–660.
9. Wan, Y.; Zhao, J.; Zeng, Y. Design and calculation of structural parameters of aviation oxygen regulator. *J. Beijing Univ. Aeronaut. Astronaut.* **2011**, *37*, 351–354.
10. Zou, N.; Zhang, B.; Wang, D.; Xiao, H.; Liu, X.; Zang, B. Performance parameter adjustment analysis and experimental verification of YTQ-7 oxygen regulator. *Aerosp. Sci. Technol.* **2011**, *5*, 45–48.
11. Yu, X.; Sun, B.; Lin, G.; Wang, H. The application of ATmega128 single chip microcomputer in electronic oxygen regulator. *Microcomput. Appl.* **2009**, *12*, 50–56.
12. Sun, C.; Cai, Y.; Long, H. Research on stepping motor fuzzy control technology application in the aircraft electronic oxygen regulator. *Meas. Control Technol.* **2013**, *32*, 78–81.
13. Jiang, Y.; Sun, Q.; Zhang, X.; Chen, Z. Pressure regulation for oxygen mask based on active disturbance rejection control. *IEEE Trans. Ind. Electron.* **2017**, *64*, 6402–6411. [CrossRef]
14. Morris-ward, M. Modification of Oxygen Regulator Functionality Tester for X-59 Regulator Testing. In *Student Poster Day*; NTRS—NASA Technical Reports Server: Washington, DC, USA, 2020.
15. Pan, R.; Lin, G.; Shi, Z.; Zeng, Y.; Yang, X. Analysis and control optimization of positive pressure fluctuation in electromechanical oxygen regulator. *Chin. J. Aeronaut.* **2021**, *34*, 205–213. [CrossRef]
16. Saha, B.K.; Songjing, L.I.; Xinbei, L.V. Analysis of pressure characteristics under laminar and turbulent flow states inside the pilot stage of a deflection flapper servo-valve: Mathematical modeling with CFD study and experimental validation. *Chin. J. Aeronaut.* **2020**, *33*, 1107–1118. [CrossRef]
17. Pan, R.; Lin, G.; Shi, Z.; Zeng, Y.; Yang, X. The application of disturbance-observer-based control in breath pressure control of aviation electronic oxygen regulator. *Energies* **2021**, *14*, 5189. [CrossRef]
18. Yinhu, L.; Shaoming, L. Single neuron PID control based on dynamic RBF neural network on-line identification. *J. Syst. Simul.* **2006**, *18* (Suppl. 2), 804–807.
19. Zhang, S.T.; Yang, F.; Hao, Q. Research and simulation of single neuron PID controller. *Mech. Eng. Autom.* **2009**, *39*, 69–70.
20. Wang, N.; Tu, J.; Chen, J. Intelligent control using a single adaptive neuron. In Proceedings of the Third Congress of Chinese Association of Automation, Beijing, China; 1991; pp. 173–177.

21. Liu, X.; Xiao, H.; Shi, Q. Evaluation of hypobaric chamber physiological test of YX-11 oxygen supply system. *Chin. J. Aerosp. Med.* **2007**, *18*, 20–25.
22. Xiao, H. Review and Prospect of Physiological Research on Aviation Oxygen Supply Equipment. In Proceedings of the Academic Conference on Environmental Control and Ergonomics of Chinese Aviation Society, Beijing, China; 2000.

energies

MDPI

Article

Research on the Modulation and Control Strategy for a Novel Single-Phase Current Source Inverter

Yi Zhang [1,*], Tao Yang [2,3] and Yiru Miao [3]

1 School of Vehicle Engineering, Chongqing University of Technology, Chongqing 400054, China
2 School of Rail Transit and Aviation Service, Chongqing Industry Polytechnic College, Chongqing 401120, China; justmuch@163.com
3 State Key Laboratory of Power Transmission Equipment & System Security and New Technology, Chongqing University, Chongqing 400044, China; miaoyiru@cqu.edu.cn
* Correspondence: zagyi81@cqut.edu.cn

Abstract: Compared to the voltage source inverter, the current source inverter (CSI) can boost voltage and improve filtering performance. However, the DC side of CSI is not a real current source, and the DC input current comprises a DC power supply and an inductor. In the switching process, the DC-link inductor is charged or discharged and is in an uncontrollable state. This paper proposes a novel CSI topology containing five switching tubes and a modulation strategy based on the hysteresis control strategy of the DC-link current. Due to the conduction and switching loss being positive to the DC-link current, the calculation method for the least reference value of the DC-link current is derived to meet power requirements. By constructing a virtual axis, we then present the control strategy of the output voltage in a two-phase rotating reference frame. Finally, we carry out the simulation and experiment are to validate the proposed topology, modulation, and control strategy.

Keywords: current source inverter; DC-link current; modulation; control strategy

Citation: Zhang, Y.; Yang, T.; Miao, Y. Research on the Modulation and Control Strategy for a Novel Single-Phase Current Source Inverter. *Energies* **2023**, *16*, 6729. https://doi.org/10.3390/en16186729

Academic Editors: Kan Liu and Wei Hu

Received: 3 September 2023
Revised: 17 September 2023
Accepted: 19 September 2023
Published: 20 September 2023

1. Introduction

Inverters can be categorized as voltage source inverters (VSI) [1] and current source inverters (CSI) [2] according to the characteristics of input power supplies. Some unique advantages of CSIs have been discovered [3], such as their boost capacity, inherent short-circuit protection capacity, and AC filtering structure. CSI, therefore, has potential applications [4] in the solar photovoltaic industry, wind energy power generation, motor drive systems, and HVDC transmission systems.

In recent years, various control and modulation strategies for CSI have been proposed [5], such as decoupling methods, low common voltage modulation, and digital vector control strategy. However, the DC-link current is considered a constant value in the above studies in which the charging or discharging process of the DC-link inductor has been ignored, leading to the DC-link current being discontinuous or continuously increasing.

Unlike VSI, the DC-link current of CSI is formed by the DC input voltage DC-link inductor, and constant DC-link current output is a necessary condition for high performance operation of CSI [6]. In [7–11], several novel CSI topologies are proposed to achieve control of the DC-link current. In [7], a three-phase current source rectifier (CSR) is introduced to adjust the DC-link current, but this topology is very complex and only suitable for AC-DC-AC applications. A buck converter is added to the DC side to regulate the DC-link current in [8]. In [6], the buck converter is replaced with a bi-directional converter that can control the current, and the buck operation can also be realized. However, the switching and conduction losses increase as the converter is introduced. In [9], a buck-boost CSI is proposed, which has the advantages of a simple structure and large voltage output range, but the control strategy is difficult to implement. The current-fed quasi-Z-source inverter is proposed in [10]. However, with the addition of many passive components, both efficiency and power density are greatly reduced.

In [11], an additional switching tube and diode are connected in parallel with the DC-side inductor of the three-phase CSI. The DC-side inductor can self-continuate when the switch is turned on. During the zero vector period, the continuous current mode is used to replace the traditional magnetizing mode, which can prevent the increase of the DC-link current. On this basis, a novel single-phase five-switch CSI topology is proposed in this paper. The modulation and control strategies are studied to obtain the constant DC-link current and high-quality AC voltage output. Finally, the simulation and experimental verification are carried out.

2. Topology and Operating Modes of the Proposed Single-Phase CSI

2.1. Analysis of the Proposed Single-Phase CSI's Operating Modes

The topology of the proposed single-phase CSI is shown in Figure 1. The DC side is composed of a DC voltage source and DC-link inductance L_{dc}. The inverter part is composed of switching tubes (S_1, S_2, S_3, S_4) and diodes (D_1, D_2, D_3, D_4). The AC side contains the filter capacitor *C* and the resistive load *R*. The switching tube S_0 and diode D_0 are parallel with L_{dc}.

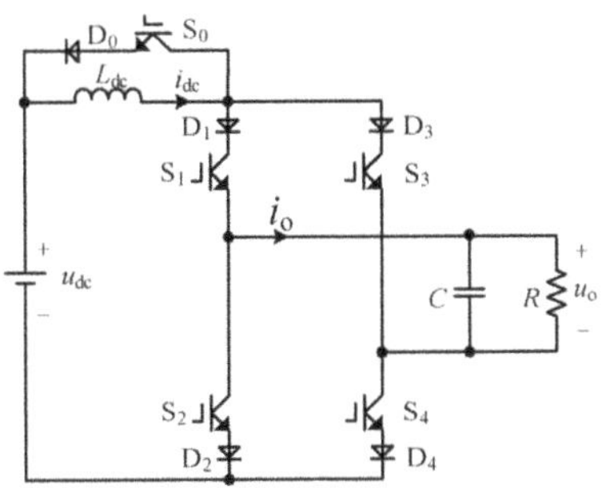

Figure 1. The topology of the proposed single-phase CSI.

There are four operating modes of the proposed single-phase CSI. The corresponding switching states are shown in Table 1.

Table 1. The operating modes, switching state, and switching function.

Operating Modes	Switching State	Switching Function *p* and *q*
magnetizing mode	S_1 and S_2 ON, S_0, S_3, and S_4 OFF	$p = 0, q = 1$
energy-supplying mode I	S_1 and S_4 ON, S_0, S_2, and S_3 OFF	$p = 1, q = 0$
energy-supplying mode II	S_2 and S_3 ON, S_0, S_1, and S_4 OFF	$p = -1, q = 0$
freewheeling model	S_0 ON, S_1, S_2, S_3, and S_4 OFF	$p = 0, q = 0$

(1) Magnetizing mode: S_1 and S_2 are turned on, and the equivalent circuit under magnetizing mode is shown in Figure 2a. At this moment, i_o is equal to 0 A, the u_{dc} is charging to L_{dc}, and *C* provides energy for the load separately.

(2) Energy-supplying mode I: S_1 and S_4 are turned on, and the equivalent circuit under this operating mode is shown in Figure 2b. At this moment, i_o is equal to i_{dc}, and the u_{dc} and L_{dc} provide energy to the AC load together.

(3) Energy-supplying mode II: S_2 and S_3 are turned on, and the equivalent circuit under this operating mode is shown in Figure 2c. Different from energy-supplying mode I, the polarity of the output current is negative; i_o is equal to $-i_{dc}$.

(4) Freewheeling model: Only S_0 is turned on, and the equivalent circuit under the freewheeling model is shown in Figure 2d. The i_{dc} freewheels through S_0, and *C* provides energy for the load separately.

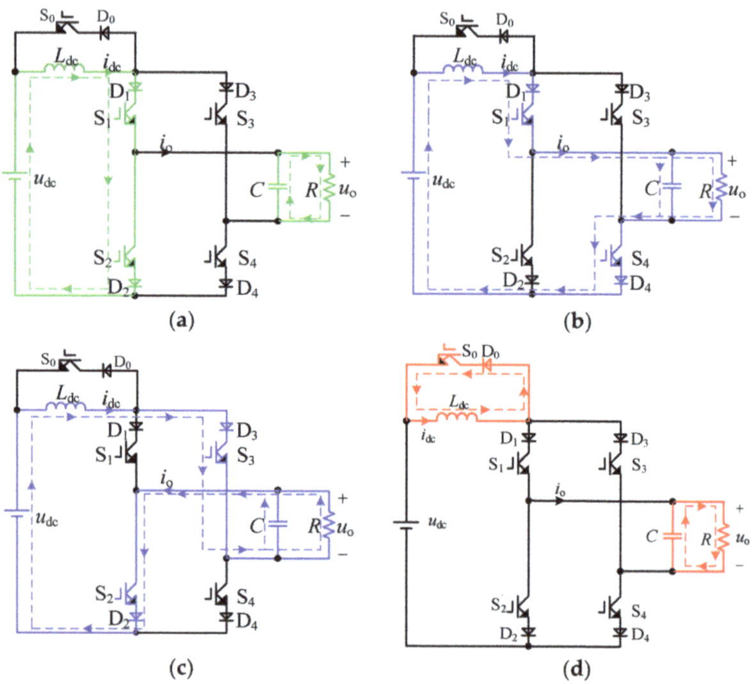

Figure 2. The operating modes of the proposed single-phase CSI: (**a**) magnetizing mode; (**b**) energy-supplying mode I; (**c**) energy-supplying mode II; (**d**) freewheeling model.

Two switching functions (p and q) are defined and shown in Table 1. The p is used to indicate whether the CSI is operating in energy-supplying mode, and the math relationship between i_{dc} and i_o can be expressed as

$$i_o = i_{dc} p \tag{1}$$

The state equation of u_o follows:

$$C\frac{du_o}{dt} = pi_{dc} - \frac{u_o}{R} \tag{2}$$

q is used to indicate whether the CSI is operating in magnetizing mode. The state equation of i_o follows:

$$L_{dc}\frac{di_{dc}}{dt} = qu_{dc} - pu_o \tag{3}$$

Since the conduction current of all switching tubes is i_{dc}, the current stresses of all switching tubes are equal to the DC-link current. The voltage stresses can be derived according to Figure 2, and the voltage stresses of all switching tubes under different operating modes are summarized in Table 2.

Table 2. Voltage stresses of all switching tubes under different operating modes.

Operating Mode	Withstand Voltage S_0	Withstand Voltage S_1	Withstand Voltage S_2	Withstand Voltage S_3	Withstand Voltage S_4
Magnetizing mode	u_{dc}	0	0	$-u_o$	u_o
Energy-supplying mode	$u_{dc} - u_o$	0	$-u_o$	$-u_o$	0
Freewheeling model	0	u_o	$-u_{dc} - u_o$	0	$-u_{dc}$

2.2. Comparison of Different CSI Topologies

The recent literature presents many CSI structures comprising distinct tradeoffs among different topologies, as shown in Figure 3. According to [12,13], a cost function (*CF*) is established in (4) to carry out a fair comparison of different CSI topologies:

$$CF = N_V(N_{US} + 2N_{BS} + N_D + N_C + N_{drv} + N_T)/N_l \quad (4)$$

where N_V is the number of DC voltages, N_{US} is the number of unidirectional switches, N_{BS} is the number of bidirectional switches, N_D is the number of diodes, N_C is the number of capacitors, N_{drv} is the number of individual drivers, N_T is the number of transformers, and N_l is the number of output current levels.

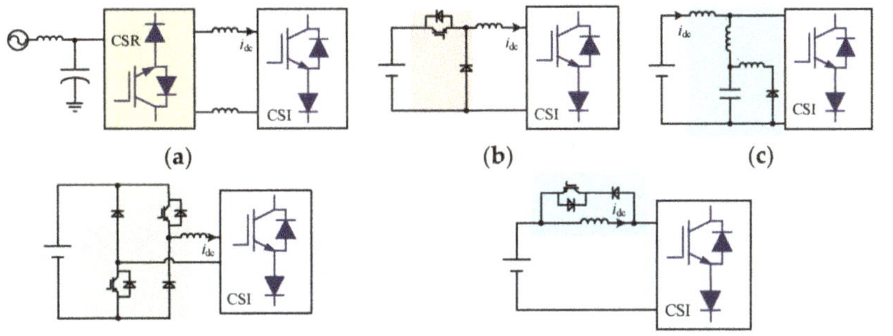

Figure 3. The topologies of different CSI: (**a**) current source rectifier; (**b**) buck converter; (**c**) quasi-Z source network; (**d**) bidirectional DC chopper; (**e**) proposed single-phase CSI.

The comparison result is presented in Table 3. Although the quasi-Z source network receives the lowest *CF*, it contains numerous passive components, and its control system is too complex. The proposed single-phase CSI and the buck converter have the same CF. However, the buck converter contains more switches than the proposed single-phase CSI. The bidirectional DC chopper's *CF* is slightly higher than the proposed one because of the additional two diodes and switching tubes. The current source rectifier rates the worst *CF* because it includes too many components.

Table 3. Comparison of different CSI topologies.

Parameter	Current Source Rectifier	Buck Converter	Quasi-Z Source Network	Bidirectional DC Chopper	Proposed Single-Phase CSI
N_V	1	1	1	1	1
N_{US}	0	0	0	0	0
N_{BS}	8	5	4	6	5
N_D	8	5	5	6	5
N_C	2	1	2	1	1
N_{drv}	8	5	4	6	5
N_T	0	0	1	0	0
N_l	3	3	3	3	3
CF	11.33	7	6.67	8.33	7
Input H Bridge	Yes	No	No	Yes	No
Modulation	PWM	PWM	PWM	PWM	PWM

3. Modulation Strategy Based on DC-Link Current Control

Unlike VSI, switching signals cannot be directly generated by comparing modulated signals with carrier waves [14]. In [15], a logic conversion circuit is adopted to convert the switching signals, which will delay them. In this section, a novel modulation strategy is presented which can be implemented by the DSP without a logic conversion circuit. Three logic signals (p_1, p_2, and p_3) are set, where p_1 represents the comparison result between the modulation wave and carrier wave, as shown in Figure 4, and where d is the duty cycle and is defined as

$$d = \frac{|i_o|}{i_{dc}} \quad (5)$$

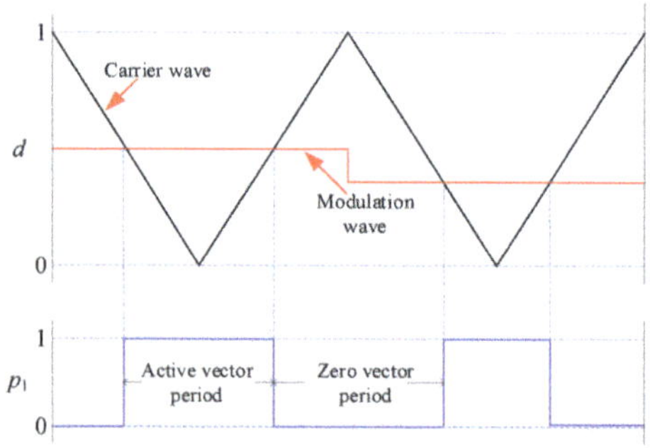

Figure 4. The diagram of p_1, carrier wave, and modulation wave.

$p_1 = 1$ represents the active vector period, and the CSI operates at energy-supplying mode I or II, which is determined by the polarity of i_o. p_2 represents the polarity of i_o. If $i_o > 0$, $p_2 = 1$; otherwise, $p_2 = 0$.

$p_1 = 0$ represents the zero vector period, and the CSI operates on either the magnetizing mode or freewheeling model, which is determined by i_{dc}. If $i_{dc} >$ the reference value i^*_{dc}, the freewheeling model will be selected to prevent i_{dc} from increasing in the zero vector period. Otherwise, the magnetizing mode will be implemented to charge L_{dc} in the zero vector period. p_3 is defined to represent the math relationship between i_{dc} and i^*_{dc}. If $i_{dc} > i^*_{dc}$, $p_3 = 1$; else, $p_3 = 0$. The changing process of i_{dc} and p_3 is described in Figure 5. i_{dc} can be adjusted according to p_3. This approach belongs to a hysteresis control method [16].

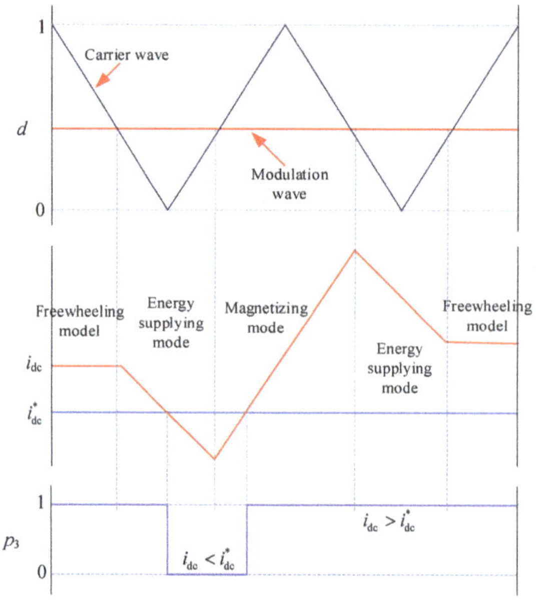

Figure 5. The changing process of i_{dc} and p_3.

Based on the above settings and analysis, the production logic of the five switching signals is shown in Figure 6. The specific logic expression for each switching is described as follows:

$$\begin{cases} p_{S1} = \overline{p}_1 \& \overline{p}_3 | p_2 \\ p_{S2} = \overline{p}_1 \& \overline{p}_3 | \overline{p}_2 \\ p_{S3} = p_1 \& \overline{p}_2 \\ p_{S4} = p_1 \& p_2 \\ p_{S0} = \overline{p}_2 \& p_3 \end{cases} \tag{6}$$

where p_{S1}, p_{S2}, p_{S3}, p_{S4}, and p_{S5} represent the driving logic of switch S_1, S_2, S_3, S_4, and S_5, respectively.

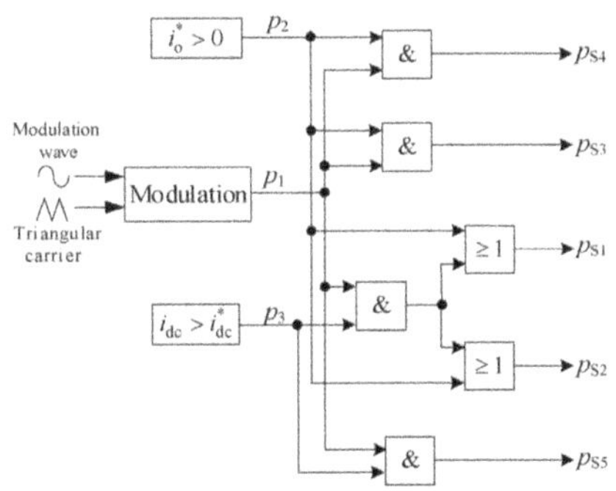

Figure 6. Logical conversion for the proposed single-phase CSI.

4. Calculation for the Optimal Reference of DC-Link Current

In Section 3, a DC-link current hysteresis control is introduced into the modulation scheme. However, the reference of the DC-link current must be determined for the following reasons [17]. If i_{dc}^* is set too small, the DC-link current is unable to provide sufficient power to the AC load. If i_{dc}^* is set too great, the conduction loss, switching loss, and harmonic distortion will increase. Thus, the DC-link current should be reduced to satisfy the current requirements. The calculation method for the optimal reference of the DC-link current is derived in this section.

It is assumed that u_o remains in a constant state in a switching period. Under energy-supplying mode, the reduction Δi_{dc_down} of i_{dc} can be expressed as follows:

$$\Delta i_{dc\_down} = \frac{T_s}{L_{dc}} d(t)[u_o(t) - u_{dc}] \tag{7}$$

The increment of Δi_{dc_up} under magnetizing mode can be expressed as

$$\Delta i_{dc\_up} = \frac{u_{dc} T_s}{L_{dc}} [1 - d(t)] \tag{8}$$

According to (7) and (8), the total reduction Δi_{dc} of i_{dc} in a switching period can be expressed as

$$\Delta i_{dc} = \Delta i_{dc\_down} - \Delta i_{dc\_up} = \frac{T_s}{L_{dc}} [u_o(t) d(t) - u_{dc}] \tag{9}$$

By substituting (5) with (9), Δi_{dc} can be rewritten as

$$\Delta i_{dc} = \frac{T_s}{L_{dc}} \left[\frac{u_o(t) i_o(t)}{i_{dc}(t)} - u_{dc} \right] \tag{10}$$

Ignoring the harmonic component and initial phase, u_o is expressed as

$$u_o(t) = U \sin \omega t \tag{11}$$

where U represents the fundamental amplitude of u_o, and ω represents the fundamental frequency. i_o is expressed as

$$i_o(t) = I \sin(\omega t + \theta) \tag{12}$$

where I represents the amplitude of i_o, and θ represents the initial phase of i_o. Due to the load of the inverter consisting of the resistance and the capacitance paralleling with the load, the following relations are satisfied:

$$\begin{cases} I = U\sqrt{1 + (\omega CR)^2}/R \\ \theta = \arctan(\omega CR) \end{cases} \tag{13}$$

Substituting (11) and (12) with (10), the following can be obtained:

$$\Delta i_{dc} = \frac{T_s}{L}\left[\frac{UI \sin(\omega t) \sin(\omega t + \theta)}{i_{dc}(t)} - u_{dc}\right] \tag{14}$$

Formula (14) is simplified as follows:

$$\Delta i_{dc} = \frac{T_s}{2L}\left\{\frac{UI[\cos\theta - \cos(2\omega t + \theta)]}{i_{dc}(t)} - 2u_{dc}\right\} \tag{15}$$

The maximum reduction Δi_{dcmax} of i_{dc} in a switching period is shown as follows:

$$\Delta i_{dcmax} = \frac{T_s}{2L}\left[\frac{UI(\cos\theta + 1)}{i_{dc}(t)} - 2u_{dc}\right] \tag{16}$$

If $\Delta i_{dcmax} < 0$, i_{dc} can continue increasing in any switching period, which consists of magnetizing mode and energy-supplying mode. Thus, i_{dc} needs to be satisfied as follows:

$$i_{dc} > \frac{UI(\cos\theta + 1)}{2u_{dc}} \tag{17}$$

Substituting (13) with (17), and replacing i_{dc} with i^*_{dc}, Formula (17) can be rewritten as follows:

$$i^*_{dc} > \frac{U^2\left[1+\sqrt{1 + (\omega CR)^2}\right]}{2u_{dc}R} \tag{18}$$

The theoretical waveforms in a fundamental period, including i_{dc}, u_o, and switching signals are presented in Figure 7. Based on the operation mode, switching signals, DC-link current optimal reference, and control strategy, the theoretical waveform is divided into eight stages:

(1) Stage 1 ($t_0 - t_1$): $0 < u_o < u_{dc}$, L_{dc} is charged in energy-supplying mode I; S_1 remains on-state; S_0 is turned on in the zero vector period to prevent i_{dc} from continuously increasing.

(2) Stage 2 ($t_1 - t_2$): $-u_{dc} < u_o < 0$, L_{dc} is charged in energy-supplying mode II; S_3 remains on-state; S_0 is turned on in the zero vector period to prevent i_{dc} from continuously increasing.

(3) Stage 3 ($t_2 - t_3$): $u_o < -u_{dc}$; L_{dc} discharges in energy-supplying mode II due to i_{dc} being still greater than i^*_{dc}; S0 is also turned on in the zero vector period; i_{dc} begins to decrease.

(4) Stage 4 ($t_3 - t_4$): $u_o < -u_{dc}$ and $i_{dc} < i^*_{dc}$; to prevent i_{dc} further decrease, S4 is turned on in the zero vector period, since i^*_{dc} is the optimal reference of the DC-link current; i_{dc} will be clamped near i^*_{dc}.

(5) Stage 5 ($t_4 - t_5$), Stage 6 ($t_5 - t_6$), Stage 7 ($t_6 - t_7$), and Stage 8 ($t_7 - t_8$) are similar to Stage 1 ($t_0 - t_1$), Stage 2 ($t_1 - t_2$), Stage 3 ($t_2 - t_3$), and Stage 4 ($t_3 - t_4$), respectively.

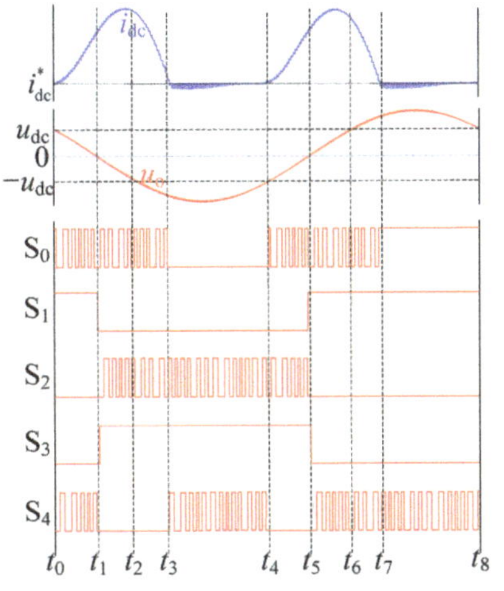

Figure 7. The theoretical waveforms in a fundamental period.

5. Control Strategy of the Output Voltage

Since the PI controller is not suitable for AC models, the math model of the singe phase CSI is established in d-q frame. The fundamental component of u_o is represented as follows:

$$u_o(t) = u_{od}\sin(\omega t) + u_{oq}\cos(\omega t) \tag{19}$$

where u_{od} = U, and u_{oq} = 0. The quadrature virtual component of u_o is introduced, which is shown as follows:

$$u'_o(t) = -U_o\sin(\omega t - \frac{\pi}{2}) \tag{20}$$

The d-q frame components u_{od} and u_{oq} can be obtained as follows:

$$\begin{cases} u_{od}(t) = u_o(t)\sin(\omega t) + u'_o(t)\cos(\omega t) \\ u_{oq}(t) = u_o(t)\cos(\omega t) - u'_o(t)\sin(\omega t) \end{cases} \tag{21}$$

The same representation for i_o is shown as follows:

$$i_o(t) = i_{od}\sin(\omega t) + i_{oq}\cos(\omega t) \tag{22}$$

where i_{od} and i_{oq} are the DC component. Substituting Equations (21) and (22) with (2), it yields the following:

$$\begin{cases} C\frac{du_{od}}{dt} + \frac{u_{od}}{R} = \omega C u_{oq} + i_{od} \\ C\frac{du_{oq}}{dt} + \frac{u_{oq}}{R} = -\omega C u_{od} + i_{oq} \end{cases} \tag{23}$$

u_{od} and u_{oq} are adjusted by the PI controller, and feedback decoupling is adopted to cancel out the coupling terms $\omega C u_{od}$ and $\omega C u_{oq}$. The control strategy of u_o is shown in Figure 8, where u^*_{od} and u^*_{oq} are the reference of u_{od} and u_{oq}, and i^*_{od} and i^*_{oq} are the reference of i_{od} and i_{oq}, respectively. i^*_{od} and i^*_{oq} are expressed as follows:

$$\begin{cases} i^*_{od} = \frac{k_p s + k_i}{s}(u^*_{od} - u_{od}) - \omega C u_{oq} \\ i^*_{oq} = \frac{k_p s + k_i}{s}(u^*_{oq} - u_{oq}) + \omega C u_{od} \end{cases} \tag{24}$$

where k_p and k_i are the proportional coefficient and integral coefficient, respectively.

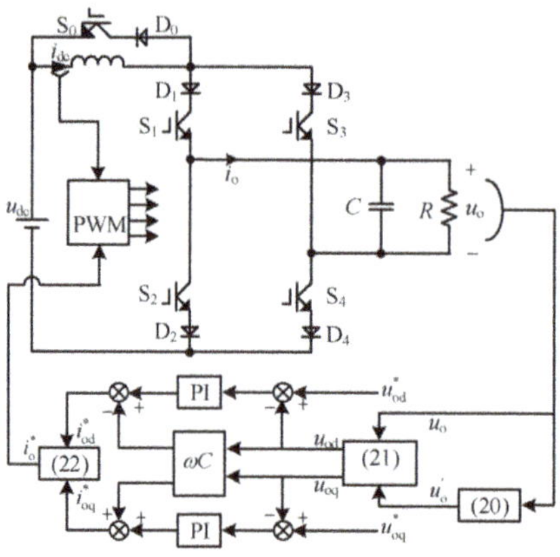

Figure 8. Control strategy diagram of output voltage.

Considering the digital delay, the open-loop transfer function $G_{open}(s)$ can be derived as follows:

$$G_{open}(s) = \frac{k_p s + k_i}{s(1 + T_s s)(1 + RCs)} \tag{25}$$

where T_s is the switching period and is equal to 100 μs. Zero point is set to offset the pole, and the cutoff frequency is set at 1 kHz. k_p and k_i are set as $2000\pi RC$ and 2000π. The Bode plot of $G_{open}(s)$ is shown in Figure 9a, and the Bode plot of the closed-loop transfer function $G_{close}(s)$ is shown in Figure 9b. Good track performance and fast dynamic response can thus be achieved.

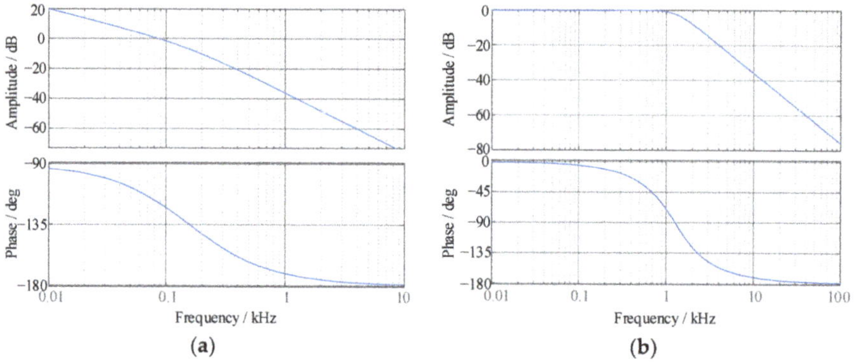

Figure 9. The Bode-plots of the output voltage control: (**a**) $G_{open}(s)$, (**b**) $G_{close}(s)$.

6. Experimental Results

An experimental prototype of a single-phase CSI is established and shown in Figure 10. The corresponding soft simulation is presented in the Supplementary Material. The algorithm of modulation and control are implemented by TMS320F28335, and the IGBT and diode are PM400HSA120 and RM300HA-24F, respectively. u_{dc} is supplied by a DC power supply. The parameters of the passive components are consistent with Table 4.

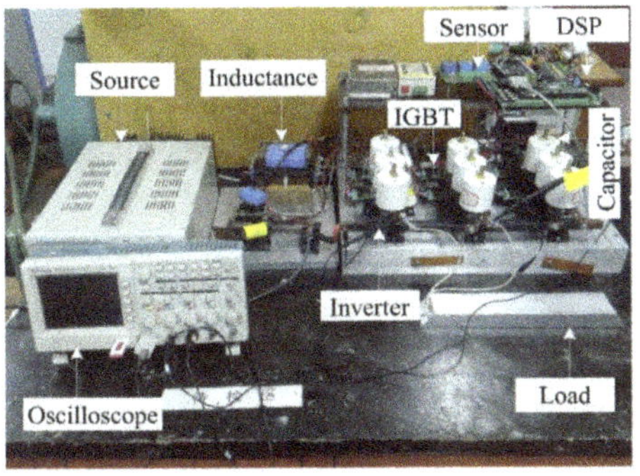

Figure 10. Experimental prototype of proposed single-phase CSI.

Table 4. Parameters of experiment.

Name	Value
DC input voltage (V)	25
AC output voltage (V)	30~150
Load power (W)	$\leq 10^3$
Output frequency (Hz)	$\leq$500 Hz
Switching frequency (Hz)	10 k
Sampling frequency (Hz)	10 k
L_{dc} (mH)	4
C (μF)	265
R (Ω)	25

6.1. Experimental Results of Steady-State

The amplitude and frequency of the reference of u_o are 50 V and 50 Hz. i^*_{dc} is set to 13.5 A. The experimental waveforms of i_{dc} and u_o are shown in Figure 11a, where the amplitude and frequency of u_o are consistent with the reference, and i_{dc} is maintained in the range of 13.5 A to 15 A. Figure 11b shows that u_o's THD is only 0.61%, and its harmonics are limited.

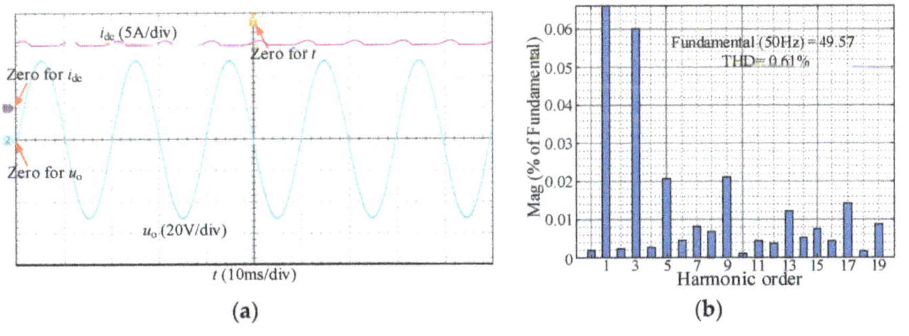

Figure 11. Experimental results under i^*_{dc} = 13.5 A: (**a**) experimental waveforms of i_{dc} and u_o; (**b**) FFT results.

In order to verify that i^*_{dc} = 13.5 A is the optimal reference of the DC-link current, a steady-state experiment is carried out in Figure 12, where i^*_{dc} is set to 11.5 A. A large fluctuation occurs in i_{dc}, which cannot maintain above 11.5 A. In some switching periods, i_{dc} cannot meet the requirement of current for the AC load. Therefore, significant low-order harmonic distortion occurs in u_o, whose THD is 23.88%, and fundamental amplitude is 38.5 V, lower than the reference 50 V.

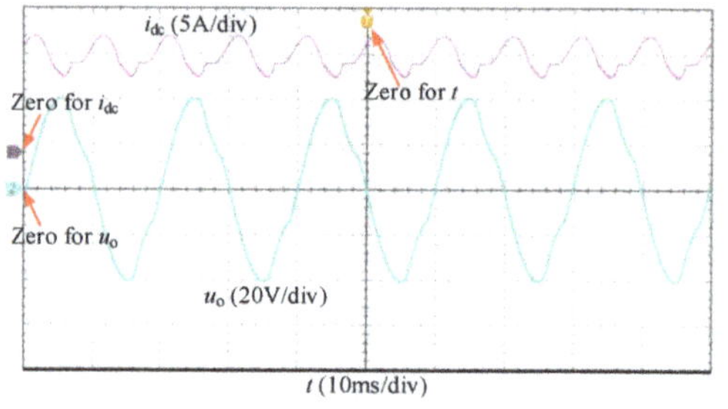

(a)

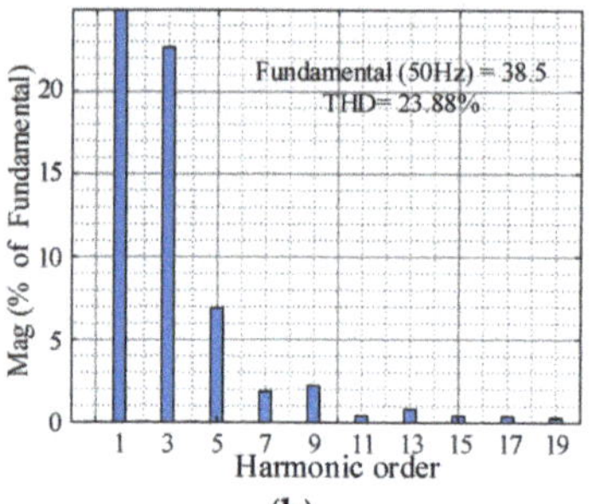

(b)

Figure 12. Experimental results under i_{dc}^{*} = 11.5 A: (**a**) experimental waveforms of i_{dc} and u_o; (**b**) FFT results.

In Figure 13, i_{dc}^{*} is set to 20 A, and i_{dc} maintains above 20 A with small fluctuation. Since i_{dc} is higher than the amplitude of i_o, u_o can track the reference. u_o's THD is 1.38%, greater than the one when i_{dc}^{*} = 13.5 A. Meanwhile, the switching loss and conduction loss of IGBT and diode are positively related to i_{dc}. The above experimental results show that the calculation method for the optimal reference of DC-link current is correct and feasible.

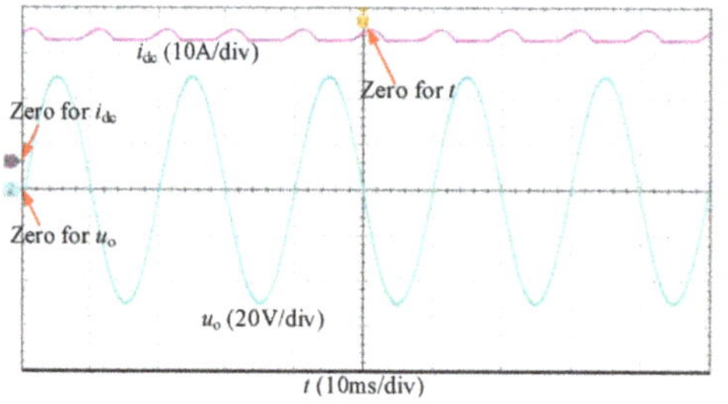

(a)

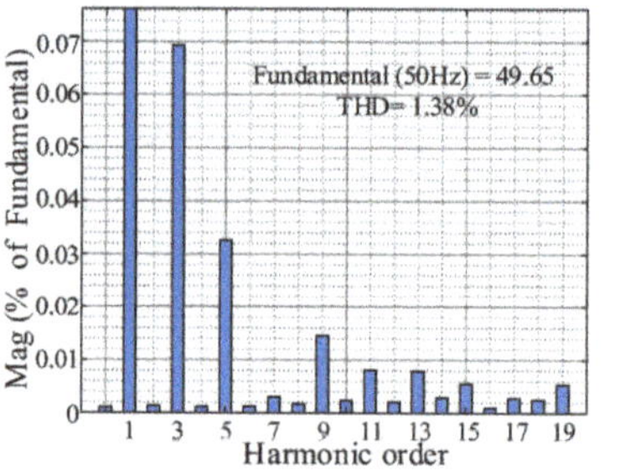

(b)

Figure 13. Experimental results under i_{dc}^{*} = 20 A: (**a**) Experimental waveforms of i_{dc} and u_o; (**b**) FFT results.

6.2. Experimental Results of Dynamic-State

The dynamic performance of the output voltage control strategy will be verified in this section. The frequency remains 50 Hz, and the amplitude of the reference voltage is adjusted from 40 V to 60 V. According to Formula (18), i_{dc}^{*} is set to 9 A and 17.5 A, respectively. The experimental waveforms of i_{dc} and u_o are shown in Figure 14. i_{dc} reaches the steady state again after 2 ms and remains above 17.5 A. In Figure 14, the u_o's amplitude is adjusted to 60 V with smooth changing.

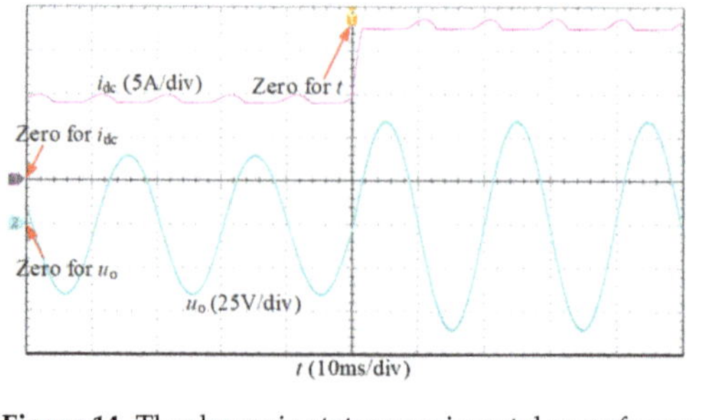

Figure 14. The dynamic-state experimental waveforms under amplitude changing.

Figure 15 shows the experimental waveforms when the amplitude is set to 50 V, and the frequency ω changes from 50 Hz to 100 Hz. According to (18), i_{dc}^{*} is related to ω, so it should be adjusted from 13.5 A to 16.5 A. In Figure 15, i_{dc} and u_o can track the reference quickly. In conclusion, the superior steady-state and dynamic-state performance of the output voltage control strategy can be fully proved, and the DC-link current reference from the proposed calculation method is the minimum value that can meet the power demand.

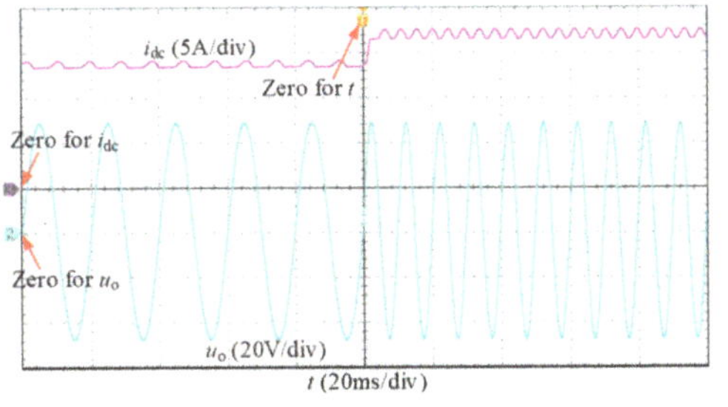

Figure 15. The dynamic-state experimental waveforms under frequency changing.

Finally, the CSI's efficiency has been tested under different load powers from 100 W to 500 W. A curve map of the CSI's efficiency is illustrated in Figure 16. When the load power increases, the power loss in the conduction and switching increases relatively more slowly than the load power, so the CSI's efficiency rises with the increased load power.

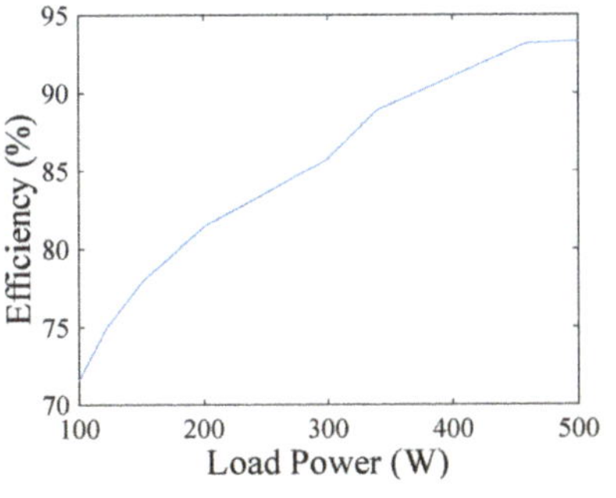

Figure 16. The efficiency map of the proposed single-phase CSI.

7. Conclusions

This paper proposes a novel PWM modulation method to control DC-link current for the improved topology of single-phase CSI; the corresponding operating modes and modulation strategy with DC-link current hysteresis control are also introduced in detail. The relationship between the optimal reference of DC-link current and output voltage is derived. A voltage control strategy based on d-q frame components is discussed. A series of simulations and experiments is set up to demonstrate the feasibility of the proposed method. The conclusions are summarized as follows:

(1) DC-link current can remain in the expected range by adopting the improved PWM modulation, and the number of switching activities is the same as the traditional modulation.

(2) The calculation method of optimal reference for DC-link current can meet the AC side load demand, improving current utilization and reducing loss.

(3) The control of DC-link current is implemented by switching magnetizing mode and freewheeling mode, which is separated from output voltage control. Thus, the DC-side model can be considered a controlled current source.

Supplementary Materials: The following supporting information can be downloaded at: https://www.mdpi.com/article/10.3390/en16186729/s1. Figure S1. Simulation waveforms in steady state: (a) DC-link current; (b) AC output volt-age. Figure S2. Simulation waveforms on the traditional single-phase CSI when the initial DC-link current is 0A: (a) DC-link current; (b) AC output voltage. Figure S3. Simulation waveforms on the traditional single-phase CSI when the initial DC-link current is 16A: (a) DC-link current; (b) AC output voltage.

Author Contributions: Methodology, Y.Z. and T.Y.; Software, T.Y.; Writing—original draft, Y.Z.; Writing—review and editing, Y.M. All authors have read and agreed to the published version of the manuscript.

Funding: This work is supported by the Youth project of science and technology research program of Chongqing Education Commission of China (No. KJQN202001105).

Data Availability Statement: The data presented in this study are available on request from the corresponding author. The data are not publicly available due to the funder's requirement.

Conflicts of Interest: The authors declare no conflict of interest.

Nomenclature

VSI	Voltage Source Inverter
CSI	Current Source Inverter
HVDC	High Voltage Direct Current
L_{dc}	DC-link inductance (mH)
C	AC side's filter capacitor (μF)
R	AC side's resistive load (Ω)
i_{dc}	DC side's current (A)
i_o	Output current on the resistive load R (A)
u_{dc}	DC side's voltage (V)
u_o	Output voltage on the resistive load R (A)
p/q	Switching functions for S_0, S_1, S_2, S_3, and S_4 (-)
d	Duty cycle (%)
i_o^*	Reference of the output current i_o (A)
i_{dc}^*	Reference of the DC side's current i_{dc} (A)
Δi_{dc_down}	Decrement of i_{dc} when discharging (A)
Δi_{dc_up}	Increment of i_{dc} when charging (A)
T_s	Switching period (s)
U	Fundamental amplitude of u_o (V)
I	Fundamental amplitude of i_o (A)
θ	Initial phase of i_o (rad)
ω	Fundamental frequency of i_o (Hz)
Δi_{dcmax}	Maximum reduction of i_{dc} (A)
u_{od}/u_{od}	d-q frame components of u_o
u_{od}^*/u_{oq}^*	Reference of u_{od}/u_{od}
i_{od}/i_{od}	d-q frame components of i_o
i_{od}^*/i_{oq}^*	Reference of i_{od}/i_{od}

References

1. Chakrabarti, A.; Sarkar, K.; Kasari, P.R.; Das, B.; Biswas, S.K. A CB-PWM Technique for Eliminating CMV in Multilevel Multiphase VSI. *IEEE Trans. Ind. Electron.* **2022**, *70*, 8666–8675. [CrossRef]
2. He, J.; Zhang, C.; Lei, M.; Han, J.; Guo, P.; Li, Y. An Active Bypass Pulse Injection-Based Low Switching Frequency PWM Approach for Harmonic Compensation of Current-Source Converters. *IEEE Trans. Power Electron.* **2021**, *36*, 1614–1625. [CrossRef]
3. Rajeev, M.; Agarwal, V. Single phase current source inverter with multiloop control for transformerless grid–PV interface. *IEEE Trans. Ind. Appl.* **2018**, *54*, 2416–2424. [CrossRef]
4. Titus, J.; Harikrishnan, P.; Hatua, K. An SCR-Based CSI-Fed Induction Motor Drive for High Power Medium Voltage Applications. *IEEE Trans. Ind. Electron.* **2020**, *68*, 4657–4666. [CrossRef]
5. Guo, X.-Q.; Sui, S.; Wang, B.; Zhang, W. A Current-Based Approach for Short-Circuit Fault Diagnosis in Closed-Loop Current Source Inverter. *IEEE Trans. Ind. Electron.* **2019**, *67*, 7941–7950. [CrossRef]

6. Miao, Y.; Liao, W.; Huang, S.; Liu, P.; Wu, X.; Song, P.; Li, G. DC-Link Current Minimization Scheme for IM Drive System Fed by Bidirectional DC Chopper-Based CSI. *IEEE Trans. Transp. Electrif.* **2023**, *9*, 2839–2850. [CrossRef]
7. Li, Y.W.; Pande, M.; Zargari, N.R.; Wu, B. DC-link current minimization for high-power current-source motor drives. *IEEE Trans. Power Electron.* **2008**, *24*, 232–240.
8. Geng, Y.; Yang, K.; Lai, Z.; Zheng, P.; Liu, H.; Deng, R. A Novel Low Voltage Ride Through Control Method for Current Source Grid-Connected Photovoltaic Inverters. *IEEE Access* **2019**, *7*, 51735–51748. [CrossRef]
9. Wang, W.; Gao, F.; Yang, Y.; Blaabjerg, F. Operation and modulation of H7 current-source inverter with hybrid SiC and Si semiconductor switches. *IEEE J. Emerg. Sel. Top. Power Electron.* **2017**, *6*, 387–399. [CrossRef]
10. Yang, S.; Peng, F.Z.; Lei, Q.; Inoshita, R.; Qian, Z. Current-fed quasi-Z-source inverter with voltage Buck–Boost and regeneration capability. *IEEE Trans. Ind. Appl.* **2010**, *47*, 882–892. [CrossRef]
11. Guo, X.; Wang, N.; Zhang, J.; Wang, B.; Nguyen, M.-K. A Novel Transformerless Current Source Inverter for Leakage Current Reduction. *IEEE Access* **2019**, *7*, 50681–50690. [CrossRef]
12. Nascimento, A.J.P.; de Menezes, B.F.; de Mesquita, S.J.; Costa, K.R.; Tofoli, F.L.; Daher, S.; Antunes, F.L.M. Bidirectional Isolated Asymmetrical Multilevel Inverter. *IEEE Trans. Circuits Syst. II Express Briefs* **2022**, *70*, 151–155. [CrossRef]
13. Sun, R.; Wang, X.; Ye, Y. Seventeen-Level Inverter Based on Switched-Capacitor and Flying-Capacitor-Fed T-type Unit. *IEEE Access* **2022**, *10*, 33561–33570. [CrossRef]
14. Ding, L.; Li, Y.W. Simultaneous DC current balance and common-mode voltage control with multilevel current source inverters. *IEEE Trans. Power Electron.* **2018**, *33*, 9188–9197. [CrossRef]
15. Guo, Q.; Zhou, C.; Li, S. A Multiple Loops Control Strategy Based on DC Link Voltage of Current Source PWM Rectifiers. *Trans. China Electrotech. Soc.* **2022**, *37*, 2051–2063.
16. Wu, F.; Feng, F.; Luo, L.; Duan, J.; Sun, L. Sampling period online adjusting-based hysteresis current control without band with constant switching frequency. *IEEE Trans. Ind. Electron.* **2014**, *62*, 270–277. [CrossRef]
17. Wang, Z.; Wu, B.; Xu, D.; Cheng, M.; Xu, L. DC-link current ripple mitigation for current-source grid-connected converters under unbalanced grid conditions. *IEEE Trans. Ind. Electron.* **2016**, *63*, 4967–4977. [CrossRef]

applied sciences

MDPI

Article

Clustering Optimization of IPMSM for Electric Vehicles: Considering Inverter Control Strategy

Jiabao Bu, Shangbin Yuan † and Jinhua Du *

The State Key Laboratory of Electrical Insulation and Power Equipment, Xi'an Jiaotong University, Xi'an 710049, China; 3122104026@stu.xjtu.edu.cn (J.B.); yuanshangbin000@163.com (S.Y.)
* Correspondence: jinhuadu@mail.xjtu.edu.cn
† Current address: Power Dispatch and Control Center of Guangxi Power Grid, Nanning 530023, China.

Abstract: The actual performance of driving motors in the electric vehicle (EV) powertrain depends not only on the electromagnetic design of the motor itself but also on the driving condition of the vehicle. The traditional motor optimization method at the rated point is difficult to deal with because of the mismatch between its high-efficiency area and the actual operation area. This paper systematically proposes an optimal design method for driving motors for EVs, considering the driving conditions and control strategy to improve motor efficiency and passengers' riding comfort. It uses cluster analysis to identify representative points and related energy weights to consider motors' comprehensive performance in different driving cycles. Three typical operation conditions are selected to implement the proposed optimization process. In the design process, by using the sensitivity analysis method, the significance of the structural parameters is effectively evaluated. Moreover, the semianalytical efficiency model and torque model of permanent magnet driving motors based on finite element analysis results are deduced to consider the influence of magnetic saturation, space harmonics, and cross-coupling between d-axis and q-axis magnetic fields. Based on the driving system demands of an A0 class pure EV, the whole optimization design is divided into four steps and three scales, including the motor scale, control scale, and system scale. By using the multi-objective optimization method, Pareto optimality of motor efficiency and torque ripple is achieved under the city driving cycle and highway driving cycle. Compared to the optimization only at the rated condition, the proportion of motor sweet region increased about 1.25 times and 3.5 times by the proposed system-scale optimization under two driving cycles, respectively. Finally, the effectiveness of the proposed optimization method is verified by the prototype experiments.

Keywords: multi-objective optimization; control strategy; driving cycle; electric vehicle

Citation: Bu, J.; Yuan, S.; Du, J. Clustering Optimization of IPMSM for Electric Vehicles: Considering Inverter Control Strategy. *Appl. Sci.* **2023**, *13*, 10792. https://doi.org/10.3390/app131910792

Academic Editors: Kan Liu and Wei Hu

Received: 15 August 2023
Revised: 25 September 2023
Accepted: 26 September 2023
Published: 28 September 2023

1. Introduction

With the deepening of vehicle electrification in the global market, the driving motor as the main power of the electrical vehicle (EV) drive system holds promising growth potential and attracts extensive attention [1,2]. At present, more than 80% of driving motors in the electrical drive system of pure and hybrid EVs on the market adopt permanent magnet synchronous motors (PMSM) for wider speed regulation range, higher efficiency, and power density [3–5], such as Toyota Prius, Chevrolet Bolt, BMW i3, BYD E6, and so on. To obtain a better driving experience and a longer driving range, it is of great significance to improve PMSM performances through optimization design technologies [6,7].

Generally, the optimization design of PMSM for an EV can be based on a single operation condition, such as the rated operation point of the torque-speed curve, which could ensure satisfactory working performance at this specific point. Yet, once the condition changes, the PMSM performance will deteriorate and even result in undesired reliability in the power train. The operating conditions of EVs are very complex and involve frequent starting and braking, frequent accelerating and decelerating, cruising, and climbing.

Therefore, the optimized PMSM at the rated operation point could not guarantee the whole operation requirements within the considered driving cycle of an EV, which puts forward inevitable demands for the PMSM design considering the driving cycle [8–11]. Different from the PMSM that operated at the rated operation point, the actual performance of the PMSM at each operating point in different driving cycles is closely related to the control strategy and control parameters. To achieve high system-level performance, the perfect cooperation of the motor and its control systems, considering the driving cycle, must be investigated synchronously.

Although both motor and control are important to the system's performance, not much work has been reported in the literature. The design and optimization are mostly at the component level of motors [10,11], which can be classified into two types of single-objective and multi-objective optimization. For single-objective optimization, any one of efficiency, torque ripple, cost, or weight can be used as the optimization objective. In [12,13], by using the finite element method (FEM)-based model and the exact subdomain model with intelligent optimization algorithms, the torque ripple and cogging torque are respectively obtained minimally in the PMSM. For multi-objective optimization, two or more design objectives are selected simultaneously [14,15]. Efficiency and cogging torque are selected as the two design objectives for optimization of a PMSM in [16], where two-stage design optimization using the Taguchi method and response surface model are used to identify the three design variables. In [17], the multi-objective genetic algorithm is adopted to achieve low cost and low torque ripple for a PMSM optimization, and the resulting motor performances meet all the given design demands. The above optimization on the component level of motors is generally conducted under the rated operation point.

When component-level-based optimization involves more operation points in the driving cycle, the simple way is to repeat the performance calculation process in rated point optimization at all operation points and then use the optimization algorithm to select the optimal scheme [18,19]. However, the optimizations with a precise model that includes all operating points will be very time-consuming, with a significant increase in computational cost and complexity. There are two main solutions to this problem. One is using approximate models for as many operating points as possible to obtain the global optimal design. The approximate models for the former solution could be the Kriging model, equivalent circuit model, and response surface model [20–22], which are generally used to replace the finite element precise model and reduce the computer cost. However, the inclusion of more points will still make the calculation more complicated. The other solution is choosing representative points (RPs) to represent the driving cycle and the application torque-speed profile [23,24]. The accuracy of this solution depends on the number and values of the RPs and the corresponding current calculation under different RPs. At present, motor optimizations using RPs to consider driving cycles mainly use the ideal current value, which ignores the influence of controllers [25,26]. However, individually optimizing motor structure parameters cannot ensure optimal performance for the entire drive system, especially when the difference in the impact of the PWM carrier harmonic on the loss is significant under different control strategies and control parameters [27,28].

On the other hand, for the control part, although there is much literature related to control strategies focusing on torque ripple reduction or motor efficiency improvements, such as advanced angle field weakening control, loss minimization direct torque control, and robust adaptive current control with disturbance observer [29–31], they usually consider the influence of harmonics or control parameters on the performance of the motor after the motor geometry structure is determined. It means that the proposed control strategies are generally designed and optimized at the control level and have not been combined with the motor design. As previously discussed, it can hardly achieve optimal system performance with this component-level-based optimization based only on motor or control. Therefore, to meet the challenging requirements and satisfy the wide operating range of EVs, it is necessary to optimize the driving motor, considering the inverter control strategy throughout the whole driving cycle in the product design phase.

To deal with the aforementioned issue, this paper presents an effort to develop a system-scale multi-objective optimization procedure for driving motors based on the clustering technique under different operating conditions. And a full-speed domain vector controller with maximum torque per current (MTPA) and flux weakening control strategies [32–34] is adopted to meet the requirements of EVs. The main contributions of this paper are shown as follows:

- The K-means clustering analysis is proposed as a preprocessing step to obtain insight into the distribution of operation points in two typical driving cycles, and the RPs of the torque-speed profile with their energy weights are identified.
- The semianalytical efficiency model and torque model of an interior PMSM (IPMSM) based on FEM results are deduced to consider the influence of magnetic saturation, space harmonics, and cross coupling between d-axis and q-axis magnetic fields.
- An optimal method to tune the motor and control parameters based on the multi-objective particle swarm optimization (MOPSO) algorithm is proposed, where the control model is added to consider the performance of IPMSMs under different control strategies in the whole driving cycle.

This paper is organized as follows: Section 2 describes the clustering optimization design procedure for the IPMSM driving motor. Section 3 presents the motor-scale, control-scale, and system-scale optimization models. Based on these models, a comparative study has been carried out under three working condition requirements: one is under the rated operation point, and the other two are under driving cycles. And the optimization results are discussed and verified by experiments in Section 4, followed by the Section 5.

2. Clustering Optimization Design Procedure

The target driving system that powers an A0 class pure EV [34] consists of a 48-slot 8-pole IPMSM and a full-speed domain vector controller with MTPA and flux weakening control strategies, whose design requirements are shown in Table 1.

Table 1. Specification of the investigated EV.

Parameter	Value	Parameter	Value
Vehicle mass	1100 kg	Radius of wheels	0.308 m
Maximum gradeability	20%	Maximum climbing speed	20 km/h
Motor rated speed	3000 rpm	Motor maximum speed	8000 rpm
Acceleration time	<10 s for 0~50 km/h, <15 s for 50~100 km/h		

2.1. Design Framework

The overall optimization procedure is represented in Figure 1, which includes the following four steps.

Step 1: Analyze the driving conditions of EVs and make a preliminary design for the motor. The K-means clustering algorithm is used to identify the RPs of the motor for evaluating the performance of the motor in the whole driving cycle of EVs.

Step 2: Establish the semianalytical models and evaluate the steady performance of the motor on the motor scale. The optimal parameters of the motor are selected by sensitivity analysis.

Step 3: Evaluate the dynamic performance of the motor on the control scale with respect to the nonlinear factors given by the motor characteristic parameters. The full-speed domain control strategy is considered to meet the requirements of low-speed starting and high-speed cruising for the proposed IPMSM driving motor.

Step 4: Performance evaluation of the system scale based on the fitness function and obtaining the optimal design with MOPSO. When the result satisfies the iterative stop condition, the Pareto solution is output. Otherwise, the motor parameters are updated based on the MOPSO and then returned to Step 2.

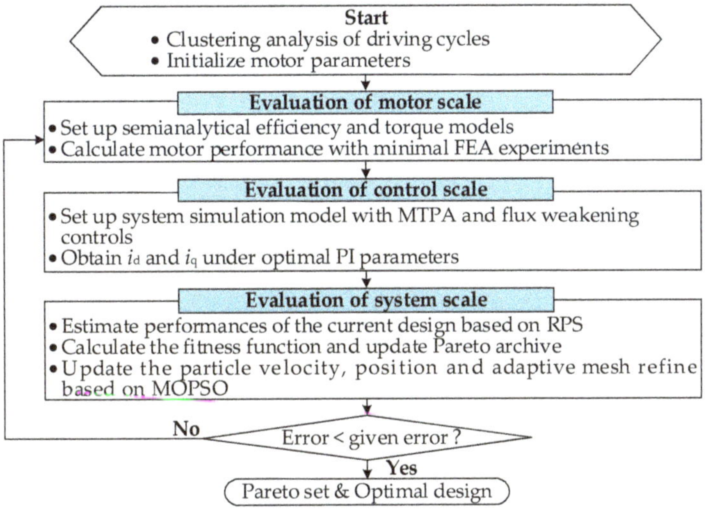

Figure 1. Framework of the proposed system-scale optimization for driving motors targeting driving cycles.

2.2. Clustering Analysis of Driving Cycles

There are three case studies proposed in this paper, which are optimizations at the rated operation point, under the urban dynamometer driving schedule (UDDS), and the highway fuel economy test (HWFET). To improve the optimization efficiency, the K-means clustering method is used to analyze driving cycles [24,26], which splits the torque-speed profile obtained from the driving cycle data into different groups considering the position and density of the operating points on the plane. Euclidean distance is used to determine the similarity between samples; therefore, the cluster S_i and its centroid l_i can be determined iteratively as follows:

$$S_i^{(t)} = \left\{ z_j : \left\| z_j - l_i^{(t)} \right\|^2 \le \left\| z_j - l_i^{(t)} \right\|^2 \forall i, i = 1,2,3\ldots k \right\} \tag{1}$$

$$l_i^{(t+1)} = \frac{1}{\left| S_i^t \right|} \sum_{x_j \in S_i^t} x_j \tag{2}$$

The iteration repeats until the centroid of each cluster no longer changes. The cluster number k is determined based on the sum of the distances of the load points to their corresponding cluster centroid, and the cluster centroid is selected as the RP of motor optimization. In this way, the two specific driving cycles shown in Figure 2a,c can be partitioned into several clusters, as shown in Figure 2b,d. Moreover, the RPs and their energy weights are listed in Table 2. It can be found that the motor operating conditions are mostly located in the low-speed or low-torque region in UDDS, and the motor runs less in the conditions of high speed or heavy load. Meanwhile, the motor operating conditions are mostly located in the high-speed and low-torque regions of HFET. Therefore, the clusters and the number of RPs determined by the proposed clustering analysis are different for the two driving cycles, which will be used for the subsequent optimization design.

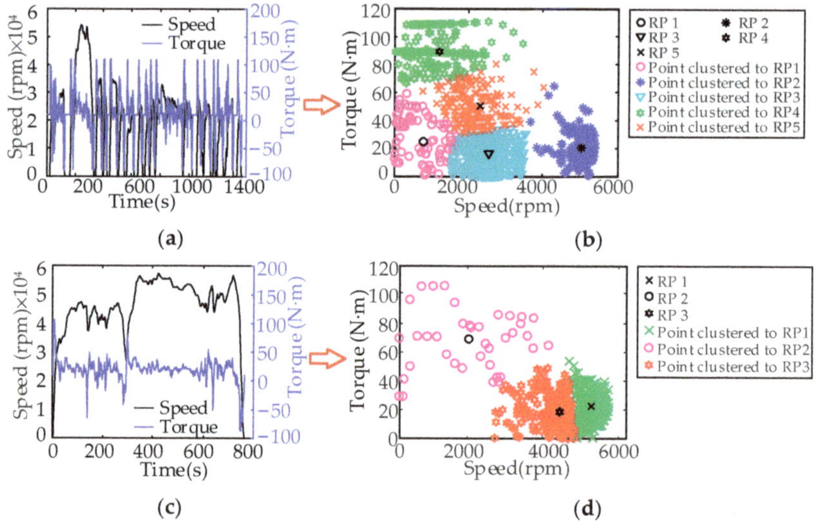

Figure 2. Driving cycle clustering results. (**a**) Speed curve of UDDS. (**b**) Torque-speed curve of UDDS. (**c**) Speed curve of HWFET. (**d**) Torque-speed curve of HWFET.

Table 2. Specification of RPs in two driving cycles.

	RPs	Speed (rpm)	Torque (N·m)	w (%)
UDDS	1	738	27	7.8
	2	4934	21.66	9.9
	3	2546	14	46.26
	4	1126	91	18.63
	5	2157	50	17.37
HWFET	1	5308	21	47
	2	2051	68	5
	3	4269	20	48

2.3. Optimization Algorithm and Fitness Function

The multi-objective PSO algorithm is utilized to obtain the optimal design, which includes adaptive mesh generation based on the roulette algorithm and global best updating as presented in [34].

In order to improve the vehicle range and driving comfort, two objectives are considered: maximize the driving cycle efficiency and minimize the torque ripple. Efficiency is the most important target for EV motors, which has a great influence on vehicle range and comfort, especially considering the current constraints of battery technology faced by vehicles. Maximum driving cycle efficiency can ensure the maximum efficiency of the drive system in the driving cycle. Likewise, the torque ripple is another key factor influencing the performance of the EV motor, which can lead to electromagnetic noise and vibration, affect vehicle comfort, and shorten the service life of the power train. Accordingly, the design objectives could be written as

$$\min f_1(x) = 1 - \frac{\sum_{j=1}^{m}(w_j T_j \Omega_j)}{\sum_{j=1}^{m}(w_j(T_j\Omega_j + P_{\text{loss},j}))} \times 100\%, \quad (3)$$

$$\min f_2(x) = \sum_{j=1}^{m}(\frac{T_{rj}w_j}{T_j}) \times 100\%, \quad (4)$$

where x is the optimization variable, the subscript j represents the RP j, and m is the total RP number.

Furthermore, the fitness function can be defined as

$$\begin{array}{c} \min F(x) = \min(f_1(x), f_2(x)) \\ s.t.\ g_1(x) = 30000 - P_{\text{out}}|_{n_{\max}} \leq 0 \\ g_2(x) = 2 - {}^{n_{\max}}/{}_{n_{\text{N}}} \leq 0 \\ g_3(x) = k_{\text{s}} - 0.8 \leq 0 \end{array}. \tag{5}$$

3. System-Scale Optimization of Drive Systems

For system-scale optimization, it is crucial to investigate the perfect cooperation of the motor and its controller synchronously. A design example with the semianalytical efficiency model and torque model of IPMSMs and a full-speed domain vector controller is investigated in this section.

3.1. Optimization Model for Motor Level

The investigated motor is parameterized as shown in Figure 3, where five independent design variables x_1, x_2, x_3, x_4, and x_5 are selected and are listed in Table 3. They are the air gap, tooth width, slot width, permanent magnet width, and yoke thickness, respectively. Therefore, $x = [x_1, x_2, x_3, x_4]^T$. It should be noted that the outer and inner diameters and the stack length are fixed [34] throughout the optimization process to meet the maximum installation space shown in Table 3.

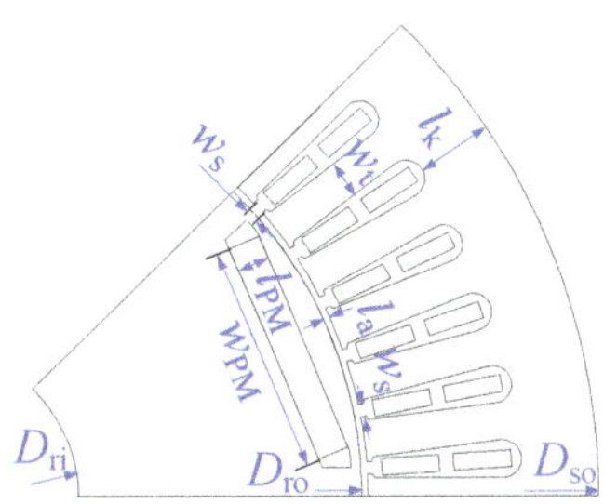

Figure 3. Parameterized model of the investigated IPMSM.

Table 3. Design parameters of the investigated IPMSM.

Parameter	Value	Parameter	Value	Parameter	Value
D_{so} (mm)	210	l_{e} (mm)	153	w_{PM} (mm)	x_4
l_{PM} (mm)	5	l_{a} (mm)	x_1	l_k (mm)	x_5
D_{ri} (mm)	48	w_{t} (mm)	x_2	α	x_6
D_{ro} (mm)	136	w_{s} (mm)	x_3	β	x_7

3.1.1. Semianalytical Torque Model and Efficacy Model of IPMSMs

For each RP with a specific torque and speed, motor performance is evaluated by semianalytical models based on FEA results to balance accuracy and computation cost. Considering the saturation, cross coupling, and spatial harmonics of IPMSMs, the torque T is derived through the co-energy method [26] as follows:

$$\begin{aligned} T &= \frac{\partial W'_{\text{m}}(i_{\text{d}}, i_{\text{q}}, \theta)}{\partial \theta} = i_{\text{s}} \frac{\partial \lambda(i_{\text{d}}, i_{\text{q}}, \theta)}{\partial \theta} - \frac{\partial W_{\text{m}}(i_{\text{d}}, i_{\text{q}}, \theta)}{\partial \theta} \\ &= \underbrace{\tfrac{3p}{2}(\lambda_{\text{d}} i_{\text{q}} - \lambda_{\text{q}} i_{\text{d}})}_{T_1} + \underbrace{\tfrac{3}{2}\left(i_{\text{q}} \tfrac{\partial \lambda_{\text{d}}}{\partial \theta} + i_{\text{d}} \tfrac{\partial \lambda_{\text{q}}}{\partial \theta}\right)}_{T_2} - \underbrace{\tfrac{\partial W_{\text{m}}(i_{\text{d}}, i_{\text{q}}, \theta)}{\partial \theta}}_{T_3} \end{aligned}, \tag{6}$$

where the subscribed d and q represent variables under the d-axis and q-axis.

It can be noted that there are three torque components in (6), which take torque ripple components T_2 and T_3 into consideration compared with the traditional torque equation, which only considers the average torque T_1.

Moreover, the magnetic density inside the driving motor core is usually non-sinusoidal and rich in harmonics. Additionally, the two magnetization modes, alternating and rotating, will make the variation trends of the magnetic density trajectory completely different. In order to consider the effects of alternating magnetization, rotating magnetization, and magnetic field harmonics, the iron losses P_{Fe} of each RP are calculated by the magnetic density orthogonal decomposition model [21] as follows:

$$P_{\mathrm{Fe}} = K_{\mathrm{y}}(fh)\sum_{h=1}^{h_{\mathrm{m}}}\left(B_{rh}^2 + B_{\mathrm{t}h}^2\right) + K_{\mathrm{c}}(fh)^2\sum_{h=1}^{h_{\mathrm{m}}}\left(B_{rh}^2 + B_{\mathrm{t}h}^2\right) + K_{\mathrm{e}}(fh)^{1.5}\sum_{h=1}^{h_{\mathrm{m}}}\left(B_{rh}^2 + B_{\mathrm{t}h}^2\right). \quad (7)$$

where h, or the subscript h, represents the number of magnetic density harmonics, and h_{m} is the total harmonic number.

The λ_{d}, λ_{q}, B_{r}, and B_{t} in analytical expressions (6) and (7) are calculated using computationally efficient FEA [26], which is the so-called semianalytical model. According to the magnetic and electric symmetry and periodicity of the considered IPMSM, when n position samples are selected and flux linkage of winding and flux density of iron core are calculated by n magnetostatics FEA of 1/8 model, there will be another $5n$ samples reconstructing by post-processing techniques, which is much more time-saving than FEA. Figure 4 shows the reconstructed results of λ_{d}, λ_{q}, B_{r}, and B_{t}, which are used to set up the efficiency model and torque model of IPMSMs. Figure 4a,b presents the variation of λ_{d} and λ_{q} with stator phase current amplitude I_{s} and rotor angle θ, where the rainbow color maps the flux linkage value. Figure 4c,d present the variation of B_{r} and B_{t} with stator current and rotor angle in the stator teeth, where the rainbow color maps the flux density value.

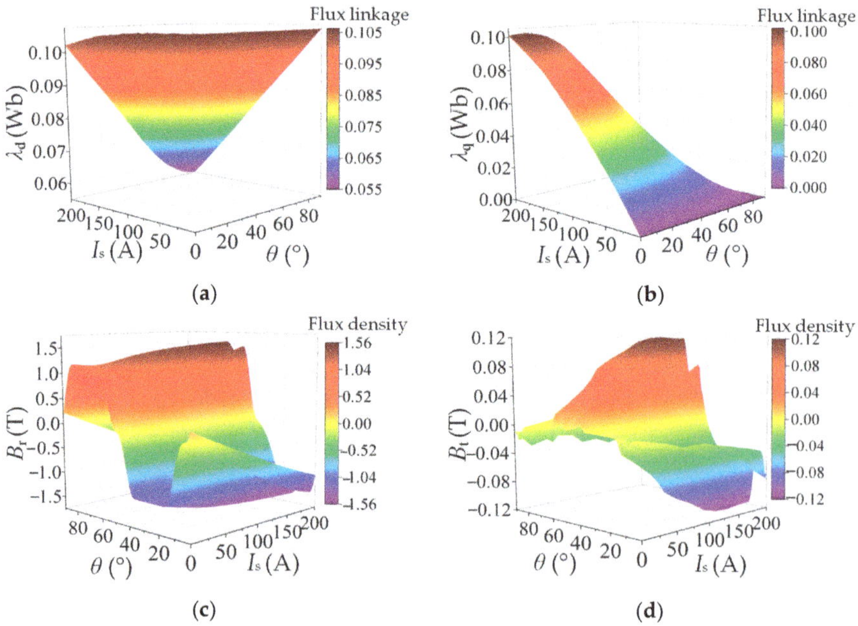

Figure 4. Computationally efficient FEA results for semianalytical models. (**a**) λ_{d}. (**b**) λ_{q}. (**c**) B_{r} in stator teeth. (**d**) B_{t} in stator teeth.

3.1.2. Sensitivity Analysis of Optimization Parameters

In this paper, the Taguchi method [35] is adopted to analyze the parameter sensitivity to save computation costs for the following global optimization. The orthogonal table selects four horizontal values for each factor as shown in Table 4, which include row number 16, factor number 5, and level number 4, and can be expressed as $L_{16}(4^5)$. Considering the geometric constraints of the motor structure, the range of values for each parameter above is specified.

Table 4. Orthogonal table of motor factors.

Number	x_1	x_2	x_3	x_4	x_5
1	0.93	5.59	1.87	32.4	11.2
2	1.2	5.59	2.13	35.87	9.6
3	1.07	5.83	2.13	34.13	14.4
4	1.07	5.83	2.13	32.4	12.8
5	0.8	5.36	1.6	32.4	9.6
6	1.2	5.36	1.87	37.6	12.8
7	0.8	5.83	1.87	35.87	14.4
8	1.07	5.36	2.4	35.87	11.2
9	1.07	6.06	1.87	34.13	9.6
10	1.07	5.59	1.6	37.6	14.4
11	1.2	5.83	1.6	34.13	11.2
12	0.8	5.59	2.4	34.13	12.8
13	0.8	6.06	2.13	37.6	11.2
14	0.93	5.83	2.4	37.6	9.6
15	1.2	6.06	2.4	32.4	14.4
16	0.93	6.06	1.87	35.87	12.8

The sensitivity analysis method [35] is considered an effective approach to analyzing the significance of design variables. And the sensitivity calculation model $S(O)$ in this paper can be expressed as

$$S(O) = \frac{\sum_{i=1}^{d} (p_{O(i)} - p_t)^2}{d}. \tag{8}$$

Based on (8), the effects of the selected five variables on the optimization objectives can be evaluated effectively. Figure 5 shows the sensitivity results. It can be seen that the factors x_4 and x_5 have great impacts on efficiency, and the factors x_3 and x_4 have great influences on torque ripple. Consequently, x_3, x_4, and x_5 are comprehensively considered as optimization variables for the following optimization design.

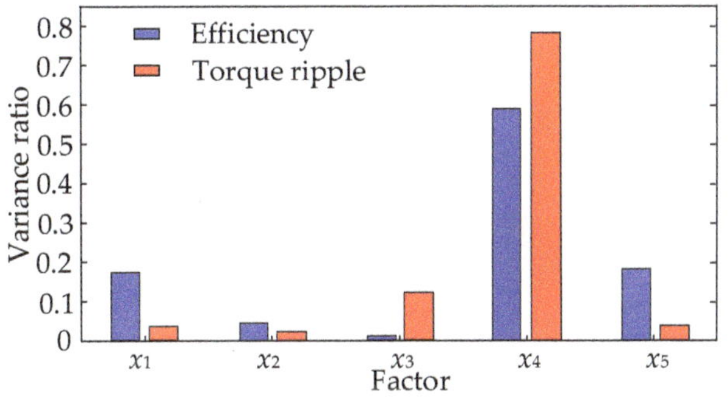

Figure 5. Variance ratio of motor parameters.

3.2. Optimization Model for Control Level

The response current of the motor under different operating conditions is related to the control strategy and the limits of the inverter bus voltage. A full-speed domain vector controller is adapted to meet the requirements of low-speed starting and high-speed cruising for the proposed IPMSM driving motor. In order to optimize the performance of the motor under different operating conditions, MTPA and flux-weakening control are used in this paper to meet the requirements of EVs. The control block is shown in Figure 6.

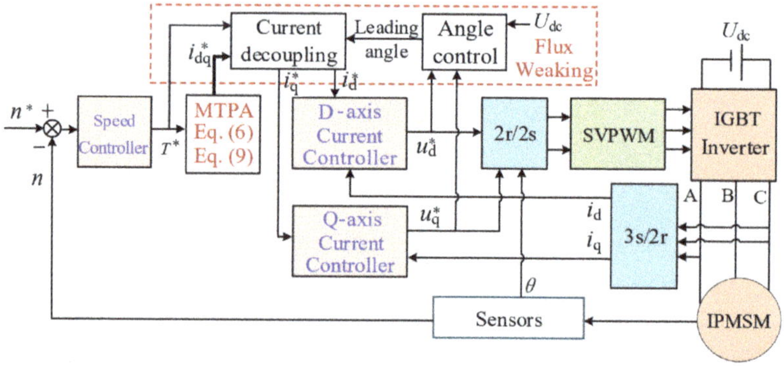

Figure 6. Control block diagram of the full-speed domain vector controller.

When the load is given and the motor terminal voltage does not reach the limit with the speed below base speed, the IPMSM adopts the MTPA control strategy and runs at the MTPA trajectory. As shown in Figure 6, in this MTPA mode, by comparing the real speed n with the speed demand n^*, the torque demand T^* could be obtained from the speed controller and input into the MTPA control part. And then, the current demand $i_{dq}{}^*$ can be solved from torque Equations (6) and (9). If the motor operates in MTPA mode, d-axis and q-axis current demand $i_d{}^*$, $i_q{}^*$ are decoupled from $i_{dq}{}^*$ and sent directly to the d-axis current controller and the q-axis current controller, respectively. Subsequently, the voltage demands $u_d{}^*$, $u_q{}^*$ are obtained, and through the 2r/2s transformation and SVPWM module, they are finally turned into the switching signal to control the IGBT inverter.

$$\frac{\partial(T/i_s)}{\partial i_d} = \frac{\partial(T/i_s)}{\partial i_q} = 0. \tag{9}$$

When the speed demand is higher, the motor provides more torque to accelerate the rotor. At the early stage of speed regulation, the motor does not reach the base speed. However, as the motor voltage reaches its limit shown in (10), the motor speed approaches the base speed, and the speed loop saturates with the maximum given current. Therefore, the flux-weakening control strategy is adopted to redistribute the current in the dq axis through the leading angle so that the speed loop exits the saturation state and the motor torque meets the load. As shown in Figure 6, in this flux weakening mode, the leading angle δ is calculated from the angle controller by comparing the bus voltage U_{dc} with the motor voltage. And the d-axis and q-axis current demand $i_d{}^*$, $i_q{}^*$ are redistributed by the current decoupling module according to δ as (11). It can be found that the flux-weakening control can increase the motor torque and expand the operating range in comparison with the MTPA control at the same motor speed.

$$u_s^2 = p\Omega\sqrt{(L_q i_q)^2 + (L_d i_d + \lambda_{pm})} \leq u_{lim}^2, \tag{10}$$

where the motor stator resistance is neglected, and u_{lim} is the limited value of motor phase voltage.

$$\begin{cases} i_{\text{d}}^{*} = i_{\text{s}} \cos(\delta + \Delta\delta) \\ i_{\text{q}}^{*} = i_{\text{s}} \sin(\delta + \Delta\delta) \end{cases}. \quad (11)$$

As described above, the speed demand is converted into torque demand by the speed controller, and the current demand is converted into voltage demand by two current controllers. In the implementation, the three PI controllers shown in Figure 6 have the form

$$\begin{cases} K_{\text{sp}} = \frac{\alpha+1}{\alpha} \frac{\pi J}{360 T_{\text{s}} p \lambda_{\text{pm}}}; K_{\text{si}} = \frac{1}{\alpha} \frac{K_{\text{sp}}}{4T_{\text{s}}} \\ K_{\text{ip\_d}} = \beta L_{\text{d}}; K_{\text{ip\_q}} = \beta L_{\text{q}}; K_{\text{ii\_d}} = K_{\text{ii\_q}} = \beta R_{\text{s}} \end{cases}. \quad (12)$$

In the whole optimization process, motor parameters and operating points change frequently, which means the fixed PI controller based on classical control theory could not satisfy the optimization demand. Therefore, to ensure that the performance evaluation of each motor scheme is carried out under the same optimal control level, two parameters α and β are optimized in the control level with the minimum control error as follows:

$$\min f(\alpha, \beta) = \int_0^{\infty} t|e(t)|\text{d}t. \quad (13)$$

where t is the calculation time, and $e(t)$ is the instantaneous response error [36].

4. Discussion and Optimization Results

4.1. Optimization Results

According to the optimization procedure shown in Figure 1, after determining the value of representative currents of RPs under different operating conditions, the system's performance could be evaluated based on the fitness function, and MOPSO is used to obtain the optimal design. To illustrate the optimistic effect, the optimized motor scheme designed by the proposed optimization method will be compared with that designed by the single operation point optimization method in this section. The results of multi-objective optimization are shown in Figure 7. Figure 7a–c shows the Pareto solution set optimized by the rated operation point optimization method and the proposed optimization method under UDDS and HWFET, respectively. The black curves in these three figures are the Pareto optimal fronts for each optimization design. Each solution in the Pareto front solution set is optimal. Hence, the optimal motor scheme could be selected from this Pareto front solution set by the decision maker. Based on the VIKOR model [37], which is a multi-attribute decision-making method, the optimal solution could be selected according to (14).

$$\max f(x) = v_1 \frac{\eta - \eta_{\min}}{\eta_{\max} - \eta_{\min}} + v_2 \frac{T_{\text{r max}} - T_{\text{r}}}{T_{\text{r max}} - T_{\text{r min}}}, \quad (14)$$

where η and T_{r} are the efficiency and torque ripple corresponding to the current scheme, and the subscript max and min represent the maximum and minimum values of the corresponding variables. v_1 and v_2 are weight coefficients of two objectives, which depend on the will and preference of decision-makers, and are 0.55 and 0.45, respectively, in this paper.

In this way, three optimal motor schemes are selected and marked as red points in Figure 7. The motor scheme designed by the rated operation point optimization method is termed L1. The motor scheme designed by the proposed optimization method under UDDS is termed L2, and that optimized under HWFET is termed L3.

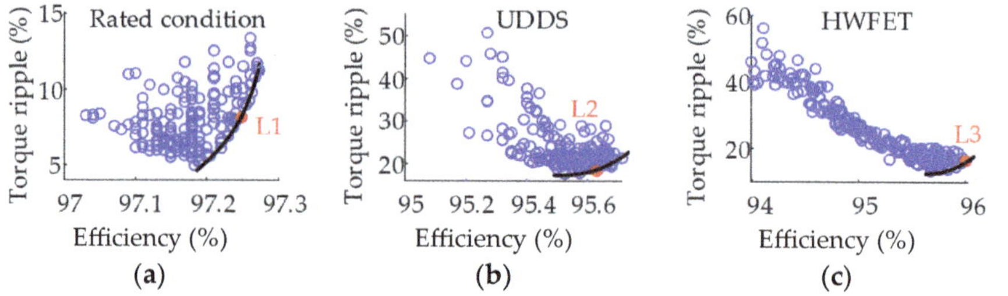

Figure 7. Motor optimization results under different working conditions. (**a**) Underrated condition. (**b**) Under UDDS. (**c**) Under HWFET. The blue dots are optimization solutions and black curves are Pareto optimal fronts.

The values of the optimal variables and objective performance of the three motor schemes are listed in Table 5. Furthermore, their performances under different driving cycles are shown in Figure 8. The area in the red dotted box in Figure 8 is defined as the sweet region with high efficiency and low torque ripple, where 92% < T_r < 98% and 5% < η < 30%. Figure 8a shows the performance evaluation of L1 and L2 under UDDS. It can be observed that the motor optimized at the rated point (L1) has about 40% of the operation points falling into the sweet region, and this ratio is about 90% for the motor optimized under UDDS (L2). Figure 8b shows the performance evaluation of L1 and L3 under HWFET. It can be seen that L1 has about 20% of the operation points falling into the sweet region, and this ratio is over 90% for the motor optimized under HWFET (L3). Therefore, it can be concluded that the sweet region can be expanded by the proposed system-scale optimization, considering driving cycles.

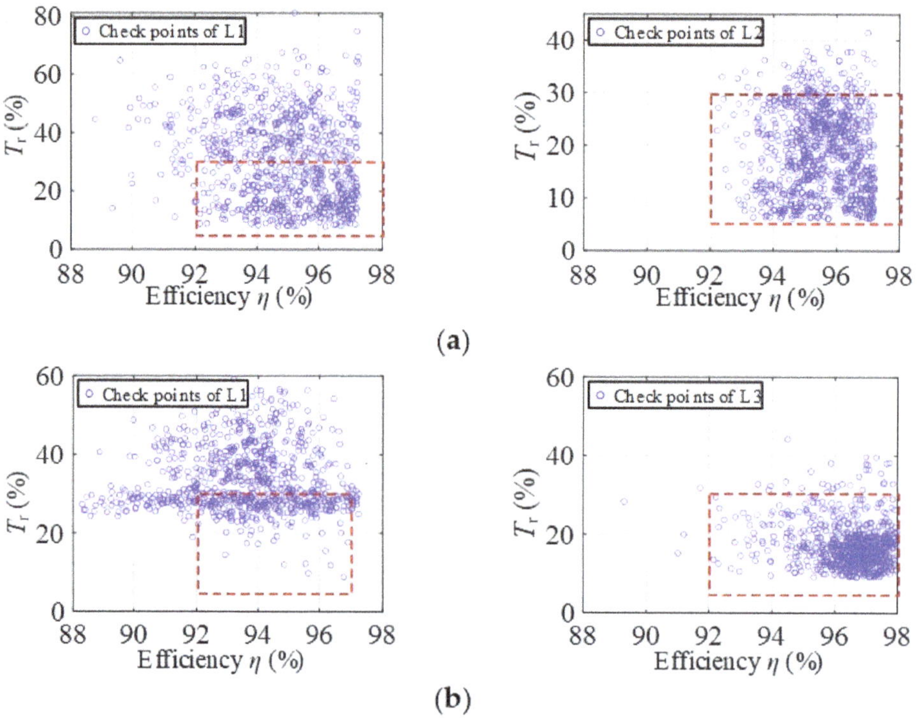

Figure 8. Performance evaluations of three motor schemes under different working conditions. (**a**) UDDS. (**b**) HWFET. The area in the red-dotted box is the sweet region.

Table 5. Objective performances and optimal variables.

	RPs	L1	L2	L3
		I_P (A), P_{Fe} (W), P_{Cu} (W), T_r (%)		
UDDS	1	29.8, 39.35, 53.25, 21	31.56, 34.24, 59.76, 14	none
	2	66.8, 355.11, 268, 35	53.37, 324.2, 171.24, 21	
	3	15.26, 178.09, 14, 42	16.32, 155.81, 16, 25	
	4	92.74, 80.5, 515.9, 10	96.55, 76.14, 559.18, 8	
	5	52.6, 154.5, 165.8, 15	55.18, 140.78, 182.7, 11	
HWFET	1	82.8, 381, 411.25, 38	none	82.75, 59.48, 325, 15
	2	71, 153.22, 303.4, 14		71.11, 76.4, 135.76, 10
	3	34.39, 327, 71.27, 28		34.39, 21.14, 296.67, 17
Rated		97.3, 284, 567.9, 7.91	100.75, 274, 608.92, 6.1	103, 268.32, 638, 9.26
Optimal results		$x_1, x_2, x_3, x_4, x_5, x_6, x_7$		
		0.8, 5.5, 1.5, 36, 14, 8750, 6500	0.8, 5.5, 1.7, 33.5, 14.5, 7854, 6280	0.8, 5.5, 1.7, 32.5, 14.5, 7525, 7500

4.2. Performance Evaluation by Experiments

To further prove the effectiveness of the optimization method described above, an IPMSM prototype is tested on the experimental platform shown in Figure 9, which is based on the dSPACE controller. The magnetic powder brake is used to provide various load torques, and a dynamic torque sensor is used to measure dynamic torque and speed. The target motor is used to power an A0 class pure EV, which adopts UDDS as the main test specification for emissions certification and fuel economy performance testing. Therefore, L2 shown in Table 5 is chosen to be tested as the prototype, which is optimized under UDDS. Meanwhile, the UDDS driving cycle is carried out to conduct the motor performance evaluation.

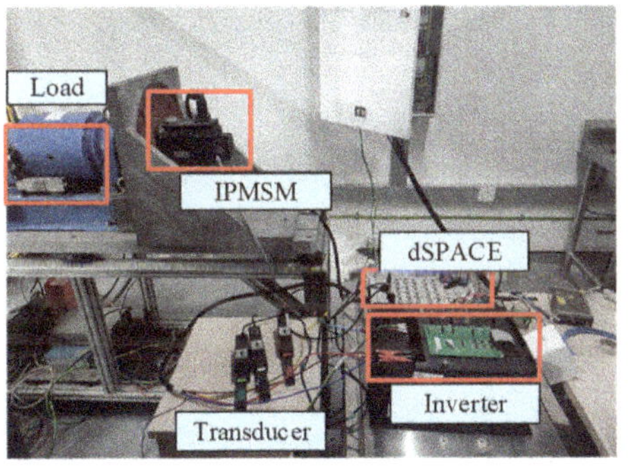

Figure 9. Experimentation platform of the IPMSM prototype.

Firstly, the no-load performance of the proposed IPMSM is tested, and the results are shown in Figure 10. In Figure 10a, it can be observed that the IPMSM can quickly follow the given speed during the starting or speed switching process. And Figure 10b shows the torque ripple at no load is less than ± 0.2 N·m, which demonstrates the good design of cogging torque. From Figure 10c, it can be seen that its three-phase back EMFs are sinusoidal and symmetric. Moreover, the no-load current is smaller than 1A to provide no-load loss, as shown in Figure 10d. The changes in the above curves are consistent with the changes in motor working conditions, and the motor speed regulation effect shows the feasibility of the proposed control system.

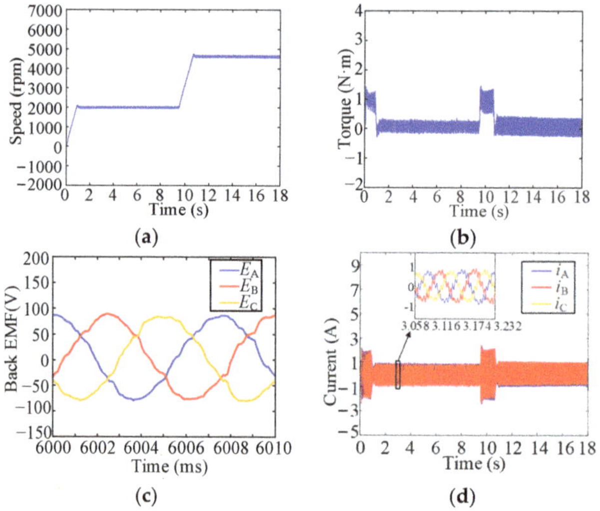

Figure 10. Experimental results at no load. (**a**) Speed. (**b**) Torque. (**c**) Back EMF. (**d**) Current.

Moreover, the loaded performance of the proposed IPMSM under MTPA and flux weakening control is tested, and the results are shown in Figure 11. It can be seen from Figure 11a,b that the motor driving modes can be switched from MTPA mode to flux weakening mode at about 5 s to expand its high-speed carrying capacity. With the flux weakening control under constant power operation, more negative i_d is applied to the IPMSM to achieve the same torque requirements as the MTPA shown in Figure 11c. In Figure 11a, it can be seen that the IPMSM can quickly follow the given speed at starting, speed switching, or load-changing working conditions. Moreover, the torque ripple at 2000 rpm with 5 N·m is about ±1 N·m; it is similar at 3500 rpm with 5 N·m; and it is about ±3 N·m at 3500 rpm with 15 N·m, which is shown in Figure 11b. Generally, torque ripples in the two modes are about 20%, which agrees with the results in Table 5. Furthermore, the three-phase currents shown in Figure 11c are sinusoidal and symmetric, which also shows the effectiveness of the proposed control system.

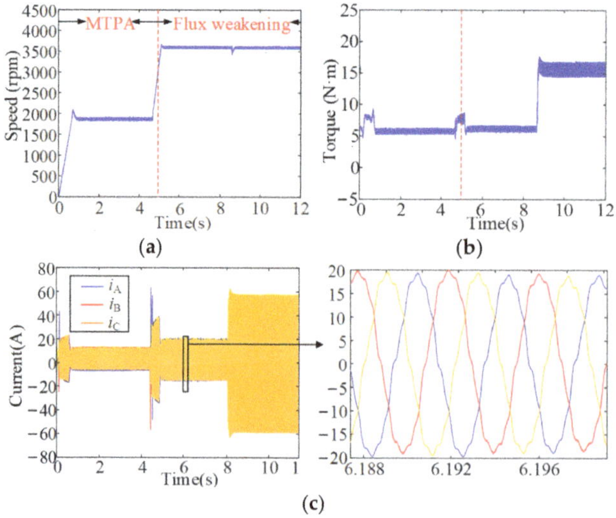

Figure 11. Experimental results are loaded. (**a**) Speed. (**b**) Torque. (**c**) Current.

In order to further verify the effectiveness of the proposed optimization method, the prototype is tested under the five RPs of UDDS. The measured efficiency and torque ripple of the prototype with the proposed control system are compared with the calculated results from the optimization process, which are listed in Table 6. The calculated results of efficiency shown in Table 6 are obtained according to the data in Tables 2 and 5, where the input power is calculated by $2\pi nT/60$ from Table 2 and the output power is approximately equal to the sum of the input power, P_{Fe}, and P_{Cu} in Table 5. Meanwhile, the calculated results of torque ripple shown in Table 6 are the same as T_{r} in Table 5. It can be observed that the measured results are almost in accordance with the optimized results. The error of efficiency is within 8%, and the maximum error of torque ripple is 20%. These errors are mainly due to the inaccurate estimation of losses and vibrations caused by mechanical factors such as wind wear and bearing in simulations. In spite of this, it can still greatly reflect the effectiveness of the proposed system-scale optimization design method, considering driving cycles.

Table 6. Optimization parameters of the investigated IPMSM.

L2	RPs	Efficiency (%)			Torque Ripple (%)		
		Calculated Result	Experimental Result	Error *	Calculated Result	Experimental Result	Error *
UDDS	1	95.7	91	5.15	14	15	6.67
	2	95.7	89	7.59	21	20	−5
	3	95.6	89	7.41	25	23	−8.7
	4	94.4	89	6.07	8	10	20
	5	97.2	91	6.83	11	13	15

* Error = (Calculated result-Experimental result)/Experimental result.

5. Conclusions

In this paper, a system-scale multi-objective optimization methodology for driving motors based on the clustering technique under different operating conditions is proposed for electric vehicle applications. The clustering approach has been introduced to identify energy weights and RPs in the torque-speed profile for two vehicle driving cycles. To evaluate the effectiveness of the system-scale optimization design of the driving motor, a comparative study has been carried out based on the semianalytical efficiency model and torque model of IPMSMs, where the motor was optimized for UDDS, HWFET, and the rated operating point. Finally, a prototype of IPMSM with an MPTA-flux weakening controller was investigated to verify the proposed method. From the discussion, the following conclusions could be drawn:

First, compared with the optimization design under the rated operation point, the solution from the proposed optimization considering driving cycles has a larger sweet region with higher efficiency and lower torque ripple.

Second, the control parameters are considered in the proposed system-scale optimization as well as the motor parameters, which can ensure the system's dynamic performances and are valuable for engineering batch production. Third, the proposed clustering optimization of driving motors considering inverter control strategy could well be applied for the rapid development of serialized motors in other applications, e.g., ship propulsion motors and robot motors.

The future work will be extended to more structural forms of motors with more diverse control strategies, e.g., axial-flux PMSM with direct torque control. Moreover, the multidisciplinary approach for system-scale optimization will be investigated in the following, which will consider the effects of thermal and mechanical factors from the perspective of practical engineering applications.

Author Contributions: Resources, S.Y.; Writing—original draft, J.B.; Writing—review & editing, J.D. All authors have read and agreed to the published version of the manuscript.

Funding: This research was funded by the Natural Science Foundation of China, grant number 52277065, and the National Key R&D Program of China, grant number 2020YFA0710500.

Institutional Review Board Statement: Not applicable.

Informed Consent Statement: Not applicable.

Data Availability Statement: Not applicable.

Conflicts of Interest: The authors declare no conflict of interest.

Nomenclature

B_r	Radial magnetic density component	P_{loss}	Motor power loss
B_t	Tangential magnetic density component	P_{out}	Motor output power
d	Total level number	P_{Fe}	Iron loss
f	Magnetic field alternates frequency	P_{Cu}	Copper loss
i_s	Stator phase current	R_s	Phase resistance
J	Moment of inertia	T_s	Sampling period
k_s	Fill factor of winding	T_r	Motor torque ripple
K_c	Eddy current loss coefficient	u_{lim}	Motor phase voltage limitation
K_e	Additional loss coefficient	w	Energy weights
K_y	Hysteresis loss coefficient	W'_m	Co-energy
K_{sp}	Proportion coefficient of speed controller	W_m	Stored energy
K_{si}	Integration coefficient of speed controller	z_i	Data point of sample i
K_{ip_d}	Proportion coefficient of d-axis current controller	$l_i^{(t)}$	Centroid of sample i in the tth iteration
K_{ii_d}	Integration coefficient of d-axis current controller	$S_i^{(t)}$	Cluster of sample i in the tth iteration
K_{ip_q}	Proportion coefficient of q-axis current controller	$p_{O(i)}$	Average motor performance of factor O at level i
K_{ii_q}	Integration coefficient of q-axis current controller	δ	Angle between the current vector and the d-axis
n_{max}	Motor maximum speed	θ	Rotor position
n_N	Motor rated speed	λ	Flux linkage
p	Pole pair number	λ_{pm}	Permanent magnet flux linkage
p_t	Overall mean of motor performance data	Ω	Rotor mechanical angular velocity

References

1. Cui, W.; Ren, L.; Zhou, J.; Zhang, Q. A new IPMSM with Hybrid rotor structure for electrical vehicle with reduced magnet loss. *IEEE Trans. Magn.* **2022**, *58*, 1. [CrossRef]
2. Geethanjali, S.; Vijayakumar, K. Testing and implementation of dual way DC-DC converter for electric vehicle power train system. *IEICE Electron. Express* **2022**, *19*, 20220343. [CrossRef]
3. Zhao, X.; Kou, B.; Huang, C.; Zhang, L. A reverse-salient permanent magnet synchronous motor for electric vehicles considering operating conditions. *IEEE Trans. Energy Convers.* **2023**, *38*, 262–272. [CrossRef]
4. Xu, Y.; Ai, M.; Xu, Z.; Liu, W.; Wang, Y. Research on interior permanent magnet synchronous motor based on performance matching of electric bus. *IEEE Trans. Appl. Supercond.* **2021**, *31*, 8. [CrossRef]
5. Liu, J.; Liang, Y.; Yang, P. Research on novel flat wire transposed winding of pmsm for electric vehicle. *IEEE Trans. Transp. Electrif.* **2023**, *9*, 771–781. [CrossRef]
6. Sun, X.; Shi, Z.; Cai, Y.; Lei, G.; Guo, Y.; Zhu, J. Driving-Cycle-oriented design optimization of a permanent magnet hub motor drive system for a Four-Wheel-Drive electric vehicle. *IEEE Trans. Transp. Electrif.* **2020**, *6*, 1115. [CrossRef]
7. Sun, X.; Hu, C.; Zhu, J.; Wang, S.; Zhou, W.; Yang, Z.; Lei, G.; Li, K.; Zhu, B.; Guo, Y. MPTC for PMSMs of EVs with multi-motor driven system considering optimal energy allocation. *IEEE Trans. Magn.* **2019**, *55*, 6. [CrossRef]
8. Chen, Q.; Fan, X.; Liu, G.; Xu, L.; Xu, M. Regulation of high-efficiency region in permanent magnet machines according to a given driving cycle. *IEEE Trans. Magn.* **2017**, *53*, 1. [CrossRef]
9. Tian, L.; Wu, L.; Huang, X.; Fang, Y. Driving range parametric analysis of electric vehicles driven by interior permanent magnet motors considering driving cycles. *CES Trans. Electr. Mach. Syst.* **2019**, *3*, 377–381. [CrossRef]
10. Fatemi, A.; Demerdash, N.A.O.; Nehl, T.W.; Ionel, D.M. Large-scale design optimization of PM machines over a target operating cycle. *IEEE Trans. Ind. Appl.* **2016**, *52*, 3772–3782. [CrossRef]
11. Sarigiannidis, A.G.; Beniakar, M.E.; Kladas, A.G. Fast adaptive evolutionary PM traction motor optimization based on electric vehicle drive cycle. *IEEE Trans. Veh. Technol.* **2017**, *66*, 5762–5774. [CrossRef]
12. Polat, M.; Yildiz, A.; Akinci, R. Performance analysis and reduction of torque ripple of axial flux permanent magnet synchronous motor manufactured for electric vehicles. *IEEE Trans. Magn.* **2021**, *57*, 1. [CrossRef]

13. Xue, Z.; Li, H.; Zhou, Y.; Ren, N.; Wen, W. Analytical prediction and optimization of cogging torque in surface-mounted permanent magnet machines with modified particle swarm optimization. *IEEE Trans. Ind. Electron.* **2017**, *64*, 9795–9805. [CrossRef]
14. Jung, S.-W.; Yoon, J.; Choi, K.; Bang, J.; Bong, U.; Hahn, S. Comparative design study of HTS synchronous motor with inner and outer rotor type based on multi-objective optimization. *IEEE Trans. Appl. Supercond.* **2022**, *32*, 6. [CrossRef]
15. Liu, X.; Hu, C.; Li, X.; Gao, J.; Huang, S. An Online data-driven multi-objective optimization of a permanent magnet linear synchronous motor. *IEEE Trans. Magn.* **2021**, *57*, 7. [CrossRef]
16. Cho, S.-K.; Jung, K.-H.; Choi, J.-Y. Design optimization of interior permanent magnet synchronous motor for electric compressors of air-conditioning systems mounted on EVs and HEVs. *IEEE Trans. Magn.* **2018**, *54*, 1–5. [CrossRef]
17. Zhao, W.; Yang, Z.; Liu, Y.; Wang, X. Analysis of a Novel surface-mounted permanent magnet motor with hybrid magnets for low cost and low torque pulsation. *IEEE Trans. Magn.* **2021**, *57*, 1. [CrossRef]
18. Diao, K.; Sun, X.; Lei, G.; Bramerdorfer, G.; Guo, Y.; Zhu, J. System-level robust design optimization of a switched reluctance motor drive system considering multiple driving cycles. *IEEE Trans. Energy Convers.* **2021**, *36*, 348–357. [CrossRef]
19. Carraro, E.; Morandin, M.; Bianchi, N. Traction PMASR motor optimization according to a given driving cycle. *IEEE Trans. Ind. Appl.* **2016**, *52*, 209. [CrossRef]
20. Bernard, N.; Dang, L.; Moreau, L.; Bourguet, S. A Pre-sizing method for salient pole synchronous reluctance machines with loss minimization control for a small urban electrical vehicle considering the driving cycle. *Energies* **2022**, *15*, 9110. [CrossRef]
21. Yamazaki, K.; Seto, Y. Iron loss analysis of interior permanent-magnet synchronous motors-variation of main loss factors due to driving condition. *IEEE Trans. Ind. Appl.* **2016**, *42*, 1045. [CrossRef]
22. Lei, G.; Wang, T.; Zhu, J.; Guo, Y.; Wang, S. System-level design optimization method for electrical drive systems—Robust approach. *IEEE Trans. Ind. Electron.* **2015**, *62*, 4702–4713. [CrossRef]
23. Zhou, X.; Zhu, X.; Wu, W.; Xiang, Z.; Liu, Y.; Quan, L. Multi-objective optimization design of variable-saliency-ratio PM motor considering driving cycles. *IEEE Trans. Ind. Electron.* **2021**, *68*, 6516. [CrossRef]
24. Qiu, H.; Cui, S.; Wang, S.; Wang, Y.; Feng, M. A Clustering-Based optimization method for the driving cycle construction: A case study in fuzhou and Putian, China. *IEEE Trans. Intell. Transp. Syst.* **2022**, *23*, 18681. [CrossRef]
25. Salameh, M.; Brown, I.P.; Krishnamurthy, M. Fundamental evaluation of data clustering approaches for driving Cycle-Based machine design optimization. *IEEE Trans. Transp. Electrif.* **2019**, *5*, 1395. [CrossRef]
26. Chen, H.; Liu, X.; Demerdash, N.A.; El-Refaie, A.M.; Chen, Z.; He, J. Computationally efficient optimization of a Five-Phase Flux-Switching PM machine under different operating conditions. *IEEE Trans. Veh. Technol.* **2019**, *68*, 6495. [CrossRef]
27. Lai, C.; Feng, G.; Iyer, K.L.V.; Mukherjee, K.; Kar, N.C. Genetic Algorithm-Based current optimization for torque ripple reduction of interior PMSMs. *IEEE Trans. Ind. Appl.* **2017**, *53*, 4493–4503. [CrossRef]
28. Jia, H.; Cheng, M.; Hua, W.; Zhao, W.; Li, W. Torque ripple suppression in flux-switching PM motor by harmonic current injection based on voltage Space-Vector modulation. *IEEE Trans. Magn.* **2020**, *46*, 1527–1530. [CrossRef]
29. Zhang, R.; Yin, Z.; Du, N.; Liu, J.; Tong, X. Robust Adaptive current control of a 1.2-MW Direct-Drive PMSM for traction drives based on internal model control with disturbance observer. *IEEE Trans. Transp. Electrif.* **2021**, *7*, 1466–1481. [CrossRef]
30. De Klerk, M.L.; Saha, A.K. A Comprehensive Review of advanced traction motor control techniques suitable for electric vehicle applications. *IEEE Access* **2021**, *9*, 125080–125108. [CrossRef]
31. Deng, T.; Su, Z.; Liu, J.; Tang, P.; Chen, X.; Liu, P. Advanced angle field weakening control strategy of permanent magnet synchronous motor. *IEEE Trans. Veh. Technol.* **2019**, *68*, 3424–3435. [CrossRef]
32. Alzayed, M.; Chaoui, H.; Farajpour, Y. Dynamic direct voltage MTPA current sensorless drives for interior PMSM-Based electric vehicles. *IEEE Trans. Veh. Technol.* **2023**, *72*, 3175–3185. [CrossRef]
33. Song, C.-H.; Song, I.-S.; Shin, H.-S.; Lee, C.-H.; Kim, K.-C. A Design of IPMSM for High-Power electric vehicles with wide-field-weakening control region. *IEEE Trans. Magn.* **2022**, *58*, 2. [CrossRef]
34. Yuan, S.; Du, J.; Wei, Y. Optimization of Permanent Magnet Machine for Electric Vehicles Considering the Control Strategy and Inverter. In Proceedings of the 2020 23rd International Conference on Electrical Machines and Systems (ICEMS), Melaka, Malaysia, 24–27 November 2020. [CrossRef]
35. He, J.; Li, G.; Zhou, R.; Wang, Q. Optimization of Permanent-Magnet spherical motor based on taguchi method. *IEEE Trans. Magn.* **2020**, *56*, 2. [CrossRef]
36. Abdo, Q.M.; Ewad, H.; Mohamed, K.A. Optimized PID Controller for Single Area Thermal Power System Based on Time Varying Acceleration Coefficients Particle Swarm optimization. In Proceedings of the 2020 International Conference on Computer, Control, Electrical, and Electronics Engineering, Khartoum, Sudan, 26 February–1 March 2021; pp. 1–5. [CrossRef]
37. Falch, L.; de Silva, C.W. Improvement of VIKOR method with application to Multi-Objective design problems. *Int. J. Inf. Technol. Decis. Mak.* **2023**, *22*, 777–802. [CrossRef]

machines MDPI

Article

Increased Dynamic Drivetrain Performance by Implementing a Modular Design with Decentralized Control Architecture

Niels Divens [1,*], Théo Tuerlinckx [1], Bernhard Westerhof [1], Kurt Stockman [2,3], David van Os [2,3] and Koen Laurijssen [1]

[1] MotionS Core Lab, Flanders Make vzw, 3001 Leuven, Belgium; theo.tuerlinckx@flandersmake.be (T.T.); bernhard.westerhof@flandersmake.be (B.W.); koen.laurijssen@flandersmake.be (K.L.)
[2] Department of Electromechanical, Systems and Metal Engineering, Ghent University, 9052 Gent, Belgium; kurt.stockman@ugent.be (K.S.); david.vanos@ugent.be (D.v.O.)
[3] MIRO Core Lab, Flanders Make vzw, 9052 Gent, Belgium
* Correspondence: niels.divens@flandersmake.be

Abstract: This paper assesses the energy consumption, control performance, and application-specific functional requirements of a modular drivetrain in comparison to a benchmark drivetrain. A decentralised control architecture has been developed and validated using mechanical plant models. Simscape models have been validated with data from an experimental setup including an equivalent modular and benchmark drivetrain. In addition, the control strategy has been implemented and validated on the experimental setup. The results prove the ability of the control strategy to synchronize the motion of the different sliders, resulting in crank position tracking errors below 0.032 radians on the setup. The model and experimental data show an increased performance of the modular drivetrain compared to the benchmark drivetrain in terms of energy consumption, control performance, and functional requirements. The modular drivetrain is especially advantageous for machines running highly dynamic motion profiles due to the reduced inertia. For such motion profiles, an increased position tracking of up to 84% has been measured. In addition, it is shown that the modular drivetrain root mean square (RMS) torque is reduced with 32% compared to the benchmark drivetrain. However, these mechanical energy savings are partly counteracted by the higher motor losses seen in the modular drivetrain, resulting in potential electrical energy savings of around 29%.

Keywords: modular drivetrain; drivetrain performance; energy consumption

Citation: Divens, N.; Tuerlinckx, T.; Westerhof, B.; Stockman, K.; van Os, D.; Laurijssen, K. Increased Dynamic Drivetrain Performance by Implementing a Modular Design with Decentralized Control Architecture. *Machines* **2023**, *11*, 1036. https://doi.org/10.3390/machines11111036

Academic Editors: Kan Liu and Wei Hu

Received: 2 October 2023
Revised: 14 November 2023
Accepted: 17 November 2023
Published: 20 November 2023

1. Introduction

In the manufacturing industry, more stringent requirements are continuously set on operating machines and systems as a way to leverage manufacturers' competitive advantages. Productivity, reliability, energy consumption, and flexibility are frequently recurring key performance indicators (KPI) for designing, developing, and selecting machines. The concept of modularity has been proposed [1–3] as a potential method to design and build machines with increased operational performance and customisation potential. The demand for easy customisation requires more flexibility in machine configuration. The literature has shown that modularity has benefits when it comes to developing more flexible machines [4,5]. A module-based approach can further reduce engineering and time costs [6], as scaling the system does not require a detailed redesign, instead relying on the addition or removal of modules built from standardized components. This paper adheres to the following definition of modularity [7]: the degree to which a system's components may be separated and recombined, often enhancing the system's scalability, flexibility, and variety of use cases. The unique set of separated and recombined components is called a module.

In [2], the authors showed that for direct drive systems a modular approach can result in increased operational efficiency. Furthermore, controller reference tracking outperformed the benchmark case, indicating the potential to push machines to higher operating speeds and thereby obtain an increased production rate. The benefits of introducing modularity have been illustrated for other applications as well, such as cranes [8], wind turbines [9], and electric vehicles [10,11]. For crane applications, ref. [8] showed that load-sharing electrical drive configurations result in a more uniform load distribution between motors and have the ability to eliminate crane skew. The grid connection of large wind turbines [9] can benefit from reduced capacitor voltage ripple when connected to a hybrid modular multilevel converter. For electric vehicle applications, ref. [11] demonstrated low harmonic distortion and high efficiency when implementing a modular multilevel topology using power electronics transformers.

This research paper aims to assess the performance of application cases for modular drivetrains typically used in tufting, weaving, and stamping machines. These machines often rely on the conversion of rotational motion to translational motion by means of bar and/or slider–crank mechanisms, which typically introduce a variable inertia. In [12], the authors demonstrated the robust performance of a slider–crank mechanism driven by a permanent magnet synchronous machine (PMSM) with an adaptive controller. However, this paper is limited to the motion control of a single slider–crank mechanism. Therefore, a modular drivetrain concept with six motors and slider-crank mechanisms (modules) is used as a test case to evaluate the performance of a decentralized control architecture for a modular drivetrain. Each independent drivetrain module executes a local control strategy without any information on the neighbouring modules. In addition, a benchmark (non-modular) drivetrain is developed in detail as a reference for a performance comparison. An experimental setup including the modular and benchmark drivetrain was designed and built, allowing the operational performance of the modular drivetrain system to be experimentally compared with the benchmark system.

Industrial weaving machines typically contain different functional components operating at high speeds, and each of their respective motion patterns must be synchronized to meet the machines' functional requirements. Traditionally, these machines have a single motor that is mechanically linked to the machine load through bar linkages, cams, etc. This is equivalent to the benchmark drivetrain presented in this paper. The sliders of this benchmark drivetrain are inherently synchronized (neglecting the non-rigidity of crankshaft) by the mechanical design; ref. [13] already showed that the application of programmable controllers and electronic drive regulators can improve the operational efficiency and reliability of multi-motor machines in the textile industry. Modularity in drivetrain systems furthermore implies modifications in the control architecture. The motions of each individual module need to be electronically synchronized with each other in order to achieve the required motion of the load without causing unnecessary internal stresses or damage to the system. This topic has been discussed elaborately in previous research [14–19] for different kind of applications. In [16], for example, the authors applied the total sliding mode control method in order to synchronize multiple asynchronous machines. However, the above research focuses on centralized control architectures for synchronizing the modules, introducing the disadvantage of a single point of failure; if the controller fails, the entire drivetrain system fails to operate. For example, ref. [19] presented a speed control method for a dual PMSM connected in parallel and regulated with a single inverter. In [20], the author investigated a decentralised control strategy for synchronized control of multi-motor drive systems; however, his research was limited to a simulation environment and was not experimentally validated. Furthermore, the case of mechanically coupled motor loads, which are often present in textile machines, was not investigated. In contrast, the contributions of the present research include a performance analysis of a drivetrain system with mechanically linked loads. In practice, the six sliders are connected with an aluminum bar as the load. With respect to drivetrain modularization, this poses more stringent requirements on drivetrain control performance in order to avoid mechanical failures during drivetrain operation. Assuming a rigid system, multiple actuators provide control with a single degree of freedom. Along with implementing a modular mechanical design of the drivetrain system, this research further aims to implement a modular and scalable masterless control architecture. This paper describes a decentralised control architecture, aiming to avoid a single point of failure and allow continuous drivetrain operation in the event that the controller of a single module fails. The local module control strategy consists of cascaded proportional–integral (PI) loops and a feed-forward term. The dynamic drivetrain model in Matlab Simscape [21] allows this control strategy to be safely validated in a virtual environment before implementing and testing it on the experimental setup. To the best of our knowledge, introducing modularity in terms of both mechanical design and control architecture on a slider–crank drivetrain system with coupled loads represents a novel research approach.

The remainder of this paper is organised as follows. Section 2 introduces the theory, models, and methods used to design, develop, and implement the modular slider–crank drivetrain on both the system and component levels. Section 2.5 elaborates the design of the experimental setup and planned experiments used to assess the drivetrain performance. The model validation and drivetrain performance assessment are then discussed in Section 3, while Section 4 provides a final summary.

2. Materials and Methods

This section first elaborates on the methodology used to compare and assess the different drivetrain architectures. Second, the the concepts underlying the two drivetrain architectures under investigation are described. Third, the modelling of the drivetrain architectures and the control architecture and strategy are explained. Lastly, the design of the experimental setup is described in detail and the drivetrain KPIs are defined.

2.1. Assessment Methodology

The comparison of the drivetrains was executed following the principles illustrated in Figure 1. In [22], this methodology was explained and applied for the assessment of modular drivetrains in a simulation environment. Multiple drivetrain architectures have been conceptually designed and modelled. An architecture can consist of Z physical models, representing, for example, a slider–crank mechanism. The detailed simulation parameters are implemented in the simulation model. These may include physical component parameters, controller parameters, environmental conditions, etc. The simulation models are run with a certain load profile that defines the motion and external forces applied to the system. Furthermore, a set of key performance indicators are defined; using the output of the simulation models, the KPI values are calculated for the different drivetrain architectures, allowing a performance assessment and comparison to be carried out. Figure 1 focuses on a model-based performance assessment approach. This paper applies a similar methodology, using data from experimental setups instead of simulation models.

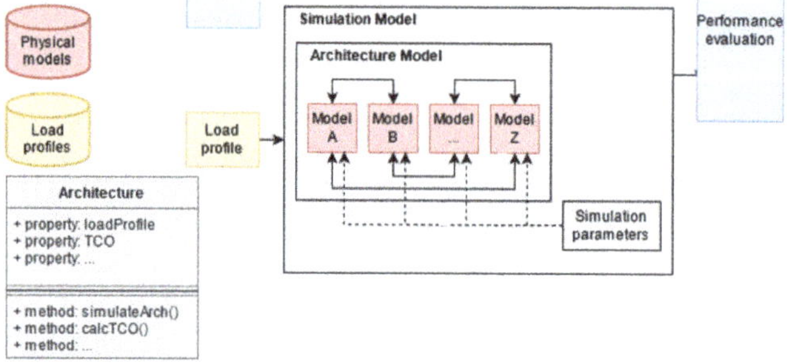

Figure 1. Assessment methodology.

2.2. Drivetrain Architectures

The definition of the modular drivetrain concept is based on the application of an industrial weaving (tufting) machine [1]. These machines make use of multiple slider–crank and/or multi-link bar mechanisms in parallel to convert rotational motion to translational motion on the load. In this case, the load is a uniformly-distributed mass connecting all of the sliders. Traditionally, these slider–crank mechanisms are connected by a crankshaft and driven by a single rotational power source, e.g., an electric motor. Figure 2 illustrates a simplified system architecture of this benchmark drivetrain with six slider–crank mechanisms. In [1], a novel modular alternative to this benchmark drivetrain was introduced, as shown schematically in Figure 2. This modular drivetrain architecture contains six slider–crank mechanisms, with the sliders all connected to a single load. This load is identical to the one in the benchmark drivetrain; however, each slider–crank mechanism is driven by a separate electric motor. Hence, the modular drivetrain system has six independently powered and controlled modules moving the load. This paper evaluates and compares the performance of both the benchmark and modular drivetrains by means of a dynamic mechanical model developed in Mathworks Simscape and an experimental test setup.

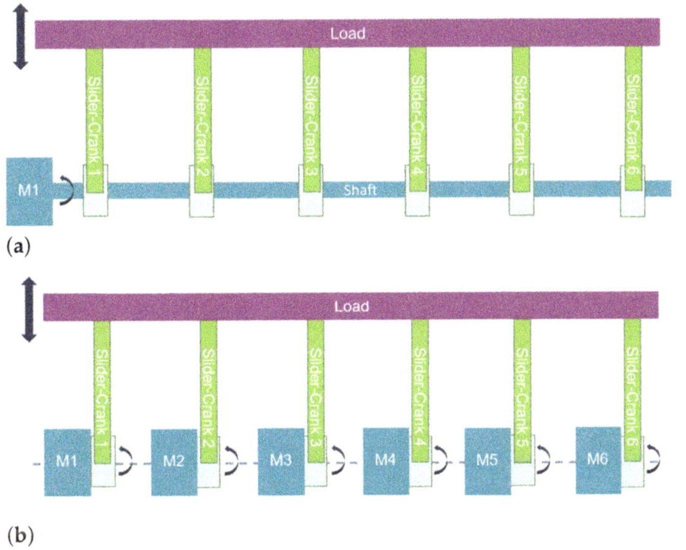

Figure 2. Drivetrain architectures: (**a**) benchmark and (**b**) modular.

2.3. Drivetrain Models

The two system architectures depicted in Figure 2 were both modelled in the Matlab Simscape (R2019a) environment. Blocks from the Simscape foundation library were used, and custom Simscape blocks were developed as well. In order to build the models of the above-mentioned drivetrain systems, component models of the electric motor, slider–crank mechanism, and inertial load are required. The benchmark system architecture additionally requires modelling of the crankshaft. The different component models are described in more detail below.

The permanent magnet synchronous machine (PMSM) and its drive were simplified as a first-order system with rotational inertia and output torque saturated at the motor peak torque curve. In order to add a minimum of motor dynamics, a discrete low-pass filter was implemented. The transfer function for this filter allows the electromagnetic dynamic performance of the modelled motor to be limited by filtering the motor torque target T_{Tgt} generated by the controller:

$$G = \frac{T_{Act}}{T_{Tgt}} = \frac{(t_s/\tau)z^{-1}}{1 + (t_s/\tau - 1)z^{-1}} \tag{1}$$

where T_{Act} is the actual torque generated by the electric motor. The time constant τ is chosen such that the filter cutoff frequency (−6 dB) equals 800 Hz. Furthermore, the motor resolver data is modelled to include a communication delay and measurement noise. The delay is simple signal time delay between the resolver and the motor controller feedback loop (Section 2.4). White noise with a constant power spectral density (PSD) was superimposed on the crank resolver signal. The delay and noise values were fitted using experimental data of the setup, and are listed in Table 1.

A separate Matlab script based on the power loss model of the PMSM, inverter, and cables was used to estimate the consumed electrical energy from the grid based on the measured (or simulated) motor speed and torque. The power loss model of the PMSM was defined by first establishing the current and voltage equations of a PMSM depending on the torque and speed setpoint [23], allowing efficiency maps to be computed by taking into account iron loss, copper loss, and mechanical loss [24] and validation to be carried out based on the efficiency maps provided by the the supplier. The motor performance characteristics were obtained from the datasheets of the PMSM motors used in the physical setup; more details can be found in Section 2.5. The cable Joule losses were directly computed from the motor currents and cable resistance. The inverter model was simplified through a load-dependent efficiency curve that was tuned to match the nominal loss from the suppliers' datasheets.

In the benchmark drivetrain, the PMSM is connected to a crankshaft, as depicted in Figure 2. The crankshaft is driven by the electric motor, and connects the cranks of the six slider–crank mechanisms. The shaft was modelled based on the rotational stiffness, damping, and inertia.

While the Simscape foundation library already collects a large set of component models, a detailed model for the slider–crank mechanism was lacking. First, the default Simscape slider–crank block does not include friction. Second, it does not include the effect of the position of the centre of gravity (Cog) of the crank and rod on the dynamic behavior. Third, gravity itself is unmodelled behaviour in the default Simscape block. Thus, the slider–crank mechanism [25] was analytically modelled and implemented in a custom Simscape block. The analytical equations included the kinematic and dynamic relationships between the rotational motion of the crank and the translational motion of the end of the rod (i.e., the slider) while taking into account the variable inertia effects. Second, the vertical and horizontal forces at the crankshaft, crank rod, and rod–load connections were calculated while taking into account the gravitational effects. All equations were derived based on the slider–crank free-body diagram depicted in Figure 3, and can be found in Appendix A.

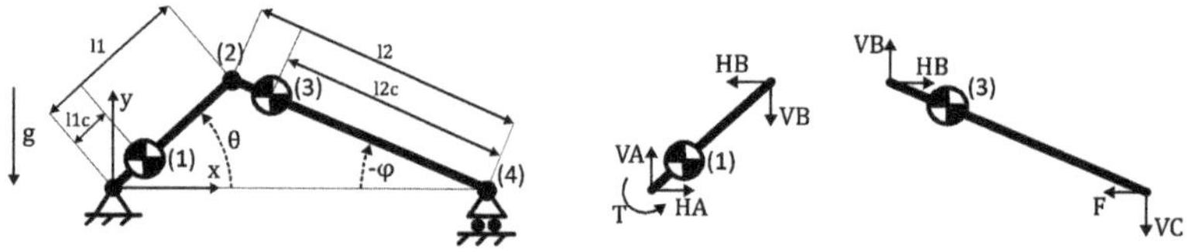

Figure 3. Slider–crank free-body diagram.

Table 1. Component properties.

Component	Parameter	Value	Units
AM8063L motor	Nominal power	4.97	kW
	Nominal torque	33.9	Nm
	Peak torque	111	Nm
	Torque constant	3.33	Nm/A
	Rotor inertia	29.0	$kg \times cm^2$
AM8051F motors	Nominal power	1.52	kW
	Nominal torque	5.8	Nm
	Peak torque	17.7	Nm
	Torque constant	1.77	Nm/A
	Rotor inertia	2.24	$kg \times cm^2$
Motor resolver	delay	1.1	ms
	Noise sample time	10^{-5}	s
	Noise PSD	10^{-20}	Mag^2/Hz
Crankshaft	Inertia	1.208×10^{-3}	$kg \times m^2$
	Length	1.96	m
	Torsional stiffness	3.4×10^3	Nm/rad
	Torsional damping	0.02	$Nm \times s/rad$
Slider-crank	Crank length	20	mm
	Rod length	95	mm
	Crank inertia	12	$kg \times cm^2$
	Rod inertia	28	$kg \times cm^2$
	Crank mass	1.04	kg
	Rod mass	0.77	kg
	Slider friction coefficient	0.45	/
	Crank Coulomb friction	0.77	Nm
	Crank viscous friction	0.03	$Nm \times s/rad$
Load	Mass	4.42	kg
	Length	1.8	m
	Segment stiffness	9.55×10^5	N/m
	Segment damping	8.29	$N \times s/m$

The load consisted of an aluminium bar connecting the six sliders of the slider–crank mechanisms. For modelling purposes, the load was divided into five segments, with each located in between two slider–cranks. The segments were modelled using the lateral bending stiffness and damping along with the translational mass between the different sliders. The load's total mass was determined on the basis of the mechanical CAD model and the lateral moment of inertia ($I_y = 51,739\ mm^4$), which is necessary in order to calculate the load segment stiffness. For each segment, the stiffness was separately calculated using the following equation:

$$K = \frac{1}{2} \times \frac{48 \times E \times Iy}{2 \times L_{segment}^3} \quad (2)$$

where E is the Young's modulus, I_y is the moment of inertia, and $L_{segment}$ is the length of the segment. The parameter values of the load are summarized in Table 1.

Additionally, static and kinetic frictions were taken into account in the model using a Simscape rotational friction element for the crank bearings and the translational friction for the slider carriage. The slider carriage friction model additionally took into account the total normal force acting on the slider carriage, including the gravitational force and the dynamic vertical force on the rod–load connection computed by the model of the slider–crank mechanism. The friction coefficients were determined by parameter optimization using experimental data from the setup (see Section 2.5), and are summarized in Table 1.

Figure 4 schematically shows how the modular drivetrain system model was built using the component models. The figure only shows two of the six modules. The benchmark drivetrain system model was built up equivalently, with the additional inclusion of a crankshaft model connecting all six cranks.

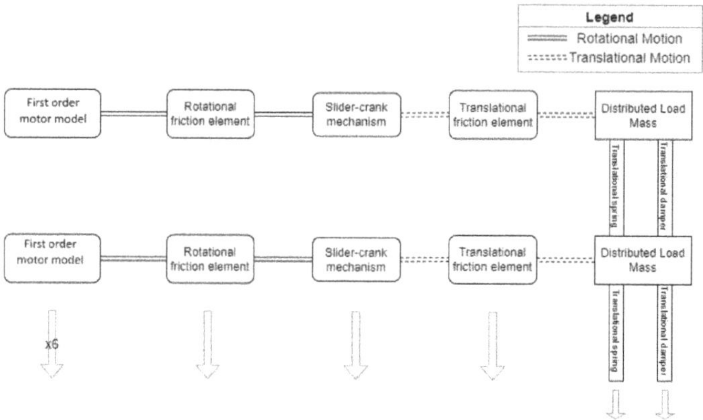

Figure 4. Modular drivetrain mechanical system model.

The main model parameters were defined from the component datasheets, and the friction parameters were tuned and validated by comparison with measurements taken on the physical setup. Table 1 provides the parameters values.

The mechanical system plant models of the benchmark and modular drivetrain were utilized to develop and validate the drivetrain control strategy, which is discussed in Section 2.4 below.

2.4. Control Architecture and Strategy

The sliders of the benchmark drivetrain are inherently synchronized by the presence of the crankshaft, which mechanically connects all six cranks. However, the modular drivetrain does not have mechanically linked slider–crank mechanisms, and as such the sliders have to be electronically synchronized through adequate control of all six motors.

Previous research on control methods has shown good performance for multi-drive systems; however, the control architectures were largely implemented in a centralized manner [8,14–18], limiting the modularity to the mechanical design of the system. In this research, a decentralised control architecture for the modular drivetrain has been developed, with each motor being controlled locally. In this way, the concept of drivetrain modularity is further extended to the drivetrain control. In addition, removing the master controller scheme eliminates it as a potential single point of failure. A schematic of this control architecture is visualized in Figure 5. Each motor is controlled by a local controller with a sample time of 4 kHz. All local controllers act towards the same reference position and speed. This decentralised control architecture allows for increased system reliability and fault tolerance while maintaining synchronization of the six sliders.

Both the benchmark and modular drivetrains were implemented with an identical local control strategy, which is visualized in Figure 6. While in the benchmark drivetrain only a single motor needs to be controlled, the modular drivetrain requires synchronized control of six independent motors. The applied control strategy is composed of the two main components depicted in Figure 6. First, a cascaded proportional integrator (PI) loop contains two PI feedback loops. This includes a positional PI controller which aims to track a rotational reference position for the motor. The output of this position controller is an input for the second PI loop, which aims to track a reference velocity. The proportional term of this velocity feedback loop is equivalent to the derivative term of a traditional PID feeback loop. Both PI loops include an anti-windup mechanism. The cascaded PI loop (position + velocity) was chosen over a traditional PID position controller because it enables sharing of the control parameter tuning between continuous and reciprocating motion (see Section 2.6 for definitions of continuous and reciprocating motion). For both drivetrains, the control parameters of the cascaded PI loop were optimized using the model-based PI tuning method. The control performance for each setup was numerically optimized for position and velocity tracking during high-speed operation.

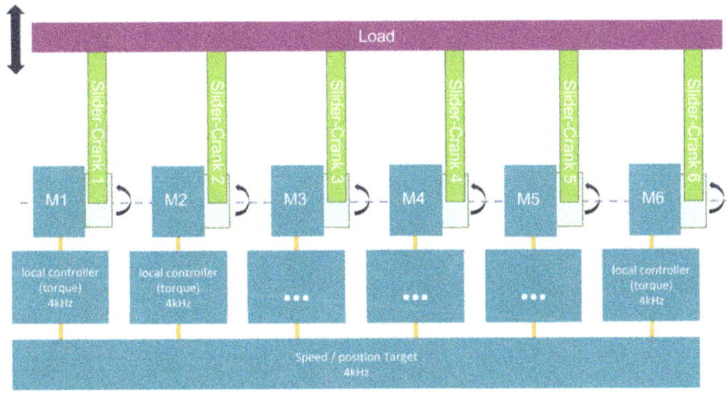

Figure 5. Modular control architecture.

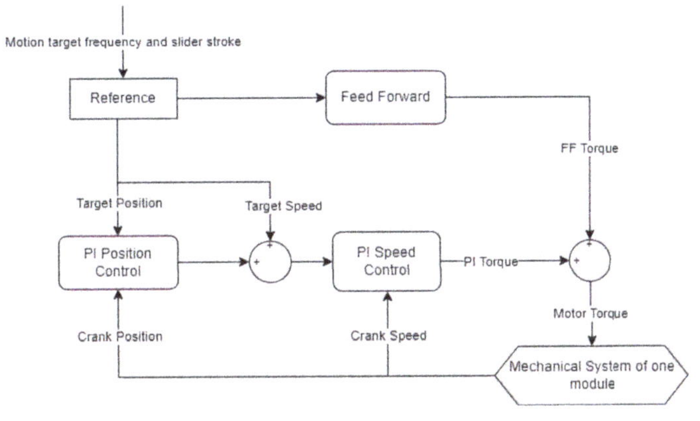

Figure 6. Local control strategy.

Second, a feed-forward controller based on the analytic equations of the slider–crank mechanism (see Appendix A) was implemented. This block outputs an estimation of the required motor torque for a certain reference crank position, velocity, acceleration, and slider–crank properties. The slider–crank properties include the dimensions, weight, inertia, and frictional characteristics of the mechanism. The exact properties and their values are listed in Table 1. The frictional characteristics were empirically defined using the experimental data from the setup, which is described below and is equivalent to the inverse of the slider–crank model described in Section 2.3.

The target motion time series is defined by a motion profile type, target slider frequency (number of slider cycles/second), and target stroke length (only applicable for the reciprocating motion type). Using the above two inputs, the reference block in Figure 6 computes the target crank position, velocity, and acceleration time traces for the different motion profile types under investigation. More information on these time-dependent motion profiles can be found in Section 2.6.

It is important to note that for the modular drivetrain, this control strategy needs to be applied in a decentralized manner; hence, each motor has its own local independent controller acting towards the same reference, resulting in a total of six active controllers. Such a decentralized control architecture eliminates the single point of failure and allows the system to remain functional (potentially with limited performance) in the event that a motor or its respective controller fails.

The combination of the mechanical plant models and control algorithm in the Matlab Simscape environment allows the performance of both the modular and non-modular (benchmark) system architectures to be virtually evaluated. Next, we aimed to further validate these findings with experimental results. Therefore, an experimental setup was developed and built, which is described in the next section.

2.5. Design of Experimental Setup

A mechanical design and component selection process was executed based on the conceptual diagrams of the system architectures visualized in Figure 2.

The setup is visualized in Figure 7 and its operation is demonstrated in the video provided in [26]. Both the modular and benchmark drivetrain were installed on the same Welda test bed table and rotated in opposite directions, resulting in the inertial forces of both setups being counterbalanced, thereby reducing the vibrations and swinging of the table. However, this has the negative consequence that a single drivetrain cannot be operated individually at high speeds, inhibiting the experimental evaluation of the maximum operating speed of the fastest drivetrain. The Welda table was installed with multiple T-slot plates, allowing for easy and flexible mounting of all components on the table. This mounting flexibility and freedom is particularly important for the modular drivetrain, as the mechanical alignment of the six modules is both challenging and critical for proper operation. For the benchmark drivetrain, the six slider–crank mechanisms were inherently aligned (within design tolerances) by the presence of the crankshaft.

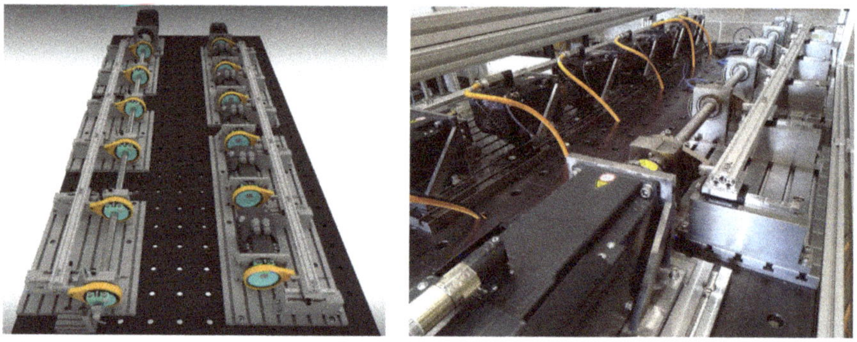

(**a**) (**b**)

Figure 7. Experimental setup: (**a**) CAD design and (**b**) setup in the lab.

To keep the research industrially relevant, component sizing and selection focused primarily on off-the-shelf components. The modular drivetrain used six AM8051F motors, while the benchmark drivetrain used a single AM8063L motor. Table 1 lists several basic properties of these motors. The motors' front faces were bolted onto a steel frame. This steel frame had slots in the bottom to affix the frame to the test bed using the T-slots, allowing for a range of movement when positioning the motors. This range of movement is particularly important for the modular drivetrain, as it has no crankshaft connecting all modules, meaning that the six slider–crank mechanisms are not inherently mechanically aligned. Custom alignment tools were developed together with a specific alignment procedure to achieve mechanical alignment within tolerances. The slider carriage was a caged ball linear guide of THK (SSR 15 X W 1 SS QZ + 160 L), and is shown in Figure 8. The load bar connecting all six slider carriages was a standard aluminum item profile with a total weight of 4.42 kg. All electronic components, such as the motor drives, IPC, IO cards, and low-voltage power supply, were stock components. A total of four type AX5000 Beckhoff drives were installed in the electrical cabinet: a single drive for the benchmark drivetrain, and three dual-channel drives for the modular drivetrain. For the latter dual-channel drives, the two channels were separated, allowing independent control of the six motors of the modular drivetrain. Each drive was operated at a switching frequency of 8 kHz. The IPC implemented and ran the control algorithm, as discussed in Section 2.4, at a sample frequency of 4 kHz. The IPC then sent the individual torque targets to each drive over an EtherCAT communication network.

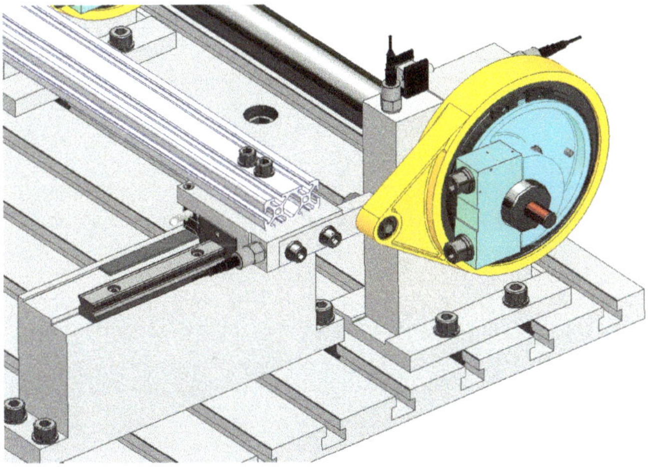

Figure 8. Slider carriage and eccentric bearing.

The main custom components of the setup were the crankshaft and the slider–crank mechanism. The crankshaft was a custom machined steel rod connecting the motor to the six slider–crank mechanisms of the benchmark drivetrain. The modular drivetrain did not have this crankshaft. All slider–crank mechanisms were identical by design. The mechanism was composed of an eccentric bearing converting the rotational motion to translational motion. The eccentricity of the bearing was obtained by offsetting the bearing's centre axis with respect to the rotational axis. The eccentric bearing is visualized in more detail in Figure 8. The slider–crank mechanism had a crank length of 20 mm, allowing a maximum full slider stroke of 40 mm to be achieved. The dimensions and properties of the main off-the-shelf and custom components are listed in Table 1.

2.6. Performance KPIs

The availability of a simulation environment and experimental environment allowed an accurate performance assessment of the benchmark and modular drivetrains to be carried out.

The following KPIs were defined for evaluating the performance of both drivetrain systems following the approach described in Section 2.1.

The maximum torque required for a certain operating condition is an indicator of the peak load of a drivetrain system. A lower peak load results in higher system reliability and lifetime. The crank torque is calculated on the basis of the motor currents reported by the respective drive. The following equation was applied for field-oriented controlled (FOC) PMSM:

$$T(t) = k_t * i_q(t) \tag{3}$$

where k_t and $i_q(t)$ are the motor torque constant and the motor q-axis current, respectively. Next, the maximum torque for a time series torque signal is calculated as follows:

$$T_{max} = \max(\|T(t)\|) \tag{4}$$

For the modular drivetrain, $T(t)$ is the total drivetrain torque, which is equal to the sum of the torques of the six independent motors.

RMS torque allows the mechanical energy consumption of the system drivetrains to be compared for a given motion profile.

$$T_{rms} = \sqrt{mean(T(t)^2)} \tag{5}$$

The crank position control error defines the control accuracy of the system drivetrain. The crank position is measured by the motor resolvers and read out from the drives. Based on these resolver signals, the KPI is calculated as follows.

$$e_{\theta}^{max} = \max\{\| \left(\theta_{act}(t) - \theta_{ref}(t)\right) \|\} \tag{6}$$

The slider error defines how accurately the slider positions are aligned and synchronized with each other during motion. For an ideal drivetrain (meaning one without disturbances such as misalignment and other design tolerances), this largely correlates with the above-mentioned crank position control error; however, for a non-ideal system such as the experimental setup, this is not necessarily the case. In practice, design tolerances, mounting misalignments, etc., are inherently present. The slider positions are measured using linear encoders (RLS LA11) attached to the slider carriage. Based on these measurements, the maximum slider error is calculated as follows:

$$e_{sliderPos}^{max} = \max_{t\epsilon[t_0, t_{end}]} \{\| \max_{i\epsilon[1,6]}\{PS_{i,t}\} - \min_{i\epsilon[1,6]}\{PS_{i,t}\} \|\} \tag{7}$$

where $PS_{i,t}$ is the position of slider number i at instant t, t_0 is the start time of the experiment with the machine at steady speed, and t_{end} is the end of the experiment.

The RMS electrical power defines the average power drawn from the electrical grid for a certain motion profile. The instantaneous electrical power $P^{elec}(t)$ is estimated using the motor, inverter, and cable loss models, which require the motor speed and torque as inputs. For accurate estimation of the electrical energy consumption, the measured speed and estimated torque signals of the setup experiments are used as inputs to the loss models. The RMS electrical power consumption for a certain motion profile is calculated as follows.

$$P_{rms}^{elec} = \sqrt{mean(P^{elec}(t)^2)} \tag{8}$$

The performance comparison using the above KPIs of the benchmark and modular system drivetrain was performed for an equal load profile (see Figure 1). The performance was investigated for three different motion profiles: continuous motion, reciprocating motion, and start/stop motion. In the continuous motion profile, the crankshaft continuously rotates in one direction at a constant speed. The start/stop motion rotates the crankshaft in one direction as well, except that the shaft stops rotating every full cycle with a duty cycle of 50%. This motion profile is typical for pick-and-place applications. In reciprocating motion, the crankshaft smoothly oscillates in two directions, which means that the motor operates in both positive and negative speed ranges. Examples of these different motion profiles are visualised in Figure 9 for the crank. It is clear that both the reciprocating and start/stop motion profiles are far more dynamic and introduce higher acceleration than the more stationary continuous motion profile. Both the continuous motion and start/stop motion profiles have a fixed stroke length, i.e., the maximum stroke is defined by the mechanical design. The reciprocating motion profile has the additional benefit of providing an electronically controllable stroke length, meaning that the stroke length is determined by the amplitude of the sinusoidal crank position time signal.

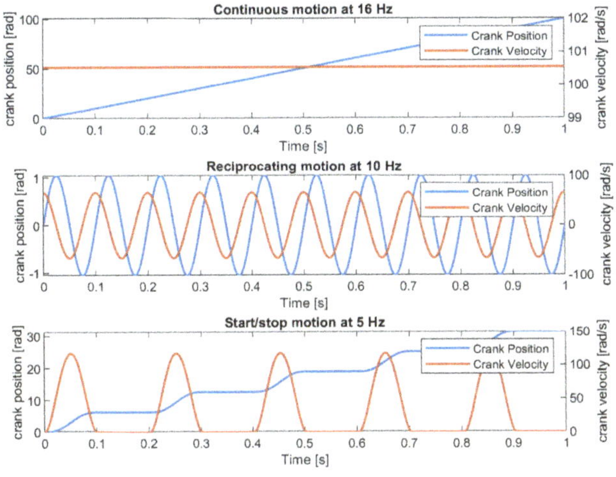

Figure 9. Crank motion profiles.

3. Results and Discussion

3.1. Simulation Versus Experimental Results

A first analysis was carried out to compare the simulation results with the experimental results for both the modular and benchmark drivetrains. For the continuous, reciprocating, and start/stop motion profiles, the above mentioned KPIs were compared using a fixed slider motion frequency of 16 Hz, 10 Hz, and 5 Hz, respectively. These operational frequencies are close to the maximum obtainable operational frequencies of the experimental benchmark drivetrain for the continuous, reciprocating and start/stop motion profiles, respectively. Figure 10 shows a comparison of the results for the modular drivetrain running the start/stop motion profile. The accuracy of the simulation estimation for each KPI is calculated as follows:

$$accuracy_{pct} = (1 + (KPI_{sim} - KPI_{exp})/KPI_{exp}) * 100 \tag{9}$$

The resulting accuracy is a percentage value, where 100 percent indicates a perfect match between simulation and experiment.

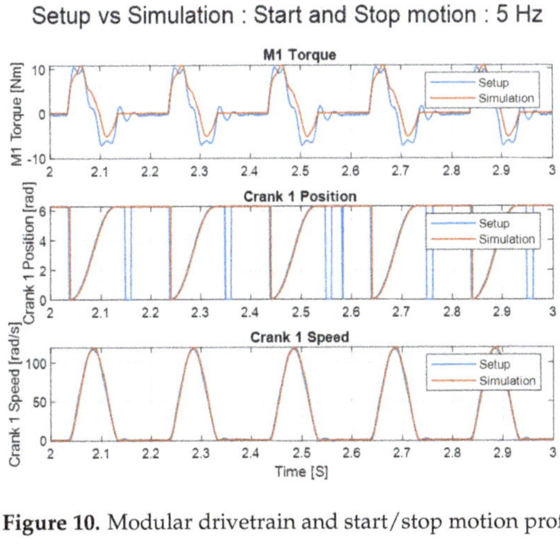

Figure 10. Modular drivetrain and start/stop motion profile: simulation versus experiment.

All analysis results are summarized in Tables 2 and 3.

Table 2. Modular drivetrain: simulation versus experiment.

KPI		Reciprocating 10 Hz	Continuous 16 Hz	Start/Stop 5 Hz
RMS torque	simulation	32 Nm	27 Nm	28 Nm
	experiment	38 Nm	13 Nm	28 Nm
	accuracy	83%	202%	98%
Max torque	simulation	51 Nm	42 Nm	67 Nm
	experiment	57 Nm	23 Nm	64 Nm
	accuracy	90%	183%	105%
Max crank error	simulation	0.018 rad	0.025 rad	0.026 rad
	experiment	0.032 rad	0.019 rad	0.022 rad
	accuracy	56%	132%	122%

Table 3. Benchmark drivetrain: simulation versus experiment.

KPI		Reciprocating 10 Hz	Continuous 16 Hz	Start/Stop 5 Hz
RMS torque	simulation	42 Nm	32 Nm	32 Nm
	experiment	57 Nm	13 Nm	41 Nm
	accuracy	74%	235%	76%
Max torque	simulation	59 Nm	40 Nm	85 Nm
	experiment	98 Nm	28 Nm	108 Nm
	accuracy	60%	145%	79%
Max crank error	simulation	0.071 rad	0.067 rad	0.111 rad
	experiment	0.160 rad	0.023 rad	0.135 rad
	accuracy	45%	286%	82%

The slider error is not analysed in the tables, as the simulation was assumed to be ideal in the sense that mechanical misalignment or design tolerances were not included in the model. The electrical power consumption is not included either, as the sensors required for electrical measurements were not installed in the experimental setup.

The results show accurate torque estimation of the models for both the start/stop motion and reciprocating motion profiles. This can be clearly seen in Figure 10 for the modular drivetrain running the start/stop motion profile, and is reflected in the comparison of RMS and maximum torque as well. For example, in Table 2, a 98% fit for the RMS torque of the modular drivetrain running the start/stop motion profile can be observed. A less accurate fit is observed for the continuous motion profile, with an accuracy of only about 200% when estimating the RMS and maximum torque. This can be explained by the need to analyse the different terms of the total required torque for the three motion profiles. The total torque amount consists of the inertial torque, gravity torque, and friction torque. The model relies on accurate analytic, kinematic, and dynamic equations for estimating the inertial and gravity torque based on known design parameters such as the geometry, mass, and inertia. These parameters are easily measured and/or calculated with a high degree of confidence. However, the friction torque is far more difficult to predict, and is a source of uncertainty. Although the friction parameters were fitted using measured data, the friction model can be further improved. For example, a large temperature dependency of the friction torque on the setup was observed. The current model does not include a thermal domain, which would help to estimate component temperatures. As our model does not include temperature-dependent friction, its estimation of the friction torque is far less accurate than that of the inertial and gravity torque. Start/stop and reciprocating motion are highly dynamic motion profiles causing high inertial torque values. As a result, the total required torque for these motion profiles is dominated by the inertial torque, meaning that the ratio of the accurately estimated inertial torque over the less accurately estimated friction torque is fairly high. This explains the better torque estimation results for the more dynamic start/stop and reciprocating motion profiles.

The same reasoning applies when comparing the torque estimation accuracy of the modular and benchmark drivetrains. The results in Tables 2 and 3 generally show a better model fit for the modular drivetrain. The benchmark drivetrain design includes a crankshaft and supporting bearings, while the modular drivetrain does not have these components. Our experimental observations using the testing setup show a significant amount of friction in these grease-lubricated supporting bearings of the crankshaft on the benchmark drivetrain. Accordingly, the frictional torque is higher for the benchmark, resulting in lower model accuracy for this drivetrain.

3.2. *Performance Comparison of the Benchmark and Modular Drivetrains*

This subsection focuses on a technical performance comparison between the modular and benchmark drivetrains. Similar to the previous subsection, the performance of both drivetrains was analyzed for the three different motion profiles with the modular and benchmark drivetrains running simultaneously at the same speed. Table 4 shows the resulting KPI values of both drivetrains for each motion profile and the KPI differences between the modular drivetrain and the benchmark drivetrain in percentages. All KPI values shown in the table were directly derived from experimental data except for the RMS electrical power.

Table 4. Performance comparison.

KPI		Reciprocating 10 Hz	Continuous 16 Hz	Start/Stop 5 Hz
RMS torque	Benchmark	57 Nm	13 Nm	41 Nm
	Modular	38 Nm	13 Nm	28 Nm
	Delta	−33%	−2%	−32%
Max torque	Benchmark	98 Nm	28 Nm	108 Nm
	Modular	57 Nm	23 Nm	64 Nm
	Delta	−42%	−18%	−41%
Max crank error	Benchmark	0.160 rad	0.023 rad	0.135 rad
	Modular	0.032 rad	0.019 rad	0.022
	Delta	−80%	−9%	−84%
Max slider error	Benchmark	0.43 mm	0.95 mm	0.46 mm
	Modular	0.39 mm	0.83 mm	0.41 mm
	Delta	−9%	−13%	−11%
RMS electrical power	Benchmark	3.58	2.01	5.17
	Modular	3.1	2.2	3.68
	Delta	−13%	+9%	−29%

For the continuous motion profile, the measurements show only a minor reduction of 2% in the RMS torque required to run the motion cycle. The maximum observed torque is reduced by about 18%. Hence, while the peak load of the modular drivetrain is clearly reduced, the average mechanical power delivered by the motors in the continuous motion profile is not significantly altered. Unfortunately, the electrical power consumption of the modular drivetrain is 9% higher than the benchmark drivetrain. Figure 11 shows that this can be explained by the increased overall motor losses of the six smaller motors compared to the motor loss of the single motor in the benchmark drivetrain. Furthermore, the experimental results shows a 19% increase in the position tracking performance of the modular drivetrain compared with the benchmark. As a result, the maximum slider error is reduced by 13%.

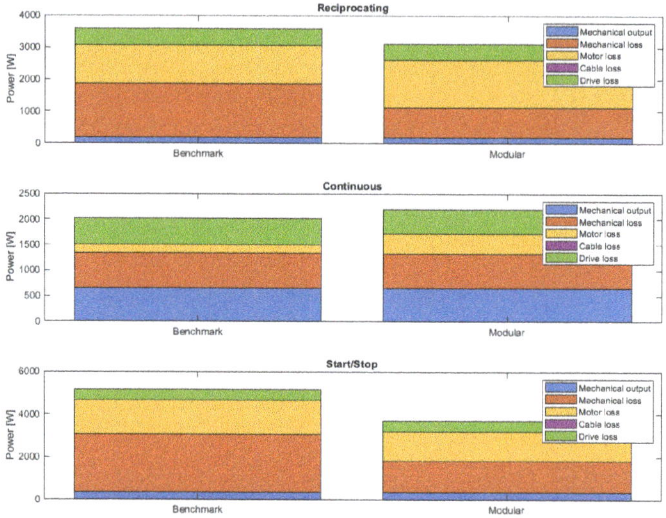

Figure 11. Electrical power consumption.

More significant performance improvements are experimentally observed for the reciprocating and start/stop motions. The total RMS torque delivered by the six modular motors is up to 33% lower compared to the RMS torque delivered by the single motor of the benchmark drivetrain. The maximum torque is decreased by up to 42%, resulting in lower fatigue stress on the modular drivetrain. As a result, for these more dynamic (high-acceleration) motion profiles, significantly less average mechanical power is required to drive the modular system compared to the benchmark system. In addition, a reduction in electrical energy consumption of up to 29% is estimated, and an 84% reduction in the tracking error is observed. Having the system load distributed over the six actuators helps to achieve better control performance. As a consequence, the slider error is reduced due to mechanical play in the pin connection between the slider and the rod, albeit less drastically, up to 11%.

These experimental results demonstrate that a modular drivetrain can result in better tracking performance, lower peak system loads, and reduced average mechanical and electrical power consumption with respect to the standard non-modular drivetrains currently used by industry. This is especially true for highly dynamic motion profiles with large acceleration values, which can benefit from a mechanical system with reduced inertia.

3.3. Outlook

Improvements could be made to both the testing setup and the model. The accuracy of the friction model could be increased by including the thermal domain. For validation of thermal behavior, temperature sensors would need to be added to the setup. In order to analyse the high-frequency system dynamics (>1 kHz), the simulation environment could benefit from a detailed model of the motor drive that includes current control.

Interesting future work could include the application of novel robust control techniques for uncertain nonlinear systems on the modular drivetrain discussed in this research. For example, a sliding mode controller such as the one described in [27] could be implemented and tested using the experimental setup described in this paper.

Future research should examine additional functionalities that modular drivetrains could potentially generate for applications in tufting, weaving, and stamping machines. We have already begun research on the potential of active damping of vibrations in the modular system. Because the modular drivetrain has six independently acting motors at different locations, there is the possibility of correcting any crank and/or slider errors. An active damping method based on a dynamic average consensus algorithm [28] has been developed and integrated in the current decentralized control architecture. Experimental validation is ongoing, and simulations already show promising results [29].

Furthermore, the impact of modular drivetrains on system reliability is being investigated. Future work could include researching safe and robust methods for handling failure. As the modular drivetrain has built-in actuator redundancy, the system can operate in the event of a single motor failure. How to implement and execute this failure handling method and what performance can be expected after such a failure are among the questions that we plan to answer in our future research.

Other future work could focus on the economics of modular drivetrains through a comparison of the total cost of ownership (TCO), including the purchase cost, cost of usage (energy), and maintenance and repair costs. Such an analysis should further build on the performance and reliability conclusions of existing research to support industrial machine manufacturers in quantifying the economical impact of modular drivetrains and making informed design decisions.

4. Conclusions

Machine manufacturers continuously aim to increase their market competitiveness by designing machines with improved productivity, reliability, energy consumption, and flexibility. The concept of drivetrain modularity was proposed in [1,2] as a method to enhance the operational performance of industrial machines. In this paper, we have introduced the concept and detailed design of a modular drivetrain architecture that implements a decentralized control architecture. A simulation plant model was set up to support the development and validation of the control strategy. The modular drivetrain performance was assessed and compared with a benchmark drivetrain using the above-mentioned simulation model as well as an experimental setup. Both drivetrains were subjected to the same load profiles, and a performance evaluation was conducted based on a set of key performance indicators: maximum torque, RMS torque, crank position tracking error, slider error, and electrical energy consumption. In addition, three different load profiles (motion cycles) were defined: continuous, reciprocation, and start/stop motion.

Comparison of the simulation KPI results with the experimental KPI results showed a good match for the more dynamic motion cycles (reciprocating and start/stop motion). For the modular drivetrain, a simulation accuracy of 98% was observed for the RMS torque KPI. The continuous motion profile, on the other hand, showed a slightly less accurate fit between the results of the simulation model and those of the experimental setup, which was due to unmodelled thermal characteristics in the friction model. The same reasoning applies to the observed discrepancies between the simulation and testing setup results for the benchmark drivetrain. As the benchmark drivetrain has more bearings, the (less accurately known) frictional force are more dominant.

Our experiments on the test setup showed significant performance improvements for the modular drivetrain as compared to the benchmark drivetrain for the highly dynamic motion profiles (reciprocating and start/stop motion). A 42% reduction in peak torque was observed and the position tracking performance was increased up to 84%, resulting in an 11% reduction in the slider error. The electrical power consumption could be reduced by up to 29%. Unfortunately, the performance increase was less pronounced for motion cycles with lower accelerations, such as the continuous motion cycle, which is largely due to the increased electrical motor losses. In closing, it can be concluded that the operating load profiles of industrial machines with high acceleration rates could benefit from a more modular drivetrain architecture.

Author Contributions: Conceptualization, N.D., T.T., B.W., K.S., D.v.O. and K.L.; methodology, N.D., T.T., B.W. and K.L.; software, N.D., T.T. and B.W.; validation, N.D. and T.T.; formal analysis, N.D., T.T. and B.W.; investigation, N.D., T.T. and B.W.; resources, K.L. and K.S.; writing—original draft preparation, N.D.; writing—review and editing, N.D., T.T., B.W., KS, D.v.O. and K.L.; visualization, N.D. and T.T.; supervision, K.L. and K.S.; project administration, K.L. and K.S.; funding acquisition, K.L. and K.S. All authors have read and agreed to the published version of the manuscript.

Funding: This research is part of the ModulAr SBO project funded and supported by Flanders Make vzw, the strategic research center for the manufacturing industry.

Data Availability Statement: The data presented in this study are available on request from the corresponding author. The data are not publicly available due to privacy restrictions.

Conflicts of Interest: The authors declare no conflict of interest.

Abbreviations

The following abbreviations are used in this manuscript:

MDPI	Multidisciplinary Digital Publishing Institute
TLA	Three-Letter Acronym
KPI	Key Performance Indicator
PI	Proportional Integrator
RMS	Root Mean Square
PMSM	Permanent Magnet Synchronous Machine
CAD	Computer-Aided Design
CoG	Center of Gravity

Appendix A. Slider–Crank Mechanism Equations

In Figure 3, (1) is the CoG of the crank, (3) is the CoG of the rod, and (4) is the slider position. The angular position of te crank is defined by θ, while the angular position of the rod is defined by ϕ; ϕ is indicated as a negative angle in order to maintain right-hand sign conventions. The respective centers of gravity (CoG) in (1) and (3) have the inertial properties $m_{1,2}$ and $I_{1,2}$, respectively, while gravity is indicated by g.

The kinematic relationships are as follows.

$$x_1 = l_{1,c}\cos(\theta) \tag{A1}$$
$$y_1 = l_{1,c}\sin(\theta) \tag{A2}$$
$$x_2 = l_1\cos(\theta) \tag{A3}$$
$$y_2 = l_1\sin(\theta) \tag{A4}$$
$$x_3 = x_2 + (l_2 - l_{2,c})\cos(-\phi) \tag{A5}$$
$$y_3 = l_{2,c}\sin(-\phi) \tag{A6}$$
$$x_4 = x_2 + l_2\cos(-\phi) \tag{A7}$$
$$\phi = -\arcsin(\frac{l_1}{l_2}\sin(\theta)) \tag{A8}$$

The equations of motion are as follows.

$$\ddot{x}_1 = \frac{HA - HB}{m_1} \tag{A9}$$
$$\ddot{y}_1 = \frac{VA - VB}{m_1} - g \tag{A10}$$
$$\ddot{x}_3 = \frac{HB - F}{m_2} \tag{A11}$$
$$\ddot{y}_3 = \frac{VB - VC}{m_2} - g \tag{A12}$$
$$\ddot{\theta} = \frac{T + (l_{1,c}HA + (l_1 - l_{1,c})HB)\sin(\theta)}{I1} + \frac{(-l_{1,c}VA - (l_1 - l_{1,c})VB)\cos(\theta)}{I1} \tag{A13}$$
$$\ddot{\phi} = \frac{(-l_{2,c}F - (l_2 - l_{2,c})HB)\sin(-\phi)}{I2}$$

$$+\frac{(-l_{2,c}VC-(l_2-l_{2,c})VB)\cos(-\phi)}{l2} \tag{A14}$$

Using the the MATLAB symbolic toolbox and differentiating (A1)–(A8) except for x_2, y_2, and θ provides the following.

$$\ddot{x}_1=-l_{1,c}\sin(\theta)\ddot{\theta}-l_{1,c}\cos(\theta)\dot{\theta}^2 \tag{A15}$$

$$\ddot{y}_1=l_{1,c}\cos(\theta)\ddot{\theta}-l_{1,c}\sin(\theta)\dot{\theta}^2 \tag{A16}$$

$$\begin{aligned}\ddot{x}_3=&\frac{l_1^2\sin(\theta)^2(l_2-l_{2,c})\dot{\theta}^2}{l_2^2\sqrt{1-\frac{l_1^2\sin(\theta)^2}{l_2^2}}}-l_1\sin(\theta)\ddot{\theta}\\&-\frac{l_1^2\cos(\theta)^2(l_2-l_{2,c})\dot{\theta}^2}{l_2^2\sqrt{1-\frac{l_1^2\sin(\theta)^2}{l_2^2}}}-l_1\cos(\theta)\dot{\theta}^2\\&-\frac{l_1^4\cos(\theta)^2\sin(\theta)^2(l_2-l_{2,c})\dot{\theta}^2}{l_2^4\left(1-\frac{l_1^2\sin(\theta)^2}{l_2^2}\right)^{3/2}}\\&-\frac{l_1^2\cos(\theta)\sin(\theta)(l_2-l_{2,c})\ddot{\theta}}{l_2^2\sqrt{1-\frac{l_1^2\sin(\theta)^2}{l_2^2}}}\end{aligned} \tag{A17}$$

$$\ddot{y}_3=\frac{(l_1l_{2,c}\cos(\theta)\ddot{\theta})-(l_1l_{2,c}\sin(\theta)\dot{\theta}^2)}{l_2} \tag{A18}$$

$$\begin{aligned}\ddot{\phi}=&\frac{l_1\sin(\theta)\dot{\theta}^2}{l_2\sqrt{1-\frac{l_1^2\sin(\theta)^2}{l_2^2}}}-\frac{l_1\cos(\theta)\ddot{\theta}}{l_2\sqrt{1-\frac{l_1^2\sin(\theta)^2}{l_2^2}}}\\&-\frac{l_1^3\cos(\theta)^2\sin(\theta)\dot{\theta}^2}{l_2^3\left(1-\frac{l_1^2\sin(\theta)^2}{l_2^2}\right)^{3/2}}\end{aligned} \tag{A19}$$

References

1. Vandenhove, T.; Diltoer, R.; Lenaerts, B.; van Os, D.; Stockman, K.; Schlimpert, S. Modular Drivetrains for Increased Performance and Reliability. In Proceedings of the 2021 IEEE International Conference on Mechatronics (ICM), Kashiwa, Japan, 7–9 March 2021; pp. 1–8. [CrossRef]
2. Lemmens, J.; Driesen, J. Synchronization and efficiency analysis of a direct-drive multi-motor application. In Proceedings of the 6th IET International Conference on Power Electronics, Machines and Drives (PEMD 2012), Bristol, UK, 27–29 March 2012; pp. 1–6. [CrossRef]
3. Otto, K.; Hölttä-Otto, K.; Simpson, T.W.; Krause, D.; Ripperda, S.; Ki Moon, S. Global Views on Modular Design Research: Linking Alternative Methods to Support Modular Product Family Concept Development. *J. Mech. Des.* **2016**, *138*, 071101. [CrossRef]
4. Schuh, G.; Dölle, C.; Barg, S.; Kuhn, M.; Breunig, S. Efficient Modular Product Platform Design of Mechatronic Systems. In Proceedings of the 2018 IEEE International Conference on Industrial Engineering and Engineering Management (IEEM), Bangkok, Thailand, 16–19 December 2018; pp. 1391–1395. [CrossRef].
5. Xu, Z.; Xi, F.; Liu, L.; Chen, L. A Method for Design of Modular Reconfigurable Machine Tools. *Machines* **2017**, *5*, 5. [CrossRef]
6. Leandro Gauss, D.P.L.; Miguel, P.A.C. Module-based product family design: Systematic literature review and meta-synthesis. *J. Intell. Manuf.* **2021**, *32*, 265–312. [CrossRef]
7. Definition of Modularity. Available online: https://www.merriam-webster.com/dictionary/modularity (accessed on 8 August 2023).
8. Mitrovic, N.; Petronijevic, M.; Kostic, V.; Jeftenic, B. Electrical Drives for Crane Application. In *Mechanical Engineering*; IntechOpen: London, UK, 2012. [CrossRef]
9. Debnath, S.; Saeedifard, M. A New Hybrid Modular Multilevel Converter for Grid Connection of Large Wind Turbines. *IEEE Trans. Sustain. Energy* **2013**, *4*, 1051–1064. [CrossRef]
10. Nemeth, T.; Bubert, A.; Becker, J.N.; De Doncker, R.W.; Sauer, D.U. A Simulation Platform for Optimization of Electric Vehicles With Modular Drivetrain Topologies. *IEEE Trans. Transp. Electrif.* **2018**, *4*, 888–900. [CrossRef]

11. Chang, F.; Lienkamp, M. A modular multilevel topology using power electronic transformers for the modular drivetrains of electric vehicles. In Proceedings of the 2016 18th European Conference on Power Electronics and Applications (EPE'16 ECCE Europe), Karlsruhe, Germany, 5–9 September 2016; pp. 1–9. [CrossRef]
12. Lin, F.J.; Fung, R.F.; Lin, Y.S. Adaptive control of slider-crank mechanism motion: Simulations and experiments. *Int. J. Syst. Sci.* **1997**, *28*, 1227–1238. [CrossRef]
13. Grumelart, A.; McKinnon, W.L. Application of Sectional Drives and PLC's to Textile Ranges. *IEEE Trans. Ind. Appl.* **1983**, *2*, 145–147. [CrossRef]
14. Perez-Pinal.; Nuñez, C.; Alvarez, R.; Cervantes, I. Comparison of multi-motor synchronization techniques. In Proceedings of the 30th Annual Conference of IEEE Industrial Electronics Society, 2004 IECON, Busan, Republic of Korea, 2–6 November 2004; Volume 2, pp. 1670–1675.
15. Zhao, D.; Li, C.; Ren, J. Speed synchronization of multiple induction motors with adjacent cross coupling control. In Proceedings of the 48h IEEE Conference on Decision and Control (CDC) held jointly with 2009 28th Chinese Control Conference, Shanghai, China, 15–18 December 2009; pp. 6805–6810. [CrossRef]
16. Peng, W.; Zhao, D. Speed Synchronization of Multi Induction Motors with Total Sliding Mode Control. In Proceedings of the 2010 Asia-Pacific Power and Energy Engineering Conference, Chengdu, China, 28–31 March 2010; pp. 1–9. [CrossRef]
17. Kong, W.; Wang, J.; Kong, D.; Cong, Y.; Feng, S. Motor shifting and torque distribution control of a multi-motor driving system in electric construction vehicles. *Adv. Mech. Eng.* **2021**, *13*, 16878140211028446. [CrossRef]
18. Xu, X.; Su, P.; Wang, F.; Chen, L.; Xie, J.; Atindana, V.A. Coordinated Control of Dual-Motor Using the Interval Type-2 Fuzzy Logic in Autonomous Steering System of AGV. *Int. J. Fuzzy Syst.* **2021**, *23*, 1070–1086. [CrossRef]
19. Asri Abd Samat, A.; Ishak, D.; Saedin, P.; Iqbal, S. Speed- Sensorless Control of Parallel- Connected PMSM fed by A Single Inverter Using MRAS. In Proceedings of the 2012 IEEE International Power Engineering and Optimization Conference (PEOCO2012), Melaka, Malaysia, 6–7 June 2012. [CrossRef]
20. Mikhov, M. An Algorithm for Synchronized Control of Multi-Motor Drive Systems. In Proceedings of the International Scientific Conference on Information, Communication and Energy Systems and Technologies, Sofia, Bulgaria, 16–18 October 2003.
21. Simscape r2019b. Available online: https://nl.mathworks.com/help/simscape/index.html (accessed on 14 July 2023).
22. Van Os, D.; Laurijssen, K.; Vansompel, H.; Sergeant, P.; Divens, N.; Stockman, K. Evaluation Framework for the Comparison of Modular Drivetrain Architectures. In Proceedings of the 2023 IEEE International Conference on Mechatronics (ICM), Loughborough, UK, 15–17 March 2023; pp. 1–6. [CrossRef]
23. MathWorks: Permanent Magnet Synchronous Motor with Sinusoidal Flux Distribution. Available online: https://nl.mathworks.com/help/sps/ref/pmsm.html (accessed on 14 July 2023).
24. Novak, M.; Novak, J.; Novak, Z. Efficiency Mapping of a Small Permanent Magnet Synchronous Motor. In *Mechatronics 2017: Recent Technological and Scientific Advances*; Březina, T., Jabłoński, R., Eds.; Springer: Cham, Switzerland, 2018; pp. 185–193.
25. Ha, J.L.; Fung, R.F.; Chen, K.Y.; Hsien, S.C. Dynamic modeling and identification of a slider-crank mechanism. *J. Sound Vib.* **2006**, *289*, 1019–1044. [CrossRef]
26. Youtube, Modular Motion System. Available online: https://www.youtube.com/watch?v=rURG46nZq28 (accessed on 14 November 2023).
27. Shao, K.; Zheng, J.; Tang, R.; Li, X.; Man, Z.; Liang, B. Barrier Function Based Adaptive Sliding Mode Control for Uncertain Systems With Input Saturation. *IEEE/ASME Trans. Mechatron.* **2022**, *27*, 4258–4268. [CrossRef]
28. Kia, S.S.; Van Scoy, B.; Cortes, J.; Freeman, R.A.; Lynch, K.M.; Martinez, S. Tutorial on Dynamic Average Consensus: The Problem, Its Applications, and the Algorithms. *IEEE Control. Syst. Mag.* **2019**, *39*, 40–72. [CrossRef]
29. Tuerlinckx, T.; Divens, N.; Laurijssen, K. Modular Drivetrain Active Damping based on Dynamic Average Consensus Algorithm in Continuous Motion. In Proceedings of the 2023 3rd International Conference on Electrical, Computer, Communications and Mechatronics Engineering (ICECCME), Tenerife, Spain, 19–21 July 2023; pp. 1–10. [CrossRef]

machines

MDPI

Article

Thermal Model and Thermal Analysis of the Dual Drive Sliding Feed System

Hui Li [1,2], Haiyang Liu [1,2,*], Xianying Feng [1,2], Yandong Liu [1,2], Ming Yao [1,2] and Anning Wang [1,2]

1 School of Mechanical Engineering, Shandong University, Jinan 250061, China; lihuifs@sdu.edu.cn (H.L.); fxying@sdu.edu.cn (X.F.); liuyandong@mail.sdu.edu.cn (Y.L.); 202014301@mail.sdu.edu.cn (M.Y.); 202014288@mail.sdu.edu.cn (A.W.)
2 Key Laboratory of High Efficiency and Clean Mechanical Manufacture of Ministry of Education, Shandong University, Jinan 250061, China
* Correspondence: haiyangliu@mail.sdu.edu.cn

Abstract: The dual drive sliding feed system can obtain a uniform and stable resolution at extremely low speeds and significantly reduce the system's nonlinear friction. However, the numerous thermal sources within the system and the multipoint sliding contact during transmission result in a significant temperature rise, leading to considerable thermal deformation and errors. Moreover, the responsive mechanism of the thermal characteristics needs to be clarified. Therefore, firstly, a frictional torque model of the engagement of the screw and nut is established, and the heat generation, heat transfer, and thermal contact resistance (TCR) are solved. Then, based on the solution, a finite element thermal simulation model of the dual drive sliding feed system is established, and experiments are performed for validation. The results show that the error in temperature at the measuring point is less than 2.1 °C, and the axial thermal elongation of the screw is less than 6.2 μm. Finally, the thermal characteristics of the feeding system under various operating conditions are analyzed. The results show that the established thermal simulated model can effectively describe the dynamic thermal characteristics of the dual drive sliding feed system during operation. The effects of the rotational speed and ambient temperature on the dynamic thermal characteristics of the dual drive sliding feed system are investigated separately. The temperature increase in each part of the screw during the operation is characterized.

Keywords: sliding screw; friction torque; feed system; thermal characteristic

Citation: Li, H.; Liu, H.; Feng, X.; Liu, Y.; Yao, M.; Wang, A. Thermal Model and Thermal Analysis of the Dual Drive Sliding Feed System. *Machines* **2023**, *11*, 1084. https://doi.org/10.3390/machines11121084

Academic Editors: Kan Liu and Wei Hu

Received: 13 November 2023
Revised: 8 December 2023
Accepted: 11 December 2023
Published: 13 December 2023

1. Introduction

Due to the rising demand for diversity and quality products in recent years, many high-end parts now need their surface form and precision to reach the micron or even nano-scale levels. As an important part of the machining process, the feed system's positional accuracy directly affects the part's final forming accuracy. However, a large amount of heat will be generated during the operation of the feed system [1,2]. Numerous studies have shown that thermal errors are difficult to control in machining due to their nonlinear and random characteristics. Thermal errors have become the most important factor affecting positioning errors [3,4]. Studying the thermal characteristics of a feed system under complex working conditions is essential for improving the forming accuracy.

Most existing feed systems use mechanical feed mechanisms; however, they are difficult to use in ultra-precision machining due to their disadvantages: mechanical clearance, friction and wear, and crawling. Therefore, Feng et al. proposed a dual drive feed system [5], which utilizes two motors rotating at approximately equal speeds in the same direction at high speeds. The worktable may get an exact micro-feed by the difference between the two speeds. The system can also adjust the worktable's speed from extremely low to very high by switching the two motors from opposite to reverse spin. The system becomes a traditional ball screw drive system when just the screw motor is powered. This type of drive significantly reduces the impact of nonlinear friction on the feed system [6]. However, because the actuator employed in the existing dual drive feed system is a ball screw pair, when the system enters nano-scale control, the unevenness of the ball and the multibody dynamic coupling characteristics have a significant impact on motion control at very low speeds, which leads to difficulties in achieving accurate and uniform displacement under nano-scale conditions. Therefore, based on previous research [7], this paper presents a novel dual drive sliding feed system that replaces the ball screw in the conventional dual drive feed system with a low-cost sliding screw pair, aiming to achieve uniform and stable displacement resolution under extremely low-speed conditions. However, due to the addition of the nut motor and the nut bearing and the multipoint contact inside the sliding screw pair, more frictional heat will be generated during operation. Therefore, the thermal field of the dual drive sliding feed system is extremely complex for obtaining accurate and uniform feed displacement and improving the forming accuracy. Clearly, it is necessary to establish a thermal estimation model of the dual drive sliding feed system and analyze the thermal characteristics of the system.

Due to its high frictional resistance and severe wear, the standard sliding screw feed system has been restricted in its application for ultra-precision machining. Therefore, studies on the frictional properties and heat generation mechanism of sliding screws are rare. However, research on planetary roller screws, ball screws, rolling bearings, and other related transmission parts is highly developed [8,9]. Zhao et al. [10] analyzed the relationship between the friction coefficient and torque balance equation of angular contact bearing between a rolling body and a raceway. They clarified the change law of the friction coefficient under changes in external factors. Helmig et al. [11] realized the dynamic calculation of the heat transfer coefficient during bearing operation using infrared thermal imaging. Ba et al. [12] established a friction model of the ball screw feed system and designed a corresponding feed-forward compensation controller. Min et al. [13] established a finite element model of a ball screw feed system and analyzed the temperature characteristics of the system during operation. Ma et al. [14] established the motion model of a planetary roller screw, analyzed the relative motion speed of the internal contact interface, and solved the distribution of the friction forces inside it. Du et al. [15] established a finite element model of a planetary roller screw and analyzed the temperature characteristics of the system at different times and positions by considering the moving heat source during operation.

In conclusion, although research on the friction characteristics, heat generation mechanism, and thermal field characteristics of rolling ball screws and other transmission components has reached a high level of maturity, the working mode of the dual drive sliding feed system still differs significantly from that of traditional transmission components. A thermal model for a dual drive sliding feed system based on existing calculation methods cannot be established. Moreover, there is no existing research on the thermal characteristics analysis of the dual drive sliding feed system. This paper establishes a thermal model of the feed system by analyzing the friction at the engagement of the screw and nut. The influence of feed speed and ambient temperature on the dynamic thermal field characteristics of the dual drive sliding feed system are explored. The framework of this paper is organized as follows. In Section 2, the thermal boundary conditions of the dual drive sliding feed system, including the heat generation, the thermal contact resistance (TCR) between components and the convective heat transfer coefficient (CHTC) are solved separately. On this basis, a finite element simulation model of the dual drive sliding feed system is established in Section 3 to analyze the thermal characteristics of the dual drive sliding feed system. The accuracy of the established thermal estimation model in terms of temperature rise characteristics and axial deformation is also verified through experiments in Section 4. Conclusions are provided in Section 5.

2. Friction Torque Modeling of the Sliding Screw

2.1. Dual Drive Sliding Feed System

The dual drive sliding feed system is illustrated in Figure 1. The main difference between this system and the traditional feed system is the addition of a motor to drive the nut. During operation, the screw and the nut rotate simultaneously, and the worktable obtains displacement through the velocity difference between the two. The distinction between the dual drive sliding feed system and the existing dual drive feed system lies in the replacement of the ball screw with a sliding screw.

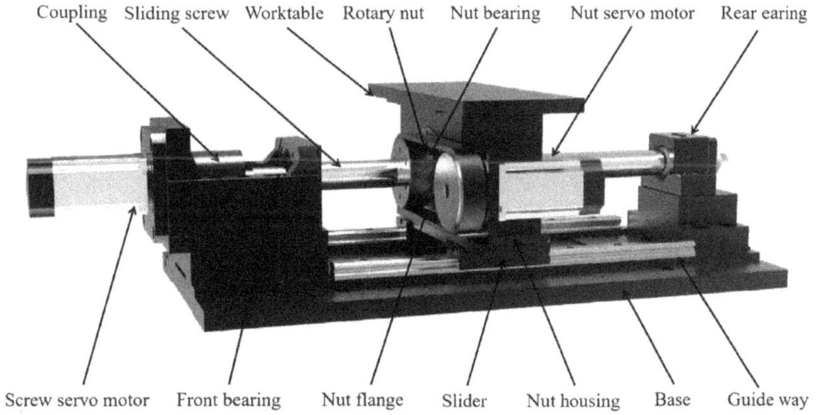

Figure 1. Structure of the dual drive sliding feed system.

The friction between components is the source of heat generation within the dual drive sliding feed system, and the friction torque is an important parameter for measuring the friction effect. Therefore, the thermal estimation model of the dual drive sliding feed system is accurately established, and the internal heat generation mechanism of the system is explored. Modeling and analyzing the frictional torque of the sliding screw are necessary. The frictional torque of a dual drive sliding feed system is affected by many factors, such as lubrication, material, and usage environment. The main reason for the frictional torque is the sliding frictional resistance that is generated by the threaded tooth contact when the relative rotation of the screw and nut occurs. Sliding friction comes from the differential sliding, elastic hysteresis, and lubrication viscosity between the screw and nut raceways. Therefore, the total friction torque of the sliding screw M_T can be expressed as:

$$M_T = \sum(M_d + M_f + M_l) \tag{1}$$

where M_d is the friction torque caused by differential sliding, M_f is the friction torque caused by elastic hysteresis, and M_l is the friction torque caused by the lubrication viscosity.

In the following, the details of each friction torque will be provided in Sections 2.2–2.4, respectively.

2.2. Friction Torque Caused by Differential Sliding

The contact part of the screw and nut is obtained as the dual drive sliding feed system is not rigid. According to the Hertz contact law, the contact surface of the screw and nut will produce a small contact area under the normal driving force, and its shape is an ellipse, as shown in Figure 2. The red area in the figure indicates the contact area between the screw and the nut, and the circumferential speed of the screw surface v_s is proportional to the distance to its axis. Similarly, the circumferential speed of the nut surface v_n is proportional to the distance to its axis. Pure rolling occurs only on two lines in the contact area. Therefore, relative sliding in opposite directions, called differential sliding, occurs at the remaining positions. The frictional moment caused by differential sliding is due to the local shear stress in the contact area. Therefore, the differential sliding friction torque M_d can be expressed as [16]:

$$M_d = \int \tau \cdot z \cdot dA = \int \mu \cdot P \cdot z \cdot dA \quad (2)$$

where z is the distance between the pure rolling line in the contact area and the local shear stress location, mm, μ is the differential sliding friction coefficient, P is the normal pressure in the contact area, MPa, and A is the area of contact, mm^2.

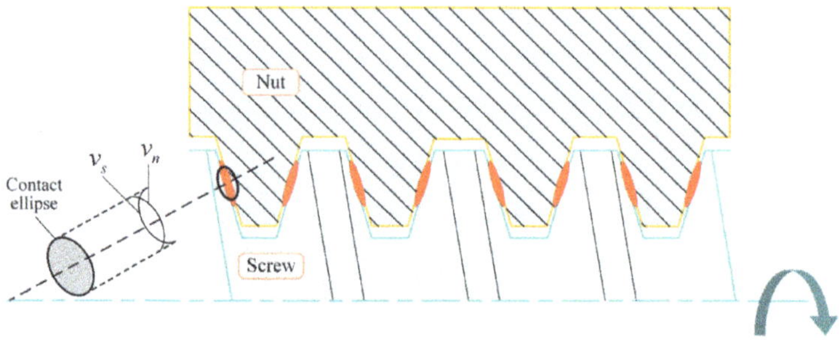

Figure 2. Schematic of differential sliding of dual drive sliding feed system.

Essentially, the actuator of the dual drive sliding feed system is similar to the driving mechanism of the planetary roller screw. Therefore, the differential sliding friction torque of the dual drive sliding screw pair is modeled and analyzed utilizing the friction torque modeling method of the planetary roller screw [17]. Taking the j contact area of the screw and nut as an example, the differential sliding friction of the contact area is integrated, and the differential sliding friction moment of the j contact area M_{dj} can be obtained as follows:

$$M_{dj} = 0.1\mu Q_j \frac{a^2}{2R}\left(\int_{-1}^{1}\left(X^2 - X^4\right)dX - 2\int_{X_1}^{X_2}\left(X^2 - X^4\right)dX\right) \quad (3)$$

where Q_j is the normal load of the j contact area, N, a is the semi-major axis of the contact ellipse, mm, R is the radius of the effective ball for Hertzian deformation (mm), and X_1 and X_2 are dimensionless parameters specified as:

$$\begin{aligned} X &= \tfrac{x}{a} \\ X_1 &= \tfrac{a_1}{a} \\ X_2 &= \tfrac{a_2}{a} \end{aligned} \quad (4)$$

where a_1 and a_2 are the distances between the centerline of the contact ellipse and the pure rolling line on both sides (mm).

Due to the direction of the differential sliding friction torque being along the long axis of the contact ellipse, there exists an angle between it and the centerline of the screw. Consequently, in the process of engagement between the screw and the nut, the resulting differential sliding friction torque can be expressed as:

$$M_d = \sum_{j=1}^{\sigma} M_{dj} \cos\alpha \quad (5)$$

where σ is the number of engaged thread teeth.

2.3. Frictional Moment Caused by Elastic Hysteresis

During the meshing and transmission process of double-drive sliding screw pairs, elastic deformation occurs at some positions on the thread teeth. Consequently, the strain lags behind the stress during loading and unloading, resulting in unbalanced elastic contact forces on both sides of the meshing surface. This imbalance leads to resistive torque during operation, referred to as frictional torque caused by elastic hysteresis. According to the Hertz contact law, the work done by the compressive stress when moving forward one unit distance is:

$$W = \frac{3bBP}{32} \tag{6}$$

where b is the short axis of the contact ellipse (mm), B is the effective axial distance of the engaging contact between balls $B = 2R$ (mm).

During the rotation of the screw and the nut, the system loses energy due to the elastic hysteresis phenomenon. Therefore, the friction torque can be expressed as:

$$M_e = \eta W \tag{7}$$

where η is the energy loss coefficient.

Since the rotation of the dual drive sliding screw is driven through the threaded teeth, the frictional moment caused by elastic hysteresis occurs between all the engaged threaded teeth. The total elastic hysteresis frictional moment is evenly distributed among the engaged parts and can therefore be expressed as [18]:

$$M_f = \sum_{i=1}^{\sigma} \frac{3}{32}\eta B\gamma \sqrt[3]{\frac{3E}{2\sum\rho}} P_i^{\frac{4}{3}} \tag{8}$$

where γ is the dimensionless coefficient of the semi-short axis of the contact ellipse, E is the Young's modulus, MPa, and $\sum\rho$ is the sum of the curvatures on the contact side.

2.4. Friction Torque Caused by Lubrication Viscosity

To enhance the transmission efficiency and minimize friction, lubricating grease must be injected at the meshing point of the screw and nut. Therefore, when the screw and the nut are driven by differential rotation, the system needs to overcome the friction caused by the shearing effect of the grease. Concerning the rolling bearing calculation method, the frictional characteristics at the engagement are analyzed by the elastic dynamics lubrication effect. The resistance caused by the viscosity of the lubrication F_l can be solved as follows [19]:

$$F_l = 2.86E' f_t {R_x}^2 k^{0.348} G^{0.022} U^{0.66} W^{0.47} \tag{9}$$

where f_t is the thermal influence coefficient, R_x is the equivalent radius of curvature of the contact ellipse in the semi-major axis direction (mm), k is the equivalent radius of curvature ratio, G is the dimensionless material coefficient, U is the dimensionless velocity coefficient, W is the dimensionless load coefficient, and E' is the equivalent modulus of elasticity of the screw and nut (Mpa), and can be expressed as:

$$E' = \left(\frac{1-\mu_{\mathrm{I}}^2}{E_{\mathrm{I}}} + \frac{1-\mu_{\mathrm{II}}^2}{E_{\mathrm{II}}}\right)^{-1} \tag{10}$$

where μ_{I} and μ_{II} are the Poisson's ratios of the screw and nut, respectively, E_{I} and E_{II} are the Young's modulus of the screw and nut, respectively (Mpa).

The j contact area between the screw and the nut is taken as an example. The frictional moment caused by the lubricating viscosity of the j contact area can be obtained as:

$$M_{lj} = F_{lj} R_s \tag{11}$$

where R_s is the nominal radius of the screw (mm). Therefore, the friction torque caused by the lubrication viscosity is as follows:

$$M_l = \sum_{j=1}^{\sigma} M_{lj} \tag{12}$$

3. Thermal Boundary Conditions of the Dual Drive Sliding Feed System

In the case of the feed system, a significant basis when performing a simulation of thermal characteristics is to establish accurate thermal boundary conditions. The heat transfer coefficient includes heat generation, the TCR between components, and the convective heat transfer coefficient. Existing studies have revealed that the thermal error of a feed system is influenced by external and internal heat sources. External heat sources mainly include space temperature and previous environmental memory. It is difficult to carry out a quantitative analysis. Precision machining processes mostly occur in a constant temperature and humidity environment, so such heat sources are not considered in this paper. The internal heat sources mainly include motors, bearings, screw nut, and guideways. To accurately establish the thermal behavior model of the system, it is necessary to clarify the heat generation and heat transfer mechanism. The heat generation and heat transfer mechanism of the dual drive sliding feed system are shown in Figure 3.

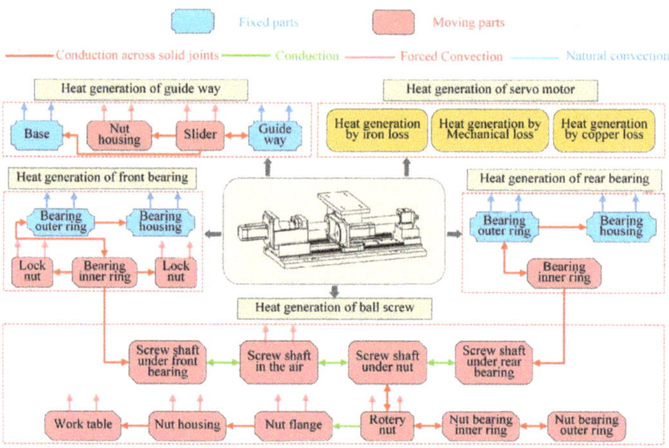

Figure 3. Heat generation and heat transfer mechanism.

3.1. *Analysis of Heat Generation*

The dual drive feed system has 4 major heat sources: (1) heat generation of the guideways, (2) heat generation of the servo motors, (3) heat generation of the bearings, and (4) heat generation of the sliding screw. The sliders move continuously on the guideways; the friction caused by the rotating motion of the ball between the sliders and guideways is the main reason for the heat generation of the guideway. The heat generated by the guideways H_g can be expressed as:

$$H_g = \lambda f_g F_g v_g \tag{13}$$

where f_g is the friction coefficient of the guideways, F_g is the vertical load carried by the guideways, N; and v_g is the moving speed of the sliders relative to the guideways (m/s).

The dual drive sliding feed system has two servo motors, which drive the screw shaft and the nut. The motor heat generation can be calculated by [20]:

$$H_m = \frac{M_m n_m}{9550}(1 - \delta) \tag{14}$$

where M_m is the output torque of the motor (N.mm), n_m is the rotation speed of the motor (rpm), and δ is the mechanical efficiency of the motor.

When bearing as a rotating part, friction is the main source of heat. The heat generation of bearings can be calculated by [21]:

$$H_b = 1.047 \times 10^{-4} n_b (M_s + M_v) \tag{15}$$

where n_b is the rotation speed of the bearing (rpm), M_s is the load friction torque caused by the elastic lag of the material and the local differential friction, (N·mm), and M_v is the viscous friction torque generated by the bearing and the lubricant (N·mm). The load friction torque can be calculated by:

$$M_s = f_1 p_1 d_b \tag{16}$$

where f_1 is the coefficient related to the type and load of the bearing, p_1 is the preload of the bearing (N), and d_b is the mean diameter of the bearing (mm).

For the front bearing and nut bearing [22]:

$$f_{1a} = 10^{-3} (P_0/C_0)^{0.33} \tag{17}$$

$$p_{1a} = 1.4Fa - 0.1Fr \tag{18}$$

where P_0, C_0, Fa, and Fr are the equivalent static load (N), the rated static load (N), axial load (N), and a radial load of the bearing (N), respectively.

For the rear bearing:

$$f_{1b} = 9 \times 10^{-4} (P_0/C_0)^{0.55} \tag{19}$$

$$p_{1b} = 3Fa - 0.1Fr \tag{20}$$

The viscous friction torque generated by the bearing and the lubricant can be calculated by [21]:

$$M_v = 10^{-7} f_0 (v_0 n_b)^{2/3} d_b. \ v_0 n_b \geq 2000 \tag{21}$$

$$M_v = 160 \times f_0 {d_b}^3. \ v_0 n_b < 2000 \tag{22}$$

where f_0 is the coefficient related to the bearing type and lubrication, and v_0 is the kinematic viscosity of the lubricant inside the bearing. In this study, Equation (21) is used for the calculation.

The friction of the dual drive sliding screw is the cause of its heat generation, which is related to the total friction torque and the speed of the sliding screw. The heat generation can be expressed as [8]:

$$H_T = \frac{M_T n_T}{9550} \tag{23}$$

where n_T is the relative speed of the screw and the nut (rpm).

3.2. Analysis of Heat Transfer

The heat transfer process of the dual drive sliding feed system can be divided into two types. One is the natural convective heat transfer between the fixed surface and the surrounding air. The other is the forced convective heat transfer due to the relative motion of the parts and the air. The CHTCs of both can be expressed as [23]:

$$h = \frac{Nu \cdot \lambda}{L_s} \tag{24}$$

where Nu is the Nusselt number, λ is the fluid thermal conductivity (W/m·K), and L_s is the component feature size (mm).

Natural convection heat transfer occurs on the surfaces of fixed components such as the bearing housing and base. The Nusselt number Nu_n can be calculated by:

$$Nu_n = C(Gr \cdot Pr)_{\varepsilon}^{\partial} \tag{25}$$

where C and ∂ are the constants determined by the shape of the heat source and fluid conditions, Gr is the Grashof number, Pr is the Prandtl number, and ε is the qualitative temperature. Gr can be calculated by:

$$Gr = \frac{g\beta L_s{}^3(T_s - T_a)}{v_a{}^2} \tag{26}$$

where g is the acceleration of gravity (m/s^2), β is the volume expansion coefficient of air, T_s is the surface temperature of the part (°C), T_a is the ambient temperature (°C), and v_a is the kinematic viscosity of the air (mm^2/s).

Forced convection heat transfer occurs on the surface of the moving components, mainly including the screw shaft, worktable, nut servo motor, and other parts. The screw shaft rotates and has a spiral groove on its surface, and its Nusselt number can be calculated as:

$$Nu_s = 0.133Re^{2/3}Pr^{1/3} \tag{27}$$

where Re is the Reynolds number. For the top and lateral surfaces of the remaining parts, the Nusselt constant Nu_p can be calculated by:

$$Nu_p = 0.332Re^{1/2}Pr^{1/3} \tag{28}$$

For the front surface perpendicular to the direction of motion, the Nusselt constant Nu_v can be calculated by:

$$Nu_v = 0.228Re^{0.731}Pr^{1/3} \tag{29}$$

The motion of the worktable of the dual drive sliding feed system is the combined motion of the screw motor and the nut motor. To analyze the thermal field characteristics of the dual drive sliding feed system, the heat generation and key heat transfer coefficients of each part of the system were calculated based on Equations (13)–(29) at an ambient temperature of 20 °C for a synthesis speed of 0.18 m/min, as shown in Tables 1 and 2.

Table 1. Partial heat production of the dual drive sliding feed system.

Speed of Synthesis (m/Min)	Speed of Screw (Rpm)	Speed of Nut (Rpm)	Heat Production (W)					
			Guideways	Screw Servo Motor	Nut Servo Motor	Front Bearing	Nut Bearing	Rear Bearing
0.18	396	360	0.39	1.87	1.70	4.95	5.31	4.36

Table 2. Partial CHTCS of the dual drive sliding feed system.

Speed of Synthesis (m/Min)	Speed of Screw (Rpm)	Speed of Nut (Rpm)	CHTCS (W/m^2·K)					
			Screw Pair	Screw Shaft	Nut Servo Motor	Nut Housing	Slider	Worktable
0.18	396	360	9.17	56.3	15.52	31.26	14.53	16.23

3.3. TCRs between Rough Contact Surfaces

The surface of the internal joint of the feed system is particularly rough at the micro level due to the processing level constraint, resulting in the so-called rough joint. When heat flows through a rough joint, it differs from heat conduction in a solid, which is hampered, and a portion of the heat is lost. The temperature will drop on both contact surfaces.

Therefore, to improve the simulation accuracy of the thermal model of the dual drive sliding feed system, it is necessary to accurately describe the TCR between the rough contact surfaces inside the system. According to the previous analysis of the TCR solution by this group, the TCRs of some key bond surfaces of the dual drive sliding feed system are shown in Table 3.

Table 3. The TCRs of the dual drive sliding feed system at key joints.

Joint Component	TCR ($m^2 \cdot K/W$)
Bearing outer ring-bearing housing	1.25×10^{-3}
Bearing inner ring-screw shaft	1.48×10^{-4}
Nut-screw	1.65×10^{-4}
Guideway-slider	6.89×10^{-4}
Nut servo motor-nut	3.68×10^{-3}

4. Establishment and Verification of the Thermal Model for Dual Drive Sliding Feed System

To accurately describe the thermal characteristics of the dual drive sliding feed system and compensate for the thermal error generated by the system, it is necessary to establish a simulation model of the thermal characteristics of the feed system. The finite element method is used to simulate the system, and an experiment is carried out to verify the comparative analysis.

4.1. Finite Element Simulation Model of Dual Drive Sliding Feed System

A finite element simulation model of the dual drive feed system was created based on the TCR, heat generation, and heat transfer coefficient calculated above to analyze the thermal characteristics of the dual drive sliding feed system. To increase the solving efficiency, the following appropriate simplifications and assumptions were made:

(1) The chamfers and some tiny parts inside the system were ignored;
(2) The screw shaft ignores the grooves on its surface and treats it as a cylinder;
(3) The parameters of heat generation and CHTCs obtained from the previous calculation do not vary with the movement or temperature rise of the components.

The hexahedral and tetrahedral meshes were used to mesh the model once it was imported into ANSYS. The heat sources, such as the screw and bearing, were coarsely meshed to reduce simulation error. The mesh created 152,924 nodes and 282,832 elements. The dual drive sliding feed system was subjected to the same thermal boundary conditions and material attributes as in the preceding section. Table 4 shows the material qualities of several of the system's essential components.

Table 4. Material properties of the dual drive sliding feed system.

Application Components	Material	Density (kg/m^3)	Modulus of Elasticity (GPa)	Poisson's Ratio	Linear Expansion Coefficient (10^{-5}/K)	Thermal Conductivity (W/m·K)	Specific Heat Capacity (J/kg·K)
Nut	Copper alloy	8400	110	0.34	1.7	115	387
Screw/bearing	GCr15	7800	200	0.28	1.2	48	729
Base/bearing housing	Q345B	7850	206	0.3	1.2	46	460
Guideways/slider	40Cr	7850	200	0.3	1.13	51	477

4.2. Experimental Verification Device and Scheme

To verify the accuracy of the solution of the simulation model, an experimental study on the temperature rise and deformation of the dual drive sliding feed system is necessary. Figure 4 depicts the experimental device. Concerning current studies, the temperature measurement locations that have a substantial impact on the total thermal field of the feed system were chosen [22–24]. The temperatures of the front bearing (T1), the rear bearing (T2), the nut flange (T3), the screw shaft (T4), the worktable (T5), the environment (T6), and the axial elongation of the screw shaft (D) were measured separately. The temperature data were acquired at T1-T3, T5, and T6 by five temperature sensors (PT100, KAIPUSEN, Xinghua, China) with a resolution of 0.1 °C. The NI PXIe-1082 was used to acquire and process the temperature data. The temperature at T4 was measured by a thermal infrared imager (E86, FLIR, Oregon, United States) with a resolution of 0.1 °C. Axial elongation (D) was measured using the laser displacement sensor (CL-3000, KEYENCE, Osaka, Japan). When the temperature rise reached 95% of the maximum temperature rise and the stable fluctuation of the value of the multiple temperature measurement points was less than 0.5 °C. The system can be considered to have reached thermal equilibrium. To minimize the influence of experimental errors on the analysis results, the values of each temperature measurement point were taken as the result of this experiment when the system reached thermal equilibrium.

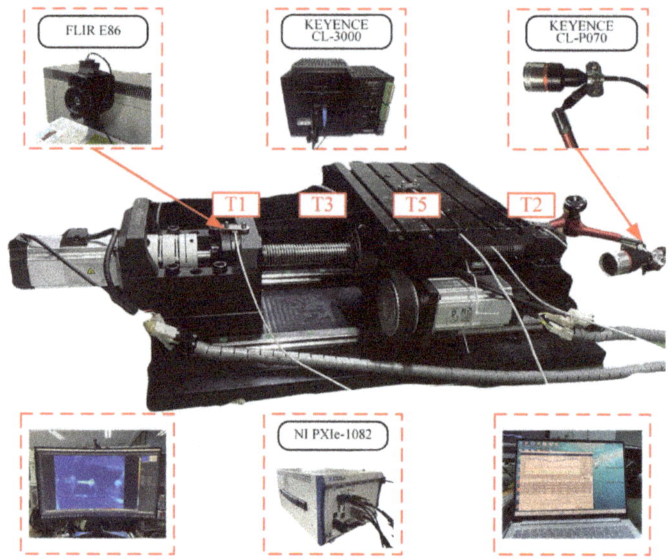

Figure 4. Temperature rise and deformation experimental device.

To clarify the thermal characteristics of the dual drive sliding feed system, the operating conditions shown in Table 5 are used for exploratory tests. The thermal characteristics of the dual drive sliding feed system at different operating speeds and different ambient temperatures are analyzed separately. Since the working temperature of the machine tool is generally 15–25 °C, the working conditions in this paper are set the ambient temperature to 15, 20, and 25 °C.

Table 5. Operating conditions of the experiments.

Operating Condition	Speed of Synthesis (m/Min)	Speed of Screw (Rpm)	Speed of Nut (Rpm)	Ambient Temperature (°C)
I	0.18	396	360	20
II	0.36	1116	1080	20
III	0.54	2214	2160	20
IV	0.36	1116	1080	15
V	0.36	1116	1080	25

4.3. Comparison of the Simulation and Experimental Results

In the simulation platform, the nut was defined to reciprocate relative to the screw, and the worktable was moved periodically throughout the entire stroke. The transient thermal analysis end time was set to 10,800 s. The corresponding simulation parameters were set according to the five operating conditions in Table 6. The temperature field and the axial deformation of the screw after the stabilization of the dual drive sliding feed system are shown in Figures 5–9.

Table 6. Comparison of the measured and simulated temperatures.

Operation Condition	Measured Position	Simulated Value (°C)	Measured Value (°C)	Deviation (°C)
I	T1	28.1	27.1	1.0
	T2	25.7	24.5	1.2
	T3	35.9	34.5	1.4
	T4	33.9	33.2	0.7
	T5	33.7	31.9	1.8
II	T1	30.8	29.6	1.2
	T2	27.4	26.2	1.2
	T3	38.5	37.1	1.4
	T4	36.4	35.8	0.6
	T5	36.2	34.2	2.0
III	T1	31.1	30.1	1.0
	T2	28.3	27.3	1.0
	T3	42.1	40.1	2.0
	T4	39.2	38.2	1.0
	T5	38.7	36.6	2.1
IV	T1	24.7	23.6	1.1
	T2	21.9	20.9	1.0
	T3	33.4	31.8	1.6
	T4	30.2	29.5	0.7
	T5	30.0	28.5	1.5
V	T1	34.5	33.2	1.3
	T2	31.7	30.5	1.2
	T3	42.4	40.8	1.6
	T4	39.9	38.8	1.1
	T5	40.0	37.9	2.1

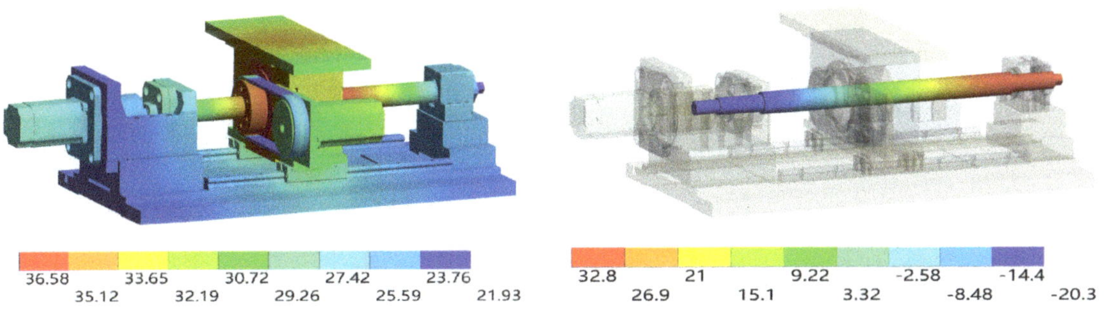

Figure 5. Temperature field and axial deformation under the operating condition I. (**a**) Temperature field distribution. (**b**) Axial deformation distribution of the screw shaft.

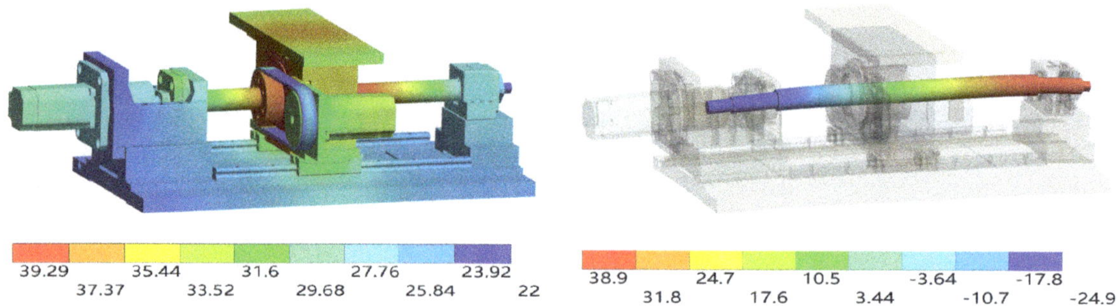

(a) (b)

Figure 6. Temperature field and axial deformation under the operating condition II. (**a**) Temperature field distribution. (**b**) Axial deformation distribution of the screw shaft.

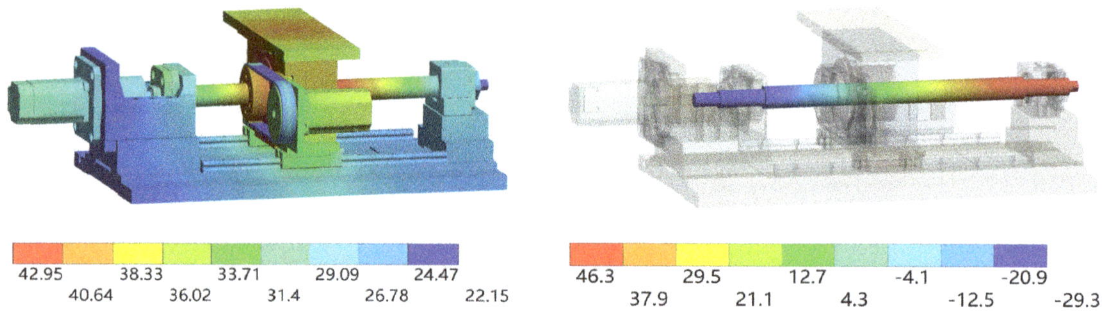

(a) (b)

Figure 7. Temperature field and axial deformation under the operating condition III. (**a**) Temperature field distribution. (**b**) Axial deformation distribution of the screw shaft.

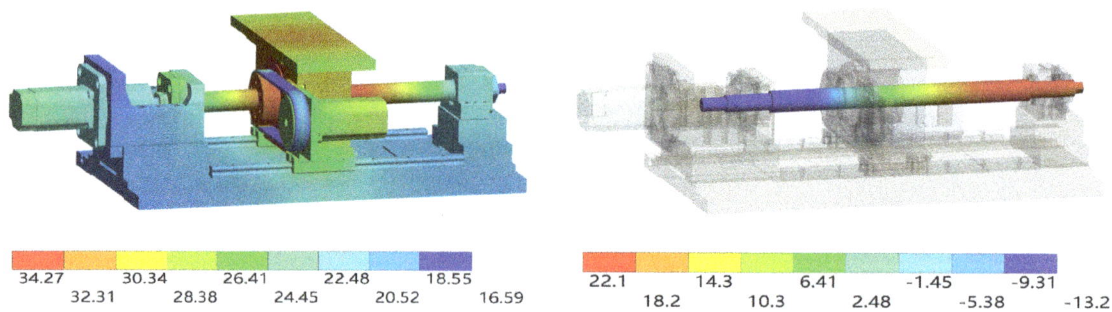

(a) (b)

Figure 8. Temperature field and axial deformation under the operating condition IV. (**a**) Temperature field distribution. (**b**) Axial deformation distribution of the screw shaft.

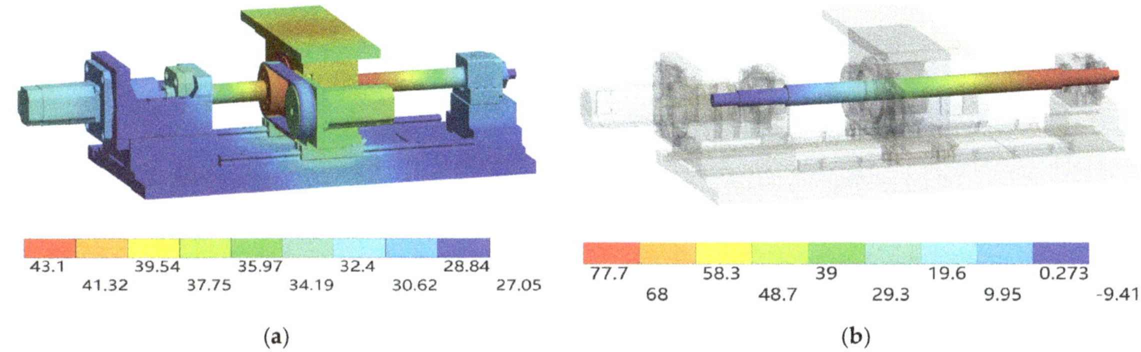

Figure 9. Temperature field and axial deformation under the operating condition V. (**a**) Temperature field distribution. (**b**) Axial deformation distribution of the screw shaft.

The solved thermal field distribution shows that the dual drive sliding feed system generates heat through the friction of the contact area, which leads to thermal deformation and thermal error. Among them, the sliding friction at the engagement of the nut and screw produces the most heat and is also affected by the heat generated by the nut bearing. Moreover, because the bearing is inside the nut housing, convective heat transfer with the surrounding air does not occur. This results in heat accumulation and local high temperatures. Therefore, the highest temperature occurs at the nut. The lowest temperature appears at the synchronous belt, which is due to the material of the synchronous belt being rubber and its low heat transfer rate. The synchronous belt experiences forced convective heat transfer with the surrounding air during operation, resulting in a lower temperature than that of the other components.

From the solved axial deformation of the screw, it can be seen that the screw generates thermal elongation in the axial direction under the action of a temperature load. The deformation of the screw is affected by multiple thermal loads, such as the nut, front bearing, and rear bearing. Thermal deformation occurs in both the forward and reverse directions of the screw axis. The axial deformation at the connection between the screw shaft and the front bearing is small. The axial deformation at the connection between the screw shaft and the rear bearing is the largest. The reason is that the rear bearing is supporting and cannot limit the axial displacement of the screw at all. Given the numerous internal heat sources and the substantial heat generation of the sliding screw in the dual drive sliding feed system, the axial deformation of the system under identical operating conditions is greater than that of the conventional feed system [6].

The minimum and maximum temperatures of the thermal field of the entire dual drive sliding feed system increased with increasing rotational speed. The thermal elongation at the rear bearing also increased. Moreover, Figures 8 and 9 show that the thermal field distribution, as well as the axial elongation, significantly changes after the ambient temperature changes. Therefore, the dynamic thermal characteristics of the dual drive sliding feed system are analyzed. The transient thermal field and axial elongation of the screw are analyzed for the dual drive sliding feed system. According to the preliminary experiment, the temperature increased rapidly during the initial period. Therefore, every 120 s was taken as a measurement cycle before 600 s, and every 600 s was taken as a measurement cycle after 600 s. All rotating axes operated at specified speeds during the experiment, and the worktable moved periodically throughout its entire stroke. The comparative analysis results are shown in Figure 10, the temperature deviation after the system reaches the thermal equilibrium state is shown in Table 6, and the axial thermal elongation deviation is shown in Table 7.

Table 7. Comparison of the measured and simulated elongations.

Operation Condition	Simulated Value (μm)	Measured Value (μm)	Deviation (μm)
I	32.8	29.9	2.9
II	38.9	35.5	3.4
III	46.3	42.5	3.8
IV	22.1	20.3	1.8
V	77.7	71.5	6.2

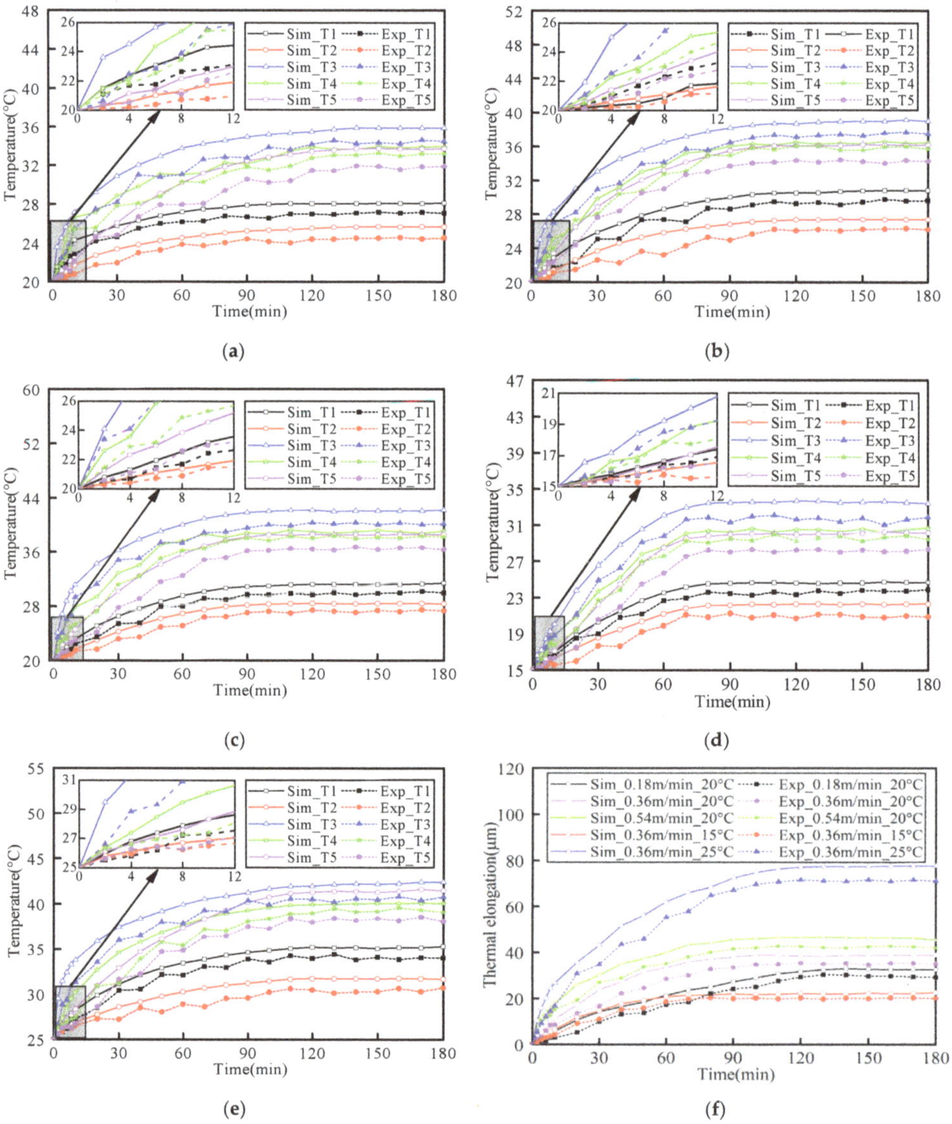

Figure 10. Comparative analysis results. (**a**) Temperature field under operating condition I. (**b**) Temperature field under operating condition II. (**c**) Temperature field under operating condition III. (**d**) Temperature field under operating condition IV. (**e**) Temperature field under operating condition V. (**f**) Axial thermal elongation under the screw.

As shown in Figure 10, along with the increase in the system running time, the temperature of each part, as well as the axial thermal elongation, will increase. Moreover, the temperature difference at T3 is large relative to the other temperature measurement points. The reason is that T3 is the result of the superposition of two major heat sources, and the sliding screw produces more heat. As shown from the experimental and simulated comparison data for each temperature measurement point in Figure 10, the values measured in the experiment are lower than the simulation values. The reason is that the temperature sensor is attached to the component surface. There is a large thermal contact resistance, which affects the accuracy of the temperature measurements. In contrast, the deviation in the temperature measured at T4 is smaller because the temperature at T4 was measured by an infrared thermometer. There was no thermal contact resistance between the sensor and the component. In addition, this experiment used air conditioning for temperature control. Moreover, the experimental environment space is small, and the airflow was disrupted during the experiment. As a result, the accuracy of temperature collection was compromised, causing fluctuations in the experimental temperature measurement data. Comparing the characteristics of the temperature rise for each operating condition in Figure 10. It can be seen that the location with a higher temperature at the end has a faster temperature rise at the beginning due to its higher heat generation. However, the temperature rise at T5 was slower than at T3 and T4. The reason was that the T5 was not a heat source, and temperature rise could only be realized through heat transfer between components. Therefore, the temperature at T5 was also easily affected by the environment and other heat sources. So, the error between the experiment and simulation was more significant.

A comparison of Figure 10a–c reveals that the temperature increase rate at each temperature measurement point increases with increasing rotational speed. The temperature at which thermal equilibrium was reached also increased, and the thermal equilibrium time gradually decreased. The system reached a thermal equilibrium state at 120, 100, and 90 min. The heat generation and heat transfer of the system reached a dynamic equilibrium state. However, the increase in each part is different. Because the temperature at the engagement point between the screw and the nut was the highest, the temperature increase in the surrounding parts was greater. The temperature increase in the regions away from the meshing was low, as shown in T1 and T2 in the figure.

Comparing Figure 10b,d,e, it can be seen that the influence of ambient temperature on the temperature characteristics of the feed system was significant. When the ambient temperature was high, the heat generated by the system was more difficult to dissipate, resulting in a faster temperature rise. On the contrary, when the ambient temperature was low, the system can more easily dissipate heat, resulting in a slower rate of temperature rise. The ambient temperature will affect the time for the feed system to reach thermal equilibrium. At an ambient temperature of 25 °C, the dual drive sliding feed system reached thermal equilibrium at 110 min. At an ambient temperature of 15 °C, the dual drive sliding feed system reached thermal equilibrium at 70 min. It can be seen that at a higher ambient temperature, the feed system needs longer to achieve thermal equilibrium due to the difficulty of heat dissipation. The feed system can achieve heat equilibrium with the environment at a lower ambient temperature more quickly.

Comparing Figure 10c,e,f, the temperature field of condition III was similar to that of condition V when the thermal equilibrium state is reached. However, due to the difference in ambient temperature, the final axial thermal elongation of the screw was 29 μm. This once again shows that the ambient temperature has an essential influence on the thermal elongation of the feed system. Therefore, the feed system's regular operation and precision stability must be ensured. Not only can the internal heat generation be controlled, but the ambient temperature should also be controlled within the appropriate range.

Tables 6 and 7 show the key temperature measurement points of the dual drive sliding feed system. The deviation in the temperature simulation of the thermal model was less than 2.1 °C, and the error in the axial elongation of the screw was less than 6.2 μm, which proves the effectiveness of the built thermal simulation model. It is shown that the developed thermal simulation model can more accurately describe the thermal characteristics of the dual drive sliding feed system. The error in the axial thermal elongation of the screw will be slightly greater than the error at each temperature measurement point. The reason is that the axial elongation of the screw is the result of the combined action of all the heat sources in the feed system and the environmental conditions. The deviation results at several key temperature measurement points do not fully represent the final axial deformation deviation of the screw. Therefore, a more accurate simulation solution model was established, laying the foundation for thermal error modeling. The key temperature measurement points of the dual drive feed system should be further optimized, and the thermal boundary conditions of the system should be optimized.

Thermal simulation can obtain more operational process data than experiments can, thus providing insight into the thermal characteristics of the feed system. Operation condition I is taken as an example based on the transient thermal analysis of the feed system. The temperature field of the screw shaft with respect to time and position changes is determined. This is shown in Figure 11.

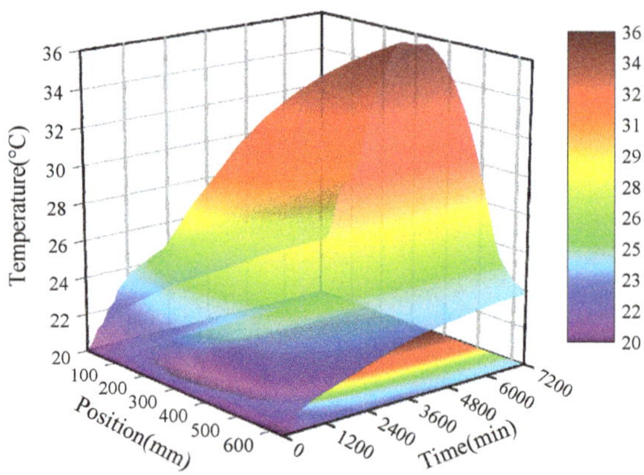

Figure 11. The overall temperature rise of the screw.

As shown in Figure 11, over time, all the parts of the screw have different degrees of temperature rise. The temperature in the middle of the screw is the highest, and the temperature at both ends is the lowest. The axial variation along the screw is from low to high and back again. The temperature at both ends is approximately symmetrical relative to the middle. The conventional screw axial temperature distribution is the opposite of the dual drive sliding feed system. The axial temperature distribution of the conventional screw is low in the middle and high at both ends [25]. The reason for this difference is that the dual drive sliding feed system utilizes a sliding screw, which results in greater heat generation. In addition, a bearing is added inside the nut, and the nut motor affects the distribution of the temperature field during operation. Therefore, the temperature rise of the screw should be controlled, and the axial deformation should be reduced. Subsequently, the structure of the dual drive sliding feed system should be further optimized to improve the temperature field distribution.

5. Conclusions

To accurately depict the thermal field characteristics of a dual drive sliding feed system under complex heat source effects and to elucidate the response mechanism of temperature rise and thermal deformation. In this study, the dynamic thermal characteristics of the system are explored by establishing a thermal simulation model and verifying the analysis with the experimental results. Based on the results and analysis, the following conclusions can be drawn:

1. The established thermal simulation model can effectively describe the dynamic thermal characteristics of the dual drive sliding feed system. By comparing the temperature rise and thermal elongation under simulation and experimental conditions, it can be concluded that the temperature rise deviation under five operating conditions is less than 2.1 °C, and the error in the axial thermal deformation of the screw is less than 6.2 μm. The established thermal characteristic simulation model can effectively describe the thermal dynamic response characteristics of the dual drive sliding feed system during operation.
2. The thermal field distribution and axial deformation of the dual drive sliding feed system differ from those of conventional feed systems. Due to the difficulty of heat dissipation and the combined effect of the screw and nut bearings, the main heat distribution region of the dual drive feed system is at the nut. Given the numerous heat sources in the system and the significant temperature increase in the sliding screw, the axial deformation of the screw in the dual drive sliding feed system is greater than that in the conventional feed system under the same operating conditions.
3. The thermal characteristics of a dual drive sliding system are significantly influenced by both rotational speed and ambient temperature. An increase in rotational speed results in a faster rate of temperature rise and a shorter time to reach thermal equilibrium. Conversely, higher ambient temperatures lead to a quicker temperature rise and a longer time to achieve thermal equilibrium. The ambient temperature also has a significant impact on the axial deformation of the screw. Even when the temperature field is similar, substantial differences in the axial thermal elongation of the screw can occur due to varying ambient temperatures.

This study investigates the temperature and deformation field distributions of a dual drive sliding system under various operating conditions and describes the dynamic thermal characteristics of the system. This research provides a foundation for temperature rise and deformation control and guidance for structural optimization and installing cooling systems. This work also lays the foundation for developing data-driven models to realize dynamic thermal error compensation. However, due to the use of empirical formulas in establishing thermal models, there are deviations in the simulation results. Future work should consider optimizing the system's thermal boundary conditions to further improve the thermal model's accuracy.

Author Contributions: H.L. (Hui Li): investigation, software, writing—original draft. H.L. (Haiyang Liu): conceptualization, data curation, formal analysis, writing—review and editing, supervision. X.F.: funding acquisition, writing—review and editing. Y.L.: methodology. M.Y.: validation. A.W.: validation. All authors have read and agreed to the published version of the manuscript.

Funding: This work was supported by the Shandong Provincial Natural Science Foundation (grant number ZR2019MEE003).

Data Availability Statement: Data supporting this study are included within the article.

Conflicts of Interest: The authors declare that they have no known competing financial interest or personal relationships that could have appeared to influence the work reported in this paper.

References

1. Li, Z.-J.; Tan, Z.; Chen, Y.; Lu, Z.-C.; Fan, Y.-C. Thermal error prediction of ball screw feed system based on inverse heat transfer analysis. *Int. J. Adv. Manuf. Technol.* **2022**, *122*, 2607–2624. [CrossRef]
2. Li, Y.; Wei, W.; Su, D.; Wu, W.; Zhang, J.; Zhao, W. Thermal characteristic analysis of ball screw feed drive system based on finite difference method considering the moving heat source. *Int. J. Adv. Manuf. Technol.* **2020**, *106*, 4533–4545. [CrossRef]
3. Lei, M.; Yang, J.; Wang, S.; Zhao, L.; Xia, P.; Jiang, G.; Mei, X. Semi-supervised modeling and compensation for the thermal error of precision feed axes. *Int. J. Adv. Manuf. Technol.* **2019**, *104*, 4629–4640. [CrossRef]
4. Liu, J.; Ma, C.; Wang, S.; Wang, S.; Yang, B.; Shi, H. Thermal boundary condition optimization of ball screw feed drive system based on response surface analysis. *Mech. Syst. Signal Process.* **2019**, *121*, 471–495. [CrossRef]
5. Du, F.; Zhang, M.; Wang, Z.; Chen, Y.; Feng, X.; Li, P. Identification and compensation of friction for a novel two-axis differential micro-feed system. *Mech. Syst. Signal Process.* **2018**, *106*, 453–465. [CrossRef]

6. Yang, H.; Xing, R.; Du, F. Thermal error modelling for a high-precision feed system in varying conditions based on an improved Elman network. *Int. J. Adv. Manuf. Technol.* **2020**, *106*, 279–288. [CrossRef]
7. Lu, Z.; Feng, X.; Su, Z.; Liu, Y.; Yao, M. Friction Parameters Dynamic Change and Compensation for a Novel Dual-Drive Micro-Feeding System. *Actuators* **2022**, *11*, 236. [CrossRef]
8. Xu, Z.Z.; Liu, X.J.; Kim, H.K.; Shin, J.H.; Lyu, S.K. Thermal error forecast and performance evaluation for an air-cooling ball screw system. *Int. J. Mach. Tools Manuf.* **2011**, *51*, 605–611. [CrossRef]
9. Qiao, G.; Liu, G.; Ma, S.; Wang, Y.; Li, P.; Lim, T.C. Thermal characteristics analysis and experimental study of the planetary roller screw mechanism. *Appl. Therm. Eng.* **2019**, *149*, 1345–1358. [CrossRef]
10. Zhao, C.J.; Yu, X.K.; Huang, Q.K.; Ge, S.D.; Gao, X. Analysis on the load characteristics and coefficient of friction of angular contact ball bearing at high speed. *Tribol. Int.* **2015**, *87*, 50–56. [CrossRef]
11. Helmig, T.; Kneer, R. A novel transient infrared-thermography based experimental method for the inverse estimation of heat transfer coefficients in rotating bearings. *Int. J. Therm. Sci.* **2021**, *167*, 107000. [CrossRef]
12. Bui, B.D.; Uchiyama, N.; Simba, K.R. Contouring control for three-axis machine tools based on nonlinear friction compensation for lead screws. *Int. J. Mach. Tools Manuf.* **2016**, *108*, 95–105. [CrossRef]
13. Min, X.; Jiang, S. A thermal model of a ball screw feed drive system for a machine tool. *Proc. Inst. Mech. Eng. Part C J. Mech. Eng. Sci.* **2011**, *225*, 186–193. [CrossRef]
14. Ma, S.; Wu, L.; Liu, G.; Fu, X. Local contact characteristics of threaded surfaces in a planetary roller screw mechanism. *J. Struct. Mech.* **2020**, *48*, 1–26. [CrossRef]
15. Du, C.; Liu, G.; Qiao, G.; Ma, S.; Cai, W. Transient thermal analysis of standard planetary roller screw mechanism based on finite element method. *Adv. Mech. Eng.* **2018**, *10*, 1687814018812305. [CrossRef]
16. Houpert, L. Numerical and analytical calculation in ball bearings. *Proc. Congr. Roulements Toulouse Toulouse* **1999**, *5–7*, 1–15.
17. Qiao, G.; Liu, G.; Ma, S.; Shi, Z.; Wang, Y.; Lim, T.C. An improved thermal estimation model of the inverted planetary roller screw mechanism. *Proc. Inst. Mech. Eng. Part C J. Mech. Eng. Sci.* **2018**, *232*, 4430–4446. [CrossRef]
18. Yang, J.J. Effect of Preload on Axial Deformation and Friction of Planetary Roller Screw. *J. Mech. Transm.* **2011**, *35*, 16–22.
19. Olaru, D.; Puiu, G.C.; Balan, L.C.; Puiu, V. A New Model to Estimate Friction Torque in a Ball Screw System. *Prod. Eng. Eco-Des. Technol. Green Energy* **2005**, 333–346. Available online: https://link.springer.com/chapter/10.1007/1-4020-2933-0_20 (accessed on 12 November 2023).
20. Kim, J.-J.; Jeong, Y.H.; Cho, D.-W. Thermal behavior of a machine tool equipped with linear motors. *Int. J. Mach. Tools Manuf.* **2004**, *44*, 749–758. [CrossRef]
21. Su, D.; Li, Y.; Zhao, W.; Zhang, H. Transient thermal error modeling of a ball screw feed system. *Int. J. Adv. Manuf. Technol.* **2023**, *124*, 2095–2107. [CrossRef]
22. Li, D.; Feng, P.; Zhang, J.; Wu, Z.; Yu, D. Method for modifying convective heat transfer coefficients used in the thermal simulation of a feed drive system based on the response surface methodology. *Numer. Heat Transf. Part A Appl.* **2015**, *69*, 51–66. [CrossRef]
23. Oyanguren, A.; Larraaga, J.; Ulacia, I. Thermo-mechanical modelling of ball screw preload force variation in different working conditions. *Int. J. Adv. Manuf. Technol.* **2018**, *97*, 723–739. [CrossRef]
24. Mao, X.; Mao, K.; Wang, F.; Yan, B.; Lei, S. A convective heat transfer coefficient algorithm for thermal analysis of machine tools considering a temperature change. *Int. J. Adv. Manuf. Technol.* **2018**, *99*, 1877–1889. [CrossRef]
25. Wu, H.; Guan, Q.; Xi, C.; Zuo, D. Construction of dynamic temperature field model of ball screw based on superposition of positive and negative temperature fields. *Numer. Heat Transf. Part A Appl. Int. J. Comput. Methodol.* **2023**, *83*, 343–360. [CrossRef]

MDPI
St. Alban-Anlage 66
4052 Basel
Switzerland
www.mdpi.com

MDPI Books Editorial Office
E-mail: books@mdpi.com
www.mdpi.com/books

www.ingramcontent.com/pod-product-compliance
Lightning Source LLC
Chambersburg PA
CBHW041934260926
51850CB00018B/132
* 9 7 8 3 7 2 5 8 1 0 7 3 4 *